INSTRUCTOR'S SOLUTIONS MANUAL

CALCULUS
ONE VARIABLE
10th Edition

INSTRUCTOR'S SOLUTIONS MANUAL

Garret Etgen
University of Houston

to accompany

CALCULUS
ONE VARIABLE

10th Edition

Saturnino Salas
Einar Hille
Garret Etgen
University of Houston

John Wiley & Sons, Inc.

Bicentennial Logo Design: Richard Pacifico

10 9 8 7 6 5 4 3 2 1

Printed and bound by Lightning Source.

CONTENTS

CHAPTER 1 1

CHAPTER 2 33

CHAPTER 3 65

CHAPTER 4 118

CHAPTER 5 229

CHAPTER 6 288

CHAPTER 7 341

CHAPTER 8 404

CHAPTER 9 481

CHAPTER 10 524

CHAPTER 11 588

CHAPTER 12 632

CHAPTER 1

SECTION 1.2

1. rational

2. rational

3. rational

4. irrational

5. rational

6. irrational

7. rational

8. rational

9. rational

10. rational

11. $\dfrac{3}{4} = 0.75$

12. $0.33 < \dfrac{1}{3}$

13. $\sqrt{2} > 1.414$

14. $4 = \sqrt{16}$

15. $-\dfrac{2}{7} < -0.285714$

16. $\pi < \dfrac{22}{7}$

17. $|6| = 6$

18. $|-4| = 4$

19. $|-3-7| = 10$

20. $|-5| - |8| = -3$

21. $|-5| + |-8| = 13$

22. $|2 - \pi| = \pi - 2$

23. $|5 - \sqrt{5}| = 5 - \sqrt{5}$

24.

25.

26.

27.

28.

29.

30.

31.

32.

33.

34.

35.

36.

37.

38.

39.

40.

41. bounded, lower bound 0, upper bound 4

42. bounded above by 0

43. not bounded

44. bounded above by 4

45. not bounded

46. bounded; lower bound 0, upper bound 1

47. bounded above, upper bound $\sqrt{2}$

48. $\sqrt{2} < \sqrt[3]{\pi} < 2^{\sqrt{\pi}} < \pi^3 < 3^{\pi}$

49. $x_0 = 2$, $x_1 \cong 2.75$, $x_2 \cong 2.58264$, $x_3 \cong 2.57133$, $x_4 \cong 2.57128$, $x_5 \cong 2.57128$; bounded; lower bound 2, upper bound 3 (the smallest upper bound $\cong 2.57128\cdots$); $x_n \cong 2.5712815907$ (10 decimal places)

50. $x_n \to 2.970...$;   bounded

51. $x^2 - 10x + 25 = (x - 5)^2$ **52.** $9(x - \frac{2}{3})(x + \frac{2}{3})$

53. $8x^6 + 64 = 8(x^2 + 2)(x^4 - 2x^2 + 4)$ **54.** $27(x - \frac{2}{3})(x^2 + \frac{2}{3}x + \frac{4}{9})$

55. $4x^2 + 12x + 9 = (2x + 3)^2$ **56.** $4(x^2 + \frac{1}{2})^2$

57. $x^2 - x - 2 = (x - 2)(x + 1) = 0$; $x = 2, -1$ **58.** $-3, 3$

59. $x^2 - 6x + 9 = (x - 3)^2$; $x = 3$ **60.** $-\frac{1}{2}, 3$

61. $x^2 - 2x + 2 = 0$;   no real zeros   **62.** -4

63.   no real zeros   **64.**   no real zeros

65. $5! = 120$ **66.** $\dfrac{5!}{8!} = \dfrac{1}{8 \cdot 7 \cdot 6} = \dfrac{1}{336}$ **67.** $\dfrac{8!}{3!5!} = \dfrac{8 \cdot 7 \cdot 6}{3 \cdot 2 \cdot 1} = 56$

68. $\dfrac{9!}{3!6!} = \dfrac{9 \cdot 8 \cdot 7}{3 \cdot 2 \cdot 1} = 84$ **69.** $\dfrac{7!}{0!7!} = \dfrac{7!}{1 \cdot 7!} = 1$

70. $\dfrac{p_1}{q_1} + \dfrac{p_2}{q_2} = \dfrac{p_1 q_2 + p_2 q_1}{q_1 q_2}$, $p_1 q_2 + p_2 q_1$ and $q_1 q_2$ are integers, and $q_1 q_2 \neq 0$

71.   Let r be a rational number and s an irrational number. Suppose $r + s$ is rational. Then $(r + s) - r = s$ is rational, a contradiction.

72. $\left(\dfrac{p_1}{q_1}\right)\left(\dfrac{p_2}{q_2}\right) = \dfrac{p_1 p_2}{q_1 q_2}$, $p_1 p_2$ and $q_1 q_2$ are integers, and $q_1 q_2 \neq 0$

73.   The product of a rational and an irrational number may either be rational or irrational; $0 \cdot \sqrt{2} = 0$ is rational, $1 \cdot \sqrt{2} = \sqrt{2}$ is irrational.

74. $\sqrt{2} + 3\sqrt{2} = 4\sqrt{2}$ irrational; $\pi + (1 - \pi) = 1$,   rational.
$(\sqrt{2})(\sqrt{3}) = \sqrt{6}$ irrational; $(\sqrt{2})(3\sqrt{2}) = 6$,   rational.

75.   Suppose that $\sqrt{2} = p/q$ where p and q are integers and $q \neq 0$. Assume that p and q have no common factors (other than ± 1). Then $p^2 = 2q^2$ and p^2 is even. This implies that $p = 2r$ is even. Therefore $2q^2 = 4r^2$ which implies that q^2 is even, and hence q is even. It now follows that p and q are both even, contradicting the assumption that p and q have no common factors.

76.   Assume $\sqrt{3} = \dfrac{p}{q}$, where p and q have no common factors. Then $3 = \dfrac{p^2}{q^2}$, so $p^2 = 3q^2$. Thus p^2 is divisible by 3, and therefore p is divisible by 3, say $p = 3a$. Then $9a^2 = 3q^2$, so $3a^2 = q^2$, where q must also be divisible by 3, contracting our assumption.

77. Let x be the length of a rectangle that has perimeter P. Then the width y of the rectangle is given by $y = (1/2)P - x$ and the area is

$$A = x\left(\frac{1}{2}P - x\right) = \left(\frac{P}{4}\right)^2 - \left(x - \frac{P}{4}\right)^2.$$

It follows that the area is a maximum when $x = P/4$. Since $y = P/4$ when $x = P/4$, the rectangle of perimeter P having the largest area is a square.

78. Circle: perimeter $2\pi r = p \implies r = \dfrac{p}{2\pi} \implies$ area $= \pi r^2 = \dfrac{p^2}{4\pi}$

square: perimeter $4x = p \implies x = \dfrac{p}{4} \implies$ area $= x^2 = \dfrac{p^2}{16} < \dfrac{p^2}{4\pi}$.

For an arbitrary rectangle, $p = 2(x+y)$, so $y = \dfrac{p}{2} - x$, and area $= xy = x(\dfrac{p}{2} - x)$. This is the equation of a parabola with vertex (hence maximum value) at $x = \dfrac{p}{4}$. Thus $y = \dfrac{p}{4}$ and the rectangle is a square. The circle still has larger area.

SECTION 1.3

1. $2 + 3x < 5$

$3x < 3$

$x < 1$

Ans: $(-\infty, 1)$

2. $\frac{1}{2}(2x + 3) < 6$

$2x + 3 < 12$

$x < \frac{9}{2}$

Ans: $(-\infty, \frac{9}{2})$

3. $16x + 64 \leq 16$

$16x \leq -48$

$x \leq -3$

Ans: $(-\infty, -3]$

4. $3x + 5 > \frac{1}{4}(x - 2)$

$12x + 20 > x - 2$

$11x > -22$

$x > -2$

Ans: $(-2, \infty)$

5. $\frac{1}{2}(1 + x) < \frac{1}{3}(1 - x)$

$3(1 + x) < 2(1 - x)$

$3 + 3x < 2 - 2x$

$5x < -1$

$x < -\frac{1}{5}$

Ans: $(-\infty, -\frac{1}{5})$

6. $3x - 2 \leq 1 + 6x$

$-3x \leq 3$

$x \geq -1$

Ans: $[-1, \infty)$

7. $\quad x^2 - 1 < 0$

$(x + 1)(x - 1) < 0$

Ans: $(-1, 1)$

8. $x^2 + 9x + 20 < 0$

$(x + 5)(x + 4) < 0$

Ans: $(-5, -4)$

9. $x^2 - x - 6 \geq 0$

$(x - 3)(x + 2) \geq 0$

Ans: $(\infty, -2] \cup [3, \infty)$

10. $x^2 - 4x - 5 > 0$

$(x - 5)(x + 1) > 0$

Ans: $(-\infty, -1) \cup (5, \infty)$

11. $2x^2 + x - 1 \leq 0$

$(2x - 1)(x + 1) \leq 0$

Ans: $[-1, 1/2]$

12. $3x^2 + 4x - 4 \geq 0$

$(3x - 2)(x + 2) \geq 0$

Ans: $(-\infty, -2] \cup [2/3, \infty)$

13. $x(x - 1)(x - 2) > 0$

Ans: $(0, 1) \cup (2, \infty)$

14. $x(2x - 1)(3x - 5) \leq 0$

Ans: $(-\infty, 0] \cup [\frac{1}{2}, \frac{5}{3}]$

15. $x^3 - 2x^2 + x \geq 0$

$x(x - 1)^2 \geq 0$

Ans: $[0, \infty)$

16. $x^2 - 4x + 4 \le 0$ **17.** $x^3(x-2)(x+3)^2 < 0$ **18.** $x^2(x-3)(x+4)^2 > 0$

$(x-2)^2 \le 0$

Ans: $\{2\}$ Ans: $(0,2)$ Ans: $(3,\infty)$

19. $x^2(x-2)(x+6) > 0$ **20.** $7x(x-4)^2 < 0$

Ans: $(-\infty, -6) \cup (2, \infty)$ Ans: $(-\infty, 0)$

21. $(-2, 2)$ **22.** $(-\infty, -1] \cup [1, \infty)$ **23.** $(-\infty, -3) \cup (3, \infty)$

24. $(0, 2)$ **25.** $(\frac{3}{2}, \frac{5}{2})$ **26.** $(-\frac{3}{2}, \frac{5}{2})$

27. $(-1, 0) \cup (0, 1)$ **28.** $(-\frac{1}{2}, 0) \cup (0, \frac{1}{2})$ **29.** $(\frac{3}{2}, 2) \cup (2, \frac{5}{2})$

30. $(-\frac{3}{2}, \frac{1}{2}) \cup (\frac{1}{2}, \frac{5}{2})$ **31.** $(-5, 3) \cup (3, 11)$ **32.** $(\frac{2}{3}, \frac{8}{3})$

33. $(-\frac{5}{8}, -\frac{3}{8})$ **34.** $(\frac{1}{2}, \frac{7}{10})$ **35.** $(-\infty, -4) \cup (-1, \infty)$

36. $(-\infty, -2) \cup (\frac{4}{3}, \infty)$ **37.** $|x-0| < 3$ or $|x| < 3$ **38.** $|x-0| < 2$ or $|x| < 2$

39. $|x-2| < 5$ **40.** $|x-2| < 2$

41. $|x-(-2)| < 5$ or $|x+2| < 5$ **42.** $\left| x - \frac{b+a}{2} \right| < \frac{b-a}{2}$

43. $|x-2| < 1 \quad \Longrightarrow \quad |2x-4| = 2|x-2| < 2$, so $|2x-4| < A$ true for $A \ge 2$.

44. $|x-2| < A \quad \Longrightarrow \quad 2|x-2| = |2x-4| < 2A \quad \Longrightarrow \quad |2x-4| < 3$

provided that $0 < A \le \frac{3}{2}$

45. $|x+1| < A \quad \Longrightarrow \quad |3x+3| = 3|x+1| < 3A \quad \Longrightarrow \quad |3x+3| < 4$

provided that $0 < A \le \frac{4}{3}$

46. $|x+1| < 2 \quad \Longrightarrow \quad 3|x+1| = |3x+3| < 6 \quad \Longrightarrow \quad |3x+3| < A$

provided that $A \ge 6$

47. (a) $\dfrac{1}{x} < \dfrac{1}{\sqrt{x}} < 1 < \sqrt{x} < x$ (b) $x < \sqrt{x} < 1 < \dfrac{1}{\sqrt{x}} < \dfrac{1}{x}$

48. $\sqrt{\dfrac{x}{x+1}} < \sqrt{\dfrac{x+1}{x+2}}$.

49. If a and b have the same sign, then $ab > 0$. Suppose that $a < b$. Then $a - b < 0$ and

$$\frac{1}{b} - \frac{1}{a} = \frac{a-b}{ab} < 0.$$

Thus, $(1/b) < (1/a)$.

50. $a^2 \leq b^2 \implies b^2 - a^2 = (b+a)(b-a) \geq 0 \implies b - a \geq 0 \implies a \leq b.$

51. With $a \geq 0$ and $b \geq 0$

$$b \geq a \implies b - a = (\sqrt{b} + \sqrt{a})(\sqrt{b} - \sqrt{a}) \geq 0 \implies \sqrt{b} - \sqrt{a} \geq 0 \implies \sqrt{b} \geq \sqrt{a}.$$

52. $|a - b| = |a + (-b)| \leq |a| + |-b| = |a| + |b|.$

53. By the hint

$$\big| |a| - |b| \big|^2 = (|a| - |b|)^2 = |a|^2 - 2|a|\,|b| + |b|^2 = a^2 - 2|ab| + b^2$$
$$\leq a^2 - 2ab + b^2 = (a - b)^2.$$
$$(ab \leq |ab|)$$

Taking the square root of the extremes, we have

$$\big| |a| - |b| \big| \leq \sqrt{(a - b)^2} = |a - b|.$$

54. If $a \geq 0$ and $b \geq 0$: $|a + b| = a + b = |a| + |b|.$
If $a < 0$ and $b < 0$: $|a + b| = -(a + b) = -a - b = |a| + |b|.$
If $a \geq 0$ and $b < 0$: If $a \geq |b|$ then $|a + b| = a - |b| < a + |b| = |a| + |b|.$
         If $a < |b|$ then $|a + b| = |b| - a < |b| + a = |a| + |b|.$
Similarly, $a < 0, b \geq 0 \implies |a + b| < |a| + |b|.$
Thus equality holds iff a and b are of the same sign.

55. With $0 \leq a \leq b$

$$a(1 + b) = a + ab \leq b + ab = b(1 + a).$$

Division by $(1 + a)(1 + b)$ gives

$$\frac{a}{1 + a} \leq \frac{b}{1 + b}.$$

56. $\dfrac{a}{1 + a} \leq \dfrac{b + c}{1 + b + c} = \dfrac{b}{1 + b + c} + \dfrac{c}{1 + b + c} \leq \dfrac{b}{1 + b} + \dfrac{c}{1 + c}.$
$\big\uparrow$
$\quad$ by Exercise 55

57. Suppose that $a < b$. Then

$$a = \frac{a + a}{2} \leq \frac{a + b}{2} \leq \frac{b + b}{2} = b.$$

$\dfrac{a + b}{2}$ is the midpoint of the line segment $\overline{ab}$.

58. First inequality: $a = (\sqrt{a})^2 \leq \sqrt{a}\sqrt{b} = \sqrt{ab}.$
Last inequality: $\frac{1}{2}(a + b) \leq \frac{1}{2}(b + b) = b.$

Middle inequality:

$$0 \le (a-b)^2 = a^2 - 2ab + b^2 = (a+b)^2 - 4ab$$

$$4ab \le (a+b)^2$$

$$2\sqrt{ab} \le (a+b) \quad \text{(by Exercise 50)}$$

$$\sqrt{ab} \le \frac{1}{2}(a+b)$$

SECTION 1.4

1. $d(P_0, P_1) = \sqrt{(6-0)^2 + (-3-5)^2} = \sqrt{36+64} = \sqrt{100} = 10$

2. $d(P_0, P_1) = \sqrt{(5-2)^2 + (5-2)^2} = 3\sqrt{2}$

3. $d(P_0, P_1) = \sqrt{[5-(-3)]^2 + (-2-2)^2} = \sqrt{64+16} = 4\sqrt{5}$

4. $d(P_0, P_1) = \sqrt{(-4-2)^2 + (7-7)^2} = 6$

5. $\left(\dfrac{2+6}{2}, \dfrac{4+8}{2}\right) = (4, 6)$ 　　　　　**6.** $\left(\dfrac{3-1}{2}, \dfrac{-1+5}{2}\right) = (1, 2)$

7. $\left(\dfrac{2+7}{2}, \dfrac{-3-3}{2}\right) = (\frac{9}{2}, -3)$ 　　　**8.** $m = \left(\dfrac{a+3}{2}, \dfrac{3+a}{2}\right)$

9. $m = \dfrac{5-1}{(-2)-4} = \dfrac{4}{-6} = -\dfrac{2}{3}$ 　　　**10.** $m = \dfrac{-3-(-7)}{4-(-2)} = \dfrac{4}{6} = \dfrac{2}{3}$

11. $m = \dfrac{b-a}{a-b} = -1$ 　　　　　　　**12.** $m = \dfrac{-1-(-1)}{4-(-3)} = \dfrac{0}{7} = 0$

13. $m = \dfrac{0-y_0}{x_0-0} = -\dfrac{y_0}{x_0}$ 　　　　**14.** $m = \dfrac{0-y_0}{0-x_0} = \dfrac{-y_0}{-x_0} = \dfrac{y_0}{x_0}$

15. Equation is in the form $y = mx + b$.   Slope is 2; y-intercept is -4.

16. Rewrite as $5x = 6$, or $x = \frac{6}{5}$, This is a vertical line with slope undefined, no y-intercept.

17. Write equation as $y = \frac{1}{3}x + 2$.   Slope is $\frac{1}{3}$; y-intercept is 2.

18. Write equation as $y = \frac{1}{2}x - \frac{4}{3}$. Slope is $\frac{1}{2}$, y-intercept is $-\frac{4}{3}$.

19. Write equation as $y = \frac{7}{3}x + \frac{4}{3}$.   Slope is $\frac{7}{3}$; y-intercept is $\frac{4}{3}$.

20. Write equation as $y = 3$; This is a horizontal line. Slope is 0, y-intercept is 3.

21. $y = 5x + 2$ **22.** $y = 5x - 2$ **23.** $y = -5x + 2$ **24.** $y = -5x - 2$

25. $y = 3$ **26.** $y = -3$ **27.** $x = -3$ **28.** $x = 3$

29. Every line parallel to the x-axis has an equation of the form $y = a$ constant. In this case $y = 7$.

30. Every line parallel to the y-axis has an equation of the form $x = a$ constant. In this case $x = 2$.

31. The line $3y - 2x + 6 = 0$ has slope $\frac{2}{3}$. Every line parallel to it has that same slope. The line through $P(2, 7)$ with slope $\frac{2}{3}$ has equation $y - 7 = \frac{2}{3}(x - 2)$, which reduces to $3y - 2x - 17 = 0$.

32. The line $y - 2x + 5 = 0$ has slope 2. Every line perpendicular to it has the slope $-\frac{1}{2}$. The line through $P(2, 7)$ with slope $-\frac{1}{2}$ has equation $y - 7 = -\frac{1}{2}(x - 2)$, which reduces to $2y + x - 16 = 0$.

33. The line $3y - 2x + 6 = 0$ has slope $\frac{2}{3}$. Every line perpendicular to it has slope $-\frac{3}{2}$.
The line through $P(2, 7)$ with slope $-\frac{3}{2}$ has equation $y - 7 = -\frac{3}{2}(x - 2)$, which reduces to
$2y + 3x - 20 = 0$.

34. The line $y - 2x + 5 = 0$ has slope 2. Every line parallel to it has slope 2.
The line through $P(2, 7)$ with slope 2 has equation $y - 7 = 2(x - 2)$, which reduces to
$y - 2x - 3 = 0$.

35. $\left(\frac{1}{2}\sqrt{2}, \frac{1}{2}\sqrt{2}\right), \left(-\frac{1}{2}\sqrt{2}, -\frac{1}{2}\sqrt{2}\right)$ [Substitute $y = x$ into $x^2 + y^2 = 1$.]

36. $\left(\dfrac{2}{\sqrt{1 + m^2}}, \dfrac{2m}{\sqrt{1 + m^2}}\right), \left(\dfrac{-2}{\sqrt{1 + m^2}}, \dfrac{-2m}{\sqrt{1 + m^2}}\right)$ [Substitute $y = mx$ into $x^2 + y^2 = 4$.]

37. $(3, 4)$ [Write $4x + 3y = 24$ as $y = \frac{4}{3}(6 - x)$ and substitute into $x^2 + y^2 = 25$.]

38. $(0, b), \left(\dfrac{-2mb}{1 + m^2}, \dfrac{b(1 - m^2)}{1 + m^2}\right)$ [Substitute $y = mx + b$ into $x^2 + y^2 = b^2$.]

39. $(1, 1)$ **40.** $\left(\frac{23}{37}, \frac{116}{37}\right)$ **41.** $\left(-\frac{2}{23}, \frac{38}{23}\right)$ **42.** $\left(-\frac{17}{73}, -\frac{2}{73}\right)$

43. We select the side joining $A(1, -2)$ and $B(-1, 3)$ as the base of the triangle.
length of side AB : $\sqrt{29}$;   equation of line through A and B : $5x + 2y - 1 = 0$
equation of line through $C(2, 4) \perp 5x + 2y - 1 = 0$: $y - 4 = \frac{2}{5}(x - 2)$
point of intersection of the two lines: $\left(\dfrac{-27}{29}, \dfrac{82}{29}\right)$
altitude of the triangle: $\sqrt{\left(2 + \frac{27}{29}\right)^2 + \left(4 - \frac{82}{29}\right)^2} = \dfrac{17}{\sqrt{29}}$
area of triangle: $\dfrac{1}{2}\left(\sqrt{29}\right)\left(\dfrac{17}{\sqrt{29}}\right) = \dfrac{17}{2}$

44. Let the side joining $A(-1, 1)$ and $B\left(3, \sqrt{2}\right)$ as the base of the triangle.

length of base AB : $\sqrt{19 - 2\sqrt{2}}$;   equation of line through A and B : $y - 1 = \dfrac{\sqrt{2} - 1}{4}(x + 1)$

equation of line through $C\left(\sqrt{2}, -1\right) \perp$ base : $y + 1 = \dfrac{-4}{\sqrt{2} - 1}\left(x - \sqrt{2}\right)$

point of intersection of the two lines: $\left(\dfrac{-5 - 10\sqrt{2}}{-19 + 2\sqrt{2}}, \dfrac{-17 - 2\sqrt{2}}{-19 + 2\sqrt{2}}\right)$

altitude of the triangle: $\dfrac{9}{\sqrt{19 - 2\sqrt{2}}}$

area of triangle: $\dfrac{1}{2}\left(\sqrt{19 - 2\sqrt{2}}\right)\left(\dfrac{9}{\sqrt{19 - 2\sqrt{2}}}\right) = \dfrac{9}{2}$

45. Substitute $y = m(x - 5) + 12$ into $x^2 + y^2 = 169$ and you get a quadratic in x that involves m. That quadratic has a unique solution iff $m = -\frac{5}{12}$. (A quadratic $ax^2 + bx + c = 0$ has a unique solution iff $b^2 - 4ac = 0$).

46. $(x - 1)^2 + (y + 3)^2 = 25$, the center is at $(1, -3)$. The radius through $(4, 1)$ is the line with slope $\frac{4}{3}$. Therefore the tangent to the circle is $(y - 1) = \frac{3}{4}(x - 4)$,   or $3x + 4y - 16 = 0$.

47. The slope of the line through the center of the circle and the point P is -2. Therefore the slope of the tangent line is $\frac{1}{2}$. The equation for the tangent line to the circle at P is

$$(y + 1) = \frac{1}{2}(x - 1), \text{ or } x - 2y - 3 = 0.$$

48. $\left(-\frac{29}{7}, -\frac{34}{7}\right)$ 　　　　　　　　　　**49.** $(2.36, -0.21)$

50. $(-0.43, -1.95)$, $(1.97, -0.35)$ 　　　　**51.** $(0.61, 2.94)$, $(2.64, 1.42)$

52. Slope of the line segment: $\dfrac{-4 - 3}{3 + 1} = -\dfrac{7}{4}$. Midpoint: $= \left(\dfrac{3 - 1}{2}, \dfrac{-4 + 3}{2}\right) = \left(1, -\frac{1}{2}\right)$.
Equation of perpendicular bisector: $y + \frac{1}{2} = \frac{4}{7}(x - 1)$.

53. Midpoint of line segment $\overline{PQ}$: $\left(\frac{5}{2}, \frac{5}{2}\right)$
Slope of line segment $\overline{PQ}$: $\frac{13}{3}$
Equation of the perpendicular bisector: $y - \frac{5}{2} = -\left(\frac{3}{13}\right)\left(x - \frac{5}{2}\right)$ or $3x + 13y - 40 = 0$

54. Length of sides: $\overline{P_0 P_1} : \sqrt{(-4 + 4)^2 + (-1 - 3)^2} = 4$, $\overline{P_0 P_2} : \sqrt{(2 + 4)^2 + (1 - 3)^2} = \sqrt{40}$

$\overline{P_1 P_2} : \sqrt{(2 + 4)^2 + (1 + 1)^2} = \sqrt{40}$.   The triangle is isosceles.

Slope of sides: $\overline{P_0 P_1} : \dfrac{-1 - 3}{-4 + 4}$; $\overline{P_0 P_2} : \dfrac{1 - 3}{2 + 4} = -\dfrac{1}{3}$

$\overline{P_1 P_2} : \dfrac{1 + 1}{2 + 4} = \dfrac{1}{3}$.   The triangle is not a right triangle.

55. $d(P_0, P_1) = \sqrt{(-2-1)^2 + (5-3)^2} = \sqrt{13}, \quad d(P_0, P_2) = \sqrt{[-2-(-1)]^2 + (5-0)^2} = \sqrt{26},$
$d(P_1, P_2) = \sqrt{[1-(-1)]^2 + (3-0)^2} = \sqrt{13}.$
Since $d(P_0, P_1) = d(P_1, P_2),$ the triangle is isosceles.
Since $[d(P_0, P_1)]^2 + [d(P_1, P_2)]^2 = [d(P_0, P_2)]^2,$ the triangle is a right triangle.

56. Length of sides: $\overline{P_0 P_1} : \sqrt{(0+2)^2 + (7+1)^2} = \sqrt{68}, \quad \overline{P_0 P_2} : \sqrt{(3+2)^2 + (2+1)^2} = \sqrt{34}$
$\overline{P_1 P_2} : \sqrt{(3-0)^2 + (2-7)^2} = \sqrt{34}.$ The triangle is isosceles.
slope of sides: $\overline{P_0 P_1} : \dfrac{7+1}{0+2} = 4, \quad \overline{P_0 P_2} : \dfrac{2+1}{3+2} = \dfrac{3}{5}, \quad \overline{P_1 P_2} : \dfrac{2-7}{3-0} = -\dfrac{5}{3}.$ The triangle is a right triangle.

57. The line l_2 through the origin perpendicular to $l_1 : Ax + By + C = 0$ has equation $y = \dfrac{B}{A}x.$ The lines l_1 and l_2 intersect at the point $P\left(\dfrac{-AC}{A^2+B^2}, \dfrac{-BC}{A^2+B^2}\right).$ The distance from P to the origin is $\dfrac{|C|}{\sqrt{A^2 + b^2}}.$

58. Length of side: $\sqrt{(4-0)^2 + (3-0)^2} = 5.$ We need a point (x, y) that is at a distance 5 from both $(0, 0)$ and $(4, 3).$ Thus $x^2 + y^2 = 25$ and $(x-4)^2 + (y-3)^2 = 25.$ From this we get $36(25 - x^2) = 25^2 - 400x + 64x^2.$ Solving gives two possibilities for the third vertex:
$$\left(2 + \frac{1}{2}\sqrt{27}, \frac{9 - 4\sqrt{27}}{6}\right), \quad \left(2 - \frac{1}{2}\sqrt{27}, \frac{9 + 4\sqrt{27}}{6}\right).$$

59. The coordinates of M are $\left(\dfrac{a}{2}, \dfrac{b}{2}\right);$ and
$d(M, (0, b)) = d(M, (0, a)) = d(M, (0, 0)) = \frac{1}{2}\sqrt{a^2 + b^2}.$

60. Let $A = (-1, -2), B = (2, 1), C = (4, -3).$
Midpoint $\overline{AB} = (\frac{1}{2}, -\frac{1}{2});$ distance to $C = \sqrt{(4 - \frac{1}{2})^2 + (-3 + \frac{1}{2})^2} = \dfrac{\sqrt{74}}{2}.$
Midpoint $\overline{AC} = (\frac{3}{2}, -\frac{5}{2});$ distance to $B = \sqrt{(2 - \frac{3}{2})^2 + (1 + \frac{5}{2})^2} = \frac{5}{2}\sqrt{2}.$
Midpoint $\overline{BC} = (3, -1);$ distance to $A = \sqrt{(3 + 1)^2 + (-2 + 1)^2} = \sqrt{17}.$

61. Denote the points $(1, 0), (3, 4)$ and $(-1, 6)$ by A, B and C, respectively. The midpoints of the line segments $\overline{AB}$, $\overline{AC}$, and $\overline{BC}$ are $P(2, 2), Q(0, 3)$ and $R(1, 5).$
An equation for the line through A and R is: $x = 1.$
An equation for the line through B and Q is: $y = \frac{1}{3}x + 3.$
An equation for the line through C and P is: $y - 2 = -\frac{4}{3}(x - 2).$
These lines intersect at the point $(1, \frac{10}{3}).$

62. The three midpoints are $\left(\dfrac{c}{2}, 0\right), \left(\dfrac{a+c}{2}, \dfrac{b}{2}\right),$ and $\left(\dfrac{a}{2}, \dfrac{b}{2}\right).$ The equations of the medians are:
$y = \dfrac{2b}{2a - c}\left(x - \dfrac{c}{2}\right), y = \dfrac{b}{a + c}x,$ and $y = \dfrac{b}{a - 2c}(x - c).$ These lines intersect at $\left(\dfrac{a+c}{3}, \dfrac{b}{3}\right).$

63. Let $A(0,0)$ and $B(a,0)$, $a > 0$, be adjacent vertices of a parallelogram. If $C(b,c)$ is the vertex opposite B, then the vertex D opposite A has coordinates $(a+b,c)$. [See the figure.]

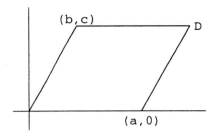

The line through A and D has equation: $y = \dfrac{c}{a+b}\,x$.

The line through B and C has equation: $y = -\dfrac{c}{a-b}\,(x-a)$.

These lines intersect at the point $\left(\dfrac{a+b}{2}, \dfrac{c}{2}\right)$ which is the midpoint of each of the line segments $\overline{AD}$ and $\overline{BC}$.

64. Midpoints: $M_1 = \left(\dfrac{x_1+x_2}{2}, \dfrac{y_1+y_2}{2}\right)$, $M_2 = \left(\dfrac{x_2+x_3}{2}, \dfrac{y_2+y_3}{2}\right)$, $M_3 = \left(\dfrac{x_3+x_4}{2}, \dfrac{y_3+y_4}{2}\right)$,

$M_4 = \left(\dfrac{x_4+x_1}{2}, \dfrac{y_4+y_1}{2}\right)$. Slope $\overline{M_1M_2} = \dfrac{y_3-y_1}{x_3-x_1}$; Slope $\overline{M_3M_4} = \dfrac{y_1-y_3}{x_1-x_3} = $ slope $\overline{M_1M_2}$.

Similarly, slope of $\overline{M_2M_3} = $ slope of $\overline{M_4M_1}$. Therefore the quadrilateral is a parallelogram.

65. Since the relation between F and C is linear, $F = mC + b$ for some constants m and C. Setting $C = 0$ and $F = 32$ gives $b = 32$. Thus $F = mC + 32$. Setting $C = 100$ and $F = 212$ gives $m = (212-32)/100 = 9/5$. Therefore

$$F = \frac{9}{5}C + 32$$

The Fahrenheit and Centigrade temperatures are equal when

$$C = F = \frac{9}{5}C + 32$$

which implies $C = F = -40°$.

66. $K - 373 = \dfrac{373 - 273}{212 - 32}(F - 212) \implies K = \dfrac{5}{9}F + \dfrac{2297}{9}$

$K - 373 = \dfrac{373 - 273}{100 - 0}(C - 100) \implies K = C + 273,$ linear

SECTION 1.5

1. (a) $f(0) = 2(0)^2 - 3(0) + 2 = 2$ (b) $f(1) = 2(1)^2 - 3(1) + 2 = 1$

(c) $f(-2) = 2(-2)^2 - 3(-2) + 2 = 16$ (d) $f(\frac{3}{2}) = 2(3/2)^2 - 3(3/2) + 2 = 2$

2. (a) $-\dfrac{1}{4}$ (b) $\dfrac{1}{5}$ (c) $-\dfrac{5}{8}$ (d) $\dfrac{8}{25}$

3. (a) $f(0) = \sqrt{0^2 + 2 \cdot 0} = 0$ (b) $f(1) = \sqrt{1^2 + 2 \cdot 1} = \sqrt{3}$

     (c) $f(-2) = \sqrt{(-2)^2 + 2(-2)} = 0$ (d) $f(\frac{3}{2}) = \sqrt{(3/2)^2 + 2(3/2)} = \frac{1}{2}\sqrt{21}$

4. (a) 3   (b) -1 (c) 11   (d) -3

5. (a) $f(0) = \dfrac{2 \cdot 0}{|0 + 2| + 0^2} = 0$ (b) $f(1) = \dfrac{2 \cdot 1}{|1 + 2| + 1^2} = \dfrac{1}{2}$

     (c) $f(-2) = \dfrac{2 \cdot (-2)}{|-2 + 2| + (-2)^2} = -1$ (d) $f(\frac{3}{2}) = \dfrac{2 \cdot (3/2)}{|(3/2) + 2| + (3/2)^2} = \dfrac{12}{23}$

6. (a) 0   (b) $\dfrac{3}{4}$ (c) 0   (d) $\dfrac{21}{25}$

7. (a) $f(-x) = (-x)^2 - 2(-x) = x^2 + 2x$ (b) $f(1/x) = (1/x)^2 - 2(1/x) = \dfrac{1 - 2x}{x^2}$

     (c) $f(a + b) = (a + b)^2 - 2(a + b) = a^2 + 2ab + b^2 - 2a - 2b$

8. (a) $f(-x) = -\dfrac{x}{x^2 + 1}$ (b) $f(\dfrac{1}{x}) = \dfrac{x}{x^2 + 1}$ (c) $f(a + b) = \dfrac{a + b}{(a + b)^2 + 1}$

9. (a) $f(-x) = \sqrt{1 + (-x)^2} = \sqrt{1 + x^2}$ (b) $f(1/x) = \sqrt{1 + (1/x)^2} = \sqrt{1 + x^2}/|x|$

     (c) $f(a + b) = \sqrt{1 + (a + b)^2} = \sqrt{a^2 + 2ab + b^2 + 1}$

10. (a) $f(-x) = -\dfrac{x}{|x^2 - 1|}$ (b) $f(\dfrac{1}{x}) = \dfrac{1}{x|\frac{1}{x^2} - 1|}$ (c) $f(a + b) = \dfrac{a + b}{|(a + b)^2 - 1|}$

11. (a) $f(a + h) = 2(a + h)^2 - 3(a + h) = 2a^2 + 4ah + 2h^2 - 3a - 3h$

     (b) $\dfrac{f(a + h) - f(a)}{h} = \dfrac{[2(a + h)^2 - 3(a + h)] - [2a^2 - 3a]}{h} = \dfrac{4ah + 2h^2 - 3h}{h} = 4a + 2h - 3$

12. (a) $f(a + h) = \dfrac{1}{a + h - 2}$

     (b) $\dfrac{f(a + h) - f(a)}{h} = \dfrac{\dfrac{1}{a + h - 2} - \dfrac{1}{a - 2}}{h} = \dfrac{-h}{h(a + h - 2)(a - 2)} = \dfrac{-1}{(a + h - 2)(a - 2)}$

13. $x = 1, 3$ **14.** $x = 0$ **15.** $x = -2$

16. $x = 5 \pm 2\sqrt{7}$ **17.** $x = -3, 3$ **18.** all $x > 0$

19. $\text{dom}(f) = (-\infty, \infty)$;   range $(f) = [0, \infty)$ **20.** $\text{dom}(g) = (-\infty, \infty)$;   range $(g) = [-1, \infty)$

21. $\text{dom}(f) = (-\infty, \infty)$;   range $(f) = (-\infty, \infty)$ **22.** $\text{dom}(g) = [0, \infty)$;   range $(g) = [5, \infty)$

23. $\text{dom}(f) = (-\infty, 0) \cup (0, \infty)$;   range $(f) = (0, \infty)$

24. $\text{dom}(g) = (-\infty, 0) \cup (0, \infty)$;   range $(g) = (-\infty, 0) \cup (0, \infty)$

25. $\text{dom}(f) = (-\infty, 1]$;   range $(f) = [0, \infty)$ **26.** $\text{dom}(g) = [3, \infty)$;   range $(g) = [0, \infty)$

27. dom $(f) = (-\infty, 7]$;   range $(f) = [-1, \infty)$

28. dom $(g) = [1, \infty)$;   range $(g) = [-1, \infty)$

29. dom $(f) = (-\infty, 2)$;   range $(f) = (0, \infty)$

30. dom $(g) = (-2, 2)$;   range $(g) = [\frac{1}{2}, \infty)$

31. horizontal line one unit above the x-axis.

32. horizontal line one unit below the x-axis.

33. line through the origin with slope 2.

34. line through $(0, 1)$ with slope 2.

35. line through $(0, 2)$ with slope $\frac{1}{2}$.

36. line through $(0, -3)$ with slope $-\frac{1}{2}$.

37. upper semicircle of radius 2 centered at the origin.

38. upper semicircle of radius 3 centered at the origin.

39. dom $(f) = (-\infty, \infty)$

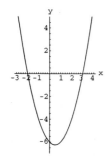

40. dom $(f) = (-\infty, \infty)$

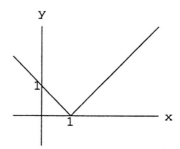

41. dom $(f) = (-\infty, 0) \cup (0, \infty)$; range $(f) = \{-1, 1\}$.

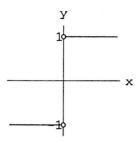

42. dom $(f) = (-\infty, \infty)$; range $f = (-\infty, \infty)$.

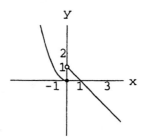

43. dom $(f) = [0, \infty)$; range $(f) = [1, \infty)$.

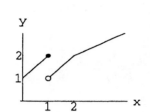

44. dom $(f) = (-\infty, 0) \cup (0, 2) \cup (2\, \infty)$. range $(f) = \{-1\} \cup (0, \infty)$.

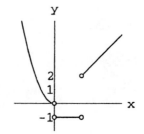

45. The curve is the graph of a function: domain $[-2, 2]$,   range $[-2, 2]$.

46. Not a function.

47. The curve is not the graph of a function; it fails the *vertical line test*.

48. Function;   domain: $(-\infty, \infty)$,   range: $(-1, 1)$

49. odd: $f(-x) = (-x)^3 = -x^3 = -f(x)$ **50.** even.

51. neither even nor odd: $g(-x) = -x(-x - 1) = x(x + 1);$ $g(-x) \neq g(x)$ and $g(-x) \neq -g(x)$

52. odd.   **53.** even.   **54.** odd.

55. odd   **56.** odd

57. (a)

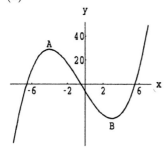

58. (a)

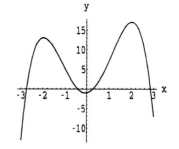

   (b) $-6.566, \ -0.493, \ 5.559$

   (c) $A(-4, 28.667), \quad B(3, -28.500)$

   (b) $-2.739, \ -0.427, \ 0.298, \ 2.868$

   (c) $A(-1.968, 13.016), \quad B(2.031, 17.015)$

59. $-5 \leq x \leq 8, \quad 0 \leq y \leq 100$

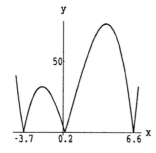

60. $2 \leq x \leq 10, 0 \leq y \leq 32$

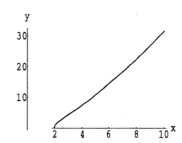

61. Range: $[-9, \infty)$.

   (a) $y = x^2 - 4x + 5 = x^2 - 4x + 4 - 9 = (x - 2)^2 - 9$. Therefore $y \geq -9$.

   (b) $x = \dfrac{4 \pm \sqrt{36 + 4y}}{2}$ which implies $y \geq -9$.

62. Range: $y \neq -2$.

(a) Divide $4 - x$ into $2x$. The result is: $y = -2 + \dfrac{8}{4 - x}$ which implies $y \neq -2$.

(b) $x = \dfrac{4y}{y + 2}$, $y \neq -2$.

63. $A = \dfrac{C^2}{4\pi}$, where C is the circumference; $\operatorname{dom}(A) = [0, \infty)$

64. $A = 4\pi r^2 \implies r = \sqrt{\dfrac{A}{4\pi}} \implies V = \dfrac{4}{3}\pi r^3 = \dfrac{4}{3}\pi \left(\dfrac{A}{4\pi}\right)^{3/2} = \dfrac{A^{3/2}}{3\sqrt{4\pi}}.$

65. $V = s^{3/2}$, where s is the area of a face; $\operatorname{dom} V = [0, \infty)$

66. $A = 6x^2 \implies V = x^3 = \left(\dfrac{A}{6}\right)^{3/2}.$

67. $S = 3d^2$, where d is the diagonal of a face; $\operatorname{dom}(S) = [0, \infty)$

68. $d = \sqrt{3}x \implies V = x^3 = \left(\dfrac{d}{\sqrt{3}}\right)^3 = \dfrac{d^3\sqrt{3}}{9}.$

69. $A = \dfrac{\sqrt{3}}{4}x^2$, where x is the length of a side; $\operatorname{dom}(A) = [0, \infty)$

70. $h = \sqrt{c^2 - x^2}$ so $V = \dfrac{1}{3}\pi r^2 h = \dfrac{1}{3}\pi x^2 \sqrt{c^2 - x^2}.$

71. Let y be the length of the rectangle. Then

$$x + 2y + \dfrac{\pi x}{2} = 15 \quad \text{and} \quad y = \dfrac{15}{2} - \dfrac{2 + \pi}{4}x, \qquad 0 \leq x \leq \dfrac{30}{2 + \pi}$$

Area: $A = xy + \tfrac{1}{2}\pi(x/2)^2 = \left(\dfrac{15}{2} - \dfrac{2 + \pi}{4}x\right)x + \dfrac{1}{8}\pi x^2 = \dfrac{15}{2}x - \dfrac{x^2}{2} - \dfrac{\pi}{8}x^2, \quad 0 < x < \dfrac{30}{2 + \pi}.$

72. $3x + 2y = 15 \implies y = \dfrac{1}{2}(15 - 3x).$ $\qquad A = xy + \dfrac{1}{2}x\left(\dfrac{\sqrt{3}}{2}x\right) = \dfrac{1}{2}x(15 - 3x) + \dfrac{\sqrt{3}}{4}x^2.$

73. The coordinates x and y are related by the equation $y = -\dfrac{b}{a}(x - a), \ 0 \leq x \leq a$.

The area A of the rectangle is given by $A = xy = x\left[-\dfrac{b}{a}(x - a)\right] = bx - \dfrac{b}{a}x^2, \ 0 \leq x \leq a.$

74. Let $x = a$ be the x-intercept. Then the line is $y = \dfrac{5}{2 - a}(x - a)$, with y-intercept $\dfrac{5a}{a - 2}$.

The area is $A = \dfrac{1}{2}xy = \dfrac{1}{2}a\dfrac{5a}{a - 2}$, or in terms of x, $A = \dfrac{5x^2}{2(x - 2)}.$

75. Let P be the perimeter of the square. Then the edge length of the square is $P/4$ and the area of the square is $A_s = (P/4)^2 = P^2/16$. The circumference of the circle is $28 - P$ which implies that the radius is $\frac{1}{2\pi}(28 - \pi)$. Thus, the area of the circle is $A_c = \pi \left[\frac{1}{2\pi}(28 - P)\right]^2 = \frac{1}{4\pi}(28 - P)^2$ and the total area is $A_s + A_c = \frac{P^2}{16} + \frac{1}{4\pi}(28 - P)^2$, $0 \le P \le 28$.

76. By similar triangles, $\frac{r}{h} = \frac{10}{20}$, so $r = \frac{1}{2}h$. Therefore $V = \frac{1}{3}\pi r^2 h = \frac{1}{3}\pi \left(\frac{h}{2}\right)^2 h = \frac{\pi}{12}h^3$.

77. Set length plus girth equal to 108. Then $l = 108 - 2\pi r$, and $V = (108 - 2\pi r)\pi r^2$.

SECTION 1.6

1. polynomial, degree 0

2. polynomial, degree 1

3. rational function

4. polynomial, degree 2

5. neither

6. polynomial, degree 4

7. neither

8. rational function.

9. neither

10. $h(x) = \dfrac{x - 4}{x^2 + 4}$, rational function

11. $\mathrm{dom}\,(f) = (-\infty, \infty)$

12. $\mathrm{dom}\,(f) = (-\infty, -1) \cup (-1, \infty)$

13. $\mathrm{dom}\,(f) = (-\infty, \infty)$

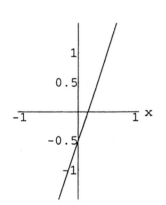

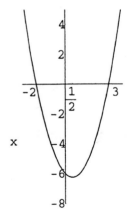

14. $\mathrm{dom}\,(f) = (-\infty, \infty)$

15. $\mathrm{dom}\,(f) = \{x : x \ne \pm 2\}$

16. $\mathrm{dom}\,(g) = (-\infty, 0) \cup (0, \infty)$

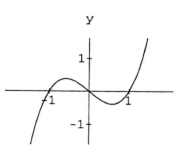

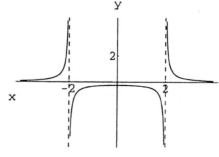

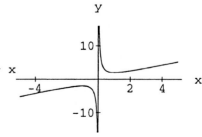

17. $225\left(\dfrac{\pi}{180}\right) = \dfrac{5\pi}{4}$

18. $-210° = -\dfrac{7\pi}{6}$ rads

19. $(-300)\left(\dfrac{\pi}{180}\right) = -\dfrac{5\pi}{3}$

20. $450° = \dfrac{5\pi}{2}$ rads

21. $15\left(\dfrac{\pi}{180}\right) = \dfrac{\pi}{12}$

22. $3° = \dfrac{\pi}{60}$ rads

23. $\left(-\dfrac{3\pi}{2}\right)\left(\dfrac{180}{\pi}\right) = -270°$

24. $225°$

25. $\left(\dfrac{5\pi}{3}\right)\left(\dfrac{180}{\pi}\right) = 300°$

26. $-330°$

27. $2\left(\dfrac{180}{\pi}\right) \cong 114.59°$

28. $-\dfrac{\sqrt{3}}{\pi}180°$

29. Let x be the arc subtended by an angle θ radians on a circle of radius r. By similarity, $\theta/1 = x/r$, which implies $x = r\theta$.

30. Let A be the area of the sector. By similarity, $\dfrac{A}{\pi r^2} = \dfrac{\theta}{2\pi}$, which implies $A = \frac{1}{2}r^2\theta$.

31. $\sin x = \frac{1}{2};\ \ x = \pi/6,\ \ 5\pi/6$

32. $2\pi/3,\ \ 4\pi/3$

33. $\tan(x/2) = 1;\ \ x = \pi/2$

34. $\pi/2,\ \ 3\pi/2$

35. $\cos x = \sqrt{2}/2;\ \ x = \pi/4,\ \ 7\pi/4$

36. $2\pi/3,\ \ 5\pi/6\ \ 5\pi/3\ \ 11\pi/6$

37. $\cos 2x = 0;\ \ x = \pi/4,\ \ 3\pi/4,\ \ 5\pi/4,\ \ 7\pi/4$

38. $\dfrac{2\pi}{3},\ \dfrac{5\pi}{3}$

39. $\sin 51° \cong 0.7771$

40. $\cos 17° \cong 0.9563$

41. $\sin(2.352) \cong 0.7101$

42. $\cos(-13.461) \cong 0.6258$

43. $\tan 72.4° \cong 3.1524$

44. $\cot(7.311) \cong 0.6035$

45. $\sin x = 0.5231;\ \ x = 0.5505,\ \ \pi - 0.5505$

46. $x = 2.5398,\ \ 2\pi - 2.5398$

47. $\tan x = 6.7192;\ \ x = 1.4231,\ \ \pi + 1.4231$

48. $x = 2.8263,\ \ \pi + 2.8263$

49. $\sec x = -4.4073;\ \ x = 1.7997,\ \ \pi + 1.7997$

50. $x = 0.0976,\ \ \pi - 0.0976$

51. The x coordinates of the points of intersection are: $x \cong 1.31,\ 1.83,\ 3.40,\ 3.93,\ 5.50,\ 6.02$

52. The x coordinate of the point of intersection is: $x \cong 1.45$

53. $\mathrm{dom}\,(f) = (-\infty, \infty);\ \mathrm{range}\,(f) = [0, 1]$

54. $\mathrm{dom}\,(g) = (-\infty, \infty);\ \mathrm{range}\,(g) = \{1\}$

55. $\mathrm{dom}\,(f) = (-\infty, \infty);\ \mathrm{range}\,(f) = [-2, 2]$

56. $\mathrm{dom}\,(F) = (-\infty, \infty);\ \mathrm{range}\,(F) = [0, 2]$

57. $\mathrm{dom}(f) = \left(k\pi - \dfrac{\pi}{2}, k\pi + \dfrac{\pi}{2}\right),\ \ k = 0,\ \pm 1,\ \pm 2,\ \ldots;\ \ \mathrm{range}\,(f) = [1, \infty)$

58. $\text{dom}(h) = (-\infty, \infty)$; $\text{range}(h) = [0, 1]$ **59.** period: $\dfrac{2\pi}{\pi} = 2$

60. period: $\dfrac{2\pi}{2} = \pi$ **61.** period: $\dfrac{2\pi}{1/3} = 6\pi$ **62.** period: $\dfrac{2\pi}{1/2} = 4\pi$

63.

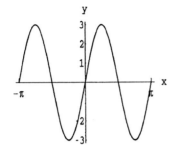

64.

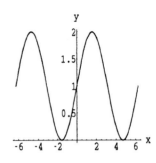

65.

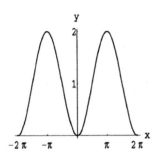

66.

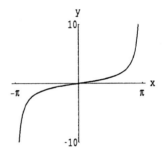

67.

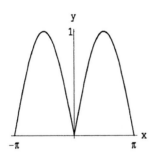

68.

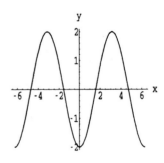

69. odd   **70.** odd   **71.** even

72. even   **73.** odd   **74.** even

75. Assume that $\theta_2 > \theta_1$. Let $m_1 = \tan\theta_1$, $m_2 = \tan\theta_2$. The angle α between l_1 and l_2 is the smaller of $\theta_2 - \theta_1$ and $180° - [\theta_2 - \theta_1]$. In the first case

$$\tan\alpha = \tan[\theta_2 - \theta_1] = \frac{\tan\theta_2 - \tan\theta_1}{1 + \tan\theta_2 \tan\theta_1} = \frac{m_2 - m_1}{1 + m_2 m_1} > 0$$

In the second case, $\tan \alpha = \tan[180° - (\theta_2 - \theta_1)] = -\tan(\theta_2 - \theta_1) = -\dfrac{m_2 - m_1}{1 + m_2 m_1} < 0$

Thus $\tan \alpha = \left| \dfrac{m_2 - m_1}{1 + m_2 m_1} \right|$

76. $(1,1);$ $\alpha \cong 39°$ $[m_1 = 4 = \tan \theta_1,\ \theta_1 \cong 76°;\quad m_2 = \frac{3}{4} = \tan \theta_2,\ \theta_2 \cong 37°]$

77. $\left(\frac{23}{37}, \frac{116}{37}\right);$ $\alpha \cong 73°$ $[m_1 = -3 = \tan \theta_1,\ \theta_1 \cong 108°;\quad m_2 = \frac{7}{10} = \tan \theta_2,\ \theta_2 \cong 35°]$

78. $\left(-\frac{2}{23}, \frac{38}{23}\right);$ $\alpha \cong 17°$ $[m_1 = 4 = \tan \theta_1,\ \theta_1 \cong 76°;\quad m_2 = -19 = \tan \theta_2,\ \theta_2 \cong 93°]$

79. $\left(-\frac{17}{13}, -\frac{2}{13}\right);$ $\alpha \cong 82°$ $[m_1 = \frac{5}{6} = \tan \theta_1,\ \theta_1 \cong 40°;\quad m_2 = -\frac{8}{5} = \tan \theta_2,\ \theta_2 \cong 122°]$

80. For each positive rational number p, $f(x + p) = f(x)$. There is no smallest such p.

81. By similar triangles, $\sin \theta = \dfrac{\sin \theta}{1} = \dfrac{\text{opp}}{\text{hyp}}$, and $\cos \theta = \dfrac{\cos \theta}{1} = \dfrac{\text{adj}}{\text{hyp}}$.

82. From the figure, area $A = \frac{1}{2} ah = \frac{1}{2} ab \sin C$.

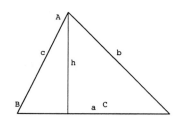

83. From the figure, $h = b \sin C = c \sin B$.

Therefore $\dfrac{\sin B}{b} = \dfrac{\sin C}{c}$

Similarly, you can show that $\dfrac{\sin A}{a} = \dfrac{\sin B}{b}$

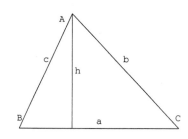

84. From the figure, $h = c \sin A, x = c \cos A,$ so

$a^2 = h^2 + (b - x)^2$

$\quad = c^2 \sin^2 A + b^2 - 2bc \cos A + c^2 \cos^2 A$

$\quad = b^2 + c^2 - 2bc \cos A$

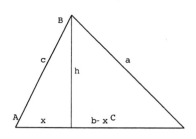

85. By the law of cosines

$1^2 + 1^2 - 2(1)(1) \cos(\alpha - \beta) = (\cos \beta - \cos \alpha)^2 + (\sin \beta - \sin \alpha)^2$

$= \cos^2 \beta - 2 \cos \beta \cos \alpha + \cos^2 \alpha + \sin^2 \beta - 2 \sin \beta \sin \alpha + \sin^2 \alpha$

$= 2 - 2 \cos \beta \cos \alpha - 2 \sin \beta \sin \alpha.$

The result follows.

86. Replace β by $-\beta$ in the identity of Exercise 85:

87. From the identities $\sin\left(\frac{1}{2}\pi + \theta\right) = \cos\theta$ and $\cos\left(\frac{1}{2}\pi + \theta\right) = -\sin\theta$ we get

$$\sin\left(\frac{1}{2}\pi - \theta\right) = \sin\left[\frac{1}{2}\pi + (-\theta)\right] = \cos(-\theta) = \cos\theta$$

and

$$\cos\left(\frac{1}{2}\pi - \theta\right) = \cos\left[\frac{1}{2}\pi + (-\theta)\right] = -\sin(-\theta) = \sin\theta.$$

88. By the Hint,

$$\sin(\alpha + \beta) = \cos\left[\left(\frac{1}{2}\pi - \alpha\right) - \beta\right] = \cos\left(\frac{1}{2}\pi - \alpha\right)\cos\beta + \sin\left(\frac{1}{2}\pi - \alpha\right)\sin\beta$$

$$= \sin\alpha\cos\beta + \cos\alpha\sin\beta.$$

89. Replace β by $-\beta$ in the identity of Exercise 88.

91. (a)

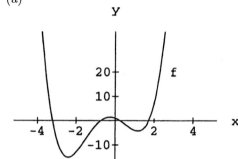 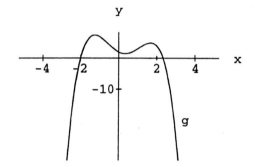

(c)

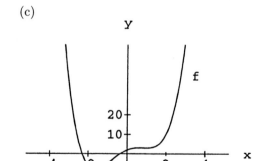 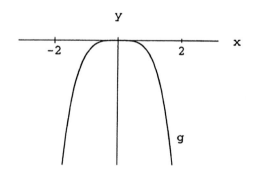

92. (a)

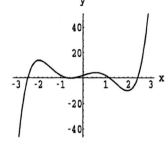 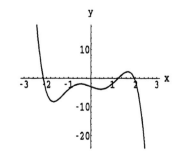

93. (a)

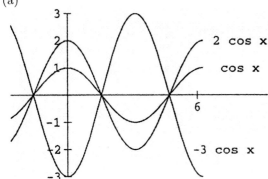

(b)

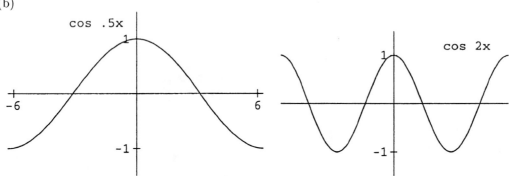

(c) *A* changes the amplitude; *B* stretches or compresses horizontally

94. (b)

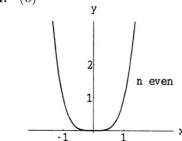
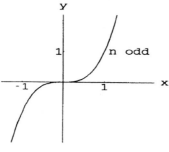

(c) $f_k(x) \geq f_{k+1}(x)$ on $[0, 1]$; $f_{k+1}(x) > f_k(x)$ on $(1, \infty)$

SECTION 1.7

1. $(f + g)(2) = f(2) + g(2) = 3 + \frac{9}{2} = \frac{15}{2}$

2. $(f - g)(-1) = 6$

3. $(f \cdot g)(-2) = f(-2)g(-2) = 15 \cdot \frac{7}{2} = \frac{105}{2}$

4. $\dfrac{f}{g}(1) = 0$

5. $(2f - 3g)(\frac{1}{2}) = 2f(\frac{1}{2}) - 3g(\frac{1}{2}) = 2 \cdot 0 - 3 \cdot \frac{9}{4} = -\frac{27}{4}$

6. $\left(\dfrac{f+2g}{f}\right)(-1) = 1$

7. $(f \circ g)(1) = f[g(1)] = f(2) = 3$ **8.** $(g \circ f)(1) = g(0)$, undefined.

9. $(f+g)(x) = f(x) + g(x) = x - 1;$ $\mathrm{dom}\,(f+g) = (-\infty, \infty)$
$(f-g)(x) = f(x) - g(x) = 3x - 5;$ $\mathrm{dom}\,(f-g) = (-\infty, \infty)$
$(f \cdot g)(x) = f(x)g(x) = -2x^2 + 7x - 6;$ $\mathrm{dom}\,(f \cdot g) = (-\infty, \infty)$
$(f/g)(x) = \dfrac{2x-3}{2-x};$ $\mathrm{dom}\,(f/g) = \{x : \ x \neq 2\}$

10. $(f+g)(x) = f(x) + g(x) = x^2 + x - 1 + \dfrac{1}{x};$ $\mathrm{dom}\,(f+g) = (-\infty, 0) \cup (0, \infty)$

$(f-g)(x) = f(x) - g(x) = x^2 - x - 1 - \dfrac{1}{x};$ $\mathrm{dom}\,(f-g) = (-\infty, 0) \cup (0, \infty)$

$(f \cdot g)(x) = f(x)g(x) = \dfrac{x^4 - 1}{x};$ $\mathrm{dom}\,(f \cdot g) = (-\infty, 0) \cup (0, \infty)$

$(f/g)(x) = \dfrac{x^3 - x}{x^2 + 1};$ $\mathrm{dom}\,(f/g) = (-\infty, 0) \cup (0, \infty)$ [$g(0)$ is undefined.]

11. $(f+g)(x) = x + \sqrt{x-1} - \sqrt{x+1};$ $\mathrm{dom}\,(f+g) = [1, \infty)$
$(f-g)(x) = \sqrt{x-1} + \sqrt{x+1} - x;$ $\mathrm{dom}\,(f-g) = [1, \infty)$
$(f \cdot g)(x) = \sqrt{x-1}\,\left(x - \sqrt{x+1}\right) = x\sqrt{x-1} - \sqrt{x^2-1};$ $\mathrm{dom}\,(f \cdot g) = [1, \infty)$

$(f/g)(x) = \dfrac{\sqrt{x-1}}{x - \sqrt{x+1}};$ $\mathrm{dom}\,(f/g) = \{x : x \geq 1 \ \text{ and } \ x \neq \tfrac{1}{2}(1 + \sqrt{5})\}$

12. $(f+g)(x) = \sin^2 x + \cos 2x;$ $\mathrm{dom}\,(f+g) = (-\infty, \infty)$
$(f-g)(x) = \sin^2 x - \cos 2x;$ $\mathrm{dom}\,(f-g) = (-\infty, \infty)$
$(f \cdot g)(x) = \sin^2 x \cos 2x;$ $\mathrm{dom}\,(f \cdot g) = (-\infty, \infty)$

$(f/g)(x) = \dfrac{\sin^2 x}{\cos 2x};$ $\mathrm{dom}\,(f/g) = \{x : x \neq \dfrac{2n+1}{4}\pi, \ n = 0, \pm 1, \pm 2, \cdots\}$

13. (a) $(6f + 3g)(x) = 6(x + 1/\sqrt{x}) + 3(\sqrt{x} - 2/\sqrt{x}) = 6x + 3\sqrt{x}; \quad x > 0$

(b) $(f - g)(x) = x + 1/\sqrt{x} - (\sqrt{x} - 2/\sqrt{x}) = x + 3/\sqrt{x} - \sqrt{x}; \quad x > 0$

(c) $(f/g)(x) = \dfrac{x\sqrt{x} + 1}{x - 2}; \quad x > 0, \ x \neq 2$

14.

$$(f+g)(x) = \begin{cases} 1 - x, \ x \leq 1 \\ 2x - 1, \ 1 < x < 2 \\ 2x - 2, \ x \geq 2 \end{cases} \qquad (f-g)(x) = \begin{cases} 1 - x, \ x \leq 1 \\ 2x - 1, \ 1 < x < 2 \\ 2x, \quad x \geq 2 \end{cases}$$

$$(f \cdot g)(x) = \begin{cases} 0, \ x < 2 \\ 1 - 2x, \ x \geq 2 \end{cases}$$

15.

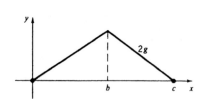

16.

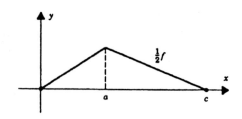

17.

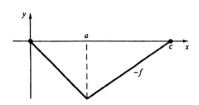

18. $y = 0$ on $(0, c]$

19.

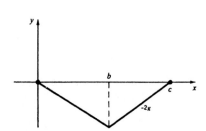

20.

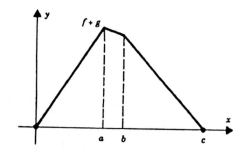

21.

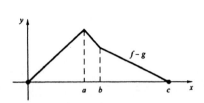

22.

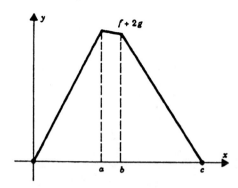

23. $(f \circ g)(x) = 2x^2 + 5$; $\text{dom}\,(f \circ g) = (-\infty, \infty)$ **24.** $(f \circ g)(x) = (2x + 5)^2$; $\text{dom}\,(f \circ g) = (-\infty, \infty)$

25. $(f \circ g)(x) = \sqrt{x^2 + 5}$; $\text{dom}\,(f \circ g) = (-\infty, \infty)$ **26.** $(f \circ g)(x) = x + \sqrt{x}$; $\text{dom}\,(f \circ g) = [0, \infty)$

27. $(f \circ g)(x) = \dfrac{x}{x-2};$ $\text{dom}\,(f \circ g) = \{x : \ x \neq 0, 2\}$

28. $(f \circ g)(x) = \dfrac{1}{x^2 - 1};$ $\text{dom}\,(f \circ g) = \{x \neq \pm 1\}$

29. $(f \circ g)(x) = \sqrt{1 - \cos^2 2x} = |\sin 2x|;$ $\text{dom}\,(f \circ g) = (-\infty, \infty)$

30. $(f \circ g)(x) = \sqrt{1 - 2\cos x};$ $\text{dom}\,(f \circ g) = [0, \pi/3] \cup [5\pi/3, 2\pi]$

31. $(f \circ g \circ h) = 4\,[g(h(x))] = 4\,[h(x) - 1] = 4(x^2 - 1);$ $\text{dom}\,(f \circ g \circ h) = (-\infty, \infty)$

32. $(f \circ g \circ h)(x) = f(g(h(x))) = f(g(x^2)) = f(4x^2) = 4x^2 - 1;$ $\text{dom}\,(f \circ g \circ h) = (-\infty, \infty)$

33. $(f \circ g \circ h)\dfrac{1}{g(h(x))} = \dfrac{1}{1/[2h(x) + 1]} = 2h(x) + 1 = 2x^2 + 1;$ $\text{dom}\,(f \circ g \circ h) = (-\infty, \infty)$

34. $(f \circ g \circ h)(x) = f(g(h(x))) = f(g(x^2)) = f\left(\dfrac{1}{2x^2 + 1}\right) = \dfrac{1/(2x^2 + 1) + 1}{1/(2x^2 + 1)} = 1 + (2x^2 + 1) = 2x^2 + 2$

35. Take $f(x) = \dfrac{1}{x}$ since $\dfrac{1 + x^4}{1 + x^2} = F(x) = f(g(x)) = f\left(\dfrac{1 + x^2}{1 + x^4}\right).$

36. Take $f(x) = ax + b$ since $f(g(x)) = f(x^2) = ax^2 + b = F(x)$

37. Take $f(x) = 2\sin x$ since $2\sin 3x = F(x) = f(g(x)) = f(3x).$

38. Take $f(x) = \sqrt{a^2 - x},$ since $f(g(x)) = f(-x^2) = \sqrt{a^2 - (-x^2)} = \sqrt{a^2 + x^2} = F(x).$

39. Take $g(x) = \left(1 - \dfrac{1}{x^4}\right)^{2/3}$ since $\left(1 - \dfrac{1}{x^4}\right)^2 = F(x) = f(g(x)) = [\,g(x)\,]^3.$

40. Take $g(x) = a^2 x^2 \,(x \neq 0),$ since $a^2 x^2 + \dfrac{1}{a^2 x^2} = F(x) = f(g(x)) = g(x) + \dfrac{1}{g(x)}.$

41. Take $g(x) = 2x^3 - 1$ (or $-(2x^3 - 1)$) since $(2x^3 - 1)^2 + 1 = F(x) = f(g(x)) = [\,g(x)\,]^2 + 1.$

42. Take $g(x) = \dfrac{1}{x}$ since $\sin\dfrac{1}{x} = F(x) = f(g(x)) = \sin(g(x)).$

43. $(f \circ g)(x) = f(g(x)) = \sqrt{g(x)} = \sqrt{x^2} = |x|;$
$(g \circ f)(x) = g(f(x)) = [f(x)]^2 = [\sqrt{x}\,]^2 = x, \quad x \geq 0$

44. $(f \circ g)(x) = f(g(x)) = 3g(x) + 1 = 3x^2 + 1, \quad (g \circ f)(x) = g(f(x)) = (f(x))^2 = (3x + 1)^2$

45. $(f \circ g)(x) = f(g(x)) = 1 - \sin^2 x = \cos^2 x;$ $(g \circ f)(x) = g(f(x)) = \sin f(x) = \sin(1 - x^2).$

46. $(f \circ g)(x) = f(g(x)) = (x-1) + 1 = x; \quad (g \circ f)(x) = g(f(x)) = \sqrt[3]{(x^3+1)-1} = x.$

47. $(f+g)(x) = f(x) + g(x) = f(x) + c; \quad quadg(x) = c.$

48. $(f \circ g)(x) = f(g(x)) = g(x) + c$ implies $f(x) = x + c.$

49. $(fg)(x) = f(x)g(x) = c\,f(x)$ implies $g(x) = c.$

50. $(f \circ g)(x) = f(g(x)) = c\,g(x)$ implies $f(x) = cx.$

51. (a) The graph of g is the graph of f shifted 3 units to the right. dom $(g) = [3, a+3]$, range $(g) = [0, b]$.

(b) The graph of g is the graph of f shifted 4 units to the left and scaled vertically by a factor of 3. dom $(g) = [-4, a-4]$, range $(g) = [0, 3b]$.

(c) The graph of g is the graph of f scaled horizontally by a factor of 2. dom $(g) = [0, a/2]$, range $(g) = [0, b]$.

(d) The graph of g is the graph of f scaled horizontally by a factor of $\frac{1}{2}$. dom $(g) = [0, 2a]$, range $(g) = [0, b]$.

52. even: $(fg)(-x) = f(-x)g(-x) = (-f(x))(-g(x)) = f(x)g(x) = (fg)(x).$

53. fg is even since $(fg)(-x) = f(-x)g(-x) = f(x)g(x) = (fg)(x).$

54. odd: $(fg)(-x) = f(-x)g(-x) = (f(x))(-g(x)) = -f(x)g(x) = -(fg)(x).$

55. (a) If f is even, then

$$f(x) = \begin{cases} -x, & -1 \le x < 0 \\ 1, & x < -1. \end{cases}$$

(b) If f is odd, then

$$f(x) = \begin{cases} x, & -1 \le x < 0 \\ -1, & x < -1. \end{cases}$$

56. (a) $f(x) = x^2 + x,$ (b) $f(x) = -x^2 - x$

57. $g(-x) = f(-x) + f[-(-x)] = f(-x) + f(x) = g(x)$

58. $h(-x) = f(-x) - f[-(-x)] = f(-x) - f(x) = -[f(x) - f(-x)] = -h(x)$

59. $f(x) = \underbrace{\frac{1}{2}[f(x) + f(-x)]}_{\text{even}} + \underbrace{\frac{1}{2}[f(x) - f(-x)]}_{\text{odd}}$

60.

	f_1	f_2	f_3	f_4	f_5	f_6
f_1	f_1	f_2	f_3	f_4	f_5	f_6
f_2	f_2	f_1	f_4	f_3	f_6	f_5
f_3	f_3	f_5	f_1	f_6	f_2	f_4
f_4	f_4	f_6	f_2	f_5	f_1	f_3
f_5	f_5	f_3	f_6	f_1	f_4	f_2
f_6	f_6	f_4	f_5	f_2	f_3	f_1

61. (a) $(f \circ g)(x) = \dfrac{5x^2 + 16x - 16}{(2 - x)^2}$ (b) $(g \circ k)(x) = x$ (c) $(f \circ k \circ g)(x) = x^2 - 4$

62. (a) $(g \circ f)(x) = \dfrac{3(x^2 - 4)}{6 - x^2}$ (b) $(k \circ g)(x) = x$ (c) $(g \circ f \circ k)(x) = -\dfrac{18(2x + 3)}{x^2 + 18x + 27}$

63. (a) For fixed a, varying b varies the y-coordinate of the vertex of the parabola.

(b) For fixed b, varying a varies the x-coordinate of the vertex of the parabola

(c) The graph of $-F$ is the reflection of the graph of F in the x-axis.

64. $a = \dfrac{1}{4}, \quad b = -\dfrac{49}{16}$

65. (a) For $c > 0$, the graph of cf is the graph of f scaled vertically by the factor c; for $c < 0$, the graph of cf is the graph of f scaled vertically by the factor $|c|$ and then reflected in the x-axis.

(b) For $c > 1$, the graph of $f(cx)$ is the graph of f compressed horizontally; for $0 < c < 1$, the graph of $f(cx)$ is the graph of f stretched horizontally; for $-1 < c < 0$, the graph of $f(cx)$ is the graph of f stretched horizontally and reflected in the y-axis; for $c < -1$, the graph of $f(cx)$ is the graph of f compressed horizontally and reflected in the y-axis.

66. (a) The graph of $f(x - c)$ is the graph of $f(x)$ shifted c units to the right if $c > 0$ and $|c|$ units to the left if $c < 0$.

(b) a changes the amplitude, b changes the period, c shifts the graph right or left $|c/b|$ units.

SECTION 1.8

1. Let S be the set of integers for which the statement is true. Since $2(1) \leq 2^1$, S contains 1. Assume now that $k \in S$. This tells us that $2k \leq 2^k$, and thus

$$2(k + 1) = 2k + 2 \leq 2^k + 2 \leq 2^k + 2^k = 2(2^k) = 2^{k+1}.$$

$$(k \geq 1)$$

This places $k + 1$ in S.

We have shown that

$$1 \in S \quad \text{and that} \quad k \in S \quad \text{implies} \quad k+1 \in S.$$

It follows that S contains all the positive integers.

2. Use $\quad 1 + 2(n+1) = 1 + 2n + 2 \leq 3^n + 2 < 3^n + 3^n = 2 \cdot 3^n < 3^{n+1}.$

3. Let S be the set of integers for which the statement is true. Since $(1)(2) = 2$ is divisible by $2, 1 \in S$. Assume now that $k \in S$. This tells us that $k(k+1)$ is divisible by 2 and therefore

$$(k+1)(k+2) = k(k+1) + 2(k+1)$$

is also divisible by 2. This places $k+1 \in S$.

We have shown that

$$1 \in S \text{ and that} \quad k \in S \text{ implies} \quad k+1 \in S.$$

It follows that S contains all the positive integers.

4. Use $\quad 1 + 3 + 5 + \cdots + (2(n+1) - 1) = n^2 + 2n + 1 = (n+1)^2$

5. Use
$$\begin{aligned}
1^2 + 2^2 + \cdots + k^2 + (k+1)^2 &= \tfrac{1}{6}k(k+1)(2k+1) + (k+1)^2 \\
&= \tfrac{1}{6}(k+1)[k(2k+1) + 6(k+1)] \\
&= \tfrac{1}{6}(k+1)(2k^2 + 7k + 6) \\
&= \tfrac{1}{6}(k+1)(k+2)(2k+3) \\
&= \tfrac{1}{6}(k+1)[(k+1) + 1][2(k+1) + 1].
\end{aligned}$$

6. Use
$$\begin{aligned}
1^3 + 2^3 + \cdots + n^3 + (n+1)^3 &= (1 + 2 + \cdots + n)^2 + (n+1)^3 \\
&= \left[\frac{n(n+1)}{2}\right]^2 + (n+1)^3 \quad \text{(by example 1)} \\
&= \frac{n^4 + 6n^3 + 13n^2 + 12n + 4}{4} \\
&= \left[\frac{(n+1)(n+2)}{2}\right]^2 \\
&= [1 + 2 + \cdots + n + (n+1)]^2
\end{aligned}$$

7. By Exercise 6 and Example 1

$$1^3 + 2^3 + \cdots + (n-1)^3 = [\tfrac{1}{2}(n-1)n]^2 = \tfrac{1}{4}(n-1)^2 n^2 < \tfrac{1}{4}n^4$$

and

$$1^3 + 2^3 + \cdots + n^3 = [\tfrac{1}{2}n(n+1)]^2 = \tfrac{1}{4}n^2(n+1)^2 > \tfrac{1}{4}n^4.$$

8. By Exercise 5,

$$1^2 + 2^2 + \cdots + (n-1)^2 = \tfrac{1}{6}(n-1)n(2n-1) < \tfrac{1}{3}n^3$$

$$1^2 + 2^2 + \cdots + n^2 = \tfrac{1}{6}n(n+1)(2n+1) > \tfrac{1}{3}n^3$$

9. Use

$$\frac{1}{\sqrt{1}} + \frac{1}{\sqrt{2}} + \frac{1}{\sqrt{3}} + \cdots + \frac{1}{\sqrt{n}} + \frac{1}{\sqrt{n+1}}$$

$$> \sqrt{n} + \frac{1}{\sqrt{n+1}+\sqrt{n}} \left(\frac{\sqrt{n+1}-\sqrt{n}}{\sqrt{n+1}-\sqrt{n}} \right) = \sqrt{n+1}.$$

10. Use $\dfrac{1}{1\cdot 2} + \dfrac{1}{2\cdot 3} + \dfrac{1}{3\cdot 4} + \cdots + \dfrac{1}{(n+1)(n+2)} = \dfrac{n}{n+1} + \dfrac{1}{(n+1)(n+2)} = \dfrac{n(n+2)+1}{(n+1)(n+2)} = \dfrac{n+1}{n+2}$

11. Let S be the set of integers for which the statement is true. Since

$$3^{2(1)+1} + 2^{1+2} = 27 + 8 = 35$$

is divisible by 7, we see that $1 \in S$.

Assume now that $k \in S$. This tells us that

$$3^{2k+1} + 2^{k+2} \text{ is divisible by 7.}$$

It follows that

$$3^{2(k+1)+1} + 2^{(k+1)+2} = 3^2 \cdot 3^{2k+1} + 2 \cdot 2^{k+2}$$

$$= 9 \cdot 3^{2k+1} + 2 \cdot 2^{k+2}$$

$$= 7 \cdot 3^{2k+1} + 2(3^{2k+1} + 2^{k+2})$$

is also divisible by 7. This places $k + 1 \in S$.

We have shown that

$$1 \in S \quad \text{and that} \quad k \in S \quad \text{implies} \quad k+1 \in S.$$

It follows that S contains all the positive integers.

12. $n \geq 1$: True for $n = 1$. For the induction step, use

$$9^{n+1} - 8(n+1) - 1 = 9 \cdot 9^n - 8n - 9 - 64n + 64n = 9(9^n - 8n - 1) + 64n$$

13. For all positive integers $n \geq 2$,

$$\left(1 - \frac{1}{2}\right)\left(1 - \frac{1}{3}\right) \cdots \left(1 - \frac{1}{n}\right) = \frac{1}{n}.$$

To see this, let S be the set of integers n for which the formula holds. Since $1 - \frac{1}{2} = \frac{1}{2}$, $2 \in S$. Suppose now that $k \in S$. This tells us that

$$\left(1 - \frac{1}{2}\right)\left(1 - \frac{1}{3}\right) \cdots \left(1 - \frac{1}{k}\right) = \frac{1}{k}$$

and therefore that

$$\left(1 - \frac{1}{2}\right)\left(1 - \frac{1}{3}\right)\cdots\left(1 - \frac{1}{k}\right)\left(1 - \frac{1}{k+1}\right) = \frac{1}{k}\left(1 - \frac{1}{k+1}\right) = \frac{1}{k}\left(\frac{k}{k+1}\right) = \frac{1}{k+1}.$$

This places $k + 1 \in S$ and verifies the formula for $n \geq 2$.

14. The product is $\dfrac{n+1}{2n}$; use $\dfrac{n+1}{2n}\left(1 - \dfrac{1}{(n+1)^2}\right) = \dfrac{n+1}{2n}\left(\dfrac{n^2+2n}{(n+1)^2}\right) = \dfrac{n+2}{2(n+1)}$

15. From the figure, observe that adding a vertex V_{N+1} to an N-sided polygon increases the number of diagonals by $(N-2) + 1 = N - 1$. Then use the identity
$\frac{1}{2}N(N-3) + (N-1) = \frac{1}{2}(N+1)(N+1-3).$

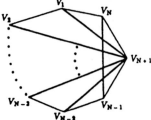

16. From the figure for Exercise 15, observe that adding a vertex (V_{N+1}) to an N-sided polygon increases the angle sum by $180°$.

17. To go from k to $k+1$, take $A = \{a_1, \ldots, a_{k+1}\}$ and $B = \{a_1, \ldots, a_k\}$. Assume that B has 2^k subsets: $B_1, B_2, \ldots B_{2^k}$. The subsets of A are then $B_1, B_2, \ldots, B_{2^k}$ together with

$$B_1 \cup \{a_{k+1}\}, \; B_2 \cup \{a_{k+1}\}, \ldots, B_{2^k} \cup \{a_{k+1}\}.$$

This gives $2(2^k) = 2^{k+1}$ subsets for A.

18. Assuming that we can construct a line segment of length $\sqrt{k}$, construct a right triangle with side lengths 1 and $\sqrt{k}$. Then the hypotenuse is a line segment of length $\sqrt{k+1}$.

19. $n = 41$

CHAPTER 1. REVIEW EXERCISES

1. rational

2. rational

3. irrational

4. rational

5. bounded below by 1

6. bounded above by 1

7. bounded; lower bound -5, upper bound 1

8. bounded; lower bound -1, upper bound $\frac{1}{4}$

9. $2x^2 + x - 1 = (2x - 1)(x + 1)$; $x = \frac{1}{2}, -1$

10. no real roots

11. $x^2 - 10x + 25 = (x-5)^2$; $x = 5$

12. $9x^3 - x = x(3x+1)(3x-1)$; $x = 0, \frac{1}{3}, -\frac{1}{3}$

13. $5x - 2 < 0$

$5x < 2$

$x < \frac{2}{5}$

Ans: $(-\infty, \frac{2}{5})$

14. $3x + 5 < \frac{1}{2}(4-x)$

$3x + 5 < 2 - \frac{x}{2}$

$\frac{7}{2}x < -3$

$x < -\frac{-6}{7}$

Ans: $(-\infty, -\frac{6}{7})$

15. $x^2 - x - 6 \geq 0$

$(x-3)(x+2) \geq 0$

Ans: $(-\infty, -2] \cup [3, \infty)$

16. $x(x^2 - 3x + 2) \leq 0$

$x(x-1)(x-2) \leq 0$

Ans: $(-\infty, 0] \cup [1, 2]$

17. $\dfrac{x+1}{(x+2)(x-2)} > 0$

Ans: $(-2, -1) \cup (2, \infty)$

18. $\dfrac{x^2 - 4x + 4}{x^2 - 2x - 3} \leq 0$

$\dfrac{(x-2)^2}{(x-3)(x+1)} \leq 0$

Ans: $(-1, 3)$

19. $|x - 2| < 1$

$-1 < x - 2 < 1$

Ans: $(1, 3)$

20. $|3x - 2| \geq 4$

$3x - 2 \geq 4$ or $3x - 2 \leq -4$

Ans: $(-\infty, -\frac{2}{3}) \cup [2, \infty)$

21. $\left|\dfrac{2}{x+4}\right| > 2$

$\dfrac{2}{x+4} > 2$ or $\dfrac{2}{x+4} < -2$

If $\dfrac{2}{x+4} > 2$

$x + 4 > 0$ and $2 > 2x + 8$

$-4 < x < -3$

If $\dfrac{2}{x+4} < -2$

$x + 4 < 0$ and $2 > -2x - 8$

$-5 < x < -4$

Ans: $(-5, -4) \cup (-4, -3)$

22. $\left|\dfrac{5}{x+1}\right| < 1$

$-1 < \dfrac{5}{x+1} < 1$

If $0 < \dfrac{5}{x+1} < 1$

$x > 4$

If $0 > \dfrac{5}{x+1} > -1$

$x < -6$

Ans: $(-\infty, -6) \cup (4, \infty)$

23. $d(P, Q) = \sqrt{(1-2)^2 + (4 - (-3))^2} = 5\sqrt{2}$;   midpoint: $\left(\dfrac{2+1}{2}, \dfrac{4-3}{2}\right) = \left(\dfrac{3}{2}, \dfrac{1}{2}\right)$

24. $d(P, Q) = \sqrt{(-1 - (-3)^2 + (6 - (-4))^2} = 2\sqrt{26}$;   midpoint: $\left(\dfrac{-3-1}{2}, \dfrac{-4+6}{2}\right) = (-2, 1)$

25. $x = 2$

26. $y = -3$

27. The line $l:\ 2x - 3y = 6$ has slope $m = 2/3$. Therefore, an equation for the line through $(2, -3)$ perpendicular to l is: $y + 3 = -\frac{3}{2}(x-2)$ or $3x + 2y = 0$

28. The line $l: 3x + 4y = 12$ has slope $m = -3/4$. Therefore, an equation for the line through $(2, -3)$ parallel to l is: $y + 3 = -\frac{3}{4}(x - 2)$ or $3x + 4y = -6$

29.

$$\begin{matrix} x - 2y = -4 \\ 3x + 4y = 3 \end{matrix} \quad \Rightarrow \quad 5x = -5 \quad \Rightarrow \quad x = -1; \quad (-1, \ \tfrac{3}{2}).$$

30.

$$\begin{matrix} 4x - y = -2 \\ 3x + 2y = 0 \end{matrix} \quad \Rightarrow \quad 11x = -4 \quad \Rightarrow \quad x = \tfrac{-4}{11}; \quad (\tfrac{-4}{11}, \tfrac{6}{11}).$$

31. Solve the equations simultaneously:

$$2x^2 = 8x - 6$$

$$2x^2 - 8x + 6 = 0$$

$$2(x - 1)(x - 3) = 0$$

the line and the parabola intersect at $(1, 2)$ and $(3, 18)$.

32. The line tangent to the circle at the point $(2, 1)$ is perpendicular to the radius at that point. The center of the circle is at $(-1, 3)$. The slope of the line through $(-1, 3)$ and $(2, 1)$ is $-2/3$. Therefore an equation for the line tangent to the circle at $(2, 1)$ is

$$y - 1 = \tfrac{3}{2}(x - 2) \quad \text{or} \quad 3x - 2y = 4.$$

33. domain: $(-\infty, \infty)$; range: $(-\infty, 4]$

34. domain: $(-\infty, \infty)$; range: $(-\infty, \infty)$

35. domain: $[4, \infty)$; range: $[0, \infty)$

36. domain: $[-\tfrac{1}{2}, \tfrac{1}{2}]$; range; $[0, \tfrac{1}{2}]$

37. domain: $(-\infty, \infty)$; range: $[1, \infty)$

38. domain: $(-\infty, \infty)$; range: $[0, \infty)$

39. domain: $(-\infty, \infty)$; range: $[0, \infty)$

40. domain: $(-\infty, \infty)$; range: $(-\infty, \infty)$

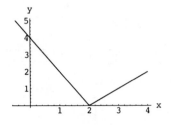

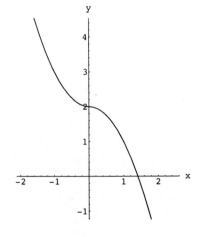

41. $x = \frac{7}{6}\pi, \frac{11}{6}\pi$

42. $x = \frac{1}{3}\pi, \frac{2}{3}\pi, \frac{4\pi}{3}, \frac{5\pi}{3}$

43. $x = \frac{3\pi}{2}$

44. $x = 0, \frac{1}{3}\pi, \frac{2}{3}\pi, \pi, \frac{4}{3}\pi, \frac{5}{3}\pi, 2\pi$

45.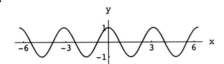

46.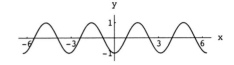

47.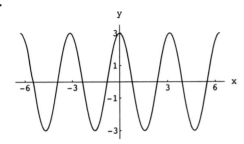

48.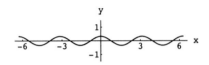

49. $(f+g)(x) = (3x+2) + (x^2-1) = x^2 + 3x + 1, \quad \mathrm{dom}\,(f+g) = (-\infty, \infty).$
$(f-g)(x) = (3x+2) - (x^2-1) = 3 + 3x - x^2, \quad \mathrm{dom}\,(f-g) = (-\infty, \infty).$
$(f \cdot g)(x) = (3x+2)(x^2-1) = 3x^3 + 2x^2 - 3x - 2, \quad \mathrm{dom}\,(f \cdot g) = (-\infty, \infty).$
$\left(\dfrac{f}{g}\right)(x) = \dfrac{3x+2}{x^2-1}, \quad \mathrm{dom}\,(f/g) = (-\infty, -1) \cup (-1, 1) \cup (1, \infty).$

50. $(f+g)(x) = x^2 - 4 + x + 1/x = x^2 + x + 1/x - 4, \quad \mathrm{dom}\,(f+g) = (-\infty, 0) \cup (0, \infty).$
$(f-g)(x) = x^2 - x - 1/x - 4, \quad \mathrm{dom}\,(f-g) = (-\infty, 0) \cup (0, \infty).$
$(f \cdot g)(x) = (x^2-4)(x + 1/x) = x^3 - 3x - 4/x, \quad \mathrm{dom}\,(f \cdot g) = (-\infty, 0) \cup (0, \infty).$
$\left(\dfrac{f}{g}\right)(x) = \dfrac{x^2-4}{x+1/x} = \dfrac{x^3-4x}{x^2+1}, \quad \mathrm{dom}\,(f/g) = (-\infty, 0) \cup (0, \infty).$

51. $(f+g)(x) = \cos^2 x + \sin 2x, \quad \mathrm{dom}\,(f+g) = [0, 2\pi].$
$(f-g)(x) = \cos^2 x - \sin 2x, \quad \mathrm{dom}\,(f-g) = [0, 2\pi].$
$(f \cdot g)(x) = \cos^2 x(\sin 2x) = 2\cos^3 x \sin x, \quad \mathrm{dom}\,(f \cdot g) = [0, 2\pi].$
$\left(\dfrac{f}{g}\right)(x) = \dfrac{\cos^2 x}{\sin 2x} = \frac{1}{2}\cot x, \quad \mathrm{dom}(f/g): x \in (0, 2\pi), x \neq \frac{1}{2}\pi, \pi, \frac{3}{2}\pi.$

52. $(f \circ g)(x) = (x+1)^2 - 2(x+1) = x^2 - 1, \quad \mathrm{dom}\,(f \circ g) = (-\infty, \infty).$
$(g \circ f)(x) = x^2 - 2x + 1 = (x-1)^2, \quad \mathrm{dom}\,(f \circ g) = (-\infty, \infty).$

53. $(f \circ g)(x) = \sqrt{(x^2-5)+1} = \sqrt{x^2-4}, \quad \mathrm{dom}\,(f \circ g) = (-\infty, -2] \cup [2, \infty).$
$(g \circ f)(x) = (\sqrt{x+1})^2 - 5 = x - 4, \quad \mathrm{dom}\,(g \circ f) = [-1, \infty).$

54. $(f \circ g)(x) = \sqrt{1 - \sin^2 2x} = |\cos 2x|, \quad \mathrm{dom}\,(f \circ g) = (-\infty, \infty).$
$(g \circ f)(x) = \sin 2\sqrt{1-x^2}, \quad \mathrm{dom}\,(g \circ f) = [-1, 1].$

55. (a) $y = kx$.

(b) If $b = ka$, then $\alpha b = \alpha ka = k(\alpha b)$. Hence, Q is a point on l.

(c) If $\alpha > 0$, P, Q are on the same side of the origin; if $\alpha < 0$, P, Q are on opposite sides of the origin.

56. (a) Set $\alpha = -\frac{a}{2}$. Then $a = -2\alpha$ and the quadratic equation can be written as

$$x^2 - 2\alpha x + b = 0 \quad \text{or} \quad (x - \alpha)^2 - (\alpha^2 - b) = 0.$$

if $\alpha^2 - b > 0$, set $\beta^2 = (\alpha^2 - b)$ and we have $(x - \alpha)^2 - \beta^2 = 0$;

if $\alpha^2 - b = 0$, we have $(x - \alpha)^2 = 0$;

if $\alpha^2 - b < 0$, set $-\beta^2 = (\alpha^2 - b)$ and we have $(x - \alpha)^2 + \beta^2 = 0$.

(b) $x = \alpha + \beta$ or $x = \alpha - \beta$.

(c) $x = \alpha$.

(d) $(x - \alpha)^2 + \beta^2 > 0$ for all x.

57. Since $|a| = |a - b + b| \leq |a - b| + |b|$ by the given inequality, we have $|a| - |b| \leq |a - b|$.

58. (a) $P = \frac{1}{2}\pi D$

(b) $A = \frac{1}{8}\pi D^2$

CHAPTER 2

SECTION 2.1

1. (a) 2   (b) −1   (c) does not exist   (d) −3

2. (a) −4   (b) −4   (c) −4   (d) 2

3. (a) does not exist   (b) −3   (c) does not exist   (d) −3

4. (a) 1   (b) does not exist   (c) does not exist   (d) 1

5. (a) does not exist   (b) does not exist   (c) does not exist   (d) 1

6. (a) 1   (b) −2   (c) does not exist   (d) −2

7. (a) 2   (b) 2   (c) 2   (d) −1

8. (a) 2   (b) 2   (c) 2   (d) 2

9. (a) 0   (b) 0   (c) 0   (d) 0

10. (a) does not exist   (b) does not exist   (c) does not exist   (d) 1

11. $c = 0, 6$ 12. $c = 3$ 13. −1   14. −3

15. 12   16. 5   17. 1   18. does not exist

19. $\frac{3}{2}$ 20. does not exist   21. does not exist   22. does not exist

23. $\displaystyle \lim_{x \to 3} \frac{2x - 6}{x - 3} = \lim_{x \to 3} 2 = 2$

24. $\displaystyle \lim_{x \to 3} \frac{x^2 - 6x + 9}{x - 3} = \lim_{x \to 3} \frac{(x - 3)^2}{x - 3} = \lim_{x \to 3}(x - 3) = 0$

25. $\displaystyle \lim_{x \to 3} \frac{x - 3}{x^2 - 6x + 9} = \lim_{x \to 3} \frac{x - 3}{(x - 3)^2} = \lim_{x \to 3} \frac{1}{x - 3};$ does not exist

26. $\displaystyle \lim_{x \to 2} \frac{x^2 - 3x + 2}{x - 2} = \lim_{x \to 2} \frac{(x - 1)(x - 2)}{x - 2} = \lim_{x \to 2}(x - 1) = 1$

27. $\displaystyle \lim_{x \to 2} \frac{x - 2}{x^2 - 3x + 2} = \lim_{x \to 2} \frac{x - 2}{(x - 1)(x - 2)} = \lim_{x \to 2} \frac{1}{x - 1} = 1$

28. does not exist   29. does not exist

30. $\displaystyle \lim_{x \to 1} \left(x + \frac{1}{x} \right) = 2$ 31. $\displaystyle \lim_{x \to 0} \frac{2x - 5x^2}{x} = \lim_{x \to 0}(2 - 5x) = 2$

32. $\lim\limits_{x\to 3}\dfrac{x-3}{6-2x}=\lim\limits_{x\to 3}\dfrac{x-3}{2(3-x)}=\lim\limits_{x\to 3}\left(-\dfrac{1}{2}\right)=-\dfrac{1}{2}$

33. $\lim\limits_{x\to 1}\dfrac{x^2-1}{x-1}=\lim\limits_{x\to 1}\dfrac{(x-1)(x+1)}{x-1}=\lim\limits_{x\to 1}(x+1)=2$

34. $\lim\limits_{x\to 1}\dfrac{x^3-1}{x-1}=\lim\limits_{x\to 1}\dfrac{(x-1)(x^2+x+1)}{(x-1)}=\lim\limits_{x\to 1}x^2+x+1=3$

35. 0	**36.** does not exist	**37.** 1
38. 3	**39.** 16	**40.** 0
41. does not exist	**42.** 2	**43.** 4
44. 0	**45.** does not exist	**46.** 2

47.

$$\lim\limits_{x\to 1}\frac{\sqrt{x^2+1}-\sqrt{2}}{x-1}=\lim\limits_{x\to 1}\frac{(\sqrt{x^2+1}-\sqrt{2})(\sqrt{x^2+1}+\sqrt{2})}{(x-1)(\sqrt{x^2+1}+\sqrt{2})}$$

$$=\lim\limits_{x\to 1}\frac{x^2-1}{(x-1)(\sqrt{x^2+1}+\sqrt{2})}=\lim\limits_{x\to 1}\frac{x+1}{\sqrt{x^2+1}+\sqrt{2}}=\frac{2}{2\sqrt{2}}=\frac{1}{\sqrt{2}}$$

48.

$$\lim\limits_{x\to 5}\frac{\sqrt{x^2+5}-\sqrt{30}}{x-5}=\lim\limits_{x\to 5}\frac{(\sqrt{x^2+5}-\sqrt{30})(\sqrt{x^2+5}+\sqrt{30})}{(x-5)(\sqrt{x^2+5}+\sqrt{30})}$$

$$=\lim\limits_{x\to 5}\frac{x^2-25}{(x-5)(\sqrt{x^2+5}+\sqrt{30})}=\lim\limits_{x\to 5}\frac{x+5}{\sqrt{x^2+5}+\sqrt{30}}=\frac{10}{2\sqrt{30}}=\frac{5}{\sqrt{30}}$$

49.

$$\lim\limits_{x\to 1}\frac{x^2-1}{\sqrt{2x+2}-2}=\lim\limits_{x\to 1}\frac{x^2-1}{\sqrt{2x+2}-2}\frac{\sqrt{2x+2}+2}{\sqrt{2x+2}+2}$$

$$=\lim\limits_{x\to 1}\frac{(x-1)(x+1)\left(\sqrt{2x+2}+2\right)}{2x+2-4}=\lim\limits_{x\to 1}\frac{(x-1)(x+1)\left(\sqrt{2x+2}+2\right)}{2(x-1)}$$

$$=\lim\limits_{x\to 1}\frac{(x+1)\left(\sqrt{2x+2}+2\right)}{2}=\frac{2\cdot 4}{2}=4$$

50. 0	**51.** 2	**52.** 0.167
53. $\frac{3}{2}$	**54.** $\frac{1}{12}$	**55.** (i) 5 (ii) does not exist
56. (i) 0 (ii) $\frac{2}{3}$	**57.** (i) $-\frac{5}{4}$ (ii) 0	**58.** (i) $-\frac{17}{32}$ (ii) $-\frac{1}{2}$
59. $c=-1$	**60.** $c=\frac{5}{2}$	**61.** $c=-2$

62. $c = 3$

63. $\lim\limits_{x \to 0} f(x) = 1$;

$\lim\limits_{x \to 0} g(x) = 0$

64. $\lim\limits_{x \to 0} f(x) = 1$;

$\lim\limits_{x \to 0} g(x) = 0$

SECTION 2.2

1. $\dfrac{1}{2}$

2. $\lim\limits_{x \to 0} \dfrac{x^2(1+x)}{2x} = \lim\limits_{x \to 0} \frac{1}{2} x(1+x) = 0$

3. $\lim\limits_{x \to 0} \dfrac{x(1+x)}{2x^2} = \lim\limits_{x \to 0} \dfrac{1+x}{2x}$;   does not exist

4. $\dfrac{4}{3}$

5. $\lim\limits_{x \to 1} \dfrac{x^4 - 1}{x - 1} = \lim\limits_{x \to 1} (x^3 + x^2 + x + 1) = 4$

6. does not exist

7. does not exist

8. $\lim\limits_{x \to 1} \dfrac{x^2 - 1}{x^2 - 2x + 1} = \lim\limits_{x \to 1} \dfrac{(x-1)(x+1)}{(x-1)^2} = \lim\limits_{x \to 1} \dfrac{x+1}{x-1}$;   does not exist

9. -1

10. does not exist

11. does not exist

12. -1

13. 0

14. 0

15. $\lim\limits_{x \to 2^+} f(x) = \lim\limits_{x \to 2^+} (x^2 - x) = 2$

16. $\lim\limits_{x \to 1^-} f(x) = \lim\limits_{x \to 1^-} 1 = 1$

17. 1

18. 9

19. 1

20. 10

21. δ_1

22. ϵ_2 and ϵ_3

23. $\delta = \frac{1}{2}\epsilon = .05$

24. $\delta = \frac{1}{5}\epsilon = 0.1$

25. $2\epsilon = .02$

26. $\delta = 5\epsilon = 0.5$

27. $\delta = 1.75$

28. $\delta = .22474$

29. for $\epsilon = 0.5$ take $\delta = 0.24$;   for $\epsilon = 0.25$ take $\delta = 0.1$

30. for $\epsilon = 0.5$ take $\delta = 0.06$;   for $\epsilon = 0.25$ take $\delta = 0.03$

31. for $\epsilon = 0.5$ take $\delta = 0.75$;   for $\epsilon = 0.25$ take $\delta = 0.43$

32. for $\epsilon = 0.5$ take $\delta = 0.125$;   for $\epsilon = 0.1$ take $\delta = 0.028$

33. for $\epsilon = 0.25$ take $\delta = 0.23$;   for $\epsilon = 0.1$ take $\delta = 0.14$

34. for $\epsilon = 0.5$ take $\delta = 0.25$;   for $\epsilon = 0.1$ take $\delta = 0.05$

35. Since

$$|(2x - 5) - 3| = |2x - 8| = 2|x - 4|,$$

we can take $\delta = \frac{1}{2}\epsilon$:

$$\text{if} \quad 0 < |x - 4| < \tfrac{1}{2}\epsilon \quad \text{then,} \quad |(2x - 5) - 3| = 2|x - 4| < \epsilon.$$

36. Since

$$|(3x - 1) - 5| = |3x - 6| = 3|x - 2|,$$

we can take $\delta = \frac{1}{3}\epsilon$:

$$\text{if} \quad 0 < |x - 2| < \tfrac{1}{3}\epsilon \quad \text{then,} \quad |(3x - 1) - 5| = 3|x - 2| < \epsilon.$$

37. Since

$$|(6x - 7) - 11| = |6x - 18| = 6|x - 3|,$$

we can take $\delta = \frac{1}{6}\epsilon$:

$$\text{if} \quad 0 < |x - 3| < \tfrac{1}{6}\epsilon \quad \text{then} \quad |(6x - 7) - 11| = 6|x - 3| < \epsilon.$$

38. Since

$$|(2 - 5x) - 2| = |-5x| = 5|x|,$$

we can take $\delta = \frac{1}{5}\epsilon$:

$$\text{if} \quad 0 < |x| < \tfrac{1}{5}\epsilon \quad \text{then,} \quad |(2 - 5x) - 2| = 5|x| < \epsilon.$$

39. Since

$$\big||1 - 3x| - 5\big| = \big||3x - 1| - 5\big| \le |3x - 6| = 3|x - 2|,$$

we can take $\delta = \frac{1}{3}\epsilon$:

$$\text{if} \quad 0 < |x - 2| < \tfrac{1}{3}\epsilon \quad \text{then} \quad \big||1 - 3x| - 5\big| \le 3|x - 2| < \epsilon.$$

40. Take $\delta = \epsilon$: $\quad$ if $\quad 0 < |x - 2| < \epsilon \quad$ then $\quad \big||x - 2| - 0\big| = |x - 2| < \epsilon$

41. Statements (b), (e), (g), and (i) are necessarily true.

42. Suppose $A \neq B$ and take $\epsilon = \frac{1}{2}|A - B|$. Then $|A - B| < \epsilon = \frac{1}{2}|A - B|$ which is impossible.

43. (i) $\displaystyle\lim_{x \to 3} \frac{1}{x - 1} = \frac{1}{2}$ $\qquad\qquad$ (ii) $\displaystyle\lim_{h \to 0} \frac{1}{(3 + h) - 1} = \frac{1}{2}$

$\quad\;$ (iii) $\displaystyle\lim_{x \to 3} \left(\frac{1}{x - 1} - \frac{1}{2} \right) = 0$ $\qquad$ (iv) $\displaystyle\lim_{x \to 3} \left| \frac{1}{x - 1} - \frac{1}{2} \right| = 0$

44. (i) $\displaystyle\lim_{x \to 1} \frac{x}{x^2 + 2} = \frac{1}{3}$ $\qquad\qquad$ (ii) $\displaystyle\lim_{h \to 0} \frac{1 + h}{(1 + h)^2 + 2} = \frac{1}{3}$

$\quad\;$ (iii) $\displaystyle\lim_{x \to 1} \left(\frac{x}{x^2 + 2} - \frac{1}{3} \right) = 0$ $\qquad$ (iv) $\displaystyle\lim_{x \to 1} \left| \frac{x}{x^2 + 2} - \frac{1}{3} \right| = 0$

45. By (2.2.6) parts (i) and (iv) with $L = 0$

46. (a) Suppose $\lim\limits_{x \to c} f(x) = L$, and let $\epsilon > 0$. Then there exists $\delta > 0$ such that

$$\text{if} \quad 0 < |x - c| < \delta, \quad \text{then} \quad |f(x) - L| < \epsilon.$$

But then we also have

$$\text{if} \quad 0 < |x - c| < \delta, \quad \text{then} \quad ||f(x)| - |L|| \le |f(x) - L| < \epsilon,$$

and therefore, $\lim\limits_{x \to c} |f(x)| = |L|$

(b) (i) Set $f(x) = -2$ for all x, and let $c = 0$. Then

$$\lim\limits_{x \to 0} |f(x)| = 2 = |2| \quad \text{but} \quad \lim\limits_{x \to 0} f(x) = -2 \neq 2.$$

(ii) Set

$$f(x) = \begin{cases} -1 & \text{if } x < 0 \\ 1 & \text{if } x > 0, \end{cases} \quad c = 0. \text{ Then}$$

$$\lim\limits_{x \to 0} |f(x)| = 1 \quad \text{but} \quad \lim\limits_{x \to 0} f(x) \quad \text{does not exist.}$$

47. Let $\epsilon > 0$. If

$$\lim\limits_{x \to c} f(x) = L,$$

then there must exist $\delta > 0$ such that

$(*)$ $\qquad\qquad$ if $\quad 0 < |x - c| < \delta \quad$ then $\quad |f(x) - L| < \epsilon.$

Suppose now that

$$0 < |h| < \delta.$$

Then

$$0 < |(c + h) - c| < \delta$$

and thus by $(*)$

$$|f(c + h) - L| < \epsilon.$$

This proves that

$$\text{if} \quad \lim\limits_{x \to c} f(x) = L \quad \text{then} \quad \lim\limits_{h \to 0} f(c + h) = L.$$

If, on the other hand,

$$\lim\limits_{h \to 0} f(c + h) = L,$$

then there must exist $\delta > 0$ such that

$(**)$ $\qquad\qquad$ if $\quad 0 < |h| < \delta \quad$ then $\quad |f(c + h) - L| < \epsilon.$

Suppose now that

$$0 < |x - c| < \delta.$$

Then by (∗∗)

$$|f(c + (x - c)) - L| < \epsilon.$$

More simply stated,

$$|f(x) - L| < \epsilon.$$

This proves that

$$\text{if} \quad \lim_{h \to 0} f(c + h) = L \quad \text{then} \quad \lim_{x \to c} f(x) = L.$$

48.   See comment after (2.2.9).

49.   (a)   Set $\delta = \epsilon\sqrt{c}$. By the hint,

$$\text{if} \quad 0 < |x - c| < \epsilon\sqrt{c} \quad \text{then} \quad |\sqrt{x} - \sqrt{c}| < \frac{1}{\sqrt{c}}|x - c| < \epsilon.$$

(b)   Set $\delta = \epsilon^2$. If $0 < x < \epsilon^2$, then $|\sqrt{x} - 0| = \sqrt{x} < \epsilon$.

50.   Take $\delta = $ minimum of 1 and $\epsilon/5$. If $0 < |x - 2| < \delta$, then $|x - 2| < \epsilon/5$, $|x + 2| < 5$ and
Therefore $|x^2 - 4| = |x - 2||x + 2| < (\epsilon/5)(5) = \epsilon$.

$$|x^3 - 1| = |x^2 + x + 1||x - 1| < 7|x - 1| < 7(\epsilon/7) = \epsilon.$$

51.   Take $\delta = $ minimum of 1 and $\epsilon/7$. If $0 < |x - 1| < \delta$, then $0 < x < 2$
and $|x - 1| < \epsilon/7$. Therefore

$$|x^3 - 1| = |x^2 + x + 1||x - 1| < 7|x - 1| < 7(\epsilon/7) = \epsilon.$$

52.   Take $\delta = 2\epsilon$. If $0 < |x - 3| < \delta$, then

$$\left|\sqrt{x + 1} - 2\right| = \frac{|x - 3|}{\sqrt{x + 1} + 2} < \frac{1}{2}|x - 3| < \epsilon$$

53.   Set $\delta = \epsilon^2$.   If $3 - \epsilon^2 < x < 3$, then $-\epsilon^2 < x - 3$, $0 < 3 - x < \epsilon^2$
and therefore $|\sqrt{3 - x} - 0| < \epsilon$.

54.   Take $\delta = \epsilon$. Suppose $0 < |x - 0| < \delta$, that is, suppose $0 < |x| < \delta$. Then

for x rational $|g(x) - 0| = |x| < \delta = \epsilon$ and for x irrational $|g(x) - 0| = 0 < \epsilon$

Thus, if $0 < |x - 0| < \delta$, then $|g(x) - 0| < \epsilon$, that is $\lim_{x \to 0} g(x) = 0$.

55.   Suppose, on the contrary, that $\lim_{x \to c} f(x) = L$ for some particular c. Taking $\epsilon = \frac{1}{2}$, there must exist
$\delta > 0$ such that

$$\text{if} \quad 0 < |x - c| < \delta, \quad \text{then} \quad |f(x) - L| < \tfrac{1}{2}.$$

Let x_1 be a rational number satisfying $0 < |x_1 - c| < \delta$ and x_2 an irrational number satisfying $0 < |x_2 - c| < \delta$. (That such numbers exist follows from the fact that every interval contains both rational and irrational numbers.) Now $f(x_1) = L$ and $f(x_2) = 0$. Thus we must have both

$$|1 - L| < \tfrac{1}{2} \quad \text{and} \quad |0 - L| < \tfrac{1}{2}.$$

From the first inequality we conclude that $L > \tfrac{1}{2}$. From the second, we conclude that $L < \tfrac{1}{2}$. Clearly no such number L exists.

56.   We begin by assuming that $\lim\limits_{x \to c^-} f(x) = L$ and showing that

$$\lim_{h \to 0} f(c - |h|) = L.$$

Let $\epsilon > 0$. Since $\lim\limits_{x \to c^-} f(x) = L$, there exists $\delta > 0$ such that

(*) $\qquad\qquad\qquad\quad$ if $\;\; c - \delta < x < c \;\;$ then $\;\; |f(x) - L| < \epsilon.$

Suppose now that $0 < |h| < \delta$. Then $c - \delta < c - |h| < c$ and, by (*),

$$|f(c - |h|) - L| < \epsilon.$$

Thus $\;\;\lim\limits_{h \to 0} f(c - |h|) = L.$

Conversely we now assume that $\lim\limits_{h \to 0} f(c - |h|) = L$. Then for $\epsilon > 0$ there exists $\delta > 0$ such that

(**) $\qquad\qquad\qquad\quad$ if $\;\; 0 < |h| < \delta \;\;$ then $\;\; |f(c - |h|) - L| < \epsilon.$

Suppose now that $c - \delta < x < c$. Then $0 < c - x < \delta$ so that, by (**),

$$|f(c - (c - x)) - L| = |f(x) - L| < \epsilon.$$

Thus $\;\;\lim\limits_{x \to c^-} f(x) = L.$

57.   We begin by assuming that $\lim\limits_{x \to c^+} f(x) = L$ and showing that

$$\lim_{h \to 0} f(c + |h|) = L.$$

Let $\epsilon > 0$. Since $\lim\limits_{x \to c^+} f(x) = L$, there exists $\delta > 0$ such that

(*) $\qquad\qquad\qquad\quad$ if $\;\; c < x < c + \delta \;\;$ then $\;\; |f(x) - L| < \epsilon.$

Suppose now that $0 < |h| < \delta$. Then $c < c + |h| < c + \delta$ and, by (*),

$$|f(c + |h|) - L| < \epsilon.$$

Thus $\;\;\lim\limits_{h \to 0} f(c + |h|) = L.$

Conversely we now assume that $\lim\limits_{h \to 0} f(c + |h|) = L$. Then for $\epsilon > 0$ there exists $\delta > 0$ such that

(**) $\qquad\qquad\qquad\quad$ if $\;\; 0 < |h| < \delta \;\;$ then $\;\; |f(c + |h|) - L| < \epsilon.$

Suppose now that $c < x < c + \delta$. Then $0 < x - c < \delta$ so that, by (**),

$$|f(c + (x - c)) - L| = |f(x) - L| < \epsilon.$$

Thus $\;\;\lim\limits_{x \to c^+} f(x) = L.$

58. Suppose that $\lim_{x \to c} f(x) = L$ and let $\epsilon > 0$. Then there exists $\delta > 0$ such that

$$\text{if} \quad 0 < |x - c| < \delta, \qquad \text{then} \qquad |f(x) - L| < \epsilon \quad \Longrightarrow \quad \lim_{x \to c}[f(x) - L] = 0.$$

Now suppose that $\lim_{x \to c}[f(x) - L] = 0$ and let $\epsilon > 0$. Then there exists $\delta > 0$ such that

$$\text{if} \quad 0 < |x - c| < \delta, \qquad \text{then} \qquad |f(x) - L - 0| < \epsilon \quad \Longrightarrow \quad \lim_{x \to c}|f(x) - L| < \epsilon$$

which implies $\lim_{x \to c} f(x) = L$

59. (a) Let $\epsilon = L$. Since $\lim_{x \to c} f(x) = L$, there exists $\delta > 0$ such that if $0 < |x - c| < \delta$ then

$$L - f(x) \le |L - f(x)| = |f(x) - L| < L$$

Therefore, $f(x) > L - L = 0$ for all $x \in (c - \delta, c + \delta)$; take $\gamma = \delta$.

(b) Let $\epsilon = -L$ and repeat the argument in part (a).

60. Counterexample: Set

$$f(x) = \begin{cases} x & \text{if } x \neq -1 \\ 1 & \text{if } x = -1. \end{cases}$$

Then $f(-1) = 1 > 0$ and $\lim_{x \to -1} f(x) = \lim_{x \to -1} x = -1$. By Exercise 51 (b), $f(x) < 0$ for all $x \neq -1$ in an interval of the form $(-1 - \gamma, -1 + \gamma)$, $\gamma > 0$.

61. (a) Let $\lim_{x \to c} f(x) = L$ and $\lim_{x \to c} g(x) = M$, and let $\epsilon > 0$. There exist positive numbers δ_1 and δ_2 such that

$$|f(x) - L| < \epsilon/2 \qquad \text{if} \qquad 0 < |x - c| < \delta_1$$

and

$$|g(x) - M| < \epsilon/2 \qquad \text{if} \qquad 0 < |x - c| < \delta_2$$

Let $\delta = \min(\delta_1, \delta_2)$. Then

$$M - L = M - g(x) + g(x) - f(x) + f(x) - L \ge [M - g(x)] + [f(x) - L]$$

$$\ge -\epsilon/2 - \epsilon/2 = -\epsilon$$

for all x such that $0 < |x - c| < \delta$. Since ϵ is arbitrary, it follows that $M \ge L$.

(b) No. For example, if $f(x) = x^2$ and $g(x) = |x|$ on $(-1, 1)$, then $f(x) < g(x)$ on $(-1, 1)$ except at $x = 0$, but $\lim_{x \to 0} x^2 = \lim_{x \to 0} |x| = 0$.

62. Let $\epsilon = 1$. Since $\lim_{x \to c} f(x) = L$, there exists a $\delta > 0$ such that if $0 < |x - c| < \delta$, then $|f(x) - L| < \epsilon = 1$. Thus,

$$-1 < f(x) - L < 1 \qquad \text{or} \qquad L - 1 < f(x) < L + 1$$

Take $B = \max\{|L - 1|, |L + 1|\}$. Then $|f(x)| < B$ for all x such that $0 < |x - c| < \delta$.

SECTION 2.3

1. (a) 3   (b) 4   (c) -2 (d) 0   (e) does not exist   (f) $\frac{1}{3}$

2. (a) 5   (b) -8 (c) does not exist   (d) 0   (e) $4/5$ (f) 9

3. $\displaystyle\lim_{x\to 4}\left(\frac{1}{x}-\frac{1}{4}\right)\left(\frac{1}{x-4}\right)=\lim_{x\to 4}\left(\frac{4-x}{4x}\right)\left(\frac{1}{x-4}\right)=\lim_{x\to 4}\frac{-1}{4x}=-\frac{1}{16};$ Theorem 2.3.2 does not apply
   since $\displaystyle\lim_{x\to 4}\frac{1}{x-4}$ does not exist.

4. Theorem 2.3.10 does not apply since $\displaystyle\lim_{x\to 3}(x^2+x-12)=0.$
   $$\lim_{x\to 3}\frac{x^2+x-12}{x-3}=\lim_{x\to 3}\frac{(x+4)(x-3)}{x-3}=\lim_{x\to 3}(x+4)=7$$

5. 3   6. 49   7. -3

8. 9   9. 5   10. 2

11. does not exist   12. $-\frac{23}{20}$ 13. -1

14. 0   15. does not exist

16. $\displaystyle\lim_{h\to 0}h\left(1-\frac{1}{h}\right)=\lim_{h\to 0}(h-1)=-1$ 17. $\displaystyle\lim_{h\to 0}h\left(1+\frac{1}{h}\right)=\lim_{h\to 0}(h+1)=1$

18. $\displaystyle\lim_{x\to 2}\frac{x-2}{x^2-4}=\lim_{x\to 2}\frac{1}{x+2}=\frac{1}{4}$ 19. $\displaystyle\lim_{x\to 2}\frac{x^2-4}{x-2}=\lim_{x\to 2}\frac{x+2}{1}=4$

20. $\displaystyle\lim_{x\to -2}\frac{(x^2-x-6)^2}{x+2}=\lim_{x\to -2}\frac{(x-3)^2(x+2)^2}{x+2}=\lim_{x\to -2}(x+3)^2(x+2)=0$

21. $\displaystyle\lim_{x\to 4}\frac{\sqrt{x}-2}{x-4}=\lim_{x\to 4}\frac{\sqrt{x}-2}{x-4}\cdot\frac{\sqrt{x}+2}{\sqrt{x}+2}=\lim_{x\to 4}\frac{x-4}{(x-4)(\sqrt{x}+2)}=\frac{1}{4}$

22. $\displaystyle\lim_{x\to 1}\frac{x-1}{\sqrt{x}-1}=\lim_{x\to 1}\frac{(\sqrt{x}-1)(\sqrt{x}+1)}{\sqrt{x}-1}=\lim_{x\to 1}(\sqrt{x}+1)=2$

23. $\displaystyle\lim_{x\to 1}\frac{x^2-x-6}{(x+2)^2}=\lim_{x\to 1}\frac{(x+2)(x-3)}{(x+2)^2}=\lim_{x\to 1}\frac{x-3}{x+2}=-\frac{2}{3}$

24. $\displaystyle\lim_{x\to -2}\frac{(x^2-x-6)}{(x+2)^2}=\lim_{x\to -2}\frac{(x-3)(x+2)}{(x+2)^2}=\lim_{x\to -2}\frac{x-3}{x+2};$ does not exist

25. $\displaystyle\lim_{h\to 0}\frac{1-1/h^2}{1-1/h}=\lim_{h\to 0}\frac{h^2-1}{h^2-h}=\lim_{h\to 0}\frac{(h+1)(h-1)}{h(h-1)}=\lim_{h\to 0}\frac{h+1}{h};$ does not exist

26. $\displaystyle\lim_{h\to 0}\frac{1-1/h^2}{1+1/h^2}=\lim_{h\to 0}\frac{h^2-1}{h^2+1}=-1$

27. $\lim\limits_{h \to 0} \dfrac{1 - 1/h}{1 + 1/h} = \lim\limits_{h \to 0} \dfrac{h - 1}{h + 1} = -1$

28. $\lim\limits_{h \to 0} \dfrac{1 + 1/h}{1 + 1/h^2} = \lim\limits_{h \to 0} \dfrac{h^2 + h}{h^2 + 1} = 0$

29. $\lim\limits_{t \to -1} \dfrac{t^2 + 6t + 5}{t^2 + 3t + 2} = \lim\limits_{t \to -1} \dfrac{(t + 1)(t + 5)}{(t + 1)(t + 2)} = \lim\limits_{t \to -1} \dfrac{t + 5}{t + 2} = 4$

30. $\lim\limits_{x \to 2^+} \dfrac{\sqrt{x^2 - 4}}{x - 2} = \lim\limits_{x \to 2^+} \dfrac{\sqrt{x + 2}\,\sqrt{x - 2}}{x - 2} = \lim\limits_{x \to 2^+} \dfrac{\sqrt{x + 2}}{\sqrt{x - 2}};$ does not exist

31. $\lim\limits_{t \to 0} \dfrac{t + a/t}{t + b/t} = \lim\limits_{t \to 0} \dfrac{t^2 + a}{t^2 + b} = \dfrac{a}{b}$

32. $\lim\limits_{x \to 1} \dfrac{x^2 - 1}{x^3 - 1} = \lim\limits_{x \to 1} \dfrac{(x - 1)(x + 1)}{(x - 1)(x^2 + x + 1)} = \lim\limits_{x \to 1} \dfrac{x + 1}{x^2 + x + 1} = \dfrac{2}{3}$

33. $\lim\limits_{x \to 1} \dfrac{x^5 - 1}{x^4 - 1} = \lim\limits_{x \to 1} \dfrac{(x - 1)(x^4 + x^3 + x^2 + x + 1)}{(x - 1)(x^3 + x^2 + x + 1)} = \lim\limits_{x \to 1} \dfrac{x^4 + x^3 + x^2 + x + 1}{x^3 + x^2 + x + 1} = \dfrac{5}{4}$

34. $\lim\limits_{h \to 0} h^2 \left(1 + \dfrac{1}{h}\right) = \lim\limits_{h \to 0} (h^2 + h) = 0$

35. $\lim\limits_{h \to 0} h \left(1 + \dfrac{1}{h^2}\right) = \lim\limits_{h \to 0} \dfrac{h^2 + 1}{h};$ does not exist

36. $\lim\limits_{x \to -4} \left(\dfrac{3x}{x + 4} + \dfrac{8}{x + 4}\right) = \lim\limits_{x \to -4} \dfrac{3x + 8}{x + 4};$ does not exist

37. $\lim\limits_{x \to -4} \left(\dfrac{2x}{x + 4} + \dfrac{8}{x + 4}\right) = \lim\limits_{x \to -4} \dfrac{2x + 8}{x + 4} = \lim\limits_{x \to -4} 2 = 2$

38. $\lim\limits_{x \to -4} \left(\dfrac{2x}{x + 4} - \dfrac{8}{x + 4}\right) = \lim\limits_{x \to -4} \dfrac{2x - 8}{x + 4};$ does not exist

39. (a) $\lim\limits_{x \to 4} \left(\dfrac{1}{x} - \dfrac{1}{4}\right) = \lim\limits_{x \to 4} \dfrac{4 - x}{4x} = 0$

   (b) $\lim\limits_{x \to 4} \left[\left(\dfrac{1}{x} - \dfrac{1}{4}\right)\left(\dfrac{1}{x - 4}\right)\right] = \lim\limits_{x \to 4} \left[\left(\dfrac{4 - x}{4x}\right)\left(\dfrac{1}{x - 4}\right)\right] = \lim\limits_{x \to 4} \left(-\dfrac{1}{4x}\right) = -\dfrac{1}{16}$

   (c) $\lim\limits_{x \to 4} \left[\left(\dfrac{1}{x} - \dfrac{1}{4}\right)(x - 2)\right] = \lim\limits_{x \to 4} \dfrac{(4 - x)(x - 2)}{4x} = 0$

   (d) $\lim\limits_{x \to 4} \left[\left(\dfrac{1}{x} - \dfrac{1}{4}\right)\left(\dfrac{1}{x - 4}\right)^2\right] = \lim\limits_{x \to 4} \dfrac{4 - x}{4x(x - 4)^2} = \lim\limits_{x \to 4} \dfrac{1}{4x(4 - x)};$ does not exist

40. (a) $\lim\limits_{x \to 3} \dfrac{x^2 + x + 12}{x - 3};$ does not exist

   (b) $\lim\limits_{x \to 3} \dfrac{x^2 + x - 12}{x - 3} = \lim\limits_{x \to 3} \dfrac{(x + 4)(x - 3)}{x - 3} = \lim\limits_{x \to 3} (x + 4) = 7$

(c) $\lim\limits_{x \to 3} \dfrac{(x^2 + x - 12)^2}{x - 3} = \lim\limits_{x \to 3} \dfrac{(x+4)^2(x-3)^2}{x - 3} = \lim\limits_{x \to 3} (x+4)^2(x-3) = 0$

(d) $\lim\limits_{x \to 3} \dfrac{x^2 + x - 12}{(x - 3)^2} = \lim\limits_{x \to 3} \dfrac{(x+4)(x-3)}{(x-3)^2} = \lim\limits_{x \to 3} \dfrac{x+4}{x-3};$ does not exist

41. (a) $\lim\limits_{x \to 4} \dfrac{f(x) - f(4)}{x - 4} = \lim\limits_{x \to 4} \dfrac{(x^2 - 4x) - (0)}{x - 4} = \lim\limits_{x \to 4} x = 4$

(b) $\lim\limits_{x \to 1} \dfrac{f(x) - f(1)}{x - 1} = \lim\limits_{x \to 1} \dfrac{x^2 - 4x + 3}{x - 1} = \lim\limits_{x \to 1} \dfrac{(x-1)(x-3)}{x - 1} = \lim\limits_{x \to 1} (x-3) = -2$

(c) $\lim\limits_{x \to 3} \dfrac{f(x) - f(1)}{x - 3} = \lim\limits_{x \to 3} \dfrac{x^2 - 4x + 3}{x - 3} = \lim\limits_{x \to 3} \dfrac{(x-1)(x-3)}{x - 3} = \lim\limits_{x \to 3} (x-1) = 2$

(d) $\lim\limits_{x \to 3} \dfrac{f(x) - f(2)}{x - 3} = \lim\limits_{x \to 3} \dfrac{x^2 - 4x + 4}{x - 3};$ does not exist

42. (a) $\lim\limits_{x \to 3} \dfrac{f(x) - f(3)}{x - 3} = \lim\limits_{x \to 3} \dfrac{x^3 - 27}{x - 3} = \lim\limits_{x \to 3} (x^2 + 3x + 9) = 27$

(b) $\lim\limits_{x \to 3} \dfrac{f(x) - f(2)}{x - 3} = \lim\limits_{x \to 3} \dfrac{x^3 - 8}{x - 3} = \lim\limits_{x \to 3} \dfrac{(x-2)(x^2 + 2x + 4)}{x - 3};$ does not exist

(c) $\lim\limits_{x \to 3} \dfrac{f(x) - f(3)}{x - 2} = \lim\limits_{x \to 3} \dfrac{x^3 - 27}{x - 2} = 0$

(d) $\lim\limits_{x \to 1} \dfrac{f(x) - f(1)}{x - 1} = \lim\limits_{x \to 1} \dfrac{x^3 - 1}{x - 1} = \lim\limits_{x \to 1} (x^2 + x + 1) = 3$

43. $f(x) = 1/x, \quad g(x) = -1/x$ with $c = 0$

44. Set, for instance,

$$f(x) = \begin{cases} 0, & x < c \\ 1, & x > c, \end{cases} \qquad g(x) = \begin{cases} 1, & x < c \\ 0, & x > c \end{cases}$$

45. True. Let $\lim\limits_{x \to c} [f(x) + g(x)] = L$. If $\lim\limits_{x \to c} g(x) = M$ exists, then $\lim\limits_{x \to c} f(x) = \lim\limits_{x \to c} [f(x) + g(x) - g(x)] = L - M$ also exists. This contradicts the fact that $\lim\limits_{x \to c} f(x)$ does not exist.

46. False, because $\lim\limits_{x \to c} g(x) = \lim\limits_{x \to c} [f(x) + g(x) - f(x)] = \lim\limits_{x \to c} [f(x) + g(x)] - \lim\limits_{x \to c} f(x).$ exists.

47. True. If $\lim\limits_{x \to c} \sqrt{f(x)} = L$ exists, then $\lim\limits_{x \to c} \sqrt{f(x)}\,\sqrt{f(x)} = L^2$ also exists.

48. False. Set $f(x) = -1$ for all $x, \quad c = 0$.

49. False;   for example set $f(x) = x$ and $c = 0$

50. False; for example, neither limit need exist: set $f(x) = \dfrac{1}{(x-3)^2}$ and $g(x) = \dfrac{2}{(x-3)^2}.$

51. False;   for example, set $f(x) = 1 - x^2, \; g(x) = 1 + x^2,$ and $c = 0$.

52. (a) If $f(x) \geq g(x)$ then $|f(x) - g(x)| = f(x) - g(x)$ and

$$\tfrac{1}{2}\{[f(x) + g(x)] + |f(x) - g(x)|\} = \tfrac{1}{2}\{f(x) + g(x) + f(x) - g(x)\}$$
$$= \tfrac{1}{2} \cdot 2f(x) = f(x) = \max\{f(x), g(x)\}.$$

If $f(x) \leq g(x)$ then $|f(x) - g(x)| = -[f(x) - g(x)] = g(x) - f(x)$ and

$$\tfrac{1}{2}\{[f(x) + g(x)] + |f(x) - g(x)|\} = \tfrac{1}{2}\{f(x) + g(x) + g(x) - f(x)\}$$
$$= \tfrac{1}{2} \cdot 2g(x) = g(x) = \max\{f(x), g(x)\}.$$

(b) $\min\{f(x), g(x)\} = \tfrac{1}{2}\{[f(x) + g(x)] - |f(x) - g(x)|\}$

53. If $\lim_{x \to c} f(x) = L$ and $\lim_{x \to c} g(x) = L,$ then

$$\lim_{x \to c} h(x) = \lim_{x \to c} \tfrac{1}{2}\{[f(x) + g(x)] - |f(x) - g(x)|\}$$

$$= \lim_{x \to c} \tfrac{1}{2}[f(x) + g(x)] - \lim_{x \to c} \tfrac{1}{f}(x) - g(x)|$$

$$= \tfrac{1}{2}(L + L) - \tfrac{1}{2}(L - L) = L.$$

A similar argument works for H.

54. (a) Let $\epsilon > 0$. Since $\lim_{x \to c} f(x) = L,$ there exists $\delta_1 > 0$ such that

$$\text{if } \quad 0 < |x - c| < \delta_1 \quad \text{then} \quad (*) \quad |f(x) - L| < \epsilon$$

Let $\delta_2 = \min\{|x_i - c| : 1 \leq i \leq n, x_i \neq c\}$. Then

$$(**) \quad f(x) = g(x) \quad \text{if } \quad 0 < |x - c| < \delta_2$$

Now, choose $\delta = \min\{\delta_1, \delta_2\}$. Then, if $\quad 0 < |x - c| < \delta$
$$|f(x) - L| < \epsilon \quad \text{by } (*), \text{ and}$$
$$f(x) = g(x) \quad \text{by } (**).$$

Therefore, if $0 < |x - c| < \delta,$ then $|g(x) - L| < \epsilon \implies \lim_{x \to c} g(x) = L.$

(b) If $\lim_{x \to c} g(x)$ exists, then $\lim_{x \to c} f(x)$ must exist by part (a).

55. (a) Suppose on the contrary that $\lim_{x \to c} g(x)$ does exist. Let $L = \lim_{x \to c} g(x)$. Then

$$\lim_{x \to c} f(x)g(x) = \lim_{x \to c} f(x) \cdot \lim_{x \to c} g(x) = 0 \cdot L = 0.$$

This contradicts the fact that $\lim_{x \to c} f(x)g(x) = 1$

(b) $\lim_{x \to c} g(x)$ exists since $\lim_{x \to c} g(x) = \lim_{x \to c} \dfrac{f(x)g(x)}{f(x)} = \dfrac{1}{L}.$

56. Suppose $\lim_{x \to c} f(x)$ does not exist. Let $g(x) = -f(x)$. Then $\lim_{x \to c} g(x)$ does not exist.
Now, $\lim_{x \to c}[f(x) + g(x)] = \lim_{x \to c}[f(x) - f(x)] = \lim_{x \to c} 0 = 0$ exists. This contradicts the hypothesis.

57. $f(x) = 2x^2 - 3x,$ $a = 2,$ $f(2) = 2$

$$\lim_{x \to a} \frac{f(x) - f(a)}{x - a} = \lim_{x \to 2} \frac{2x^2 - 3x - 2}{x - 2}$$

$$= \lim_{x \to 2} \frac{(x - 2)(2x + 1)}{x - 2} = \lim_{x \to 2} (2x + 1) = 5$$

58. $f(x) = x^3 + 1,$ $a = -1,$ $f(-1) = 0$

$$\lim_{x \to a} \frac{f(x) - f(a)}{x - a} = \lim_{x \to -1} \frac{x^3 + 1}{x + 1}$$

$$= \lim_{x \to -1} \frac{(x + 1)(x^2 - x + 1)}{x + 1} = \lim_{x \to -1} (x^2 - x + 1) = 3$$

59. $f(x) = \sqrt{x},$ $a = 4,$ $f(4) = 2$

$$\lim_{x \to a} \frac{f(x) - f(a)}{x - a} = \lim_{x \to 4} \frac{\sqrt{x} - 2}{x - 4} = \lim_{x \to 4} \frac{\sqrt{x} - 2}{x - 2} \cdot \frac{\sqrt{x} + 2}{\sqrt{x} + 2}$$

$$= \lim_{x \to 4} \frac{x - 4}{(x - 4)(\sqrt{x} + 2)} = \lim_{x \to 4} \frac{1}{\sqrt{x} + 2} = \frac{1}{4}$$

60. $f(x) = 1/(x + 1),$ $a = 1,$ $f(1) = 1/2$

$$\lim_{x \to a} \frac{f(x) - f(a)}{x - a} = \lim_{x \to 1} \frac{\dfrac{1}{x + 1} - \dfrac{1}{2}}{x - 1}$$

$$= \lim_{x \to 1} \frac{\dfrac{-(x - 1)}{2(x + 1)}}{x - 1} = \lim_{x \to 1} \frac{-1}{2(x + 1)} = -\frac{1}{4}$$

61. (a) $\displaystyle\lim_{h \to 0} \frac{f(x + h) - f(x)}{h} = \lim_{h \to 0} \frac{x + h - x}{h} = 1$

(b) $\displaystyle\lim_{h \to 0} \frac{f(x + h) - f(x)}{h} = \lim_{h \to 0} \frac{(x + h)^2 - x^2}{h} = \lim_{h \to 0} \frac{x^2 + 2xh + h^2 - x^2}{h}$

$$= \lim_{h \to 0} (2x + h) = 2x$$

(c) $\displaystyle\lim_{h \to 0} \frac{f(x + h) - f(x)}{h} = \lim_{h \to 0} \frac{(x + h)^3 - x^3}{h} = \lim_{h \to 0} \frac{x^3 + 3x^2 h + 3xh^2 + h^3 - x^3}{h}$

$$= \lim_{h \to 0} (3x^2 + 3xh + h^2) = 3x^2$$

(d) $\displaystyle\lim_{h \to 0} \frac{f(x + h) - f(x)}{h} = \lim_{h \to 0} \frac{(x + h)^4 - x^4}{h} = \lim_{h \to 0} \frac{x^4 + 4x^3 h + 6x^2 h^2 + 4xh^3 + h^4 - x^4}{h}$

$$= \lim_{h \to 0} (4x^3 + 6x^2 h + 4xh^2 + h^3) = 4x^3$$

(e) $\displaystyle\lim_{h \to 0} \frac{f(x + h) - f(x)}{h} = \lim_{h \to 0} \frac{(x + h)^n - x^n}{h} = nx^{n-1}$ for any positive integer n.

SECTION 2.4

1. (a) f is discontinuous at $x = -3,\ 0,\ 2,\ 6$

(b)   at -3, neither; f is continuous from the right at 0; at 2 and 6, neither

2. g is continuous on $(-4, -1), (-1, 3], (3, 5), (5, 8]$

3. continuous

4. continuous

5. continuous

6. continuous

7. continuous

8. jump discontinuity

9. removable discontinuity

10. removable discontinuity

11. jump discontinuity

12. removable discontinuity

13. continuous

14. indefinite discontinuity

15. infinite discontinuity

16. infinite discontinuity

17.

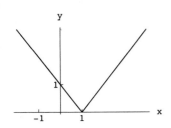

no discontinuities

18.

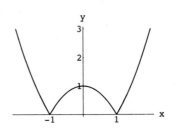

no discontinuities

19.

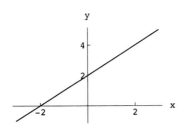

no discontinuities

20.

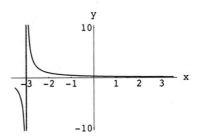

infinite discontinuity at -3

21.

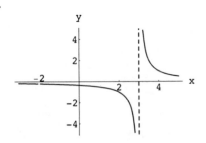

continuous at -2

infinite discontinuity at 3

22.

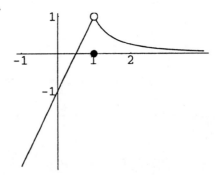

removable discontinuity at 1

23.

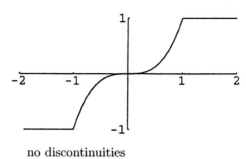

no discontinuities

24.

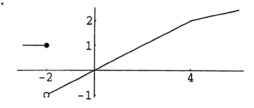

jump discontinuity at -2

25.

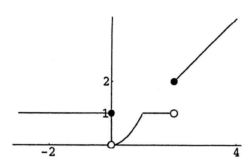

jump discontinuities at 0 and 2

26.

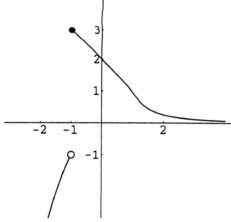
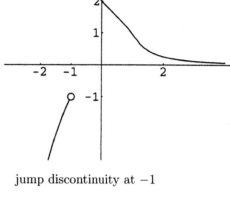

jump discontinuity at -1

27.

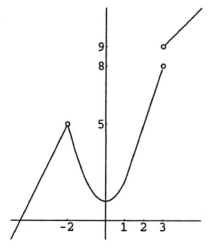

removable discontinuity at -2;
jump discontinuity at 3

28.

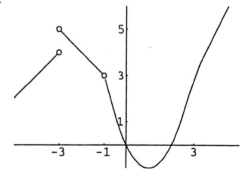

jump discontinuity at -3
removable discontinuity at -1

29.

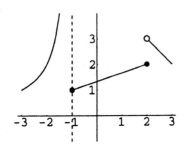

(One possibility)

30.

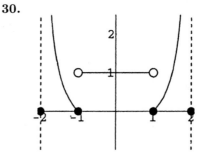

(One possibility)

31. $f(1) = 2$

32. impossible

33. impossible;

$$\lim_{x \to 1^-} f(x) = -1; \quad \lim_{x \to 1^+} f(x) = 1$$

34. f(1)=0

35. Since $\lim_{x \to 1^-} f(x) = 1$ and $\lim_{x \to 1^+} f(x) = A - 3 = f(1)$, take $A = 4$.

36. Since $\lim_{x \to 2^-} f(x) = 4A^2 = f(2)$ and $\lim_{x \to 2^+} f(x) = 2(1 - A)$, we need $4A^2 = 2(1 - A)$ or $2A^2 + A - 1 = 0$. This gives $A = \frac{1}{2}, -1$.

37. The function f is continuous at $x = 1$ iff

$$f(1) = \lim_{x \to 1^-} f(x) = A - B \quad \text{and} \quad \lim_{x \to 1^+} f(x) = 3$$

are equal; that is, $A - B = 3$. The function f is discontinuous at $x = 2$ iff

$$\lim_{x \to 2^-} f(x) = 6 \quad \text{and} \quad \lim_{x \to 2^+} f(x) = f(2) = 4B - A$$

are unequal; that is, iff $4B - A \neq 6$. More simply we have $A - B = 3$ with $B \neq 3$:

$$A - B = 3, 4B - A \neq 6 \implies A - B = 3, 3B - 3 \neq 6 \implies A - B = 3, B \neq 3.$$

38. Discontinuous at $x = 1$: $\lim_{x \to 1^-} f(x) \neq \lim_{x \to 1^+} f(x) \implies A - B \neq 3$.

Continuous at $x = 2$: $\lim_{x \to 2^-} f(x) = \lim_{x \to 2^+} f(x) \implies 4B - A = 6$.

Now,

$$4B - A = 6, \text{ and } A - B \neq 3 \implies 4B - A = 6 \text{ and } B \neq 3.$$

39. $c = -3$ **40.** $c = 2, d = -3$ **41.** $f(5) = \frac{1}{6}$

42. $f(5) = 0$ **43.** $f(5) = \frac{1}{3}$ **44.** $f(5) = \frac{1}{4}$

45. nowhere; see Figure 2.1.8   **46.** continuous only at $x = 0$.

47. $x = 0$, $x = 2$, and all non-integral values of x

48. (a)

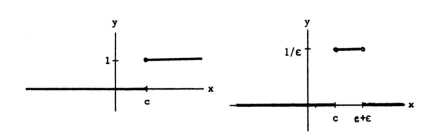

(b) H_c is continuous on $(-\infty, c) \cup [c, \infty)$.

$P_{\epsilon,c}$ is continuous on $(-\infty, c) \cup [c, c+\epsilon) \cup [c+\epsilon, \infty)$.

δ_c is continuous on $(-\infty, c) \cup (c, \infty)$.

49. Refer to (2.2.6). Use the equivalence of (i) and (ii) setting $L = f(c)$.

50. (a) Let $\epsilon = f(c) > 0$. By the continuity of f at c, there exists $\delta > 0$ such that

$$|f(x) - f(c)| < f(c) \quad \text{for all } x \in (c-\delta, c+\delta).$$

This implies that $f(x) > 0$ for all $x \in (c-\delta, c+\delta)$.

(b) Apply the result in part (a) to $h(x) = -f(x)$.

(c) Apply the result in part (a) to $h(x) = g(x) - f(x)$.

51. Suppose that g does not have a non-removable discontinuity at c. Then either g is continuous at c or it has a removable discontinuity at c. In either case, $\lim g(x)$ as $x \to c$ exists. Since $g(x) = f(x)$ except at a finite set of points $x_1, x_2, \ldots, x_n$, $\lim f(x)$ exists as $x \to c$ by Exercise 54, Section 2.3.

52. (a) Choose any point c and let $\epsilon > 0$. Since f is continuous at c, there exists $\delta > 0$ such that

$$\text{if } |x - c| < \delta \quad \text{then} \quad |f(x) - f(c)| < \epsilon.$$

Now, since

$$\big||f(x)| - |f(c)|\big| \leq |f(x) - f(c)|,$$

it follows that

$$\big||f(x)| - |f(c)|\big| < \epsilon \quad \text{if} \quad |x - c| < \delta.$$

Thus, $|f|$ is continuous at c.

(b)

$$\text{Let} \quad f(x) = \begin{cases} 1, & x \neq 1 \\ -1, & x = 1. \end{cases} \quad \text{Then } f \text{ is not continuous at } x = 1.$$

However, $|f(x)| = 1$ for all x is continuous everywhere.

(c)

$$\text{Let} \quad f(x) = \begin{cases} 1, & x \text{ rational} \\ -1, & x \text{ irratiional.} \end{cases} \quad \text{Then } f \text{ is nowhere continuous.}$$

However, $|f(x)| = 1$ for all x is continuous everywhere.

53. By implication, f is defined on $(c - p, c + p)$. The given inequality implies that $B \geq 0$. If $B = 0$, then $f \equiv f(c)$ is a constant function and hence is continuous. Now assume that $B > 0$. Let $\epsilon > 0$ and let $\delta = \min\{\epsilon/B, p\}$. If $|x - c| < \delta$ then $x \in (c - p, c + p)$ and

$$|f(x) - f(c)| \leq B|x - c| < B \cdot \delta \leq B \cdot \frac{\epsilon}{B} = \epsilon$$

Thus, f is continuous at c.

54. Choose any point $c \in (a, b)$, and let $\epsilon > 0$. Now choosing $\delta = \epsilon$, we have

$$\text{if } |x - c| < \delta, \ x \in (a, b), \quad \text{then} \quad |f(x) - f(c)| \leq |x - c| < \delta = \epsilon.$$

Thus, f is continuous at c, and it follows that f is continuous on (a, b).

55. $\lim_{h \to 0} [f(c + h) - f(c)] = \lim_{h \to 0} \left[\frac{f(c + h) - f(c)}{h} \cdot h \right] = \lim_{h \to 0} \left[\frac{f(c + h) - f(c)}{h} \right] \cdot \lim_{h \to 0} h = L \cdot 0 = 0.$
Therefore f is continuous at c by Exercise 47.

56. Let $f_e(x) = \frac{1}{2}[f(x) + f(-x)]$ and $f_o(x) = \frac{1}{2}[f(x) - f(-x)]$. Then f_e is an even function, f_o is an odd function, each function is continuous on $(-\infty, \infty)$, and $f = f_e + f_o$.

57. $\frac{5}{2}$ **58.** $\frac{1}{2}$

59. $\lim_{x \to 0^-} f(x) = -1$; $\lim_{x \to 0^+} f(x) = 1$; f is discontinuous for all k.

60. 2

SECTION 2.5

1. $\lim_{x \to 0} \frac{\sin 3x}{x} = \lim_{x \to 0} 3\left(\frac{\sin 3x}{3x} \right) = 3(1) = 3$ **2.** $\lim_{x \to 0} \frac{2x}{\sin x} = 2 \lim_{x \to 0} \frac{\sin x}{x} = 2(1) = 2$

3. $\lim_{x \to 0} \frac{3x}{\sin 5x} = \lim_{x \to 0} \frac{3}{5}\left(\frac{5x}{\sin 5x} \right) = \frac{3}{5}(1) = \frac{3}{5}$ **4.** $\lim_{x \to 0} \frac{\sin 3x}{2x} = \frac{3}{2} \lim_{x \to 0} \frac{\sin 3x}{3x} = \frac{3}{2}$

5. $\lim_{x \to 0} \frac{\sin 4x}{\sin 2x} = \lim_{x \to 0} \frac{4x}{2x} \cdot \frac{\sin 4x}{4x} \cdot \frac{2x}{\sin 2x} = 2(1)(1) = 2$

6. $\lim_{x \to 0} \frac{\sin 3x}{5x} = \frac{3}{5} \lim_{x \to 0} \frac{\sin 3x}{3x} = \frac{3}{5}$

7. $\lim\limits_{x\to 0}\dfrac{\sin x^2}{x}=\lim\limits_{x\to 0}x\left(\dfrac{\sin x^2}{x^2}\right)=\lim\limits_{x\to 0}x\cdot\lim\limits_{x\to 0}\dfrac{\sin x^2}{x^2}=0(1)=0$

8. $\lim\limits_{x\to 0}\dfrac{\sin x^2}{x^2}=1$ **9.** $\lim\limits_{x\to 0}\dfrac{\sin x}{x^2}=\lim\limits_{x\to 0}\dfrac{(\sin x)/x}{x};$ does not exist

10. $\lim\limits_{x\to 0}\dfrac{\sin^2 x^2}{x^2}=\lim\limits_{x\to 0}\left(\sin x^2\right)\left(\dfrac{\sin x^2}{x^2}\right)=0(1)=0$

11. $\lim\limits_{x\to 0}\dfrac{\sin^2 3x}{5x^2}=\lim\limits_{x\to 0}\dfrac{9}{5}\left(\dfrac{\sin 3x}{3x}\right)^2=\dfrac{9}{5}(1)=\dfrac{9}{5}$

12. $\lim\limits_{x\to 0}\dfrac{\tan^2 3x}{4x^2}=\lim\limits_{x\to 0}\dfrac{9}{4}\cdot\dfrac{1}{\cos^2 3x}\cdot\dfrac{\sin^2 3x}{(3x)^2}=\dfrac{9}{4}(1)(1)=\dfrac{9}{4}$

13. $\lim\limits_{x\to 0}\dfrac{2x}{\tan 3x}=\lim\limits_{x\to 0}\dfrac{2x\cos 3x}{\sin 3x}=\lim\limits_{x\to 0}\dfrac{2}{3}\left(\dfrac{3x}{\sin 3x}\right)\cos 3x=\dfrac{2}{3}(1)(1)=\dfrac{2}{3}$

14. $\lim\limits_{x\to 0}\dfrac{4x}{\cot 3x}=\lim\limits_{x\to 0}\dfrac{4x\sin 3x}{\cos 3x}=0$ **15.** $\lim\limits_{x\to 0}x\csc x=\lim\limits_{x\to 0}\dfrac{x}{\sin x}=1$

16. $\lim\limits_{x\to 0}\dfrac{\cos x-1}{2x}=-\dfrac{1}{2}\lim\limits_{x\to 0}\dfrac{1-\cos x}{x}=0$

17. $\lim\limits_{x\to 0}\dfrac{x^2}{1-\cos 2x}=\lim\limits_{x\to 0}\dfrac{x^2}{1-\cos 2x}\cdot\left(\dfrac{1+\cos 2x}{1+\cos 2x}\right)=\lim\limits_{x\to 0}\dfrac{x^2(1+\cos 2x)}{\sin^2 2x}$

$\qquad\qquad =\lim\limits_{x\to 0}\dfrac{1}{4}\left(\dfrac{2x}{\sin 2x}\right)^2(1+\cos 2x)=\dfrac{1}{4}(1)(2)=\dfrac{1}{2}$

18. $\lim\limits_{x\to 0}\dfrac{x^2-2x}{\sin 3x}=\lim\limits_{x\to 0}\left(\dfrac{x-2}{3}\right)\left(\dfrac{3x}{\sin 3x}\right)=\left(-\dfrac{2}{3}\right)(1)=-\dfrac{2}{3}$

19. $\lim\limits_{x\to 0}\dfrac{1-\sec^2 2x}{x^2}=\lim\limits_{x\to 0}\dfrac{-\tan^2 2x}{x^2}=\lim\limits_{x\to 0}\dfrac{-\sin^2 2x}{x^2\cos^2 2x}=\lim\limits_{x\to 0}\left[-4\left(\dfrac{\sin 2x}{2x}\right)^2\dfrac{1}{\cos^2 2x}\right]=-4$

20. $\lim\limits_{x\to 0}\dfrac{1}{2x\csc x}=\dfrac{1}{2}\lim\limits_{x\to 0}\dfrac{\sin x}{x}=\dfrac{1}{2}$ **21.** $\lim\limits_{x\to 0}\dfrac{2x^2+x}{\sin x}=\lim\limits_{x\to 0}(2x+1)\dfrac{x}{\sin x}=1$

22. $\lim\limits_{x\to 0}\dfrac{1-\cos 4x}{9x^2}=\lim\limits_{x\to 0}\left(\dfrac{1-\cos 4x}{9x^2}\right)\left(\dfrac{1+\cos 4x}{1+\cos 4x}\right)=\lim\limits_{x\to 0}\dfrac{1-\cos^2 4x}{9x^2(1+\cos 4x)}$

$\qquad =\lim\limits_{x\to 0}\dfrac{\sin^2 4x}{9x^2(1+\cos 4x)}=\dfrac{16}{9}\lim\limits_{x\to 0}\dfrac{\sin^2 4x}{(4x)^2}\cdot\dfrac{1}{1+\cos 4x}=\dfrac{16}{9}\cdot 1\cdot\dfrac{1}{2}=\dfrac{8}{9}$

23. $\lim\limits_{x\to 0}\dfrac{\tan 3x}{2x^2+5x}=\lim\limits_{x\to 0}\dfrac{1}{x(2x+5)}\dfrac{\sin 3x}{\cos 3x}=\lim\limits_{x\to 0}\dfrac{3}{2x+5}\left(\dfrac{\sin 3x}{3x}\right)\dfrac{1}{\cos 3x}=\dfrac{3}{5}(1)(1)=\dfrac{3}{5}$

24. $\lim\limits_{x\to 0}x^2(1+\cot^2 3x)=\lim\limits_{x\to 0}\left(x^2+\dfrac{x^2\cos^2 3x}{\sin^2 3x}\right)=\lim\limits_{x\to 0}\left(x^2+\dfrac{(3x)^2}{\sin^2 3x}\cdot\dfrac{\cos^2 3x}{9}\right)=0+(1)\dfrac{1}{9}=\dfrac{1}{9}$

25. $\lim\limits_{x \to 0} \dfrac{\sec x - 1}{x \sec x} = \lim\limits_{x \to 0} \dfrac{\dfrac{1}{\cos x} - 1}{x\left(\dfrac{1}{\cos x}\right)} = \lim\limits_{x \to 0} \dfrac{1 - \cos x}{x} = 0$

26. $\lim\limits_{x \to \pi/4} \dfrac{1 - \cos x}{x} = \dfrac{1 - \cos(\pi/4)}{\pi/4} = \dfrac{4 - 2\sqrt{2}}{\pi}$ **27.** $\dfrac{2\sqrt{2}}{\pi}$

28. $\lim\limits_{x \to 0} \dfrac{\sin^2 x}{x(1 - \cos x)} = \lim\limits_{x \to 0} \dfrac{x}{1 - \cos x} \cdot \dfrac{\sin^2 x}{x^2}$; does not exist

29. $\lim\limits_{x \to \pi/2} \dfrac{\cos x}{x - \pi/2} = \lim\limits_{h \to 0} \dfrac{\cos(h + \pi/2)}{h} = \lim\limits_{h \to 0} \dfrac{-\sin h}{h} = -1$

$\qquad\qquad h = x - \pi/2 \qquad\qquad \cos(h + \pi/2) = \cos h \cos \pi/2 - \sin h \sin \pi/2$

30. $\lim\limits_{x \to \pi} \dfrac{\sin x}{x - \pi} = \lim\limits_{x \to \pi} \dfrac{-\sin(x - \pi)}{x - \pi} = -1$

31. $\lim\limits_{x \to \pi/4} \dfrac{\sin(x + \pi/4) - 1}{x - \pi/4} = \lim\limits_{h \to 0} \dfrac{\sin(h + \pi/2) - 1}{h} = \lim\limits_{h \to 0} \dfrac{\cos h - 1}{h} = 0$

$\qquad\qquad\qquad\qquad \overset{\uparrow}{\underset{\textstyle\quad h = x - \pi/4}{\rule{0pt}{1em}}}$

32. $\lim\limits_{x \to \pi/6} \dfrac{\sin[x + (\pi/3)] - 1}{x - (\pi/6)} = \lim\limits_{x \to \pi/6} \dfrac{\sin[x - (\pi/6) + (\pi/2)] - 1}{x - (\pi/6)} = \lim\limits_{x \to \pi/6} \dfrac{\cos[x - (\pi/6)] - 1}{x - (\pi/6)} = 0$

33. Equivalently we will show that $\lim\limits_{h \to 0} \cos(c + h) = \cos c$. The identity

$$\cos(c + h) = \cos c \cos h - \sin c \sin h$$

gives

$$\lim_{h \to 0} \cos(c + h) = \cos c \left(\lim_{h \to 0} \cos h\right) - \sin c \left(\lim_{h \to 0} \sin h\right) = (\cos c)(1) - (\sin c)(0) = \cos c.$$

34. $\dfrac{a}{b}$ **35.** 0   **36.** $\dfrac{a}{b}$ **37.** 1

38. Let $\epsilon > 0$. There exists $\delta_1 > 0$ such that

$$\text{if} \quad 0 < |x| < \delta_1, \quad \text{then} \quad |f(x) - L| < \epsilon.$$

Take $\delta = \delta_1/|a|$:

$$\text{if} \quad 0 < |x| < \delta, \quad \text{then} \quad 0 < |ax| = |a|\,|x| < \delta_1 \quad \text{and} \quad |f(ax) - L| < \epsilon.$$

39. $f(x) = \sin x; \quad a = \pi/4$

$$\lim_{h\to 0} \frac{f(a+h) - f(a)}{h} = \lim_{h\to 0} \frac{\sin\left(\frac{\pi}{4} + h\right) - \sin\left(\frac{\pi}{4}\right)}{h} = \lim_{h\to 0} \frac{\sin(\pi/4)\cos h + \cos(\pi/4)\sin h - \sin(\pi/4)}{h}$$

$$= \lim_{h\to 0} \frac{-\sin(\pi/4)(1 - \cos h) + \cos(\pi/4)\sin h}{h}$$

$$= -\sin(\pi/4) \lim_{h\to 0} \frac{1 - \cos h}{h} + \cos(\pi/4) \lim_{h\to 0} \frac{\sin h}{h} = \cos(\pi/4) = \frac{\sqrt{2}}{2}$$

tangent line: $\quad y - \dfrac{\sqrt{2}}{2} = \dfrac{\sqrt{2}}{2}\left(x - \dfrac{\pi}{4}\right)$

40. $f(x) = \cos x; \quad a = \pi/3$

$$\lim_{h\to 0} \frac{f(a+h) - f(a)}{h} = \lim_{h\to 0} \frac{\cos\left(\frac{\pi}{3} + h\right) - \cos\left(\frac{\pi}{3}\right)}{h} = \lim_{h\to 0} \frac{\cos(\pi/3)\cos h - \sin(\pi/3)\sin h - \cos(\pi/3)}{h}$$

$$= \lim_{h\to 0} \frac{-\cos(\pi/3)(1 - \cos h) - sin(\pi/3)\sin h}{h}$$

$$= -\cos(\pi/3) \lim_{h\to 0} \frac{1 - \cos h}{h} - \sin(\pi/3) \lim_{h\to 0} \frac{\sin h}{h} = -\sin(\pi/3) = -\frac{\sqrt{3}}{2}$$

tangent line: $\quad y - \dfrac{1}{2} = -\dfrac{\sqrt{3}}{2}\left(x - \dfrac{\pi}{3}\right)$

41. $f(x) = \cos 2x; \quad a = \pi/6$

$$\lim_{h\to 0} \frac{f(a+h) - f(a)}{h} = \lim_{h\to 0} \frac{\cos 2\left(\frac{\pi}{6} + h\right) - \cos\left(2\frac{\pi}{6}\right)}{h} = \lim_{h\to 0} \frac{\cos(2h + \pi/3) - \cos(\pi/3)}{h}$$

$$= \lim_{h\to 0} \frac{\cos(\pi/3)\cos 2h - \sin(\pi/3)\sin 2h - \cos(\pi/3)}{h}$$

$$= -\cos(\pi/3) \lim_{h\to 0} 2\frac{1 - \cos 2h}{h} - \sin(\pi/3) \lim_{h\to 0} 2\frac{\sin 2h}{h}$$

$$= -\cos(\pi/3) \cdot 2 \cdot 0 - \sin(\pi/3) \cdot 2 \cdot 1 = -\sqrt{3}$$

tangent line: $\quad y - \dfrac{1}{2} = -\sqrt{3}\left(x - \dfrac{\pi}{6}\right)$

42. $f(x) = \sin 3x; \quad a = \pi/2$

$$\lim_{h\to 0} \frac{f(a+h) - f(a)}{h} = \lim_{h\to 0} \frac{\sin 3\left(\frac{\pi}{2} + h\right) - \sin\left(3\frac{\pi}{2}\right)}{h} = \lim_{h\to 0} \frac{\sin(3h + 3\pi/2) - \sin(3\pi/2)}{h}$$

$$= \lim_{h\to 0} \frac{\sin(3\pi/2)\cos 3h + \cos(3\pi/2)\cos 3h - \sin(3\pi/2)}{h}$$

$$= \lim_{h\to 0} \frac{1 - \cos 3h}{h} = 3 \lim_{h\to 0} \frac{1 - \cos 3h}{3h} = 3 \cdot 0 = 0$$

tangent line: $\quad y - (-1) = 0\left[x - (3\pi/2)\right] \quad \text{or} \quad y = -1$

43. For $x \neq 0$, $\quad |x\sin(1/x)| = |x||\sin(1/x)| \leq |x|$. Thus,

$$-|x| \leq |x\sin(1/x)| \leq |x|$$

Since $\lim_{x\to 0}(-|x|) = \lim_{x\to 0}|x| = 0$, the result follows by the pinching theorem.

44. Since $0 \le \cos^2[1/(x-\pi)] \le 1$ for all $x \ne \pi$, we have $|(x-\pi)\cos^2[1/(x-\pi)]| \le |x-\pi|$. Thus,

$$-|x-\pi| \le |(x-\pi)\cos[1/(x-\pi)]| \le |x-\pi|$$

Since $\lim\limits_{x\to\pi}(-|x-\pi|) = \lim\limits_{x\to\pi}|x-\pi| = 0$, the result follows by the pinching theorem.

45. For x close to 1(radian), $0 < \sin x \le 1$. Thus,

$$0 < |x-1|\sin x \le |x-1|$$

and the result follows by the pinching theorem.

46. Clearly, $|f(x)| \le 1$ for all x. Therefore,

$$|x\,f(x)| \le |x| \quad \text{which implies} \quad -|x| \le xf(x) \le |x| \quad \text{for all } x.$$

The result follows by the pinching theorem.

47. Suppose that there is a number B such that $|f(x)| \le B$ for all $x \ne 0$. Then $|x\,f(x)| \le B|x|$ and

$$-B|x| \le x\,f(x) \le B|x|$$

The result follows by the pinching theorem.

48. We have

$$f(x) = x \cdot \frac{f(x)}{x} \quad \text{and} \quad \left|\frac{f(x)}{x}\right| \le B,\ x \ne 0.$$

Therefore, the result follows from Exercise 47.

49. Suppose that there is a number B such that $\left|\dfrac{f(x)-L}{x-c}\right| \le B$ for $x \ne c$. Then

$$0 \le |f(x)-L| = \left|(x-c)\frac{f(x)-L}{x-c}\right| \le B|x-c|$$

By the pinching theorem, $\lim\limits_{x\to c}|f(x)-L| = 0$ which implies $\lim\limits_{x\to c}f(x) = L.$

50. Let $\epsilon > 0$. Let $\delta > 0$ be such that for $|\delta - c| < p$,

$$\text{if } |x-c| < \delta \text{ then } |f(x)| < \frac{\epsilon}{B}.$$

In other words, $B|f(x)| < \epsilon$.

But $|g(x)| \le B$, so $|f(x)g(x)| \le B|f(x)|$. Thus $|f(x)g(x)| < \epsilon$.

51. $\lim\limits_{x\to 0}\dfrac{20x-15x^2}{\sin 2x} = 10$

52. $\lim\limits_{x\to 0}\dfrac{\tan x}{x^2}$ does not exist

53. $\frac{1}{3}$

54. $\frac{1}{2}$

SECTION 2.6

1. Set $f(x) = 2x^3 - 4x^2 + 5x - 4$. Then f is continuous on $[1, 2]$ and $f(1) = -1 < 0$. $f(2) = 6 > 0$.
   By the intermediate-value theorem there is a c in $[1, 2]$ such that $f(c) = 0$.

2. Set $f(x) = x^4 - x - 1$. Then f is continuous on $[-1, 1]$ and $f(-1) = 1 > 0$, $f(1) = -1 < 0$.
   By the intermediate-value theorem there is a c in $[-1, 1]$ such that $f(c) = 0$.

3. Set $f(x) = \sin x + 2\cos x - x^2$ Then f is continuous on $[0, \frac{\pi}{2}]$ and $f(0) = 2 > 0$, $f(\frac{\pi}{2}) = 1 - \frac{\pi^2}{4} < 0$.
   By the intermediate-value theorem there is a c in $[0, \frac{\pi}{2}]$ such that $f(c) = 0$.

4. Set $f(x) = 2\tan x - x$. Then f is continuous on $[0, \frac{\pi}{4}]$ and $f(0) = 0 < 1$, $f(\frac{\pi}{4}) = 2 - \frac{\pi}{4} > 1$.
   By the intermediate-value theorem there is a c in $[0, \frac{\pi}{4}]$ such that $f(c) = 1$.

5. Set $f(x) = x^2 - 2 + \frac{1}{2x}$. Then f is continuous on $[\frac{1}{4}, 1]$ and $f(\frac{1}{4}) = \frac{1}{16} > 0$, $f(1) = -\frac{1}{2} < 0$.
   By the intermediate-value theorem there is a c in $[\frac{1}{4}, 1]$ such that $f(c) = 0$.

6. Set $f(x) = x^{\frac{5}{3}} + x^{\frac{1}{3}}$. Then f is continuous on $[-1, 1]$ and $f(-1) = -2 < 1$, $f(1) = 2 > 1$.
   By the intermediate-value theorem there is a c in $[-1, 1]$ such that $f(c) = 1$.

7. Set $f(x) = x^3 - \sqrt{x + 2}$. Then f is continuous on $[1, 2]$ and $f(1) = 1 - \sqrt{3} < 0$, $f(2) = 6 > 0$.
   By the intermediate-value theorem there is a c in $[1, 2]$ such that $f(c) = 0$.
   i.e. $c^3 = \sqrt{c + 2}$.

8. Set $f(x) = \sqrt{x^2 - 3x} - 2$. Then f is continuous on $[3, 5]$ and $f(3) = -2 < 0$, $f(5) = \sqrt{10} + 2 > 0$.
   By the intermediate-value theorem there is a c in $[3, 5]$ such that $f(c) = 0$.

9. $f(x)$ is continuous on $[0, 1]$; $f(0) = 0 < 1$ and $f(1) = 4 > 1$.
   By the intermediate value theorem there is a c in $(0, 1)$ such that $f(c) = 1$.

10. $f(x)$ is continuous on $[2, 3]$; $f(2) = \frac{1}{2} > 0$ and $f(3) = -\frac{1}{2} < 0$,
    By the intermediate value theorem there is a c in $(2, 3)$ (hence in $(1, 4)$) such that $f(c) = 0$.

11. Set $f(x) = x^3 - 4x + 2$.   Then $f(x)$ is continuous on $[-3, 3]$.
    Checking the integer values on this interval,

    $$f(-3) = -13 < 0, \quad f(-2) = 2 > 0, \quad f(0) = 2 > 0, \quad f(1) = -1 < 0, \text{ and } f(2) = 2 > 0.$$

    By the intermediate value theorem there are roots in $(-3, -2), (0, 1)$ and $(1, 2)$.

12. Set $f(x) = x^2$.   Then $f(x)$ is continuous on $[1, 2]$, $f(1) = 1 < 2$ and $f(2) = 4 > 2$.
    By the intermediate value theorem there is a c in $(1, 2)$ such that $f(c) = 2$.

13.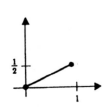

14.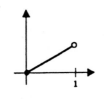

15.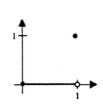

16. Impossible

17.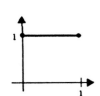

18.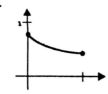

19.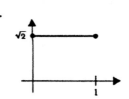

20.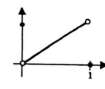

21. Impossible

22.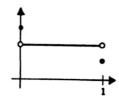

23.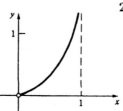

24. Impossible

25. If $f(0) = 0$ or if $f(1) = 1$, we have a fixed point. If $f(0) \neq 0$ and $f(1) \neq 1$, then set $g(x) = x - f(x)$. g is continuous on $[0, 1]$ and $g(0) < 0 < g(1)$. Therefore there exists a number $c \in (0, 1)$ such that $g(c) = c - f(c) = 0$.

26. Set $h(x) = f(x) - g(x)$. Then h is continuous on $[a, b]$, and $h(a) = f(a) - g(a) < 0$, $h(b) = f(b) - g(b) > 0$. By the intermediate value theorem there exists a number $c \in (a, b)$ such that $h(c) = 0$. Thus, $f(c) = g(c)$.

27. (a) If $f(0) = 1$ or if $f(1) = 0$, we're done. Suppose $f(0) \neq 1$ and $f(1) \neq 0$. The diagonal from $(0, 1)$ to $(1, 0)$ has equation $y = 1 - x$, $0 \leq x \leq 1$. Set $g(x) = (1 - x) - f(x)$. g is continuous on $[0, 1]$ and $g(0) = 1 - f(0) > 0$; $g(1) = -f(1) < 0$. Therefore there exists a number $c \in (0, 1)$ such that $g(c) = (c - 1) - f(c) = 0$.
(b) Suppose $g(0) = 0$ and $g(1) = 1$. Set $h(x) = g(x) - f(x)$; h is continuous on $[0, 1]$. If $h(0) = 0$ or if $h(1) = 0$, we're done. If not, then $h(0) < 0$ and $h(1) > 0$. Therefore there exists a number $c \in (0, 1)$ such that $h(c) = 0$. The same argument works for the other case.

28. Let $f(x) = x^3$ and repeat the argument given in Exercise 37.

29. The cubic polynomial $P(x) = x^3 + ax^2 + bx + c$ is continuous on $(-\infty, \infty)$.. Writing P as

$$P(x) = x^3 \left(1 + \frac{a}{x} + \frac{b}{x^2} + \frac{c}{x^3} \right) \quad x \neq 0$$

it follows that $P(x) < 0$ for large negative values of x and $P(x) > 0$ for large positive values of x. Thus there exists a negative number N such that $P(x) < 0$ for $x < N$, and a positive number M such that $P(x) > 0$ for $x > M$. By the intermediate-value theorem, P has a zero in $[N, M]$.

30. (a) Let S be the set of positive integers for which the statement is true. Then $1 \in S$ by hypothesis. Now assume that $k \in S$. Then $a^k < b^k$ and

$$a^{k+1} = (a)a^k < (a)b^k < (b)b^k = b^{k+1}.$$

   Therefore, $k + 1 \in S$ and S is the set of positive integers.

   (b) Clearly 0 is the unique nth root of 0. Choose any positive number x and let $f(t) = t^n - x$. Since $f(0) = -x < 0$ and $f(t) \to \infty$ as $t \to \infty$, there exists a number $c > 0$ such that $f(c) = 0$. The number c is an nth root of x. The uniqueness follows from part (a).

31. The function T is continuous on $[4000, 4500]]$; and $T(4000) \cong 98.0995$, $T(4500) \cong 97.9478$. Thus, by the intermediate-value theorem, there is an elevation h between 4000 and 4500 meters such that $T(h) = 98$.

32. Think of the equator as being a circle and choose a reference point P and a positive direction. For example, choose P to be $0°$ longitude and let "eastward" be the positive direction. Using radian measure, let x, $0 \le x \le 2\pi$ denote the coordinate of a point x radians from P. Then, x and $x + \pi$ are diametrically opposite points on the equator. Let $T(x)$ be the temperature at the point x, and let $f(x) = T(x) - T(x + \pi)$. If $f(0) = 0$, then the temperatures at the points 0 and π are equal. If $f(x) \ne 0$, then $f(0) = T(0) - T(\pi)$ and $f(\pi) = T(\pi) - T(2\pi) = t(\pi) - T(0)$ have opposite sign. Thus, there exists a point $c \in (0, \pi)$ at which $f(c) = 0$, and $T(c) = T(c + \pi)$.

33. Let $A(r)$ denote the area of a circle with radius r, $r \in [0, 10]$. Then $A(r) = \pi r^2$ is continuous on $[0, 10]$, and $A(0) = 0$ and $A(10) = 100\pi \cong 314$. Since $0 < 250 < 314$ it follows from the intermediate value theorem that there exists a number $c \in (0, 10)$ such that $A(c) = 250$.

34. Let x and y be the dimensions of a rectangle in $\mathcal{R}$. Then, $2x + 2y = P$ and $y = \dfrac{P}{2} - x$. The area function

$$A(x) = xy = x\left(\frac{P}{2} - x\right) = \frac{P}{2}x - x^2, \quad x \in [0, P/2]$$

   is continuous. Therefore, A has a maximum value on $[0, P/2]$. Since

$$A(x) = \frac{P}{2}x - x^2 = \frac{P^2}{16} - \left(x - \frac{P}{4}\right)^2$$

   it is clear that the rectangle with maximum area has dimensions $x = y = \dfrac{P}{4}$.

35. Inscribe a rectangle in a circle of radius R. Introduce a coordinate system with the origin at the center of the circle and the sides of the rectangle parallel to the coordinate axes. Then the area of the rectangle

is given by

$$A(x) = 4x\sqrt{R^2 - x^2}, \quad x \in [0, R].$$

Since A is continuous on $[0, R]$, A has a maximum value.

36. $f(0) = -4$, $f(1) = 2$.   Thus, f has a zero in $(0, 1)$ at $r = 0.771$.

37. $f(-3) = -9$, $f(-2) = 5$; $f(0) = 3$, $f(1) = -1$; $f(1) = -1$, $f(2) = 1$ Thus, f has a zero in $(-3, -2,)$ in $(0, 1)$ and in $(1, 2)$.
$r_1 = -2.4909$, $r_2 = 0.6566$,   and $r_3 = 1.8343$

38. $f(-2) = -25$, $f(-1) = 1$; $f(0) = 1$, $f(1) = -1$; $f(1) = -1$, $f(2) = 27$ Thus, f has a zero in $(-2, -1,)$ in $(0, 1)$ and in $(1, 2)$.
$r_1 = -1.3888$, $r_2 = 0.3345$,   and $r_3 = 1.2146$

39. $f(-2) = -5.6814$, $f(-1) = 1.1829$; $f(0) = 0.5$, $f(1) = -0.1829$; $f(1) = -0.1829$, $f(2) = 6.681$
Thus, f has a zero in $(-2, -1)$, in $(0, 1)$ and in $(1, 2)$.
$r_1 = -1.3482$, $r_2 = 0.2620$,   and $r_3 = 1.0816$

40. f satisfies the hypothesis of the intermediate-value theorem: $c \cong -0.83$

41. f is not continuous at $x = 1$. Therefore f does not satisfy the hypothesis of the intermediate-value theorem. However,

$$\frac{f(-3) + f(2)}{2} \neq \frac{1}{2} = f(c) \text{ for } c \cong 0.163$$

42. f is not continuous at $x = -\pi/2, \pi/2, 3\pi/2$; $\frac{1}{2}\left[f(-\pi) + f(2\pi)\right] = 0$; $f(x) \neq 0$ for all $x \in [-\pi, 2\pi]$.

43. f satisfies the hypothesis of the intermediate-value theorem:

$$\frac{f(\pi/2) + f(2\pi)}{2} = \frac{1}{2} = f(c) \text{ for } c \cong 2.38, \; 4.16, \; 5.25$$

44. f is bounded.
    $\max(f) = 6 \quad [f(0) = 6]$
    $\min(f) \cong -0.376 \quad [f(1.46) \cong -0.376]$

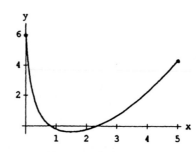

45. f is bounded.

$\max(f) = 1 \quad [f(1) = 1]$

$\min(f) = -1 \quad [f(-1) = -1]$

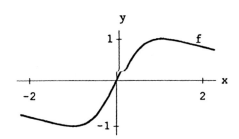

46. f is unbounded

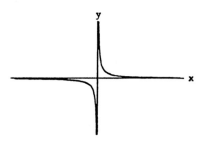

47. f is bounded.

no max f is not defined at $x = 0$

$\min(f) \cong 0.3540$

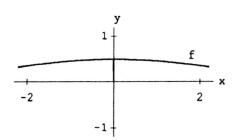

PROJECT 2.6

1. (a) $\dfrac{2-1}{2^n} < 0.001 \quad \Longrightarrow \quad 2^n > 1000 \quad \Longrightarrow \quad n > 9.$

The minimum number of iterations required is $n = 10$.

$$\sqrt{2} \simeq \frac{1449}{1024} \simeq 1.415034.$$

(b) $\dfrac{2-1}{2^n} < 0.0001 \quad \Longrightarrow \quad 2^n > 10,000 \quad \Longrightarrow \quad n > 13.$

The minimum number of iterations required is $n = 14$.

$$\frac{2-1}{2^n} < 0.00001 \quad \Longrightarrow \quad 2^n > 100,000 \quad \Longrightarrow \quad n > 16.$$

The minimum number of iterations required is $n = 17$.

2. $f(x) = x^3 + x - 9;\ f(1) = -7$ and $f(2) = 1.$ Therefore, $f(c) = 0$ for some $c \in (1, 2)$.

(a) A minimum of 7 iterations are required; $c \simeq \dfrac{245}{128} \simeq 1.9140625.$

Accurate to 7 decimal places, the root is $c \simeq 1.9201751$.

(b) $\dfrac{2-1}{2^n} < 0.001 \quad \Longrightarrow \quad 2^n > 1,000 \quad \Longrightarrow \quad n > 9.$

The minimum number of iterations required is $n = 10$.

$$\frac{2-1}{2^n} < 0.0001 \quad \Longrightarrow \quad 2^n > 10,000 \quad \Longrightarrow \quad n > 13.$$

The minimum number of iterations required is $n = 14$.

3. $f(x) = \sin x + x + 3$; $f(-3) \approx -0.142$ and $f(-2) \approx 0.091$.   Therefore, $f(c) = 0$ for some $c \in (-3, -2)$.

(a) A minimum of 7 iterations are required; $c \simeq \dfrac{-279}{128} \simeq -2.17969$.

Accurate to 6 decimal places, the root is $c \simeq -2.179727$.

(b) $\dfrac{2-1}{2^n} < 0.00001 \quad \Longrightarrow \quad 2^n > 100,000 \quad \Longrightarrow \quad n > 16$.

The minimum number of iterations required is $n = 17$.

$$\frac{2-1}{2^n} < 0.000001 \quad \Longrightarrow \quad 2^n > 1,000,000 \quad \Longrightarrow \quad n > 19.$$

The minimum number of iterations required is $n = 20$.

4. For $f(x) = x^2 - 2$, the first three iterations are:

$c_1 = \frac{4}{3}$, $c_2 = \frac{5}{3} \approx 1.6667$, $c_3 = \frac{3}{2} = 1.5$

For $f(x) = x^3 + x - 9$, the first three iterations are:

$c_1 = \frac{15}{8} = 1.875$, $c_2 = \frac{31}{16} = 1.9375$, $c_3 = \frac{61}{32} = 1.90625$.

For $f(x) = \sin x + x + 3$, the first three iterations are:

$c_1 \approx -2.39056$, $c_2 \approx -2.19528$, $c_3 \approx -2.09764$.

CHAPTER 2. REVIEW EXERCISES

1. $\displaystyle\lim_{x \to 3} \frac{x^2 - 3}{x + 3} = \frac{3^2 - 3}{3 + 3} = 1.$

2. $\displaystyle\lim_{x \to 2} \frac{x^2 + 4}{x^2 + 2x + 1} = \frac{2^2 + 4}{2^2 + 2(2) + 1} = \frac{8}{9}.$

3. $\displaystyle\lim_{x \to 3} \frac{(x - 3)^2}{x + 3} = \frac{0}{6} = 0.$

4. $\displaystyle\lim_{x \to 3} \frac{x^2 - 9}{x^2 - 5x + 6} = \lim_{x \to 3} \frac{(x - 3)(x + 3)}{(x - 2)(x - 3)} = \lim_{x \to 3} \frac{x + 3}{x - 2} = 6.$

5. $\displaystyle\lim_{x \to 2^+} \frac{x - 2}{|x - 2|} = \lim_{x \to 2^+} \frac{x - 2}{x - 2} = 1.$

6. $\displaystyle\lim_{x \to -2} \frac{|x|}{x - 2} = \lim_{x \to -2} \frac{-x}{x - 2} = -\frac{1}{2}.$

7. $\displaystyle\lim_{x \to 0} \left(\frac{1}{x} - \frac{1 - x}{x} \right) = \lim_{x \to 0} \frac{x}{x} = 1.$

8. $\lim\limits_{x\to 3^+} \dfrac{\sqrt{x-3}}{|x-3|} = \lim\limits_{x\to 3^+} \dfrac{\sqrt{x-3}}{x-3} = \lim\limits_{x\to 3^+} \dfrac{1}{\sqrt{x-3}}$; does not exist.

9. $\lim\limits_{x\to 1^+} \dfrac{|x-1|}{x} = \lim\limits_{x\to 1^+} \dfrac{x-1}{x} = 0$ and $\lim_{x\to 1^-} \dfrac{|x-1|}{x} = \lim_{x\to 1^-} \dfrac{-x+1}{x} = 0$. $\lim_{x\to 1} \dfrac{|x-1|}{x} = 0$.

10. $\lim\limits_{x\to 1^+} \dfrac{x^3-1}{|x^3-1|} = \lim\limits_{x\to 1^+} \dfrac{x^3-1}{x^3-1} = 1$ and $\lim_{x\to 1^-} \dfrac{x^3-1}{|x^3-1|} = \lim_{x\to 1^-} \dfrac{x^3-1}{-x^3+1} = -1$; does not exist.

11. $\lim\limits_{x\to 1} \dfrac{\sqrt{x}-1}{x-1} = \lim\limits_{x\to 1} \dfrac{\sqrt{x}-1}{(\sqrt{x}-1)(\sqrt{x}+1)} = \lim\limits_{x\to 1} \dfrac{1}{\sqrt{x}+1} = \dfrac{1}{2}$.

12. $\lim\limits_{x\to 3^+} \dfrac{x^2-2x-3}{\sqrt{x-3}} = \lim\limits_{x\to 3^+} \dfrac{(x-3)(x+1)}{\sqrt{x-3}} = \lim\limits_{x\to 3^+} \sqrt{x-3}(x+1) = 0$.

13. $\lim\limits_{x\to 3^+} \dfrac{\sqrt{x^2-2x-3}}{x-3} = \lim\limits_{x\to 3^+} \dfrac{\sqrt{x+1}}{\sqrt{x-3}}$, does not exist.

14. $\lim\limits_{x\to 4} \dfrac{\sqrt{x+5}-3}{x-4} = \lim\limits_{x\to 4} \dfrac{(\sqrt{x+5}-3)(\sqrt{x+5}+3)}{(x-4)(\sqrt{x+5}+3)} = \lim\limits_{x\to 4} \dfrac{x-4}{(x-4)(\sqrt{x+5}+3)} = \dfrac{1}{6}$.

15. $\lim\limits_{x\to 2} \dfrac{x^3-8}{x^4-3x^2-4} = \lim\limits_{x\to 2} \dfrac{(x-2)(x^2+2x+4)}{(x-2)(x+2)(x^2+1)} = \lim\limits_{x\to 2} \dfrac{x^2+2x+4}{(x+2)(x^2+1)} = \dfrac{3}{5}$.

16. $\lim\limits_{x\to 0} \dfrac{5x}{\sin 2x} = \lim\limits_{x\to 0} \dfrac{5}{2}\dfrac{2x}{\sin 2x} = \dfrac{5}{2}$.

17. $\lim\limits_{x\to 0} \dfrac{\tan^2 2x}{3x^2} = \lim\limits_{x\to 0} \dfrac{4}{3\cos^2 2x}\left(\dfrac{\sin 2x}{2x}\right)^2 = \dfrac{4}{3}$.

18. $\lim\limits_{x\to 0} x\csc 4x = \lim\limits_{x\to 0} \dfrac{4x}{4\sin 4x} = \dfrac{1}{4}$.

19. $\lim\limits_{x\to 0} \dfrac{x^2-3x}{\tan x} = \lim\limits_{x\to 0} \dfrac{x(x-3)\cos x}{\sin x} = -3$.

20. $\lim\limits_{x\to \pi/2} \dfrac{\cos x}{2x-\pi} = \lim\limits_{x\to \pi/2} \dfrac{\sin(\pi/2-x)}{-2(\pi/2-x)} = -\dfrac{1}{2}$.

21. $\lim\limits_{x\to 0} \dfrac{\sin 3x}{5x^2-4x} = \lim\limits_{x\to 0} \dfrac{\sin 3x}{3x\left(\dfrac{5}{3}x - \dfrac{4}{3}\right)} = -\dfrac{3}{4}$.

22. $\lim\limits_{x\to 0} \dfrac{5x^2}{1-\cos 2x} = \lim\limits_{x\to 0} \dfrac{5x^2}{2\sin^2 x} = \dfrac{5}{2}$.

23. $\lim\limits_{x\to -\pi} \dfrac{x+\pi}{\sin x} = \lim\limits_{x\to -\pi} \dfrac{x+\pi}{-\sin(x+\pi)} = -1$.

24. $\lim\limits_{x\to 2} \dfrac{x^2-4}{x^3-8} = \lim\limits_{x\to 2} \dfrac{(x-2)(x+2)}{(x-2)(x^2+2x+4)} = \lim\limits_{x\to 2} \dfrac{x+2}{x^2+2x+4} = \dfrac{1}{3}$.

25. $\displaystyle\lim_{x\to 2^-}\frac{x-2}{|x^2-4|} = \lim_{x\to 2^-} -\frac{x-2}{(x-2)(x+2)} = \lim_{x\to 2^-} -\frac{1}{x+2} = -\frac{1}{4}.$

26. $\displaystyle\lim_{x\to 2}\frac{1-2/x}{1-4/x^2} = \lim_{x\to 2}\frac{x^2-2x}{x^2-4} = \lim_{x\to 2}\frac{x}{x+2} = \frac{1}{2}.$

27. $\displaystyle\lim_{x\to 1^+}\frac{x^2-3x+2}{\sqrt{x-1}} = \lim_{x\to 1^+}\frac{(x-2)(x-1)}{\sqrt{x-1}} = \lim_{x\to 1^+}(x-2)\sqrt{x-1} = 0.$

28. $\displaystyle\lim_{x\to 3}\frac{1-9/x^2}{1+3/x} = \frac{1-9/9}{1+3/3} = 0.$

29. $\displaystyle\lim_{x\to 2} f(x) = 3(2) - 2^2 = 2.$

30. $\displaystyle\lim_{x\to -2^+} f(x) = (-2)^2 - 3 = 1$ and $\displaystyle\lim_{x\to -2^-} f(x) = 3 - 2 = 1.$ Hence $\displaystyle\lim_{x\to -2} f(x) = 1.$

31. (a) False   (b) False   (c) True   (d) False   (e) True

32. Since $\displaystyle\lim_{x\to 3} x^2 - 2x - 3 = 0,$ we must have $\displaystyle\lim_{x\to 3}(2x^2 - 3ax + x - a - 1) = 0.$

$\displaystyle\lim_{x\to 3}(2x^2 - 3ax + x - a - 1) = 18 - 9a + 3 - a - 1 = 0;\quad -10a + 20 = 0;\quad a = 2.$

$\displaystyle\lim_{x\to 3}\frac{2x^2-5x-3}{x^2-2x-3} = \lim_{x\to 3}\frac{(2x+1)(x-3)}{(x+1)(x-3)} = \frac{7}{4}.$

33. (a)

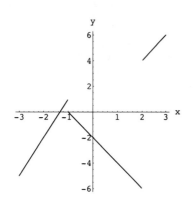

(b) (i) 1   (ii) 0   (iii) does not exist   (iv) -6 (v) 4   (vi) does not exist

(c) (i) f is continuous from the left at -1; f is not continuous from the right at -1.

   (ii) f is not continuous from the left at 2; f is continuous from the right at 2.

34. (a) $\displaystyle\lim_{x\to 0}\cos\left(\frac{1-\cos x}{2x}\right) = \cos\left(\lim_{x\to 0}\frac{1-\cos x}{2x}\right) = \cos 0 = 1$

(b) $\displaystyle\lim_{x\to 0}\cos\left(\frac{\pi\sin x}{2x}\right) = \cos\left(\lim_{x\to 0}\frac{\pi\sin x}{2x}\right) = \cos\tfrac{1}{2}\pi = 0$

35. For f to be continuous at 2, we must have

$$\lim_{x \to 2^-} (2x^2 - 1) = 7 = A = \lim_{x \to 2^+} (x^3 - 2Bx) = 8 - 4B.$$

These equations imply that $A = 7$ and $B = \dfrac{1}{4}$.

36. For f to continuous at $x = -1$, we must have

$$\lim_{x \to -1^-} (Ax + B) = B - A = -2 \quad \text{or} \quad A = B + 2.$$

For f to be discontinuous at $x = 2$, we must have

$$\lim_{x \to 2^+} (2Bx - A) = 4B - A \neq f(2) = 4.$$

$A = B + 2$ and $4B - A \neq 4$ imply that $B \neq 2$. Necessary and sufficient conditions for f to be continuous at $x = -1$ and discontinuous at $x = 2$ are: $A = B + 2$ with $B \neq 2$.

37. $\displaystyle\lim_{x \to -3} \frac{x^2 - 2x - 15}{x + 3} = \lim_{x \to -3} \frac{(x - 5)(x + 3)}{x + 3} = -8;$ set $f(-3) = -8.$

38. $\displaystyle\lim_{x \to 3} \frac{\sqrt{x + 1} - 2}{x - 3} = \lim_{x \to 3} \frac{(\sqrt{x + 1} - 2)(\sqrt{x + 1} + 2)}{(x - 3)(\sqrt{x + 1} + 2)} = \lim_{x \to 3} \frac{x - 3}{(x - 3)(\sqrt{x + 1} + 2)} = \frac{1}{4};$ set $f(3) = \frac{1}{4}.$

39. $\displaystyle\lim_{x \to 0} \frac{\sin \pi x}{x} = \lim_{x \to 0} \frac{\pi \sin \pi x}{\pi x} = \pi;$ set $f(0) = \pi.$

40. $\dfrac{1 - \cos x}{x^2} = \dfrac{1 - \cos x}{x^2} \cdot \dfrac{1 + \cos x}{1 + \cos x} = \dfrac{1 - \cos^2 x}{x^2} \cdot \dfrac{1}{1 + \cos x} = \dfrac{\sin^2 x}{x^2} \cdot \dfrac{1}{1 + \cos x}.$

Therefore, $\displaystyle\lim_{x \to 0} \frac{1 - \cos x}{x^2} = \lim_{x \to 0} \frac{\sin^2 x}{x^2} \cdot \frac{1}{1 + \cos x} = \frac{1}{2};$ set $f(0) = \frac{1}{2}.$

41. (a) False; need f continuous   (b) True; intermediate-value theorem

(c) False; need f continuous on $[a, b]$ (d) False; consider $f(x) = x^2$ on $[-1.1]$.

42. Set $f(x) = x^3 - 3x - 4.$ Then $f(2) = -2$ and $f(3) = 14.$ By the intermediate-value theorem, $f(x)$ has zero in $[2, 3]$.

43. Set $f(x) = 2 \cos x - x + 1.$ Then $f(1) = 2 \cos 1 > 0$ and $f(2) = 2 \cos 2 - 1 < 0.$ By the intermediate-value theorem, $f(x)$ as zero in $[1, 2]$.

44. Let $\epsilon > 0.$

$$|(5x - 4) - 6| = |5x - 10| = 5|x - 2|.$$

If $0 < |x - 2| < \delta = \epsilon/5,$ then $|(5x - 4) - 6| < \epsilon.$ Therefore, $\displaystyle\lim_{x \to 2}(5x - 4) = 6.$

45. If $0 < |x - (-4)| = |x + 4| < 1$, then $2x + 5 < 0$ and $|2x + 5| = -2x - 5$.

Let $\epsilon > 0$.

$$\big|\,|2x + 5| - 3\big| = |-2x - 5 - 3| = |-2x - 8| = 2|x + 4|.$$

If $\delta = \min\{1, \epsilon/2\}$, then $\big|\,|2x + 5| - 3\big| < \epsilon$. Therefore, $\lim\limits_{x \to -4} |2x + 5| = 3$.

46. Let $\epsilon > 0$.

$$\left|\sqrt{x - 5} - 2\right| = \left|\sqrt{x - 5} - 2 \cdot \frac{\sqrt{x - 5} + 2}{\sqrt{x - 5} + 2}\right| = \left|\frac{x - 9}{\sqrt{x - 5} + 2}\right| < |x - 9|.$$

If $0 < |x - 9| < \delta = \epsilon$, then $\left|\sqrt{x - 5} - 2\right| < \epsilon$. Therefore, $\lim\limits_{x \to 9} \sqrt{x - 5} = 2$.

47. Suppose that $\lim\limits_{x \to 0} \dfrac{f(x)}{x} = L$ exists. Then

$$\lim_{x \to 0} f(x) = \lim_{x \to 0} \frac{f(x)\, x}{x} = \lim_{x \to 0} \frac{f(x)}{x}\, x = L \lim_{x \to 0} x = 0.$$

48. Suppose that $\lim\limits_{x \to c} g(x) = l$ and that f is continuous at l. Let $\epsilon > 0$. By the continuity of f at l, there is a number $\delta_1 > 0$ such that $|f(y) - f(l)| < \epsilon$ whenever $|y - l| < \delta_1$. Since $\lim\limits_{x \to c} g(x) = l$, there is a positive number δ such that $|g(x) - l| < \delta_1$ whenever $0 < |x - c| < \delta$. Therefore, if $0 < |x - c| < \delta$, then $|g(x) - l| < \delta_1$ and $|f(g(x) - f(l)| < \epsilon$. Therefore, $\lim\limits_{x \to c} f(g(x)) = l$.

49. (a) Yes. For example, set $f(x) = \begin{cases} \dfrac{\sin \pi x}{x}, & x > 0 \\[2mm] \pi, & x = 0 \end{cases}$

(b) No. By the continuity of f (from the right) at 0, $f(1/n) = 0$ for each postive integer implies that $f(0) = 0$.

50. (a) No. Suppose that f and g are everywhere continuous. If $f(c) \neq g(c)$, then there is an interval $(c - p, c + p)$ on which $f(c) \neq g(c)$.

(b) No. Suppose that f and g are everywhere continuous. If $f(x) \neq g(x)$ on $[a, b]$, then $f(a) \neq g(a)$ and $f(b) \neq g(b)$. Therefore, by the continuity of f and g, there exist postive numbers p and q such that $f(x) \neq g(x)$ on $(a - p, b + q)$.

(c) Yes. For example, set $f(x) = 0$ for all x and $g(x) = \begin{cases} x(1 - x), & 0 \leq x \leq 1 \\[2mm] 0, & \text{otherwise.} \end{cases}$

f and g are everywhere continuous and $f(x) \neq g(x)$ only on $(0, 1)$.

CHAPTER 3

SECTION 3.1

1. $f'(x) = \lim\limits_{h \to 0} \dfrac{f(x+h) - f(x)}{h} = \lim\limits_{h \to 0} \dfrac{[2 - 3(x+h)] - [2 - 3x]}{h} = \lim\limits_{h \to 0} \dfrac{-3h}{h} = \lim\limits_{h \to 0} -3 = -3$

2. $f'(x) = \lim\limits_{h \to 0} \dfrac{f(x+h) - f(x)}{h} = \lim\limits_{h \to 0} \dfrac{k - k}{h} = \lim\limits_{h \to 0} 0 = 0$

3. $f'(x) = \lim\limits_{h \to 0} \dfrac{f(x+h) - f(x)}{h} = \lim\limits_{h \to 0} \dfrac{[5(x+h) - (x+h)^2] - (5x - x^2)}{h}$

$= \lim\limits_{h \to 0} \dfrac{5h - 2xh - h^2}{h} = \lim\limits_{h \to 0} (5 - 2x - h) = 5 - 2x$

4. $f'(x) = \lim\limits_{h \to 0} \dfrac{f(x+h) - f(x)}{h} = \lim\limits_{h \to 0} \dfrac{[2(x+h)^3 + 1] - [2x^3 + 1]}{h}$

$= \lim\limits_{h \to 0} \dfrac{2(x^3 + 3x^2h + 3xh^2 + h^3) - 2x^3}{h} = \lim\limits_{h \to 0} \dfrac{6x^2h + 6xh^2 + 2h^3}{h}$

$= \lim\limits_{h \to 0} (6x^2 + 6xh + 2h^2) = 6x^2$

5. $f'(x) = \lim\limits_{h \to 0} \dfrac{f(x+h) - f(x)}{h} = \lim\limits_{h \to 0} \dfrac{(x+h)^4 - x^4}{h}$

$= \lim\limits_{h \to 0} \dfrac{(x^4 + 4x^3h + 6x^2h^2 + 4xh^3 + h^4) - x^4}{h}$

$= \lim\limits_{h \to 0} (4x^3 + 6x^2h + 4xh^2 + h^3) = 4x^3$

6. $f'(x) = \lim\limits_{h \to 0} \dfrac{f(x+h) - f(x)}{h} = \lim\limits_{h \to 0} \dfrac{\dfrac{1}{x+h+3} - \dfrac{1}{x+3}}{h}$

$\lim\limits_{h \to 0} \dfrac{(x+3) - (x+h+3)}{h(x+h+3)(x+3)} = \lim\limits_{h \to 0} \dfrac{-h}{h(x+h+3)(x+3)}$

$\lim\limits_{h \to 0} \dfrac{-1}{(x+h+3)(x+3)} = \dfrac{-1}{(x+3)^2}$

7. $f'(x) = \lim\limits_{h \to 0} \dfrac{f(x+h) - f(x)}{h} = \lim\limits_{h \to 0} \dfrac{\sqrt{x+h-1} - \sqrt{x-1}}{h}$

$= \lim\limits_{h \to 0} \dfrac{(x+h-1) - (x-1)}{h(\sqrt{x+h-1} + \sqrt{x-1})} = \lim\limits_{h \to 0} \dfrac{1}{\sqrt{x+h-1} + \sqrt{x-1}} = \dfrac{1}{2\sqrt{x-1}}$

8. $f'(x) = \lim\limits_{h \to 0} \dfrac{f(x+h) - f(x)}{h} = \lim\limits_{h \to 0} \dfrac{[(x+h)^3 - 4(x+h)] - [x^3 - 4x]}{h}$

$= \lim\limits_{h \to 0} \dfrac{3x^2h + 3xh^2 + h^3 - 4h}{h} = \lim\limits_{h \to 0} (3x^2 + 3xh + h^2 - 4) = 3x^2 - 4$

9. $f'(x) = \lim\limits_{h \to 0} \dfrac{f(x+h) - f(x)}{h} = \lim\limits_{h \to 0} \dfrac{\dfrac{1}{(x+h)^2} - \dfrac{1}{x^2}}{h}$

$\qquad = \lim\limits_{h \to 0} \dfrac{x^2 - (x^2 + 2hx + h^2)}{hx^2(x+h)^2} = \lim\limits_{h \to 0} \dfrac{-2x - h}{x^2(x+h)^2} = -\dfrac{2}{x^3}$

10. $f'(x) = \lim\limits_{h \to 0} \dfrac{f(x+h) - f(x)}{h} = \lim\limits_{h \to 0} \dfrac{\dfrac{1}{\sqrt{x+h}} - \dfrac{1}{\sqrt{x}}}{h}$

$\qquad \lim\limits_{h \to 0} \dfrac{\sqrt{x} - \sqrt{x+h}}{h\sqrt{x}\sqrt{x+h}} = \lim\limits_{h \to 0} \dfrac{\left(\sqrt{x} - \sqrt{x+h}\right)\left(\sqrt{x} + \sqrt{x+h}\right)}{h\sqrt{x}\sqrt{x+h}\left(\sqrt{x} + \sqrt{x+h}\right)} = \lim\limits_{h \to 0} \dfrac{x - (x+h)}{h\sqrt{x}\sqrt{x+h}\left(\sqrt{x} + \sqrt{x+h}\right)}$

$\qquad \lim\limits_{h \to 0} \dfrac{-h}{h\sqrt{x}\sqrt{x+h}\left(\sqrt{x} + \sqrt{x+h}\right)} = \dfrac{-1}{2x\sqrt{x}}$

11. $f(x) = x^2 - 4x; \quad c = 3$:

difference quotient:

$$\dfrac{f(3+h) - f(3)}{h} = \dfrac{(3+h)^2 - 4(3+h) - (-3)}{h}$$

$$= \dfrac{9 + 6h + h^2 - 12 - 4h + 3}{h} = \dfrac{2h + h^2}{h} = 2 + h$$

Therefore, $\quad f'(3) = \lim\limits_{h \to 0} \dfrac{f(3+h) - f(3)}{h} = \lim\limits_{h \to 0}(2 + h) = 2$

12. $f(x) = 7x - x^2; \quad c = 2$:

difference quotient:

$$\dfrac{f(2+h) - f(2)}{h} = \dfrac{7(2+h) - (2+h)^2 - (10)}{h}$$

$$= \dfrac{14 + 7h - 4 - 4h - h^2 - 10}{h} = \dfrac{3h - h^2}{h} = 3 - h$$

Therefore, $\quad f'(2) = \lim\limits_{h \to 0} \dfrac{f(2+h) - f(2)}{h} = \lim\limits_{h \to 0}(3 - h) = 3$

13. $f(x) = 2x^3 + 1; \quad c = 1$:

difference quotient:

$$\dfrac{f(-1+h) - f(-1)}{h} = \dfrac{2(-1+h)^3 + 1 - (-1)}{h}$$

$$= \dfrac{2\left[-1 + 3h - 3h^2 + h^3\right] + 2}{h} = \dfrac{6h - 6h^2 + 2h^3}{h} = 6 - 6h + 2^2$$

Therefore, $\quad f'(-1) = \lim\limits_{h \to 0} \dfrac{f(-1+h) - f(-1)}{h} = \lim\limits_{h \to 0}(6 - 6h + 2h^2) = 6$

14. $f(x) = 5 - x^4; \quad c = -1$:

difference quotient:

$$\dfrac{f(1+h) - f(1)}{h} = \dfrac{5 - (1+h)^4 - (4)}{h}$$

$$= \frac{5 - 1 - 4h - 6h^2 - 4h^3 - h^4 - 4}{h}$$

$$= \frac{-4h - 6h^2 - 4h^3 - h^4}{h} = -4 - 6h - 4h^2 - h^3$$

Therefore, $\quad f'(3) = \lim_{h \to 0} \frac{f(1+h) - f(1)}{h} = \lim_{h \to 0} \left(-4 - 6h - 4h^2 - h^3\right) = -4$

15. $f(x) = \dfrac{8}{x+4}; \quad c = -2:$

difference quotient:

$$\frac{f(-2+h) - f(-2)}{h} = \frac{\dfrac{8}{(-2+h)+4} - 4}{h} = \frac{\dfrac{8}{h+2} - 4}{h}$$

$$= \frac{8 - 4h - 8}{h(h+2)} = \frac{-4}{h+2}$$

Therefore, $\quad f'(-2) = \lim_{h \to 0} \frac{f(-2+h) - f(-2)}{h} = \lim_{h \to 0} \frac{-4}{h+2} = -2$

16. $f(x) = \sqrt{6-x}; \quad c = 2:$

difference quotient:

$$\frac{f(2+h) - f(2)}{h} = \frac{\sqrt{6 - (2+h)} - (2)}{h}$$

$$= \frac{\sqrt{4-h} - 2}{h}$$

$$= \frac{\sqrt{4-h} - 2}{h} \cdot \frac{\sqrt{4-h} + 2}{\sqrt{4-h} + 2}$$

$$= \frac{-1}{\sqrt{4-h} + 2}$$

Therefore, $\quad f'(2) = \lim_{h \to 0} \frac{f(2+h) - f(2)}{h} = \lim_{h \to 0} \frac{-1}{\sqrt{4-h} + 2} = -\frac{1}{4}$

17. $f(4) = 4$. Slope of tangent at $(4,4)$ is $f'(4) = -3$. Tangent $\quad y - 4 = -3(x - 4)$.

18. $f(4) = 2$. Slope of tangent at $(4,2)$ is $f'(4) = \dfrac{1}{2\sqrt{4}} = \dfrac{1}{4}$. Tangent $y - 2 = \dfrac{1}{4}(x - 4)$.

19. $f(-2) = \dfrac{1}{4}$. Slope of tangent at $\left(-2, \dfrac{1}{4}\right)$ is $\dfrac{1}{4}$. Tangent: $\quad y - \dfrac{1}{4} = \dfrac{1}{4}(x + 2)$.

20. $f(2) = -3$. Slope of tangent at $(2, -3)$ is $f'(2) = -3(2)^2 = -12$. Tangent: $\quad y + 3 = -12(x - 2)$;

21. (a) f is not continuous at $c = -1$ and $c = 1$; f has a removable discontinuity at $c = -1$ and a jump discontinuity at $c = 1$.

   (b) f is continuous but not differentiable at $c = 0$ and $c = 3$.

22. (a) f is not continuous at $c = 2$; f has a jump discontinuity at 2

   (b) f is continuous but not differentiable at $c = -2$ and $c = 3$.

23.   at $x = -1$

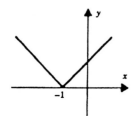

24.   at $x = \frac{5}{2}$

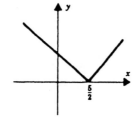

25.   at $x = 0$

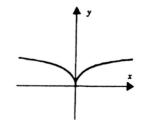

26.   at $x = -2, 2$

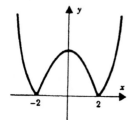

27.   at $x = 1$

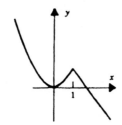

28.   at $x = 2$

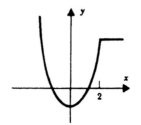

29. $f'(1) = 4$

$$\lim_{h \to 0^-} \frac{f(1+h) - f(1)}{h} = \lim_{h \to 0^-} \frac{4(1+h) - 4}{h} = 4$$

$$\lim_{h \to 0^+} \frac{f(1+h) - f(1)}{h} = \lim_{h \to 0^+} \frac{2(1+h)^2 + 2 - 4}{h} = 4$$

30. $f'(1) = 6$

$$\lim_{h \to 0^-} \frac{f(1+h) - f(1)}{h} = \lim_{h \to 0^-} \frac{3(1+h)^2 - 3}{h} = 6$$

$$\lim_{h \to 0^+} \frac{f(1+h) - f(1)}{h} = \lim_{h \to 0^+} \frac{[2(1+h)^3 + 1] - 3}{h} = 6$$

31. $f'(-1)$ does not exist

$$\lim_{h \to 0^-} \frac{f(-1+h) - f(-1)}{h} = \lim_{h \to 0^-} \frac{h - 0}{h} = 1$$

$$\lim_{h \to 0^+} \frac{f(-1+h) - f(-1)}{h} = \lim_{h \to 0^+} \frac{h^2 - 0}{h} = 0$$

32. $f'(3)$ does not exist because f is not continuous at $x = 3$.

33.

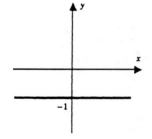

34.

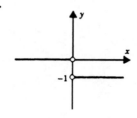

35.

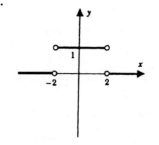

36. 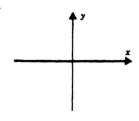 **37.** 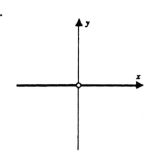 **38.**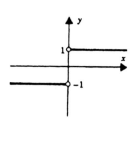

39. Since $f(1) = 1$ and $\lim\limits_{x \to 1^+} f(x) = 2$, f is not continuous at 1. Therefore, by (3.1.3), f is not differentiable at 1.

40. (a) $f'(x) = \begin{cases} 2(x+1), & x < 0 \\ 2(x-1), & x > 0. \end{cases}$

(b) $\lim\limits_{h \to 0^-} \dfrac{f(0+h) - f(0)}{h} = \lim\limits_{h \to 0^-} \dfrac{(h+1)^2 - 1}{h} = \lim\limits_{h \to 0^-} (h+2) = 2,$

$\lim\limits_{h \to 0^+} \dfrac{f(0+h) - f(0)}{h} = \lim\limits_{h \to 0^+} \dfrac{(h-1)^2 - 1}{h} = \lim\limits_{h \to 0^+} (h-2) = -2.$

41. Continuity at $x = 1$: $\lim\limits_{x \to 1^-} f(x) = 1 = \lim\limits_{x \to 1^+} f(x) = A + B.$ Thus $A + B = 1$.
Differentiability at $x = 1$:

$$\lim\limits_{h \to 0^-} \dfrac{f(1+h) - f(1)}{h} = \lim\limits_{h \to 0^-} \dfrac{(1+h)^3 - 1}{h} = 3 = \lim\limits_{h \to 0^+} \dfrac{f(1+h) - f(1)}{h} = A$$

Therefore, $A = 3, \implies B = -2$.

42. Continuity at $x = 2 \implies 4B + 2A = 2$;   differentiability at $x = 2 \implies 4B + A = 4$;
$A = -2, \ B = \frac{3}{2}$

43. $f(x) = c, \ c$ any constant.   **44.** $f(x) = \begin{cases} 1, & x \neq 0 \\ 0, & x = 0. \end{cases}$

45. $f(x) = |x + 1|$;   or $f(x) = \begin{cases} 0, & x \neq -1 \\ 1, & x = -1. \end{cases}$

46. $f(x) = |x^2 - 1|$

47. $f(x) = 2x + 5$ **48.** $f(x) = -|x|$

49. (a) $\lim\limits_{x \to 2^+} f(x) = \lim\limits_{x \to 2^-} f(x) = f(2) = 2$ Thus, f is continuous at $x = 2$.

(b) $\lim\limits_{h \to 0^-} \dfrac{f(2+h) - f(2)}{h} = \lim\limits_{h \to 0^-} \dfrac{(2+h)^2 - (2+h) - 2}{h} = 3$

$\lim\limits_{h \to 0^+} \dfrac{f(2+h) - f(2)}{h} = \lim\limits_{h \to 0^+} \dfrac{2(2+h) - 2 - 2}{h} = 2$ f is not differentiable at 2.

50.

$$f'(x) = \lim_{h \to 0} \frac{(x+h)\sqrt{x+h} - x\sqrt{x}}{h}$$

$$= \lim_{h \to 0} \frac{x\left(\sqrt{x+h} - \sqrt{x}\right) + h\sqrt{x+h}}{h}$$

$$= \lim_{h \to 0} \frac{xh}{h\left(\sqrt{x+h} + \sqrt{x}\right)} + \lim_{h \to 0} \sqrt{x+h}$$

$$= \tfrac{1}{2}\sqrt{x} + \sqrt{x} = \tfrac{3}{2}\sqrt{x}$$

51. (a) f is not continuous at 0 since $\lim_{x \to 0^-} f(x) = 1$ and $\lim_{x \to 0^+} f(x) = 0$. Therefore f is not differentiable at 0.

(b)

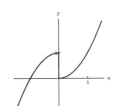

52. (a) $\lim_{h \to 0} \dfrac{f(h) - 0}{h} = \begin{cases} \lim_{h \to 0} \dfrac{h}{h} = 1, & h \text{ rational} \\[2mm] \lim_{h \to 0} \dfrac{0}{h} = 0, & h \text{ irrational.} \end{cases}$

Therefore, $\lim_{h \to 0} \dfrac{f(0+h) - f(0)}{h}$ does not exist.

(b) $\lim_{h \to 0} \dfrac{g(h) - 0}{h} = \begin{cases} \lim_{h \to 0} \dfrac{h^2}{h} = 0, & h \text{ rational} \\[2mm] \lim_{h \to 0} \dfrac{0}{h} = 0, & h \text{ irrational} \end{cases}$

Therefore, g is differentiable at 0 and $g'(0) = 0$.

53. (a) If the tangent line is horizontal, then the normal line is vertical: $x = c$.

(b) If the tangent line has slope $f'(c) \neq 0$, then the normal line has slope $-\dfrac{1}{f'(c)}$:

$$y - f(c) = -\frac{1}{f'(c)}(x - c).$$

(c) If the tangent line is vertical, then the normal line is horizontal: $y = f(c)$.

54. The center of the circle.

55. The normal line has slope -4 : $y - 2 = -4(x - 4)$.

56.

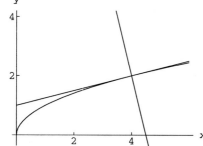

57. $f(x) = x^2,\ f'(x) = 2x$

The tangent line at $(3,9)$ has equation $y - 9 = 6(x - 3)$ and x-intercept $x = \dfrac{3}{2}$.

The normal line at $(3,9)$ has equation $y - 9 = -\frac{1}{6}(x - 3)$ and x-intercept $x = 57$.

$s = 57 - \dfrac{3}{2} = \dfrac{111}{2}$.

58. (a) The slope of the normal at P is y/x.

(b) The slope of the tangent at P is $-x/y$.

(c)
$$\frac{\sqrt{1 - (x+h)^2} - \sqrt{1 - x^2}}{h} = \frac{\sqrt{1 - (x+h)^2} - \sqrt{1 - x^2}}{h} \cdot \frac{\sqrt{1 - (x+h)^2} + \sqrt{1 - x^2}}{\sqrt{1 - (x+h)^2} + \sqrt{1 - x^2}}$$

$$= \frac{1 - (x+h)^2 - (1 - x^2)}{h\left[\sqrt{1 - (x+h)^2} + \sqrt{1 - x^2}\right]}$$

$$= \frac{-2xh - h^2}{h\left[\sqrt{1 - (x+h)^2} + \sqrt{1 - x^2}\right]} = \frac{-2x - h}{\left[\sqrt{1 - (x+h)^2} + \sqrt{1 - x^2}\right]}$$

$$\lim_{h \to 0} \frac{f(x+h) - f(x)}{h} = \lim_{h \to 0} \frac{-2x - h}{\sqrt{1 - (x+h)^2} + \sqrt{1 - x^2}} = -\frac{2x}{2\sqrt{1 - x^2}} = -\frac{x}{y}$$

where $y = \sqrt{1 - x^2}$.

59. (a) Since $\quad |\sin(1/x)| \le 1 \quad$ it follows that

$$-x \le f(x) \le x \qquad \text{and} \qquad -x^2 \le g(x) \le x^2$$

Thus $\quad \lim_{x \to 0} f(x) = f(0) = 0 \quad$ and $\quad \lim_{x \to 0} g(x) = g(0) = 0, \quad$ which implies that f and g are continuous at 0.

(b) $\displaystyle \lim_{h \to 0} \frac{h \sin(1/h) - 0}{h} = \lim_{h \to 0} \sin(1/h) \quad$ does not exist.

(c) $\displaystyle \lim_{h \to 0} \frac{h^2 \sin(1/h) - 0}{h} = \lim_{h \to 0} h \sin(1/h) = 0.$ Thus g is differentiable at 0 and $g'(0) = 0$.

60. $f'(2) = \lim\limits_{h \to 0} \dfrac{f(2+h) - f(2)}{h} = \lim\limits_{h \to 0} \dfrac{[(2+h)^3 + 1] - (2^3 + 1)}{h} = \lim\limits_{h \to 0} \dfrac{12h + 6h^2 + h^3}{h} = 12$

$f'(2) = \lim\limits_{x \to 2} \dfrac{f(x) - f(2)}{x-2} = \lim\limits_{x \to 2} \dfrac{(x^3 + 1) - 9}{x-2} = \lim\limits_{x \to 2} \dfrac{(x-2)(x^2 + 2x + 4)}{x-2} = \lim\limits_{x \to 2}(x^2 + 2x + 4) = 12$

61. $f'(1) = \lim\limits_{h \to 0} \dfrac{f(1+h) - f(1)}{h} = \lim\limits_{h \to 0} \dfrac{[(1+h)^2 - 3(1+h)] - (-2)]}{h} = \lim\limits_{h \to 0} \dfrac{-h + h^2}{h} = -1$

$f'(1) = \lim\limits_{x \to 1} \dfrac{f(x) - f(1)}{x-1} = \lim\limits_{x \to 1} \dfrac{(x^2 - 3x) - (-2)}{x-1} = \lim\limits_{x \to 1} \dfrac{(x-2)(x-1)}{x-1} = \lim\limits_{x \to 1}(x - 2) = -1$

62. $f'(3) = \lim\limits_{h \to 0} \dfrac{f(3+h) - f(3)}{h} = \lim\limits_{h \to 0} \dfrac{\sqrt{1 + (3+h)} - 2}{h} = \lim\limits_{h \to 0} \dfrac{\sqrt{4+h} - 2}{h} = \lim\limits_{h \to 0} \dfrac{h}{h\sqrt{4+h} + 2} = \dfrac{1}{4}$

$f'(3) = \lim\limits_{x \to 3} \dfrac{f(x) - f(3)}{x-3} = \lim\limits_{x \to 1} \dfrac{\sqrt{1+x} - 2}{x-3} = \lim\limits_{x \to 3} \dfrac{x-3}{(x-3)(\sqrt{1+x} + 2)} = \lim\limits_{x \to 3} \dfrac{1}{\sqrt{1+x} + 2} = \dfrac{1}{4}$

63. $f'(-1) = \lim\limits_{h \to 0} \dfrac{f(-1+h) - f(-1)}{h} = \lim\limits_{h \to 0} \dfrac{(-1+h)^{1/3} + 1}{h}$

$= \lim\limits_{h \to 0} \dfrac{(-1+h)^{1/3} + 1}{h} \cdot \dfrac{(-1+h)^{2/3} - (-1+h)^{1/3} + 1}{(-1+h)^{2/3} - (-1+h)^{1/3} + 1}$

$= \lim\limits_{h \to 0} \dfrac{h}{h\left[-1+h)^{2/3} - (-1+h)^{1/3} + 1\right]} = \dfrac{1}{3}$

$f'(-1) = \lim\limits_{x \to -1} \dfrac{f(x) - f(-1)}{x - (-1)} = \lim\limits_{x \to -1} \dfrac{x^{1/3} + 1}{x+1} = \lim\limits_{x \to -1} \dfrac{x^{1/3} + 1}{x+1} \cdot \dfrac{x^{2/3} - x^{1/3} + 1}{x^{2/3} - x^{1/3} + 1}$

$= \lim\limits_{x \to -1} \dfrac{x+1}{(x+1)(x^{2/3} - x^{1/3} + 1)} = \dfrac{1}{3}$

64. $f'(0) = \lim\limits_{h \to 0} \dfrac{f(0+h) - f(0)}{h} = \lim\limits_{h \to 0} \dfrac{\dfrac{1}{h+2} - \dfrac{1}{2}}{h} = \lim\limits_{h \to 0} \dfrac{-h}{2h(h+2)} = \lim\limits_{h \to 0} \dfrac{-1}{2(h+2)} = -\dfrac{1}{4}$

$f'(0) = \lim\limits_{x \to 0} \dfrac{f(x) - f(0)}{x} = \lim\limits_{x \to 0} \dfrac{\dfrac{1}{x+2} - \dfrac{1}{2}}{x} = \lim\limits_{x \to 0} \dfrac{-x}{2x(x+2)} = \lim\limits_{x \to 0} \dfrac{-1}{2(x+2)} = -\dfrac{1}{4}$

65. (a) $D = \dfrac{(2+h)^{5/2} - 2^{5/2}}{h} \qquad -1 \le h \le 1$

(b) $f'(2) \cong 7.071$

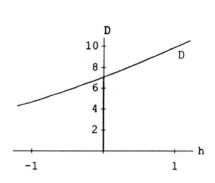

66. (a) $D = \dfrac{(2+h)^{2/3} - 2^{2/3}}{h} \qquad -1 \le h \le 1$

(b) $f'(2) \cong 0.529$

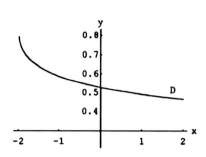

67. (a) $f'(x) = \dfrac{5}{2\sqrt{5x-4}}$; $f'(3) = \dfrac{5}{2\sqrt{11}}$

(b) $f'(x) = -2x + 16x^3 - 6x^5$; $f'(-2) = 68$

(c) $f'(x) = -\dfrac{3(3-2x)}{2+3x)^2} - \dfrac{2}{2+3x}$; $f'(-1) = 13$

68. (a) $f'(1)$ does not exist   (b) $f'(-2) = 0$ (c) $f'(3)$ does not exist

69. (c) $f'(x) = 10x - 21x^2$ (d) $f'(x) = 0$ at $x = 0,\ \frac{10}{21}$

70. (c) $f'(x) = 3x^2 + 2x - 4$ (d) $f'(x) = 0$ at $x \cong -1.5352,\ 0.8685$

71. (a) $f'(\frac{3}{2}) = -\frac{11}{4}$;   tangent line: $y - \frac{21}{8} = -\frac{11}{4}\left(x - \frac{3}{2}\right)$,   normal line: $y - \frac{21}{8} = \frac{4}{11}\left(x - \frac{3}{2}\right)$

(c) $(1.453,\ 1.547)$

72. (a) Set $f(x) = x^3$. Then

$$\frac{f(x+h) - f(x)}{h} = \frac{(x+h)^3 - x^3}{h}$$

$$= \frac{x^3 + 3x^2h + 3xh^2 + h^3 - x^3}{h}$$

$$= \frac{3x^2h + 3xh^2 + h^3}{h} = 3x^2 + 3xh + h^2$$

$$\lim_{h \to 0} \frac{f(x+h) - f(x)}{h} = \lim_{h \to 0}(3x^2 + 3xh + h^2) = 3x^2.$$

(b) Let S be the set of integers for which the statement is true. We have shown that $1, 2, 3 \in S$. Assume that $k \in S$. This tells us that if $f(x) = x^k$, then $f'(x) = \lim_{h \to 0} \dfrac{(x+h)^k - x^k}{h} = kx^{k-1}$. Set $f(x) = x^{k+1}$. Then, by the hint,

$$f'(x) = \lim_{h \to 0} \frac{(x+h)^{k+1} - x^{k+1}}{h} = \lim_{h \to 0} \frac{x(x+h)^k - x \cdot x^k + h(x+h)^k}{h}$$

$$= \lim_{h \to 0} x \left[\frac{(x+h)^k - x^k}{h}\right] + \lim_{h \to 0} (x+h)^k$$

$$= x \cdot kx^{k-1} + x^k = (k+1)x^k.$$

Therefore, $k + 1 \in S$. Thus S contains the set of positive integers.

SECTION 3.2

1. $F'(x) = -1$

2. $F'(x) = 2$

3. $F'(x) = 55x^4 - 18x^2$

4. $F'(x) = \dfrac{-6}{x^3}$

5. $F'(x) = 2ax + b$

6. $F'(x) = x^3 - x^2 + x - 1$

7. $F'(x) = 2x^{-3}$

8. $F'(x) = \dfrac{x^3(2x) - (x^2 + 2)3x^2}{x^6} = -\dfrac{x^2 + 6}{x^4}$

9. $G'(x) = (x^2 - 1)(1) + (x - 3)(2x) = 3x^2 - 6x - 1$

10. $F'(x) = 1 + \dfrac{1}{x^2}$

11. $G'(x) = \dfrac{(1 - x)(3x^2) - x^3(-1)}{(1 - x)^2} = \dfrac{3x^2 - 2x^3}{(1 - x)^2}$

12. $F'(x) = \dfrac{(cx - d)a - (ax - b)c}{(cx - d)^2} = \dfrac{bc - ad}{(cx - d)^2}$

13. $G'(x) = \dfrac{(2x + 3)(2x) - (x^2 - 1)(2)}{(2x + 3)^2} = \dfrac{2(x^2 + 3x + 1)}{(2x + 3)^2}$

14. $G'(x) = \dfrac{(x + 1)(28x^3) - (7x^4 + 11)(1)}{(x + 1)^2} = \dfrac{21x^4 + 28x^3 - 11}{(x + 1)^2}$

15. $G'(x) = (x^3 - 2x)(2) + (2x + 5)(3x^2 - 2) = 8x^3 + 15x^2 - 8x - 10$

16. $G'(x) = \dfrac{(x^2 - 1)(3x^2 + 3) - (x^3 + 3x)2x}{(x^2 - 1)^2} = \dfrac{x^4 - 6x^2 - 3}{(x^2 - 1)^2}$

17. $G'(x) = \dfrac{(x - 2)(1/x^2) - (6 - 1/x)(1)}{(x - 2)^2} = \dfrac{-2(3x^2 - x + 1)}{x^2(x - 2)^2}$

18. $G'(x) = \dfrac{x^2(4x^3) - (1 + x^4)2x}{x^4} = \dfrac{2(x^4 - 1)}{x^3}$

19. $G'(x) = (9x^8 - 8x^9)\left(1 - \dfrac{1}{x^2}\right) + \left(x + \dfrac{1}{x}\right)(72x^7 - 72x^8) = -80x^9 + 81x^8 - 64x^7 + 63x^6$

20. $G'(x) = \left(\dfrac{-1}{x^2}\right)\left(1 + \dfrac{1}{x^2}\right) + \left(1 + \dfrac{1}{x}\right)\left(\dfrac{-2}{x^3}\right) = -\dfrac{1}{x^2} - \dfrac{2}{x^3} - \dfrac{3}{x^4}$

21. $f'(x) = -(x - 2)^{-2};\quad f'(0) = -\frac{1}{4}, \quad f'(1) = -1$

22. $f'(x) = 3x^3 + 2x;\quad f'(0) = 0, \quad f'(1) = 5$

23. $f'(x) = \dfrac{(1 + x^2)(-2x) - (1 - x^2)(2x)}{(1 + x^2)^2} = \dfrac{-4x}{(1 + x^2)^2};\quad f'(0) = 0, \quad f'(1) = -1$

24. $f'(x) = \dfrac{(x^2 + 2x + 1)(4x + 1) - (2x^2 + x + 1)(2x + 2)}{(x^2 + 2x + 1)^2} = \dfrac{(x + 1)(4x + 1) - 2(2x^2 + x + 1)}{(x + 1)^3};$

$f'(0) = -1, \quad f'(1) = \frac{1}{4}$

25. $f'(x) = \dfrac{(cx+d)a - (ax+b)c}{(cx+d)^2} = \dfrac{ad-bc}{(cx+d)^2},$ $f'(0) = \dfrac{ad-bc}{d^2},$ $f'(1) = \dfrac{ad-bc}{(c+d)^2}$

26. $f'(x) = \dfrac{(cx^2+bx+a)(2ax+b) - (ax^2+bx+c)(2cx+b)}{(cx^2+bx+a)^2};$ $f'(0) = \dfrac{b(a-c)}{a^2},$ $f'(1) = \dfrac{2(a-c)}{a+b+c}$

27. $f'(x) = xh'(x) + h(x);$ $f'(0) = 0h'(0) + h(0) = 0(2) + 3 = 3$

28. $f'(x) = 6xh(x) + 3x^2h'(x) - 5;$ $f'(0) = -5$

29. $f'(x) = h'(x) + \dfrac{h'(x)}{[h(x)]^2},$ $f'(0) = h'(0) + \dfrac{h'(0)}{[h(0)]^2} = 2 + \dfrac{2}{3^2} = \dfrac{20}{9}$

30. $f'(x) = h'(x) + \dfrac{h(x) - xh'(x)}{h^2(x)};$ $f'(0) = h'(0) + \dfrac{h(0)}{h^2(0)} = 2 + \dfrac{1}{3} = \dfrac{7}{3}$

31. $f'(x) = \dfrac{(x+2)(1) - x(1)}{(x+2)^2} = \dfrac{2}{(x+2)^2},$
slope of tangent at $(-4, 2) : f'(-4) = \frac{1}{2};$ equation for tangent: $y - 2 = \frac{1}{2}(x+4)$

32. $f'(x) = (x^3 - 2x + 1)(4) + (4x - 5)(3x^2 - 2) = 16x^3 - 15x^2 - 16x + 14$
slope of tangent at $(2, 15) : f'(2) = 50;$ equation for tangent: $y - 15 = 50(x-2)$

33. $f'(x) = (x^2 - 3)(5 - 3x^2) + (5x - x^3)(2x);$
slope of tangent at $(1, -8) : f'(1) = (-2)(2) + (4)(2) = 4;$ equation for tangent: $y + 8 = 4(x-1)$

34. $f'(x) = 2x + \dfrac{10}{x^2};$
slope of tangent at $(-2, 9) : f'(-2) = -\frac{3}{2};$ equation for tangent: $y - 9 = -\frac{3}{2}(x+2)$

35. $f'(x) = (x-2)(2x-1) + (x^2 - x - 11)(1) = 3(x-3)(x+1);$
$f'(x) = 0$ at $x = -1, 3;$ $(-1, 27),$ $(3, -5)$

36. $f'(x) = 2x + \dfrac{16}{x^2} = \dfrac{2(x^3 + 8)}{x^2};$ $f'(x) = 0$ at $x = -2;$ $(-2, 12)$

37. $f'(x) = \dfrac{(x^2+1)(5) - 5x(2x)}{(x^2+1)^2} = \dfrac{5(1-x^2)}{(x^2+1)^2},$ $f'(x) = 0$ at $x = \pm1;$ $(-1, -5/2),$ $(1, 5/2)$

38. $f'(x) = (x+2)(2x-2) + (x^2 - 2x - 8)(1) = 3x^2 - 12 = 3(x^2 - 4);$
$f'(x) = 0$ at $x = \pm2;$ $(-2, 0),$ $(2, -32)$

39. $f(x) = x^4 - 8x^2 + 3;$ $f'(x) = 4x^3 - 16x = 4x(x^2 - 4) = 4x(x-2)(x+2)$
  (a) $f'(x) = 0$ at $x = 0,$ ±2
  (b) $f'(x) > 0$ on $(-2, 0) \cup (2, \infty)$
  (c) $f'(x) < 0$ on $(-\infty, -2) \cup (0, 2)$

40. $f(x) = 3x^4 - 4x^3 - 2; \quad f'(x) = 12x^3 - 12x^2 = 12x^2(x-1)$

(a) $f'(x) = 0$ at $x = 0, 1$.

(b) $f'(x) > 0$ on $(1, \infty)$.

(c) $f'(x) < 0$ on $(-\infty, 0) \cup (0, 1)$

41. $f(x) = x + \dfrac{4}{x^2}; \quad f'(x) = 1 - \dfrac{8}{x^3} = \dfrac{x^3 - 8}{x^3} = \dfrac{(x-2)(x^2 + 2x + 4)}{x^3}$

(a) $f'(x) = 0$ at $x = 2$.

(b) $f'(x) > 0$ on $(-\infty, 0) \cup (2, \infty)$

(c) $f'(x) < 0$ on $(0, 2)$

42. $f(x) = \dfrac{x^2 - 2x + 4}{x^2 + 4}; \quad f'(x) = \dfrac{(x^2 + 4)(2x - 2) - (x^2 - 2x + 4)(2x)}{(x^2 + 4)^2} = \dfrac{2(x^2 - 4)}{(x^2 + 4)^2}$

(a) $f'(x) = 0$ at $x = 2, -2$.

(b) $f'(x) > 0$ on $(-\infty, -2) \cup (2, \infty)$

(c) $f'(x) < 0$ on $(-2, 2)$

43. slope of line 4; slope of tangent $f'(x) = -2x; \quad -2x = 4$ at $x = -2; \quad (-2, -10)$

44. slope of line 3/5; slope of tangent $f'(x) = 3x^2 - 3;$
perpendicular when $3x^2 - 3 = -\frac{5}{3}; \quad x = \pm\frac{2}{3}; \quad \left(-\frac{2}{3}, \frac{46}{27}\right), \left(\frac{2}{3}, -\frac{46}{27}\right)$

45. $f(x) = x^3 + x^2 + x + C$ **46.** $f(x) = x^4 - x^2 + 4x + C$

47. $f(x) = \dfrac{2x^3}{3} - \dfrac{3x^2}{2} + \dfrac{1}{x} + C$ **48.** $f(x) = \dfrac{x^5}{5} + \dfrac{x^4}{2} + \sqrt{x} + C$

49. We want f to be continuous at $x = 2$. That is, we want

$$\lim_{x \to 2^-} f(x) = f(2) = \lim_{x \to 2^+} f(x).$$

This gives

(1) $8A + 2B + 2 = 4B - A.$

We also want

$$\lim_{x \to 2^-} f'(x) = \lim_{x \to 2^+} f'(x).$$

This gives

(2) $12A + B = 4B.$

Equations (1) and (2) together imply that $A = -2$ and $B = -8$.

50. First we need, $\lim_{x \to -1^-} f(x) = \lim_{x \to -1^+} f(x), \quad \Longrightarrow \quad A + B = -B - A + 4 \text{ or } A + B = 2.$

Next we need, $\lim_{x \to -1^-} f'(x) = \lim_{x \to -1^+} f'(x) \quad \Longrightarrow \quad -2A = 5B + A \text{ or } 3A + 5B = 0.$

Solving these equations gives $A = 5, B = -3.$

51.

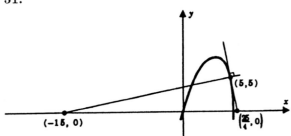

slope of tangent at $(5,5)$ is $f'(5) = -4$

tangent $y - 5 = -4(x-5)$ intersects
  x-axis at $\left(\frac{25}{4}, 0\right)$

normal $y - 5 = \frac{1}{4}(x-5)$ intersects
  x-axis at $(-15, 0)$

area of triangle is
$$\tfrac{1}{2}(5)(15 + \tfrac{25}{4}) = \tfrac{425}{8}$$

52.

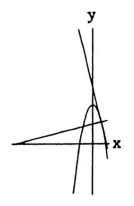

slope of tangent at $(2,5)$ is $f'(2) = -4$

tangent $y - 5 = -4(x-2)$ intersects
  x-axis at $\left(\frac{13}{4}, 0\right)$

normal $y - 5 = \frac{1}{4}(x-2)$ intersects
  x-axis at $(-18, 0)$

area of triangle is
$$\tfrac{1}{2}(5)(18 + \tfrac{13}{4}) = \tfrac{425}{8}$$

53. If the point $(1, 3)$ lies on the graph, we have $f(1) = 3$ and thus

$(*)$ $\qquad\qquad\qquad\qquad\qquad A + B + C = 3.$

If the line $4x + y = 8$ (slope -4) is tangent to the graph at $(2, 0)$, then
$f(2) = 0$ and $f'(2) = -4$. Thus,

$(**)$ $\qquad\qquad\qquad 4A + 2B + C = 0 \quad\text{and}\quad 4A + B = -4.$

Solving the equations in $(*)$ and $(**)$, we find that $A = -1, \quad B = 0, \quad C = 4$.

54. First, $f(1) = 0$ and $f'(1) = 3 \quad\Longrightarrow\quad A + B + C + D = 0 \ \text{ and } \ 3A + 2B + C = 3$
Next, $f(2) = 9$ and $f'(2) = 18 \quad\Longrightarrow\quad 8A + 4B + 2C + D = 9 \ \text{ and } \ 12A + 4B + C = 18$
Solving these equations gives $A = 3, \ B = -6, \ C = 6, \ D = -3$.

55. Let $f(x) = ax^2 + bx + c$. Then $f'(x) = 2ax + b$ and $f'(x) = 0$ at $x = -b/2a$.

56. The derivative of p is the quadratic $p'(x) = 3ax^2 + 2bx + c$. Its discriminant is

$$D = (2b)^2 - 4(3a)(c) = 4b^2 - 12ac$$

(a) p has two horizontal tangents iff p' has two real roots iff $D > 0$.

(b) p has exactly one horizontal tangent iff p has only one real root iff $D = 0$.

(c) p has no horizontal tangent iff p has no real roots iff $D < 0$.

57. Let $f(x) = x^3 - x$. The secant line through $(-1, f(-1)) = (-1, 0)$ and $(2, f(2)) = (2, 6)$ has slope

$m = \dfrac{6 - 0}{2 - (-1)} = 2$. Now, $f'(x) = 3x^2 - 1$ and $3c^2 - 1 = 2$ implies $c = -1,\ 1$.

58. $m_{sec} = \dfrac{f(3) - f(1)}{3 - 1} = \dfrac{\frac{3}{4} - \frac{1}{2}}{2} = \frac{1}{8}$

$f'(x) = \dfrac{(x + 1)(1) - x(1)}{(x + 1)^2} = \dfrac{1}{(x + 1)^2};\quad f'(c) = \frac{1}{8} \implies c = -1 \pm 2\sqrt{2}$

59. Let $f(x) = 1/x$, $x > 0$. Then $f'(x) = -1/x^2$. An equation for the tangent line to the graph of f at the point $(a, f(a))$, $a > 0$, is $y = (-1/a^2)x + 2/a$. The y-intercept is $2/a$ and the x-intercept is $2a$. The area of the triangle formed by this line and the coordinate axes is: $A = \frac{1}{2}(2/a)(2a) = 2$ square units.

60. Let (x, y) be the point on the graph that the tangent line passes through. $f'(x) = 3x^2$, so $x^3 - 8 = 3x^2(x - 2)$. Thus $x = 2$ or $x = -1$. The lines are $y - 8 = 12(x - 2)$ and $y + 1 = 3(x + 1)$.

61. Let (x, y) be the point on the graph that the tangent line passes through. $f'(x) = 3x^2 - 1$, so $x^3 - x - 2 = (3x^2 - 1)(x + 2)$. Thus $x = 0$ or $x = -3$. The lines are $y = -x$ and $y + 24 = 26(x + 3)$.

62. (a) $f(c) = c^3$; $f'(x) = 3x^2$ and $f'(c) = 3c^2$. Tangent line: $y - c^3 = 3c^2(x - c)$ or $y = 3c^2x - 2c^3$.

(b) We solve the equation $3c^2x - 2c^3 = x^3$:

$$x^3 - 3c^2x + 2c^3 = 0 \implies (x - c)(x^2 + cx - 2c^2) = 0 \implies (x - c)^2(x + 2c) = 0$$

Thus, the tangent line at $x = c$, $c \neq 0$ intersects the graph at $x = -2c$.

63. Since f and $f + g$ are differentiable, $g = (f + g) - f$ is differentiable. The functions $f(x) = |x|$ and $g(x) = -|x|$ are not differentiable at $x = 0$ yet their sum $f(x) + g(x) \equiv 0$ is differentiable for all x.

64. No. If f and fg are differentiable, then $g = \dfrac{fg}{f}$ will be differentiable where $f(x) \neq 0$.

65. Since

$$\left(\frac{f}{g}\right)(x) = \frac{f(x)}{g(x)} = f(x) \cdot \frac{1}{g(x)},$$

it follows from the product and reciprocal rules that

$$\left(\frac{f}{g}\right)'(x) = \left(f \cdot \frac{1}{g}\right)'(x) = f(x)\left(-\frac{g'(x)}{[g(x)]^2}\right) + f'(x) \cdot \frac{1}{g(x)} = \frac{g(x)f'(x) - f(x)g'(x)}{[g(x)]^2}.$$

66.
$$(fgh)'(x) = [(fg)(x) \cdot h(x)]' = (fg)(x)h'(x) + h(x)[(fg)(x)]'$$

$$= f(x)g(x)h'(x) + h(x)[f(x)g'(x) + g(x)f'(x)]$$

$$= f'(x)g(x)h(x) + f(x)g'(x)h(x) + f(x)g(x)h'(x)$$

67. $F'(x) = 2x\left(1 + \dfrac{1}{x}\right)(2x^3 - x + 1) + (x^2 + 1)\left(\dfrac{-1}{x^2}\right)(2x^3 - x + 1) + (x^2 + 1)\left(1 + \dfrac{1}{x}\right)(6x^2 - 1)$

68. $G'(x) = \dfrac{1}{2\sqrt{x}}\left(\dfrac{1}{1+2x}\right)(x^2+x-1) + \sqrt{x}\left(\dfrac{-2}{(1+2x)^2}\right)(x^2+x-1) + \sqrt{x}\left(\dfrac{1}{1+2x}\right)(2x+1)$

69. $g(x) = [f(x)]^2 = f(x)\cdot f(x)$

$g'(x) = f(x)f'(x) + f(x)f'(x) = 2f(x)f'(x)$

70. $g'(x) = 2(x^3 - 2x^2 + x + 2)(3x^2 - 4x + 1)$

71. (a) $f'(x) = 0$ at $x = 0,\ -2$ (b) $f'(x) > 0$ on $(-\infty, -2)\cup(0,\infty)$

    (c) $f'(x) < 0$ on $(-2,-1)\cup(-1,0)$

72. (a) $f'(x) = 0$ at $x = 0,\ 1,\ \frac{5}{2}$ (b) $f'(x) > 0$ on $(-\infty,0)\cup\left(1,\frac{5}{2}\right)\cup(5/2,\infty)$

    (c) $f'(x) < 0$ on $(0,1)$

73. (a) $f'(x) \neq 0$ for all $x \neq 0$ (b) $f'(x) > 0$ on $(0,\infty)$

    (c) $f'(x) < 0$ on $(-\infty,0)$

74. (a) $f'(x) = 0$ at $x = -\sqrt[3]{4} \cong -1.587$ (b) $f'(x) > 0$ on $\left(-\sqrt[3]{4},0\right)$

    (c) $f'(x) < 0$ on $\left(-\infty,-\sqrt[3]{4}\right)\cup(0,\infty)$

75. (a) $\dfrac{\sin(0+0.001)-\sin 0}{0.001} \cong 0.99999$ $\dfrac{\sin(0-0.001)-\sin 0}{-0.001} \cong 0.99999$

$\dfrac{\sin[(\pi/6)+0.001]-\sin(\pi/6)}{0.001} \cong 0.86578$ $\dfrac{\sin[(\pi/6)-0.001]-\sin(\pi/6)}{-0.001} \cong 0.86628$

$\dfrac{\sin[(\pi/4)+0.001]-\sin(\pi/4)}{0.001} \cong 0.70675$ $\dfrac{\sin[(\pi/4)-0.001]-\sin(\pi/4)}{-0.001} \cong 0.70746$

$\dfrac{\sin[(\pi/3)+0.001]-\sin(\pi/3)}{0.001} \cong 0.49957$ $\dfrac{\sin[(\pi/3)-0.001]-\sin(\pi/3)}{-0.001} \cong 0.50043$

$\dfrac{\sin[(\pi/2)+0.001]-\sin(\pi/2)}{0.001} \cong -0.0005$ $\dfrac{\sin[(\pi/2)-0.001]-\sin(\pi/2)}{-0.001} \cong 0.0005$

(b) $\cos 0 = 1,\quad \cos(\pi/6) \cong 0.866025,\quad \cos(\pi/4) \cong 0.707107,\quad \cos(\pi/3) = 0.5,\quad \cos(\pi/2) = 0$

(c) If $f(x) = \sin x$ then $f'(x) = \cos x$.

76. (a)

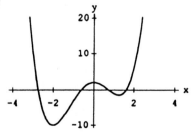

$x = -2,\ 0,\ \frac{5}{4}$

(b)

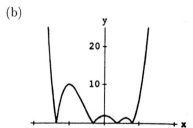

$x_1 = -2.732,\quad x_2 = -0.618,$

$x_3 = 0.732,\quad x_4 = 1.618$

SECTION 3.3

1. $\dfrac{dy}{dx} = 12x^3 - 2x$

2. $\dfrac{dy}{dx} = 2x - 8x^{-5}$

3. $\dfrac{dy}{dx} = 1 + \dfrac{1}{x^2}$

4. $\dfrac{dy}{dx} = \dfrac{(1-x)2 - 2x(-1)}{(1-x)^2} = \dfrac{2}{(1-x)^2}$

5. $\dfrac{dy}{dx} = \dfrac{(1+x^2)(1) - x(2x)}{(1+x^2)^2} = \dfrac{1-x^2}{(1+x^2)^2}$

6. $y = x^3 - x^2 - 2x; \ \dfrac{dy}{dx} = 3x^2 - 2x - 2$

7. $\dfrac{dy}{dx} = \dfrac{(1-x)2x - x^2(-1)}{(1-x)^2} = \dfrac{x(2-x)}{(1-x)^2}$

8. $y = \dfrac{2x - x^2}{3 + 3x}; \ \dfrac{dy}{dx} = \dfrac{(3+3x)(2-2x) - (2x-x^2)(3)}{(3+3x)^2} = \dfrac{2 - 2x - x^2}{3(1+x)^2}$

9. $\dfrac{dy}{dx} = \dfrac{(x^3-1)3x^2 - (x^3+1)3x^2}{(x^3-1)^2} = \dfrac{-6x^2}{(x^3-1)^2}$

10. $\dfrac{dy}{dx} = \dfrac{(1+x)2x - x^2(1)}{(1+x)^2} = \dfrac{x^2 + 2x}{(1+x)^2}$

11. $\dfrac{d}{dx}(2x - 5) = 2$

12. $\dfrac{d}{dx}(5x + 2) = 5$

13. $\dfrac{d}{dx}\left[(3x^2 - x^{-1})(2x + 5)\right] = (3x^2 - x^{-1})2 + (2x + 5)(6x + x^{-2}) = 18x^2 + 30x + 5x^{-2}$

14. $\dfrac{d}{dx}\left[(2x^2 + 3x^{-1})(2x - 3x^{-2})\right] = (2x^2 + 3x^{-1})(2 + 6x^{-3}) + (2x - 3x^{-2})(4x - 3x^{-2}) = 12x^2 + 27x^{-4}$

15. $\dfrac{d}{dt}\left(\dfrac{t^4}{2t^3 - 1}\right) = \dfrac{(2t^3 - 1)4t^3 - t^4(6t^2)}{(2t^3 - 1)^2} = \dfrac{2t^3(t^3 - 2)}{(2t^3 - 1)^2}$

16. $\dfrac{d}{dt}\left(\dfrac{2t^3 + 1}{t^4}\right) = \dfrac{d}{dt}\left(\dfrac{2}{t} + \dfrac{1}{t^4}\right) = -\dfrac{2}{t^2} - \dfrac{4}{t^5} = -\dfrac{2(t^3 + 2)}{t^5}$

17. $\dfrac{d}{du}\left(\dfrac{2u}{1 - 2u}\right) = \dfrac{(1-2u)2 - 2u(-2)}{(1-2u)^2} = \dfrac{2}{(1-2u)^2}$

18. $\dfrac{d}{du}\left(\dfrac{u^2}{u^3 + 1}\right) = \dfrac{(u^3+1)(2u) - u^2(3u^2)}{(u^3+1)^2} = \dfrac{u(2-u^3)}{(u^3+1)^2}$

19. $\dfrac{d}{du}\left(\dfrac{u}{u-1} - \dfrac{u}{u+1}\right) = \dfrac{(u-1)(1) - u}{(u-1)^2} - \dfrac{(u+1)(1) - u}{(u+1)^2}$
$$= -\dfrac{1}{(u-1)^2} - \dfrac{1}{(u+1)^2} = -\dfrac{2(1+u^2)}{(u^2-1)^2}$$

20. $\dfrac{d}{du}\left[u^2(1 - u^2)(1 - u^3)\right] = \dfrac{d}{du}[u^2 - u^4 - u^5 + u^7] = 2u - 4u^3 - 5u^4 + 7u^6$

21. $\dfrac{d}{dx}\left(\dfrac{x^3 + x^2 + x + 1}{x^3 - x^2 + x - 1}\right) = \dfrac{(x^3 - x^2 + x - 1)(3x^2 + 2x + 1) - (x^3 + x^2 + x + 1)(3x^2 - 2x + 1)}{(x^3 - x^2 + x - 1)^2}$

$\qquad\qquad\qquad\qquad\quad = \dfrac{-2(x^4 + 2x^2 + 1)}{(x^2 + 1)^2(x - 1)^2} = \dfrac{-2}{(x - 1)^2}$

22. $\dfrac{d}{dx}\left(\dfrac{x^3 + x^2 + x - 1}{x^3 - x^2 + x + 1}\right) = \dfrac{(x^3 - x^2 + x + 1)(3x^2 + 2x + 1) - (x^3 + x^2 + x - 1)(3x^2 - 2x + 1)}{(x^3 - x^2 + x + 1)^2}$

$\qquad\qquad\qquad\qquad\quad = \dfrac{-2x^4 + 8x^2 + 2}{(x^3 - x^2 + x + 1^2)}$

23. $\dfrac{dy}{dx} = (x + 1)\dfrac{d}{dx}[(x + 2)(x + 3)] + (x + 2)(x + 3)\dfrac{d}{dx}(x + 1) = (x + 1)(2x + 5) + (x + 2)(x + 3)$

At $x = 2$, $\dfrac{dy}{dx} = (3)(9) + (4)(5) = 47$.

24. $\dfrac{dy}{dx} = (x + 1)(x^2 + 2)(3x^2) + (x + 1)(x^3 + 3)(2x) + (x^2 + 2)(x^3 + 3)(1)$

At $x = 2$, $\dfrac{dy}{dx} = 3(6)(12) + 3(11)(4) + (6)(11) = 414$.

25. $\dfrac{dy}{dx} = \dfrac{(x + 2)\dfrac{d}{dx}[(x - 1)(x - 2)] - (x - 1)(x - 2)(1)}{(x + 2)^2}$

$\qquad = \dfrac{(x + 2)(2x - 3) - (x - 1)(x - 2)}{(x + 2)^2}$

At $x = 2$, $\quad \dfrac{dy}{dx} = \dfrac{4(1) - 1(0)}{16} = \dfrac{1}{4}$.

26. $y = \dfrac{x^4 - x^2 - 2}{x^2 + 2}; \quad \dfrac{dy}{dx} = \dfrac{(x^2 + 2)(4x^3 - 2x) - (x^4 - x^2 - 2)(2x)}{(x^2 + 2)^2}$

At $x = 2$, $\quad \dfrac{dy}{dx} = \dfrac{6(28) - (10)4}{36} = \dfrac{32}{9}$

27. $f'(x) = 21x^2 - 30x^4$, $f''(x) = 42x - 120x^3$ $\qquad$ **28.** $f'(x) = 10x^4 - 24x^3 + 2$, $f''(x) = 40x^3 - 72x^2$

29. $f'(x) = 1 + 3x^{-2}$, $f''(x) = -6x^{-3}$ $\qquad\qquad$ **30.** $f'(x) = 2x + 2x^{-3}$, $f''(x) = 2 - 6x^{-4}$

31. $f(x) = 2x^2 - 2x^{-2} - 3$, $\quad f'(x) = 4x + 4x^{-3}$, $\quad f''(x) = 4 - 12x^{-4}$

32. $f(x) = 4x - 9x^{-1}$, $\quad f'(x) = 4 + 9x^{-2}$, $\quad f''(x) = -18x^{-3}$

33. $\dfrac{dy}{dx} = x^2 + x + 1$ $\qquad\qquad$ **34.** $\dfrac{dy}{dx} = 10 + 50x$ $\qquad\qquad$ **35.** $\dfrac{dy}{dx} = 8x - 20$

$\qquad \dfrac{d^2y}{dx^2} = 2x + 1$ $\qquad\qquad\qquad \dfrac{d^2y}{dx^2} = 50$ $\qquad\qquad\qquad\qquad \dfrac{d^2y}{dx^2} = 8$

$\qquad \dfrac{d^3y}{dx^3} = 2$ $\qquad\qquad\qquad\quad\; \dfrac{d^3y}{dx^3} = 0$ $\qquad\qquad\qquad\qquad\; \dfrac{d^3y}{dx^3} = 0$

36. $\dfrac{dy}{dx} = \frac{1}{2}\,x^2 - \frac{1}{2}\,x + 1$

$\dfrac{d^2y}{dx^2} = x - \frac{1}{2}$

$\dfrac{d^3y}{dx^3} = 1$

37. $\dfrac{dy}{dx} = 3x^2 + 3x^{-4}$

$\dfrac{d^2y}{dx^2} = 6x - 12x^{-5}$

$\dfrac{d^3y}{dx^3} = 6 + 60x^{-6}$

38. $\dfrac{dy}{dx} = 3x^2 - 2x^{-2}$

$\dfrac{d^2y}{dx^2} = 6x + 4x^{-3}$

$\dfrac{d^3y}{dx^3} = 6 - 12x^{-4}$

39. $\dfrac{d}{dx}\left[x\dfrac{d}{dx}\left(x - x^2\right)\right] = \dfrac{d}{dx}\left[x(1 - 2x)\right] = \dfrac{d}{dx}\left[x - 2x^2\right] = 1 - 4x$

40.
$$\dfrac{d^2}{dx^2}\left[(x^2 - 3x)\dfrac{d}{dx}\left(x + x^{-1}\right)\right] = \dfrac{d^2}{dx^2}\left[(x^2 - 3x)(1 - x^{-2})\right]$$
$$= \dfrac{d^2}{dx^2}[x^2 - 3x - 1 + 3x^{-1}]$$
$$= \dfrac{d}{dx}\left(2x - 3 - 3x^{-2}\right) = 2 + 6x^{-3}$$

41. $\dfrac{d^4}{dx^4}[3x - x^4] = \dfrac{d^3}{dx^3}[3 - 4x^3] = \dfrac{d^2}{dx^2}[-12x^2] = \dfrac{d}{dx}[-24x] = -24$

42.
$$\dfrac{d^5}{dx^5}[ax^4 + bx^3 + cx^2 + dx + e] = \dfrac{d^4}{dx^4}[4ax^3 + 3bx^2 + 2cx + d]$$
$$= \dfrac{d^3}{dx^3}[12ax^2 + 6bx + 2c]$$
$$= \dfrac{d^2}{dx^2}[24ax + 6b] = \dfrac{d}{dx}[24a] = 0$$

43. $\dfrac{d^2}{dx^2}\left[(1 + 2x)\dfrac{d^2}{dx^2}\left(5 - x^3\right)\right] = \dfrac{d^2}{dx^2}\left[(1 + 2x)(-6x)\right] = \dfrac{d^2}{dx^2}\left[-6x - 12x^2\right] = -24$

44.
$$\dfrac{d^3}{dx^3}\left[\dfrac{1}{x}\dfrac{d^2}{dx^2}\left[x^4 - 5x^2\right]\right] = \dfrac{d^3}{dx^3}\left[\dfrac{1}{x}\left(12x^2 - 10\right)\right]$$
$$= \dfrac{d^3}{dx^3}[12x - 10x^{-1}] = \dfrac{d^2}{dx^2}[12 + 10x^{-2}]$$
$$= \dfrac{d}{dx}[-20x^{-3}] = 60x^{-4}$$

45. $y = x^4 - \dfrac{x^3}{3} + 2x^2 + C$

46. $y = \dfrac{x^2}{2} + \dfrac{1}{x^2} + 3x + C$

47. $y = x^5 - \dfrac{1}{x^4} + C$

48. $y = \dfrac{2x^6}{3} + \dfrac{5}{3x^3} - 2x + C$

49. Let $p(x) = ax^2 + bx + c$. Then $p'(x) = 2ax + b$ and $p''(x) = 2a$. Now

$$p''(1) = 2a = 4 \implies a = 2; \quad p'(1) = 2(2)(1) + b = -2 \implies b = -6;$$
$$p(1) = 2(1)^2 - 6(1) + c = 3 \implies c = 7$$

Thus $p(x) = 2x^2 - 6x + 7$.

50. $p(x) = ax^3 + bx^2 + cx + d$ $p'''(-1) = 6 \quad \Longrightarrow \quad a = 1$

$p'(x) = 3ax^2 + 2bx + c$ $p''(-1) = -2 \quad \Longrightarrow \quad b = 2$

$p''(x) = 6ax + 2b$ $p'(-1) = 3 \quad \Longrightarrow \quad c = 4$

$p'''(x) = 6a$ $p(-1) = 0 \quad \Longrightarrow \quad d = 3$

Therefore, $p(x) = x^3 + 2x^2 + 4x + 3$.

51.   (a) If $k = n$, $f^{(n)}(x) = n!$ (b) If $k > n$, $f^{(k)}(x) = 0$.

      (c) If $k < n$, $f^{(k)}(x) = n(n-1)(n-2)\cdots(n-k+1)x^{n-k}$.

52.   (a) $\dfrac{d^n}{dx^n} = n!\, a_n$ (b) $\dfrac{d^k}{dx^k} = 0$ if $k > n$.

53.   Let $f(x) = \begin{cases} x^2 & x \geq 0 \\ 0 & x \leq 0 \end{cases}$

      (a) $f'_+(0) = \lim\limits_{h \to 0^+} \dfrac{f(0+h) - f(0)}{h} = \lim\limits_{h \to 0^+} \dfrac{h^2 - 0}{h} = 0,$

          $f'_-(0) = \lim\limits_{h \to 0^-} \dfrac{f(0+h) - f(0)}{h} = \lim\limits_{h \to 0^-} \dfrac{0}{h} = 0$

      Therefore, f is differentiable at 0 and $f'(0) = 0$.

      (b) $f'(x) = \begin{cases} 2x & x \geq 0 \\ 0 & x \leq 0 \end{cases}$ (d)

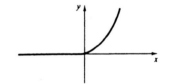

      (c) $f''_+(0) = \lim\limits_{h \to 0^+} \dfrac{f'(0+h) - f'(0)}{h} = \lim\limits_{h \to 0^+} \dfrac{2h - 0}{h} = 2,$

          $f''_-(0) = \lim\limits_{h \to 0^-} \dfrac{f'(0+h) - f'(0)}{h} = \lim\limits_{h \to 0^-} \dfrac{0}{h} = 0$

      Since $f''_+(0) \neq f''_-(0)$, $f''(0)$ does not exist.

54.   Let $g(x) = \begin{cases} x^3 & x \geq 0 \\ 0 & x < 0 \end{cases}$

      (a) $g'_+(0) = \lim\limits_{h \to 0^+} \dfrac{g(0+h) - g(0)}{h} = \lim\limits_{h \to 0^+} \dfrac{h^3 - 0}{h} = 0,$

          $g'_-(0) = \lim\limits_{h \to 0^-} \dfrac{g(0+h) - g(0)}{h} = \lim\limits_{h \to 0^-} \dfrac{0}{h} = 0$

      Therefore, g is differentiable at 0 and $g'(0) = 0$.

      $g'(x) = \begin{cases} 3x^2 & x \geq 0 \\ 0 & x < 0 \end{cases}$

      $g''_+(0) = \lim\limits_{h \to 0^+} \dfrac{g'(0+h) - g,(0)}{h} = \lim\limits_{h \to 0^+} \dfrac{3h^2 - 0}{h} = 0,$

      $g''_-(0) = \lim\limits_{h \to 0^-} \dfrac{g'(0+h) - g'(0)}{h} = \lim\limits_{h \to 0^-} \dfrac{0}{h} = 0$

      Therefore, g' is differentiable at 0 and $g''(0) = 0$.

(b) $g'(x) = \begin{cases} 3x^2 & x \geq 0 \\ 0 & x < 0 \end{cases}$

$g''(x) = \begin{cases} 6x & x \geq 0 \\ 0 & x < 0 \end{cases}$ (d)

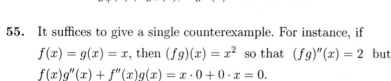

(c) $g'''_+(0) = \lim\limits_{h \to 0^+} \dfrac{g''(0+h) - g''(0)}{h} = \lim\limits_{h \to 0^+} \dfrac{6h - 0}{h} = 6$

$g'''_-(0) = \lim\limits_{h \to 0^-} \dfrac{g''(0+h) - g''(0)}{h} = \lim\limits_{h \to 0^-} \dfrac{0}{h} = 0$

Since $g'''_+(0) \neq g'''_-(0)$, $g'''(0)$ does not exist.

55. It suffices to give a single counterexample. For instance, if
$f(x) = g(x) = x$, then $(fg)(x) = x^2$ so that $(fg)''(x) = 2$ but
$f(x)g''(x) + f''(x)g(x) = x \cdot 0 + 0 \cdot x = 0$.

56. $\dfrac{d}{dx}\left[f(x)g'(x) - f'(x)g(x)\right] = \left[f(x)g''(x) + f'(x)g'(x)\right] - \left[f'(x)g'(x) + f''(x)g(x)\right]$

$= f(x)g''(x) - f''(x)g(x)$

57. $f''(x) = 6x$; (a) $x = 0$ (b) $x > 0$ (c) $x < 0$

58. $f''(x) = 12x^2$; (a) $x = 0$ (b) all $x \neq 0$ (c) none

59. $f''(x) = 12x^2 + 12x - 24$; (a) $x = -2,\ 1$ (b) $x < -2,\ \ x > 1$ (c) $-2 < x < 1$

60. $f''(x) = 12x^2 + 18x - 12$; (a) $x = -2,\ \frac{1}{2}$ (b) $x < -2,\ \ x > \frac{1}{2}$ (c) $-2 < x < \frac{1}{2}$

61. The result is true for $n = 1$:
$$\frac{d^1 y}{dx^1} = \frac{dy}{dx} = -x^{-2} = (-1)^1 1!\, x^{-1-1}.$$

If the result is true for $n = k$:
$$\frac{d^k y}{dx^k} = (-1)^k k!\, x^{-k-1}$$

then the result is true for $n = k + 1$:
$$\frac{d^{k+1} y}{dx^{k+1}} = \frac{d}{dx}\left[\frac{d^k y}{dx^k}\right] = \frac{d}{dx}\left[(-1)^k k!\, x^{-(k+1)}\right] = (-1)^{(k+1)}(k+1)!\, x^{-(k+1)-1}.$$

62. $y' = -2x^{-3}$, $y'' = 6x^{-4}$, $y''' = -24x^{-5}$; $y^{(n)} = (-1)^n(n+1)!\, x^{-(n+2)}$

Let S be the set of positive integers for which the result holds. Then $1 \in S$. Assume that the positive integer $k \in S$. Now,
$$y^{(k+1)} = \frac{d}{dx}\, y^{(k)} = \frac{d}{dx}\left[(-1)^k(k+1)!\, x^{-(k+2)}\right]$$

$$= -(-1)^k(k+2)(k+1)!\, x^{-(k+2)-1} = (-1)^{k+1}(k+2)!\, x^{-(k+3)}$$

Thus, $k + 1 \in S$, and S is the set of positive integers.

63. $\dfrac{d}{dx}(uvw) = uv\dfrac{dw}{dx} + uw\dfrac{dv}{dx} + vw\dfrac{du}{dx}$

64. (a) $\dfrac{d^n}{dx^n}[x^n] = n!$ (b) $\dfrac{d^{n+1}}{dx^{n+1}}[x^n] = 0$

65. By the reciprocal rule, $\dfrac{d}{dx}\left[\dfrac{1}{1-x}\right] = \dfrac{-(-1)}{(1-x)^2} = \dfrac{1}{(1-x)^2}.$

$$\dfrac{d^2}{dx^2}\left[\dfrac{1}{1-x}\right] = \dfrac{d}{dx}\left[\dfrac{1}{(1-x)^2}\right]$$

$$= \dfrac{d}{dx}\left[\dfrac{1}{1-x}\cdot\dfrac{1}{1-x}\right] = \dfrac{1}{1-x}\cdot\dfrac{1}{(1-x)^2} + \dfrac{1}{1-x}\cdot\dfrac{1}{(1-x)^2}$$

$$= \dfrac{2}{(1-x)^3}$$

$$\dfrac{d^3}{dx^3}\left[\dfrac{1}{1-x}\right] = \dfrac{d}{dx}\left[\dfrac{2}{(1-x)^3}\right]$$

$$= 2\dfrac{d}{dx}\left[\dfrac{1}{1-x}\cdot\dfrac{1}{(1-x)^2}\right] = 2\dfrac{1}{1-x}\cdot\dfrac{2}{(1-x)^3} + 2\dfrac{1}{(1-x)^2}\cdot\dfrac{1}{(1-x)^2}$$

$$= \dfrac{6}{(1-x)^4} = \dfrac{3!}{(1-x)^4}$$

and so on.

In general, $\dfrac{d^n}{dx^n}\left[\dfrac{1}{1-x}\right] = \dfrac{n!}{(1-x)^{n+1}}.$ You can use induction to prove this result.

66. $\dfrac{d^n}{dx^n}\left[\dfrac{1-x}{1+x}\right] = \dfrac{(-1)^n\, 2\cdot n!}{(1+x)^{n+1}}$

67. (b) The lines tangent to the graph of f are parallel to the line $x - 2y + 12 = 0$ at the points $\left(\dfrac{-1}{\sqrt{2}}, \dfrac{1}{2\sqrt{2}}\right)$ and $\left(\dfrac{1}{\sqrt{2}}, \dfrac{-1}{2\sqrt{2}}\right).$

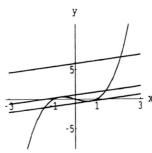

68. (b) The normals to the graph of f are perpendicular to l at the points where

$$x = -\frac{1}{2}, \ x = \frac{1-\sqrt{5}}{4}, \ x = \frac{1+\sqrt{5}}{4}$$

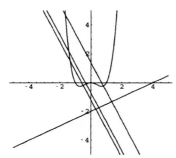

69. (a) $f(x) = x^3 + x^2 - 4x + 1; \quad f'(x) = 3x^2 + 2x - 4.$

(b)

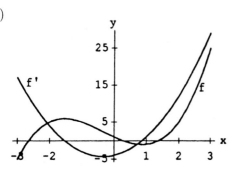

(c) The graph is "falling" when $f'(x) < 0$;
    the graph is "rising" when $f'(x) > 0$.

70. (a) $f(x) = x^4 - x^3 - 5x^2 - x - 2; \quad f'(x) = 4x^3 - 3x^2 - 10x - 1.$

(b)

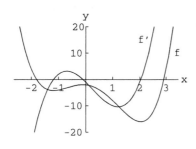

(c) The graph is "falling" when $f'(x) < 0$;
    the graph is "rising" when $f'(x) > 0$.

71. (a) $f(x) = \frac{1}{2}x^3 - 3x^2 + 3x + 3; \quad f'(x) = \frac{3}{2}x^2 - 6x + 3$

(b)

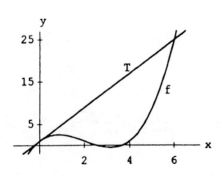

(c) The line tangent to the graph is horizontal; the
    graph turns from rising to falling, or from
    falling to rising.

(d) $x_1 \cong 0.586, \quad x_2 \cong 3.414$

72. (a) $f(x) = \frac{1}{2}x^3 - 3x^2 + 4x + 1$; $f'(x) = \frac{3}{2}x^2 - 6x + 4$.   (b)

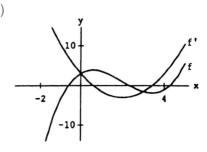

(c) The line tangent to the graph is horizontal; the graph turns from rising to falling, or from falling to rising.

(d) $x_1 \cong 0.845$, $x_2 \cong 3.155$

SECTION 3.4

1. $A = \pi r^2$, $\dfrac{dA}{dr} = 2\pi r$.   When $r = 2$, $\dfrac{dA}{dr} = 4\pi$.   **2.** $V = s^3$, $\dfrac{dV}{ds} = 3s^2$.   When $s = 4$, $\dfrac{dV}{ds} = 48$.

3. $A = \dfrac{1}{2}z^2$, $\dfrac{dA}{dz} = z$.   When $z = 4$, $\dfrac{dA}{dz} = 4$.   **4.** $\dfrac{dy}{dx} = -x^{-2}$.   When $x = -1$, $\dfrac{dy}{dx} = -1$.

5. $y = \dfrac{1}{x(1+x)}$, $\dfrac{dy}{dx} = \dfrac{-(2x+1)}{x^2(1+x)^2}$.   At $x = 2$, $\dfrac{dy}{dx} = -\dfrac{5}{36}$.

6. $\dfrac{dy}{dx} = 3x^2 - 24x + 45 = 3(x-3)(x-5)$; $\dfrac{dy}{dx} = 0$ at $x = 3, 5$.

7. $V = \dfrac{4}{3}\pi r^3$, $\dfrac{dV}{dr} = 4\pi r^2 = $ the surface area of the ball.

8. $S = 4\pi r^2$; $\dfrac{dS}{dr} = 8\pi r$ and $\dfrac{dS}{dr} = 8\pi r_0$ at $r = r_0$. $\dfrac{dS}{dr} = 1 \implies r_0 = \dfrac{1}{8\pi}$.

9. $y = 2x^2 + x - 1$, $\dfrac{dy}{dx} = 4x + 1$; $\dfrac{dy}{dx} = 4$ at $x = \frac{3}{4}$.   Therefore $x_0 = \frac{3}{4}$.

10.   (a) $A = \pi \left(\dfrac{d}{2}\right)^2 = \dfrac{\pi}{4}d^2$; $A' = \dfrac{\pi}{2}d$ (b) $A = \pi \left(\dfrac{C}{2\pi}\right)^2 = \dfrac{C^2}{4\pi}$; $\dfrac{dA}{dC} = \dfrac{C}{2\pi}$

11.   (a) $w = s\sqrt{2}$, $V = s^3 = \left(\dfrac{w}{\sqrt{2}}\right)^3 = \dfrac{\sqrt{2}}{4}w^3$, $\dfrac{dV}{dw} = \dfrac{3\sqrt{2}}{4}w^2$.

   (b) $z^2 = s^2 + w^2 = 3s^2$, $z = s\sqrt{3}$. $V = s^3 = \left(\dfrac{z}{\sqrt{3}}\right)^3 = \dfrac{\sqrt{3}}{9}z^3$, $\dfrac{dV}{dz} = \dfrac{\sqrt{3}}{3}z^2$.

12. $A = bh = $ (constant) $\implies h = \dfrac{A}{b}$; $\dfrac{dh}{db} = -\dfrac{A}{b^2} = -\dfrac{bh}{b^2} = -\dfrac{h}{b}$

13.   (a) $\dfrac{dA}{d\theta} = \dfrac{1}{2}r^2$ (b) $\dfrac{dA}{dr} = r\theta$

   (c) $\theta = \dfrac{2A}{r^2}$ so $\dfrac{d\theta}{dr} = \dfrac{-4A}{r^3} = \dfrac{-4}{r^3}\left(\dfrac{1}{2}r^2\theta\right) = \dfrac{-2\theta}{r}$

14.   (a) $\dfrac{dA}{dh} = 2\pi r$ (b) $\dfrac{dA}{dr} = 2\pi(2r + h)$

   (c) $h = \dfrac{A}{2\pi r} - r$; $\dfrac{dh}{dr} = -\dfrac{A}{2\pi r^2} - 1 = \dfrac{-2\pi r(r+h) - 2\pi r^2}{2\pi r^2} = -\dfrac{2r+h}{r}$

15. $y = ax^2 + bx + c,$ $\qquad\qquad\qquad\qquad z = bx^2 + ax + c.$

$\dfrac{dy}{dx} = 2ax + b,$ $\qquad\qquad\qquad\qquad\quad \dfrac{dz}{dx} = 2bx + a.$

$\dfrac{dy}{dx} = \dfrac{dz}{dx}$ iff $2ax + b = 2bx + a.$ With $a \ne b$, this occurs only at $x = \dfrac{1}{2}$.

16. Set $k(x) = f(x)g(x)h(x).$ Then $k'(x) = f(x)g(x)h'(x) + f(x)g'(x)h(x) + f'(x)g(x)h(x)$ and

$$k'(1) = (0)(2)(0) + (0)(-1)(-2) + (1)(2)(-2) = -4$$

SECTION 3.5

1. $y = x^4 + 2x^2 + 1,\quad y' = 4x^3 + 4x = 4x(x^2 + 1)$

$\quad y = (x^2 + 1)^2,\quad y' = 2(x^2 + 1)(2x) = 4x(x^2 + 1)$

2. $y = x^6 - 2x^3 + 1,\quad y' = 6x^5 - 6x^2 = 6x^2(x^3 - 1)$

$\quad y = (x^3 - 1)^2,\quad y' = 2(x^3 - 1)(3x^2) = 6x^2(x^3 - 1)$

3. $y = 8x^3 + 12x^2 + 6x + 1,\quad y' = 24x^2 + 24x + 6 = 6(2x + 1)^2$

$\quad y = (2x + 1)^3,\quad y' = 3(2x + 1)^2(2) = 6(2x + 1)^2$

4. $y = x^6 + 3x^4 + 3x^2 + 1,\quad f'(x) = 6x^5 + 12x^3 + 6x = 6x(x^2 + 1)^2$

$\quad y = (x^2 + 1)^3,\quad y' = 3(x^2 + 1)^2(2x) = 6x(x^2 + 1)^2$

5. $y = x^2 + 2 + x^{-2},\quad y' = 2x - 2x^{-3} = 2x(1 - x^{-4})$

$\quad y = (x + x^{-1})^2,\quad y' = 2(x + x^{-1})(1 - x^{-2}) = 2x(1 + x^{-2})(1 - x^{-2}) = 2x(1 - x^{-4})$

6. $y = 9x^4 - 12x^3 + 4x^2,\quad y' = 36x^3 - 36x^2 + 8x = 4x(3x - 2)(3x - 1)$

$\quad y = (3x^2 - 2x)^2,\quad y' = 2(3x^2 - 2x)(6x - 2) = 4x(3x - 2)(3x - 1)$

7. $f'(x) = -1(1 - 2x)^{-2} \dfrac{d}{dx}(1 - 2x) = 2(1 - 2x)^{-2}$

8. $f'(x) = 5(1 + 2x)^4 \dfrac{d}{dx}(1 + 2x) = 10(1 + 2x)^4$

9. $f'(x) = 20(x^5 - x^{10})^{19} \dfrac{d}{dx}(x^5 - x^{10}) = 20(x^5 - x^{10})^{19}(5x^4 - 10x^9)$

10. $f'(x) = 3(x^2 + x^{-2})^2 \dfrac{d}{dx}(x^2 + x^{-2}) = 6(x^2 + x^{-2})^2(x - x^{-3})$

11. $f'(x) = 4\left(x - \dfrac{1}{x}\right)^3 \dfrac{d}{dx}\left(x - \dfrac{1}{x}\right) = 4\left(x - \dfrac{1}{x}\right)^3\left(1 + \dfrac{1}{x^2}\right)$

12. $f(t) = (1 + t)^{-4};\quad f'(t) = -4(1 + t)^{-5} \dfrac{d}{dt}(1 + t) = -4(1 + t)^{-5}$

13. $f'(x) = 4(x - x^3 + x^5)^3 \dfrac{d}{dx}(x - x^3 + x^5) = 4(x - x^3 + x^5)^3(1 - 3x^2 + 5x^4)$

14. $f'(t) = 3(t - t^2)^2 \dfrac{d}{dt}(t - t^2) = 3(t - t^2)^2(1 - 2t)$

15. $f'(t) = 4(t^{-1} + t^{-2})^3 \dfrac{d}{dt}(t^{-1} + t^{-2}) = 4(t^{-1} + t^{-2})^3(-t^{-2} - 2t^{-3})$

16. $f'(x) = 3\left(\dfrac{4x+3}{5x-2}\right)^2 \dfrac{d}{dx}\left(\dfrac{4x+3}{5x-2}\right) = 3\left(\dfrac{4x+3}{5x-2}\right)^2\left[\dfrac{(5x-2)4-(4x+3)5}{(5x-2)^2}\right] = -\dfrac{69(4x+3)^2}{(5x-2)^4}$

17. $f'(x) = 4\left(\dfrac{3x}{x^2+1}\right)^3 \dfrac{d}{dx}\left(\dfrac{3x}{x^2+1}\right) = 4\left(\dfrac{3x}{x^2+1}\right)^3\left[\dfrac{(x^2+1)3-3x(2x)}{(x^2+1)^2}\right] = \dfrac{324x^3(1-x^2)}{(x^2+1)^5}$

18. $f'(x) = 3\left[(2x+1)^2 + (x+1)^2\right]^2 \dfrac{d}{dx}\left[(2x+1)^2 + (x+1)^2\right]$

$\qquad = 3\left[(2x+1)^2 + (x+1)^2\right]^2\left[2(2x+1)(2) + 2(x+1)(1)\right]$

$\qquad = 6\left[(2x+1)^2 + (x+1)^2\right]^2(5x+3)$

19. $f'(x) = -\left(\dfrac{x^3}{3} + \dfrac{x^2}{2} + \dfrac{x}{1}\right)^{-2}\dfrac{d}{dx}\left(\dfrac{x^3}{3} + \dfrac{x^2}{2} + \dfrac{x}{1}\right) = -\left(\dfrac{x^3}{3} + \dfrac{x^2}{2} + x\right)^{-2}(x^2 + x + 1)$

20. $f'(x) = 2[(6x + x^5)^{-1} + x]\dfrac{d}{dx}[(6x + x^5)^{-1} + x] = 2[(6x + x^5)^{-1} + x][1 - (6x + x^5)^{-2}(6 + 5x^4)]$

21. $\dfrac{dy}{dx} = \dfrac{dy}{du}\dfrac{du}{dx} = \dfrac{-2u}{(1+u^2)^2}\cdot(2)$

At $x = 0$, we have $u = 1$ and thus $\quad \dfrac{dy}{dx} = \dfrac{-4}{4} = -1.$

22. $\dfrac{dy}{dx} = \dfrac{dy}{du}\dfrac{du}{dx} = (1 - u^{-2})\cdot 4(3x+1)^3(3)$

At $x = 0$, we have $u = 1$ and thus $\quad \dfrac{dy}{dx} = 0.$

23. $\dfrac{dy}{dx} = \dfrac{dy}{du}\dfrac{du}{dx} = \dfrac{(1-4u)2 - 2u(-4)}{(1-4u)^2}\cdot 4(5x^2+1)^3(10x) = \dfrac{2}{(1-4u)^2}\cdot 40x(5x^2+1)^3$

At $x = 0$, we have $u = 1$ and thus $\dfrac{dy}{dx} = \dfrac{2}{9}(0) = 0.$

24. $\dfrac{dy}{dx} = \dfrac{dy}{du}\dfrac{du}{dx} = (3u^2 - 1)\cdot\left(\dfrac{-2}{(1+x)^2}\right)$

At $x = 0$, we have $u = 1$ and thus $\quad \dfrac{dy}{dx} = 2(-2) = -4.$

25. $\dfrac{dy}{dt} = \dfrac{dy}{du}\dfrac{du}{dx}\dfrac{dx}{dt} = \dfrac{(1+u^2)(-7) - (1-7u)(2u)}{(1+u^2)^2}(2x)(2)$

$\qquad = \dfrac{7u^2 - 2u - 7}{(1+u^2)^2}(4x) = \dfrac{4x(7x^4 + 12x^2 - 2)}{(x^4 + 2x^2 + 2)^2} = \dfrac{4(2t-5)[7(2t-5)^4 + 12(2t-5)^2 - 2]}{[(2t-5)^4 + 2(2t-5)^2 + 2]^2}$

26. $\dfrac{dy}{dt} = \dfrac{dy}{du}\dfrac{du}{dx}\dfrac{dx}{dt} = 2u\left(\dfrac{(1+x^2)(-7) - (1-7x)(2x)}{(1+x^2)^2}\right)(5)$

$\qquad = 10u\cdot\dfrac{7x^2 - 2x - 7}{(1+x^2)^2} = \dfrac{10[1 - 7(5t+2)][7(5t+2)^2 - 2(5t+2) - 7]}{[1 + (5t+2)^2]^2}$

27. $\dfrac{dy}{dx} = \dfrac{dy}{ds}\dfrac{ds}{dt}\dfrac{dt}{dx} = 2(s+3)\cdot\dfrac{1}{2\sqrt{t-3}}\cdot(2x)$

At $x=2$, we have $t=4$ so that $s=1$ and thus $\dfrac{dy}{dx} = 2(4)\dfrac{1}{2\cdot 1}(4) = 16$.

28. $\dfrac{dy}{dx} = \dfrac{dy}{ds}\dfrac{ds}{dt}\dfrac{dt}{dx} = \dfrac{2}{(1-s)^2}\left(1+\dfrac{1}{t^2}\right)\dfrac{1}{2\sqrt{x}}$

At $x=2$, we have $t=\sqrt{2}$ and $s=\sqrt{2}/2$. Thus $\dfrac{dy}{dx} = \dfrac{2}{(1-\sqrt{2}/2)^2}\left(1+\dfrac{1}{2}\right)\dfrac{1}{2\sqrt{2}} = \dfrac{3}{3\sqrt{2}-4}$.

29. $(f\circ g)'(0) = f'(g(0))g'(0) = f'(2)g'(0) = (1)(1) = 1$

30. $(f\circ g)'(1) = f'(g(1))g'(1) = f'(1)g'(1) = (1)(0) = 0$

31. $(f\circ g)'(2) = f'(g(2))g'(2) = f'(2)g'(2) = (1)(1) = 1$

32. $(g\circ f)'(0) = g'(f(0))f'(0) = g'(1)f'(0) = (0)(2) = 0$

33. $(g\circ f)'(1) = g'(f(1))f'(1) = g'(0)f'(1) = (1)(1) = 1$

34. $(g\circ f)'(2) = g'(f(2))f'(2) = g'(1)f'(2) = (0)(1) = 0$

35. $(f\circ h)'(0) = f'(h(0))h'(0) = f'(1)h'(0) = (1)(2) = 2$

36. $(f\circ h\circ g)'(1) = f'(h(g(1)))\ h'(g(1))g'(1) = f'(2)h'(1)g'(1) = (1)(1)(0) = 0$

37. $(g\circ f\circ h)'(2) = g'(f(h(2)))\ f'(h(2))h'(2) = g'(1)f'(0)h'(2) = (0)(2)(2) = 0$

38. $(g\circ h\circ f)'(0) = g'(h(f(0)))\ h'(f(0))f'(0) = g'(2)h'(1)f'(0) = (1)(1)(2) = 2$

39. $f'(x) = 4(x^3+x)^3(3x^2+1)$

$f''(x) = 3(4)(x^3+x)^2(3x^2+1)^2 + 4(x^3+x)^3(6x) = 12(x^3+x)^2[(3x^2+1)^2 + 2x(x^3+x)]$

40. $f'(x) = 10(x^2-5x+2)^9(2x-5)$

$f''(x) = 9(10)(x^2-5x+2)^8(2x-5)^2 + 10(x^2-5x+2)^9(2)$

$= (10)(x^2-5x+2)^8\left[9(2x-5)^2 + 2(x^2-5x+2)\right]$

41. $f'(x) = 3\left(\dfrac{x}{1-x}\right)^2\cdot\dfrac{1}{(1-x)^2} = \dfrac{3x^2}{(1-x)^4}$

$f''(x) = \dfrac{6x(1-x)^4 - 3x^2(4)(1-x)^3(-1)}{(1-x)^8} = \dfrac{6x(1+x)}{(1-x)^5}$

42. $f'(x) = \dfrac{1}{2\sqrt{x^2+1}}(2x) = \dfrac{x}{\sqrt{x^2+1}}$

$$f''(x) = \frac{\sqrt{x^2+1}(1) - x\dfrac{x}{\sqrt{x^2+1}}}{\left(\sqrt{x^2+1}\right)^2} = \frac{1}{(x^2+1)^{3/2}}$$

43. $2xf'(x^2+1)$

44. $f'\left(\dfrac{x-1}{x+1}\right)\dfrac{d}{dx}\left(\dfrac{x-1}{x+1}\right) = \dfrac{2}{(x+1)^2}\,f'\left(\dfrac{x-1}{x+1}\right)$

45. $2f(x)f'(x)$

46. $\dfrac{[f(x)+1]f'(x) - [f(x)-1]f'(x)}{[f(x)+1]^2} = \dfrac{2f'(x)}{[f(x)+1]^2}$

47. $f'(x) = -4x(1+x^2)^{-3};$ (a) $x = 0$ (b) $x < 0$ (c) $x > 0$

48. $f'(x) = 2(1-x^2)(-2x) = -4x(1-x^2);$ (a) $x = -1, 0, 1$ (b) $-1 < x < 0,\ x > 1$
(c) $x < -1,\ 0 < x < 1$

49. $f'(x) = \dfrac{1-x^2}{(1+x^2)^2};$ (a) $x = \pm 1$ (b) $-1 < x < 1$ (c) $x < -1,\ x > 1$

50. $f'(x) = (1-x^2)^3 + x(3)(1-x^2)^2(-2x) = (1-x^2)^2(1-7x^2);$
(a) $x = \pm 1,\ x = \pm\frac{1}{7}\sqrt{7}$ (b) $-\frac{1}{7}\sqrt{7} < x < \frac{1}{7}\sqrt{7}$
(c) $x < -1,\ \ -1 < x < -\frac{1}{7}\sqrt{7},\ \ \frac{1}{7}\sqrt{7} < x < 1,\ \ x > 1$

51. $\dfrac{n!}{(1-x)^{n+1}}$

52. $\dfrac{(-1)^{n+1}n!}{(1+x)^{n+1}}$

53. $n!\,b^n$

54. $\dfrac{(-1)^n n!\,ab^n}{(bx+c)^{n+1}}$

55. $y = (x^2+1)^3 + C$

56. $y = \dfrac{(x^2-1)^2}{2} + C$

57. $y = (x^3-2)^2 + C$

58. $y = \dfrac{(x^3+2)^3}{3} + C$

59. $L'(x) = \dfrac{1}{x^2+1}\cdot 2x = \dfrac{2x}{x^2+1}$

60. $H'(x) = 2f(x)f'(x) - 2g(x)g'(x) = 2f(x)g(x) - 2g(x)f(x) = 0$

61. $T'(x) = 2f(x)\cdot f'(x) + 2g(x)\cdot g'(x) = 2f(x)\cdot g(x) - 2g(x)\cdot f(x) = 0$

62. (a) Suppose f is even: $[f(x)]' = [f(-x)]' = f'(-x)(-1) = -f'(-x);$ thus $f'(-x) = -f'(x).$
(b) Suppose f is odd: $[f(x)]' = -[f(-x)]' = -f'(-x)(-1) = f'(-x);$ thus $f'(-x) = f'(x).$

63. Suppose $P(x) = (x-a)^2 q(x),$ where $q(a) \neq 0.$ Then

$$P'(x) = 2(x-a)q(x) + (x-a)^2 q'(x) \quad \text{and} \quad P''(x) = 2q(x) + 4(x-a)q'(x) + (x-a)^2 q''(x),$$

and it follows that $P(a) = P'(a) = 0,$ and $P''(a) \neq 0.$

Now suppose that $P(a) = P'(a) = 0$ and $P''(a) \neq 0$.

$$P(a) = 0 \quad \Longrightarrow \quad P(x) = (x-a)g(x) \quad \text{for some polynomial } g.$$

Then $P'(x) = g(x) + (x-a)g'(x)$ and

$$P'(a) = 0 \quad \Longrightarrow \quad g(a) = 0 \text{ and so } g(x) = (x-a)q(x) \text{ for some polynomial } q.$$

Therefore, $P(x) = (x-a)^2 q(x)$. Finally, $P''(a) \neq 0$ implies $q(a) \neq 0$.

64. Suppose $P(x) = (x-a)^3 q(x)$, where $q(a) \neq 0$. Then

$$P'(x) = 3(x-a)^2 q(x) + (x-a)^3 q'(x)$$

$$P''(x) = 6(x-a)q(x) + 6(x-a)^2 q'(x) + (x-a)^3 q''(x)$$

$$P'''(x) = 6q(x) + 18(x-a)q'(x) + 9(x-a)^2 q''(x) + (x-a)^3 q'''(x)$$

and it follows that $P(a) = P'(a) = P''(a) = 0$, $P'''(a) \neq 0$.

Now suppose that $P(a) = P'(a) = P''(a) = 0$ and $P'''(a) \neq 0$.

$$P(a) = 0 \quad \Longrightarrow \quad P(x) = (x-a)g(x) \quad \text{for some polynomial } g.$$

Then $P'(x) = g(x) + (x-a)g'(x)$ and

$$P'(a) = 0 \quad \Longrightarrow \quad g(a) = 0 \text{ and so } g(x) = (x-a)h(x) \text{ for some polynomial } h.$$

Therefore, $P(x) = (x-a)^2 h(x)$. Now $P''(x) = 2h(x) + 4(x-a)h'(x) + (x-a)^2 h''(x)$ and

$$P''(a) = 0 \quad \Longrightarrow \quad h(a) = 0 \text{ and so } h(x) = (x-a)q(x) \text{ for some polynomial } q.$$

Therefore, $P(x) = (x-a)^3 q(x)$. Finally, $P'''(a) \neq 0$ implies $q(a) \neq 0$.

65. Let P be a polynomial function of degree n. The number a is a root of P of multiplicity k, $(k < n)$ if and only if $P(a) = P'(a) = \cdots = P^{(k-1)}(a) = 0$ and $P^{(k)}(a) \neq 0$.

66. $A = \dfrac{\sqrt{3}}{4} x^2$, where $x = \dfrac{2\sqrt{3}}{3} h$. Now

$$\frac{dA}{dh} = \frac{dA}{dx}\frac{dx}{dh} = \frac{\sqrt{3}}{2} x \cdot \frac{2\sqrt{3}}{3} = \frac{2\sqrt{3}}{3} h; \quad \frac{dA}{dh} = 4 \text{ when } h = 2\sqrt{3}$$

67. $V = \frac{4}{3}\pi r^3$ and $\dfrac{dr}{dt} = 2$ cm/sec. By the chain rule, $\dfrac{dV}{dt} = \dfrac{dV}{dr}\dfrac{dr}{dt} = 4\pi r^2 \dfrac{dr}{dt} = 8\pi r^2$.

At the instant the radius is 10 centimeters, the volume is increasing at the rate

$$\frac{dV}{dt} = 8\pi (10)^2 = 800\pi \text{ cm}^3/\text{sec}.$$

68. $V = \frac{4}{3}\pi r^3$, $S = 4\pi r^2$, and $\dfrac{dV}{dt} = 200$.

$$\frac{dS}{dt} = \frac{dS}{dr} \cdot \frac{dr}{dV} \cdot \frac{dV}{dt}$$

$$= 8\pi r \cdot \frac{1}{4\pi r^2} \cdot 200$$

$$= \frac{400}{r}$$

$$= 80 \quad \text{when } r = 5$$

The surface area is increasing 80 cm^2/sec. at the instant the radius is 5 centimeters.

69. (a) $\dfrac{dF}{dt} = \dfrac{dF}{dr} \cdot \dfrac{dr}{dt} = \dfrac{2k}{r^3} \cdot (49 - 9.8t) = \dfrac{2k}{(49t - 4.9t^2)^3}(49 - 9.8t), \ 0 \le t \le 10$

     (b) $\dfrac{dF}{dt}(3) = \dfrac{2k}{(102.9)^3}(19.6); \quad \dfrac{dF}{dt}(7) = -\dfrac{2k}{(102.9)^3}(19.6).$

70. (a) $f(9) = -2, \ f'(9) = -\frac{1}{12}$; tangent $T : y = -\frac{1}{12}x - \frac{5}{4}$.

     (b) 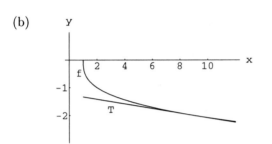   (c) $(7.4, 10.8)$

71. (a) $f(1) = \frac{1}{2}, \ f'(1) = -\frac{1}{2}$; tangent $T : y = -\frac{1}{2}x + 1.$

     (b) 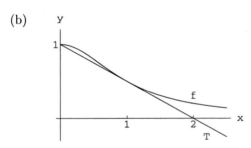   (c) $(0.8, 1.2)$

72. $\dfrac{d}{dx}\left[x^2 \dfrac{d^4}{dx^4}\left(x^2 + 1\right)^4\right] = 288\left(35x^5 + 20x^3 + x\right)$

73. (a) $\dfrac{d}{dx}\left[f\left(\dfrac{1}{x}\right)\right] = -\dfrac{f'(1/x)}{x^2}$ (b) $\dfrac{d}{dx}\left[f\left(\dfrac{x^2 - 1}{x^2 + 1}\right)\right] = \dfrac{4x\, f'\left[(x^2 - 1)/(x^2 + 1)\right]}{(1 + x^2)^2}$

     (c) $\dfrac{d}{dx}\left[\dfrac{f(x)}{1 + f(x)}\right] = \dfrac{f'(x)}{[1 + f(x)]^2}$

74. (a) $\dfrac{d}{dx}\left[u_1\left(u_2(x)\right)\right] = u_1'\left(u_2(x)\right) u_2'(x)$ (b) $\dfrac{d}{dx}\left[u_1\left(u_2(u_3(x))\right)\right] = u_1'\left[u_2(u_3(x))\right] u_2'\left(u_3(x)\right)$

                                                                            $u_3'(x)$

     (c) $\dfrac{d}{dx}\left[u_1\left(u_2[u_3(u_4(x))]\right)\right] = u_1'\left[u_2(u_3[u_4(x)])\right] u_2'\left(u_3[u_4(x)]\right) u_3'[u_4(x)]u_4'(x)$

75. $\dfrac{d^2}{dx^2}\left[f(g(x))\right] = [g'(x)]^2 f''[g(x)] + f'[g(x)]g''(x)$

SECTION 3.6

1. $\dfrac{dy}{dx} = -3\sin x - 4\sec x \tan x$

2. $\dfrac{dy}{dx} = 2x\sec x + x^2\sec x\tan x$

3. $\dfrac{dy}{dx} = 3x^2\csc x - x^3\csc x\cot x$

4. $\dfrac{dy}{dx} = 2\sin x\,\cos x$

5. $\dfrac{dy}{dt} = -2\cos t\sin t$

6. $\dfrac{dy}{dt} = 6t\,\tan t + 3t^2\sec^2 t$

7. $\dfrac{dy}{du} = 4\sin^3\sqrt{u}\,\dfrac{d}{du}\left(\sin\sqrt{u}\,\right) = 4\sin^3\sqrt{u}\,\cos\sqrt{u}\,\dfrac{d}{du}\left(\sqrt{u}\,\right) = 2u^{-1/2}\sin^3\sqrt{u}\,\cos\sqrt{u}$

8. $\dfrac{dy}{du} = \csc u^2 - 2u^2\csc u^2\cot u^2$

9. $\dfrac{dy}{dx} = \sec^2 x^2\,\dfrac{d}{dx}\left(x^2\right) = 2x\sec^2 x^2$

10. $\dfrac{dy}{dx} = -\dfrac{1}{2\sqrt{x}}\,\sin\sqrt{x}$

11. $\dfrac{dy}{dx} = 4[x + \cot\pi x]^3[1 - \pi\csc^2\pi x]$

12. $\dfrac{dy}{dx} = 3(x^2 - \sec 2x)^2(2x - 2\sec 2x\tan 2x) = 6(x^2 - \sec 2x)^2(x - \sec 2x\tan 2x)$

13. $\dfrac{dy}{dx} = \cos x,\ \dfrac{d^2y}{dx^2} = -\sin x$

14. $\dfrac{dy}{dx} = -\sin x,\ \dfrac{d^2y}{dx^2} = -\cos x$

15. $\dfrac{dy}{dx} = \dfrac{(1 + \sin x)(-\sin x) - \cos x\,(\cos x)}{(1 + \sin x)^2} = \dfrac{-\sin x - (\sin^2 x + \cos^2 x)}{(1 + \sin x)^2} = -(1 + \sin x)^{-1}$

$\dfrac{d^2y}{dx^2} = (1 + \sin x)^{-2}\dfrac{d}{dx}(1 + \sin x) = \cos x\,(1 + \sin x)^{-2}$

16. $\dfrac{dy}{dx} = 3\tan^2(2\pi x)\sec^2(2\pi x)(2\pi) = 6\pi\tan^2(2\pi x)\sec^2(2\pi x)$

$\dfrac{d^2y}{dx^2} = 6\pi(2)\tan(2\pi x)\sec^2(2\pi x)\sec^2(2\pi x)(2\pi) + 6\pi\tan^2(2\pi x)(2)\sec(2\pi x)[\sec(2\pi x)\tan(2\pi x)](2\pi)$

$= 24\pi^2\tan(2\pi x)\sec^2(2\pi x)[\sec^2(2\pi x) + \tan^2(2\pi x)]$

17. $\dfrac{dy}{du} = 3\cos^2 2u\,\dfrac{d}{du}(\cos 2u) = -6\cos^2 2u\sin 2u$

$\dfrac{d^2y}{du^2} = -6[\cos^2 2u\,\dfrac{d}{du}(\sin 2u) + \sin 2u\,\dfrac{d}{du}(\cos^2 2u)]$

$= -6[2\cos^3 2u + \sin 2u\,(-4\cos 2u\sin 2u)] = 12\cos 2u\,[2\sin^2 2u - \cos^2 2u]$

18. $\dfrac{dy}{dt} = 5\sin^4(3t)\,\cos(3t)(3) = 15\sin^4(3t)\,\cos(3t)$

$\dfrac{d^2y}{dt^2} = 15(4)\sin^3(3t)\,3\cos^2(3t) + 15\sin^4(3t)[-3\sin(3t)] = 45\sin^3(3t)[4\cos^2(3t) - \sin^2(3t)]$

19. $\dfrac{dy}{dt} = 2\sec^2 2t,\quad \dfrac{d^2y}{dt^2} = 4\sec 2t\,\dfrac{d}{dt}(\sec 2t) = 8\sec^2 2t\tan 2t$

20. $\dfrac{dy}{du} = -4\,\csc^2 4u; \quad \dfrac{d^2y}{du^2} = -4(2)\,\csc(4u)[-\csc(4u)\,\cot(4u)(4)] = 32\,\csc^2(4u)\,\cot(4u)$

21. $\dfrac{dy}{dx} = x^2(3\cos 3x) + 2x\sin 3x$

$\dfrac{d^2y}{dx^2} = [x^2(-9\sin 3x) + 2x(3\cos 3x)] + [2x(3\cos 3x) + 2(\sin 3x)]$

$\qquad = (2 - 9x^2)\sin 3x + 12x\cos 3x$

22. $\dfrac{dy}{dx} = \dfrac{(1-\cos x)\cos x - \sin x\,(-[-\sin x])}{(1-\cos x)^2} = \dfrac{\cos x - 1}{(1-\cos x)^2} = \dfrac{-1}{1-\cos x}$

$\dfrac{d^2y}{dx^2} = \sin x(1-\cos x)^{-2}$

23. $y = \sin^2 x + \cos^2 x = 1 \;\; \text{so} \;\; \dfrac{dy}{dx} = \dfrac{d^2y}{dx^2} = 0$

24. $y = \sec^2 x - \tan^2 x = 1 \;\; \text{so} \;\; \dfrac{dy}{dx} = \dfrac{d^2y}{dx^2} = 0$

25. $\dfrac{d^4}{dx^4}(\sin x) = \dfrac{d^3}{dx^3}(\cos x) = \dfrac{d^2}{dx^2}(-\sin x) = \dfrac{d}{dx}(-\cos x) = \sin x$

26. $\dfrac{d^4}{dx^4}(\cos x) = \dfrac{d^3}{dx^3}(-\sin x) = \dfrac{d^2}{dx^2}(-\cos x) = \dfrac{d}{dx}(\sin x) = \cos x$

27. $\dfrac{d}{dt}\left[t^2\dfrac{d^2}{dt^2}(t\cos 3t)\right] = \dfrac{d}{dt}\left[t^2\dfrac{d}{dt}(\cos 3t - 3t\sin 3t)\right]$

$\qquad\qquad = \dfrac{d}{dt}[t^2(-3\sin 3t - 3\sin 3t - 9t\cos 3t)]$

$\qquad\qquad = \dfrac{d}{dt}[-6t^2\sin 3t - 9t^3\cos 3t]$

$\qquad\qquad = (-18t^2\cos 3t - 12t\sin 3t) + (27t^3\sin 3t - 27t^2\cos 3t)$

$\qquad\qquad = (27t^3 - 12t)\sin 3t - 45t^2\cos 3t$

28. $\dfrac{d}{dt}\left[t\dfrac{d}{dt}(\cos t^2)\right] = \dfrac{d}{dt}\left[-t\,\sin t^2(2t)\right] = \dfrac{d}{dt}\left[-2t^2\,\sin t^2\right]$

$\qquad\qquad = -4t\,\sin t^2 - 2t^2\cos t^2(2t) = -4t(\sin t^2 + t^2\cos t^2)$

29. $\dfrac{d}{dx}[f(\sin 3x)] = f'(\sin 3x)\dfrac{d}{dx}(\sin 3x) = 3\cos 3x\,f'(\sin 3x)$

30. $\dfrac{d}{dx}[\sin f(3x)] = \cos[f(3x)]f'(3x)(3) = 3f'(3x)\cos[f(3x)]$

31. $\dfrac{dy}{dx} = \cos x; \;\;$ slope of tangent at $(0,0)$ is 1; tangent: $\;\; y = x$.

32. $\dfrac{dy}{dx} = \sec^2 x;$ slope of tangent at $(\pi/6, \sqrt{3}/3)$ is $\sec^2(\pi/6) = 4/3;$

tangent: $y - \frac{1}{3}\sqrt{3} = \frac{4}{3}\left(x - \frac{1}{6}\pi\right)$

33. $\dfrac{dy}{dx} = -\csc^2 x;$ slope of tangent at $(\dfrac{\pi}{6}, \sqrt{3})$ is $-4,$ an equation for

tangent: $y - \sqrt{3} = -4\left(x - \dfrac{\pi}{6}\right).$

34. $\dfrac{dy}{dx} = -\sin x;$ slope of tangent at $(0, 1)$ is $0;$ tangent: $y = 1.$

35. $\dfrac{dy}{dx} = \sec x \tan x,$ slope of tangent at $(\dfrac{\pi}{4}, \sqrt{2})$ is $\sqrt{2},$ an equation for

tangent is $y - \sqrt{2} = \sqrt{2}\left(x - \dfrac{\pi}{4}\right).$

36. $\dfrac{dy}{dx} = -\csc x \cot x,$ slope of tangent at $(\pi/3, 2\sqrt{3}/3)$ is $-2/3;$

tangent: $y - \frac{2}{3}\sqrt{3} = -\frac{2}{3}\left(x - \frac{1}{3}\pi\right).$

37. $\dfrac{dy}{dx} = -\sin x;$ $x = \pi$ **38.** $\dfrac{dy}{dx} = \cos x;$ $x = \frac{1}{2}\pi,\ x = \frac{3}{2}\pi$

39. $\dfrac{dy}{dx} = \cos x - \sqrt{3}\sin x;$ $\dfrac{dy}{dx} = 0$ gives $\tan x = \dfrac{1}{\sqrt{3}};$ $x = \dfrac{\pi}{6}, \dfrac{7\pi}{6}$

40. $\dfrac{dy}{dx} = -\sin x - \sqrt{3}\cos x;$ $\dfrac{dy}{dx} = 0$ gives $\tan x = -\sqrt{3};$ $x = \dfrac{2\pi}{3}, \dfrac{5\pi}{3}$

41. $\dfrac{dy}{dx} = 2\sin x \cos x = \sin 2x;$ $x = \dfrac{\pi}{2},\ \pi,\ \dfrac{3\pi}{2}$

42. $\dfrac{dy}{dx} = -2\sin x \cos x = -\sin 2x;$ $x = \dfrac{\pi}{2},\ \pi,\ \dfrac{3\pi}{2}$

43. $\dfrac{dy}{dx} = \sec^2 x - 2;$ $\dfrac{dy}{dx} = 0$ gives $\sec x = \pm\sqrt{2};$ $x = \dfrac{\pi}{4}, \dfrac{3\pi}{4}, \dfrac{5\pi}{4}, \dfrac{7\pi}{4}$

44. $\dfrac{dy}{dx} = -3\csc^2 x + 4;$ $\dfrac{dy}{dx} = 0$ gives $\csc x = \pm\dfrac{2}{\sqrt{3}};$ $x = \dfrac{\pi}{3}, \dfrac{2\pi}{3}, \dfrac{4\pi}{3}, \dfrac{5\pi}{3}$

45. $\dfrac{dy}{dx} = 2\sec x \tan x + \sec^2 x;$ since $\sec x$ is never zero, $\dfrac{dy}{dx} = 0$ gives

$2\tan x + \sec x = 0$ so that $\sin x = -1/2;$ $x = \dfrac{7\pi}{6}, \dfrac{11\pi}{6}$

46. $\dfrac{dy}{dx} = -\csc^2 x + 2\csc x \cot x;$ since $\csc x$ is never zero, $\dfrac{dy}{dx} = 0$ gives

$2\cot x - \csc x = 0$ so that $\cos x = 1/2;$ $x = \dfrac{\pi}{3}, \dfrac{5\pi}{3}$

47. $f(x) = x + 2\cos x,$ $f'(x) = 1 - 2\sin x.$

(a) $xf'(x) = 0:$ $1 - 2\sin x = 0,$ $\sin x = 1/2,$ $x = \pi/6,\ 5\pi/6.$

$f':$ +   −   +   ; 0 $\dfrac{\pi}{6}$ $\dfrac{5\pi}{6}$ 2π

(b) $f'(x) > 0$ on $(0, \pi/6) \cup (5\pi/6, 2\pi),$ (c) $f'(x) < 0$ on $(\pi/6, 5\pi/6).$

48. $f(x) = x - \sqrt{2}\sin x,$ $f'(x) = 1 - \sqrt{2}\cos x.$

(a) $f'(x) = 0:$ $1 - \sqrt{2}\cos x = 0,$ $\cos x = \sqrt{2}/2,$ $x = \pi/4,\ 7\pi/4.$

$f':$ −   +   ; 0 $\dfrac{\pi}{4}$ $\dfrac{7\pi}{4}$ 2π

(b) $f'(x) > 0$ on $(\pi/4, 7\pi/4),$ (c) $f'(x) < 0$ on $(0, \pi/4) \cup (7\pi/4, 2\pi).$

49. $f(x) = \sin x + \cos x,$ $f'(x) = \cos x - \sin x.$

(a) $f'(x) = 0:$ $\cos x - \sin x = 0,$ $\cos x = \sin x,$ $x = \pi/4,\ 5\pi/4.$

$f':$ +   −   +   ; 0 $\dfrac{\pi}{4}$ $\dfrac{5\pi}{4}$ 2π

(b) $f'(x) > 0$ on $(0, \pi/4) \cup (5\pi/4, 2\pi),$ (c) $f'(x) < 0$ on $(\pi/4, 5\pi/4).$

50. $f(x) = \sin x - \cos x,$ $f'(x) = \cos x + \sin x.$

(a) $f'(x) = 0:$ $\cos x + \sin x = 0,$ $\cos x = -\sin x,$ $x = 3\pi/4,\ 7\pi/4.$

$f':$ +   −   +   ; 0 $\dfrac{3\pi}{4}$ $\dfrac{7\pi}{4}$ 2π

(b) $f'(x) > 0$ on $(0, 3\pi/4) \cup (7\pi/4, 2\pi),$ (c) $f'(x) < 0$ on $(3\pi/4, 7\pi/4).$

51. (a) $\dfrac{dy}{dt} = \dfrac{dy}{du}\dfrac{du}{dx}\dfrac{dx}{dt} = (2u)(\sec x \tan x)\pi = 2\pi \sec^2 \pi t \tan \pi t$

(b) $y = \sec^2 \pi t - 1,$ $\dfrac{dy}{dt} = 2\sec \pi t\,(\sec \pi t \tan \pi t)\pi = 2\pi \sec^2 \pi t \tan \pi t$

52. (a) $\dfrac{dy}{dt} = \dfrac{dy}{du}\dfrac{du}{dx}\dfrac{dx}{dt} = 3\left[\dfrac{1}{2}(1+u)\right]^2\left(\dfrac{1}{2}\right)(-\sin x)(2) = 3\left[\dfrac{1}{2}(1+\cos 2t)\right]^2(-\sin 2t)$

$\qquad = 3(\cos^4 t)(-2\sin t \cos t) = -6\cos^5 t \sin t$

(b) $y = \left[\dfrac{1}{2}(1+\cos 2t)\right]^3 = \cos^6 t;$ $\dfrac{dy}{dt} = 6\cos^5 t(-\sin t) = -6\cos^5 t \sin t$

53. (a) $\dfrac{dy}{dt} = \dfrac{dy}{du}\dfrac{du}{dx}\dfrac{dx}{dt} = 4\left[\dfrac{1}{2}(1-u)\right]^3\left(-\dfrac{1}{2}\right)(-\sin x)(2) = 4\left[\dfrac{1}{2}(1-\cos 2t)\right]^3\sin 2t$

$\qquad = 4\sin^6 t\,(2\sin t \cos t) = 8\sin^7 t \cos t$

(b) $y = \left[\dfrac{1}{2}(1 - \cos 2t)\right]^4 = \sin^8 t$, $\dfrac{dy}{dt} = 8\sin^7 t \cos t$

54. (a) $\dfrac{dy}{dt} = \dfrac{dy}{du}\dfrac{du}{dx}\dfrac{dx}{dt} = (-2u)(-\csc x \cot x)(3) = (-2\csc(3t)(-\csc(3t)\cot(3t)3$

$\qquad\qquad = 6\csc^2(3t)\cot(3t)$

$\quad$ (b) $y = 1 - csc^2(3t)$; $\dfrac{dy}{dt} = -2\csc(3t)[-\csc(3t)\cot(3t)](3) = 6\csc^2(3t)\cot(3t)$

55. $\dfrac{d^n}{dx^n}(\cos x) = \begin{cases} (-1)^{(n+1)/2}\sin x, & n \text{ odd} \\ (-1)^{n/2}\cos x, & n \text{ even} \end{cases}$

56. (a) $\dfrac{d}{dx}(\cot x) = \dfrac{d}{dx}\left(\dfrac{\cos x}{\sin x}\right) = \dfrac{\sin x(-\sin x) - \cos x(\cos x)}{\sin^2 x} = \dfrac{-1}{\sin^2 x} = -\csc^2 x$

$\quad$ (b) $\dfrac{d}{dx}(\sec x) = \dfrac{d}{dx}\left(\dfrac{1}{\cos x}\right) = \dfrac{-1}{\cos^2 x}(-\sin x) = \dfrac{1}{\cos x}\cdot\dfrac{\sin x}{\cos x} = \sec x \tan x$

$\quad$ (c) $\dfrac{d}{dx}(\csc x) = \dfrac{d}{dx}\left(\dfrac{1}{\sin x}\right) = \dfrac{-1}{\sin^2 x}(\cos x) = \dfrac{-1}{\sin x}\cdot\dfrac{\cos x}{\sin x} = -\csc x \cot x$

57. $\dfrac{d}{dx}(\cos x) = \dfrac{d}{dx}\left[\sin\left(\dfrac{\pi}{2} - x\right)\right] = -\cos\left(\dfrac{\pi}{2} - x\right) = -\sin x$

58. Differentiating both sides, $2\cos 2x = 2(\cos^2 x - \sin^2 x)$.   Thus $\cos 2x = \cos^2 x - \sin^2 x$.

59. $f'(0) = \lim\limits_{h\to 0}\dfrac{\sin(0+h) - \sin 0}{h} = \lim\limits_{h\to 0}\dfrac{\sin h}{h} = \lim\limits_{x\to 0}\dfrac{\sin x}{x}$

60. $f'(0) = \lim\limits_{h\to 0}\dfrac{\cos(0+h) - \cos 0}{h} = \lim\limits_{h\to 0}\dfrac{\cos h - 1}{h} = \lim\limits_{x\to 0}\dfrac{\cos x - 1}{x}$

61. $f(x) = 2\sin x + 3\cos x + C$ $\qquad\qquad$ **62.** $f(x) = \tan x + \cot x + C$

63. $f(x) = \sin 2x + \sec x + C$ $\qquad\qquad$ **64.** $f(x) = \dfrac{-\cos 3x}{3} + \dfrac{\csc 2x}{2} + C$

65. $f(x) = \sin(x^2) + \cos 2x + C$ $\qquad\qquad$ **66.** $f(x) = \dfrac{\tan(x^3)}{3} + \sec 2x + C$

67. (a) $f'(x) = \sin(1/x) + x\,\cos(1/x)(-1/x^2)$

$\qquad\qquad = \sin(1/x) - (1/x)\cos(1/x)$

$\qquad g'(x) = 2x\,\sin(1/x) + x^2\,\cos(1/x)(-1/x^2)$

$\qquad\qquad = 2x\,\sin(1/x) - \cos(1/x)$

$\quad$ (b) $\lim\limits_{x\to 0} g'(x) = \lim\limits_{x\to 0}[2x\,\sin(1/x) - \cos(1/x)] = -\lim\limits_{x\to 0}\cos(1/x)$ does not exist

68. (a) f must be continuous at 0:

$$\lim_{x \to 0^+} f(x) = \lim_{x \to 0^+} \cos x = 1, \quad \lim_{x \to 0^-} f(x) = \lim_{x \to 0^-} (ax + b) = b; \text{ thus } b = 1$$

Differentiable at 0 :

$$\lim_{h \to 0^+} \frac{f(h) - f(0)}{h} = \lim_{h \to 0^+} \frac{\cos h - 1}{h} = 0,$$

$$\lim_{h \to 0^-} \frac{f(h) - f(0)}{h} = \lim_{h \to 0^+} \frac{ah + 1 - 1}{h} = a$$

Therefore, f is differentiable at 0 if $a = 0$ and $b = 1$.

(b)

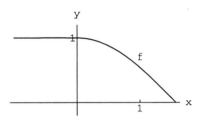

69. (a) Continuity:

$$\lim_{x \to 2\pi/3^-} \sin x = \frac{\sqrt{3}}{2}, \quad \lim_{x \to 2\pi/3^+} (ax + b) = \frac{2\pi a}{3} + b; \text{ thus } \frac{2\pi a}{3} + b = \frac{\sqrt{3}}{2}$$

Differentiability:

$$\lim_{x \to 2\pi/3^-} \cos x = -\frac{1}{2}, \quad \lim_{x \to 2\pi/3^+} (a) = a; \text{ thus } a = -\frac{1}{2}$$

Therefore, f is differentiable at $2\pi/3$ if $a = -\frac{1}{2}$ and $b = \frac{1}{2}\sqrt{3} + \frac{1}{3}\pi$

(b)

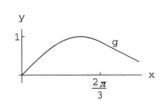

70. (a) Continuity:

$$\lim_{x \to \pi/3^-} (1 + a \cos x) = 1 + \frac{1}{2}a, \quad \lim_{x \to \pi/3^+} [b + \sin(x/2)] = b + \frac{1}{2} \text{ which implies } b = \frac{1}{2} + \frac{1}{2}a.$$

Differentiability:

$$\lim_{x \to \pi/3^-} (-a \sin x) = -\frac{1}{2}\sqrt{3}\,a, \quad \lim_{x \to \pi/3^+} [\frac{1}{2}\cos(x/2)] = \frac{1}{4}\sqrt{3} \text{ which implies } a = -\frac{1}{2}.$$

Therefore, f is differentiable at $\pi/3$ if $a = -\frac{1}{2}$ and $b = \frac{1}{4}$.

(b)

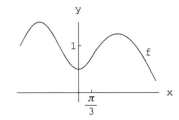

71. Let $y(t) = A \sin \omega t + B \cos \omega t$. Then

$$y'(t) = \omega A \cos \omega t - \omega B \sin \omega t \quad \text{and} \quad y''(t) = -\omega^2 A \sin \omega t - \omega^2 B \cos \omega t$$

Thus,

$$\frac{d^2 y}{dt^2} + \omega^2 y = 0.$$

72. (a) $\theta = a \sin(\omega t + \phi); \quad \theta' = a\omega \cos(\omega t + \phi); \quad \theta'' = -a\omega^2 \sin(\omega t + \phi)$

Thus, θ satisfies the equation.

(b)

$$\theta = a \sin(\omega t + \phi_0)$$

$$= a \sin(\omega t) \cos \phi_0 - a \cos(\omega t) \sin \phi_0$$

$$= A \sin(\omega t) + B \cos(\omega t) \text{ where } A = -a \sin \phi_0, \quad B = a \cos \phi_0$$

73. $A = \frac{1}{2} c^2 \sin x; \quad \dfrac{dA}{dx} = \dfrac{1}{2} c^2 \cos x$

74. $c = \sqrt{a^2 + b^2 - 2ab \cos x}; \quad \dfrac{dc}{dx} = \dfrac{1}{2\sqrt{a^2 + b^2 - 2ab \cos x}} (2ab \sin x) = \dfrac{ab \sin x}{\sqrt{a^2 + b^2 - 2ab \cos x}}$

75. (a) $f^{(4p)}(x) = k^{4p} \cos kx, \quad f^{(4p+1)}(x) = -k^{4p+1} \sin kx, \quad f^{(4p+2)}(x) = -k^{4p+2} \cos kx,$

$f^{(4p+3)}(x) = k^{4p+3} \sin kx, \ p = 0, 1, 2, \ldots$

(b) $m = k^2, \ k = 1, 2, 3, \ldots$

76. $f(x) = A \cos \sqrt{2} x + B \sin \sqrt{2} x; \qquad f'(x) = -A\sqrt{2} \sin \sqrt{2} x + B\sqrt{2} \cos \sqrt{2} x$

$f(0) = 2 \quad \Longrightarrow \quad A = 2; \qquad f'(0) = -3 \quad \Longrightarrow \quad B = -\dfrac{3}{\sqrt{2}} = -\dfrac{3\sqrt{2}}{2}$

77. f has horizontal at the points with x-coordinate $\quad \frac{1}{2}\pi, \ \frac{3}{2}\pi, \ \cong 3.39, \ \cong 6.03$

78. $f(x) = \sin x - \sin^2 x$ has horizontal tangents at: $\quad \left(\frac{\pi}{6}, \frac{1}{4}\right), \ \left(\frac{\pi}{2}, 0\right), \ \left(\frac{5\pi}{6}, \frac{1}{4}\right), \ \left(\frac{3\pi}{2}, -2\right)$

79. $f(x) = \sin x, \ f'(x) = \cos x; \quad f(0) = 0, \ f'(0) = 1$. Therefore an equation for the tangent line T is $y = x$.

The graphs of f and T are:

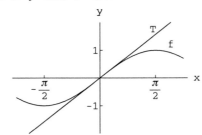

$|\sin x - x| < 0.01$ on $(-0.39, 0.39)$.

80. $f(x) = \tan x,\ f'(x) = \sec^2 x;\quad f(\pi/4) = 1,\ f'(\pi/4) = 2$. Therefore an equation for the tangent line T is $y = 2x + 1 - \frac{1}{2}\pi$.

The graphs of f and T are:

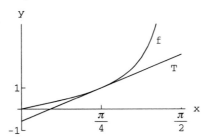

$|\tan x - (2x + 1 - \frac{1}{2}\pi)| < 0.01$ on $(0.712, 0.852)$.

SECTION 3.7

1.
$$x^2 + y^2 = 4$$
$$2x + 2y\frac{dy}{dx} = 0$$
$$\frac{dy}{dx} = \frac{-x}{y}$$

2.
$$x^3 + y^3 - 3xy = 0$$
$$3x^2 + 3y^2\frac{dy}{dx} - 3\left(y + x\frac{dy}{dx}\right) = 0$$
$$\frac{dy}{dx} = \frac{y - x^2}{y^2 - x}$$

3.
$$4x^2 + 9y^2 = 36$$
$$8x + 18y\frac{dy}{dx} = 0$$
$$\frac{dy}{dx} = \frac{-4x}{9y}$$

4.
$$\sqrt{x} + \sqrt{y} = 4$$
$$\frac{1}{2\sqrt{x}} + \frac{1}{2\sqrt{y}}\frac{dy}{dx} = 0$$
$$\frac{dy}{dx} = -\frac{\sqrt{y}}{\sqrt{x}}$$

5.
$$x^4 + 4x^3y + y^4 = 1$$
$$4x^3 + 12x^2y + 4x^3\frac{dy}{dx} + 4y^3\frac{dy}{dx} = 0$$
$$\frac{dy}{dx} = -\frac{x^3 + 3x^2y}{x^3 + y^3}$$

6.
$$x^2 - x^2 y + xy^2 + y^2 = 1$$

$$2x - 2xy - x^2 \frac{dy}{dx} + y^2 + 2xy\frac{dy}{dx} + 2y\frac{dy}{dx} = 0$$

$$\frac{dy}{dx} = \frac{2xy - 2x - y^2}{2xy + 2y - x^2}$$

7.
$$(x - y)^2 - y = 0$$

$$2(x - y)\left(1 - \frac{dy}{dx}\right) - \frac{dy}{dx} = 0$$

$$\frac{dy}{dx} = \frac{2(x - y)}{2(x - y) + 1}$$

8.
$$(y + 3x)^2 - 4x = 0$$

$$2(y + 3x)\left(\frac{dy}{dx} + 3\right) - 4 = 0$$

$$\frac{dy}{dx} = -3 + \frac{2}{y + 3x}$$

9.
$$\sin(x + y) = xy$$

$$\cos(x + y)\left(1 + \frac{dy}{dx}\right) = x\frac{dy}{dx} + y$$

$$\frac{dy}{dx} = \frac{y - \cos(x + y)}{\cos(x + y) - x}$$

10. $\tan xy = xy;$ $\sec^2(xy)\left(y + x\frac{dy}{dx}\right) = y + x\frac{dy}{dx};$ $\frac{dy}{dx} = -\frac{y}{x}$

11.
$$y^2 + 2xy = 16$$

$$2y\frac{dy}{dx} + 2x\frac{dy}{dx} + 2y = 0$$

$$(x + y)\frac{dy}{dx} + y = 0.$$

Differentiating a second time, we have

$$(x + y)\frac{d^2y}{dx^2} + \frac{dy}{dx}\left(2 + \frac{dy}{dx}\right) = 0.$$

Substituting $\dfrac{dy}{dx} = \dfrac{-y}{x + y},$ we have

$$(x + y)\frac{d^2y}{dx^2} - \frac{y}{(x + y)}\left(\frac{2x + y}{x + y}\right) = 0, \quad \frac{d^2y}{dx^2} = \frac{2xy + y^2}{(x + y)^3} = \frac{16}{(x + y)^3}.$$

12.
$$x^2 - 2xy + 4y^2 = 3$$

$$2x - 2y - 2x\frac{dy}{dx} + 8y\frac{dy}{dx} = 0$$

$$x - y + (4y - x)\frac{dy}{dx} = 0.$$

Differentiating a second time, we have

$$1 - \frac{dy}{dx} + \left(4\frac{dy}{dx} - 1\right)\frac{dy}{dx} + (4y - x)\frac{d^2y}{dx^2} = 0.$$

Substituting $\quad \frac{dy}{dx} = \frac{y - x}{4y - x}, \quad$ we have

$$\frac{d^2y}{dx^2} = -\frac{3(x^2 - 2xy + 4y^2)}{(4y - x)^3} = \frac{-9}{(4y - x)^3}.$$

13.
$$y^2 + xy - x^2 = 9$$
$$2y\frac{dy}{dx} + x\frac{dy}{dx} + y - 2x = 0.$$

Differentiating a second time, we have

$$\left[2\left(\frac{dy}{dx}\right)^2 + 2y\frac{d^2y}{dx^2}\right] + \left[x\frac{d^2y}{dx^2} + \frac{dy}{dx}\right] + \frac{dy}{dx} - 2 = 0$$

$$(2y + x)\frac{d^2y}{dx^2} + 2\left[\left(\frac{dy}{dx}\right)^2 + \frac{dy}{dx} - 1\right] = 0.$$

Substituting $\quad \frac{dy}{dx} = \frac{2x - y}{2y + x}, \quad$ we have

$$(2y + x)\frac{d^2y}{dx^2} + 2\left[\frac{(2x - y)^2 + (2x - y)(2y + x) - (2y + x)^2}{(2y + x)^2}\right] = 0$$

$$\frac{d^2y}{dx^2} = \frac{10(y^2 + xy - x^2)}{(2y + x)^3} = \frac{90}{(2y + x)^3}.$$

14.
$$x^2 - 3xy = 18$$
$$2x - 3y - 3x\frac{dy}{dx} = 0.$$

Differentiating a second time, we have

$$2 - 3\frac{dy}{dx} - 3\frac{dy}{dx} - 3x\frac{d^2y}{dx^2} = 0$$

Substituting $\quad \frac{dy}{dx} = \frac{2x - 3y}{3x}, \quad$ we have

$$\frac{d^2y}{dx^2} = -\frac{6(x - 3y)}{9x^2} = -\frac{6(x^2 - 3xy)}{9x^3} = -\frac{6(18)}{9x^3} = -\frac{12}{x^3}$$

15.
$$4\tan y = x^3$$
$$4\sec^2 y\,\frac{dy}{dx} = 3x^2$$
$$\frac{dy}{dx} = \frac{3}{4}x^2\cos^2 y$$
$$\frac{d^2y}{dx^2} = \frac{3}{2}x\cos^2 y + \frac{3}{4}x^2\left(2\,\cos y(-\sin y)\frac{dy}{dx}\right)$$
$$= \frac{3}{2}x\cos^2 y - \frac{9}{8}x^4\sin y\cos^3 y$$

16.
$$\sin^2 x + \cos^2 y = 1$$

$$2\sin x \cos x - 2\cos y \sin y \frac{dy}{dx} = 0$$

$$\sin 2x - \sin 2y \frac{dy}{dx} = 0$$

$$\frac{dy}{dx} = \frac{\sin 2x}{\sin 2y}$$

$$\frac{d^2y}{dx^2} = \frac{\sin 2y(\cos 2x)2 - \sin 2x(\cos 2y)2\frac{dy}{dx}}{\sin^2 2y}$$

Substituting $\dfrac{dy}{dx} = \dfrac{\sin 2x}{\sin 2y}$ and using the double angle formulas, we find that

$$\frac{d^2y}{dx^2} = \frac{8\left[\cos^2 x \cos^2 y(\sin^2 y - \sin^2 x) - \sin^2 x \sin^2 y(\cos^2 y - \cos^2 x)\right]}{\sin^3 2y} = 0$$

since $\sin^2 x = \sin^2 y$ and $\cos^2 x = \cos^2 y$ from the original equation.

17. $x^2 - 4y^2 = 9, \qquad 2x - 8y\dfrac{dy}{dx} = 0.$

At $(5, 2)$, we get $\dfrac{dy}{dx} = \dfrac{5}{8}.$ Then,

$$2 - 8\left[y\frac{d^2y}{dx^2} + \left(\frac{dy}{dx}\right)^2\right] = 0.$$

At $(5, 2)$ we get

$$2 - 8\left[2\frac{d^2y}{dx^2} + \frac{25}{64}\right] = 0 \quad \text{so that} \quad \frac{d^2y}{dx^2} = -\frac{9}{128}.$$

18. $x^2 + 4xy + y^3 + 5 = 0, \qquad 2x + 4y + 4x\dfrac{dy}{dx} + 3y^2\dfrac{dy}{dx} = 0.$

At $(2, -1)$, we get $4 - 4 + 8\dfrac{dy}{dx} + 3\dfrac{dy}{dx} = 0$ so $\dfrac{dy}{dx} = 0.$ Differentiating again,

$$2 + 4\frac{dy}{dx} + 4\frac{dy}{dx} + 4x\frac{d^2y}{dx^2} + 6y\left(\frac{dy}{dx}\right)^2 + 3y^2\frac{d^2y}{dx^2} = 0$$

At $(2, -1)$ we get $2 + 11\dfrac{d^2y}{dx^2} = 0$ so $\dfrac{d^2y}{dx^2} = -\dfrac{2}{11}$

19. $\cos(x + 2y) = 0 \qquad -\sin(x + 2y)\left(1 + 2\dfrac{dy}{dx}\right) = 0.$

At $(\pi/6, \pi/6)$, we get $\dfrac{dy}{dx} = -1/2.$ Then,

$$-\cos(x + 2y)\left(1 + 2\frac{dy}{dx}\right)^2 - \sin(x + 2y)\left(2\frac{d^2y}{dx^2}\right) = 0.$$

At $(\pi/6, \pi/6)$, we get

$$-\cos\frac{\pi}{2}(0)^2 - \sin\frac{\pi}{2}\left(2\frac{d^2y}{dx^2}\right) = 0 \quad \text{so that} \quad \frac{d^2y}{dx^2} = 0.$$

20. $\quad x = \sin^2 y \qquad 1 = 2\sin y \cos y \dfrac{dy}{dx} = \sin 2y \dfrac{dy}{dx}.$

At $(1/2, \pi/4)$, we get $\dfrac{dy}{dx} = 1.$ Differentiating again,

$$0 = 2\cos 2y \left(\dfrac{dy}{dx}\right)^2 + \sin 2y \dfrac{d^2y}{dx^2}.$$

At $(1/2, \pi/4)$, we get $\dfrac{d^2y}{dx^2} = 0.$

21.
$$2x + 3y = 5$$
$$2 + 3\dfrac{dy}{dx} = 0$$
slope of tangent at $(-2, 3):\quad -2/3$
tangent: $\quad y - 3 = -\frac{2}{3}(x + 2)$
normal: $\quad y - 3 = \frac{3}{2}(x + 2)$

22.
$$9x^2 + 4y^2 = 72$$
$$18x + 8y\dfrac{dy}{dx} = 0$$
slope of tangent at $(2, 3):\quad -\frac{3}{2}$
tangent: $\quad y - 3 = -\frac{3}{2}(x - 2)$
normal: $\quad y - 3 = \frac{2}{3}(x - 2)$

23.
$$x^2 + xy + 2y^2 = 28$$
$$2x + x\dfrac{dy}{dx} + y + 4y\dfrac{dy}{dx} = 0$$
slope of tangent at $(-2, -3):\quad -1/2$
tangent: $\quad y + 3 = -\frac{1}{2}(x + 2)$
normal: $\quad y + 3 = 2(x + 2)$

24.
$$x^3 - axy + 3ay^2 = 3a^3$$
$$3x^2 - ay - ax\dfrac{dy}{dx} + 6ay\dfrac{dy}{dx} = 0$$
slope of tangent at $(a, a):\quad -\frac{2}{5}$
tangent: $\quad y - a = -\frac{2}{5}(x - a)$
normal: $\quad y - a = \frac{5}{2}(x - a)$

25.
$$x = \cos y \dfrac{dy}{dx}$$
$$1 = -\sin y \dfrac{dy}{dx}$$
slope of tangent at $\left(\dfrac{1}{2}, \dfrac{\pi}{3}\right):\quad \dfrac{-2}{\sqrt{3}}$
tangent: $\quad y - \dfrac{\pi}{3} = -\dfrac{2}{\sqrt{3}}\left(x - \dfrac{1}{2}\right)$
normal: $\quad y - \dfrac{\pi}{3} = \dfrac{\sqrt{3}}{2}\left(x - \dfrac{1}{2}\right)$

26.
$$\tan xy = x$$
$$\sec^2(xy)\left(y + x\dfrac{dy}{dx}\right) = 1$$
slope of tangent at $(1, \pi/4):\quad \dfrac{2 - \pi}{4}$
tangent: $\quad y - \dfrac{\pi}{4} = \dfrac{2 - \pi}{4}(x - 1)$
normal: $\quad y - \dfrac{\pi}{4} = -\dfrac{4}{2 - \pi}(x - 1)$

27. $\dfrac{dy}{dx} = \dfrac{1}{2}(x^3 + 1)^{-1/2}\dfrac{d}{dx}(x^3 + 1) = \dfrac{3}{2}x^2(x^3 + 1)^{-1/2}$ $\qquad$ **28.** $\dfrac{dy}{dx} = \frac{1}{3}(x + 1)^{-2/3}$

29. $\dfrac{dy}{dx} = \dfrac{1}{4}(2x^2 + 1)^{-3/4}\dfrac{d}{dx}(2x^2 + 1) = x(2x^2 + 1)^{-3/4}$

30. $\dfrac{dy}{dx} = \frac{1}{3}(x + 1)^{-2/3}(x + 2)^{2/3} + (x + 1)^{1/3}\left(\frac{2}{3}\right)(x + 2)^{-1/3} = \dfrac{3x + 4}{3(x + 1)^{2/3}(x + 2)^{1/3}}$

31. $\dfrac{dy}{dx} = \sqrt{2 - x^2}\left[\dfrac{-x}{\sqrt{3 - x^2}}\right] + \sqrt{3 - x^2}\left[\dfrac{-x}{\sqrt{2 - x^2}}\right] = \dfrac{x(2x^2 - 5)}{\sqrt{2 - x^2}\,\sqrt{3 - x^2}}$

32. $\dfrac{dy}{dx} = \frac{3}{2}(x^4 - x + 1)^{1/2}(4x^3 - 1) = \frac{3}{2}\sqrt{x^4 - x + 1}\,(4x^3 - 1)$

33. $\dfrac{d}{dx}\left(\sqrt{x}+\dfrac{1}{\sqrt{x}}\right) = \dfrac{d}{dx}\left(x^{1/2}+x^{-1/2}\right) = \dfrac{1}{2}x^{-1/2} - \dfrac{1}{2}x^{-3/2} = \dfrac{1}{2}x^{-3/2}(x-1)$

34. $\dfrac{d}{dx}\left(\sqrt{\dfrac{3x+1}{2x+5}}\right) = \dfrac{1}{2}\left(\dfrac{3x+1}{2x+5}\right)^{-1/2}\left(\dfrac{(2x+5)3-(3x+1)2}{(2x+5)^2}\right) = \dfrac{13}{2(2x+5)^2}\sqrt{\dfrac{2x+5}{3x+1}}$

35. $\dfrac{d}{dx}\left(\dfrac{x}{\sqrt{x^2+1}}\right) = \dfrac{d}{dx}\left(x(x^2+1)^{-1/2}\right)$

$\qquad = x\left(-\dfrac{1}{2}(x^2+1)^{-3/2}(2x)\right) + (x^2+1)^{-1/2} = (x^2+1)^{-3/2}$

36. $\dfrac{d}{dx}\left(\dfrac{\sqrt{x^2+1}}{x}\right) = \dfrac{x\left(\frac{1}{2}\right)(x^2+1)^{-1/2}(2x)-(x^2+1)^{1/2}}{x^2} = \dfrac{\dfrac{x^2}{\sqrt{x^2+1}}-\sqrt{x^2+1}}{x^2} = \dfrac{-1}{x^2\sqrt{x^2+1}}$

37. (a)   (b)   (c)

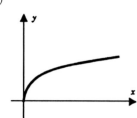
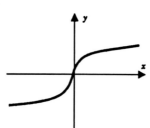
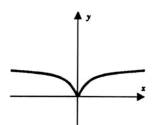

38. $y = (a^2+x^2)^{1/2}; \quad \dfrac{dy}{dx} = \dfrac{1}{2}(a^2+x^2)^{-1/2}(2x) = x(a^2+x^2)^{-1/2};$

$\qquad \dfrac{d^2y}{dx^2} = (a^2+x^2)^{-1/2} - \dfrac{1}{2}x(a^2+x^2)^{-3/2}(2x) = \dfrac{a^2}{(a^2+x^2)^{3/2}}$

39. $y = (a+bx)^{1/3}; \quad \dfrac{dy}{dx} = \dfrac{b}{3}(a+bx)^{-2/3}; \quad \dfrac{d^2y}{dx^2} = \dfrac{-2b^2}{9}(a+bx)^{-5/3}$

40. $y = x(a^2-x^2)^{1/2}; \quad \dfrac{dy}{dx} = (a^2-x^2)^{1/2} + \frac{1}{2}x(a^2-x^2)^{-1/2}(-2x) = (a^2-x^2)^{1/2} - x^2(a^2-x^2)^{-1/2}$

$\qquad \dfrac{d^2y}{dx^2} = \frac{1}{2}(a^2-x^2)^{-1/2}(-2x) - 2x(a^2-x^2)^{-1/2} - x^2\left(-\frac{1}{2}\right)(a^2-x^2)^{-3/2}(-2x) = \dfrac{x(2x^2-3a^2)}{(a^2-x^2)^{3/2}}$

41. $y = \sqrt{x}\,\tan\sqrt{x}; \quad \dfrac{dy}{dx} = \dfrac{1}{2\sqrt{x}}\tan\sqrt{x} + \sqrt{x}\,\sec^2\sqrt{x}\left(\dfrac{1}{2\sqrt{x}}\right) = \dfrac{1}{2\sqrt{x}}\tan\sqrt{x} + \dfrac{1}{2}\sec^2\sqrt{x}$

$\qquad \dfrac{d^2y}{dx^2} = \dfrac{2\sqrt{x}\,\sec^2\sqrt{x}\,(1/2\sqrt{x}) - \tan\sqrt{x}\,(1/\sqrt{x})}{4x} + \sec\sqrt{x}\,\sec\sqrt{x}\,\tan\sqrt{x}\,(1/2\sqrt{x})$

$\qquad = \dfrac{\sqrt{x}\,\sec^2\sqrt{x} - \tan\sqrt{x} + 2x\,\sec^2\sqrt{x}\,\tan\sqrt{x}}{4x\sqrt{x}}$

42. $y = \sqrt{x}\,\sin\sqrt{x}; \quad \dfrac{dy}{dx} = \dfrac{1}{2\sqrt{x}}\sin\sqrt{x} + \sqrt{x}\,\cos\sqrt{x}\left(\dfrac{1}{2\sqrt{x}}\right) = \dfrac{1}{2\sqrt{x}}\sin\sqrt{x} + \dfrac{1}{2}\cos\sqrt{x}$

$$\frac{d^2y}{dx^2} = \frac{2\sqrt{x}\cos\sqrt{x}(1/2\sqrt{x}) - \sin\sqrt{x}(1/\sqrt{x})}{4x} - \frac{1}{2}\sin\sqrt{x}(1/2\sqrt{x})$$

$$= \frac{\sqrt{x}\cos\sqrt{x} - \sin\sqrt{x}}{4x^{3/2}} - \frac{\sin\sqrt{x}}{4\sqrt{x}}$$

43. Differentiation of $x^2 + y^2 = r^2$ gives $2x + 2y\dfrac{dy}{dx} = 0$ so that the slope of the normal line is

$$\frac{-1}{dy/dx} = \frac{y}{x} \quad (x \neq 0).$$

Let (x_0, y_0) be a point on the circle. Clearly, if $x_0 = 0$, the normal line, $x = 0$, passes through the origin. If $x_0 \neq 0$, the normal line is

$$y - y_0 = \frac{y_0}{x_0}(x - x_0), \quad \text{which simplifies to} \quad y = \frac{y_0}{x_0}x,$$

a line through the origin.

44. $y^2 = x; \quad 2y\dfrac{dy}{dx} = 1; \quad \dfrac{dy}{dx} = \dfrac{1}{2y}$

When $x = a$, $y = \pm\sqrt{a}$.

Slope of tangent at $(a, \sqrt{a})$: $\quad \dfrac{1}{2\sqrt{a}}$

Equation of tangent line: $\quad y - \sqrt{a} = \dfrac{1}{2\sqrt{a}}(x - a)$

x-intercept: $\quad -\sqrt{a} = \dfrac{1}{2\sqrt{a}}(x - a) \quad \Longrightarrow \quad x = -a$

Slope of tangent at $(a, -\sqrt{a})$: $\quad -\dfrac{1}{2\sqrt{a}}$

Equation of tangent line: $\quad y + \sqrt{a} = -\dfrac{1}{2\sqrt{a}}(x - a)$

x-intercept: $\quad \sqrt{a} = -\dfrac{1}{2\sqrt{a}}(x - a) \quad \Longrightarrow \quad x = -a$

45. For the parabola $y^2 = 2px + p^2$, we have $2y\dfrac{dy}{dx} = 2p$ and the slope of a tangent is given by $m_1 = p/y$. For the parabola $y^2 = p^2 - 2px$, we obtain $m_2 = -p/y$ as the slope of a tangent. The parabolas intersect at the points $(0, \pm p)$. At each of these points $m_1 m_2 = -1$; the parabolas intersect at right angles.

46. For $y = 2x$, the slope is $m_1 = 2$. For $x^2 - xy + 2y^2 = 28$, we have

$$2x - y - x\frac{dy}{dx} + 4y\frac{dy}{dx} = 0 \quad \text{so} \quad \frac{dy}{dx} = m_2 = \frac{y - 2x}{4y - x}$$

At a point of intersection of the line and the curve, we have $m_2 = 0$ since $y = 2x$. Thus

$$\tan\alpha = |-m_1| = 2 \quad \Longrightarrow \quad \alpha \cong 1.107(\text{radians}) \cong 63.4°$$

47. For $y = x^2$ we have $m_1 = \dfrac{dy}{dx} = 2x$; for $x = y^3$ we have $3y^2\dfrac{dy}{dx} = 1$ or $m_2 = \dfrac{dy}{dx} = 1/3y^2$.

At $(1,1)$, $m_1 = 2$, $m_2 = 1/3$ and

$$\tan\alpha = \left|\frac{m_1 - m_2}{1 - m_1 m_2}\right| = \left|\frac{2 - (1/3)}{1 + 2(1/3)}\right| = 1 \quad \Rightarrow \quad \alpha = \frac{\pi}{4}$$

At $(0,0)$, $m_1 = 0$ and m_2 is undefined. Thus $\alpha = \pi/2$.

48. $(x-1)^2 + y^2 = 10$; $2(x-1) + 2y\dfrac{dy}{dx} = 0$, $m_1 = \dfrac{dy}{dx} = -\dfrac{x-1}{y}$

$x^2 + (y-2)^2 = 5$; $2x + 2(y-2)\dfrac{dy}{dx} = 0$, $m_2 = \dfrac{dy}{dx} = -\dfrac{x}{y-2}$

The circles intersect at the points $(-2,1)$ and $(2,3)$.

At $(-2,1)$: $m_1 = 3$, $m_2 = -2$ and $\tan\alpha = \left|\dfrac{3+2}{1+(3)(-2)}\right| = 1$. Thus $\alpha = \dfrac{\pi}{4}$.

At $(2,3)$: $m_1 = -1/3$, $m_2 = -2$ and $\tan\alpha = \left|\dfrac{(-1/3)+2}{1+(-1/3)(-2)}\right| = 1$. Thus $\alpha = \dfrac{\pi}{4}$.

49. The hyperbola and the ellipse intersect at the points $(\pm 3, \pm 2)$. For the hyperbola, $\dfrac{dy}{dx} = \dfrac{x}{y}$ and for the ellipse $\dfrac{dy}{dx} = -\dfrac{4x}{9y}$. The product of these slopes is $-\dfrac{4x^2}{9y^2}$. This product is -1 at each of the points of intersection. Therefore the hyperbola and ellipse are orthogonal.

50. The curves intersect at the points $(\pm 1, 1)$. For the ellipse, $\dfrac{dy}{dx} = -\dfrac{3x}{2y}$ and for $y^3 = x^2$ we have $\dfrac{dy}{dx} = \dfrac{2x}{3y^2}$. The product of these slopes is $-\dfrac{6x^2}{6y^3} = -\dfrac{x^2}{y^3}$. This product is -1 at each of the points of intersection. Therefore the curves are orthogonal.

51. For the circles, $\dfrac{dy}{dx} = -\dfrac{x}{y}$, $y \neq 0$, and for the straight lines, $\dfrac{dy}{dx} = m = \dfrac{y}{x}$, $x \neq 0$. Since the product of the slopes is -1, it follows that the two families are orthogonal trajectories.

52. For the parabolas, $m_1 = \dfrac{dy}{dx} = \dfrac{1}{2ay}$, $y \neq 0$, and for the ellipses, $m_2 = \dfrac{dy}{dx} = -\dfrac{2x}{y}$, $y \neq 0$. Let (x_0, y_0) be a point of intersection of a parabola and an ellipse. Then

$$m_1 \cdot m_2 = \frac{1}{2ay_0} \cdot \left(-\frac{2x_0}{y_0}\right) = -\frac{x_0}{ay_0^2} = -1 \text{ since } x_0 = ay_0^2.$$

53. The line $x + 2y + 3 = 0$ has slope $m = -1/2$. Thus, a line perpendicular to this line will have slope 2. A tangent line to the ellipse $4x^2 + y^2 = 72$ has slope $m = \dfrac{dy}{dx} = -\dfrac{4x}{y}$. Setting $-\dfrac{4x}{y} = 2$ gives $y = -2x$.

Substituting into the equation for the ellipse, we have

$$4x^2 + 4x^2 = 72 \quad \Rightarrow \quad 8x^2 = 72 \quad \Rightarrow \quad x = \pm 3$$

It now follows that $y = \mp 6$ and the equations of the tangents are:

at $(3, -6):$ $\quad y + 6 = 2(x - 3)$ $\quad$ or $\quad y = 2x - 12;$

at $(-3, 6):$ $\quad y - 6 = 2(x + 3)$ $\quad$ or $\quad y = 2x + 12.$

54. The line $2x + 5y - 4 = 0$ has slope $m = -2/5$. A tangent line to the hyperbola $4x^2 - y^2 = 36$ has slope $\dfrac{dy}{dx} = \dfrac{4x}{y}$. Therefore a normal line to the hyperbola will have slope $m = -\dfrac{y}{4x}$. Setting $-\dfrac{y}{4x} = -\dfrac{2}{5}$ gives $y = 8x/5$. Substituting this into the equation for the hyperbola, we have

$$4x^2 - \frac{64}{25}x^2 = 36 \quad \Longrightarrow \quad x = \pm 5$$

It now follows that $y = 8$ when $x = 5$ and $y = -8$ when $x = -5$. The equations of the normals are:

at $(5, 8):$ $\quad y - 8 = -\tfrac{2}{5}(x - 5)$ $\quad$ or $\quad y = -\tfrac{2}{5}x + 10;$

at $(-5, -8):$ $\quad y + 8 = -\tfrac{2}{5}(x + 5)$ $\quad$ or $\quad y = -\tfrac{2}{5}x - 10.$

55. Differentiate the equation $(x^2 + y^2)^2 = x^2 - y^2$ implicitly with respect to x:

$$2(x^2 + y^2)\left(2x + 2y\frac{dy}{dx}\right) = 2x - 2y\frac{dy}{dx}$$

Now set $dy/dx = 0$. This gives

$$2x(x^2 + y^2) = x$$

$$x^2 + y^2 = \frac{1}{2} \quad (x \neq 0)$$

Substituting this result into the original equation, we get

$$x^2 - y^2 = \frac{1}{4}$$

Now

$$\begin{matrix} x^2 + y^2 = 1/2 \\ x^2 - y^2 = 1/4 \end{matrix} \Rightarrow x = \pm\frac{\sqrt{6}}{4}, \quad y = \pm\frac{\sqrt{2}}{4}$$

Thus, the points on the curve at which the tangent line is horizontal are:

$$(\sqrt{6}/4, \sqrt{2}/4), \quad (\sqrt{6}/4, -\sqrt{2}/4), \quad (-\sqrt{6}/4, \sqrt{2}/4), \quad (-\sqrt{6}/4, -\sqrt{2}/4).$$

56. (a) $x^{2/3} + y^{2/3} = a^{2/3};$ $\quad \tfrac{2}{3}x^{-1/3} + \tfrac{2}{3}y^{-1/3}\dfrac{dy}{dx} = 0,$ and $\dfrac{dy}{dx} = -\left(\dfrac{y}{x}\right)^{1/3}$

Thus, the slope at $(x_1, y_1),$ $x_1 \neq 0$ is: $m = -\left(\dfrac{y_1}{x_1}\right)^{1/3}$

(b) $m = 0:$ $\quad -\left(\dfrac{y_1}{x_1}\right)^{1/3} = 0 \quad \Longrightarrow \quad y_1 = 0$ and $x_1 = \pm a;$ $\quad (a, 0), (-a, 0)$

$$m = 1: \quad -\left(\frac{y_1}{x_1}\right)^{1/3} = 1 \quad \Longrightarrow \quad y_1 = -x_1 \text{ and } x_1 = \pm \tfrac{1}{4} a\sqrt{2};$$

$$\left(\tfrac{1}{4} a\sqrt{2}, -\tfrac{1}{4} a\sqrt{2}\right), \; \left(-\tfrac{1}{4} a\sqrt{2}, \tfrac{1}{4} a\sqrt{2}\right)$$

$$m = -1: \quad -\left(\frac{y_1}{x_1}\right)^{1/3} = -1 \quad \Longrightarrow \quad y_1 = x_1 \text{ and } x_1 = \pm \tfrac{1}{4} a\sqrt{2};$$

$$\left(\tfrac{1}{4} a\sqrt{2}, \tfrac{1}{4} a\sqrt{2}\right), \; \left(-\tfrac{1}{4} a\sqrt{2}, -\tfrac{1}{4} a\sqrt{2}\right)$$

57. Differentiate the equation $x^{1/2} + y^{1/2} = c^{1/2}$ implicitly with respect to x :

$$\frac{1}{2} x^{-1/2} + \frac{1}{2} y^{-1/2} \frac{dy}{dx} = 0 \quad \text{which implies} \quad \frac{dy}{dx} = -\left(\frac{y}{x}\right)^{1/2}$$

An equation for the tangent line to the graph at the point (x_0, y_0) is

$$y - y_0 = -\left(\frac{y_0}{x_0}\right)^{1/2} (x - x_0)$$

The x- and y-intercepts of this line are

$$a = (x_0 y_0)^{1/2} + x_0 \quad \text{and} \quad b = (x_0 y_0)^{1/2} + y_0 \quad \text{respectively.}$$

Now

$$a + b = 2(x_0 y_0)^{1/2} + x_0 + y_0 = \left(x_0^{1/2} + y_0^{1/2}\right)^2 = c.$$

58. The circle has equation $x^2 + (y - a)^2 = 1$ and $2x + 2(y - a)\dfrac{dy}{dx} = 0$. Thus $\dfrac{dy}{dx} = -\dfrac{x}{y - a}$.

A tangent line to the parabola has slope $\dfrac{dy}{dx} = 4x$. Now

$$4x = -\frac{x}{y - a} \quad \Longrightarrow \quad x(4y - 4a + 1) = 0 \quad \Longrightarrow \quad 4y - 4a + 1 = 0 \text{ since } x \neq 0$$

It follows that

$$y = a - \frac{1}{4} \quad \Longrightarrow \quad x = \pm\frac{\sqrt{15}}{4} \quad \Longrightarrow \quad y = \frac{15}{8}$$

Points of intersection: $\left(\pm\dfrac{\sqrt{15}}{4}, \dfrac{15}{8}\right)$

59. (a) $d(h) = \dfrac{3\sqrt[3]{h}}{h} = \dfrac{3}{h^{2/3}}$

(b) $d(h) \to \infty$ as $h \to 0^-$ and as $h \to 0^+$

(c) The graph of f has a vertical tangent at $(0, 0)$.

60. (a) $d(h) = \dfrac{3\sqrt[3]{h^2}}{h} = \dfrac{3}{h^{1/3}}$

(b) $d(h) \to -\infty$ as $h \to 0^-$; and $d(h) \to \infty$ as $h \to 0^+$

(c) The graph is said to have a "cusp" at $(0, 0)$.

61. $f'(x) > 0$ on $(-\infty, \infty)$

62. (a) $f'(x) = 0$ at $x = \dfrac{\sqrt{3}}{3}$;   (b) $f'(x) > 0$ on $(\sqrt{3}/3, \infty)$;   (c) $f'(x) < 0$ on $(0, \sqrt{3}/3)$

63. $x = t, \quad y = \sqrt{4 - t^2}$ $x = t, \quad y = -\sqrt{4 - t^2}$

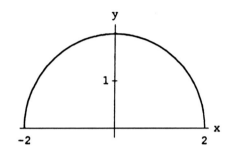   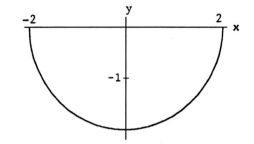

64. $\left.\dfrac{dy}{dx}\right|_{(2,1)} = -\dfrac{3}{2}$ **65.** $\left.\dfrac{dy}{dx}\right|_{(3,4)} = 3$ **66.** $\left.\dfrac{dy}{dx}\right|_{(0,\pi/6)} = 0$

67. $\left.\dfrac{dy}{dx}\right|_{(1,3\sqrt{3})} = -\sqrt{3}$

68. (b) $x = 3$ implies $y = 3$; $\left.\dfrac{dy}{dx}\right|_{(3,3)} = -1$;   tangent line: $y - 3 = -(x - 3)$ or $x + y = 6$.

   (c)

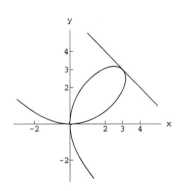

69. (a) The graph of $x^4 = x^2 - y^2$ is:

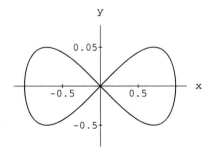

   (b) Differentiate the equation $x^4 = x^2 - y^2$
        implicitly with respect to x :

$$4x^3 = 2x - 2y\frac{dy}{dx}$$

   Now set $dy/dx = 0$. This gives $4x^3 = 2x$
   which implies $x = \pm\dfrac{\sqrt{2}}{2}$.

70. The graph of $(2-x)y^2 = x^3$ is:

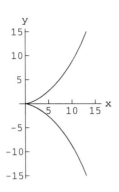

REVIEW EXERCISES

1.
$$f'(x) = \lim_{h \to 0} \frac{f(x+h) - f(x)}{h} = \lim_{h \to 0} \frac{[(x+h)^3 - 4(x+h) + 3] - [x^3 - 4x + 3]}{h}$$

$$= \lim_{h \to 0} \frac{3x^2 h + 3xh^2 + h^3 - 4h}{h} = \lim_{h \to 0} (3x^2 + 3xh + h^2 - 4) = 3x^2 - 4.$$

2.
$$f'(x) = \lim_{h \to 0} \frac{f(x+h) - f(x)}{h} = \lim_{h \to 0} \frac{\sqrt{1 + 2(x+h)} - \sqrt{1 + 2x}}{h}$$

$$= \lim_{h \to 0} \frac{\sqrt{1 + 2(x+h)} - \sqrt{1 + 2x}}{h} \cdot \frac{\sqrt{1 + 2(x+h)} + \sqrt{1 + 2x}}{\sqrt{1 + 2(x+h)} + \sqrt{1 + 2x}}$$

$$= \lim_{h \to 0} \frac{2h}{h \left[\sqrt{1 + 2(x+h)} + \sqrt{1 + 2x} \right]} = \frac{1}{\sqrt{1 + 2x}}$$

3.
$$g'(x) = \lim_{h \to 0} \frac{f(x+h) - f(x)}{h} = \lim_{h \to 0} \frac{\dfrac{1}{x+h-2} - \dfrac{1}{x-2}}{h}$$

$$= \lim_{h \to 0} \frac{(x-2) - (x+h-2)}{h(x+h-2)(x-2)} = \lim_{h \to 0} \frac{-1}{(x+h-2)(x-2)} = \frac{-1}{(x-2)^2}.$$

4.
$$F'(x) = \lim_{h \to 0} \frac{F(x+h) - F(x)}{h} = \lim_{h \to 0} \frac{(x+h)\sin(x+h) - x\sin x}{h}$$

$$= \lim_{h \to 0} \frac{(x+h)(\sin x \cos h + \cos x \sin h) - x \sin x}{h}$$

$$= \lim_{h \to 0} \frac{x \sin x(\cos h - 1) + h \sin x \cos h + x \cos x \sin h + h \cos x \sin h}{h}$$

$$= x \sin x \lim_{h \to 0} \frac{\cos h - 1}{h} + \lim_{h \to 0} \sin x \cos h + x \cos x \lim_{h \to 0} \frac{\sin h}{h} + \lim_{h \to 0} \cos x \sin h$$

$$= \sin x + x \cos x.$$

5. $y' = \frac{2}{3}x^{-1/3}$

6. $y' = 2(\frac{3}{4})x^{-1/4} - 4(-\frac{1}{4})x^{-5/4} = \frac{3}{2}x^{-1/4} + x^{-5/4}.$

7. $y' = \dfrac{(x^3-1)(2x+2) - (1+2x+x^2)3x^2}{(x^3-1)^2} = -\dfrac{x^4 + 4x^3 + 3x^2 + 2x + 2}{(x^3-1)^2}.$

8. $f'(t) = 3(2-3t^2)^2(-6t) = -18t(2-3t^2)^2.$

9. $f(x) = (a^2 - x^2)^{-1/2}; \quad f'(x) = -\dfrac{1}{2}(a^2-x^2)^{-3/2}(-2x) = \dfrac{x}{(a^2-x^2)^{3/2}}.$

10. $y' = 2\left(a - \dfrac{b}{x}\right)\left(\dfrac{b}{x^2}\right) = \dfrac{2b}{x^2}\left(a - \dfrac{b}{x}\right).$

11. $y' = 3\left(a + \dfrac{b}{x^2}\right)^2\left(-\dfrac{2b}{x^3}\right) = -\dfrac{6b}{x^3}\left(a + \dfrac{b}{x^2}\right)^2.$

12. $y' = \sqrt{2+3x} + \dfrac{x}{2}(2+3x)^{-1/2}(3) = \sqrt{2+3x} + \dfrac{3x}{2\sqrt{2+3x}}.$

13. $y' = \sec^2\sqrt{2x+1}\left[\frac{1}{2}(2x+1)^{-1/2}(2)\right] = \dfrac{\sec^2\sqrt{2x+1}}{\sqrt{2x+1}}.$

14. $g'(x) = x^2\left[-\sin(2x-1)(2)\right] + 2x\cos(2x-1) = 2x\cos(2x-1) - 2x^2\sin(2x-1).$

15. $F'(x) = (x+2)^2\frac{1}{2}(x^2+2)^{-1/2}(2x) + \sqrt{x^2+2}\,(2)(x+2) = 2(x+2)\sqrt{x^2+2} + \dfrac{x(x+2)^2}{\sqrt{x^2+2}}.$

16. $y' = \dfrac{(a^2-x^2)2x - (a^2+x^2)(-2x)}{(a^2-x^2)^2} = \dfrac{4a^2x}{(a^2-x^2)^2}.$

17. $h'(t) = \sec t^2 + t(2t)\sec t^2\tan t^2 + 6t^2 = \sec t^2 + 2t^2\tan t^2\sec t^2 + 6t^2.$

18. $y' = \dfrac{(1+\cos x)2\cos 2x - \sin 2x(-\sin x)}{(1+\cos x)^2} = \dfrac{2\cos 2x}{1+\cos x} + \dfrac{\sin 2x\sin x}{(1+\cos x)^2}.$

19. $s' = \dfrac{1}{3}\left(\dfrac{2-3t}{2+3t}\right)^{-2/3}\cdot\dfrac{(2+3t)(-3) - (2-3t)(3)}{(2+3t)^2} = -\dfrac{4}{(2+3t)^{4/3}(2-3t)^{2/3}}.$

20. $r' = \theta^2(\frac{1}{2})(3-4\theta)^{-1/2}(-4) + 2\theta\sqrt{3-4\theta} = 2\theta\sqrt{3-4\theta} - \dfrac{2\theta^2}{\sqrt{3-4\theta}}.$

21. $f'(\theta) = -3\csc^2(3\theta + \pi).$

22. $y' = \dfrac{(1+x^2)(\sin 2x + 2x\cos 2x) - x\sin 2x(2x)}{(1+x^2)^2} = \dfrac{\sin 2x + 2x\cos 2x + 2x^3\cos 2x - x^2\sin 2x}{(1+x^2)^2}.$

23. $f'(x) = \frac{1}{3}x^{-2/3} + \frac{1}{2}x^{-1/2} = \dfrac{1}{3x^{2/3}} + \dfrac{1}{2x^{1/2}}; \quad f'(64) = \dfrac{1}{48} + \dfrac{1}{16} = \dfrac{1}{12}.$

24. $f'(x) = \dfrac{x}{2}\dfrac{1}{\sqrt{8-x^2}}(-2x) + \sqrt{8-x^2} = \sqrt{8-x^2} - \dfrac{x^2}{\sqrt{8-x^2}}; \quad f'(2) = 0.$

25. $f'(x) = x^2(2)(\pi)\sin\pi x\cos\pi x + 2x\sin^2\pi x = 2x\sin^2\pi x + 2\pi x^2\sin\pi x\cos\pi x$; $f'(1/6) = \dfrac{1}{12} + \dfrac{\pi\sqrt{3}}{72}$.

26. $f' = -3\csc^2 3x$; $f'(\pi/9) = -4$.

27. $f(x) = 2x^3 - x^2 + 3$, $f'(x) = 6x^2 - 2x$; $f'(1) = 4$.
Tangent line: $y - 4 = 4(x - 1)$ or $y = 4x$;   normal line: $y - 4 = -\frac{1}{4}(x - 1)$ or $y = -\frac{1}{4}x + \frac{17}{4}$.

28. $f(x) = \dfrac{2x - 3}{3x + 4}$, $f'(x) = \dfrac{(3x + 4)2 - (2x - 3)3}{(3x + 4)^2} = \dfrac{17}{(3x + 4)^2}$; $f'(-1) = 17$.
Tangent line: $y + 5 = 17(x + 1)$;   normal line: $y + 5 = -\frac{1}{17}(x + 1)$.

29. $f(x) = (x + 1)\sin 2x$, $f'(x) = 2(x + 1)\cos 2x + \sin 2x$; $f'(0) = 2$.
Tangent line: $y = 2x$;   normal line: $y = -\frac{1}{2}x$.

30. $f(x) = x\sqrt{1 + x^2}$, $f'(x) = \sqrt{1 + x^2} + \dfrac{x^2}{\sqrt{1 + x^2}}$; $f'(1) = \frac{3}{2}\sqrt{2}$.
Tangent line: $y - \sqrt{2} = \frac{3}{2}\sqrt{2}\,(x - 1)$;   normal line: $y - \sqrt{2} = -\frac{1}{3}\sqrt{2}\,(x - 1)$

31. $f'(x) = -\sin(2 - x)(-1) = \sin(2 - x)$, $f''(x) = \cos(2 - x)(-1) = -\cos(2 - x)$.

32. $f'(x) = \frac{3}{2}(x^2 + 4)^{1/2}(2x) = 3x(x^2 + 4)^{1/2}$,
$f'' = 3x(\frac{1}{2})(x^2 + 4)^{-1/2}(2x) + 3(x^2 + 4)^{1/2} = \dfrac{3x^2}{(x^2 + 4)^{1/2}} + 3(x^2 + 4)^{1/2} = \dfrac{6x^2 + 12}{\sqrt{x^2 + 4}}$.

33. $y' = x\cos x + \sin x$, $y'' = 2\cos x - x\sin x$.

34. $g'(u) = 2\tan u\sec^2 u$, $g''(u) = 2\tan u \cdot 2\sec u\cdot\sec u\tan u + 2\sec^4 u = 4\sec^2 u\tan^2 u + 2\sec^4 u$.

35. $(-1)^n\, n!\, b^n$ **36.** $\dfrac{(-1)^n\, n!\, ab^n}{(bx + c)^{n+1}}$

37. $3x^2 y + x^3 y' + y^3 + 3xy^2 y' = 0$, $y' = -\dfrac{y^3 + 3x^2 y}{x^3 + 3xy^2}$.

38. $(1 + 2y')\sec^2(x + 2y) = 2xy + x^2 y'$, $y' = \dfrac{2xy - \sec^2(x + 2y)}{2\sec^2(x + 2y) - x^2}$.

39. $6x^2 + 3\cos y - 3xy'\sin y = 2y + 2xy'$, $y' = \dfrac{6x^2 + 3\cos y - 2y}{2x + 3x\sin y}$.

40. $2x + 3\sqrt{y} + \dfrac{3xy'}{2\sqrt{y}} = \dfrac{1}{y} - \dfrac{xy'}{y^2}$, $y' = \dfrac{2y^2 - 4xy^3 - 6y^{7/2}}{2xy + 3xy^{5/2}}$.

41. $2x + 2y + 2xy' - 6yy' = 0$, $y' = \dfrac{x + y}{3y - x}$;   at $(3, 2)$ $y' = \dfrac{5}{3}$.
Tangent line: $y - 2 = \frac{5}{3}(x - 3)$ or $5x - 3y = 9$; normal line: $y - 2 = -\frac{3}{5}(x - 3)$ or $3x + 5y = 19$.

42. $2y \cos 2x + y' \sin 2x - xy' \cos y - \sin y = 0$, $y' = \dfrac{\sin y - 2t \cos 2x}{\sin 2x - x \cos y}$;   at $(\frac{1}{4}\pi, \frac{1}{2}\pi)$ $y' = 1$.

Tangent line: $y - \frac{1}{2}\pi = x - \frac{1}{4}\pi$ or $y = x + \frac{1}{4}\pi$;   normal line: $y - \frac{1}{2}\pi = -(x - \frac{1}{4}\pi)$ or $y = -x + \frac{3}{4}\pi$.

43. $f'(x) = 3(x-4)(x-2)$.

f':

(a) $f'(x) = 0$ at $x = 2,\ 4$,   (b) $f'(x) > 0$ on $(-\infty, 2) \cup (4, \infty)$,

(c) $f'(x) < 0$ on $(2, 4)$.

44. $f'(x) = \dfrac{2 - 4x^2}{(1 + 2x^2)^2}$.

f':

(a) $f'(x) = 0$ at $x = \pm\sqrt{2}/2$,   (b) $f'(x) > 0$ on $(-\sqrt{2}/2, \sqrt{2}/2)$,

(c) $f'(x) < 0$ on $(-\infty, -\sqrt{2}/2) \cup (\sqrt{2}/2, \infty)$.

45. $f'(x) = 1 + 2\cos 2x$.

f':

(a) $f'(x) = 0$ at $x = \pi/3,\ 2\pi/3,\ 4\pi/3,\ 5\pi/3$,

(b) $f'(x) > 0$ on $(0, \pi/3) \cup (2\pi/3, 4\pi/3) \cup (5\pi/3, 2\pi)$,

(c) $f'(x) < 0$ on $(\pi/3, 2\pi/3) \cup (4\pi/3, 5\pi/3)$.

46. $f'(x) = \sqrt{3} + 2\sin x$.

f':

(a) $f'(x) = 0$ at $x = 4\pi/3,\ 5\pi/3$,   (b) $f'(x) > 0$ on $(0, 4\pi/3) \cup (5\pi/3, 2\pi)$,

(c) $f'(x) < 0$ on $(4\pi/3, 5\pi/3)$.

47. $y' = \sqrt{x}$.   (a) 1,   (b) 3,   (c) $\frac{1}{3}$.

48. $y' = 3x^2$. The tangent line to the curve $y = x^3$ at the point (a, a^3) has equation

$$y - a^3 = 3a^2(x - a) \quad \text{or} \quad y = 3a^2 x - 2a^3.$$

Since this line must pass through $(0, 2)$, we have

$$2 = -2a^3 \quad \text{which implies} \quad a = -1.$$

There is one tangent line that passes through $(0, 2)$, namely $y = 3x + 2$.

49. $y' = 3x^2 - 1$. The tangent line to the curve $y = x^3 - x$ at the point $(a, a^3 - a)$ has equation

$$y - (a^3 - a) = (3a^2 - 1)(x - a) \quad \text{or} \quad y = (3a^2 - 1)x - 2a^3.$$

Since this line must past through $(-2, 2)$, we have

$$2 = -6a^2 + 2 - 2a^3 \quad \text{which gives} \quad 6a^2 + 2a^3 = 0 \quad \text{and} \quad 2a^2(3 + a) = 0.$$

Thus, $a = 0, -3$. There are two tangent lines that pass through $(-2, 2)$: $y = -x$, $y = 26x + 54$.

50. The curve $y = Ax^2 + Bx + C$ passes through the points $(1, 3)$ and $(2, 3)$. This implies that $A + B + C = 3$ and $4A + 2B + C = 3$.

$y' = 2Ax + B$. The line $x - y + 1 = 0$ has slope 1. At $x = 2$, $y' = 4A + B$. Therefore we have $4A + B = 1$. The solution set of the system of equations

$$A + B + C = 3$$

$$4A + 2B + C = 3$$

$$4A + B = 1$$

is: $A = 1$, $B = -3$, $C = 5$; $\quad y = x^2 - 3x + 5$.

51. The curve $y = Ax^3 + Bx^2 + Cx + D$ passes through the points $(1, 1)$ and $(-1, -9)$. This implies that $A + B + C + D = 1$ and $-A + B - C + D = -9$.

$y' = 3Ax^2 + 2Bx + C$. The line $y = 5x - 4$ has slope 5; the line $y = 9x$ has slope 9. At $x = 1$, $y' = 3A + 2B + C$; at $x = -1$, $y' = 3A - 2B + C$. Therefore we have $3A + 2B + C = 5$ and $3A - 2B + C = 9$. The solution set of the system of equations

$$A + B + C + D = 1$$

$$-A + B - C + D = -9$$

$$3A + 2B + C = 5$$

$$3A - 2B + C = 9$$

is: $A = 1$, $B = -1$, $C = 4$, $D = -3$; $\quad y = x^3 - x^2 + 4x - 3$.

52.

$$\frac{1}{h}\left[\frac{1}{(x+h)^n} - \frac{1}{x^n}\right] = \frac{1}{h}\left[\frac{x^n - (x+h)^n}{(x+h)^n x^n}\right]$$

$$= \frac{1}{h}\left[\frac{x^n - x^n - nx^{n-1}h - (\text{terms of the form } Cx^k h^m, m \geq 2)}{(x+h)^n x^n}\right]$$

$$= \frac{-nx^{n-1} - (\text{terms of the form } Cx^k h^m, m \geq 1)}{(x+h)^n x^n}.$$

Therefore,

$$\lim_{h \to 0} \frac{1}{h} \left[\frac{1}{(x+h)^n} - \frac{1}{x^n} \right] = \lim_{h \to 0} \frac{-nx^{n-1} - (\text{terms of the form } Cx^k h^m, m \geq 1)}{(x+h)^n x^n}$$

$$= \frac{-nx^{n-1}}{x^{2n}} = \frac{-n}{x^{n+1}}.$$

53. Let $f(x) = x^2 - 2x + 1$; $f'(x) = 2x - 2$.

$$\lim_{h \to 0} \frac{(1+h)^2 - 2(1+h) + 1}{h} = \lim_{h \to 0} \frac{(1+h)^2 - 2(1+h) + 1 - 0}{h} = f'(1) = 0$$

54. Let $f(x) = \sqrt{x}$; $f'(x) = 1/2\sqrt{x}$.

$$\lim_{h \to 0} \frac{\sqrt{9+h} - 3}{h} = f'(9) = \frac{1}{6}$$

55. Let $f(x) = \sin x$; $f'(x) = \cos x$.

$$\lim_{h \to 0} \frac{\sin(\frac{1}{6}\pi + h) - \frac{1}{2}}{h} = f'(\frac{1}{6}\pi) = \frac{\sqrt{3}}{2}$$

56. Let $f(x) = x^5$; $f'(x) = 5x^4$.

$$\lim_{x \to 2} \frac{x^5 - 32}{x - 2} = f'(2) = 80$$

57. Let $f(x) = \sin x$; $f'(x) = \cos x$.

$$\lim_{x \to \pi} \frac{\sin x}{x - \pi} = f'(\pi) = -1$$

58. (a) From Figure A, M increases on $[a, b]$, is constant on $[b, e]$, and increases on $[e, \infty)$; M is differentiable at b $(f'(b) = 0)$, M is not differentiable at e $(f'(e) \neq 0)$.

(b) From Figure B, m is constant on $[a, c]$, decreases on $[c, d]$, and is constant on $[d, \infty)$; m is not differentiable at c $(f'(c) \neq 0)$, m is differentiable at d $(f'(d) = 0)$.

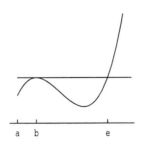

Figure A

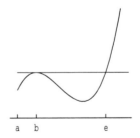

Figure B

CHAPTER 4

SECTION 4.1

1. f is differentiable on $(0,1)$, continuous on $[0,1]$; and $f(0) = f(1) = 0$.

$f'(c) = 3c^2 - 1; \quad 3c^2 - 1 = 0 \Longrightarrow c = \dfrac{\sqrt{3}}{3} \quad \left(-\dfrac{\sqrt{3}}{3} \notin (0,1) \right)$

2. f is differentiable on $(-2,2)$, continuous on $[-2,2]$; and $f(-2) = f(2) = 0$.

$f'(c) = 4c^3 - 4c; \quad 4c(c^2 - 1) = 0 \Longrightarrow c = 0, \pm 1$

3. f is differentiable on $(0,2\pi)$, continuous on $[0,2\pi]$; and $f(0) = f(2\pi) = 0$.

$f'(c) = 2\cos 2c; \quad 2\cos 2c = 0 \Longrightarrow 2c = \dfrac{\pi}{2} + n\pi, \quad \text{and} \quad c = \dfrac{\pi}{4} + \dfrac{n\pi}{2}, \quad n = 0, \pm 1, \pm 2 \ldots$

Thus, $c = \dfrac{\pi}{4}, \dfrac{3\pi}{4}, \dfrac{5\pi}{4}, \dfrac{7\pi}{4}$

4. f is differentiable on $(0,8)$, continuous on $[0,8]$; and $f(0) = f(8) = 0$.

$f'(c) = \frac{2}{3}c^{-1/3} - \frac{2}{3}c^{-2/3} = \dfrac{2}{3} \dfrac{c^{1/3} - 1}{c^{2/3}} \quad f'(c) = 0 \quad \Longrightarrow \quad c = 1.$

5. $f'(c) = 2c, \quad \dfrac{f(b) - f(a)}{b - a} = \dfrac{4 - 1}{2 - 1} = 3; \quad 2c = 3 \quad \Longrightarrow \quad c = 3/2$

6. $f'(c) = \dfrac{3}{2\sqrt{c}} - 4, \quad \dfrac{f(b) - f(a)}{b - a} = \dfrac{-10 - (-1)}{4 - 1} = -3; \quad \dfrac{3}{2\sqrt{c}} - 4 = -3 \quad \Longrightarrow \quad c = 9/4$

7. $f'(c) = 3c^2, \quad \dfrac{f(b) - f(a)}{b - a} = \dfrac{27 - 1}{3 - 1} = 13; \quad 3c^2 = 13 \Longrightarrow c = \dfrac{1}{3}\sqrt{39} \quad \left(-\dfrac{1}{3}\sqrt{39} \text{ is not in } [a,b] \right)$

8. $f'(c) = \frac{2}{3}c^{-1/3}, \quad \dfrac{f(b) - f(a)}{b - a} = \dfrac{4 - 1}{8 - 1} = \dfrac{3}{7}; \quad \frac{2}{3}c^{-1/3} = \dfrac{3}{7} \quad \Longrightarrow \quad c = \dfrac{(14)^3}{9^3}$

9. $f'(c) = \dfrac{-c}{\sqrt{1 - c^2}}, \quad \dfrac{f(b) - f(a)}{b - a} = \dfrac{0 - 1}{1 - 0} = -1; \quad \dfrac{-c}{\sqrt{1 - c^2}} = -1 \quad \Longrightarrow \quad c = \dfrac{1}{2}\sqrt{2}$

$(-\dfrac{1}{2}\sqrt{2}$ is not in $[a,b])$

10. $f'(c) = 3c^2 - 3, \quad \dfrac{f(b) - f(a)}{b - a} = \dfrac{-2 - 2}{1 - (-1)} = -2; \quad 3c^2 - 3 = -2 \quad \Longrightarrow \quad c = \pm \dfrac{\sqrt{3}}{3}$

11. f is continuous on $[-1,1]$, differentiable on $(-1,1)$ and $f(-1) = f(1) = 0$.

$f'(x) = \dfrac{-x(5 - x^2)}{(3 + x^2)^2 \sqrt{1 - x^2}}, \quad f'(c) = 0$ for c in $(-1,1)$ implies $c = 0$.

12. (a) $f'(x) = \frac{2}{3}x^{-1/3} = \dfrac{2}{3x^{1/3}} \neq 0$ for all $x \in (-1,1)$.

(b) $f'(0)$ does not exist. Therefore, f is not differentiable on $(-1,1)$.

13. No. By the mean-value theorem there exists at least one number $c \in (0, 2)$ such that

$$f'(c) = \frac{f(2) - f(0)}{2 - 0} = \frac{3}{2} > 1.$$

14. No, by Rolle's theorem: $f(2) = f(3) = 1$ but there is no value $c \in (2, 3)$ such that $f'(c) = 0$.

15. By the mean-value theorem there is a number $c \in (2, 6)$ such that

$$f(6) - f(2) = f'(c)(6 - 2) = f'(c)4.$$

Since $1 \le f'(x) \le 3$ for all $x \in (2, 6)$, it follows that

$$4 \le f(6) - f(2) \le 12.$$

16. $f(x) = x^2 + x + 3$, $f'(x) = 2x + 1$.

The slope of the line through $(-1, 3)$ and $(2, 9)$ is 2. Setting $2x + 1 = 2$, we get $x = \frac{1}{2}$. The point on the graph of f where the tangent line is parallel to the line through $(-1, 3)$ and $(2, 9)$ is: $(1/2, 15/4)$.

17. f is everywhere continuous and everywhere differentiable except possibly at $x = -1$.

f is continuous at $x = -1$: as you can check,

$$\lim_{x \to -1^-} f(x) = 0, \quad \lim_{x \to -1^+} f(x) = 0, \quad \text{and} \quad f(-1) = 0.$$

f is differentiable at $x = -1$ and $f'(-1) = 2$: as you can check,

$$\lim_{h \to 0^-} \frac{f(-1 + h) - f(-1)}{h} = 2 \quad \text{and} \quad \lim_{h \to 0^+} \frac{f(-1 + h) - f(-1)}{h} = 2.$$

Thus f satisfies the conditions of the mean-value theorem on every closed interval $[a, b]$.

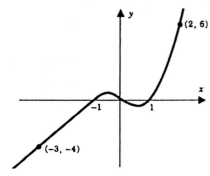

$$f'(x) = \begin{cases} 2, & x \le -1 \\ 3x^2 - 1, & x > -1 \end{cases}$$

$$\frac{f(2) - f(-3)}{2 - (-3)} = \frac{6 - (-4)}{2 - (-3)} = 2.$$

$f'(c) = 2$ with $c \in (-3, 2)$ iff $c = 1$ or $-3 < c \le -1$.

18. f is continuous and differentiable everywhere;

$$f'(x) = \begin{cases} 3x^2, & x \le 1 \\ 3, & x > 1 \end{cases}$$

$$\frac{f(2) - f(-1)}{2 - (-1)} = \frac{6 - 1}{3} = \frac{5}{3}$$

For $c \le 1$, $f'(c) = 3c^2 = \frac{5}{3} \implies c = \pm\frac{\sqrt{5}}{3}$

For $c > 1$, $f'(c) = 3 \ne \frac{5}{3}$

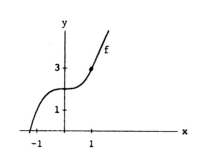

19. Let $f(x) = Ax^2 + Bx + C$. Then $f'(x) = 2Ax + B$. By the mean-value theorem

$$f'(c) = \frac{f(b) - f(a)}{b - a} = \frac{(Ab^2 + Bb + C) - (Aa^2 + Ba + C)}{b - a}$$

$$= \frac{A(b^2 - a^2) + B(b - a)}{b - a} = A(b + a) + B$$

Therefore, we have

$$2Ac + B = A(b + a) + B \implies c = \frac{a + b}{2}$$

20. $\dfrac{f(1) - f(-1)}{1 - (-1)} = 1$ and $f'(x) = -1/x^2 < 0$; f is not continuous at 0.

21. $\dfrac{f(1) - f(-1)}{1 - (-1)} = 0$ and $f'(x)$ is never zero. This result does not violate the mean-value theorem

since f is not differentiable at 0; the theorem does not apply.

22. $f(x) = \begin{cases} 2x - 4, & x \geq 1/2 \\ -2x - 2, & x < 1/2 \end{cases}$

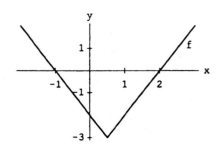

$f'(x) = \begin{cases} 2, & x > 1/2 \\ -2, & x < 1/2 \end{cases}$

$f'(x) \neq 0$ for all $x \neq \frac{1}{2}$.

Rolle's theorem is not violated because f is

not differentiable at $x = \frac{1}{2}$.

23. Set $P(x) = 6x^4 - 7x + 1$. If there existed three numbers $a < b < c$ at which $P(x) = 0$, then by Rolle's theorem $P'(x)$ would have to be zero for some x in (a, b) and also for some x in (b, c). This is not the case: $P'(x) = 24x^3 - 7$ is zero only at $x = (7/24)^{1/3}$.

24. Set $P(x) = 6x^5 + 13x + 1$. Note that $P(-1) < 0$ and $P(0) > 0$. By the intermediate-value theorem, the equation $P(x) = 0$ has at least one real root c. If this equation had another real root d, then by Rolle's theorem $P'(x)$ would have to be zero for some x between c and d. This is not the case: $P'(x) = 30x^4 + 13$ is never zero.

25. Set $P(x) = x^3 + 9x^2 + 33x - 8$. Note that $P(0) < 0$ and $P(1) > 0$. Thus, by the intermediate-value theorem, there exists some number c between 0 and 1 at which $P(x) = 0$. If the equation $P(x) = 0$ had an additional real root, then by Rolle's theorem there would have to be some real number at which $P'(x) = 0$. This is not the case: $P'(x) = 3x^2 + 18x + 33$ is never zero since the discriminant $b^2 - 4ac = (18)^2 - 12(33) < 0$.

26. (a)   Suppose that f has two zeros, $x_1, x_2 \in (a, b)$. Then, f is differentiable on (x_1, x_2) and continuous on $[x_1, x_2]$. By Rolle's theorem, f' has a zero in (x_1, x_2) which contradicts the hypothesis.

(b) If f had three zeros in (a, b), then, by Rolle's' theorem, f' would have at least two zeros in (a, b) and f'' would have at least one zero in (a, b) which contradicts the hypothesis.

27. Let c and d be two consecutive roots of the equation $P'(x) = 0$. The equation $P(x) = 0$ cannot have two or more roots between c and d for then, by Rolle's theorem, $P'(x)$ would have to be zero somewhere between these two roots and thus between c and d. In this case c and d would no longer be consecutive roots of $P'(x) = 0$.

28. If $f(x) = 0$ at $a_1,\ a_2,\ \ldots, a_n$ then by Rolle's theorem, $f'(x)$ is zero at some number $b_1 \in (a_1, a_2)$, at some number $b_2 \in (a_2, a_3),\ \ldots$, at some number $b_{n-1} \in (a_{n-1}, a_n)$; $f''(x)$, in turn, must be zero at some number $c_1 \in (b_1, b_2)$, at some number $b_2 \in (b_2, b_3),\ \ldots$, at some number $c_{n-2} \in (b_{n-2}, b_{n-1})$.

29. Suppose that f has two fixed points $a,\ b \in I$, with $a < b$. Let $g(x) = f(x) - x$. Then $g(a) = f(a) - a = 0$ and $g(b) = f(b) - b = 0$. Since f is differentiable on I, we can conclude that g is differentiable on (a, b) and continuous on $[a, b]$. By Rolle's theorem, there exists a number $c \in (a, b)$ such that $g'(c) = f'(c) - 1 = 0$ or $f'(c) = 1$. This contradicts the assumption that $f'(x) < 1$ on I.

30. Set $P(x) = x^3 + ax + b$. It is obvious that for x sufficiently large, $P(x) > 0$ and for x sufficiently large negative, $P(x) < 0$. Thus, by the intermediate-value theorem, the equation $P(x) = 0$ has at least one real root.

If $a \geq 0$, then $P'(x) = 3x^2 + a$ is positive, except possibly at 0, where it remains nonnegative. It follows that P is everywhere increasing and therefore it cannot take on the value 0 more than once.

Suppose now that $a < 0$. Then $-\frac{1}{3}\sqrt{3}\,|a|$ and $\frac{1}{3}\sqrt{3}\,|a|$ are consecutive roots of the equation $P'(x) = 0$ and thus, by Exercise 27, P cannot take on the value zero more than once between these two numbers.

31. (a) $f'(x) = 3x^2 - 3 < 0$ for all x in $(-1, 1)$. Also, f is differentiable on $(-1, 1)$ and continuous on $[-1, 1]$. Thus there cannot be a and b in $(-1, 1)$ such that $f(a) = f(b) = 0$, or they would contradict Rolle's theorem.
   (b) When $f(x) = 0$, $b = 3x - x^3 = x(3 - x^2)$. When x is in $(-1, 1)$, then $|x(3 - x^2)| < 2$.
   Thus $|b| < 2$.

32. $f'(x) = 3x^2 - 3a^2 > 0$ for all x in $(-a, a)$. Also, f is differentiable on $(-a, a)$ and continuous on $[-a, a]$. Thus there cannot be b and c in $(-a, a)$ such that $f(b) = f(c) = 0$, or they would contradict Rolle's theorem.

33. For $p(x) = x^n + ax + b$, $p'(x) = nx^{n-1} + a$, which has at most one real zero for n even $\left(x = -\frac{a}{n}^{\frac{1}{n-1}}\right)$. If there were more than two distinct real roots of $p(x)$, then by Rolle's theorem there would be more than one zero of $p'(x)$. Thus there are at most two distinct real roots of $p(x)$.

34. For $p(x) = x^n + ax + b$, $p'(x) = nx^{n-1} + a$, which has at most two real zeros for n odd. If there were more than three distinct real roots of $p(x)$, then by Rolle's theorem there would be more than two zeros of $p'(x)$. Thus there are at most three distinct real roots of $p(x)$.

35. If $x_1 = x_2$, then $|f(x_1) - f(x_2)|$ and $|x_1 - x_2|$ are both 0 and the inequality holds. If $x_1 \neq x_2$, then by the mean-value theorem

$$\frac{f(x_1) - f(x_2)}{x_1 - x_2} = f'(c)$$

for some number c between x_1 and x_2. Since $|f'(c)| \leq 1$:

$$\left| \frac{f(x_1) - f(x_2)}{x_1 - x_2} \right| \leq 1 \quad \text{and thus} \quad |f(x_1) - f(x_2)| \leq |x_1 - x_2|.$$

36. See the proof of Theorem 4.2.2.

37. Set, for instance, $f(x) = \begin{cases} 1, & a < x < b \\ 0, & x = a, b \end{cases}$

38. (a) Let $f(x) = \cos x$. Choose any numbers x and y, (assume $x < y$). By the mean-value theorem, there is a number c between x and y such that

$$\frac{f(y) - f(x)}{y - x} = f'(c) \quad \Rightarrow \quad \frac{|\cos y - \cos x|}{|y - x|} = |-\sin c| \leq 1 \quad \Rightarrow \quad |\cos x - \cos y| \leq |x - y|$$

(b) Repeat the in part (a) with $f(x) = \sin x$.

39. (a) By the mean-value theorem, there exists a number $c \in (a, b)$ such that $f(b) - f(a) = f'(c)(b - a)$. If $f'(x) \leq M$ for all $x \in (a, b)$, then it follows that

$$f(b) \leq f(a) + M(b - a)$$

(b) If $f'(x) \geq m$ for all $x \in (a, b)$, then it follows that

$$f(b) \geq f(a) + m(b - a)$$

(c) If $|f'(x)| \leq K$ on (a, b), then $-K \leq f'(x) \leq K$ on (a, b) and the result follows from parts (a) and (b).

40. Assume that $g(x) \neq 0$ for all $x \in [a, b]$ and let $h(x) = \dfrac{f(x)}{g(x)}$. Then h is defined on $[a, b]$ and $h(a) = h(b) = 0$. Therefore, by Rolle's theorem, there exists a number $c \in (a, b)$ such that

$$h'(c) = \frac{g(c)f'(c) - f(c)g'(c)}{g^2(c)} = 0$$

Thus $g(c)f'(c) - f(c)g'(c) = 0$ which contradicts the given condition $f(x)g'(x) - g(x)f'(x) \neq 0$ for all $x \in I$. Thus, g has at least one zero in (a, b).

By reversing the roles of f and g, the same argument can be used to show that g cannot have two (or more) zeros on (a, b).

41. We show first that the conditions on f and g imply that f and g cannot be simultaneously 0. Set $h(x) = f^2(x) + g(x)$. Then

$h'(x) = 2f(x)f'(x) + 2g(x)g'(x) = 2f(x)[g(x)] + 2g(x)[-f(x)] = 0 \implies f^2(x) + g^2(x) = C$ constant.

If $C = 0$, then $f^2(x) = -g^2(x) \implies f(x) \equiv 0$, contradicting the assumptions on f. Therefore, $f^2(x) + g^2(x) = C$, $C > 0$. If $f(\alpha) = 0$, then $g(\alpha \neq 0$.

Assume that $f(a) = f(b) = 0$ and $f(x) \neq 0$ on (a, b). Suppose that $g(x) \neq 0$ on (a, b). We know also that $g(a) \neq 0$ and $g(b) \neq 0$. Set $h(x) = f(x)/g(x)$. Then h is continuous on the closed interval $[a, b]$ and differentiable on the open interval (a, b). Therefore, by Rolle's theorem, there exists at least one point $c \, \epsilon (a, b)$ such that $h'(c) = 0$. But,

$$h'(x) = \frac{g(x)f'(x) - f(x)g'(x)}{g^2(x)} = \frac{g^2(x) + f^2(x)}{g^2(x)} = \frac{C}{g^2(x)} > 0$$

for all $x \, \epsilon (a, b)$, and we have a contradiction. Thus g must have at least one zero in (a, b). Since $g'(x) = -f(x) \neq 0$ in (a, b), g has exactly one zero in (a, b).

Simply reverse the roles of f and g to show that f has exactly one zero between two consecutive zeros of g.

42. We prove the result for $h > 0$. The proof for $h < 0$ is similar. If f is differentiable on $(x, x + h)$, it is continuous there and thus, by the hypothesis at x and $x + h$ continuous on $[x, x + h]$. By the mean-value theorem, there exists c in $(x, x + h)$ for which

$$\frac{f(x + h) - f(x)}{x + h - x} = f'(c).$$

Multiplying through by $(x + h) - x = h$, we have

$$f(x + h) - f(x) = f'(c)h.$$

Since c is between x and $x + h$, c can be written

$$c = x + \theta h \quad \text{with} \quad 0 < \theta < 1.$$

43.
$$f'(x_0) = \lim_{h \to 0} \frac{f(x_0 + h) - f(x_0)}{h} = \lim_{h \to 0} \frac{f'(x_0 + \theta h)h}{h} = \lim_{h \to 0} f'(x_0 + \theta h)$$

$$\overset{\big\uparrow}{\rule{0pt}{0pt}} \text{(by the hint)}$$

$$= \lim_{x \to x_0} f'(x) = L$$

$$\overset{\big\uparrow}{\rule{0pt}{0pt}} \text{(by 2.2.6)}$$

44. Suppose that $f(a) = f(b) = k$, and let $g(x) = f(x) - k$. Then g is differentiable on (a, b), continuous on $[a, b]$, and $g(a) = g(b) = 0$. Therefore, by Rolle's theorem, there exists at least one number $c \in (a, b)$ such that $g'(c) = 0$. Since $g'(x) = f'(x)$, it follows that $f'(c) = 0$.

45. Using the hint, F is continuous on $[a, b]$, differentiable on (a, b), and $F(a) = F(b)$. By Exercise 44, there is a number c in (a, b) such that $F'(c) = 0$.

Therefore $[f(b) - f(a)]g'(c) - [g(b) - g(a)]f'(c) = 0$ and $\dfrac{f(b) - f(a)}{g(b) - g(a)} = \dfrac{f'(c)}{g'(c)}$.

46. $f(x) = 2x^3 + 3x^2 - 3x - 2$ is differentiable on $(-2, 1)$, continuous on $[-2, 1]$, and $f(-2) = f(1) = 0$.

$f'(x) = 6x^2 + 6x - 3$

$f'(c) = 0$ at $c_1 \cong -1.366$, $c_2 \cong 0.366$

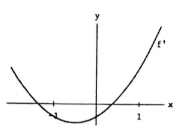

47. $f(x) = 1 - x^3 - \cos(\pi x/2)$ is differentiable on $(0, 1)$, continuous on $[0, 1]$, and $f(0) = f(1) = 0$.

$f'(x) = -3x^2 + \dfrac{\pi}{2}\sin(\pi x/2)$

$f'(c) = 0$ at $c \cong 0.676$

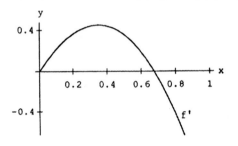

48. $f(1) = 0$; f satisfies Rolle's theorem on $[0, 1]$; $f'(c) = 0$ at $c \cong 0.6058$.

49. $b \cong 0.5437$, $\quad c \cong 0.3045$,

$f(0.3045) \cong -0.1749$

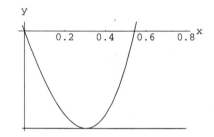

50. x-intercepts: $x = 0$, $x = 1$; $\quad f'(x) = 0$ at $x \cong 0.3874$

51. $0\ c = 0$

52. x-intercepts: $x = -2$, $x = \frac{4}{5}\ x = 2$; $\quad f'(x) = 0$ at $x \cong -1.1005$ and $x \cong 1.5577$

53. $f(x) = x^4 - 7x^2 + 2$; $\quad f'(x) = 4x^3 - 14x$

$g(x) = 4x^3 - 14x - \dfrac{f(3) - f(1)}{3 - 1} = 4x^3 - 14x - 12$

$g(c) = 0$ at $c \cong 2.205$

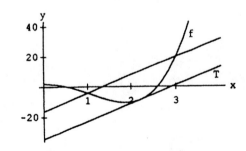

54. $f(x) = x \cos x + 4 \sin x; \quad f'(x) = \cos x - x \sin x + 4 \cos x = 5 \cos x - x \sin x$ and

$$g(x) = 5 \cos x - x \sin x - \frac{f(\pi/2) - f(-\pi/2)}{\pi} = 5 \cos x - x \sin x - \frac{8}{\pi}$$

$g(c) = 0$ at $c_1 \cong -0.872, \; c_2 \cong 0.872$

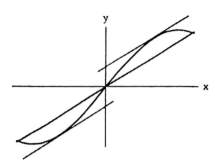

55. $f(x) = x^3 - x^2 + x - 1; \; f'(x) = 3x^2 - 2x + 1; \quad c = 8/3.$

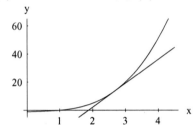

56. $f(x) = x^4 - 2x^3 - x^2 - x + 1; \; f'(x) = 4x^3 - 6x^2 - 2x - 1; \quad c = 1/2, \; -0.6180, \; 1.6180.$

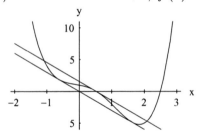

SECTION 4.2

1. $f'(x) = 3x^2 - 3 = 3\left(x^2 - 1\right) = 3(x+1)(x-1)$

f increases on $(-\infty, -1]$ and $[1, \infty)$, decreases on $[-1, 1]$

2. $f'(x) = 3x^2 - 6x = 3x(x - 2)$

f increases on $(-\infty, 0]$ and $[2, \infty)$, decreases on $[0, 2]$

3. $f'(x) = 1 - \dfrac{1}{x^2} = \dfrac{x^2 - 1}{x^2} = \dfrac{(x+1)(x-1)}{x^2}$

f increases on $(-\infty, -1]$ and $[1, \infty)$, decreases on $[-1, 0)$ and $(0, 1]$ (f is not defined at 0)

4. $f'(x) = 3(x-3)^2; \quad f$ increases on $(-\infty, \infty)$

5. $f'(x) = 3x^2 + 4x^3 = x^2(3 + 4x)$

f increases on $\left[-\frac{3}{4}, \infty\right)$,   decreases on $\left(-\infty, -\frac{3}{4}\right]$

6. $f'(x) = 3x^2 + 6x + 2$

f increases on $\left(-\infty, -1 - \frac{1}{3}\sqrt{3}\right]$ and $\left[-1 + \frac{1}{3}\sqrt{3}, \infty\right)$,   decreases on $\left[-1 - \frac{1}{3}\sqrt{3}, -1 + \frac{1}{3}\sqrt{3}\right]$

7. $f'(x) = 4(x + 1)^3$

f increases on $[-1, \infty)$,   decreases on $(-\infty, -1]$

8. $f'(x) = \dfrac{2(x^3 + 1)}{x^3}$

f increases on $[-\infty, -1]$ and $(0, \infty)$,   decreases on $[-1, 0)$

9. $f(x) = \begin{cases} \dfrac{1}{2-x}, & x < 2 \\ \dfrac{1}{x-2}, & x > 2 \end{cases}$ $f'(x) = \begin{cases} \dfrac{1}{(2-x)^2}, & x < 2 \\ \dfrac{-1}{(x-2)^2}, & x > 2 \end{cases}$

f increases on $(-\infty, 2)$,   decreases on $(2, \infty)$ (f is not defined at 2)

10. $f'(x) = \dfrac{(1 + x^2) - x(2x)}{(1 + x^2)^2}$; f increases on $[-1, 1]$,   decreases on $(-\infty, -1]$ and $[1, \infty)$

11. $f'(x) = -\dfrac{4x}{(x^2 - 1)^2}$

f increases on $(-\infty, -1)$ and $(-1, 0]$, decreases on $[0, 1)$ and $(1, \infty)$ (f is not defined at ± 1)

12. $f'(x) = \dfrac{(x^2 + 1)(2x) - x^2(2x)}{(x^2 + 1)^2} = \dfrac{2x}{(x^2 + 1)^2}$

f increases on $[0, \infty)$,   decreases on $(-\infty, 0]$

13. $f(x) = \begin{cases} x^2 - 5, & x < -\sqrt{5} \\ -\left(x^2 - 5\right), & -\sqrt{5} \le x \le \sqrt{5} \\ x^2 - 5, & \sqrt{5} < x \end{cases}$ $f'(x) = \begin{cases} 2x, & x < -\sqrt{5} \\ -2x, & -\sqrt{5} < x < \sqrt{5} \\ 2x, & \sqrt{5} < x \end{cases}$

f increases on $[-\sqrt{5}, 0]$ and $[\sqrt{5}, \infty)$, decreases on $(-\infty, -\sqrt{5}]$ and $[0, \sqrt{5}]$

14. $f'(x) = x^2(2)(1 + x) + (x + 1)^2(2x) = 2x(x + 1)(2x + 1)$

f increases on $[-1, -1/2]$ and $[0, \infty)$,   decreases on $(-\infty, -1]$ and $[-1/2, 0]$

15. $f'(x) = \dfrac{2}{(x + 1)^2}$; f increases on $(-\infty, -1)$ and $(-1, \infty)$ (f is not defined at -1)

16. $f'(x) = 2x - \dfrac{32}{x^3} = \dfrac{2(x^4 - 16)}{x^3} = \dfrac{2(x - 2)(x + 2)(x^2 + 4)}{x^3}$

f increases on $[-2, 0)$ and $[2, \infty)$,   decreases on $(-\infty, -2]$ and $(0, 2]$

17. $f'(x) = \dfrac{x}{(2 + x^2)^2}\sqrt{\dfrac{2 + x^2}{1 + x^2}}$ f increases on $[0, \infty)$,   decreases on $(-\infty, 0]$

18. $f(x) = \begin{cases} x^2 - x - 2, & x \le -1 \\ -x^2 + x + 2, & -1 < x < 2 \\ x^2 - x - 2, & x \ge 2 \end{cases}$ $f'(x) = \begin{cases} 2x - 1, & x \le -1 \\ -2x + 1, & -1 < x < 2 \\ 2x - 1, & x \ge 2 \end{cases}$

f increases on $\left[-1, \frac{1}{2}\right]$ and $[2, \infty)$,   decreases on $(-\infty, -1]$ and $\left[\frac{1}{2}, 2\right]$

19. $f'(x) = 1 + \sin x \ge 0$; f increases on $[0, 2\pi]$

20. $f'(x) = 1 + \cos x \ge 0$; f increases on $[0, 2\pi]$

21. $f'(x) = -2\sin 2x - 2\sin x = -2\sin x (2\cos x + 1)$; f increases on $\left[\frac{2}{3}\pi, \pi\right]$, decreases on $\left[0, \frac{2}{3}\pi\right]$

22. $f'(x) = -2\cos x \sin x = -2\sin 2x$; f increases on $[\pi/2, \pi]$,   decreases on $[0, \pi/2]$

23. $f'(x) = \sqrt{3} + 2\sin 2x$; f increases on $\left[0, \frac{2}{3}\pi\right]$ and $\left[\frac{5}{6}\pi, \pi\right]$, decreases on $\left[\frac{2}{3}\pi, \frac{5}{6}\pi\right]$

24. $f'(x) = 2\sin x \cos x - \sqrt{3}\cos x = \cos x \left(2\sin x - \sqrt{3}\right)$

f increases on $\left[\frac{1}{3}\pi, \frac{1}{2}\pi\right]$ and $\left[\frac{2}{3}\pi, \pi\right]$,   decreases on $\left[0, \frac{1}{3}\pi\right]$ and $\left[\frac{1}{2}\pi, \frac{2}{3}\pi\right]$

25. $\dfrac{d}{dx}\left(\dfrac{x^3}{3} - x\right) = f'(x) \implies f(x) = \dfrac{x^3}{3} - x + C$

$f(1) = 2 \implies 2 = \frac{1}{3} - 1 + C$, so $C = \frac{8}{3}$. Thus, $f(x) = \frac{1}{3}x^3 - x + \frac{8}{3}$.

26. $\dfrac{d}{dx}\left(x^2 - 5x\right) = f'(x) \implies f(x) = x^2 - 5x + C$

$f(2) = 4 \implies 4 = 4 - 10 + C$, so $C = 10$. Thus, $f(x) = x^2 - 5x + 10$.

27. $\dfrac{d}{dx}\left(x^5 + x^4 + x^3 + x^2 + x\right) = f'(x) \implies f(x) = x^5 + x^4 + x^3 + x^2 + x + C$

$f(0) = 5 \implies 5 = 0 + C$, so $C = 5$. Thus, $f(x) = x^5 + x^4 + x^3 + x^2 + x + 5$.

28. $\dfrac{d}{dx}\left(-2x^{-2}\right) = f'(x) \implies f(x) = -2x^{-2} + C$

$f(1) = 0 \implies 0 = -2 + C$, so $C = 2$. Thus, $f(x) = -2x^{-2} + 2, \ x > 0$.

29. $\dfrac{d}{dx}\left(\dfrac{3}{4}x^{4/3} - \dfrac{2}{3}x^{3/2}\right) = f'(x) \implies f(x) = \dfrac{3}{4}x^{4/3} - \dfrac{2}{3}x^{3/2} + C$

$f(0) = 1 \implies 1 = 0 + C$, so $C = 1$. Thus, $f(x) = \frac{3}{4}x^{4/3} - \frac{2}{3}x^{3/2} + 1, \ x \ge 0$.

30. $\dfrac{d}{dx}\left(-\dfrac{1}{4}x^{-4} - \dfrac{25}{4}x^{4/5}\right) = f'(x) \implies f(x) = -\dfrac{1}{4}x^{-4} - \dfrac{25}{4}x^{4/5} + C$

$f(1) = 0 \implies 0 = -\frac{1}{4} - \frac{25}{2} + C$, so $C = \frac{13}{2}$. Thus, $f(x) = -\frac{1}{4}x^{-4} - \frac{25}{4}x^{4/5} + \frac{13}{2}, \ x > 0$.

31. $\dfrac{d}{dx}\left(2x - \cos x\right) = f'(x) \implies f(x) = 2x - \cos x + C$

$f(0) = 3 \implies 3 = 0 - 1 + C$, so $C = 4$. Thus, $f(x) = 2x - \cos x + 4$.

32. $\dfrac{d}{dx}\left(2x^2 + \sin x\right) = f'(x) \implies f(x) = 2x^2 + \sin x + C$

$f(0) = 1 \implies 1 = 0 + C$, so $C = 1$. Thus, $f(x) = 2x^2 + \sin x + 1$.

33. $f'(x) = \begin{cases} 1, & x < -3 \\ -1, & -3 < x < -1 \\ 1, & -1 < x < 1 \\ -2, & 1 < x \end{cases}$

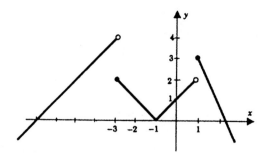

f increases on $(-\infty, -3)$ and $[-1, 1]$;

decreases on $[-3, -1]$ and $[1, \infty)$

34. $f'(x) = \begin{cases} 2(x-1), & x < 1 \\ -1, & 1 < x < 3 \\ -2, & x > 3 \end{cases}$

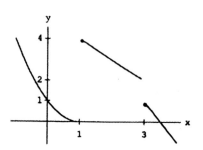

f decreases on $(-\infty, 1)$, $[1, 3)$ and $[3, \infty)$.

35. $f'(x) = \begin{cases} -2x, & x < 1 \\ -2, & 1 < x < 3 \\ 3, & 3 < x \end{cases}$

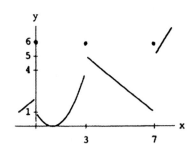

f increases on $(-\infty, 0]$ and $[3, \infty)$;

decreases on $[0, 1)$ and $[1, 3]$

36. $f'(x) = \begin{cases} 1, & x < 0 \\ 2(x-1), & 0 < x < 3 \\ -1, & 3 < x < 7 \\ 2, & x > 7 \end{cases}$

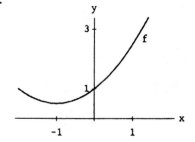

f increases on $(-\infty, 0]$, $[1, 3]$, $(7, \infty)$;

decreases on $[0, 1]$ and $[3, 7)$

37.

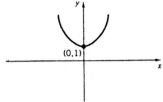

38.

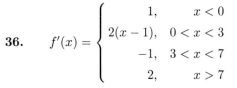

39.

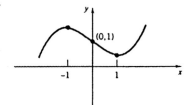

40.

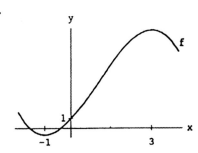

41.

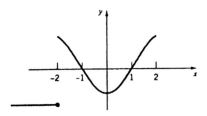

42.

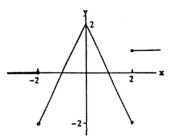

43.

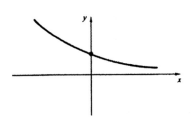

44.

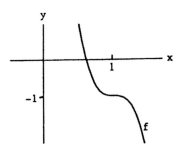

45.   Not possible; f is increasing, so $f(2)$ must be greater than $f(-1)$.

46.   Not possible; by the intermediate-value theorem, f must have a zero in $(3, 5)$.

47.   (a)   True. Let x_1, $x_2 \in [a, c]$, $x_1 < x_2$. If x_1, $x_2 \in [a, b]$, or if x_1, $x_2 \in [b, c]$, then $f(x_1) < f(x_2)$.
   If $x_1 \in [a, b)$ and $x_2 \in [b, c]$, then $f(x_1) < f(b) \le f(x_2)$. Therefore f increases on $[a, c]$

   (b)   False. A slight modification of Example 6 is a counterexample. Let $g(x) = \begin{cases} \frac{1}{2}x + 2, & x \le 1 \\ x^3, & x > 1. \end{cases}$

48.   (a)   True. Use the obvious modification of the argument in Exercise 47(a).

   (b)   False. An example like 47(b): $f(x) = \begin{cases} -x + 2, & x \le 1 \\ -x + 3, & x > 1. \end{cases}$

49.   (a)   True. If $f'(c) < 0$ at some number $c \in (a, b)$, then there exists a number δ such that
   $f(x) > f(c) > f(z)$ for $x \in (c - \delta, c)$ and $z \in (c, c + \delta)$ (Theorem 4.1.2).

   (b)   False. $f(x) = x^3$ increases on $(-1, 1)$ and $f'(0) = 0$.

50.   False. $f(x) = x + \dfrac{1}{2\pi} \sin (2x - 1)\pi$ is increasing on $[0, 4]$ and $f'(x) = 1 + \cos (2x - 1)\pi\, x = 0$ at
   $x = 1,\ x = 2.\ x = 3.$

51. Let $f(x) = x - \sin x$. Then $f'(x) = 1 - \cos x$.

(a) $f'(x) \geq 0$ for all $x \in (-\infty, \infty)$ and $f'(x) = 0$ only at $x = \dfrac{\pi}{2} + n\pi$, $n = 0, \pm 1, \pm 2, \ldots$
It follows from Theorem 4.2.3 that f is increasing on $(-\infty, \infty)$.

(b) Since f is increasing on $(-\infty, \infty)$ and $f(0) = 0 - \sin 0 = 0$, we have:

$$f(x) > 0 \text{ for all } x > 0 \Rightarrow x > \sin x \text{ on } (0, \infty);$$

$$f(x) < 0 \text{ for all } x < 0 \Rightarrow x < \sin x \text{ on } (-\infty, 0).$$

52. A proof is outlined just below the statement of the theorem.

53. $f'(x) = 2 \sec x (\sec x \tan x) = 2 \sec^2 x \tan x$ and $g'(x) = 2 \tan x \sec^2 x$.

Therefore, $f'(x) = g'(x)$ for all $x \in I$.

54. Evaluating $\sec^2 x - \tan^2 x = C$ at $x = 0$ gives $C = 1$.

55. Let f and g be functions such that $f'(x) = -g(x)$ and $g'(x) = f(x)$. Then:

(a) Differentiating $f^2(x) + g^2(x)$ with respect to x, we have

$$2f(x)f'(x) + 2g(x)g'(x) = -2f(x)g(x) + 2g(x)f(x) = 0.$$

Thus, $f^2(x) + g^2(x) = C$ (constant).

(b) $f(0) = 0$ and $g(0) = 1$ implies $C = 1$.

(c) The functions $f(x) = \sin x$, $g(x) = \cos x$ have these properties.

56. (a) Let $h(x) = f(x) - g(x)$. Then $h'(x) = f'(x) - g'(x) > 0$ on $(0, c)$, and h is increasing on $(0, c)$. Since $h(0) = f(0) - g(0) = 0$, it follows that $h(x) > 0$ on $(0, c)$. Thus, $f(x) > g(x)$ on $(0, c)$.

(b) Again let $h(x) = f(x) - g(x)$. Then h is increasing on $(-c, 0)$ which implies that $h(x) < 0$ on this interval since $h(0) = 0$. Therefore, $f(x) < g(x)$ on $(-c, 0)$.

57. Let $f(x) = \tan x$ and $g(x) = x$ for $x \in [0, \pi/2)$. Then $f(0) = g(0) = 0$ and $f'(x) = \sec^2 x > g'(x) = 1$ for $x \in (0, \pi/2)$. Thus, $\tan x > x$ for $x \in (0, \pi/2)$ by Exercise 56(a).

58. Let $f(x) = \cos x - \left(1 - \frac{1}{2}x^2\right)$ for $x \in [0, \infty)$. Then $f(0) = 0$ and $f'(x) = -\sin x + x = x - \sin x > 0$ for $x \in (0, \infty)$ by Exercise 51 (b). Thus, $f(x) > 0$ for $x \in (0, \infty)$ which implies $\cos x > 1 - \frac{1}{2}x^2$ on $(0, \infty)$.

59. Choose an integer $n > 1$. Let $f(x) = (1 + x)^n$ and $g(x) = 1 + nx$, $x > 0$. Then, $f(0) = g(0) = 1$ and $f'(x) = n(1 + x)^{n-1} > g'(x) = n$ since $(1 + x)^{n-1} > 1$ for $x > 0$. The result follows from Exercise 56(a).

60. Let $f(x) = \sin x - \left(x - \frac{1}{6}x^3\right)$. Then $f(0) = 0$ and $f'(x) = \cos x - \left(1 - \frac{1}{2}x^2\right) > 0$ by Exercise 58. Therefore, $f(x) > f(0) = 0$ for all $x \in (0, \infty)$ which implies $\sin x > x - \frac{1}{6}x^3$ on $(0, \infty)$.

61. $4° \cong 0.06981$ radians. By Exercises 51 and 60,

$$0.6981 - \frac{(0.6981)^3}{6} = 0.06975 < \sin 4° < 0.6981$$

62. (a) Let $f(x) = \cos x - (1 - \frac{1}{2}x^2 + \frac{1}{24}x^4)$. Then $f(0) = 0$ and $f'(x) = -\sin x + x - \frac{x^3}{6} < 0$ by
Exercise 60. Therefore, $f(x) < f(0) = 0$ on all $x \in (0, \infty)$, which implies $\cos x < 1 - \frac{1}{2}x^2 + \frac{1}{24}x^4$
on $(0, \infty)$.

(b) $6° = \frac{\pi}{30}$. Using this for x in $1 - \frac{1}{2}x^2 < \cos x < 1 - \frac{1}{2}x^2 + \frac{1}{24}x^4$,

$\implies 0.994517 < \cos 6° < 0.994522$.

63. Let $f(x) = 3x^4 - 10x^3 - 4x^2 + 10x + 9$, $x \in [-2, 5]$. Then $f'(x) = 12x^3 - 30x^2 - 8x + 10$.

$f'(x) = 0$ at $x \cong -0.633, \ 0.5, \ 2.633$

f is decreasing on $[-2, -0.633]$

and $[0.5, 2.633]$

f is increasing on $[-0.633, 0.5]$

and $[2.633, 5]$

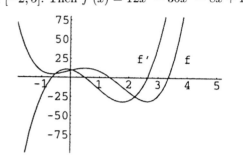

64. Let $f(x) = 2x^3 - x^2 - 13x - 6$, $x \in [-3, 4]$. Then $f'(x) = 6x^2 - 2x - 13$.

$f'(x) = 0$ at $x \cong -1.315, \ 1.648$

f is decreasing on $[-1.315, 1.648]$

f is increasing on $[-3, -1.315]$ and $[1.648, 4]$

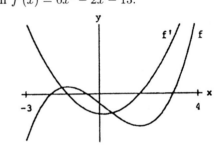

65. Let $f(x) = x \cos x - 3 \sin 2x$, $x \in [0, 6]$. Then $f'(x) = \cos x - x \sin x - 6 \cos 2x$.

$f'(x) = 0$ at $x \cong 0.770, \ 2.155, \ 3.798, \ 5.812$

f is decreasing on $[0, 0.770], \ [2.155, 3.798]$

and $[5.812, 6]$

f is increasing on $[0.770, 2.155]$

and $[3.798, 5.812]$

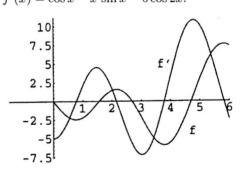

66. Let $f(x) = x^4 + 3x^3 - 2x^2 + 4x + 4$, $x \in [-5, 3]$. Then $f'(x) = 4x^3 + 9x^2 - 4x + 4$.

$f'(x) = 0$ at $x \cong -2.747$

f is decreasing on $[-5, -2.747]$

f is increasing on $[-2.747, 3]$

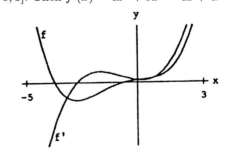

67. (a) $f'(x) = 0$ at $x = 0, \frac{\pi}{2}, \pi, \frac{3\pi}{2}, 2\pi$ (b) $f'(x) > 0$ on $\left(\pi, \frac{3\pi}{2}\right) \cup \left(\frac{3\pi}{2}, 2\pi\right)$

     (c) $f'(x) < 0$ on $\left(0, \frac{\pi}{2}\right) \cup \left(\frac{\pi}{2}, \pi\right)$

68. (b) $f'(x) > 0$ on $(-\infty, \infty)$

69. (a) $f'(x) = 0$ at $x = 0$ (b) $f'(x) > 0$ on $(0, \infty)$

     (c) $f'(x) < 0$ on $(-\infty, 0)$

70. (a) $f'(x) = 0$ at $x = -\frac{1}{2}, \frac{8}{5}, 3$ (b) $f'(x) > 0$ on $\left(-\infty, -\frac{1}{2}\right) \cup \left(-\frac{1}{2}, \frac{8}{5}\right) \cup (3, \infty)$

     (c) $f'(x) < 0$ on $\left(\frac{8}{5}, 3\right)$

71. $f = C$, constant; $f'(x) \equiv 0$

SECTION 4.3

1. $f'(x) = 3x^2 + 3 > 0$;   no critical pts, no local extreme values

2. $f'(x) = 8x^3 - 8x = 8x(x^2 - 1)$;   critical pts $-1, 0, 1$

     $f''(x) = 24x^2 - 8$; $f''(-1) = f''(1) = 16 > 0$, $f''(0) = -8 < 0$;

     $f(0) = 6$ local max, $f(-1) = 4$ local min, $f(1) = 4$ local min

3. $f'(x) = 1 - \dfrac{1}{x^2}$;   critical pts $-1, 1$

     $f''(x) = \dfrac{2}{x^3}$, $f''(-1) = -2$, $f''(1) = 2$ $f(-1) = -2$ local max, $f(1) = 2$ local min

4. $f'(x) = 2x + \dfrac{6}{x^3} = \dfrac{2x^4 + 6}{x^3}$;   no critical pts (note: 0 is not in the domain of f),
no local extreme values

5. $f'(x) = 2x - 3x^2 = x(2 - 3x)$;   critical pts $0, \frac{2}{3}$

     $f''(x) = 2 - 6x$; $f''(0) = 2$, $f''(\frac{2}{3}) = -2$

     $f(0) = 0$ local min, $f(\frac{2}{3}) = \frac{4}{27}$ local max

6. $f'(x) = -2(1 - x)(1 + x) + (1 - x)^2 = (x - 1)(3x + 1)$;   critical pts $-\frac{1}{3}, 1$

     $f''(x) = (1 + 3x) + 3(x - 1) = 2(3x - 1)$; $f''\left(-\frac{1}{3}\right) = -4$, $f''(1) = 4$

     $f\left(-\frac{1}{3}\right) = \frac{32}{27}$ local max, $f(1) = 0$ local min

7. $f'(x) = \dfrac{2}{(1 - x)^2}$;   no critical pts, no local extreme values

8. $f'(x) = \dfrac{(2 + x)(-3) - (2 - 3x)(1)}{(2 + x)^2} = -\dfrac{8}{(2 + x)^2}$;   no critical pts (note: -2 is not
in the domain of f), no local extreme values

9. $f'(x) = -\dfrac{2(2x+1)}{x^2(x+1)^2}$;   critical pt $-\dfrac{1}{2}$

$f\left(-\dfrac{1}{2}\right) = -8$ local max

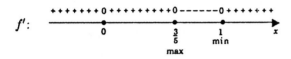

10. $f(x) = \begin{cases} x^2 - 16, & x < -4 \\ 16 - x^2, & -4 \le x < 4 \\ x^2 - 16, & x \ge 4 \end{cases}$ $f'(x) = \begin{cases} 2x, & x < -4 \\ -2x, & -4 < x < 4 \\ 2x, & x > 4 \end{cases}$

critical pts $-4, 0, 4$; $f(-4) = f(4) = 0$ local minima, $f(0) = 16$ local max

11. $f'(x) = x^2(5x - 3)(x - 1)$;   critical pts $0, \dfrac{3}{5}, 1$

$f\left(\dfrac{3}{5}\right) = \dfrac{2^2 3^3}{5^5}$ local max

$f(1) = 0$ local min

no local extreme at 0

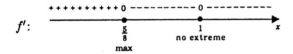

12. $f'(x) = 3\left(\dfrac{x-2}{x+2}\right)^2 \dfrac{4}{(x+2)^2} \ge 0$;   critical pt 2, no local extreme values

13. $f'(x) = (5 - 8x)(x - 1)^2$;   critical pts $\dfrac{5}{8}, 1$

$f\left(\dfrac{5}{8}\right) = \dfrac{27}{2048}$ local max

no local extreme at 1

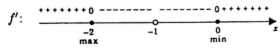

14. $f'(x) = -(1+x)^3 + (1-x)(3)(1+x)^2 = 2(1+x)^2(1 - 2x)$;   critical pts $-1, \dfrac{1}{2}$

$f\left(\dfrac{1}{2}\right) = \dfrac{27}{16}$ local max

no local extreme at -1

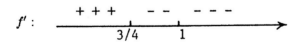

15. $f'(x) = \dfrac{x(2+x)}{(1+x)^2}$;   critical pts $-2, 0$

$f(-2) = -4$ local max

$f(0) = 0$ local min

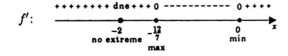

16. $f'(x) = (1-x)^{1/3} - \dfrac{1}{3}x(1-x)^{-2/3} = \dfrac{3 - 4x}{3(1-x)^{2/3}}$;   critical pts $\dfrac{3}{4}, 1$

$f(3/4) = \dfrac{3}{4^{4/3}}$ local max

no local extreme at 1

f' : $\begin{array}{ccc} +\,+\,+ & -\;- & -\;-\;- \\ \hline & & \\ 3/4 & 1 & \end{array}$

17. $f'(x) = \dfrac{1}{3}x(7x + 12)(x + 2)^{-2/3}$;   critical pts $-2, -\dfrac{12}{7}, 0$

$f\left(-\dfrac{12}{7}\right) = \dfrac{144}{49}\left(\dfrac{2}{7}\right)^{1/3}$ local max

$f(0) = 0$ local min

18. $f'(x) = \dfrac{-1}{(x+1)^2} + \dfrac{1}{(x-2)^2} = \dfrac{3(2x-1)}{(x+1)^2(x-2)^2};$ critical pt $\frac{1}{2}$

f' :

$f\left(\frac{1}{2}\right) = \frac{4}{3}$ local min

19. $f(x) = \begin{cases} 2 - 3x, & x \le -\frac{1}{2} \\ x + 4, & -\frac{1}{2} < x < 3 \\ 3x - 2, & 3 \le x \end{cases}$ $f'(x) = \begin{cases} -3, & x < -\frac{1}{2} \\ 1, & -\frac{1}{2} < x < 3 \\ 3, & 3 < x \end{cases}$

critical pts $-\frac{1}{2}, 3$

f' :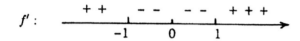

$f\left(-\frac{1}{2}\right) = \frac{7}{2}$ local min

no local extreme at 3

20. $f'(x) = \frac{7}{3}x^{4/3} - \frac{7}{3}x^{-2/3} = \frac{7}{3}\dfrac{x^2 - 1}{x^{2/3}};$ critical pts $-1, 0, 1$

f' :

$f(-1) = 6$ local max,

$f(1) = -6$ local min

no local extreme at 0

21. $f'(x) = \frac{2}{3}x^{-4/3}(x - 1);$ critical pt 1

f' :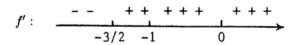

$f(1) = 3$ local min

no local extreme at 0

22. $f'(x) = \dfrac{(x+1)3x^2 - x^3}{(x+1)^2} = \dfrac{x^2(2x+3)}{(x+1)^2};$ critical pts $-\frac{3}{2}, 0$

f' :

$$
\begin{array}{c}
-\ - \quad +\ + \quad +\ +\ + \qquad +\ +\ + \\
\hline
-3/2\quad -1\qquad\quad 0
\end{array}
$$

$f\left(-\frac{3}{2}\right) = \frac{27}{4}$ local min

no local extreme at 0

23. $f'(x) = \cos x - \sin x;$ critical pts $\frac{1}{4}\pi, \frac{5}{4}\pi$

$f''(x) = -\sin x - \cos x,$ $f''\left(\frac{1}{4}\pi\right) = -\sqrt{2},$ $f''\left(\frac{5}{4}\pi\right) = \sqrt{2}$

$f(\frac{1}{4}\pi) = \sqrt{2}$ local max, $f(\frac{5}{4}\pi) = -\sqrt{2}$ local min

24. $f'(x) = 1 - 2\sin 2x;$ critical pts $\frac{\pi}{12}, \frac{5\pi}{12}$

f':

$$
\begin{array}{c}
+\ + \quad -\ -\ -\ -\ -\ -\ - \quad +\ + \\
\hline
0\quad \pi/12 \qquad\qquad 5\pi/12
\end{array}
$$

$f(\frac{1}{12}\pi) = \dfrac{\pi}{12} + \dfrac{\sqrt{3}}{2}$ local max

$f(\dfrac{5}{12}\pi) = \dfrac{5\pi}{12} - \dfrac{\sqrt{3}}{2}$ local min

25. $f'(x) = \cos x \, (2 \sin x - \sqrt{3}\,)$;   critical pts $\frac{1}{3}\pi$, $\frac{1}{2}\pi$, $\frac{2}{3}\pi$

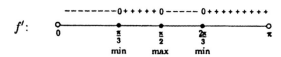

$f': $

$f(\frac{1}{3}\pi) = f(\frac{2}{3}\pi) = -\frac{3}{4}$ local mins

$f(\frac{1}{2}\pi) = 1 - \sqrt{3}$ local max

26. $f'(x) = 2 \sin x \cos x$;   critical pts $\frac{1}{2}\pi$, π, $\frac{3}{2}\pi$

$f': $ $+\,+$ $-\,-$ $+\,+$ $-\,-$

   0 $\pi/2$ π $3\pi/2$

$f\left(\frac{1}{2}\pi\right) = 1 = f\left(\frac{3}{2}\pi\right)$ local max

$f(\pi) = 0$ local min

27. $f'(x) = \cos^2 x - \sin^2 x - 3 \cos x + 2 = (2\cos x - 1)(\cos x - 1)$ critical pts $\frac{1}{3}\pi$, $\frac{5}{3}\pi$

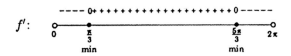

$f': $

$f\left(\frac{1}{3}\pi\right) = \frac{2}{3}\pi - \frac{5}{4}\sqrt{3}$ local min

$f\left(\frac{5}{3}\pi\right) = \frac{10}{3}\pi + \frac{5}{4}\sqrt{3}$ local max

28. $f'(x) = 6 \sin^2 x \cos x - 3 \cos x = 3 \cos x (2 \sin^2 x - 1)$;   critical pts $\frac{1}{4}\pi$, $\frac{1}{2}\pi$, $\frac{3}{4}\pi$

$f': $ $-\,-$ $+\,+$ $-\,-$ $+\,+$

   0 $\pi/4$ $\pi/2$ $3\pi/4$

$f(\frac{1}{4}\pi) = f(\frac{3}{4}\pi) = -\sqrt{2}$ local mins

$f(\frac{1}{2}\pi) = -1$ local max

29. (a) f increases on $[-2, 0]$ and $[3, \infty)$; f decreases on $(-\infty, -2]$ and $[0, 3]$.

     (b) $f(-2)$ and $f(3)$ are local minima; $f(0) = 1$ is a local maximum.

30. (a) f increases on $(-\infty, -1]$ and $[0, \infty)$; f decreases on $[-1, 0]$.

     (b) $f(-1)$ is a local maximum; $f(0) = 1$ is a local minimum.

31. Let $h(x) = f(x) - g(x)$. Then $h(x)$ gives the vertical separation between graphs of f and g at x. If h has a maximum at c, then $h'(c) = 0$. Since $h'(x) = f'(x) - g'(x)$, $h'(c) = 0$ implies $f'(c) = g'(c)$. Thus the lines tangent to the graphs of f and g are parallel at $x = c$.

32. Set $g(x) = -f(-x)$ and apply the proof of the second derivative test already given.

33. Solving $f'(x) = 2ax + b = 0$ gives a critical point at $x = -\dfrac{b}{2a}$.   Since $f''(x) = 2a$, f has a local maximum at $-\dfrac{b}{2a}$ if $a < 0$ and a local minimum at $-\dfrac{b}{2a}$ if $a > 0$.

34. Setting $f'(x) = 3ax^2 + 2bx + c = 0$ and checking the discriminant, we get

     (1)   2 local extrema if $b^2 > 3ac$

     (2)   1 local extrema if $b^2 = 3ac$

     (3)   0 local extrema if $b^2 < 3ac$

35.

$$P(x) = x^4 - 8x^3 + 22x^2 - 24x + 4$$

$$P'(x) = 4x^3 - 24x^2 + 44x - 24$$

$$P''(x) = 12x^2 - 48x + 44$$

Since $P'(1) = 0$, $P'(x)$ is divisible by $x - 1$. Division by $x - 1$ gives

$$P'(x) = (x - 1)\left(4x^2 - 20x + 24\right) = 4(x - 1)(x - 2)(x - 3).$$

The critical pts are 1, 2, 3. Since

$$P''(1) > 0, \quad P''(2) < 0, \quad P''(3) > 0,$$

$P(1) = -5$ is a local min, $P(2) = -4$ is a local max, and $P(3) = -5$ is a local min.

Since $P'(x) < 0$ for $x < 0$, P decreases on $(-\infty, 0]$. Since $P(0) > 0$, P does not take on the value 0 on $(-\infty, 0]$.

Since $P(0) > 0$ and $P(1) < 0$, P takes on the value 0 at least once on $(0, 1)$. Since $P'(x) < 0$ on $(0, 1)$, P decreases on $[0, 1]$. It follows that P takes on the value zero only once on $[0, 1]$.

Since $P'(x) > 0$ on $(1, 2)$ and $P'(x) < 0$ on $(2, 3)$, P increases on $[1, 2]$ and decreases on $[2, 3]$. Since $P(1), P(2), P(3)$ are all negative, P cannot take on the value 0 between 1 and 3.

Since $P(3) < 0$ and $P(100) > 0$, P takes on the value 0 at least once on $(3, 100)$. Since $P'(x) > 0$ on $(3, 100)$, P increases on $[3, 100]$. It follows that P takes on the value zero only once on $[3, 100]$.

Since $P'(x) > 0$ on $(100, \infty)$, P increases on $[100, \infty)$. Since $P(100) > 0$, P does not take on the value 0 on $[100, \infty)$.

36. f has a local maximum at $x = 0$; f has a local minimum at $x = -1$ and $x = 2$.

37. (a)   **(b)**

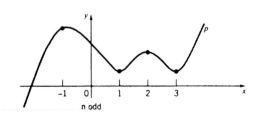   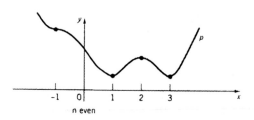

38. Let $f(x) = Ax^2 + Bx + C$. Then $f'(x) = 2Ax + B$.

$f(-1) = 3 \implies A - B + C = 3$; $f(3) = -1 \implies 9A + 3B + C = -1$

Since f has a minimum at $x = 2$, $f'(2) = 4A + B = 0$

Solving for A, B, C, we get $A = \frac{1}{2}$, $B = -2$, $C = \frac{1}{2}$.

39. Let $f(x) = \dfrac{ax}{x^2 + b^2}$. Then $f'(x) = \dfrac{a\left(b^2 - x^2\right)}{\left(b^2 + x^2\right)^2}$. Now

$$f'(0) = \frac{a}{b^2} = 1 \Rightarrow a = b^2 \quad \text{and} \quad f'(x) = \frac{b^2\left(b^2 - x^2\right)}{\left(b^2 + x^2\right)^2}$$

$$f'(-2) = \frac{b^2\left(b^2 - 4\right)}{\left(b^2 + 4\right)^2} = 0 \Rightarrow b = \pm 2$$

Thus, $a = 4$ and $b = \pm 2$.

40. (a) $f(x) = x^p(1 - x)^q$, $p, q \geq 2$; $f'(x) = x^{p-1}(1 - x)^{q-1}[p - (p + q)x]$

$$f'(x) = 0 \quad \Longrightarrow \quad x = 0, \ x = 1, \ x = \frac{p}{p + q}$$

(b) p even, $p - 1$ odd:

f' : $\quad$ -- -- $\quad$ + +

$\qquad\qquad\quad$ 0 $\qquad\qquad$ 1 $\qquad\qquad\qquad$ f has a local min at $x = 0$

(c) q even, $q - 1$ odd:

f' : $\qquad\qquad\qquad$ -- -- $\quad$ + +

$\qquad\qquad\quad$ 0 $\qquad\qquad$ 1 $\qquad\qquad\qquad$ f has a local min at $x = 1$

(d) $f''\left(\dfrac{p}{p+q}\right) = -(p+q)\left(\dfrac{p}{p+q}\right)^{p-1}\left(\dfrac{q}{p+q}\right)^{q-1} < 0 \quad \Rightarrow \quad f$ has a local max at $x = \dfrac{p}{p+q}$.

41. If p is a polynomial of degree n, then p' has degree $n - 1$. This implies that p' has at most $n - 1$ zeros, and it follows that p has at most $n - 1$ local extreme values.

42. The function $D(x) = \sqrt{x^2 + [f(x)]^2}$ gives the distance from the origin to the point $(x, f(x))$ on the graph of f. Since the graph of f does not pass through the origin,

$$D'(x) = \frac{x + f(x)f'(x)}{\sqrt{x^2 + [f(x)]^2}}$$

is defined for all $x \in \text{dom}\,(f)$. Suppose that D has a local extreme value at c. Then

$$D'(c) = \frac{c + f(c)f'(c)}{\sqrt{c^2 + [f(c)]^2}} = 0 \Rightarrow c + f(c)f'(c) = 0 \quad \text{and} \quad f'(c) = -\frac{c}{f(c)}$$

Suppose that $c \neq 0$. The slope of the line through $(0, 0)$ and $(c, f(c))$ is given by $m_1 = \dfrac{f(c)}{c}$ and the slope of the tangent line to the graph of f at $x = c$ is given by $m_2 = f'(c) = -\dfrac{c}{f(c)}$. Since $m_1 m_2 = -1$, these two lines are perpendicular. If $c = 0$, then the tangent line to the graph of f is horizontal and the line through $(0, 0)$ and $(0, f(0))$ is vertical.

43. If $f(x) = x^4 - 7x^2 - 8x - 3$, then $f'(x) = 4x^3 - 14x - 8$ and $f''(x) = 12x^2 - 14$. Since $f'(2) = -4 < 0$ and $f'(3) = 58 > 0$, f' has at least one zero in $(2, 3)$. Since $f''(x) > 0$ for $x \in (2, 3)$, f' is increasing on this interval and so it has exactly one zero. Thus, f has exactly one critical point c in $(2, 3)$.

44. If $f(x) = \sin x + \dfrac{x^2}{2} - 2x$, then $f'(x) = \cos x + x - 2$ and $f''(x) = -\sin x + 1$. Since $f'(2) = -0.4161 < 0$ and $f'(3) = 0.01 > 0$, f' has at least one zero in $(2, 3)$. Since $f''(x) > 0$ for $x \in (2, 3)$, f' is increasing on this interval and so it has exactly one zero. Thus, f has exactly one critical point c in $(2, 3)$.

45. $f(x) = \dfrac{ax^2 + b}{cx^2 + d}$ and $f'(x) = \dfrac{2(ad - bc)x}{(cx^2 + d)^2}$; $x = 0$ is a critical number.

$f''(x) = \dfrac{2(ad - bc)(cx^2 - 4cx + d)}{(cx^2 + d)^3}$; $f''(0) = \dfrac{2(ad - bc)}{d^2}$.

Therefore, $ad - bc > 0$ implies that $f(0)$ is a local minimum; $ad - bc < 0$ implies that $f(0)$ is a local maximum.

46. Let δ be any positive number and consider f on the interval $(-\delta, \delta)$. Let n be a positive integer such that

$$0 < \frac{1}{\frac{\pi}{2} + 2n\pi} < \delta \text{ and } 0 < \frac{1}{\frac{-\pi}{2} + 2n\pi} < \delta.$$

Then

$$f\left(\frac{1}{\frac{\pi}{2} + 2n\pi}\right) > 0 \text{ and } f\left(\frac{1}{\frac{-\pi}{2} + 2n\pi}\right) < 0.$$

Thus f takes on both positive and negative values in every interval centered at 0 and it follows that f cannot have a local maximum or minimum at 0.

47. (a)

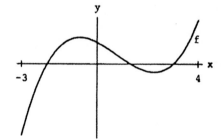

critical points: $x_1 \cong -0.692$, $x_2 \cong 2.248$

local extreme values: $f(-0.692) \cong 29.342$, $f(2.248) \cong -8.766$

(b) f is increasing on $[-3, -0.692]$, and $[2.248, 4]$; f is decreasing on $[-0.692, 2.248]$

48. (a)

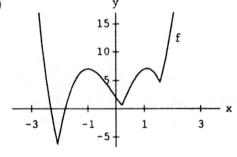

critical points: $\quad x_1 \cong -2.085, \ x_2 \cong -1, \ x_3 \cong 0.207, \ x_4 \cong 1.096, \ x_5 = 1.544$

local extreme values: $\quad f(-2.085) \cong -6.255, \ f(-1) = 7, \ f(0.207) \cong 0.621, \ f(1.096) \cong 7.097,$

$f(1.544) \cong 4.635$

(b) f is increasing on $[-2.085, -1], \ [0.207, 1.096],$ and $[1.544, 4]$

f is decreasing on $[-4, -2.085], \ [-1, 0.207],$ and $[1.096, 1.544]$

49. (a)

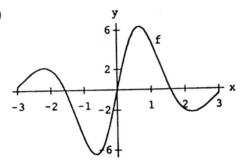

critical points: $\quad x_1 \cong -2.201, \ x_2 \cong -0.654, \ x_3 \cong 0.654, \ x_4 \cong 2.201$

local extreme values: $\quad f(-2.204) \cong 2.226, \ f(-0.654) \cong -6.634, \ f(0.654) \cong 6.634,$

$f(2.204) \cong -2.226$

(b) f is increasing on $[-3. -2.204], \ [-0.654, 0.654],$ and $[2.204, 3]$

f is decreasing on $[-2.204, -0.654],$ and $[0.654, 2.204]$

50. $\quad f'(x) = 0$ at $x = 2, \ 3; \quad f(2) = 0$ is a local minimum.

51. $\quad f'(x) > 0$ on $\left(\frac{2}{3}, \infty\right); \quad f$ has no local extrema.

52. $\quad f'(x) = 0$ at the multiples of $\frac{1}{4}\pi; \quad f(0) = 1$ is a local maximum, $\ f(\pi/4) = 0$ is a local minimum, $f(\pi/2) = 1$ is a local maximum, and so on.

53.

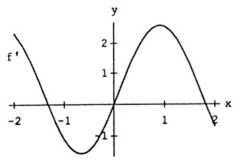

critical points of f : $\quad x_1 \cong -1.326, \ x_2 = 0, \ x_3 \cong 1.816$

$f''(-1.326) \cong -4 < 0 \quad \Rightarrow \ f$ has a local maximum at $x = -1.326$

$f''(0) = 4 > 0 \quad \Rightarrow \ f$ has a local minimum at $x = 0$

$f''(1.816) \cong -4 \quad \Rightarrow \ f$ has a local maximum at $x = 1.816$

54.

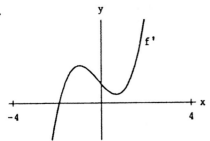

critical number of f : $x_1 \cong -1.935$

$f''(-1.935) \cong 14.60 > 0$ $\Rightarrow$ f has a local minimum at $x = -1.935$

SECTION 4.4

1. $f'(x) = \frac{1}{2}(x+2)^{-1/2}, \;\; x > -2;$

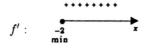

$f(-2) = 0$ endpt and abs min;   as $x \to \infty, \;\; f(x) \to \infty;$ so no abs max

2. $f'(x) = 2x - 3;$ critical pt. $\frac{3}{2};$ $f\left(\frac{3}{2}\right) = -\frac{1}{4}$ local and abs min

3. $f'(x) = 2x - 4, \;\; x \in (0,3);$

critical pt. 2;

$f(0) = 1$ endpt and abs max, $f(2) = -3$ local and abs min, $f(3) = -2$ endpt max

4. $f'(x) = 4x + 5, \;\; x \in (-2, 0);$

critical pt. $-\frac{5}{4};$

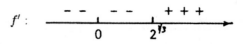

$f(-2) = -3$ endpt max, $f\left(-\frac{5}{4}\right) = -\frac{33}{8}$ local and abs min, $f(0) = -1$ endpt and abs max

5. $f'(x) = 2x - \dfrac{1}{x^2} = \dfrac{2x^3 - 1}{x^2}, \;\; x \neq 0; \;\; f'(x) = 0$ at $x = 2^{-1/3}$

critical pt. $2^{-1/3}$; $f''(x) = 2 + \dfrac{2}{x^3}, \;\; f''\left(2^{-1/3}\right) = 6$

$f\left(2^{-1/3}\right) = 2^{-2/3} + 2^{1/3} = 2^{-2/3} + 2 \cdot 2^{-2/3} = 3 \cdot 2^{-2/3}$ local min

6. $f'(x) = 1 - \dfrac{2}{x^3}$

critical pt. $2^{1/3};$ $f\left(2^{1/3}\right) = 3(2)^{-2/3}$ local min

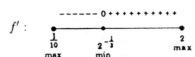

7. $f'(x) = \dfrac{2x^3 - 1}{x^2}, \;\; x \in \left(\dfrac{1}{10}, 2\right);$

critical pt. $2^{-1/3};$

$f\left(\frac{1}{10}\right) = 10\frac{1}{100}$ endpt and abs max, $f\left(2^{-1/3}\right) = 3 \cdot 2^{-2/3}$ local and abs min,

$f(2) = 4\frac{1}{2}$ endpt max

8. $f'(x) = 1 - \frac{2}{x^3}, \quad x \in (1, \sqrt{2})$

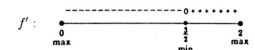

critical pt. $2^{1/3}$; $f(1) = 2$ endpt and abs max

$f\left(2^{1/3}\right) = 3(2)^{-2/3}$ local and abs min, $f\left(\sqrt{2}\right) = \sqrt{2} + \frac{1}{2}$ endpt max

9. $f'(x) = 2x - 3, \quad x \in (0, 2);$

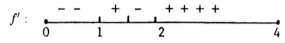

critical pt. $\frac{3}{2}$;

$f(0) = 2$ endpt and abs max, $f\left(\frac{3}{2}\right) = -\frac{1}{4}$ local and abs min,

$f(2) = 0$ endpt max

10. $f'(x) = 2(x-1)(x-2)(2x-3); \quad x \in (0, 4)$

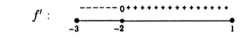

critical pts. $1, \frac{3}{2}, 2,;$ $f(0) = 4$ endpt max, $f(1) = 0$ local and abs min,

$f\left(\frac{3}{2}\right) = \frac{1}{16}$ local max, $f(2) = 0$ local and abs min, $f(4) = 36$ endpt and abs max

11. $f'(x) = \frac{(2-x)(2+x)}{(4+x^2)^2}, \quad x \in (-3, 1);$

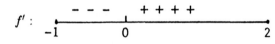

critical pt. -2;

$f(-3) = -\frac{3}{13}$ endpt max, $f(-2) = -\frac{1}{4}$ local and abs min,

$f(1) = \frac{1}{5}$ endpt and abs max

12. $f'(x) = \frac{2x}{(1+x^2)^2}, \quad x \in (-1, 2)$

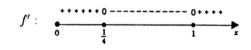

critical pt. 0; $f(-1) = \frac{1}{2}$ endpt max, $f(0) = 0$ local and abs min,

$f(2) = \frac{4}{5}$ endpt and abs max

13. $f'(x) = 2\left(x - \sqrt{x}\right)\left(1 - \frac{1}{2\sqrt{x}}\right), \quad x > 0;$

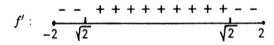

critical pts. $\frac{1}{4}, 1$;

$f(0) = 0$ endpt and abs min, $f\left(\frac{1}{4}\right) = \frac{1}{16}$ local max, $f(1) = 0$ local and abs min;

as $x \to \infty, \quad f(x) \to \infty;$ so no abs max

14. $f'(x) = \frac{2(2 - x^2)}{(4 - x^2)^{1/2}}, \quad x \in (-2, 2)$

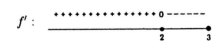

critical pts. $-\sqrt{2}, \sqrt{2}$; $f(-2) = 0$ endpt max, $f\left(-\sqrt{2}\right) = -2$ local and abs min,

$f\left(\sqrt{2}\right) = 2$ local and abs max, $f(2) = 0$ endpt min

15. $f'(x) = \frac{3(2-x)}{2\sqrt{3-x}}, \quad x < 3$

critical pt. 2;

$f(2) = 2$ local and abs max, $f(3) = 0$ endpt min;

as $x \to -\infty, \quad f(x) \to -\infty;$ so no abs min

16. $f'(x) = \frac{1}{2}\left(x^{-1/2} + x^{-3/2}\right), \quad x > 0$

no critical pts;   no extreme values.

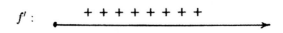

17. $f'(x) = -\frac{1}{3}(x-1)^{-2/3}, \quad x \neq 1;$

critical pt. 1;

no local extremes; $\left.\begin{array}{ll} \text{as} & x \to \infty, \quad f(x) \to -\infty \\ \text{as} & x \to -\infty, \quad f(x) \to \infty \end{array}\right\}$ no abs extremes

18. $f'(x) = \frac{8}{3}\frac{3x-1}{(4x-1)^{2/3}(2x-1)^{1/3}};$

critical pts. $\frac{1}{4}, \frac{1}{3}, \frac{1}{2}$;   no extreme value at $\frac{1}{4}$

$f\left(\frac{1}{3}\right) = \frac{1}{3}$ local max, $f\left(\frac{1}{2}\right) = 0$ local min

19. $f'(x) = \sin x \left(2\cos x + \sqrt{3}\right), x \in (0, \pi);$

critical pt. $\frac{5}{6}\pi$;

$f(0) = -\sqrt{3}$ endpt and abs min, $f\left(\frac{5}{6}\pi\right) = \frac{7}{4}$ local and abs max, $f(\pi) = \sqrt{3}$ endpt min

20. $f'(x) = -\csc^2 x + 1, \quad x \in (0, 2\pi/3);$

critical pt. $\frac{1}{2}\pi$ (note: 0 is not in the domain)

No extreme value at $\frac{1}{2}\pi$, $f\left(\frac{2}{3}\pi\right) = \frac{1}{3}\left(2\pi - \sqrt{3}\right)$ endpt and abs max

21. $f'(x) = -3\sin x \left(2\cos^2 x + 1\right) < 0, \quad x \in (0, \pi);$ no critical pts.

$f(0) = 5$ endpt and abs max, $f(\pi) = -5$ endpt and abs min

22. $f'(x) = 2\cos 2x - 1, \quad x \in (0, \pi):$

critical pts. $\frac{1}{6}\pi, \frac{5}{6}\pi$; $f(0) = 0$ endpt min, $f\left(\frac{1}{6}\pi\right) = \frac{1}{2}\sqrt{3} - \frac{1}{6}\pi$ local and abs max,

$f\left(\frac{5}{6}\pi\right) = -\frac{1}{2}\sqrt{3} - \frac{5}{6}\pi$ local and abs min, $f(\pi) = -\pi$ endpt max

23. $f'(x) = \sec^2 x - 1 \geq 0, \quad x \in \left(-\frac{1}{3}\pi, \frac{1}{2}\pi\right);$ critical pt. 0;

$f\left(-\frac{1}{3}\pi\right) = \frac{1}{3}\pi - \sqrt{3}$ endpt and abs min, no abs max

24. $f'(x) = 2\sin x \cos x(2\sin^2 x - 1), \quad x \in \left(0, \frac{2}{3}\pi\right);$

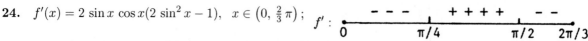

critical pts. $\frac{1}{4}\pi, \frac{1}{2}\pi$; $f(0) = 0$ endpt and abs max, $f\left(\frac{1}{4}\pi\right) = -\frac{1}{4}$ local and abs min,

$f\left(\frac{1}{2}\pi\right) = 0$ local and abs max, $f\left(\frac{2}{3}\pi\right) = -\frac{3}{16}$ endpt min

25.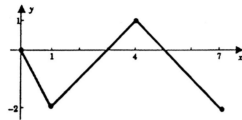

$$f'(x) = \begin{cases} -2, & 0 < x < 1 \\ 1, & 1 < x < 4 \\ -1, & 4 < x < 7 \end{cases}$$

critical pts. 1, 4;

$f(0) = 0$ endpt max, $f(1) = -2$ local and abs min,

$f(4) = 1$ local and absolute max, $f(7) = -2$ endpt and abs min

26. $f'(x) = \begin{cases} 1, & -8 < x < -3 \\ 2x + 1, & -3 < x \le 2 \\ 5, & 2 < x < 5 \end{cases}$

critical pts. $-3, -\frac{1}{2}$;

$f(-8) = 1$ endpt min, $f(-3) = 6$ local max

$f\left(-\frac{1}{2}\right) = -\frac{1}{4}$ local and abs min

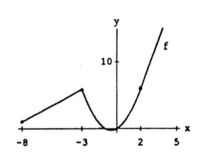

27.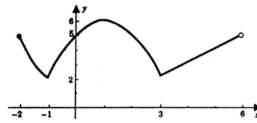

$$f'(x) = \begin{cases} 2x, & -2 < x < -1 \\ 2 - 2x, & -1 < x < 3 \\ 1, & 3 < x < 6 \end{cases}$$

critical pts. $-1, 1, 3$

$f(-2) = 5$ endpt max, $f(-1) = 2$ local and abs min,

$f(1) = 6$ local and abs max, $f(3) = 2$ local and abs min

28. $f'(x) = \begin{cases} -2x - 2, & -2 < x < 0 \\ -1, & 0 < x < 2 \\ 1, & 2 < x < 3 \\ (x - 2)^2, & 3 < x < 4 \end{cases}$

critical pts. $-1, 0, 2, 3$;

$f(-2) = 2$ endpt min, $f(-1) = 3$ local and abs max,

$f(0)$ not an extreme value, $f(2) = 0$ local and abs min,

$f(3) = \frac{1}{3}$ local min, $f(4) = \frac{8}{3}$ endpt max

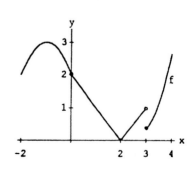

29.

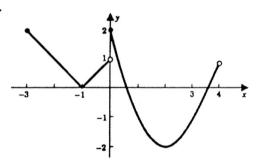

$$f'(x) = \begin{cases} -1, & -3 < x < -1 \\ 1, & -1 < x < 0 \\ 2x - 4, & 0 < x < 3 \\ 2, & 3 \leq x < 4 \end{cases}$$

critical pts. $-1, 0, 2$

$f(-3) = 2$ endpt and abs max, $f(-1) = 0$ local min,

$f(0) = 2$ local and abs max, $f(2) = -2$ local and abs min

30. $f'(x) = \begin{cases} -2x, & 0 < x < 1 \\ -2, & 1 < x < 2 \\ -x, & 2 < x < 3 \end{cases}$

Note: 1 is not in the domain of f.

critical pts. 2;

$f(0) = 0$ endpt and abs max, $f(2) = -2$ local max,

$f(3) = -\frac{9}{2}$ endpt and abs min

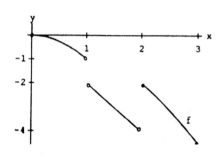

31.

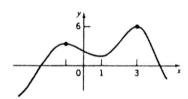

32.

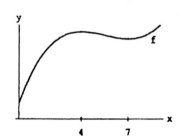

33. Not possible: $f(1) = f(3) = 0$ implies $f'(c) = 0$ for some $c \in (1, 3)$ (Rolle's theorem).

34. $f(x) = x - \dfrac{1}{2\pi} \sin 2\pi x$

35. Let $p(x) = x^3 + ax^2 + bx + c$. Then $p'(x) = 3x^2 + 2ax + b$ is a quadratic with discriminant $\Delta = 4a^2 - 12b = 4(a^2 - 3b)$. If $a^2 \leq 3b$, then $\Delta \leq 0$. This implies that $p'(x)$ does not change sign on $(-\infty, \infty)$. On the other hand, if $a^2 - 3b > 0$, then $\Delta > 0$ and p' has two real zeros, c_1 and c_2, from which it follows that p has extreme values at c_1 and c_2. Therefore, if p has no extreme values, then we must have $a^2 - 3b \leq 0$.

36. $f(x) = (1+x)^r - (1+rx), \quad x \geq -1.$

$f'(x) = r\left[(1+x)^{r-1} - 1\right]; \quad f'(x) = 0 \implies x = 0$

$f''(x) = r(r-1)(1+x)^{r-2}; \quad f''(0) = r(r-1) > 0 \implies f$ has a local minimum at $x = 0$

Since 0 is the only critical number of f, the local minimum must be an absolute minimum.

37. By contradiction. If f is continuous at c, then, by the first-derivative test (4.3.4), $f(c)$ is not a local maximum.

38. Since $f(c) \geq f(x)$ for all x in some open interval around c, and likewise $f(c) \leq f(x)$ for all x in some open interval around c, it follows that f must be constant on some open interval containing c.

39. If f is not differentiable on (a, b), then f has a critical point at each point c in (a, b) where $f'(c)$ does not exist. If f is differentiable on (a, b), then by the mean-value theorem there exists c in (a, b) where $f'(c) = [f(b) - f(a)]/(b-a) = 0$. This means c is a critical point of f.

40. We give a proof by contradiction. Suppose for no c in (c_1, c_2) is $f(c)$ a local minimum. By Theorem 2.6.2, f has a minimum on $[c_1, c_2]$ and therefore, this minimum must occur at c_1 or c_2. Suppose that $f(c_1)$ is an endpoint minimum. Then for some $\delta_1 > 0$,

(*)
$$f(x) \geq f(c), \quad x \in [c_1, c_1 + \delta_1).$$

Since $f(c_1)$ is a local maximum, there exists $\delta_2 > 0$ such that

(**)
$$f(x) \leq f(c_1), \quad x \in (c_1 - \delta_2, c_1 + \delta_2).$$

Set $\delta = \min[\delta_1, \delta_2]$. From (*) and (**), it follows that

$$f(x) = f(c_1), \quad x \in (c_1, c_1 + \delta).$$

This means that f has a local minimum on (c_1, c_2). The argument at c_2 is similar.

41. Let $f(x) = \begin{cases} 1, & \text{if } x \text{ is a rational number} \\ 0, & \text{if } x \text{ is an irrational number} \end{cases}$

42. $f(x) = \sin x$ and $f(x) = \cos x$ each have an infinite number of local maxima and local minima occurring at distinct points. However, the local maximum values are all the same, 1, and the local minimum values are all the same, -1. The function $f(x) = \frac{1}{2}x + \sin x$ has an infinite number of local maxima and local minima, and the local maximum values and the local minimum values are all different.

43. Let M be a positive number. Then

$$P(x) - M \geq a_n x^n - \left(|a_{n-1}|x^{n-1} + \cdots + |a_1|x + |a_0| + M\right) \quad \text{for} \quad x > 0$$

$$\geq a_n x^n - x^{n-1}\left(|a_{n-1}| + \cdots + |a_1| + |a_0| + M\right) \quad \text{for} \quad x > 1$$

$$\geq x^{n-1}\left[a_n x - \left(|a_{n-1}| + \cdots + |a_1| + |a_0| + M\right)\right]$$

It now follows that

$$P(x) - M \geq 0 \quad \text{for} \quad x \geq K = \frac{|a_{n-1}| + \cdots + |a_1| + |a_0| + M}{a_n}^{1/n}.$$

44. Let R be a rectangle with its diagonals having length c, and let x be the length of one of its sides. Then the length of the other side is $y = \sqrt{c^2 - x^2}$ and the area of R is given by

$$A(x) = x \sqrt{c^2 - x^2}$$

Now

$$A'(x) = \sqrt{c^2 - x^2} - \frac{x^2}{\sqrt{c^2 - x^2}} = \frac{c^2 - 2x^2}{\sqrt{c^2 - x^2}},$$

and

$$A'(x) = 0 \implies x = \frac{\sqrt{2}}{2} c$$

It is easy to verify that A has a maximum at $x = \dfrac{\sqrt{2}}{2} c$. Since $y = \dfrac{\sqrt{2}}{2} c$ when $x = \dfrac{\sqrt{2}}{2} c$, it follows that the rectangle of maximum area is a square.

45. $f(0) = f(1) = 0$ and $f(x) > 0$ on $(0, 1)$.

$f'(x) = -qx^p(1-x)^{q-1} + px^{p-1}(1-x)^q$; $f'(x) = 0$ implies $x = \dfrac{p}{p+q}$. The absolute maximum value of f is $f(p/(p+q)) = \left(\dfrac{p}{p+q}\right)^p \cdot \left(\dfrac{q}{p+q}\right)^q$

46. Let $S = x^3 + y^3$ where $x + y = 16$. Then $S(x) = x^3 + (16 - x)^3$ and

$S'(x) = 3x^2 - 3(16 - x)^2 = 96(x - 8)$; $S'(x) = 0 \implies x = 8 \implies y = 8$;

$S''(x) = 96$; $S''(8) = 96 > 0 \implies S$ has a local minimum at $x = 8$.

It now follows that $S(8)$ is the absolute minimum of S.

47. Setting $R'(\theta) = \dfrac{v^2 \cos 2\theta}{16} = 0$, gives $\theta = \dfrac{\pi}{4}$.   Since $R''\left(\dfrac{\pi}{4}\right) = -\dfrac{v^2}{8} < 0$, $\theta = \dfrac{\pi}{4}$ is a maximum.

48. Cut the wire into two pieces, one of length x and the other of length $L - x$. Suppose that the wire of length x is used to form the equilateral triangle, and the other piece is used to form the square. Then the area of the triangle is $\sqrt{3}\, x^2/36$, and the area of the square is $(L - x)^2/16$. Now, let

$$S(x) = \frac{\sqrt{3}}{36} x^2 + \frac{1}{16} (L - x)^2$$

Then

$$S'(x) = \frac{\sqrt{3}}{18} x - \frac{1}{8} (L - x)$$
$$= \frac{4\sqrt{3} + 9}{72} x - \frac{1}{8} L$$

Setting $S'(x) = 0$ we find that

$$x = \frac{9}{4\sqrt{3} + 9} L. \cong 0.5650\, L$$

Now,

$$S(0) = \frac{1}{16} L^2 = 0.0625 L^2 \quad \text{(absolute maximum)}$$

$$S\left(\frac{9}{4\sqrt{3}+9}L\right) \cong 0.0390L^2 \quad \text{(absolute minimum)}$$

$$S(L) = \frac{\sqrt{3}}{36}L^2 = 0.0481L^2$$

To maximize the sum of the areas, use the wire to form a square; to minimize the sum, use $x \cong 0.5650\,L$ to form the triangle and the remainder to form the square.

49.

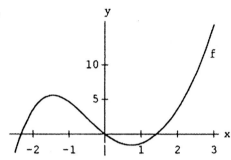

critical pts: $x_1 = -1.452,\ x_2 = 0.760$

$f(-1.452)$ local maximum

$f(0.727)$ local minimum

$f(3)$ absolute maximum

$f(-2.5)$ absolute minimum

50.

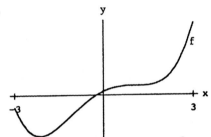

critical pts: $x_1 = -2.179,\ x_2 = 1,\ x_3 = 1.158$

$f(-3)$ endpoint maximum

$f(1)$ local maximum

$f(-2.158),\ f(1.158)$ local minima

$f(3)$ absolute maximum

$f(-2.179)$ absolute minimum

51.

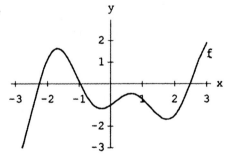

critical points: $x_1 = -1.683,\ x_2 = -0.284,$

$x_3 = 0.645,\ x_4 = 1.760$

$f(-1.683), f(0.645)$ local maxima

$f(-0.284), f(1.760)$ local minima

$f(\pi)$ absolute maximum

$f(-\pi)$ absolute minimum

52.

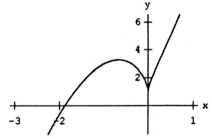

critical pts: $x_1 = -0.667,\ x_2 = 0$

$f(-0.667)$ local maximum

$f(0)$ local minimum

$f(1)$ absolute maximum

$f(-3)$ absolute minimum

53. Yes; $M = f(2) = 1$; $m = f(1) = f(3) = 0$

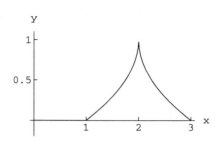

54. No; f is discontinuous at $x = 3$; $M = f(3) = \frac{7}{2}$; $m = f(0) = -\frac{19}{4}$

55. Yes; $M = f(6) = 2 + \sqrt{3}$; $m = f(1) = \frac{3}{2}$

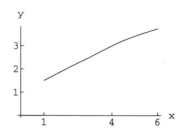

SECTION 4.5

1. Set $P = xy$ and $y = 40 - x$. We want to maximize

$$P(x) = x(40 - x), \quad 0 \le x \le 40.$$

$$P'(x) = 40 - 2x, \quad P'(x) = 0 \implies x = 20.$$

Since P increases on $(0, 20]$ and decreases on $[20, 40)$, the abs max of P occurs when $x = 20$. Then, $y = 20$ and $xy = 400$.

The maximal value of xy is 400.

2. Set $A = xy$ and $2x + 2y = 24$ or $y = 12 - x$. We want to maximize

$$A(x) = x(12 - x), \quad 0 \le x \le 12.$$

$$A'(x) = 12 - 2x, \quad P'(x) = 0 \implies x = 6.$$

Since A increases on $[0, 6]$ and decreases on $[6, 12]$, the abs max of A occurs when $x = 6$. Then, $y = 6$.

The dimensions of the rectangle having perimeter 24 and maximum area are: 6×6.

3.

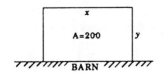

Minimize P

$P = x + 2y$, $\quad 200 = xy$, $\quad y = 200/x$

$$P(x) = x + \frac{400}{x}, \quad x > 0.$$

$P'(x) = 1 - \dfrac{400}{x^2}, \quad P'(x) = 0 \quad \Longrightarrow \quad x = 20.$

Since P decreases on $(0, 20]$ and increases on $[20, \infty)$, the abs min of P occurs when $x = 20$.

To minimize the fencing, make the garden 20 ft (parallel to barn) by 10 ft.

4.

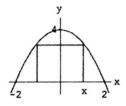

Maximize A

$A = 2xy, \quad y = 4 - x^2$

$A(x) = 2x(4 - x^2) = 8x - 2x^3, \quad 0 \le x \le 2.$

$A'(x) = 8 - 6x^2, \quad P'(x) = 0 \quad \Longrightarrow \quad x = \dfrac{2}{\sqrt{3}}.$

Since A increases on $[0, 2/\sqrt{3}]$ and decreases on $[2/\sqrt{3}, 2]$, the abs max of A occurs when $x = 2/\sqrt{3}$.

The maximal area is $\frac{32}{9}\sqrt{3}$.

5.

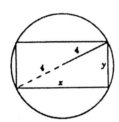

Maximize A

$A = xy, \quad x^2 + y^2 = 8^2, \quad y = \sqrt{64 - x^2}$

$A(x) = x\sqrt{64 - x^2}, \quad 0 \le x \le 8.$

$A'(x) = \sqrt{64 - x^2} + x\left(\dfrac{-x}{\sqrt{64 - x^2}}\right) = \dfrac{64 - 2x^2}{\sqrt{64 - x^2}}, \qquad A'(x) = 0 \quad \Longrightarrow \quad x = 4\sqrt{2}.$

Since A increases on $(0, 4\sqrt{2}]$ and decreases on $[4\sqrt{2}, 8)$, the abs max of A occurs when $x = 4\sqrt{2}$.
Then, $y = 4\sqrt{2}$ and $xy = 32$.
The maximal area is 32.

6. Maximize $P = 2x + 2y; \quad xy = A$ (constant) $\quad \Longrightarrow \quad y = \dfrac{A}{x}$

$P(x) = 2x + \dfrac{2A}{x}, \quad x > 0; \qquad P'(x) = 2 - \dfrac{2A}{x^2};$

$$P'(x) = 0 \quad \Longrightarrow \quad x = \sqrt{A} \quad \Longrightarrow \quad y = \sqrt{A}$$

The rectangle having a given area A and minimum perimeter is a square of side length $\sqrt{A}$.

7.

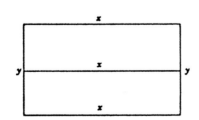

Minimize $\quad P = 3x + 2y$

$A = xy = 15,000, \quad y = \dfrac{15,000}{x}$

$$P(x) = 3x + \frac{30,000}{x}, \quad 0 < x < \infty.$$

$$P'(x) = 3 - \frac{30,000}{x^2} = \frac{3(x^2 - 10,000)}{x^2}; \quad P'(x) = 0 \implies x = 100.$$

Since P decreases on $(0, 100]$ and increases on $[100, \infty)$, the abs min of P occurs when $x = 100$. When $x = 100$, $y = 150$; at least 600 feet of fencing is needed.

8. Minimize $C = 300y + 400x$

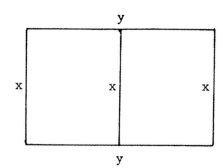

$$A = 5000 = xy \implies y = \frac{5000}{x}; p$$

$$C(x) = \frac{1,500,000}{x} + 400x, \quad x > 0;$$

$$C'(x) = -\frac{1,500,000}{x^2} + 400;$$

$$C'(x) = 0 \implies x \cong 61.24.$$

$$C''(x) = \frac{3,000,000}{x^3}; \quad C''(61.24) > 0.$$

C has an abs min at $x = 61.24$. The dimensions that will minimize the cost are: $x = 61.24$, $y = 81.65$.

9.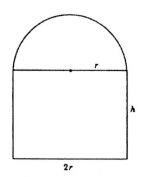

Maximize L

To account for the semi-circular portion admitting less light per square foot, we multiply its area by $1/3$.

$$L = 2xy + \frac{1}{3}\left(\frac{\pi x^2}{2}\right),$$
$$2x + 2y + \pi x = p, \quad y = \tfrac{1}{2}(p - 2x - \pi x)$$

$$L = 2x\left(\frac{p - 2x - \pi x}{2}\right) + \frac{1}{6}\pi x^2$$

$$L(x) = px - \left(2 + \frac{5}{6}\pi\right)x^2, \quad 0 \le x \le \frac{p}{2 + \pi}.$$

$$L'(x) = p - \left(4 + \frac{5}{3}\pi\right)x; \quad L'(x) = 0 \implies x = \frac{3p}{12 + 5\pi}.$$

Since $L''(x) < 0$ for all x in the domain of L, the local max at $x = 3p/(12 + 5\pi)$ is the abs max.

For the window that admits the most light, take the radius of the semicircle as $\dfrac{3p}{12 + 5\pi}$ ft.

10.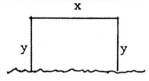

Maximize A

$$A = xy, \quad x + 2y = 800$$

$$A(y) = (800 - 2y)y = 800y - 2y^2, \quad 0 \le y \le 400.$$

$$A'(y) = 800 - 4y, \quad A'(y) = 0 \implies y = 200.$$

Since A increases on $[0, 200]$ and decreases on $[200, 400]$, the abs max of A occurs when $y = 200$.

The dimensions of the field of maximum area are: 200×400.

11.

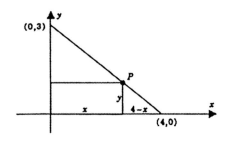

<u>Maximize A</u>

$$A = xy, \quad \frac{3}{4} = \frac{y}{4-x} \qquad \text{(similar triangles)}$$

$$y = \tfrac{3}{4}(4-x)$$

$A(x) = \dfrac{3x}{4}(4-x), \; 0 \le x \le 4.$

$A'(x) = 3 - \dfrac{3x}{2}, \; A'(x) = 0 \implies x = 2.$

Since A increases on $(0, 2]$ and decreases on $[2, 4)$, the abs max of A occurs when $x = 2$.

To maximize the area of the rectangle, take P as the point $\left(2, \tfrac{3}{2}\right)$.

12. The equation of the third side is: $y = mx + (1-m)$.

The base of the triangle is: $b = \dfrac{m-1}{m}$.

The two lines intersect when $3x = mx + (1-m)$;

$\implies \quad x = \dfrac{1-m}{3-m} \implies h = \dfrac{3(1-m)}{3-m}.$

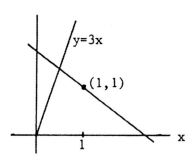

We want to minimize

$$A(m) = \frac{1}{2}\frac{m-1}{m}\frac{3(1-m)}{3-m} = -\frac{3}{2}\frac{(1-m)^2}{3m - m^2}, \quad m < 0.$$

$$A'(m) = -\frac{3}{2}\frac{(m+3)(m-1)}{(3m - m^2)^2}, \quad A'(m) = 0 \implies m = -3.$$

The area of the triangle is a minimum when the slope of the line is -3.

13.

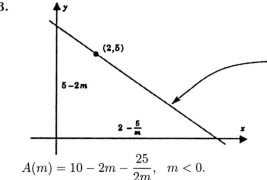

<u>Minimize A</u>

$A = \tfrac{1}{2}(x\text{-intercept})\,(y\text{-intercept})$

Equation of line: $\quad y - 5 = m(x-2)$

x-intercept: $\; 2 - \dfrac{5}{m}$

y-intercept: $\; 5 - 2m$

$A = \dfrac{1}{2}\left(2 - \dfrac{5}{m}\right)(5 - 2m) = 10 - 2m - \dfrac{25}{2m}$

$A(m) = 10 - 2m - \dfrac{25}{2m}, \; m < 0.$

$A'(m) = -2 + \dfrac{25}{2m^2}, \; A'(m) = 0 \implies m = -\dfrac{5}{2}.$

Since $A''(m) = -25/m^3 > 0$ for $m < 0$, the local min at $m = -5/2$ is the abs min.

The triangle of minimal area is formed by the line of slope $-5/2$.

14. Since $\lim\limits_{m \to 0^-} A(m) = +\infty$, no minimum exists.

15.

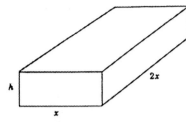

Maximize V

$$V = 2x^2h, \quad 2\left(2x^2 + xh + 2xh\right) = 100, \quad h = \frac{50 - 2x^2}{3x}$$

$$V = 2x^2 \left(\frac{50 - 2x^2}{3x}\right)$$

$V(x) = \frac{100}{3}x - \frac{4}{3}x^3, \quad 0 \le x \le 5.$

$V'(x) = \frac{100}{3} - 4x^2, \quad V'(x) = 0 \implies x = \frac{5}{3}\sqrt{3}.$

Since $V''(x) = -8x < 0$ on $(0, 5)$, the local max at $x = \frac{5}{3}\sqrt{3}$ is the abs max.

The base of the box of greatest volume measures $\frac{5}{3}\sqrt{3}$ in. by $\frac{10}{3}\sqrt{3}$ in.

16. With no top, we have $2x^2 + 2xh + 4xh = 100$, or $h = \dfrac{50 - x^2}{3x}$.

Maximize $V(x) = 2x^2 \left(\dfrac{50 - x^2}{3x}\right) = \dfrac{2x}{3}(50 - x^2), \quad 0 \le x \le 5\sqrt{2}.$

$V'(x) = \dfrac{100}{3} - 2x^2, \quad V'(x) = 0 \implies x = \dfrac{5}{3}\sqrt{6}.$

Since $V''(x) = -4x < 0$ on $(0, 5\sqrt{2})$, the local max at $x = \frac{5}{3}\sqrt{6}$ is the abs max.

The base of the box of greatest volume measures $\frac{5}{3}\sqrt{6}$ in. by $\frac{10}{3}\sqrt{6}$ in.

17.

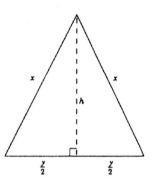

Maximize A

$A = \frac{1}{2}hy$

$2x + y = 12 \implies y = 12 - 2x$

Pythagorean Theorem:

$$h^2 + \left(\frac{y}{2}\right)^2 = x^2 \implies h = \sqrt{x^2 - \left(\frac{y}{2}\right)^2}$$

Thus, $h = \sqrt{x^2 - (6 - x)^2} = \sqrt{12x - 36}.$

$A(x) = (6 - x)\sqrt{12x - 36}, \quad 3 \le x \le 6.$

$A'(x) = -\sqrt{12x - 36} + (6 - x)\left(\dfrac{6}{\sqrt{12x - 36}}\right) = \dfrac{72 - 18x}{\sqrt{12x - 36}},$

$A'(x) = 0 \implies x = 4.$

Since A increases on $(3, 4]$ and decreases on $[4, 6)$, the abs max of A occurs at $x = 4$.

The triangle of maximal area is equilateral with side of length 4.

18. It is sufficient to minimize the square of the distance:

$$S = (x - 0)^2 + (y - 6)^2 = 8y + (y - 6)^2 \text{ since } x^2 = 8y \text{ where } y \ge 0.$$

$$S'(y) = 2y - 4, \quad S'(y) = 0 \implies y = 2.$$

The points on the parabola that are closest to $(0, 6)$ are: $(4, 2)$ and $(-4, 2)$.

19.

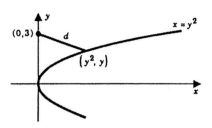

Minimize d

$$d = \sqrt{(y^2 - 0)^2 + (y - 3)^2}$$

The square-root function is increasing;

d is minimal when $D = d^2$ is minimal.

$D(y) = y^4 + (y - 3)^2$, y real.

$D'(y) = 4y^3 + 2(y - 3) = (y - 1)\left(4y^2 + 4y + 6\right)$, $D'(y) = 0$ at $y = 1$.

Since $D''(y) = 12y^2 + 2 > 0$, the local min at $y = 1$ is the abs min.

The point $(1, 1)$ is the point on the parabola closest to $(0, 3)$.

20. $f(x) = Ax^{-1/2} + Bx^{1/2}$, $f(9) = 6$ $\implies$ $\frac{1}{3}A + 3B = 6$.

$f'(x) = \dfrac{-A}{2x^{3/2}} + \dfrac{B}{2x^{1/2}}$; $f'(9) = 0$ $\implies$ $\dfrac{-A}{54} + \dfrac{B}{6} = 0$.

Solving the two equations gives: $A = 9$, $B = 1$.

21. The figure shows a rectangle inscribed in the ellipse $16x^2 + 9y^2 = 144$. The area of the rectangle is:

$A = (2x)(2y) = 4xy$. Solving the equation of the ellipse for y, we get $y = \frac{4}{3}\sqrt{9 - x^2}$.

Maximize $A = \frac{16}{3}x\sqrt{9 - x^2}$, $0 \leq x \leq 3$.

$A'(x) = \dfrac{16}{3}\left[\sqrt{9 - x^2} - \dfrac{x^2}{\sqrt{9 - x^2}}\right] = \dfrac{16}{3}\left(\dfrac{9 - 2x^2}{\sqrt{9 - x^2}}\right)$

$A'(x) = 0$ $\implies$ $x = 3/\sqrt{2}$.

Since $A(0) = A(3) = 0$, we can conclude that $A(3/\sqrt{2})$

is the absolute maximum value of A.

At $x = 3/\sqrt{2}$, $y = 4/\sqrt{2}$ and $A = 24$; the maximum possible

area for a rectangle inscribed in the ellipse is 24 sq. units.

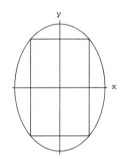

22. Simply repeat the solution of Exercise 21, replacing 9 by a^2 and 16 by b^2. The result is:

$x = a/\sqrt{2}$, $y = b/\sqrt{2}$; the maximum possible area of a rectangle inscribed in the ellipse is $A = 4(a/\sqrt{2})(b/\sqrt{2}) = 2ab$ square units.

23.

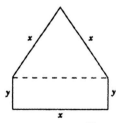

Maximize A

$$A = xy + \frac{\sqrt{3}}{4}x^2, \quad 30 = 3x + 2y, \quad y = \frac{30 - 3x}{2}$$

$A(x) = 15x - \dfrac{3}{2}x^2 + \dfrac{\sqrt{3}}{4}x^2$, $0 \leq x \leq 10$.

$A'(x) = 15 - 3x + \dfrac{\sqrt{3}}{2}x$, $A'(x) = 0$ $\implies$ $x = \dfrac{30}{6 - \sqrt{3}} = \dfrac{10}{11}(6 + \sqrt{3})$.

Since $A''(x) = -3 + \dfrac{\sqrt{3}}{2} < 0$ on $(0, 10)$, the local max at $x = \dfrac{10}{11}\left(6 + \sqrt{3}\right)$ is the abs max.

The pentagon of greatest area is composed of an equilateral triangle with side $\dfrac{10}{11}\left(6 + \sqrt{3}\right) \cong 7.03$ in. and rectangle with height $\dfrac{15}{11}\left(5 - \sqrt{3}\right) \cong 4.46$ in.

24. To maximize the area, use the cross-section that is wider at the top.

$A(h) = 4h + h\sqrt{16 - h^2}, \quad 0 \le h \le 4;$

$A'(h) = 4 + \dfrac{16 - 2h^2}{\sqrt{16 - h^2}};$

$A'(h) = 0 \implies h = 2\sqrt{3}.$

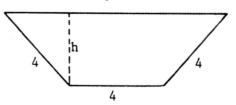

The depth of the gutter that has maximum carrying capacity is: $2\sqrt{3}$ inches.

25.

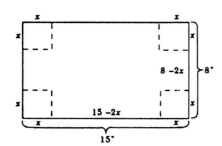

Maximize V

$V = x(8 - 2x)(15 - 2x)$

$\left.\begin{array}{r} x \ge 0 \\ 8 - 2x \ge 0 \\ 15 - 2x \ge 0 \end{array}\right\} \implies 0 \le x \le 4$

$V(x) = 120x - 46x^2 + 4x^3, \quad 0 \le x \le 4.$

$V'(x) = 120 - 92x + 12x^2 = 4(3x - 5)(x - 6), \quad V'(x) = 0 \text{ at } x = \tfrac{5}{3}.$

Since V increases on $\left(0, \tfrac{5}{3}\right)$ and decreases on $\left[\tfrac{5}{3}, 4\right)$, the abs max of V occurs when $x = \tfrac{5}{3}$.

The box of maximal volume is made by cutting out squares $5/3$ inches on a side.

26.

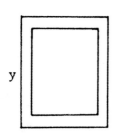

Minimize P

$P = 2x + 2y;$

$(x - 4)(y - 6) = 81; \quad y = \dfrac{81}{x - 4} + 6.$

$P(x) = 2x + 2\left(\dfrac{81}{x - 4} + 6\right), \quad x > 4.$

$P'(x) = 2 - \dfrac{162}{(x - 4)^2}, \quad P'(x) = 0 \implies x = 13.$

Since $P''(x) = \dfrac{324}{(x - 4)^3} > 0$ when $x > 4$, $x = 13$ is the abs min.

The most economical page has dimensions: width 13 cm, length 15 cm.

27.

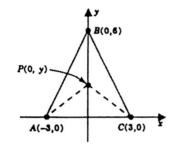

$\underline{\text{Minimize}\ \overline{AP} + \overline{BP} + \overline{CP} = S}$

length $AP = \sqrt{9 + y^2}$

length $BP = 6 - y$

length $CP = \sqrt{9 + y^2}$

$S(y) = 6 - y + 2\sqrt{9 + y^2}, \ \ 0 \le y \le 6.$

$S'(y) = -1 + \dfrac{2y}{\sqrt{9 + y^2}}, \ \ S'(y) = 0 \ \implies \ y = \sqrt{3}.$

Since

$$S(0) = 12, \quad S\left(\sqrt{3}\right) = 6 + 3\sqrt{3} \cong 11.2, \quad \text{and} \quad S(6) = 6\sqrt{5} \cong 13.4,$$

the abs min of S occurs when $y = \sqrt{3}$.

To minimize the sum of the distances, take P as the point $\left(0, \sqrt{3}\right)$.

28. Refer to Exercise 27. Here we want to minimize

$S(y) = 3 - y + 2\sqrt{36 + y^2}, \ \ 0 \le y \le 3.$

$S'(y) = -1 + \dfrac{2y}{\sqrt{36 + y^2}}, \ \ S'(y) = 0 \ \implies \ y = \sqrt{12} > 3.$

Thus, the minimum must occur at one of the endpoints: $S(0) = 15, \ \ S(3) = 2\sqrt{45} < S(0).$

To minimize the sum of the distances, take $P = (0, 3)$.

29.

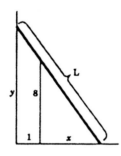

$\underline{\text{Minimize}\ L}$

$L^2 = y^2 + (x + 1)^2.$

By similar triangles $\dfrac{y}{x + 1} = \dfrac{8}{x}, \ \ y = \dfrac{8}{x}(x + 1).$

$L^2 = \left[\left(\dfrac{8}{x}\right)(x + 1)\right]^2 + (x + 1)^2 = (x + 1)^2 \left(\dfrac{64}{x^2} + 1\right)$

Since L is minimal when L^2 is minimal, we consider the function

$$f(x) = (x + 1)^2 \left(\dfrac{64}{x^2} + 1\right), \ \ x > 0.$$

$$f'(x) = 2(x + 1)\left(\dfrac{64}{x^2} + 1\right) + (x + 1)^2 \left(\dfrac{-128}{x^3}\right)$$

$$= \dfrac{2(x + 1)}{x^3}\left[x^3 - 64\right], \ \ f'(x) = 0 \ \implies \ x = 4.$$

Since f decreases on $(0, 4]$ and increases on $[4, \infty)$, the abs min of f occurs when $x = 4$.

The shortest ladder is $5\sqrt{5}$ ft long.

30. Maximize $L = x + y$.

By similar triangles, $\dfrac{y}{6} = \dfrac{x}{\sqrt{x^2 - 64}}$

$L(x) = x + \dfrac{6x}{\sqrt{x^2 - 64}}, \; x > 8$

$L'(x) = 1 - \dfrac{384}{(x^2 - 64)^{3/2}}$

$L'(x) = 0 \implies x = \sqrt{64 + (384)^{2/3}} \cong 10.81$

$L(10.81) \cong 19.73$; the longest ladder is approximately 19.7 ft.

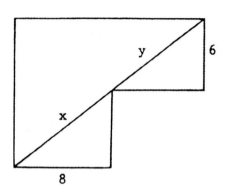

31.

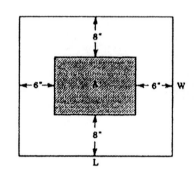

Maximize A

(We use feet rather than inches to reduce arithmetic.)

$A = (L - 1)\left(W - \dfrac{4}{3}\right)$

$LW = 27 \implies W = \dfrac{27}{L}$

$A = (L - 1)\left(\dfrac{27}{L} - \dfrac{4}{3}\right) = \dfrac{85}{3} - \dfrac{27}{L} - \dfrac{4}{3}L$

$A(L) = \dfrac{85}{3} - \dfrac{27}{L} - \dfrac{4}{3}L, \; 1 \le L \le \dfrac{81}{4}$.

$A'(L) = \dfrac{27}{L^2} - \dfrac{4}{3}, \; A'(L) = 0 \implies L = \dfrac{9}{2}$.

Since $A'(L) = -54/L^3 < 0$ for $1 < L < \dfrac{81}{4}$, the max at $L = \dfrac{9}{2}$ is the abs max.

The banner has length $9/2$ ft = 54 in. and height 6 ft = 72 in.

32. Assume $V = \dfrac{1}{3}\pi r^2 h = 1$, or $h = \dfrac{3}{\pi r^2}$.

Minimize surface area $S = \pi r \sqrt{r^2 + h^2} = \pi r \sqrt{r^2 + \dfrac{9}{\pi^2 r^4}} = \dfrac{\sqrt{\pi^2 r^6 + 9}}{r}$.

$\dfrac{dS}{dr} = \dfrac{2\pi^2 r^6 - 9}{r^2 \sqrt{\pi^2 r^6 + 9}} = 0 \implies r = \left(\dfrac{9}{2\pi^2}\right)^{1/6} = \dfrac{3^{1/3}}{2^{1/6}\pi^{1/3}}$

$\implies h = \dfrac{3}{\pi\left(\dfrac{9}{2\pi^2}\right)^{1/3}} = \dfrac{3^{1/3}2^{1/3}}{\pi^{1/3}} = r\sqrt{2}$.

33.

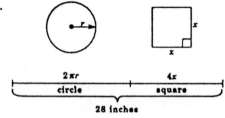

Find the extreme values of A

$A = \pi r^2 + x^2$

$2\pi r + 4x = 28 \implies x = 7 - \dfrac{1}{2}\pi r$.

$A(r) = \pi r^2 + \left(7 - \dfrac{1}{2}\pi r\right)^2, \; 0 \le r \le \dfrac{14}{\pi}$.

Note: the endpoints of the domain correspond to the instances when the string is not cut: $r = 0$
when no circle is formed, $r = 14/\pi$ when no square is formed.

$$A'(r) = 2\pi r - \pi \left(7 - \frac{1}{2}\pi r \right), \quad A'(r) = 0 \implies r = \frac{14}{4 + \pi}.$$

Since $A''(r) = 2\pi + \pi^2/2 > 0$ on $(0, 14/\pi)$, the abs min of A occurs when $r = 14/(4 + \pi)$ and
the abs max of A occurs at one of the endpts: $A(0) = 49$, $A(14/\pi) = 196/\pi > 49$.

(a) To maximize the sum of the two areas, use all of the string to form the circle.

(b) To minimize the sum of the two areas, use $2\pi r = 28\pi/(4 + \pi) \cong 12.32$ inches of string for the
circle.

34. Maximize $V = x^2 h$ given that $x^2 + 4xh = 12 \implies h = \dfrac{12 - x^2}{4x}$.

$$V(x) = x^2 \left(\frac{12 - x^2}{4x} \right) = 3x - \frac{1}{4}x^3, \quad 0 < x \le \sqrt{12}.$$

$V'(x) = 3 - \frac{3}{4}x^2$, $V'(x) = 0 \implies x = 2$. Since V increases on $(0, 2]$ and decreases on $[2, \sqrt{12}]$,

V has an abs max at $x = 2$; the maximum volume is $V(2) = 4$ cu ft.

35.

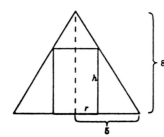

Maximize V

$V = \pi r^2 h$

By similar triangles

$$\frac{8}{5} = \frac{h}{5 - r} \quad \text{or} \quad h = \frac{8}{5}(5 - r).$$

$$V(r) = \frac{8\pi}{5}r^2(5 - r), \quad 0 \le r \le 5.$$

$$V'(r) = \frac{8\pi}{5}\left(10r - 3r^2 \right), \quad V'(r) = 0 \implies r = 10/3.$$

Since V increases on $(0, 10/3]$ and decreases on $[10/3, 5)$, the abs max of V occurs when $r = 10/3$.

The cylinder with maximal volume has radius $10/3$ and height $8/3$.

36. Maximize $A = 2\pi r h = \dfrac{16\pi}{5} r(5 - r), \quad 0 \le r \le 5, \quad \left(h = \frac{8}{5}(5 - r) \text{ from Exercise 33} \right).$

$$A'(r) = \frac{16\pi}{5}(5 - 2r), \quad A'(r) = 0 \implies r = \frac{5}{2}$$

The curved surface is a maximum when $r = \frac{5}{2}$, $h = 4$.

37.

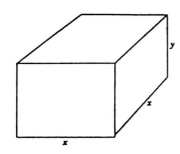

Minimize C

In dollars,

$C = $ cost base + cost top + cost sides

$$= .35 \left(x^2 \right) + .15 \left(x^2 \right) + .20(4xy)$$

$$= \frac{1}{2}x^2 + \frac{4}{5}xy$$

Volume $= x^2 y = 1250 \quad y = \dfrac{1250}{x^2}$

$$C(x) = \frac{1}{2}x^2 + \frac{1000}{x}, \quad x > 0.$$

$$C'(x) = x - \frac{1000}{x^2}, \quad C'(x) = 0 \implies x = 10.$$

Since $C''(x) = 1 + 2000/x^3 > 0$ for $x > 0$, the local min of C at $x = 10$ is the abs min.

The least expensive box is 12.5 ft tall with a square base 10 ft on a side.

38. Maximize $A = xy$. By similar triangles

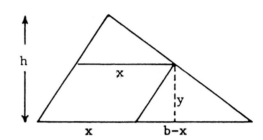

$$\frac{y}{b-x} = \frac{h}{b}, \quad \text{so}$$

$$A(x) = \frac{h}{b}x(b-x), \quad 0 \le x \le b.$$

$$A'(x) = \frac{h}{b}(b - 2x), \quad A'(x) = 0 \implies x = \frac{b}{2}.$$

Since A is increasing on $[0, b/2]$ and decreasing on $[b/2, b]$, A has an abs max at $x = b/2$

$A(b/2) = \frac{1}{4}hb = \frac{1}{2}$ area of triangle ABC.

39.

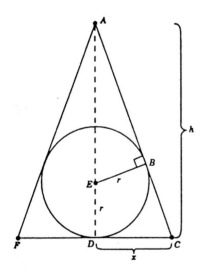

<u>Minimize A</u>

$$A = \frac{1}{2}(h)(2x) = hx$$

Triangles ADC and ABE are similar:

$$\frac{AD}{DC} = \frac{AB}{BE} \quad \text{or} \quad \frac{h}{x} = \frac{AB}{r}.$$

Pythagorean Theorem:

$$r^2 + (AB)^2 = (h - r)^2.$$

Thus

$$r^2 + \left(\frac{hr}{x}\right)^2 = (h - r)^2.$$

Solving this equation for h we find that

$$h = \frac{2x^2 r}{x^2 - r^2}.$$

$$A(x) = \frac{2x^3 r}{x^2 - r^2}, \quad x > r.$$

$$A'(x) = \frac{(x^2 - r^2)(6x^2 r) - 2x^3 r(2x)}{(x^2 - r^2)^2} = \frac{2x^2 r (x^2 - 3r^2)}{(x^2 - r^2)^2},$$

$$A'(x) = 0 \implies x = r\sqrt{3}.$$

Since A decreases on $(r, r\sqrt{3}\,]$ and increases on $[r\sqrt{3}, \infty)$, the local min at $x = r\sqrt{3}$ is the abs min of A. When $x = r\sqrt{3}$, we get $h = 3r$ so that $FC = 2r\sqrt{3}$ and $AF = FC = \sqrt{h^2 + x^2} = 2r\sqrt{3}$. The triangle of least area is equilateral with side of length $2r\sqrt{3}$.

40. Maximize $A(x) = \frac{1}{2}(r+x)2\sqrt{r^2 - x^2}$

$= (r+x)\sqrt{r^2 - x^2},\ 0 \le x \le r.$

$A'(x) = \dfrac{r^2 - rx - 2x^2}{\sqrt{r^2 - x^2}};$

$A'(x) = 0 \implies x = \dfrac{r}{2}.$

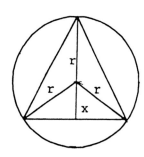

Since A increases on $[0, r/2]$ and decreases on $[r/2, r]$, A has an abs max at $x = r/2$;

$A(r/2) = \dfrac{3\sqrt{3}}{4}r^2.$

41.

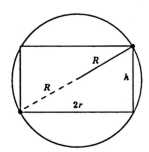

Maximize V

$V = \pi r^2 h$

By the Pythagorean Theorem,

$(2r)^2 + h^2 = (2R)^2$

so

$h = 2\sqrt{R^2 - r^2}.$

$V(r) = 2\pi r^2\sqrt{R^2 - r^2},\ 0 \le r \le R.$

$V'(r) = 2\pi \left[2r\sqrt{R^2 - r^2} - \dfrac{r^3}{\sqrt{R^2 - r^2}} \right] = \dfrac{2\pi r\left(2R^2 - 3r^2\right)}{\sqrt{R^2 - r^2}}$

$V'(r) = 0 \implies r = \frac{1}{3}R\sqrt{6}.$

Since V increases on $\left(0, \frac{1}{3}R\sqrt{6}\,\right]$ and decreases on $\left[\frac{1}{3}R\sqrt{6}, R\right)$, the local max at $r = \frac{1}{3}R\sqrt{6}$ is the abs max.

The cylinder of maximal volume has base radius $\frac{1}{3}R\sqrt{6}$ and height $\frac{2}{3}R\sqrt{3}$.

42. Maximize $A = 2\pi r h = 4\pi r\sqrt{R^2 - r^2},\ 0 \le r \le R,\ \left(h = 2\sqrt{R^2 - r^2}\text{ from Exercise 39}\right).$

$A'(r) = \dfrac{4\pi(R^2 - 2r^2)}{\sqrt{R^2 - r^2}},\quad A'(r) = 0 \implies r = \dfrac{R}{\sqrt{2}}.$

The curved surface is a maximum when $r = \dfrac{R}{\sqrt{2}},\ h = R\sqrt{2}.$

43.

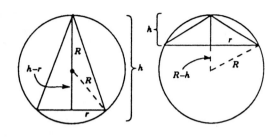

Case 1 : $h \geq R$ Case 2 : $h \leq R$

Maximize V

$$V = \tfrac{1}{3}\pi r^2 h$$

Pythagorean Theorem

Case 1 : $(h - R)^2 + r^2 = R^2$

Case 2 : $(R - h)^2 + r^2 = R^2$

In both cases

$$r^2 = R^2 - (R - h)^2 = 2hR - h^2.$$

$V(h) = \tfrac{1}{3}\pi \left(2h^2 R - h^3\right), \quad 0 \leq h \leq 2R.$

$V'(h) = \tfrac{1}{3}\pi \left(4hR - 3h^2\right), \quad V'(h) = 0 \quad \text{at} \quad h = \dfrac{4R}{3}.$

Since V increases on $\left(0, \tfrac{4}{3}R\right]$ and decreases on $\left[\tfrac{4}{3}R, 2R\right)$, the local max at $h = \tfrac{4}{3}R$ is the abs max.

The cone of maximal volume has height $\tfrac{4}{3}R$ and radius $\tfrac{2}{3}R\sqrt{2}$.

44.   Maximize $V = \tfrac{1}{3}\pi r^2 h$, where $r^2 + h^2 = a^2$.

$V(h) = \tfrac{1}{3}\pi (a^2 - h^2)h, \quad 0 \leq h \leq a,$

$V'(h) = \tfrac{1}{3}\pi (a^2 - 3h^2), \quad V'(h) = 0 \quad \Rightarrow \quad h = \dfrac{a}{\sqrt{3}}.$

Maximum volume $V\left(a/\sqrt{3}\right) = \tfrac{2}{27}\pi a^3 \sqrt{3}.$

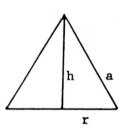

45.

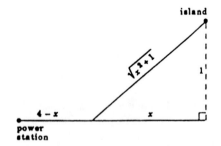

Minimize C

In units of $\$10{,}000$,

$$C = \frac{\text{cost of cable}}{\text{underground}} + \frac{\text{cost of cable}}{\text{under water}}$$

$$= 3(4 - x) \qquad + \quad 5\sqrt{x^2 + 1}.$$

Clearly, the cost is unnecessarily high if

$$x > 4 \quad \text{or} \quad x < 0.$$

$C(x) = 12 - 3x + 5\sqrt{x^2 + 1}, \quad 0 \leq x \leq 4.$

$C'(x) = -3 + \dfrac{5x}{\sqrt{x^2 + 1}}, \quad C'(x) = 0 \quad \Longrightarrow \quad x = 3/4.$

Since the domain of C is closed, the abs min can be identified by evaluating C at each critical point:

$$C(0) = 17, \quad C\left(\tfrac{3}{4}\right) = 16, \quad C(4) = 5\sqrt{17} \cong 20.6.$$

The minimum cost is $\$160,000$.

46. Maximize $\alpha - \theta$.

Since the tangent function is an increasing function

on $[0, \pi/2)$, it suffices to maximize $\tan(\alpha - \theta)$.

$$\tan(\alpha - \theta) = \frac{\tan\alpha - \tan\theta}{1 + \tan\alpha\,\tan\theta}; \quad \tan\alpha = \frac{16}{x}, \quad \tan\theta = \frac{9}{x}.$$

Thus, we maximize $T(x) = \dfrac{\dfrac{16}{x} - \dfrac{9}{x}}{1 + \dfrac{16}{x}\cdot\dfrac{9}{x}} = \dfrac{7x}{x^2 + 144}, \quad x > 0.$

$$T'(x) = \frac{7(144 - x^2)}{(x^2 + 144)^2}, \quad T'(x) = 0 \implies x = 12.$$

Stand 12 ft from the wall for the most favorable view.

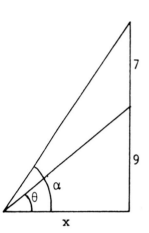

47. $P'(\theta) = \dfrac{-mW(m\cos\theta - \sin\theta)}{(m\sin\theta + \cos\theta)^2}; \quad P$ is minimized when $\tan\theta = m$.

48. $R(\theta) = \dfrac{2v^2}{g\cos^2\alpha}\cos\theta\,\sin(\theta - \alpha), \quad 0 < \theta < \tfrac{1}{2}\pi$

$$R'(\theta) = \frac{2v^2}{g\cos^2\alpha}\left[-\sin\theta\,\sin(\theta - \alpha) + \cos\theta\,\cos(\theta - \alpha)\right]$$

$$= \frac{2v^2}{g\cos^2\alpha}\cos(2\theta - \alpha)$$

$$R'(\theta) = 0 \implies 2\theta - \alpha = \tfrac{1}{2}\pi \implies \theta = \tfrac{1}{4}\pi + \tfrac{1}{2}\alpha.$$

49. Minimize $I = \dfrac{a}{x^2} + \dfrac{b}{(s - x)^2}$.

$$I'(x) = -\frac{2a}{x^3} + \frac{2b}{(s - x)^3}, \quad I'(x) = 0 \implies x = \frac{a^{\frac{1}{3}}s}{a^{\frac{1}{3}} + b^{\frac{1}{3}}}.$$

50. Minimize $D = (y - y_1)^2 + (x - x_1)^2$, where $y = -\dfrac{1}{B}(Ax + C)$.

$$D' = 2\left(-\frac{1}{B}(Ax + C) - y_1\right)\left(-\frac{A}{B}\right) + 2(x - x_1) = 0$$

$$\implies x = \frac{B^2 x_1 - AC - ABy_1}{A^2 + B^2} \text{ and thus } y = \frac{A^2 y_1 - BC - ABx_1}{A^2 + B^2}.$$

Thus $d = \sqrt{D} = \sqrt{\left(\dfrac{A^2 y_1 - BC - ABx_1}{A^2 + B^2} - y_1\right)^2 + \left(\dfrac{B^2 x_1 - AC - ABy_1}{A^2 + B^2} - x_1\right)^2}$

$$= \frac{|Ax_1 + By_1 + C|}{\sqrt{A^2 + B^2}}.$$

51. The slope of the line through (a, b) and $(x, f(x))$ is $\dfrac{f(x) - b}{x - a}$.

Let $D(x) = [x - a]^2 + [b - f(x)]^2$.   Then $D'(x) = 0$

$$\implies \quad 2[x - a] - 2[b - f(x)]f'(x) = 0$$

$$\implies \quad f'(x) = \frac{x - a}{b - f(x)}.$$

52. Let $P = (x_1, x_1^2)$ and $Q = (x_2, x_2^2)$.   The slope of the line PQ is $\dfrac{x_1^2 - x_2^2}{x_1 - x_2} = x_1 + x_2$.   This is

perpendicular to the slope of the tangent to the parabola through point P, which is $2x_1$.   Thus

$x_1 + x_2 = -\dfrac{1}{2x_1} \implies x_2 = -x_1 - \dfrac{1}{2x_1}$.   Now minimize:

$$D = (x_1^2 - x_2^2)^2 + (x_1 - x_2)^2 = (x_1^2 - x_2^2)^2(1 + (x_1 + x_2)^2)$$

$$= \left(2x_1 + \frac{1}{2x_1}\right)^2 \left(1 + \frac{1}{4x_1^2}\right)$$

Thus $D' = \left(2x_1 + \dfrac{1}{2x_1}\right)^2 \left(-\dfrac{1}{2x_1^3}\right) + 2\left(2x_1 + \dfrac{1}{2x_1}\right)\left(2 - \dfrac{1}{2x_1^2}\right)\left(1 + \dfrac{1}{4x_1^2}\right) = 0$

$$\implies \quad \left(2x_1 + \frac{1}{2x_1}\right)\left(\frac{1}{2x_1^3}\right) = 2\left(2 - \frac{1}{2x_1^2}\right)\left(1 + \frac{1}{4x_1^2}\right)$$

$$\implies \quad x_1 = \pm\frac{\sqrt{2}}{2}, \text{ and } y_1 = \frac{1}{2}.$$

53. Set $F(x) = 6x^4 - 16x^3 + 9x^2$. For integral values of x, $F(x) = f(x)$.

$F'(x) = 24x^3 - 48x^2 + 18x = 6x(4x^2 - 8x + 3) = 6x(2x - 1)(2x - 3)$

$F'(x) = 0$ at $x = 0,\ 1/2,\ 3/2$.

$F'(x) < 0$ on $(-\infty, 0) \cup (1/2, 3/2)$; $F'(x) > 0$ on $(0, 1/2) \cup (3/2, \infty)$. F has local minima at $x = 0$

and $x = 3/2$; F has a local maximum at $x = 1/2$. $F(0) = f(0) = 0$, $F(1) = f(1) = -1$, $F(2) = $

$f(2) = 4$; $n = 1$ minimizes $f(n)$.

54. Let x be the number of passengers and R the revenue in dollars.

$$R(x) = \begin{cases} 37x, & 16 \le x \le 35 \\ [37 - \frac{1}{2}(x - 35)]\,x, & 35 < x \le 48; \end{cases}$$

$$R'(x) = \begin{cases} 37, & 16 < x < 35 \\ \frac{109}{2} - x, & 35 < x < 48. \end{cases}$$

The critical number is 35. From $R(16) = 592$, $R(35) = 1295$, and $R(48) = 1464$ we conclude that

the revenue is maximized by taking a full load of 48 passengers.

55. Let x be the number of customers and P the net profit in dollars. Then $0 \leq x \leq 250$ and

$$P(x) = \begin{cases} 12x, & 0 \leq x \leq 50 \\ [12 - 0.06(x - 50)]\,x, & 50 < x \leq 250; \end{cases}$$

$$P'(x) = \begin{cases} 12, & 0 \leq x \leq 50 \\ 15 - 0.12x, & 50 < x \leq 250. \end{cases}$$

The critical points are: $x = 50$, $x = 125$. From $P(0) = 0$, $P(50) = 600$, $P(125) = 937.50$, and $P(250) = 0$, we conclude that the net profit is maximized by servicing 125 customers.

56. Maximize $P(x) = 2cy + cx$, where c is the price of low-grade steel and $y = \left(\dfrac{40 - 5x}{10 - x}\right)$.

$$P(x) = 2c(\frac{40 - 5x}{10 - x}) + cx$$

$$P'(x) = 2c\left[\frac{(10 - x)(-5) - (40 - 5x)(-1)}{(10 - x)^2}\right] + c,$$

$$P'(x) = 0 \implies x = 10 - \sqrt{20} \text{ or about } 5\frac{1}{2} \text{ tons.}$$

57. $y = mx - \dfrac{1}{400}(m^2 + 1)x^2$. When $y = 0$, $x = \dfrac{800m}{m^2 + 1}$.

Differentiating x with respect to m, $x' = \dfrac{800 - 800m^2}{(m^2 + 1)^2} = 0$

$\implies m = 1$.

58. When $x = 300$, $y = 300m - \dfrac{225}{2}(m^2 + 1)$.

Differentiating y with respect to m, $y' = 300 - 225m = 0$

$\implies m = \dfrac{4}{3}$.

59. Driving at ν mph, the trip takes $\dfrac{300}{\nu}$ hours and uses $\left(1 + \dfrac{1}{400}\nu^2\right)\dfrac{300}{\nu}$ gallons of fuel.

Thus, the expenses are:

$$E(\nu) = 2.60\left(1 + \frac{1}{400}\nu^2\right)\frac{300}{\nu} + 20\left(\frac{300}{\nu}\right) = 1.95\nu + \frac{6780}{\nu}, \quad 35 \leq x \leq 70.$$

Differentiating, we get

$$E'(\nu) = 1.95 - \frac{6780}{\nu^2}, \quad \text{and} \quad E'(\nu) = 0 \text{ at } \nu \cong 59$$

E is decreasing on $[35, 59]$ and increasing on $[59, 70]$; the minimal expenses occur when the truck is driven at 59 mph.

60. Let ν be the speed of the boat measured in kilometers per hour. A 100 kilometer trip will take $100/\nu$ hours.

Fixed costs: $F(\nu) = 2500 \left(\dfrac{100}{\nu} \right)$ dollars.

Fuel costs: $G(\nu) = k\nu^2$; $\quad 400 = k(10)^2$ implies $k = 4$. Therefore, $G(\nu) = 4\nu^2 \left(\dfrac{100}{\nu} \right)$ dollars.

Total cost: $C(\nu) = 2500 \left(\dfrac{100}{\nu} \right) + 4\nu^2 \left(\dfrac{100}{\nu} \right) = 100 \left(\dfrac{2500}{\nu} + 4\nu \right)$, $0 < \nu < \infty$. Differentiating, we get

$$C'(\nu) = 100 \left(-\frac{2500}{\nu^2} + 4 \right) = 100 \left(\frac{4\nu^2 - 2500}{\nu^2} \right) \text{ and } C'(\nu) = 0 \text{ at } \nu = 25.$$

C is decreasing on $(0, 25]$ and increasing on $[25, \infty)$; the speed that minimizes the expenses is 25 kilometers per hour. If we replace the 100 kilometers by M kilometers, the total cost will be

$$C(\nu) = M \left(\frac{2500}{\nu} + 4\nu \right)$$

and we will get exactly the same result. The minimizing speed is independent of the length of the trip.

61. Minimize $SA = 2\pi rh + 2\pi r^2$, where $2r \le h < 6$ and $\pi r^2 h = 16\pi$ (hence $h = \dfrac{16}{r^2}$).

Thus $SA = \dfrac{32\pi}{r} + 2\pi r^2$. Differentiating, $\quad SA' = -\dfrac{32\pi}{r^2} + 4\pi r = 0$

$\implies \quad r^3 = 8$, so $r = 2$ feet and $h = 4$ feet. Thus no minimum exists.

62. We want to maximize the ratio $\dfrac{\text{income}}{\text{cost}} = \dfrac{200,000n}{1,000,000n + 100,000(1 + 2 + \cdots + n - 1) + 5,000,000}$

$= \dfrac{2n}{10n + \frac{1}{2}(n-1)n + 50} = \dfrac{4n}{n^2 + 19n + 100}$.

Let $f(x) = \dfrac{4x}{x^2 + 19x + 100}$, $x > 0$

Then $f'(x) = \dfrac{(x^2 + 19x + 100)4 - 4x(2x + 19)}{(x^2 + 19x + 100)^2} = \dfrac{4(100 - x^2)}{(x^2 + 19x + 100)^2} = 0$

$\implies \quad x = 10$.

Since $f'(x) > 0$ for $x < 0$, $\quad f'(x) < 0$ for $x > 10$, f has an absolute maximum at $x = 10$.

A ten story building provides the greatest return on investment.

63. Swimming at 2 miles per hour, Maggie will reach point B in 1 hour; walking along the shore (a distance of π miles), she will reach point B in $\pi/5 \cong 0.63$ hours. Suppose that she swims to a point C and then walks to B. Let θ be the central angle (measured in radians) determined by the points B and C. See the figure. By the law of cosines, the square of the distance from A to C is given by

$$d^2 = 1^2 + 1^2 - 2(1)(1) \cos (\pi - \theta) = 2 + 2 \cos \theta.$$

Therefore the distance d from A to C is $d = \sqrt{2 + 2 \cos \theta}$. The distance from C to B is θ.

Now, the total length of time to swim to C and then walk to B is

$$T(\theta) = 2\sqrt{2 + 2\cos\theta} + 5\theta$$

Maggie wants to minimize T.

$$T'(\theta) = \frac{-2\sin\theta}{2\sqrt{2 + 2\cos\theta}} + 5.$$

Setting $T'(\theta) = 0$, we get

$$\frac{-2\sin\theta}{2\sqrt{2 + 2\cos\theta}} + 5 = 0$$

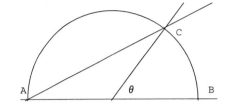

which reduces to $2\cos^2\theta + 25\cos\theta + 23 = 0$

and factors into $(\cos\theta + 1)(2\cos\theta + 23) = 0$.

This implies $\theta = \pi$. Maggie should walk the entire distance!

64. We modify the solution of Exercise 63, replacing the walking rate of 2 miles per hour by the rowing rate of 3 miles per hour.

The total length of time to row to C and then walk to B is

$$T(\theta) = 3\sqrt{2 + 2\cos\theta} + 5\theta$$

The derivative of T is

$$T'(\theta) = \frac{-3\sin\theta}{2\sqrt{2 + 2\cos\theta}} + 5.$$

Setting $T'(\theta) = 0$, we get

$$\frac{-3\sin\theta}{2\sqrt{2 + 2\cos\theta}} + 5 = 0,$$

which reduces to

$$9\cos^2\theta + 50\cos\theta + 41 = (\cos\theta + 1)(9\cos\theta + 41) = 0.$$

This implies $\theta = \pi$. Again, Maggie should walk the entire distance!

65. (b) The point on the graph of f that is closest to P is: $\left(1 + \sqrt{2},\, 2 + \sqrt{2}\right)$

   (c) $l_{PQ}: \quad y - \left(2 + \sqrt{2}\right) = \dfrac{1 - \sqrt{2}}{3 - \sqrt{2}}\left(x - \left[1 + \sqrt{2}\right]\right)$

   (d) and (e) $l_{PQ} = l_N$.

66. The point on the graph of $f(x) = x - x^3$ that is closest to $(1, 8)$ is (approximately) $(-2.1474, 7.7548)$.

67. $D(x) = \sqrt{x^2 + (7 - 3x)^2}$;   the point on the line that is closest to the origin is: $\left(\frac{21}{10}, \frac{7}{10}\right)$

68. The point on the graph of f that minimizes the distance to the point $(4, 3)$ is (approximately)

$(1.3918, 2.0629)$.

PROJECT 4.5

1. Distance over water: $\sqrt{36 + x^2}$.

Distance over land: $12 - x$.

Total energy: $E(x) = W\sqrt{36 + x^2} + L(12 - x), \quad 0 \le x \le 12$.

2. $W = 1.5L$, so $E(x) = 1.5L\sqrt{36 + x^2} + L(12 - x)$, for $0 \le x \le 12$.

$E'(x) = \dfrac{1.5Lx}{\sqrt{36 + x^2}} - L = 0 \implies x = \dfrac{12}{\sqrt{5}} \simeq 5.36$.

$E'(x) < 0$ on $(0, \dfrac{12}{\sqrt{5}})$ and $E'(x) > 0$ on $(\dfrac{12}{\sqrt{5}}, 12)$, so E has an absolute minimum at $\dfrac{12}{\sqrt{5}}$.

3. (a) $W = kL, \ k > 1$, so $E(x) = kL\sqrt{36 + x^2} + L(12 - x)$, for $0 \le x \le 12$.

$E'(x) = \dfrac{kLx}{\sqrt{36 + x^2}} - L = 0 \implies x = \dfrac{6}{\sqrt{k^2 - 1}}$.

$E'(x) < 0$ on $(0, \dfrac{6}{\sqrt{k^2 - 1}})$ and $E'(x) > 0$ on $(\dfrac{6}{\sqrt{k^2 - 1}}, 12)$, so E has an absolute minimum

at $\dfrac{6}{\sqrt{k^2 - 1}}$.

(b) As k increases, x decreases. As $k \to 1^+$, x increases.

(c) $x = 12 \implies k = \dfrac{\sqrt{5}}{2} \simeq 1.12$.

(d) No

SECTION 4.6

1. (a) f is increasing on $[a, b], \ [d, n]$; f is decreasing on $[b, d], \ [n, p]$.

(b) The graph of f is concave up on $(c, k), \ (l, m)$;

The graph of f is concave down on $(a, c), \ (k, l), \ (m, p)$.

The x-coordinates of the points of inflection are: $x = c, \ x = k, \ x = l, \ x = m$.

2. (a) g is increasing on $[a, b], \ [c, e], \ [m, n]$; g is decreasing on $[b, c], \ [e, m]$.

(b) The graph of g is concave up on $(a, b), \ (b, d)$;

The graph of g is concave down on $(d, m), \ (m, n)$.

The x-coordinate of the point of inflection is: $x = d$.

3. $(i) \ f', \quad (ii) \ f, \quad (iii) \ f''$.

4. (a) $x = -2$ local max,   no local minima.

(b) points of inflection at $x = 0, \ 2$.

(c)

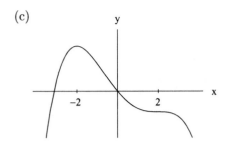

5. $f'(x) = -x^{-2}, \quad f''(x) = 2x^{-3};$

    concave down on $(-\infty, 0)$, concave up on $(0, \infty)$;   no pts of inflection

6. $f'(x) = 1 - x^{-2}, \quad f''(x) = 2x^{-3};$

    concave down on $(-\infty, 0)$, concave up on $(0, \infty)$;   no pts of inflection

7. $f'(x) = 3x^2 - 3, \quad f''(x) = 6x;$

    concave down on $(-\infty, 0)$, concave up on $(0, \infty)$;   pt of inflection $(0, 2)$

8. $f'(x) = 4x - 5, \quad f''(x) = 4;$ concave up on $(-\infty, \infty)$

9. $f'(x) = x^3 - x, \quad f''(x) = 3x^2 - 1;$

    concave up on $\left(-\infty, -\frac{1}{3}\sqrt{3}\right)$ and $\left(\frac{1}{3}\sqrt{3}, \infty\right)$, concave down on $\left(-\frac{1}{3}\sqrt{3}, \frac{1}{3}\sqrt{3}\right)$;

    pts of inflection $\left(-\frac{1}{3}\sqrt{3}, -\frac{5}{36}\right)$ and $\left(\frac{1}{3}\sqrt{3}, -\frac{5}{36}\right)$

10. $f'(x) = 3x^2 - 4x^3, \quad f''(x) = 6x - 12x^2 = 6x(1 - 2x);$

    concave down on $(-\infty, 0)$, and $\left(\frac{1}{2}, \infty\right)$, concave up on $\left(0, \frac{1}{2}\right)$;

    pts of inflection $(0,0)$, $\left(\frac{1}{2}, \frac{1}{16}\right)$

11. $f'(x) = -\dfrac{x^2 + 1}{(x^2 - 1)^2}, \quad f''(x) = \dfrac{2x\left(x^2 + 3\right)}{(x^2 - 1)^3};$

    concave down on $(-\infty, -1)$ and $(0, 1)$, concave up on $(-1, 0)$ and on $(1, \infty)$;

    pts of inflection $(0, 0)$

12. $f'(x) = \dfrac{-4}{(x - 2)^2}, \quad f''(x) = \dfrac{8}{(x - 2)^3};$

    concave down on $(-\infty, 2)$, concave up on $(2, \infty)$;   no pts of inflection

13. $f'(x) = 4x^3 - 4x, \quad f''(x) = 12x^2 - 4;$

    concave up on $\left(-\infty, -\frac{1}{3}\sqrt{3}\right)$ and $\left(\frac{1}{3}\sqrt{3}, \infty\right)$, concave down on $\left(-\frac{1}{3}\sqrt{3}, \frac{1}{3}\sqrt{3}\right)$;

    pts of inflection $\left(-\frac{1}{3}\sqrt{3}, \frac{4}{9}\right)$ and $\left(\frac{1}{3}\sqrt{3}, \frac{4}{9}\right)$

14. $f'(x) = \dfrac{6(1-x^2)}{(x^2+1)^2}, \quad f''(x) = \dfrac{12x(x^2-3)}{(x^2+1)^3}$;

concave down on $\left(-\infty, -\sqrt{3}\right)$ and $\left(0, \sqrt{3}\right)$, concave up on $\left(-\sqrt{3}, 0\right)$ and $\left(\sqrt{3}, \infty\right)$;

pts of inflection $\left(-\sqrt{3}, -\frac{3}{2}\sqrt{3}\right)$, $(0,0)$, $\left(\sqrt{3}, \frac{3}{2}\sqrt{3}\right)$

15. $f'(x) = \dfrac{-1}{\sqrt{x}\,(1+\sqrt{x}\,)^2}, \quad f''(x) = \dfrac{1+3\sqrt{x}}{2x\sqrt{x}\,(1+\sqrt{x}\,)^3}$;

concave up on $(0, \infty)$;   no pts of inflection

16. $f'(x) = \frac{1}{5}(x-3)^{-4/5}, \quad f''(x) = -\frac{4}{25}(x-3)^{-9/5}$;

concave up on $(-\infty, 3)$, concave down on $(3, \infty)$;   pt of inflection $(3, 0)$

17. $f'(x) = \frac{5}{3}(x+2)^{2/3}, \quad f''(x) = \frac{10}{9}(x+2)^{-1/3}$;

concave down on $(-\infty, -2)$, concave up on $(-2, \infty)$;   pt of inflection $(-2, 0)$

18. $f'(x) = \dfrac{4-2x^2}{(4-x^2)^{1/2}}, \quad f''(x) = \dfrac{2x(x^2-6)}{(4-x^2)^{3/2}} \qquad$ Note: $\text{dom}\,(f) = [-2, 2]$

concave up on $(-2, 0)$, concave down on $(0, 2)$;   pt of inflection $(0, 0)$

19. $f'(x) = 2\sin x \cos x = \sin 2x, \quad f''(x) = 2\cos 2x$;

concave up on $\left(0, \frac{1}{4}\pi\right)$ and $\left(\frac{3}{4}\pi, \pi\right)$, concave down on $\left(\frac{1}{4}\pi, \frac{3}{4}\pi\right)$;

pts of inflection $\left(\frac{1}{4}\pi, \frac{1}{2}\right)$ and $\left(\frac{3}{4}\pi, \frac{1}{2}\right)$

20. $f'(x) = -4\cos x \sin x - 2x, \quad f''(x) = -4(\cos^2 x - \sin^2 x) - 2 = -4\cos 2x - 2$;

concave down on $\left(0, \frac{1}{3}\pi\right)$ and $\left(\frac{2}{3}\pi, \pi\right)$, concave up on $\left(\frac{1}{3}\pi, \frac{2}{3}\pi\right)$;

pts of inflection $\left(\dfrac{1}{3}\pi, \dfrac{9-2\pi^2}{18}\right)$ and $\left(\dfrac{2}{3}\pi, \dfrac{9-8\pi^2}{18}\right)$

21. $f'(x) = 2x + 2\cos 2x, \quad f''(x) = 2 - 4\sin 2x$;

concave up on $\left(0, \frac{1}{12}\pi\right)$ and on $\left(\frac{5}{12}\pi, \pi\right)$, concave down on $\left(\frac{1}{12}\pi, \frac{5}{12}\pi\right)$;

pts of inflection $\left(\dfrac{1}{12}\pi, \dfrac{72+\pi^2}{144}\right)$ and $\left(\dfrac{5}{12}\pi, \dfrac{72+25\pi^2}{144}\right)$

22. $f'(x) = 4\sin^3 x \cos x, \quad f''(x) = 4\sin^2 x[3\cos^2 x - \sin^2 x]$;

concave up on $\left(0, \frac{1}{3}\pi\right)$ and $\left(\frac{2}{3}\pi, \pi\right)$, concave down on $\left(\frac{1}{3}\pi, \frac{2}{3}\pi\right)$;

pts of inflection $\left(\frac{1}{3}\pi, \frac{9}{16}\right)$ and $\left(\frac{2}{3}\pi, \frac{9}{16}\right)$

23. points of inflection: $(\pm 3.94822, 10.39228)$

24. $f''(x) = 0$ at $x \cong \pm 0.94, \ \pm 2.57, \ \pm 3.71, \ \pm 5.35$

25. points of inflection: $(-3, 0), \ (-2.11652, 2, 39953), \ (-0.28349, -18.43523)$

26. $f''(x) > 0$ for all $x \in \text{dom}\, f$

27. $f(x) = x^3 - 9x$

(a) $f'(x) = 3x^2 - 9 = 3(x^2 - 3)$

$f'(x) \geq 0 \Rightarrow x \leq -\sqrt{3}$ or $x \geq \sqrt{3}$;

$f'(x) \leq 0 \Rightarrow -\sqrt{3} \leq x \leq \sqrt{3}$.

Thus, f is increasing on $(-\infty, -\sqrt{3}] \cup [\sqrt{3}, \infty)$

and decreasing on $[-\sqrt{3}, \sqrt{3}]$.

(b) $f(-\sqrt{3}) \cong 10.39$ is a local maximum;

$f(\sqrt{3}) \cong -10.39$ is a local minimum.

(c) $f''(x) = 6x$;

The graph of f is concave up on $(0, \infty)$ and concave down on $(-\infty, 0)$.

(d) point of inflection: $(0, 0)$

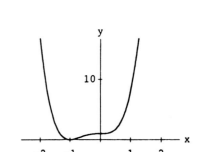

28. $f(x) = 3x^4 + 4x^3 + 1$

(a) $f'(x) = 12x^3 + 12x^2 = 12x^2(x + 1)$

$f'(x) \geq 0 \Rightarrow x \geq -1$;

$f'(x) \leq 0 \Rightarrow x \leq -1$.

Thus, f is increasing on $[-1, \infty)$

and decreasing on $(-\infty, -1]$.

(b) $f(-1) = 0$ is a local minimum;

no local maxima.

(c) $f''(x) = 36x^2 + 24x = 36x \left(x + \frac{2}{3}\right)$;

The graph of f is concave up on $\left(-\infty, -\frac{2}{3}\right)$ and $(0, \infty)$; and concave down on $\left(-\frac{2}{3}, 0\right)$.

(d) points of inflection: $\left(-\frac{2}{3}, \frac{11}{27}\right)$, $(0, 1)$

29. $f(x) = \dfrac{2x}{x^2 + 1}$

(a) $f'(x) = -\dfrac{2(x + 1)(x - 1)}{(x^2 + 1)^2}$

$f'(x) \geq 0 \Rightarrow -1 \leq x \leq 1$;

$f'(x) \leq 0 \Rightarrow x \leq -1$ or $x \geq 1$.

Thus, f is increasing on $[-1, 1]$;

and decreasing on $(-\infty, -1] \cup [1, \infty)$.

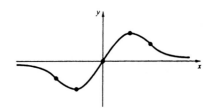

(b) $f(-1) = -1$ is a local minimum;

$f(1) = 1$ is a local maximum.

(c) $f''(x) = \dfrac{4x(x + \sqrt{3})(x - \sqrt{3})}{(x^2 + 1)^3}$

$f''(x) > 0 \Rightarrow x \leq -\sqrt{3}$ or $x \geq \sqrt{3}$;

$f''(x) < 0 \Rightarrow -\sqrt{3} < x < \sqrt{3}$.

The graph of f is concave up on $(-\sqrt{3}, 0) \cup (\sqrt{3}, \infty)$ and concave down on $(-\infty, -\sqrt{3}) \cup (0, \sqrt{3})$.

(d) points of inflection: $(-\sqrt{3}, -\sqrt{3}/2)$, $(0, 0)$, $(\sqrt{3}, \sqrt{3}/2)$

30. $f(x) = x^{1/3}(x-6)^{2/3}$

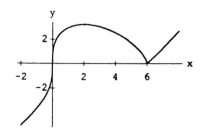

(a) $f'(x) = \dfrac{x-2}{x^{2/3}(x-6)^{1/3}}$

$f'(x) \geq 0 \Rightarrow x \leq 2, \ x \neq 0, \ \text{or} \ x > 6;$

$f'(x) \leq 0 \Rightarrow 2 \leq x < 6.$

Thus, f is increasing on $(-\infty, 2] \cup [6, \infty)$
and decreasing on $[2, 6]$.

(b) $f(2) = 2(4)^{1/3}$ is a local maximum;

$f(6) = 0$ is a local minimum.

(c) $f''(x) = \dfrac{-8}{x^{5/3}(x-6)^{4/3}};$

The graph of f is concave down on $(0, \infty)$ and concave down up $(-\infty, 0)$.

(d) point of inflection: $(0, 0)$

31. $f(x) = x + \sin x, \quad x \in [-\pi, \pi]$

(a) $f'(x) = 1 + \cos x$

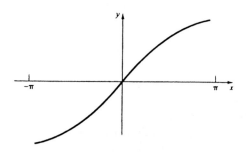

$f'(x) > 0$ on $(-\pi, \pi)$

Thus, f is increasing on $[-\pi, \pi]$.

(b) No local extrema

(c) $f''(x) = -\sin x$

$f''(x) > 0$ for $x \in (-\pi, 0);$

$f''(x) < 0$ for $x \in (0, \pi).$

The graph of f is concave up on $(-\pi, 0)$ and concave down on $(0, \pi)$.

(d) point of inflection: $(0, 0)$

32. $f(x) = \sin x + \cos x, \quad x \in [0, 2\pi]$

(a) $f'(x) = \cos x - \sin x$

$f'(x) \geq 0 \quad \Rightarrow \quad 0 \leq x \leq \frac{1}{4}\pi \ \text{or} \ \frac{5}{4}\pi \leq x \leq 2\pi$

$f'(x) \leq 0 \quad \Rightarrow \quad \frac{1}{4}\pi \leq x \leq \frac{5}{4}\pi$

Thus, f is increasing on $\left[0, \frac{1}{4}\pi\right] \cup \left[\frac{5}{4}\pi, 2\pi\right];$

f is decreasing on $\left[\frac{1}{4}\pi, \frac{5}{4}\pi\right].$

(b) $f(\pi/4) = \sqrt{2}$ is a local maximum;

$f(5\pi/4) = -\sqrt{2}$ is a local minimum.

(c) $f''(x) = -\sin x - \cos x$

$f''(x) > 0 \quad \Rightarrow \quad \frac{3}{4}\pi < x < \frac{7}{4}\pi;$

$f''(x) < 0 \quad \Rightarrow \quad 0 < x < \frac{3}{4}\pi \ \text{or} \ \frac{7}{4}\pi < x < 2\pi.$

The graph of f is concave up on $\left(\frac{3}{4}\pi, \frac{7}{4}\pi\right)$ and concave down on $\left(0, \frac{3}{4}\pi\right) \cup \left(\frac{7}{4}\pi, 2\pi\right).$

(d) points of inflection: $\left(\frac{3}{4}\pi, 0\right), \ \left(\frac{7}{4}\pi, 0\right)$

33. $f(x) = \begin{cases} x^3, & x < 1 \\ 3x - 2, & x \geq 1. \end{cases}$

(a) $f'(x) = \begin{cases} 3x^2, & x < 1 \\ 3, & x \geq 1; \end{cases}$

$f'(x) > 0$ on $(-\infty, 0) \cup (0, \infty)$

Thus, f is increasing on $(-\infty, \infty)$.

(b) No local extrema

(c) $f''(x) = \begin{cases} 6x, & x < 1 \\ 0, & x \geq 1; \end{cases}$

$f''(x) > 0$ for $x \in (0, 1);$ $f''(x) < 0$ for $x \in (-\infty, 0)$.

Thus, the graph of f is concave up on $(0, 1)$ and concave down on $(-\infty, 0)$.

The graph of f is a straight line for $x \geq 1$.

(d) point of inflection: $(0, 0)$

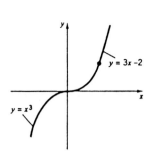

34. $f(x) = \begin{cases} 2x + 4, & x \leq -1 \\ 3 - x^2, & x > -1. \end{cases}$

(a) $f'(x) = \begin{cases} 2, & x \leq -1 \\ -2x, & x > -1; \end{cases}$

$f'(x) \geq 0$ on $(-\infty, 0];$

$f'(x) \leq 0$ on $[0, \infty)$.

Thus, f is increasing on $(-\infty, 0]$ and decreasing on $[0, \infty)$.

(b) $f(0) = 3$ is a local maximum.

(c) $f''(x) = \begin{cases} 0, & x < -1 \\ -2, & x > -1; \end{cases}$

$f''(x) < 0$ for $x \in (-1, \infty)$.

Thus, the graph of f is concave down on $(-1, \infty)$.

The graph of f is a straight line for $x \leq -1$.

(d) There are no points of inflection.

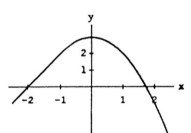

35.

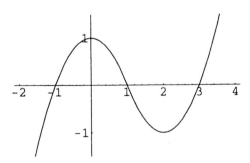

36.

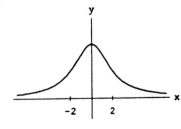

37.

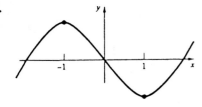

38.

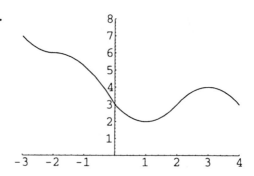

39. Since $f''(x) = 6x - 2(a + b + c)$, set $d = \frac{1}{3}(a + b + c)$. Note that $f''(d) = 0$ and that f is concave down on $(-\infty, d)$ and concave up on (d, ∞); $(d, f(d))$ is a point of inflection.

40. $f'(x) = 2cx - 2x^{-3}$, $f''(x) = 2c + 6x^{-4}$. To have a point of inflection at 1 we need
$$f''(1) = 0 \implies 2c + 6 = 0 \implies c = -3$$

41. Since $(-1, 1)$ lies on the graph, $1 = -a + b$.
Since $f''(x)$ exists for all x and there is a pt of inflection at $x = \frac{1}{3}$, we must have $f''\left(\frac{1}{3}\right) = 0$.
Therefore
$$0 = 2a + 2b.$$
Solving these two equations, we find $a = -\frac{1}{2}$ and $b = \frac{1}{2}$.
Verification: the function
$$f(x) = -\frac{1}{2}x^3 + \frac{1}{2}x^2$$
has second derivative $f''(x) = -3x + 1$. This does change sign at $x = \frac{1}{3}$.

42. $f(x) = Ax^{1/2} + Bx^{-1/2}$; $f(1) = 4 \implies A + B = 4$.
$f'(x) = \frac{1}{2}Ax^{-1/2} - \frac{1}{2}Bx^{-3/2}$, $f''(x) = -\frac{1}{4}Ax^{-3/2} + \frac{3}{4}Bx^{-5/2}$.
To have a point of inflection at $(1, 4)$, we need $f''(1) = 0 \implies -\frac{1}{4}A + \frac{3}{4}B = 0$.
Solving the two equations gives $A = 3$, $B = 1$.

43. First, we require that $\left(\frac{1}{6}\pi, 5\right)$ lie on the curve:
$$5 = \frac{1}{2}A + B.$$
Next we require that $\dfrac{d^2y}{dx^2} = -4A\cos 2x - 9B\sin 3x$ be zero (and change sign) at $x = \frac{1}{6}\pi$:
$$0 = -2A - 9B.$$
Solving these two equations, we find $A = 18$, $B = -4$.
Verification: the function
$$f(x) = 18\cos 2x - 4\sin 3x$$
has second derivative $f''(x) = -72\cos 2x + 36\sin 3x$. This does change sign at $x = \frac{1}{6}\pi$.

44. $f(x) = Ax^2 + Bx + C;$ $f'(x) = 2Ax + B;$ $f''(x) = 2A.$

(a) Concave up $\implies$ $f''(x) > 0$ $\implies$ $A > 0;$ to decrease between A and B we need

$$f'(x) < 0, \text{ for } x \text{ between } A \text{ and } B \implies B \le -2A^2.$$

(b) Concave down $\implies$ $f''(x) < 0$ $\implies$ $A < 0;$ to have $f'(x) > 0$ for x between A and B we need $2A^2 + B \ge 0$ and $2AB + B \ge 0$, that is, $B \ge -2A^2$ and $B(2A + 1) \ge 0$. If $A > -\frac{1}{2}$, then we need $B \ge 0$ (and automatically $B \ge -2A^2$). If $A \le -\frac{1}{2}$, then we need $B \le 0$ and $B \ge -2A^2$. The conditions are: $-\frac{1}{2} < A < 0,\ B \ge 0,$ or $A \le -\frac{1}{2},\ -2A^2 \le B \le 0.$

45. Let $f'(x) = 3x^2 - 6x + 3$. Then we must have $f(x) = x^3 - 3x^2 + 3x + c$ for some constant c. Note that $f''(x) = 6x - 6$ and $f''(1) = 0$. Since $(1, -2)$ is a point of inflection of the graph of f, $(1, -2)$ must lie on the graph. Therefore,

$$1^3 - 3(1)^2 + 3(1) + c = -2 \text{ which implies } c = -3$$

and so $f(x) = x^3 - 3x^2 + 3x - 3.$

46. $f'(x) = \cos x$ and $f''(x) = -\sin x = -f(x).$
Thus f is concave down when $f''(x) < 0$ $\implies$ $f(x) > 0.$
Similarly, f is concave up when $f''(x) > 0$ $\implies$ $f(x) < 0.$
$g'(x) = -\sin x$ and $g''(x) = -\cos x = -g(x)$, so $g(x)$ has the same property.

47. (a) $p''(x) = 6x + 2a$ is negative for $x < -a/3$, and positive for $x > -a/3$. Therefore, the graph of p has a point of inflection at $x = -a/3$.

(b) $p'(x) = 3x^2 + 2ax + b$. The discriminant of this quadratic is $4a^2 - 12b = 4(a^2 - 3b)$. Thus, p' has two real zeros iff $a^2 > 3b$.

(c) If $a^2 \le 3b$, then $b \ge a^2/3$ and $p'(x) = 3x^2 + 2ax + b \ge 3x^2 + 2ax + a^2/3 = 3(x + \frac{1}{3}a)^2 \ge 0$ for all x; p is increasing in this case.

48. It is sufficient to show that the x-coordinate of the point of inflection is the x-coordinate of the mid-point of the line segment connecting the local extrema. It is easy to show that the x-coordinate of the point of inflection is $x_0 = -\frac{1}{3} a$. Now suppose that p has local extrema at x_1 and x_2, $x_1 \ne x_2$. Then

$$p'(x_1) = p'(x_2) = 0 \quad \Rightarrow \quad 3x_1^2 + 2ax_1 + b - (3x_2^2 + 2ax_2 + b) = 0 \quad \Rightarrow \quad x_1 + x_2 = -\frac{2}{3} a.$$

Thus, $\dfrac{x_1 + x_2}{2} = -\frac{1}{3} a = x_0.$

49. (a)

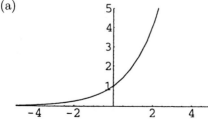

(b) No.   If $f''(x) < 0$ and $f'(x) < 0$ for all x, then
$f(x) < f'(0)x + f(0)$ on $(0, \infty)$, which implies that
$f(x) \to -\infty$ as $x \to \infty.$

50. Let $f(x) = a_n x^n + a_{n-1} x^{n-1} + \cdots + a_2 x^2 + a_1 x + a_0$.

Then $f''(x) = n(n-1)a_n x^{n-2} + \cdots + 2a_2$ is a degree $n-2$ polynomial which can have at most $n-2$ roots.   Hence f has at most $n-2$ points of inflection.

51.

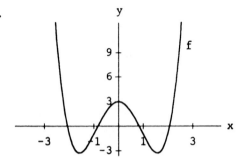

(a) concave up on $(-4, -0.913) \cup (0.913, 4)$

concave down on $(-0.913, 0.913)$

(b) pts of inflection at $x = -0.913, \ 0.913$

52.

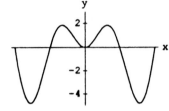

(a) up: $(-2\pi, -3.64), \ (-1.077, 1.077), \ (3.64, 2\pi)$

down: $(-3.64, -1.077) \cup (1.077, 3.64)$

(b) pts of inflection at $x = \pm 3.64, \ \pm 1.077$

53.

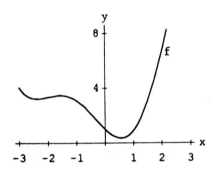

(a) concave up on:

$(-\pi, -1.996) \cup (-.0345, 2.550)$

concave down on:

$(-1.996, -0.345) \cup (2.550, \pi)$

(b) pts of inflection at:

$x = -1.996, \ -0.345, \ 2,550$

54.

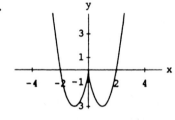

(a) concave up on $(-5, 5)$

(b) no pts of inflection

55. (a) $f''(x) = 0$ at $x = 0.68824, \ 2.27492, \ 4.00827, \ 5.59494$

(b) $f''(x) > 0$ on $(0.68824, 2.27492) \cup (4.00827, 5.59494)$

(c) $f''(x) < 0$ on $(0, 0.68824) \cup (2.27492, 4.00827) \cup (5.59494, 2\pi)$

56. (a) $f''(x) \neq 0$ for all x (b) $f''(x) > 0$ on $(-\infty, -1) \cup (1, \infty)$

(c) $f''(x) < 0$ on $(-1, 1)$

57. (a) $f''(x) = 0$ at $x = \pm 1, \ 0, \ \pm 0.32654, \ \pm 0.71523$

(b) $f''(x) > 0$ on $(-0.71523, -0.32654) \cup (0, 0.32654) \cup (0.71523, 1) \cup (1, \infty)$

(c) $f''(x) < 0$ on $(-\infty, -1) \cup (-1, -0.71523) \cup (-0.32654, 0) \cup (0, 32654, 0.71523)$

58. (a) $f''(x) = 0$ at $x = 0$ (b) $f''(x) > 0$ on $(-4, 0)$
(c) $f''(x) < 0$ on $(0, 4)$

SECTION 4.7

1. (a) ∞ (b) $-\infty$ (c) ∞ (d) 1
(e) 0 (f) $x = -1$, $x = 1$ (g) $y = 0$, $y = 1$

2. (a) d (b) c (c) $x = a$, $x = b$
(d) $y = d$ (e) p (f) q

3. vertical: $x = \frac{1}{3}$; horizontal: $y = \frac{1}{3}$

4. vertical: $x = -2$; horizontal: none

5. vertical: $x = 2$; horizontal: none

6. vertical: none; horizontal: $y = 0$

7. vertical: $x = \pm 3$; horizontal: $y = 0$

8. vertical: $x = 16$; horizontal: $y = 0$

9. vertical: $x = -\frac{4}{3}$; horizontal: $y = \frac{4}{9}$

10. vertical: $x = \frac{1}{3}$; horizontal: $y = \frac{4}{9}$

11. vertical: $x = \frac{5}{2}$; horizontal: $y = 0$

12. vertical: $x = \frac{1}{2}$; horizontal: $y = -\frac{1}{8}$

13. vertical: none; horizontal: $y = \pm \frac{3}{2}$

14. vertical: $x = 8$; horizontal: $y = 0$

15. vertical: $x = 1$; horizontal: $y = 0$

16. vertical: $x = \pm 1$; horizontal: $y = \pm 2$

17. vertical: none; horizontal: $y = 0$

18. vertical: none; horizontal: $y = 0$

19. vertical: $x = \left(2n + \frac{1}{2}\right)\pi$; horizontal: none

20. vertical: $x = 2n\pi$; horizontal: none

21. $f'(x) = \frac{4}{3}(x + 3)^{1/3}$; neither

22. $f'(x) = \frac{2}{5}x^{-3/5}$; cusp

23. $f'(x) = -\frac{4}{5}(2 - x)^{-1/5}$; cusp

24. neither; $f(-1)$ is not defined

25. $f'(x) = \frac{6}{5}x^{-2/5}\left(1 - x^{3/5}\right)$; tangent

26. $f'(x) = \frac{7}{5}(x - 5)^{2/5}$; neither

27. $f(-2)$ undefined; neither

28. $f'(x) = \frac{3}{7}(2 - x)^{-4/7}$; tangent

29. $f'(x) = \begin{cases} \frac{1}{2}(x - 1)^{-1/2}, & x > 1 \\ -\frac{1}{2}(1 - x)^{-1/2}, & x < 1; \end{cases}$ cusp

30. $f'(x) = (4x - 3)(x - 1)^{-2/3}$;   tangent

31. $f'(x) = \begin{cases} \frac{1}{3}(x+8)^{-23}, & x > -8 \\ -\frac{1}{3}(x+8)^{-2/3}, & x < -8; \end{cases}$ cusp

32. f is not defined for $x > 2$; neither

33. f not continuous at 0; neither

34. f is not continuous at 0; neither

35.

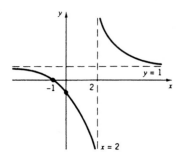

36.

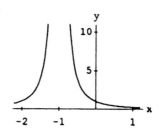

37.

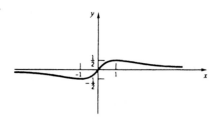

38.

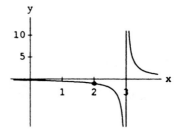

39. $f(x) = x - 3x^{1/3}$

(a) $f'(x) = 1 - \dfrac{1}{x^{2/3}}$

f is increasing on $(-\infty, -1] \cup [1, \infty)$

f is decreasing on $[-1, 1]$

(b) $f''(x) = \frac{2}{3}x^{-5/3}$

concave up on $(0, \infty)$; concave down on $(-\infty, 0)$

vertical tangent at $(0,0)$

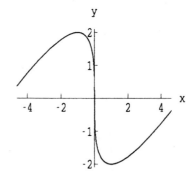

40. $f(x) = x^{2/3} - x^{1/3}$

(a) $f'(x) = \frac{2}{3}x^{-1/3} - \frac{1}{3}x^{-2/3} = \dfrac{2x^{1/3} - 1}{3x^{2/3}}$

f is increasing on $\left[\frac{1}{8}, \infty\right)$

f is decreasing on $\left(-\infty, \frac{1}{8}\right]$

(b) $f''(x) = -\frac{2}{9}x^{-4/3} + \frac{2}{9}x^{-5/3} = \dfrac{2(1 - x^{1/3})}{9x^{5/3}}$

concave up on $(0, 1)$

concave down on $(-\infty, 0) \cup (1, \infty)$

vertical tangent at $(0,0)$

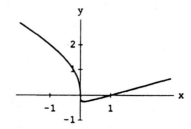

41. $f(x) = \frac{3}{5} x^{5/3} - 3x^{2/3}$

    (a) $f'(x) = x^{2/3} - 2x^{-1/3} = \dfrac{x-2}{x^{1/3}}$

       f is increasing on $(-\infty, 0] \cup [2, \infty)$

       f is decreasing on $[0, 2]$

    (b) $f''(x) = \frac{2}{3} x^{-1/3} + \frac{2}{3} x^{-4/3} = \dfrac{2x+2}{3x^{4/3}}$

       concave up on $(-1, 0) \cup (0, \infty)$; concave down on

       $(-\infty, -1)$

       vertical cusp at $(0, 0)$

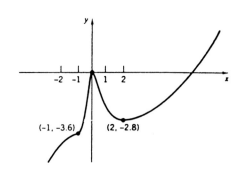

42. $f(x) = \sqrt{|x|}$

$$= \begin{cases} x^{1/2}, & x \geq 0 \\ (-x)^{1/2}, & x < 0. \end{cases}$$

    (a) $f'(x) = \begin{cases} \frac{1}{2} x^{-1/2}, & x > 0 \\ -\frac{1}{2} (-x)^{-1/2}, & x < 0; \end{cases}$

       f is increasing on $[0, \infty)$

       f is decreasing on $(-\infty, 0]$

    (b) $f''(x) = \begin{cases} -\frac{1}{4} x^{-3/2}, & x > 0 \\ -\frac{1}{4} (-x)^{-3/2}, & x < 0; \end{cases}$

       concave down on $(-\infty, 0) \cup (0, \infty)$

       vertical cusp at $(0, 0)$

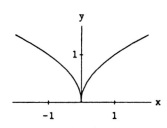

43.

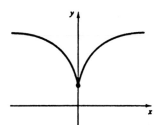

no asymptotes

vertical cusp at $(0, 1)$

44.

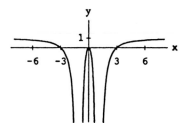

vertical asymptotes: $x = 1$, $x = -1$

horizontal asymptote: $y = 1$

no vertical tangents or cusps

45.

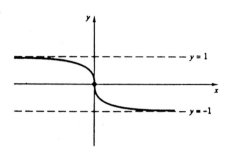

horizontal asymptotes: $y = -1$, $y = 1$

vertical tangent at $(0,0)$

46.

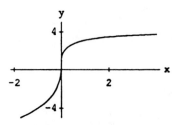

horizontal asymptote: $y = 4$

vertical tangent at $(0,1)$

47. (a) p odd; (b) p even.

48. $[r(x) - (ax + b)] = \dfrac{Q(x)}{q(x)} \to 0$ as $x \to \pm\infty$, since deg $[Q(x)] <$ deg $[q(x)]$.

49.

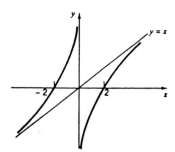

vertical asymptote: $x = 0$

oblique asymptote: $y = x$

50.

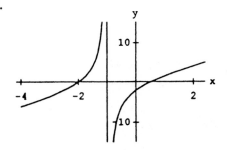

vertical asymptote: $x = -1$

oblique asymptote: $y = 2x + 1$

51.

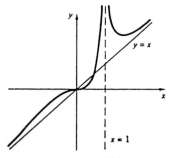

vertical asymptote: $x = 1$

oblique asymptote: $y = x$

52.

vertical asymptote: $x = 0$

oblique asymptote: $y = -3x + 1$

53. oblique asymptote: $y = 3x - 4$

54. oblique asymptote: $y = 5x - 3$

55. $y = 1$ is a horizontal asymptote.

$$f(x) = \sqrt{x^2 + 2x} - x = \left(\sqrt{x^2 + 2x} - x\right) \cdot \frac{\sqrt{x^2 + 2x} + x}{\sqrt{x^2 + 2x} + x} = \frac{2x}{\sqrt{x^2 + 2x} + x} \to 1.$$

56. $y = -\frac{1}{2}$ is a horizontal asymptote.

$$f(x) = \sqrt{x^4 - x^2} - x^2 = \left(\sqrt{x^4 - x^2} - x^2\right) \cdot \frac{\sqrt{x^4 - x^2} + x^2}{\sqrt{x^4 - x^2} + x^2} = \frac{-x^2}{\sqrt{x^4 - x^2} + x^2} \to -\frac{1}{2}.$$

SECTION 4.8

[Rough sketches; not scale drawings]

1. $f(x) = (x-2)^2$

$f'(x) = 2(x-2)$

$f''(x) = 2$

$f':$

$f'':$

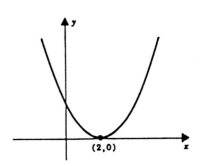

2. $f(x) = 1 - (x-2)^2$

$f'(x) = -2(x-2)$

$f''(x) = -2$

$f':$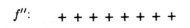

$f'':$

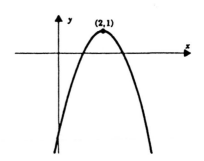

3. $f(x) = x^3 - 2x^2 + x + 1$

$f'(x) = (3x-1)(x-1)$

$f''(x) = 6x - 4$

$f':$

$f'':$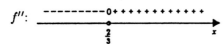

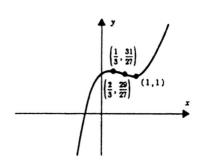

4. $f(x) = x^3 - 9x^2 + 24x - 7$

$f'(x) = 3x^2 - 18x + 24$

$f''(x) = 6x - 18$

f':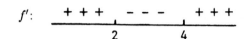

f'':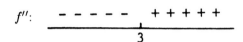

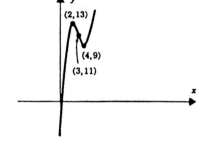

5. $f(x) = x^3 + 6x^2, \quad x \in [-4, 4]$

$f'(x) = 3x(x + 4)$

$f''(x) = 6x + 12$

f':

f'':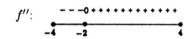

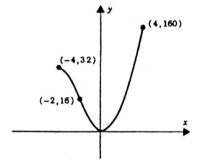

6. $f(x) = x^4 - 8x^2, \quad x \in [0, \infty)$

$f'(x) = 4x^3 - 16x$

$f''(x) = 12x^2 - 16$

f':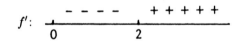

f'':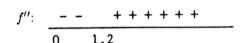

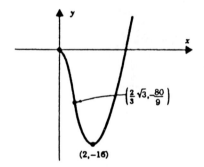

7. $f(x) = \frac{2}{3}x^3 - \frac{1}{2}x^2 - 10x - 1$

$f'(x) = (2x - 5)(x + 2)$

$f''(x) = 4x - 1$

f':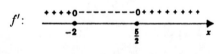

f'':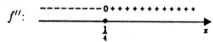

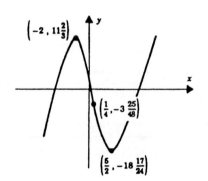

8. $f(x) = x(x^2 + 4)^2 = x^5 + 8x^3 + 16x$

$f'(x) = 5x^4 + 24x^2 + 16$

$f''(x) = 20x^3 + 48x = 4x(5x^2 + 12)$

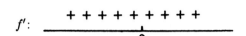

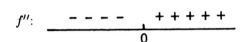

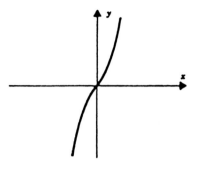

9. $f(x) = x^2 + 2x^{-1}$

$f'(x) = 2x - 2x^{-2} = 2\left(x^3 - 1\right)/x^2$

$f''(x) = 2 + 4x^{-3}$

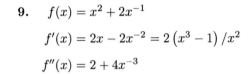

asymptote: $x = 0$

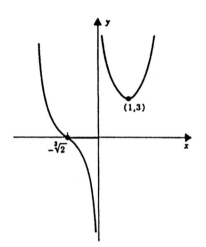

10. $f(x) = x - x^{-1},$

$f'(x) = 1 + x^{-2}$

$f''(x) = -x^{-3}$

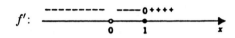

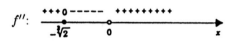

asymptotes: $x = 0, \; y = x$

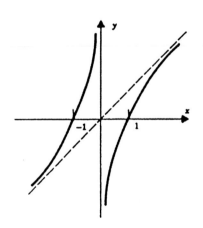

11. $\quad f(x) = (x-4)/x^2$

$\quad\quad f'(x) = (8-x)/x^3$

$\quad\quad f''(x) = (2x-24)/x^4$

$\quad f'$: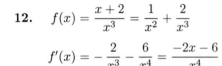

$\quad f''$:

asymptotes: $\quad x=0, \; y=0$

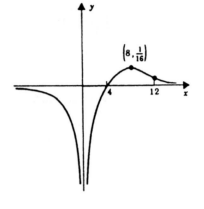

12. $\quad f(x) = \dfrac{x+2}{x^3} = \dfrac{1}{x^2} + \dfrac{2}{x^3}$

$\quad\quad f'(x) = -\dfrac{2}{x^3} - \dfrac{6}{x^4} = \dfrac{-2x-6}{x^4}$

$\quad\quad f''(x) = \dfrac{6}{x^4} + \dfrac{24}{x^5} = \dfrac{6(x+4)}{x^5}$

$\quad f'$:

$\quad f''$:

asymptotes: $\quad x=0, \; y=0$

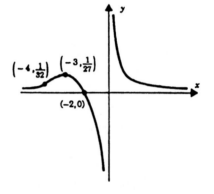

13. $\quad f(x) = 2x^{1/2} - x, \quad x \in [0,4]$

$\quad\quad f'(x) = x^{-1/2}\left(1 - x^{1/2}\right)$

$\quad\quad f''(x) = -\tfrac{1}{2}x^{-3/2}$

$\quad f'$: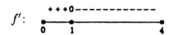

$\quad f''$:

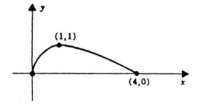

14. $\quad f(x) = \tfrac{1}{4}x - \sqrt{x}, \quad x \in [0,9]$

$\quad\quad f'(x) = \tfrac{1}{4} - \tfrac{1}{2}x^{-1/2}$

$\quad\quad f''(x) = \tfrac{1}{4}x^{-3/2}$

$\quad f'$: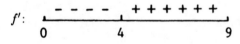

$\quad f''$: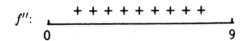

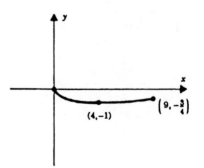

15. $f(x) = 2 + (x+1)^{6/5}$

$f'(x) = \frac{6}{5}(x+1)^{1/5}$

$f''(x) = \frac{6}{25}(x+1)^{-4/5}$

f':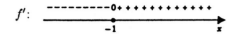

f'':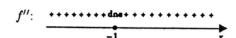

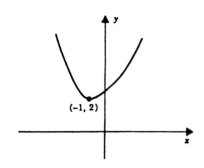

16. $f(x) = 2 + (x+1)^{7/5}$

$f'(x) = \frac{7}{5}(x+1)^{2/5}$

$f''(x) = \frac{14}{25}(x+1)^{-3/5}$

f':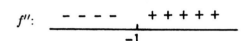

f'':

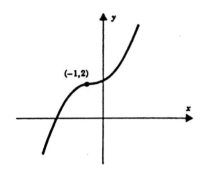

17. $f(x) = 3x^5 + 5x^3$

$f'(x) = 15x^2\left(x^2+1\right)$

$f''(x) = 30x\left(2x^2+1\right)$

f':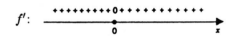

f'':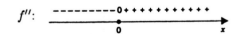

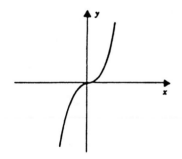

18. $f(x) = 3x^4 + 4x^3$

$f'(x) = 12x^3 + 12x^2 = 12x^2(x+1)$

$f''(x) = 36x^2 + 24x = 12x(3x+2)$

f':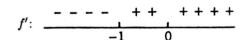

f'':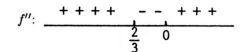

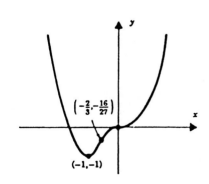

19. $f(x) = 1 + (x-2)^{5/3}$

$f'(x) = \frac{5}{3}(x-2)^{2/3}$

$f''(x) = \frac{10}{9}(x-2)^{-1/3}$

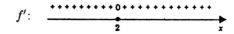

f':

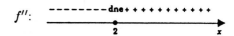

f'':

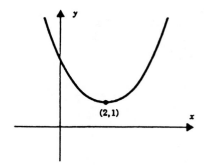

20. $f(x) = 1 + (x-2)^{4/3}$

$f'(x) = \frac{4}{3}(x-2)^{1/3}$

$f''(x) = \frac{4}{9}(x-2)^{-2/3}$

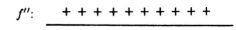

f':

f'': $+ + + + + + + + + + +$

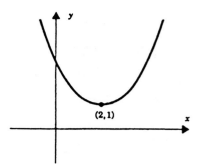

21. $f'(x) = \dfrac{8x}{(x^2+4)^2}$

$f''(x) = 48\dfrac{8(4-3x^2)}{(x^2+4)^3}$

$f'(x) < 0$ on $(-\infty, 0)$

$f'(x) > 0$ on $(0, \infty)$

$f''(x) < 0$ on $\left(-\infty, -2/\sqrt{3}\right) \cup \left(2/\sqrt{3}, \infty\right)$;

$f''(x) > 0$ on $\left(-2/\sqrt{3}, 2/\sqrt{3}\right)$;

asymptote: $y = 1$

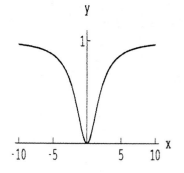

22. $f(x) = \dfrac{2x^2}{x+1}$

$f'(x) = \dfrac{2x(x+2)}{(x+1)^2}$

$f''(x) = \dfrac{4}{(x+1)^3}$

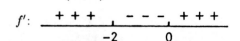

f':

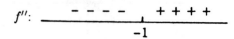

f'':

asymptote: $y = 2x - 2$

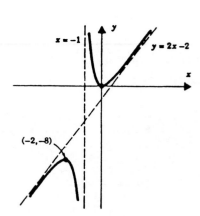

23. $f(x) = \dfrac{x}{(x+3)^2}$

$f'(x) = \dfrac{3-x}{(x+3)^3}$

$f''(x) = \dfrac{2x-12}{(x+3)^4}$

f':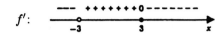

f'':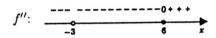

asymptotes: $x = -3, \ y = 0$

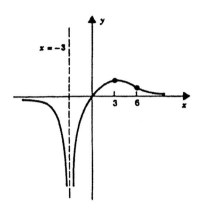

24. $f(x) = \dfrac{x}{x^2+1}$

$f'(x) = \dfrac{1-x^2}{(x^2+1)^2}$

$f''(x) = \dfrac{2x(x^2-3)}{(x^2+1)^3}$

f':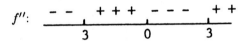

f'':

asymptote: $y = 0$

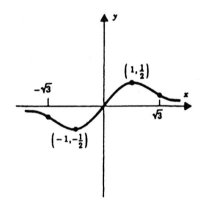

25. $f(x) = \dfrac{x^2}{x^2-4}$

$f'(x) = \dfrac{-8x}{(x^2-4)^2}$

$f''(x) = \dfrac{8(3x^2+4)}{(x^2-4)^3}$

f':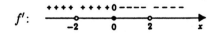

f'':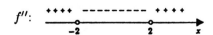

asymptotes: $x = -2, \ x = 2, \ y = 1$

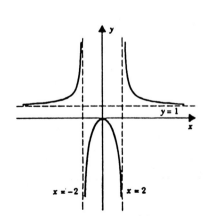

26. $f(x) = \dfrac{1}{x^3 - x}$

$f'(x) = \dfrac{1 - 3x^2}{(x^3 - x)^2}$

increasing on $\left(-1/\sqrt{3}, 0\right) \cup \left(0, 1/\sqrt{3}\right)$;

decreasing otherwise

$f''(x) = \dfrac{2\left(6x^4 - 3x^2 + 1\right)}{(x^3 - x)^3}$

concave up on $(-1, 0) \cup (1, \infty)$

concave down on $(-\infty, 0) \cup (0, 1)$

asymptotes: $x = -1, x = 0, x = 1, y = 0$

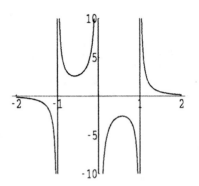

27. $f(x) = x(1 - x)^{1/2}$

$f'(x) = \tfrac{1}{2}(1 - x)^{-1/2}(2 - 3x)$

$f''(x) = \tfrac{1}{4}(1 - x)^{-3/2}(3x - 4)$

f':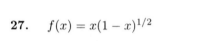

f'':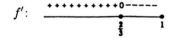

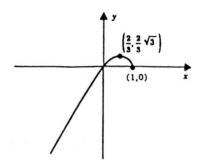

28. $f(x) = (x - 1)^4 - 2(x - 1)^2$

$f'(x) = 4(x - 1)^3 - 4(x - 1)$

$f''(x) = 12(x - 1)^2 - 4$

f':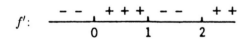

f'':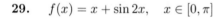

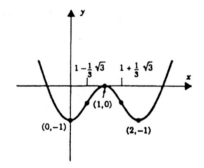

29. $f(x) = x + \sin 2x, \quad x \in [0, \pi]$

$f'(x) = 1 + 2\cos 2x$

$f''(x) = -4\sin 2x$

f':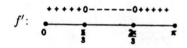

f'':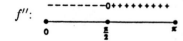

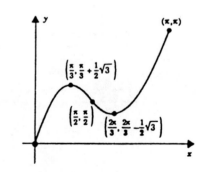

30. $f(x) = \cos^3 x + 6\cos x, \quad x \in [0, \pi]$

$f'(x) = -3\sin x(\cos^2 x + 2)$

$f''(x) = -9\cos^3 x$

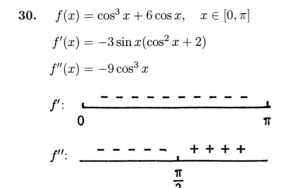

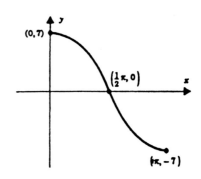

31. $f(x) = \cos^4 x, \quad x \in [0, \pi]$

$f'(x) = -4\cos^3 x \sin x$

$f''(x) = 4\cos^2 x\left(3\sin^2 x - \cos^2 x\right)$

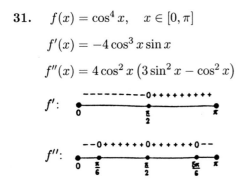

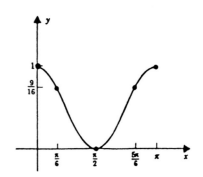

32. $f(x) = \sqrt{3}\,x - \cos 2x, \quad x \in [0, \pi]$

$f'(x) = \sqrt{3} + 2\sin 2x$

$f''(x) = 4\cos 2x$

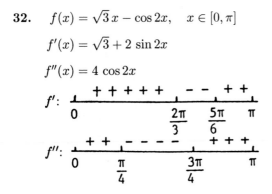

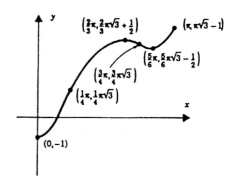

33. $f(x) = 2\sin^3 x + 3\sin x, \quad x \in [0, \pi]$

$f'(x) = 3\cos x\left(2\sin^2 x + 1\right)$

$f''(x) = 9\sin x\left(1 - 2\sin^2 x\right)$

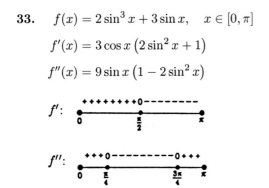

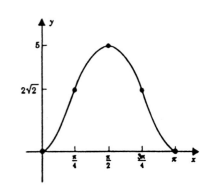

34. $f(x) = \sin^4 x, \quad x \in [0, \pi]$

$f'(x) = 4\sin^3 x \cos x$

$f''(x) = 4\sin^2 x \left(3\cos^2 x - \sin^2 x\right)$

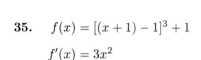

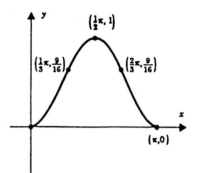

35. $f(x) = [(x+1) - 1]^3 + 1$

$f'(x) = 3x^2$

$f''(x) = 6x$

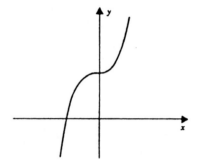

36. $f(x) = x^3(x+5)^2$

$f'(x) = 5x^2(x+3)(x+5)$

$f''(x) = 10x(2x^2 + 12x + 15)$

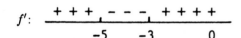

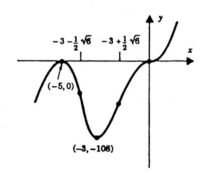

37. $f(x) = x^2(5-x)^3$

$f'(x) = 5x(2-x)(5-x)^2$

$f''(x) = 10(5-x)\left(2x^2 - 8x + 5\right)$

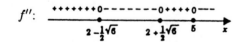

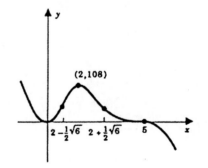

38. $f(x) = \begin{cases} 4 - 2x + x^2, & 0 < x < 2 \\ 4 + 2x - x^2, & x \le 0, \ x \ge 2 \end{cases}$

$f'(x) = \begin{cases} -2 + 2x, & 0 < x < 2 \\ 2 - 2x, & x < 0, \ x > 2 \end{cases}$

$f''(x) = \begin{cases} 2, & 0 < x < 2 \\ -2, & x < 0, \ x > 2 \end{cases}$

f': $\begin{array}{cccc} + + & - - - & + + + & - - \\ \hline & 0 & 1 & 2 \end{array}$

f'': $\begin{array}{ccc} - - & + + + + + & - - \\ \hline & 0 & 2 \end{array}$

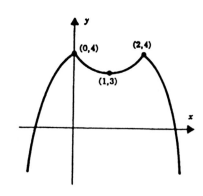

39. $f(x) = \begin{cases} 4 - x^2, & |x| > 1 \\ x^2 + 2, & -1 \le x \le 1 \end{cases}$

$f'(x) = \begin{cases} -2x, & |x| > 1 \\ 2x, & -1 < x < 1 \end{cases}$

$f''(x) = \begin{cases} -2, & |x| > 1 \\ 2, & -1 < x < 1 \end{cases}$

f': $\begin{array}{ccccc} + + + + + \ \text{dne} & - - - - & 0 & + + + + \ \text{dne} & - - - - - \\ \hline & -1 & 0 & 1 & x \end{array}$

f'': $\begin{array}{ccc} - - - - - \ \text{dne} & + + + + + + + + + \ \text{dne} & - - - - - \\ \hline & -1 & 1 & x \end{array}$

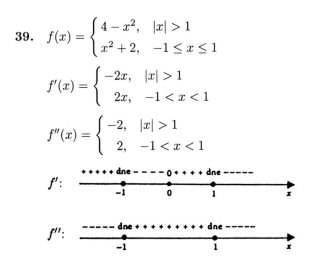

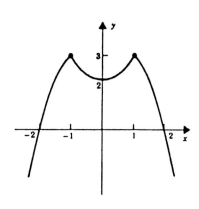

40. $f(x) = x - x^{1/3}$

$f'(x) = 1 - \frac{1}{3}\,x^{-2/3}$

$f''(x) = \frac{2}{9}\,x^{-5/3}$

f': $\begin{array}{ccc} + + + & - - - & + + + \\ \hline & 0 & 1 \end{array}$

f'': $\begin{array}{cc} - - - - - & + + + + \\ \hline & 0 \end{array}$

vertical tangent at $(0,0)$

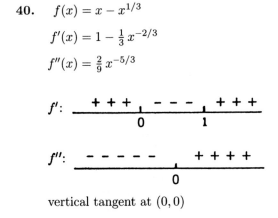

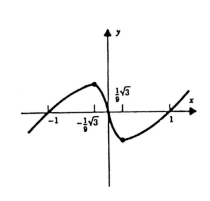

41. $f(x) = x(x-1)^{1/5}$

$f'(x) = \frac{1}{5}(x-1)^{-4/5}(6x-5)$

$f''(x) = \frac{2}{25}(x-1)^{-9/5}(3x-5)$

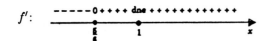

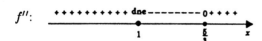

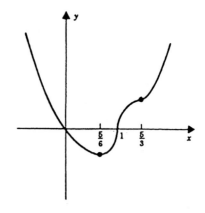

vertical tangent at $(1,0)$

42. $f(x) = x^2(x-7)^{1/3}$

$f'(x) = \dfrac{7x(x-6)}{3(x-7)^{2/3}}$

$f''(x) = \dfrac{14(2x^2 - 24x + 63)}{9(x-7)^{5/3}}$

f':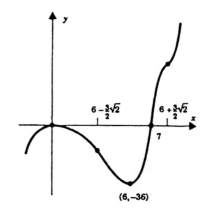

f'':

vertical tangent at $(7,0)$

43. $f(x) = x^2 - 6x^{1/3}$

$f'(x) = 2x^{-2/3}\left(x^{5/3} - 1\right)$

$f''(x) = \frac{2}{3}x^{-5/3}\left(3x^{5/3} + 2\right)$

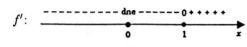

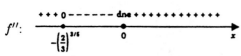

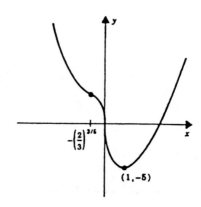

vertical tangent at $(0,0)$

44. $f(x) = \dfrac{2x}{\sqrt{x^2+1}}$

$f'(x) = \dfrac{2}{(x^2+1)^{3/2}}$

$f''(x) = \dfrac{-6x}{(x^2+1)^{5/2}}$

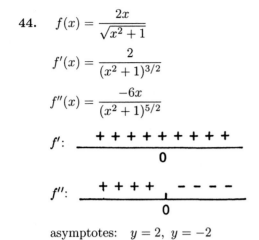

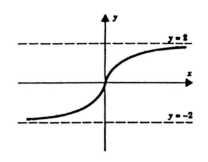

asymptotes: $y = 2,\; y = -2$

45. $f(x) = \left(\dfrac{x}{x-2}\right)^{1/2}$; $x \le 0,\; x > 2$

$f'(x) = -\left(\dfrac{x}{x-2}\right)^{-1/2}(x-2)^{-2}$

$f''(x) = (2x-1)\left(\dfrac{x}{x-2}\right)^{-3/2}(x-2)^{-4}$

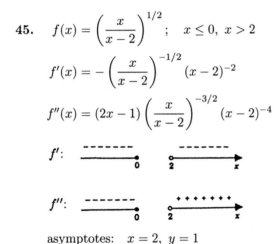

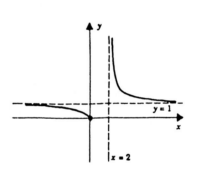

asymptotes: $x = 2,\; y = 1$

46. $f(x) = \left(\dfrac{x}{x+4}\right)^{1/2}$; $x < -4,\; x \ge 0$

$f'(x) = 2\left(\dfrac{x}{x+4}\right)^{-1/2}(x+4)^{-2}$

$f''(x) = \dfrac{-4(x+1)}{(x+4)^{5/2}x^{3/2}}$

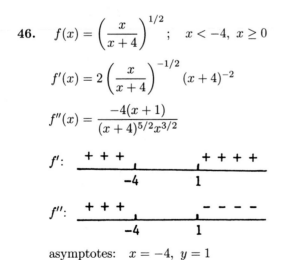

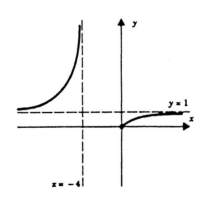

asymptotes: $x = -4,\; y = 1$

47. $f(x) = x^2 \left(x^2 - 2\right)^{-1/2}, \quad |x| > \sqrt{2}$

   $f'(x) = x \left(x^2 - 4\right) \left(x^2 - 2\right)^{-3/2}$

   $f''(x) = 2 \left(x^2 + 4\right) \left(x^2 - 2\right)^{-5/2}$

f':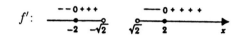

f'':

asymptotes: $x = -\sqrt{2}, \; x = \sqrt{2}$

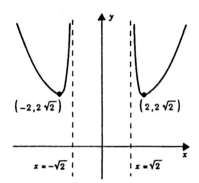

48. $f(x) = 3\cos 4x, \quad x \in [0, \pi]$

   $f'(x) = -12\sin 4x$

   $f''(x) = -48\cos 4x$

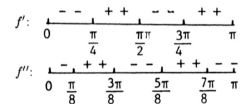

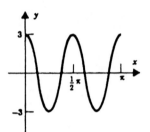

49. $f(x) = 2\sin 3x, \quad x \in [0, \pi]$

   $f'(x) = 6\cos 3x$

   $f''(x) = -18\sin 3x$

f':

f'':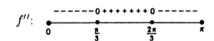

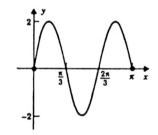

50. $f(x) = 3 + 2\cot x + \csc^2 x, \quad x \in (0, \pi/2)$

   $f'(x) = -2\csc^2 x \left(1 + \cot x\right)$

   $f''(x) = 2\csc^2 x \left(3\cot^2 x + 2\cot x + 1\right)$

f':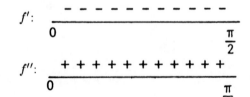

asymptote: $x = 0$

asymptote: $x = 0$

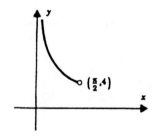

51. $f(x) = 2\tan x - \sec^2 x, \quad x \in (0, \pi/2)$

$\qquad = -(1 - \tan x)^2$

$f'(x) = 2\sec^2 x\,(1 - \tan x)$

$f''(x) = -2\sec^2 x\,\left(3\tan^2 x - 2\tan x + 1\right)$

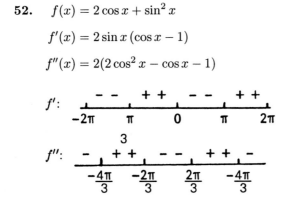

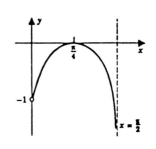

asymptote: $x = \frac{1}{2}\pi$

52. $f(x) = 2\cos x + \sin^2 x$

$f'(x) = 2\sin x\,(\cos x - 1)$

$f''(x) = 2(2\cos^2 x - \cos x - 1)$

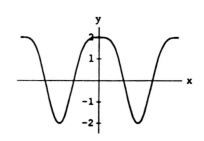

53. $f(x) = \dfrac{\sin x}{1 - \sin x}, \quad x \in (-\pi, \pi)$

$f'(x) = \dfrac{\cos x}{(1 - \sin x)^2}$

$f''(x) = \dfrac{1 - \sin x + \cos^2 x}{(1 - \sin x)^3}$

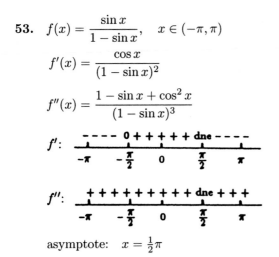

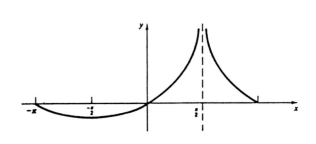

asymptote: $x = \frac{1}{2}\pi$

54. $f(x) = \dfrac{1}{1 - \cos x}, \quad x \in (-\pi, \pi)$

$f'(x) = \dfrac{-\sin x}{(1 - \cos x)^2}$

$f''(x) = \dfrac{2 - \cos x - \cos^2 x}{(1 - \cos x)^3}$

f':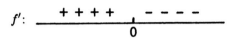

f'':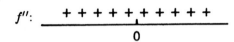

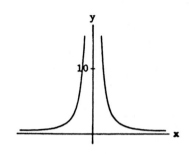

asymptote: $x = 0$

55. (a) f increases on $(-\infty, -1] \cup (0, 1] \cup [3, \infty)$;

f decreases on $[-1, 0) \cup [1, 3]$;   critical points: $x = -1, 0, 1, 3$.

(b)

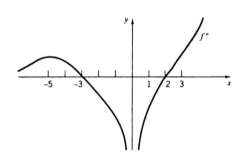

concave up on $(-\infty, -3) \cup (2, \infty)$

concave down on $(-3, 0) \cup (0, 2)$.

(c)

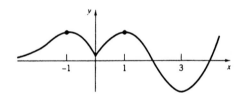

The graph does not necessarily have a horizontal asymptote.

56. (a)

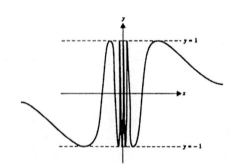

(b)   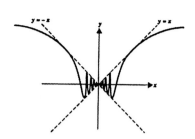   (c)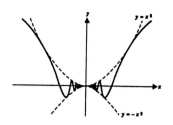

(d) F is discontinuous at 0; $\lim\limits_{x\to 0} \sin(1/x)$ does not exist

$\quad$ G is continuous at 0: $|x \sin(1/x)| = |x|\,|\sin(1/x)| \le |x| \to 0,$

$\qquad$ so that $\lim\limits_{x\to 0} G(x) = 0 = G(0).$

$\quad$ H is continuous at 0: $\left|x^2 \sin(1/x)\right| = |x|^2\,|\sin(1/x)| \le |x|^2 \to 0,$

$\qquad$ so that $\lim\limits_{x\to 0} H(x) = 0 = H(0).$

(e) F is not differentiable at 0 since it is not continuous at 0.

$\quad$ G is not differentiable at 0: $\lim\limits_{h\to 0} \dfrac{G(h) - G(0)}{h} = \lim\limits_{h\to 0} \sin(1/h)$ does not exist.

$\quad$ H is differentiable at 0: $H'(0) = \lim\limits_{h\to 0} \dfrac{H(h) - H(0)}{h} = \lim\limits_{h\to 0} h \sin(1/h) = 0.$

57. $f(x) - x^2 = \dfrac{x^3 - x^{1/3}}{x} - x^2 = \dfrac{x^3 - x^{1/3} - x^3}{x} = -\dfrac{1}{x^{2/3}} \to 0$ as $x \to \pm\infty.$

The figure shows the graphs of f and $g(x) = x^2$ for $x > 0$. Since f and g are even functions,
the graphs are symmetric about the y-axis.

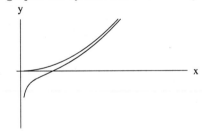

58. (a)

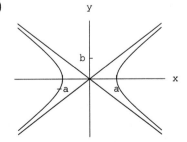

(b) $\dfrac{b}{a}\sqrt{x^2 - a^2} = \dfrac{b}{a}x\sqrt{1 - \dfrac{a^2}{x^2}} \to \dfrac{b}{a}x$ as $x \to \infty$. Therefore, $\dfrac{b}{a}\sqrt{x^2 - a^2} - \dfrac{b}{a}x \to 0$ as $x \to \infty.$

(c) If $x < 0,$ then the second quadrant arc is given by $y = -\dfrac{b}{a}x\sqrt{1 - \dfrac{a^2}{x^2}} \to -\dfrac{b}{a}x$ as $x \to -\infty.$

SECTION 4.9

1. $x(5) = -6$; $v(t) = 3 - 2t$ so $v(5) = -7$ and speed $= 7$; $a(t) = -2$ so $a(5) = -2$.

2. $x(3) = -12$; $v(t) = 5 - 3t^2$ so $v(3) = -22$ and speed $= 22$; $a(t) = -6t$ so $a(3) = -18$.

3. $x(1) = 6$; $v(t) = -18/(t+2)^2$ so $v(1) = -2$ and speed $= 2$,
   $a(t) = 36/(t+2)^3$ so $a(1) = 4/3$.

4. $x(3) = 1$; $v(t) = \dfrac{6}{(t+3)^2}$ so $v(3) = 1/6 = $ speed; $a(t) = -12/(t+3)^3$ so $a(3) = -1/18$.

5. $x(1) = 0$, $v(t) = 4t^3 + 18t^2 + 6t - 10$ so $v(1) = 18$ and speed $= 18$,
   $a(t) = 12t^2 + 36t + 6$ so $a(1) = 54$.

6. $x(2) = -20$; $v(t) = 4t^3 - 18t$ so $v(2) = -4$ and speed $= 4$,
   $a(t) = 12t^2 - 18$ so $a(2) = 30$.

7. $x(t) = 5t + 1$, $\quad x'(t) = v(t) = 5$, $\quad x''(t) = a(t) = 0$.
   $v(t) \neq 0$, $a(t) = 0$ for all t.

8. $x(t) = 4t^2 - t + 3$, $\quad x'(t) = v(t) = 8t - 1$, $\quad x''(t) = a(t) = 8$.
   $v(t) = 0$ at $t = 1/8$; $\quad a(t) \neq 0$ for all t.

9. $x(t) = t^3 - 6t^2 + 9t - 1$, $\quad x'(t) = v(t) = 3t^2 - 12t + 9$, $\quad x''(t) = a(t) = 6t - 12$.
   $v(t) = 3(t-1)(t-3)$, $v(t) = 0$ at $t = 1, 3$; $\quad a(t) = 0$ at $t = 2$.

10. $x(t) = t^4 - 4t^3 + 4t^2 + 2$, $\quad x'(t) = v(t) = 4t^3 - 12t^2 + 8t$, $\quad x''(t) = a(t) = 12t^2 - 24t + 8$.
    $v(t) = 4t(t-1)(t-2)$, $v(t) = 0$ at $t = 0, 1, 2$; $\quad a(t) = 0$ at $t = 1 + \sqrt{3}/3$.

11. A	**12.** C	**13.** A	**14.** C	**15.** A and B
16. B	**17.** A	**18.** A	**19.** A and C	**20.** B

21. The object is moving right when $v(t) > 0$. Here,
    $v(t) = 4t^3 - 36t^2 + 56t = 4t(t-2)(t-7)$; $\quad v(t) > 0$ when $0 < t < 2$ and $7 < t$.

22. The object is moving left when $v(t) < 0$. Here,
    $v(t) = 3t^2 - 24t + 21 = 3(t-7)(t-1)$; $\quad v(t) < 0$ when $1 < t < 7$.

23. The object is speeding up when $v(t)$ and $a(t)$ have the same sign.
    $v(t) = 5t^3(4-t)$ $\qquad$ sign of $v(t)$:
    $a(t) = 20t^2(3-t)$ $\qquad$ sign of $a(t)$:
    Thus, $\quad 0 < t < 3$ and $\quad 4 < t$.

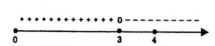

24. The object is slowing down when $v(t)$ and $a(t)$ have opposite sign.

$v(t) = 12t - 4t^3 = 4t(3 - t^2)$ sign of $v(t)$:

$a(t) = 12(1 - t)$ sign of $a(t)$: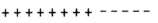

Thus, $1 < t < \sqrt{3}$.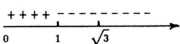

25. The object is moving left and slowing down when $v(t) < 0$ and $a(t) > 0$.

$v(t) = 3(t - 5)(t + 1)$ sign of $v(t)$:

$a(t) = 6(t - 2)$ sign of $a(t)$:

Thus, $2 < t < 5$.

26. The object is moving right and slowing down when $v(t) > 0$ and $a(t) < 0$.

$v(t) = 3(t - 5)(t + 1)$ sign of $v(t)$:

$a(t) = 6(t - 2)$ sign of $a(t)$:

This never happens.

27. The object is moving right and speeding up when $v(t) > 0$ and $a(t) > 0$.

$v(t) = 4t(t - 2)(t - 4)$ sign of $v(t)$:

$a(t) = 4(3t^2 - 12t + 8)$ sign of $a(t)$: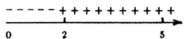

Thus, $0 < t < 2 - \frac{2}{3}\sqrt{3}$ and $4 < t$.

28. The object is moving left and speeding up when $v(t) < 0$ and $a(t) < 0$.

$v(t) = 4t(t - 2)(t - 4)$ sign of $v(t)$: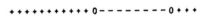

$a(t) = 4(3t^2 - 12t + 8)$ sign of $a(t)$: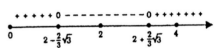

Thus, $2 < t < 2 + \frac{2}{3}\sqrt{3}$.

29. $x(t) = (t + 1)^2(t - 9)^3$, $x'(t) = v(t) = 3(t + 1)^2(t - 9)^2 + 2(t + 1)(t - 9)^3 = 5(t + 1)(t - 9)^2(t - 3)$.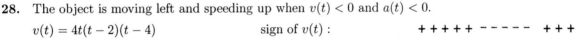
The object changes direction at time $t = 3$.

30. $x(t) = t(t - 8)^3$, $x'(t) = v(t) = (t - 8)^3 + 3t(t - 8)^2 = 4(t - 8)^2(t - 2)$.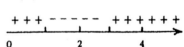
The object changes direction at time $t = 2$.

31. $x(t) = (t^3 - 12t)^4$, $x/(t) = v(t) = 4(t^3 - 12t)^3(3t^2 - 12) = 12t^3(t + 2\sqrt{3})^3(t - 2\sqrt{3})^3(t + 2)(t - 2)$.
The object changes direction at times $t = 2, 2\sqrt{3}$.

32. $x(t) = (t^2 - 8t + 15)^3$, $x'(t) = v(t) = 3(t^2 - 8t + 15)^2(2t - 8) = 6(t^2 - 8t + 15)^2(t - 4)$.
The object changes direction at time $t = 4$.

Exercises 33–38. The object is moving right with increasing speed when both $v(t)$ and $a(t)$ are positive.

33. $x(t) = \sin 3t$, $x'(t) = v(t) = 3\cos 3t$, $x''(t) = a(t) = -9\sin 3t$.

$v(t) > 0$ on $(0, \pi/6) \cup (\pi/2, 5\pi/6) \cup (7\pi/6, 3\pi/2) \cup (11\pi/6, 2\pi)$.

$a(t) > 0$ on $(\pi/3, 2\pi/3) \cup (\pi, 4\pi/3) \cup (5\pi/3, 2\pi)$.

$v(t)$ and $a(t)$ are positive on $(\pi/2, 2\pi/3) \cup (7\pi/6, 4\pi/3) \cup (11\pi/6, 2\pi)$.

34. $x(t) = \cos 2t, \quad x'(t) = v(t) = -2\sin 2t, \quad x''(t) = a(t) = -4\cos 2t.$

$v(t) > 0$ on $(\pi/2, \pi) \cup (3\pi/2, 2\pi); \quad a(t) > 0$ on $(\pi/4, 3\pi/4) \cup (5\pi/4, 7\pi/4).$

$v(t)$ and $a(t)$ are positive on $(\pi/2, 3\pi/4) \cup (3\pi/2, 7\pi/4).$

35. $x(t) = \sin t - \cos t, \quad x'(t) = v(t) = \cos t + \sin t, \quad x''(t) = a(t) = -\sin t + \cos t.$

$v(t) > 0$ on $(0, 3\pi/4) \cup (7\pi/4, 2\pi); \quad a(t) > 0$ on $(0, \pi/4) \cup (5\pi/4, 2\pi).$

$v(t)$ and $a(t)$ are positive on $(0, \pi/4) \cup (7\pi/4, 2\pi).$

36. $x(t) = \sin t + \cos t, \quad x'(t) = v(t) = \cos t - \sin t, \quad x''(t) = a(t) = -\sin t - \cos t.$

$v(t) > 0$ on $(0, \pi/4) \cup (5\pi/4, 2\pi); \quad a(t) > 0$ on $(3\pi/4, 7\pi/4).$

$v(t)$ and $a(t)$ are positive on $(5\pi/4, 7\pi/4).$

37. $x(t) = t + 2\cos t, \quad x'(t) = v(t) = 1 - 2\sin t, \quad x''(t) = a(t) = -2\cos t.$

$v(t) > 0$ on $(0, \pi/6) \cup (5\pi/6, 2\pi); \quad a(t) > 0$ on $(\pi/2, 3\pi/2).$

$v(t)$ and $a(t)$ are positive on $(5\pi/6, 3\pi/2).$

38. $x(t) = t - \sqrt{2}\sin t, \quad x'(t) = v(t) = 1 - \sqrt{2}\cos t, \quad x''(t) = a(t) = \sqrt{2}\sin t.$

$v(t) > 0$ on $(\pi/4, 7\pi/4); \quad a(t) > 0$ on $(0, \pi).$

$v(t)$ and $a(t)$ are positive on $(\pi/4, \pi).$

39. Since $v_0 = 0$ the equation of motion is

$$y(t) = -16t^2 + y_0.$$

We want to find y_0 so that $y(6) = 0$. From

$$0 = -16(6)^2 + y_0$$

we get $y_0 = 576$ feet.

40. The equation of motion is: $y(t) = -4.9t^2 + y_0$. Therefore, the velocity is given by $v(t) = -9.8t$.

Since the object hits the ground at $98 \, \text{m/sec.}$, we have $-9.8t = -98$, and $t = 10$.

Therefore, $y(10) = 0 = -4.9(10)^2 + y_0$ and $y_0 = 490$ meters.

41. The object's height and velocity at time t are given by

$$y(t) = -\frac{1}{2}gt^2 + v_0 t \quad \text{and} \quad v(t) = -gt + v_0$$

Since the object's velocity at its maximum height is 0, it takes v_0/g seconds to reach maximum height, and

$$y(v_0/g) = -\tfrac{1}{2}g(v_0/g)^2 + v_0(v_0/g) = v_0^2/2g \quad \text{or} \quad v_0^2/19.6 \quad \text{(meters)}$$

42. Since $y_0 = 0$, we have $y(t) = -16t^2 + v_0 t = t(-16t + v_0)$. Now,

$$y(8) = 0 \implies v_0 = (16)8 = 128 \implies \text{the initial velocity was 128 ft/sec.}$$

43. At time t, the object's height is $y(t) = -\frac{1}{2} gt^2 + v_0 t + y_0$, and its velocity is $v(t) = -gt + v_0$. Suppose that $y(t_1) = y(t_2)$, $t_1 \neq t_2$. Then

$$-\tfrac{1}{2} gt_1^2 + v_0 t_1 + y_0 = -\tfrac{1}{2} gt_2^2 + v_0 t_2 + y_0$$

$$\tfrac{1}{2} g(t_2^2 - t_1^2) = v_0(t_2 - t_1)$$

$$gt_2 + gt_1 = 2v_0$$

From this equation, we get $-(-gt_1 + v_0) = -gt_2 + v_0$ and so $|v(t_1)| = |v(t_2)|$.

44. Since $y_0 = 0$, we have $y(t) = -4.9t^2 + v_0 t = t(v_0 - 4.9t)$ The object hits the ground at $t = v_0/4.9$ sec., that is, the object is in the air for $v_0/4.9$ sec. At its maximum height, the velocity of the object is 0. Since $v(t) = -9.8t + v_0$, we have $-9.8t + v_0 = 0$ and $t = v_0/9.8 = \frac{1}{2}(v_0/4.9)$. The result follows from this.

45. In the equation

$$y(t) = -16t^2 + v_0 t + y_0$$

we take $v_0 = -80$ and $y_0 = 224$. The ball first strikes the ground when

$$-16t^2 - 80t + 224 = 0;$$

that is, at $t = 2$. Since

$$v(t) = y'(t) = -32t - 80,$$

we have $v(2) = -144$ so that the speed of the ball the first time it strikes the ground is 144 ft/sec. Thus, the speed of the ball the third time it strikes the ground is $\dfrac{1}{4}\left[\dfrac{1}{4}(144)\right] = 9$ ft/sec.

46. Since $y_0 = 0$, we have $y(t) = -16t^2 + v_0 t$.

$$y(2) = 64 \implies -16(2)^2 + 2v_0 = 64 \implies v_0 = 64 \quad \text{and} \quad y(t) = -16t^2 + 64t$$

Now, at the maximum height, $v(t) = -32t + 64 = 0 \implies t = 2$. We already know the height at $t = 2$, namely 64 ft.

47. The equation is $y(t) = -16t^2 + 32t$.   (Here $y_0 = 0$ and $v_0 = 32$.)

(a) We solve $y(t) = 0$ to find that the stone strikes the ground at $t = 2$ seconds.

(b) The stone attains its maximum height when $v(t) = 0$. Solving

$$v(t) = -32t + 32 = 0, \quad \text{we get } t = 1 \quad \text{and, thus, the maximum height is } y(1) = 16 \text{ feet.}$$

(c) We want to choose v_0 in

$$y(t) = -16t^2 + v_0 t$$

so that $y(t_0) = 36$ when $v(t_0) = 0$ for some time t_0.

From $v(t) = -32t + v_0 = 0$ we get $t_0 = v_0/32$ so that

$$-16\left(\frac{v_0}{32}\right)^2 + v_0\left(\frac{v_0}{32}\right) = 36, \quad \text{or} \quad \frac{v_0{}^2}{64} = 36.$$

Thus, $v_0 = 48$ ft/sec.

48. (a) Measuring height from the water surface, we have $y(t) = -16t^2 + y_0$, since $v_0(0) = 0$.
     If the stone hits the water 3 seconds later, then $y(3) = -16(3)^2 + y_0 = 0$. so $y_0 = 144$.

(b) It takes $y_0/1080$ seconds for the sound of the splash to reach the man so the stone hits the
    at time $t = 3 - y_0/1080$. Thus,

$$y(t) = -16\left(3 - \frac{y_0}{1080}\right)^2 + y_0 = 0 \quad \Longrightarrow \quad y_0 \cong 132.47 \text{ ft.}$$

49. For all three parts of the problem the basic equation is

$$y(t) = -16t^2 + v_0 t + y_0$$

with

(*) $y(t_0) = 100$ and $y(t_0 + 2) = 16$

for some time $t_0 > 0$.

We are asked to find y_0 for a given value of v_0.

From (*) we get

$$16 - 100 = y(t_0 + 2) - y(t_0)$$

$$= [-16(t_0 + 2)^2 + v_0(t_0 + 2) + y_0] - [-16t_0{}^2 + v_0 t_0 + y_0]$$

$$= -64t_0 - 64 + 2v_0$$

so that

$$t_0 = \tfrac{1}{32}(v_0 + 10).$$

Substituting this result in the basic equation and noting that $y(t_0) = 100$, we have

$$-16\left(\frac{v_0 + 10}{32}\right)^2 + v_0\left(\frac{v_0 + 10}{32}\right) + y_0 = 100$$

and therefore

(**) $y_0 = 100 - \dfrac{v_0{}^2}{64} + \dfrac{25}{16}.$

We use (**) to find the answer to each part of the problem.

(a) $v_0 = 0$ so $y_0 = \frac{1625}{16}$ ft   (b) $v_0 = -5$ so $y_0 = \frac{6475}{64}$ ft   (c) $v_0 = 10$ so $y_0 = 100$ ft

50. Let $v_0 > 0$ be the initial velocity. The equation of motion prior to the impact is: $y(t) = -16t^2 - v_0 t + 4$.

The ball hits the ground at time $t = \dfrac{\sqrt{v_0^2 + 256} - v_0}{32}$ with velocity $v = \sqrt{v_0^2 + 256}$. The equation of

motion following the impact is: $y(t) = -16t^2 + \dfrac{\sqrt{v_0^2 + 256}}{2} t$. It reaches its maximum height at time

$T = \dfrac{\sqrt{v_0^2 + 256}}{64}$. Now, $y(T) = 4 \implies v_0 = 16\sqrt{3}$.

51. Let $y_0 > 0$ be the initial height. The equation of motion becomes:
$$0 = -16(8)^2 + 5(8) + y_0, \quad \text{so } y_0 = 984 \text{ ft}.$$

52. Using $0 = -16t^2 - 5t + 984$, yields $t = \dfrac{123}{16}$ or about 7.7 sec.

53. Let $f_1(t)$ and $f_2(t)$ be the positions of the horses at time t. Consider $f(t) = f_1(t) - f_2(t)$. Let T be the time the horses finish the race. Then $f(0) = f(T) = 0$. By the mean-value theorem there is a c in $(0, T)$ such that $f'(c) = 0$. Hence $f_1'(t) = f_2'(t)$, so the horses had the same speed at time c.

54. Apply the argument given in Exercise 53 to the velocity functions $v_1(t)$, $v_2(t)$.

55. The driver must have exceeded the speed limit at some time during the trip. Let $s(t)$ denote the car's position at time t, with $s(0) = 0$ and $s(5/3) = 120$. Then, by the mean-value theorem, there exists at least one number (time) c such that
$$v(c) = s'(c) = \frac{s(5/3) - s(0)}{\frac{5}{3} - 0} = \frac{120}{\frac{5}{3}} = \frac{360}{5} = 72 \text{ mi./hr.}$$

56. Let $f(t)$ be the function that gives the car's velocity after t hours. Then $f(0) = 30$ and $f(\frac{1}{4}) = 60$. f is differentiable on $(1, \frac{1}{4})$ and continuous on $[1, \frac{1}{4}]$, so by the mean-value theorem there is a c in $(1, \frac{1}{4})$ such that $f'(c) = \dfrac{60 - 30}{0 - t_1} = 120$. i.e. The acceleration at time c was 120 mph.

57. If the speed $s(t)$ of the car is less than 60 mi/hr= 1 mi/min, then the distance traveled in 20 minutes is less than 20 miles. Therefore, the car must have gone at least 1 mi/min at some time $t < 20$. Let t_1 be the first instant the car's speed is 1 mi/min. Then the speed $s(t)$ is less than 1 mi/min on the interval $[0, t_1)$ and the distance r traveled in t_1 minutes is less than t_1 miles. Now, by the mean-value theorem, there is a time $c \in [t_1, 20]$ such that
$$s'(c) = \frac{20 - r}{20 - t_1} > \frac{20 - t_1}{20 - t_1} = 1 \ (= 60 \text{ mi/hr}).$$

58. Let $s(t)$ denote the distance that the car has traveled in t seconds since applying the brakes, $0 \le t \le 6$. Then $s(0) = 0$ and $s(6) = 280$. Assume that s is differentiable on $(0, 6)$ and continuous on $[0, 6]$. Then, by the mean-value theorem, there exists a time $c \in (0, 6)$ such that
$$s'(c) = v(c) = \frac{s(6) - s(0)}{6 - 0} = \frac{280}{6} \cong 46.67 \text{ ft/sec}$$

Now $v(0) \ge v(c) = 46.7$ ft/sec. Thus, the driver must have been exceeding the speed limit (44 ft/sec) at the instant he applied his brakes.

59. $y(t) = A\sin(\omega t + \phi_0), \quad y'(t) = v(t) = \omega A\cos(\omega t + \phi_0), \quad y''(t) = a(t) = -\omega^2 A\sin(\omega t + \phi_0)$

(a) $y''(t) + \omega^2 y(t) = -\omega^2 A\sin(\omega t + \phi_0) + \omega^2 A\sin(\omega t + \phi_0) = 0.$

(b) $v'(t) = a(t) = -\omega^2 A\sin(\omega t + \phi_0) = 0$ when $\omega t + \phi_0 = n\pi; \ t = \dfrac{n\pi - \phi_0}{\omega}.$

The extreme values of v occur at these times. Now

$$y\left(\frac{n\pi - \phi_0}{\omega}\right) = A\sin\left(\omega\,\frac{n\pi - \phi_0}{\omega} + \phi_0\right) = A\sin(n\pi) = 0.$$

The bob attains maximum speed at the equilibrium position.

(c) $a'(t) - \omega^3 A\cos(\omega t + \phi_0) = 0$ when $\omega t + \phi_0 = (2n-1)\pi/2; \ t = \dfrac{(2n-1)\pi/2 - \phi_0}{\omega}.$

The extreme values of a occur at these times. Now

$$a\left(\frac{(2n-1)\pi/2 - \phi_0}{\omega}\right) = -\omega^2 A\sin\left(\frac{(2n-1)\pi/2 - \phi_0}{\omega} + \phi_0\right) = \pm\omega^2 A.$$

The bob attains these values at

$$y\left(\frac{(2n-1)\pi/2 - \phi_0}{\omega}\right) = A\sin\left(\omega\,\frac{(2n-1)\pi/2 - \phi_0}{\omega} + \phi_0\right) = A\sin(2n-1)\pi/2 = \pm A.$$

60. (a) $x(t) = t^3 - 7t^2 + 10t + 5, \quad v(t) = 3t^2 - 14t + 10, \ 0 \le t \le 5$

(b) The object is moving to the right when $0 < t < 0.88$ and when $3.79 < t < 5$.
The object is moving to the left when $0.88 < t < 3.79$

(c) The object stops at times $t \cong 0.88$ and $t \cong 3.79$.
The maximum speed is $v \cong 6.33$ at $t \cong 2.33$.

(d) $a(t) = 6t - 14$
The object is speeding up when $v(t)$ and $a(t)$ have
the same sign: $0.88 < t < 2.33$ and $3.79 < t < 5$.
The object is slowing down when $v(t)$ and $a(t)$ have
opposite sign: $0 < t < 0.88$ and $2.33 < t < 3.79$.

PROJECT 4.9a

1. length of arc $= r\theta$, speed $= \dfrac{d}{dt}[r\theta] = r\dfrac{d\theta}{dt} = r\omega$

2. $v = r\omega$ so $KE = \frac{1}{2}mr^2\omega^2$.

3. We know that $d\theta/dt = \omega$ and, at time $t = 0, \ \theta = \theta_0$.   Therefore $\theta = \omega t + \theta_0$.   It follows that

$$x(t) = r\cos(\omega t + \theta_0) \quad and \quad y(t) = r\sin(\omega t + \theta_0).$$

$x(t) = r\cos(\omega t + \theta_0), \quad y(t) = r\sin(\omega t + \theta_0)$
$v(t) = x'(t) = -r\omega\sin(\omega t + \theta_0) = -\omega\,y(t)$
$a(t) = -r\omega^2\cos(\omega t + \theta_0) = -\omega^2\,x(t)$
$v(t) = y'(t) = r\omega\cos(\omega t + \theta_0) = \omega\,x(t)$
$a(t) = -r\omega^2\sin(\omega t + \theta_0) = -\omega^2\,y(t)$

4. For the sector $\quad A = \frac{1}{2}r^2\theta, \quad \dfrac{dA}{dt} = \frac{1}{2}r^2\dfrac{d\theta}{dt} = \frac{1}{2}r^2\omega \quad$ is constant.

For triangle T

$$A = \tfrac{1}{2}(2r\sin\tfrac{1}{2}\theta)(r\cos\tfrac{1}{2}\theta)$$

$$= \tfrac{1}{2}r^2(2\sin\tfrac{1}{2}\theta\cos\tfrac{1}{2}\theta) = \tfrac{1}{2}r^2\sin\theta,$$

$$\dfrac{dA}{dt} = \frac{1}{2}r^2\cos\theta\,\dfrac{d\theta}{dt} = \frac{1}{2}r^2\omega\cos\theta \quad \text{varies with } \theta.$$

For segment S

$$A = \tfrac{1}{2}r^2\theta - \tfrac{1}{2}r^2\sin\theta = \tfrac{1}{2}r^2(\theta - \sin\theta),$$

$$\dfrac{dA}{dt} = \frac{1}{2}r^2\left(\dfrac{d\theta}{dt} - \cos\dfrac{d\theta}{dt}\right) = \frac{1}{2}r^2\omega(1 - \cos\theta) \quad \text{varies with } \theta.$$

5. From Exercise 4, $\quad \dfrac{dA_T}{dt} = \tfrac{1}{2}r^2\omega\cos\theta$ and $\dfrac{dA_S}{dt} = \tfrac{1}{2}r^2\omega - \tfrac{1}{2}r^2\omega\cos\theta$

Now,

$$\frac{1}{2}r^2\omega\cos\theta = \frac{1}{2}r^2\omega - \frac{1}{2}r^2\omega\cos\theta \quad\Longrightarrow\quad \cos\theta = \frac{1}{2} \quad\Longrightarrow\quad \theta = \frac{\pi}{3}.$$

PROJECT 4.9B

1.

$$\frac{d}{dt}\left[mgy + \tfrac{1}{2}mv^2\right] = mg\frac{dy}{dt} + \frac{1}{2}m\frac{d}{dt}(v^2)$$

$$= mgv + \frac{1}{2}m\left[2v\frac{dv}{dt}\right]$$

$$= mgv + mv\frac{dv}{dt}$$

$$= mgv + mv(-g) \quad (\text{since} \quad dv/dt = a = -g)$$

$$= mgv - mgv = 0$$

2. By Problem 1, $mgy + \tfrac{1}{2}mv^2 = C$ (constant). Since $v = 0$ at height $y = y_0$, we have $C = mgy_0$. Thus,

$$mgy_0 = mgy + \tfrac{1}{2}mv^2 \quad \text{and} \quad |v| = \sqrt{2g(y_0 - y)}$$

3. $y(t) = -\tfrac{1}{2}gt^2 + y_0 \quad\Longrightarrow\quad gt = \sqrt{2g(y_0 - y)}$

$v(t) = y'(t) = -gt \quad$ Therefore, $\quad |v(t)| = \sqrt{2g(y_0 - y)}.$

SECTION 4.10

1. $x + 2y = 2, \quad \dfrac{dx}{dt} + 2\dfrac{dy}{dt} = 0$

   (a) If $\dfrac{dx}{dt} = 4$, then $\dfrac{dy}{dt} = -2$ units/sec.   (b) If $\dfrac{dy}{dt} = -2$, then $\dfrac{dx}{dt} = 4$ units/sec.

2. $x^2 + y^2 = 25$, $\quad 2x\dfrac{dx}{dt} + 2y\dfrac{dy}{dt} = 0$ $\quad$ and $\quad \dfrac{dx}{dt} = -\dfrac{y}{x}\dfrac{dy}{dt}$.

At the point $(3,4)$, $\dfrac{dy}{dt} = -2$. Therefore, $\dfrac{dx}{dt} = \dfrac{8}{3}$; $\quad$ the x-coordinate is increasing at the rate of 8/3 units per second.

3. $y^2 = 4(x+2)$, $\quad 2y\dfrac{dy}{dt} = 4\dfrac{dx}{dt}$ $\quad$ and $\quad \dfrac{dx}{dt} = \frac{1}{2}\,y\dfrac{dy}{dt}$

At the point $(7,6)$, $\dfrac{dy}{dt} = 3$. Therefore $\dfrac{dx}{dt} = \frac{1}{2}\cdot 6 \cdot 3 = 9$ units/sec.

4. We are given $\dfrac{dx}{dt} = 2$. Also $4y = (x+2)^2$ so $4\dfrac{dy}{dt} = 2(x+2)\dfrac{dx}{dt}$ or $\dfrac{dy}{dt} = \frac{1}{2}(x+2)\dfrac{dx}{dt}$.

At $x = 2$, $\dfrac{dy}{dt} = 4$. The distance from a point on the parabola to the point $(-2,0)$ is given by

$$S = \sqrt{(x+2)^2 + y^2} = \sqrt{4y + y^2} \quad \text{since } (x+2)^2 = 4y. \quad \text{Now}$$

$$\frac{dS}{dt} = \frac{1}{2}(4y + y^2)^{-1/2}(4 + 2y)\frac{dy}{dt} = \frac{2+y}{\sqrt{4y+y^2}}\frac{dy}{dt}.$$

Therefore, at the point $(2,4)$, $\quad \dfrac{dS}{dt} = \dfrac{6}{\sqrt{32}}4 = 3\sqrt{2}$.

5. Let $s = \sqrt{x^2 + y^2}$ denote the distance to the origin at time t. Since $x = 4\cos t$ and $y = 2\sin t$, we have

$$s(t) = \sqrt{16\cos^2 t + 4\sin^2 t} = \sqrt{12\cos^2 t + 4}$$

$$\frac{ds}{dt} = \frac{1}{2}(12\cos^2 t + 4)^{-1/2}(-24\cos t \sin t)$$

$$= \frac{-12\cos t \sin t}{\sqrt{12\cos^2 t + 4}}$$

At $t = \pi/4$, $\quad \dfrac{ds}{dt} = \dfrac{-12\cos(\pi/4)\sin(\pi/4)}{\sqrt{12\cos^2(\pi/4) + 4}} = -\frac{3}{5}\sqrt{10}$.

6. $y = x\sqrt{x} = x^{3/2}$; $\quad \dfrac{dy}{dt} = \dfrac{3}{2}x^{1/2}\dfrac{dx}{dt}$.

Now $\dfrac{dx}{dt} = \dfrac{dy}{dt} = z \neq 0$, $\quad \Longrightarrow \quad \frac{3}{2}x^{1/2} = 1 \quad \Longrightarrow \quad x = \frac{4}{9}$ and $y = \frac{8}{27}$.

Both coordinates are changing at the same rate at the point $(4/9, 8/27)$.

7. Find $\quad \dfrac{dx}{dt}$ and $\dfrac{dS}{dt}$ $\quad$ when $\quad V = 27\text{m}^3$

given that $\quad \dfrac{dV}{dt} = -2\text{m}^3/\text{min}.$

(*) $\quad V = x^3, \quad S = 6x^2$

Differentiation of equations (*) gives

$$\frac{dV}{dt} = 3x^2\frac{dx}{dt} \quad and \quad \frac{dS}{dt} = 12x\frac{dx}{dt}.$$

When $V = 27$, $x = 3$. Substituting $x = 3$ and $dV/dt = -2$, we get

$$-2 = 27\frac{dx}{dt} \quad \text{so that} \quad \frac{dx}{dt} = -2/27 \quad \text{and} \quad \frac{dS}{dt} = 12(3)\left(\frac{-2}{27}\right) = -8/3.$$

The rate of change of an edge is $-2/27$ m/min; the rate of change of the surface area is $-8/3$ m²/min.

8.　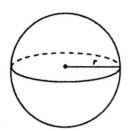
Find $\dfrac{dr}{dt}$ and $\dfrac{dS}{dt}$ when $r = 10$ ft

given that $\dfrac{dV}{dt} = 8$ ft³/min.

(∗)　$V = \frac{4}{3}\pi r^3$, $\quad S = 4\pi r^2$

Differentiation of equations (∗) with respect to t gives

$$\frac{dV}{dt} = 4\pi r^2\frac{dr}{dt} \quad \text{and} \quad \frac{dS}{dt} = 8\pi r\frac{dr}{dt}.$$

Substituting $r = 10$ and $dV/dt = 8$, we get

$$8 = 4\pi(10)^2\frac{dr}{dt} \quad \text{so that} \quad \frac{dr}{dt} = \frac{1}{50\pi} \quad \text{and} \quad \frac{dS}{dt} = 8\pi(10)\frac{1}{50\pi} = \frac{8}{5}.$$

The radius is increasing $\dfrac{1}{50\pi}$ ft/min; the surface area is increasing $\dfrac{8}{5}$ ft²/min.

9. The area of an equilateral triangle of side x is given by

$$A = \frac{1}{2}x\left(\frac{x\sqrt{3}}{2}\right) = \frac{\sqrt{3}}{4}x^2. \quad \text{Thus} \quad \frac{dA}{dt} = \frac{\sqrt{3}}{2}x\frac{dx}{dt}.$$

When $x = \alpha$, $\dfrac{dx}{dt} = k$, and $\dfrac{dA}{dt} = \frac{\sqrt{3}}{2}\alpha k$ cm²/min.

10. We have $2x + 2y = 24$, or $x + y = 12$. Thus, $A = xy = x(12 - x) = 12x - x^2$.
When $A = 32$, $x = 4$ or $x = 8$, and it follows that $\dfrac{dA}{dt} = (12 - 2x)\dfrac{dx}{dt} = \pm 4\dfrac{dx}{dt}$.

11.　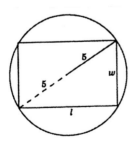
Find $\dfrac{dA}{dt}$ when $l = 6$ in.

given that $\dfrac{dl}{dt} = -2$ in./sec.

By the Pythagorean theorem

$$l^2 + w^2 = 100.$$

Also, $A = lw$. Thus, $A = l\sqrt{100 - l^2}$. Differentiation with respect to t gives

$$\frac{dA}{dt} = l\left(\frac{-l}{\sqrt{100 - l^2}}\right)\frac{dl}{dt} + \sqrt{100 - l^2}\,\frac{dl}{dt}.$$

Substituting $l = 6$ and $dl/dt = -2$, we get

$$\frac{dA}{dt} = 6\left(\frac{-6}{8}\right)(-2) + (8)(-2) = -7.$$

The area is decreasing at the rate of 7 in.²/sec.

12.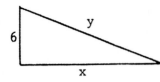

$$x^2 + 6^2 = y^2; \quad 2x\frac{dx}{dt} = 2y\frac{dy}{dt}$$

If $y = 30$ ft and $\frac{dx}{dt} = 8$ ft/min, then

$$\frac{dy}{dt} = \frac{x}{y}\frac{dy}{dt} = \frac{\sqrt{(30)^2 - 36}}{30}8 = \frac{16}{5}\sqrt{6} \text{ ft/min}$$

13.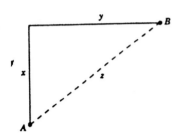

Compare $\frac{dy}{dt}$ to $\frac{dx}{dt} = -13$ mph

given that $z = 16$ and $\frac{dz}{dt} = -17$

when $x = y$.

By the Pythagorean theorem $x^2 + y^2 = z^2$. Thus,

$$2x\frac{dx}{dt} + 2y\frac{dy}{dt} = 2z\frac{dz}{dt}.$$

Since $x = y$ when $z = 16$, we have $x = y = 8\sqrt{2}$ and

$$2(8\sqrt{2})(-13) + 2(8\sqrt{2})\frac{dy}{dt} = 2(16)(-17).$$

Solving for dy/dt, we get

$$-13\sqrt{2} + \sqrt{2}\frac{dy}{dt} = -34 \quad \text{or} \quad \frac{dy}{dt} = \frac{1}{\sqrt{2}}(13\sqrt{2} - 34) \cong -11.$$

Thus, boat A wins the race.

14. $V = \frac{4}{3}\pi r^3$, so $\frac{dV}{dt} = 4\pi r^2\frac{dr}{dt} = 4\pi r^2(-\frac{1}{5})$. Thus at $r = 12$ we have $\frac{dV}{dt} = -\frac{576}{5}\pi$ cm^3/min.

15. We want to find $\frac{dA}{dt}$ when $\frac{dx}{dt} = 2$ and $x = 12$.

$A = \frac{1}{2}x\sqrt{169 - x^2}$, so $\frac{dA}{dt} = \left[\frac{1}{2}\sqrt{169 - x^2} - \frac{x^2}{2\sqrt{169 - x^2}}\right]\frac{dx}{dt}$

When $\frac{dx}{dt} = 2$ and $x = 12$, $\frac{dA}{dt} = -\frac{119}{5}$ ft^2/sec.

16.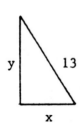

$x^2 + y^2 = (13)^2; \quad 2x\frac{dx}{dt} + 2y\frac{dy}{dt} = 0$ and $\frac{dy}{dt} = -\frac{x}{2y}$

since $\frac{dx}{dt} = 0.5$. When $x = 5$, $y = 12$ and

$\frac{dy}{dt} = -\frac{5}{24}$; the top of the ladder is

dropping 5/24 ft/sec.

17. We want to find dV/dt when $V = 1000$ ft^3 and $P = 5$ lb/in.^{2} given that $dP/dt = -0.05$ lb/in.^{2}/hr.

Differentiating $PV = C$ with respect to t, we get

$$P\frac{dV}{dt} + V\frac{dP}{dt} = 0 \quad \text{so that} \quad 5\frac{dV}{dt} + 1000(-0.05) = 0. \quad \text{Thus,} \quad \frac{dV}{dt} = 10.$$

The volume increases at the rate of 10 ft^3/hr.

18. $PV^{1.4} = C;$ $V^{1.4}\dfrac{dP}{dt} + (1.4)PV^{0.4}\dfrac{dV}{dt} = 0$ and $\dfrac{dP}{dt} = -\dfrac{1.4P}{V}\dfrac{dV}{dt}.$

With $V = 10$, $P = 50$ and $\dfrac{dV}{dt} = -1$, we have $\dfrac{dP}{dt} = -\dfrac{(1.4)50}{10}(-1) = 7$

The pressure is increasing $7\,\text{lb/in}^2/\text{sec}.$

19.

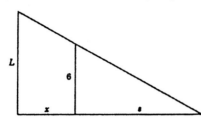

Find $\dfrac{ds}{dt}$ when $x = 3$ ft (and $s = 4$ ft)

given that $\dfrac{dx}{dt} = 400$ ft/min.

By similar triangles

$$\frac{L}{x+s} = \frac{6}{s}.$$

Substitution of $x = 3$ and $s = 4$ gives us $\dfrac{L}{7} = \dfrac{6}{4}$ so that the lamp post is $L = 10.5$ ft tall. Rewriting

$$\frac{10.5}{x+s} = \frac{6}{s} \quad \text{as} \quad s = \frac{4}{3}x$$

and differentiating with respect to t, we find that

$$\frac{ds}{dt} = \frac{4}{3}\frac{dx}{dt} = \frac{1600}{3}.$$

The shadow lengthens at the rate of $1600/3$ ft/min.

If the tip of his shadow is at the point z, then

$$z = x + s \quad \text{and} \quad \frac{dz}{dt} = \frac{dx}{dt} + \frac{ds}{dt} = 400 + \frac{1600}{3} = \frac{2800}{3}.$$

The tip of his shadow is moving at the rate of $2800/3$ ft./min.

20.

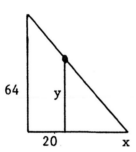

$y(t) = 64 - 16t^2;$ $\dfrac{dy}{dt} = -32t$

By similar triangles, $\dfrac{y}{x} = \dfrac{64}{20+x}.$

Thus, $y = \dfrac{64x}{20+x}.$

Now, $\dfrac{dy}{dt} = \dfrac{(20+x)(64) - 64x}{(20+x)^2}\dfrac{dx}{dt} = \dfrac{1280}{(20+x)^2} \implies \dfrac{dx}{dt} = \dfrac{(20+x)^2}{1280}\dfrac{dy}{dt}.$

At $t = 1$, $y = 48$, $x = 60$, and $\dfrac{dy}{dt} = -32$. Therefore, $\dfrac{dx}{dt} = \dfrac{(80)^2}{1280}(-32) = -160$ ft/sec.

21.   Let $W(t) = 150\left(1 + \frac{1}{4000}r\right)^{-2}$. We want to find dW/dt when $r = 400$ given that $dr/dt = 10$ mi/sec. Differentiating with respect to t, we get

$$\frac{dW}{dt} = -300\left(1 + \frac{1}{4000}r\right)^{-3}\left(\frac{1}{4000}\right)\frac{dr}{dt}$$

Now set $r = 400$ and $dr/dt = 10$. Then

$$\frac{dW}{dt} = -300\left(1 + \frac{400}{4000}\right)^{-3}\frac{10}{4000} \cong -0.5634\,\text{lbs/sec}$$

22. $M = \dfrac{m}{\sqrt{1 - v^2/c^2}}$; $\dfrac{dM}{dt} = m\left(-\tfrac{1}{2}\right)\left(1 - v^2/c^2\right)^{-3/2}\left(-2v/c^2\right)\dfrac{dv}{dt} = \dfrac{mv}{c^2\left(1 - v^2/c^2\right)^{3/2}}\dfrac{dv}{dt}.$

If $v = \dfrac{c}{2}$ and $\dfrac{dv}{dt} = \dfrac{c}{100}$, then $\dfrac{dM}{dt} = \dfrac{m(c/2)}{c^2\left(1 - c^2/4c^2\right)^{3/2}}\dfrac{c}{100} = \dfrac{\sqrt{3}}{225}\,m.$

23.

Find $\dfrac{dh}{dt}$ when $h = 3$ in.

given that $\dfrac{dV}{dt} = -\dfrac{1}{2}$ cu in./min.

By similar triangles

$\quad r = \tfrac{1}{3}h.$

Thus $V = \tfrac{1}{3}\pi r^2 h = \tfrac{1}{27}\pi h^3$. Differentiating with respect to t, we get

$$\dfrac{dV}{dt} = \dfrac{1}{9}\pi h^2 \dfrac{dh}{dt}.$$

When $h = 3$,

$$-\dfrac{1}{2} = \dfrac{1}{9}\pi(9)\dfrac{dh}{dt} \quad\text{and}\quad \dfrac{dh}{dt} = -\dfrac{1}{2\pi}.$$

The water level is dropping at the rate of $1/2\pi$ inches per minute.

24.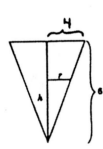

$V = \dfrac{1}{3}\pi r^2 h$ and $\dfrac{r}{h} = \dfrac{4}{6}$ (similar triangles)

so $V = \dfrac{4}{27}\pi h^3$.   Thus $\dfrac{dV}{dt} = \dfrac{4}{9}\pi h^2 \dfrac{dh}{dt}$,

so at $\dfrac{dh}{dt} = 0.5$ and $h = 2$,

$\dfrac{dV}{dt} = \dfrac{8\pi}{9}$ cubic ft per sec.

25. $\dfrac{dV}{dt} = 4\pi r^2 \dfrac{dr}{dt}$ and $\dfrac{dSA}{dt} = 8\pi r \dfrac{dr}{dt}$.   Thus when $\dfrac{dSA}{dt} = 4$ and $\dfrac{dr}{dt} = 0.1$

we get $r = \dfrac{5}{\pi}$ and $\dfrac{dV}{dt} = \dfrac{10}{\pi}$ cm^3/min.

26. $V = \pi r h^2 - \tfrac{1}{3}\pi h^3$; $\dfrac{dV}{dt} = 2\pi r h \dfrac{dh}{dt} - \pi h^2 \dfrac{dh}{dt}$, and $\dfrac{dh}{dt} = \dfrac{2}{\pi(14h - h^2)}$ since $r = 7$ and $\dfrac{dV}{dt} = 2.$

(a) When $h = 7/2$, $\dfrac{dh}{dt} = \dfrac{8}{147\pi}$ in./sec.   (b) When $h = 7$, $\dfrac{dh}{dt} = \dfrac{2}{49\pi}$ in./sec.

27.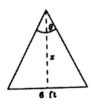

Find $\dfrac{d\theta}{dt}$ when $x = 4$ ft

given that $\dfrac{dx}{dt} = 2$ in./min.

$(*)\quad \tan\dfrac{\theta}{2} = \dfrac{3}{x}$

Differentiation of $(*)$ with respect to t gives

$$\dfrac{1}{2}\sec^2\dfrac{\theta}{2}\dfrac{d\theta}{dt} = -\dfrac{3}{x^2}\dfrac{dx}{dt} \quad\text{or}\quad \dfrac{d\theta}{dt} = -\dfrac{6}{x^2}\cos^2\dfrac{\theta}{2}\dfrac{dx}{dt}.$$

Note that $dx/dt = 2$ in./min$=1/6$ ft/min. When $x = 4$, we have $\cos\theta/2 = 4/5$ and thus

$$\frac{d\theta}{dt} = -\frac{6}{16}\left(\frac{4}{5}\right)^2\left(\frac{1}{6}\right) = -\frac{1}{25}.$$

The vertex angle decreases at the rate of 0.04 rad/min.

28.

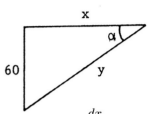

$$\tan\alpha = \frac{60}{x}; \quad \sec^2\alpha\frac{d\alpha}{dt} = -\frac{60}{x^2}\frac{dx}{dt}$$

Now, $\sec^2\alpha = \left(\dfrac{y}{x}\right)^2$ so $\dfrac{d\alpha}{dt} = -\dfrac{60}{y^2}\dfrac{dx}{dt}.$

With $y = 100$, and $\dfrac{dx}{dt} = -10$, we have $\dfrac{d\alpha}{dt} = -\dfrac{60}{(100)^2}(-10) = 0.06$ rad/min.

29.

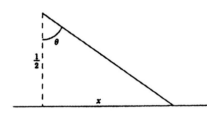

Find $\dfrac{dx}{dt}$ when $x = 1$ mi

given that $\dfrac{d\theta}{dt} = 2\pi$ rad/min.

$(*)\qquad \tan\theta = \dfrac{x}{1/2} = 2x$

Differentiation of $(*)$ with respect to t gives

$$\sec^2\theta\frac{d\theta}{dt} = 2\frac{dx}{dt}.$$

When $x = 1$, we get $\sec\theta = \sqrt{5}$ and thus $\dfrac{dx}{dt} = 5\pi.$

The light is traveling at 5π mi/min.

30. (a) We have $\tan\theta = x$, so $\sec^2\theta\dfrac{d\theta}{dt} = \dfrac{dx}{dt}.$ Switching to radians, $\dfrac{d\theta}{dt} = 4\pi$

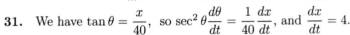

Thus at $\theta = \dfrac{\pi}{4}$, $\dfrac{dx}{dt} = 8\pi$ mi/min.

(b) At $\theta = 0$, $\dfrac{dx}{dt} = 4\pi$ mi/min.

31. We have $\tan\theta = \dfrac{x}{40}$, so $\sec^2\theta\dfrac{d\theta}{dt} = \dfrac{1}{40}\dfrac{dx}{dt}$, and $\dfrac{dx}{dt} = 4.$

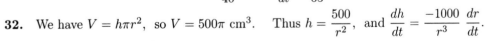

At $t = 15, x = 60$ and $\sec\theta = \dfrac{\sqrt{5200}}{40}$, so $\dfrac{d\theta}{dt} = \dfrac{2}{65} \cong 0.031$ rad/sec.

32. We have $V = h\pi r^2$, so $V = 500\pi$ cm^3.   Thus $h = \dfrac{500}{r^2}$, and $\dfrac{dh}{dt} = \dfrac{-1000}{r^3}\dfrac{dr}{dt}.$

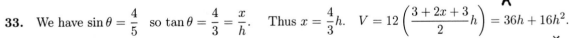

At $r = 5$ and $\dfrac{dr}{dt} = \dfrac{1}{2}$, we have $\dfrac{dh}{dt} = -4$ cm/sec.

33. We have $\sin\theta = \dfrac{4}{5}$ so $\tan\theta = \dfrac{4}{3} = \dfrac{x}{h}.$ Thus $x = \dfrac{4}{3}h.$ $V = 12\left(\dfrac{3+2x+3}{2}h\right) = 36h + 16h^2.$

Thus $\dfrac{dV}{dt} = (36 + 32h)\dfrac{dh}{dt}$, so at $\dfrac{dV}{dt} = 10$ and $h = 2,$

$\dfrac{dh}{dt} = \dfrac{1}{10}$ ft/min.

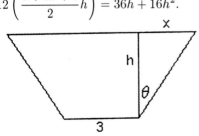

34. $\tan\alpha = \dfrac{y}{x}; \quad \sec^2\alpha\dfrac{d\alpha}{dt} = \dfrac{x\dfrac{dy}{dt} - y\dfrac{dx}{dt}}{x^2}$

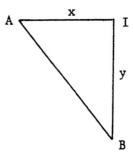

$$\dfrac{d\alpha}{dt} = \dfrac{x\dfrac{dy}{dt} - y\dfrac{dx}{dt}}{x^2} \cdot \dfrac{x^2}{x^2 + y^2}$$

$$= \dfrac{x\dfrac{dy}{dt} - y\dfrac{dx}{dt}}{x^2 + y^2}$$

Now $\dfrac{dx}{dt} = -30\,\text{mph} = -44\,\text{ft/sec}$ and $\dfrac{dy}{dt} = -22.5\,\text{mph} = -33\,\text{ft/sec}.$

At $x = 300,\ y = 400,$ we have

$$\dfrac{d\alpha}{dt} = \dfrac{300(-33) - 400(-44)}{(300)^2 + (400)^2} = 0.0308 \text{ the angle is increasing } 0.0308 \text{ rad/sec.}$$

35. Find $\dfrac{d\theta}{dt}$ when $y = 4$ ft

given that $\dfrac{dx}{dt} = 3$ ft/sec.

$\tan\theta = \dfrac{16}{x}, \quad x^2 + (16)^2 = (16 + y)^2$

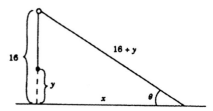

Differentiating $\tan\theta = 16/x$ with respect to t, we obtain

$$\sec^2\theta\dfrac{d\theta}{dt} = \dfrac{-16}{x^2}\dfrac{dx}{dt} \quad\text{and thus}\quad \dfrac{d\theta}{dt} = \dfrac{-16}{x^2}\cos^2\theta\dfrac{dx}{dt}.$$

From $x^2 + (16)^2 = (16 + y)^2$ we conclude that $x = 12$, when $y = 4$. Thus

$$\cos\theta = \dfrac{x}{16 + y} = \dfrac{12}{20} = \dfrac{3}{5} \quad\text{and}\quad \dfrac{d\theta}{dt} = \dfrac{-16}{(12)^2}\left(\dfrac{3}{5}\right)^2(3) = \dfrac{-3}{25}.$$

The angle decreases at the rate of 0.12 rad/sec.

36. $\tan(\alpha/2) = \dfrac{3}{x}; \quad \sec^2(\alpha/2)\dfrac{1}{2}\dfrac{d\alpha}{dt} = -\dfrac{3}{x^2}\dfrac{dx}{dt},$

and $\dfrac{d\alpha}{dt} = -\dfrac{6}{y^2}\dfrac{dx}{dt}$

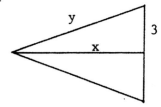

Initially, $y = 5$ so $x = 4$. 8 seconds later $x = 12$ so $y = \sqrt{(12)^2 + (3)^2} = \sqrt{153}$. Therefore,

$$\dfrac{d\alpha}{dt} = -\dfrac{6}{y^2}\dfrac{dx}{dt} = -\dfrac{6}{153}(1) = -\dfrac{6}{153}; \text{ the angle is decreasing } -6/153 \text{ rad/sec.}$$

37. Find $\dfrac{d\theta}{dt}$ when $t = 6$ min.

$\tan\theta = \dfrac{100t}{500 + 75t} = \dfrac{4t}{20 + 3t}$

Differentiation with respect to t gives

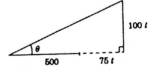

$$\sec^2\theta\dfrac{d\theta}{dt} = \dfrac{(20 + 3t)4 - 4t(3)}{(20 + 3t)^2} = \dfrac{80}{(20 + 3t)^2}.$$

When $t = 6$

$$\tan\theta = \dfrac{24}{38} = \dfrac{12}{19} \quad\text{and}\quad \sec^2\theta = 1 + \left(\dfrac{12}{19}\right)^2 = \dfrac{505}{361}$$

so that

$$\frac{d\theta}{dt} = \frac{80}{(20 + 3t)^2} \cdot \frac{1}{\sec^2 \theta} = \frac{80}{(38)^2} \cdot \frac{361}{505} = \frac{4}{101}.$$

The angle increases at the rate of 4/101 rad/min.

38. $\tan \alpha = \dfrac{x}{H}$; $\sec^2 \alpha \dfrac{d\alpha}{dt} = \dfrac{1}{H} \dfrac{dx}{dt}$

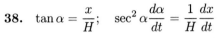

$\dfrac{d\alpha}{dt} = \dfrac{1}{H} \dfrac{dx}{dt} \cos^2 \alpha = \dfrac{H}{H^2 + x^2} \dfrac{dx}{dt}$

We have $H = 2$ mi. and $\dfrac{dx}{dt} = 400$ mph. After 2 seconds, $x = 400 \left(\dfrac{2}{3600} \right) = \dfrac{2}{9}$ miles.

$$\frac{d\alpha}{dt} = \frac{2}{2^2 + (2/9)^2}(400) = \frac{200(81)}{82} \text{ rad/hr} = \frac{9}{164} \text{ rad/sec.}$$

39. Let x be the distance from third base to the player. Then the distance from home plate to the player is given by: $y = \sqrt{(90)^2 + x^2}$. Differentiation with respect to t gives

$$\frac{dy}{dt} = \frac{x}{\sqrt{(90)^2 + x^2}} \frac{dx}{dt}$$

We are given that $dx/dt = -15$. Therefore,

$$\frac{dy}{dt} = -\frac{15x}{\sqrt{(90)^2 + x^2}} \quad \text{and} \quad \frac{dy}{dt}\bigg|_{x=10} = \frac{-15}{\sqrt{82}} \cong 1.66$$

The distance between home plate to the player is decreasing 1.66 ft./sec.

40. The plane is flying at an altitude 6 miles. If y denotes the horizontal distance between the radar station and the plane, then

$$\tan \theta = \frac{6}{y} \quad \Longrightarrow \quad \sec^2 \theta \frac{d\theta}{dt} = \frac{-6}{y} \frac{dy}{dt}$$

Therefore,

$$\frac{dy}{dt} = \frac{y^2 \sec^2 \theta}{-6} \frac{d\theta}{dt}$$

At the instant the plane is 12 miles from the station, $y = \sqrt{108}$, $\theta = \frac{1}{6}\pi$ and the speed is

$$\left| \frac{dy}{dt} \right| = \left| \frac{108}{-6} \frac{4}{3} \frac{\pi}{360} \right| = \frac{4\pi}{60} \text{ mi/sec.} \cong 754 \text{ mi/hr.}$$

41. Let x be the position of the runner on the track, let y be the distance between the runner and the spectator, and let θ be the central angle indicated in the figure. The runner's speed is 5 meters per second so

$50 \dfrac{d\theta}{dt} = 5$ and $\dfrac{d\theta}{dt} = \dfrac{1}{10}$.

By the law of cosines:

$y^2 = (50)^2 + (200)^2 - 2(50)(200) \cos \theta.$

Differentiating with respect to t, we get

$2y \dfrac{dy}{dt} = 20,000 \sin \theta \, \dfrac{d\theta}{dt}$

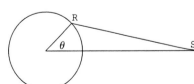

$$\frac{dy}{dt} = \frac{10,000 \sin\theta}{y} \frac{1}{10}$$

$$\frac{dy}{dt} = \frac{1,000}{y} \sin\theta$$

Now, $\left.\dfrac{dy}{dt}\right|_{y=200} = 5\dfrac{\sqrt{(200)^2 - (25)^2}}{200} = 5\dfrac{\sqrt{39375}}{200} = 4.96.$

The runner is approaching the spectator at the rate of 4.96 meters per second.

42. $\dfrac{dx}{dt} = \dfrac{23}{4}$ **43.** $\dfrac{dx}{dt} = 4$ **44.** $\dfrac{dy}{dt} = -\dfrac{1}{2\sqrt{3}}$

SECTION 4.11

1. $\Delta V = (x + h)^3 - x^3$
$\qquad = (x^3 + 3x^2h + 3xh^2 + h^3) - x^3$
$\qquad = 3x^2h + 3xh^2 + h^3,$
$\quad dV = 3x^2h,$
$\Delta V - dV = 3xh^2 + h^3 \qquad$ (see figure)

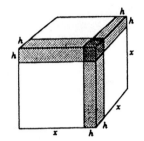

2.   The area of the ring can be thought of as the increase of the area of a disk as the radius increases from r to $r + h$.

$$A = \pi r^2; \quad \text{so} \quad dA = 2\pi r\,dr = 2\pi rh$$

The exact area is: $\pi(r + h)^2 - \pi r^2 = \pi(r^2 + 2rh + h^2) - \pi r^2 = 2\pi rh + \pi h^2.$

3. $f(x) = x^{1/3}, \quad x = 1000, \quad h = 2, \quad f'(x) = \frac{1}{3}x^{-2/3}$

$\sqrt[3]{1002} = f(x + h) \cong f(x) + hf'(x) = \sqrt[3]{1000} + 2\left[\frac{1}{3}(1000)^{-2/3}\right] = 10 + \frac{1}{150} \cong 10.0067$

4. $f(x) = \dfrac{1}{\sqrt{x}}, \quad x = 25, \quad h = -0.5, \quad f'(x) = \dfrac{-1}{2x^{3/2}}$

$\dfrac{1}{\sqrt{24.5}} = f(x + h) \cong f(x) + hf'(x) = \dfrac{1}{5} + (-\dfrac{1}{2})\dfrac{-1}{2(5)^3} = \dfrac{101}{500} = 0.202$

5. $f(x) = x^{1/4}, \quad x = 16, \quad h = -0.5, \quad f'(x) = \frac{1}{4}x^{-3/4}$

$(15.5)^{1/4} = f(x + h) \cong f(x) + hf'(x) = (16)^{1/4} + (-0.5)\left[\frac{1}{4}(16)^{-3/4}\right] = 1\frac{63}{64} \cong 1.9844$

6. $f(x) = x^{2/3}, \quad x = 27, \quad h = -1, \quad f'(x) = \frac{2}{3}x^{-1/3}$

$(26)^{2/3} = f(x + h) \cong f(x) + hf'(x) = (27)^{2/3} + (-1)\left[\frac{2}{3}(27)^{-1/3}\right] = \dfrac{79}{9} \cong 8.778$

7. $f(x) = x^{3/5}, \quad x = 32, \quad h = 1, \quad f'(x) = \frac{3}{5}x^{-2/5}$

$\quad (33)^{3/5} = f(x+h) \cong f(x) + hf'(x) = (32)^{3/5} + (1)\left[\frac{3}{5}(32)^{-2/5}\right] = 8.15$

8. $f(x) = x^{-1/5}, \quad x = 32, \quad h = 1, \quad f'(x) = -\frac{1}{5}x^{-6/5}$

$\quad (33)^{-1/5} = f(x+h) \cong f(x) + hf'(x) = (32)^{-1/5} - (1)\left[\frac{1}{5}(32)^{-6/5}\right] = \frac{159}{320} \cong 0.497$

9. $f(x) = \sin x, \quad x = \frac{\pi}{4}, \quad h = \frac{\pi}{180}, \quad f'(x) = \cos x$

$\quad \sin 46° = f(x+h) \cong f(x) + hf'(x) = \sin\frac{\pi}{4} + \frac{\pi}{180}\cos\frac{\pi}{4} = \frac{\sqrt{2}}{2}\left(1 + \frac{\pi}{180}\right) \cong 0.719$

10. $f(x) = \cos x, \quad x = \frac{\pi}{3}, \quad h = \frac{\pi}{90}, \quad f'(x) = -\sin x$

$\quad \cos 62° = f(x+h) \cong f(x) + hf'(x) = \cos\frac{\pi}{3} + \frac{\pi}{90}\left(-\sin\frac{\pi}{3}\right) = \frac{1}{2} - \left(\frac{\pi}{90}\right)\left(\frac{\sqrt{3}}{2}\right) \cong 0.470$

11. $f(x) = \tan x, \quad x = \frac{\pi}{6}, \quad h = \frac{-\pi}{90}, \quad f'(x) = \sec^2 x$

$\quad \tan 28° = f(x+h) \cong f(x) + hf'(x) = \tan\frac{\pi}{6} + \left(\frac{-\pi}{90}\right)\left(\frac{4}{3}\right) = \frac{\sqrt{3}}{3} - \frac{2\pi}{135} \cong 0.531$

12. $f(x) = \sin x, \quad x = \frac{\pi}{4}, \quad h = -\frac{\pi}{90}, \quad f'(x) = \cos x$

$\quad \sin 43° = f(x+h) \cong f(x) + hf'(x) = \sin\frac{\pi}{4} + \left(-\frac{\pi}{180}\right)\cos\frac{\pi}{4} = \frac{\sqrt{2}}{2} - \left(\frac{\pi}{180}\right)\left(\frac{\sqrt{2}}{2}\right) \cong 0.682$

13. $f(2.8) \cong f(3) + (-0.2)f'(3) = 2 + (-0.2)(2) = 1.6$

14. $f(5.4) \cong f(5) + (0.4)f'(5) = 1 + (0.4)(3) = 2.2$

15. $V(r) = \pi r^2 h; \quad \text{volume} = V(r+t) - V(r) \cong tV'(r) = 2\pi rht$

16.   Error in diameter $= 0.3 \quad \Longrightarrow \quad$ error in radius $= 0.15$.

     (a) $dS = 8\pi rh = 8\pi(8)(0.15) \cong 9.6\pi \text{ cm}^2$

     (b) $dV = 4\pi r^2 h = 4\pi(8)^2(0.15) \cong 38.4 \text{ cm}^3$

17. $V(x) = x^3, \quad V'(x) = 3x^2, \quad \Delta V \cong dV = V'(10)h = 300h$

$\quad |dV| \leq 3 \quad \Longrightarrow \quad |300h| \leq 3 \quad \Longrightarrow \quad |h| \leq 0.01, \quad \text{error} \leq 0.01 \text{ feet}$

18.   (a) Let $f(x) = \sqrt{x}$. Then $f'(x) = \frac{1}{2\sqrt{x}}$ and $\sqrt{x+1} - \sqrt{x} \cong (1)f'(x) = \frac{1}{2\sqrt{x}}$.

      Now, $\frac{1}{2\sqrt{x}} < 0.01 = \frac{1}{100} \quad \Longrightarrow \quad \sqrt{x} > 50 \quad \Longrightarrow \quad x > 2500$.

     (b) Let $f(x) = x^{1/4}$. Then $f'(x) = \frac{1}{4}x^{-3/4}$ and $\sqrt[4]{x+1} - \sqrt[4]{x} \cong (1)f'(x) = \frac{1}{4}x^{-3/4}$.

      Now, $\frac{1}{4}x^{-3/4} < 0.002 = \frac{2}{1000} \quad \Longrightarrow \quad x^{3/4} > 125 \quad \Longrightarrow \quad x > 625$.

19. $V(r) = \frac{2}{3}\pi r^3$ and $dr = 0.01$.

$$V(r + 0.01) - V(r) \cong V'(r)(0.01) = 2\pi r^2 (0.01)$$

$$= 2\pi (600)^2 (0.01) \quad (50 \text{ ft} = 600 \text{ in})$$

$$= 22619.5 \text{ in}^3 \quad \text{or} \quad 98 \text{ gallons (approx.)}$$

20. $V(r) = \frac{4}{3}\pi r^3$; $\dfrac{dV}{dr} = 4\pi r^2$.

Now, $V(r + h) - V(r) = 8 \times (10)6 \cong \dfrac{dV}{dr} h = 4\pi r^2 h$. Therefore,

$$h = \frac{8 \times (10)^6}{4\pi(4000)^2} \cong 0.0398 \text{ (miles)} \cong 210 \text{ (feet)}$$

21. $P = 2\pi \sqrt{\dfrac{L}{g}}$ implies $P^2 = 4\pi^2 \dfrac{L}{g}$

Differentiating with respect to t, we have

$$2P\frac{dP}{dt} = \frac{4\pi^2}{g} \cdot \frac{dL}{dt} = \frac{P^2}{L} \cdot \frac{dL}{dt} \quad \text{since} \quad \frac{P^2}{L} = \frac{4\pi^2}{g}.$$

Thus $\dfrac{dP}{P} = \dfrac{1}{2} \cdot \dfrac{dL}{L}$

22. $\dfrac{dP}{P} = -15 \text{ sec/hour} = -\dfrac{15}{3600}$. By Exercise 21,

$$\frac{1}{2}\frac{dL}{L} = -\frac{15}{3600} \quad \text{and} \quad dL = -\frac{30}{3600} L = -\frac{1}{120} L$$

With $L = 90$, we have $dL = -90/120 = -0.75$; the pendulum should be shortened

0.75 cm to 89.25 cm.

23. $L = 3.26 \text{ ft}$, $P = 2 \text{ sec}$, and $dL = 0.01 \text{ ft}$

$$\frac{dP}{P} = \frac{1}{2} \cdot \frac{dL}{L}$$

$$dP = \frac{1}{2} \cdot \frac{dL}{L} \cdot P = \frac{1}{2} \cdot \frac{0.01}{3.26} \cdot 2 \quad dP \cong 0.00307 \text{ sec}$$

24. Each edge increases by 0.1%; take $h = 0.001x$.

$S = 6x^2$, $dS = 12xh$, and $\dfrac{dS}{S} = \dfrac{12x(0.001x)}{6x^2} = 0.002 = 0.2\%$.

$A = x^3$, $dA = 3x^2 h$, and $\dfrac{dA}{A} = \dfrac{3x^2(0.001x)}{x^3} = 0.003 = 0.3\%$.

25. $A(x) = \dfrac{1}{4}\pi x^2$, $dA = \dfrac{1}{2}\pi x h$, $\dfrac{dA}{A} = 2\dfrac{h}{x}$

$$\frac{dA}{A} \le 0.01 \quad \Longleftrightarrow \quad 2\frac{h}{x} \le 0.01 \quad \Longleftrightarrow \quad \frac{h}{x} \le 0.005 \quad \text{within } \frac{1}{2}\%$$

26. (a) Let $y = x^n$. Then $dy = nx^{n-1}h$.

To get $\dfrac{dy}{y} = \dfrac{nx^{n-1}h}{x^n} < 0.01$, we need $\dfrac{h}{x} < \dfrac{0.01}{n}$, that is, within $\tfrac{1}{n}\%$.

(b) Let $y = x^{1/n}$. Then $dy = \tfrac{1}{n}x^{(1-n)/n}h$.

To get $\dfrac{dy}{y} = \dfrac{\tfrac{1}{n}x^{(1-n)/n}h}{x^{1/n}} < 0.01$, we need $\dfrac{h}{x} < (0.01)n$, that is, within $n\%$.

27. (a) and (b)   **28.** $\displaystyle\lim_{h\to 0} g(h) = \lim_{h\to 0}\frac{g(h)}{h}\cdot h = \left(\lim_{h\to 0}\frac{g(h)}{h}\right)\left(\lim_{h\to 0} h\right) = 0$

29. $\displaystyle\lim_{h\to 0}\frac{g_1(h)+g_2(h)}{h} = \lim_{h\to 0}\frac{g_1(h)}{h} + \lim_{h\to 0}\frac{g_2(h)}{h} = 0 + 0 = 0$

$\displaystyle\lim_{h\to 0}\frac{g_1(h)g_2(h)}{h} = \lim_{h\to 0} h\,\frac{g_1(h)g_2(h)}{h^2} = \left(\lim_{h\to 0} h\right)\left(\lim_{h\to 0}\frac{g_1(h)}{h}\right)\left(\lim_{h\to 0}\frac{g_2(h)}{h}\right) = (0)(0)(0) = 0$

30. (a) $g(h) = f(x+h) - f(x) - mh$.

(b) $\displaystyle\lim_{h\to 0}\frac{g(h)}{h} = \lim_{h\to 0}\left[\frac{f(x+h)-f(x)}{h} - m\right] = f'(x) - m = 0$ iff $m = f'(x)$.

31. Suppose that f is differentiable at x. Then there is a number m such that

$$\lim_{h\to 0}\frac{f(x+h)-f(x)}{h} = m \qquad \text{or} \qquad \lim_{h\to 0}\frac{f(x+h)-f(x)}{h} - m = 0.$$

Let $g(h) = f(x+h) - f(x) - mh$. Then

$$\frac{g(h)}{h} = \frac{f(x+h)-f(x)}{h} - m \qquad \text{and} \qquad \lim_{h\to 0}\frac{g(h)}{h} = \lim_{h\to 0}\frac{f(x+h)-f(x)}{h} - m = 0.$$

Now suppose that $f(x+h) - f(x) - mh = g(h)$ where $g(h) = o(h)$. Then

$$\lim_{h\to 0}\left(\frac{f(x+h)-f(x)}{h} - m\right) = \lim_{h\to 0}\frac{g(h)}{h} = 0.$$

Therefore

$$\lim_{h\to 0}\frac{f(x+h)-f(x)}{h} = m$$

and f is differentiable at x. $m = f'(x)$.

PROJECT 4.11

1. $C(x) = 2000 + 50x - \dfrac{x^2}{20}$.

    Marginal cost: $C'(x) = 50 - \dfrac{x}{10}$; $C'(20) = 48$

    Exact cost of the 21st component: $C(21) - C(20) = 3027.95 - 2980 = 47.95$.

2. Profit function: $P(x) = R(x) - C(x) = 650x - 5x^2 - (12,000 + 30x) = 620x - 5x^2 - 12000$.

    Breakeven points: $P(x) = 0$, implies $x^2 - 124x + 2400 = (x - 24)(x - 100) = 0$; $x = 24$, $x = 100$
    units.

    Maximum profit: $P'(x) = 620 - 10x$; $P'(x) = 0$ at $x = 62$ units.

3. (a) Profit function: $P(x) = R(x) - C(x) = 20x - \dfrac{x^2}{50} - (4x + 1400) = 16x - \dfrac{x^2}{50} - 1400$.

    Break-even points: $16x - \dfrac{x^2}{50} - 1400 = 0$ or $x^2 - 800x + 70,000 = 0$

    Thus $x = 100$, or $x = 700$ units.

    (b) $P'(x) = 16 - \dfrac{x}{25}$; $P'(x) = 0$ at $x = 400$ units.

    (c)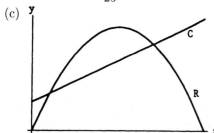

4. (a)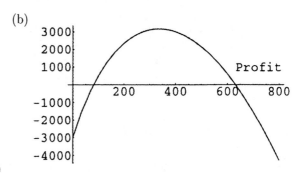

    Break-even points at $x = 81.11$ and $x = 631.19$

    (b) Maximum profit at $x = 336.11$ units

5. (a)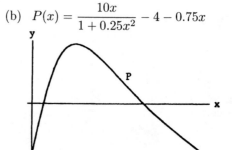

    Breakeven points at $x \cong 0.46$, and $x \cong 4.53$

    (b) $P(x) = \dfrac{10x}{1 + 0.25x^2} - 4 - 0.75x$

    Maximum profit at 175 units

6. $A(x) = \dfrac{C(x)}{x}, \quad A'(x) = \dfrac{xC'(x) - C(x)}{x^2}.$

$A'(x) = 0$ implies $xC'(x) - C(x) = 0$ or $C'(x) = \dfrac{C(x)}{x} = A(x)$. The critical points of A are the points where $C'(x) = C(x)/x$.

$A''(x) = \dfrac{x^2[xC''(x)] - [xC'(x) - C(x)](2x)}{x^4}$

At the points where $xC'(x) - C(x) = 0$, $A''(x) = \dfrac{C''(x)}{x}$. If $C''(x) > 0$ at such a point, then $A(x)$ is a minimum.

7. $B(x) = \dfrac{R(x)}{x}, \quad B'(x) = \dfrac{xP'(x) - P(x)}{x^2}.$

$B'(x) = 0$ implies $xP'(x) - P(x) = 0$ or $P'(x) = \dfrac{P(x)}{x} = B(x)$. The critical points of B are the points where $P'(x) = P(x)/x$.

$B''(x) = \dfrac{x^2[xP''(x)] - [xP'(x) - P(x)](2x)}{x^4}$

At the points where $xP'(x) - P(x) = 0$, $B''(x) = \dfrac{P''(x)}{x}$. If $P''(x) < 0$ at such a point, then $B(x)$ is a maximum.

SECTION 4.12

1. (a) $x_{n+1} = \dfrac{1}{2}x_n + 12\left(\dfrac{1}{x_n}\right)$ 　　　　　(b) $x_4 \cong 4.89898$

2. (a) $x_{n+1} = \dfrac{2x_n^3 - 1}{3x_n^2 - 4}$ 　　　　　(b) $x_4 \cong 1.86081$

3. (a) $x_{n+1} = \dfrac{2}{3}x_n + \dfrac{25}{3}\left(\dfrac{1}{x_n}\right)^2$ 　　　　　(b) $x_4 \cong 2.92402$

4. (a) $x_{n+1} = \dfrac{4}{5}x_n + 6\left(\dfrac{1}{x_n}\right)^4$ 　　　　　(b) $x_4 \cong 1.97435$

5. (a) $x_{n+1} = \dfrac{x_n \sin x_n + \cos x_n}{\sin x_n + 1}$ 　　　　　(b) $x_4 \cong 0.73909$

6. (a) $x_{n+1} = \dfrac{x_n \cos x_n - \sin x_n - x_n^2}{\cos x_n - 2x_n}$ 　　　　　(b) $x_4 \cong 0.87673$

7. (a) $x_{n+1} = \dfrac{6 + x_n}{2\sqrt{x_n + 3} - 1}$ 　　　　　(b) $x_4 \cong 2.30278$

8. (a) $x_{n+1} = \dfrac{x_n \sec^2 x_n - \tan x_n}{1 + \sec^2 x_n}$ 　　　　　(b) $x_4 \cong 2.30278$

9. Let $f(x) = x^{1/3}$. Then $f'(x) = \frac{1}{3}x^{-2/3}$. The Newton-Raphson method applied to this function gives:

$$x_{n+1} = x_n - \frac{f(x_n)}{f'(x_n)} = x_n - \frac{x_n^{1/3}}{\frac{1}{3}x_n^{-2/3}} = -2x_n.$$

Choose any $x_1 \neq 0$. Then $x_2 = -2x_1$, $x_3 = -2x_2 = 4x_1$, $\cdots$,

$x_n = -2x_{n-1} = (-1)^{n-1}2^n x_1, \cdots$.

10. $x_{n+1} = x_n$ for all n.

11. (a) f is a continuous function, and $f(1) = -2 < 0$, $f(2) = 3 > 0$. Thus, f has a root in $(1, 2)$.

(b) $f'(x) = 6x^2 - 6x$ and $f'(1) = 0$. Therefore, $x_1 = 1$ will fail to generate values that will approach the root in $(1, 2)$.

(c) $x_{n+1} = x_n - \dfrac{2x_n^3 - 3x_n^2 - 1}{6x_n^2 - 6x_n}$;

$x_1 = 2$, $x_2 = 1.75$, $x_3 = 1.68254$, $x_4 = 1.67768$, $f(x_4) \cong 0.00020$.

12. (a) Let $f(x) = x^4 - 2x^2 - \frac{17}{16}$. Then $f'(x) = 4x^3 - 4x$. The Newton-Raphson method applied to this function gives:

$$x_{n+1} = x_n - \frac{x_n^4 - 2x_n^2 - \frac{17}{16}}{4x_n^3 - 4x_n}$$

If $x_1 = \frac{1}{2}$, then $x_2 = -\frac{1}{2}$, $x_3 = \frac{1}{2}$, $\cdots x_n = (-1)^{n-1} \frac{1}{2}$, $\cdots$.

(b) $x_1 = 2$, $x_2 = 1.71094$, $x_3 = 1.58569$, $x_4 = 1.56165$; $f(x_4) = 0.00748$

13. (a) $f(x) = x^2 - a$; $f'(x) = 2x$. Substituting into (4.12.1), we have

$$x_{n+1} = x_n - \frac{x_n^2 - a}{2x_n} = \frac{x_n^2 + a}{2x_n} = \frac{1}{2} \left(x_n + \frac{a}{x_n} \right), \quad n \geq 1$$

(b) Let $a = 5$, $x_1 = 2$, and $x_{n+1} = \frac{1}{2} \left(x_n + \frac{a}{x_n} \right)$, $n \geq 1$.

Then $x_2 = 2.25$, $x_3 = 2.23611$, $x_4 = 2.23607$, and $f(x_4) \cong 0.000009045$.

14. (a) Let $f(x) = x^k - a$. Then $f'(x) = kx^{k-1}$. The Newton-Raphson method applied to this function gives:

$$x_{n+1} = x_n - \frac{x_n^k - a}{kx_n^{k-1}} = x_n - \frac{1}{k}x_n + \frac{1}{k}\frac{a}{x_n^{k-1}} = \frac{1}{k}\left[(k-1)x_n + \frac{a}{x_n^{k-1}} \right]$$

(b) Let $a = 23$, $k = 3$ and $x_1 = 3$. Then

$x_1 = 3$, $x_2 = 2.85185$, $x_3 = 2.84389$, $x_4 = 2.84382$; $f(x_4) = -0.00114$

15. (a) Let $f(x) = \frac{1}{x} - a$. Then $f'(x) = -\frac{1}{x^2}$. The Newton-Raphson method applied to this function gives:

$$x_{n+1} = x_n - \frac{\frac{1}{x_n} - a}{-\frac{1}{x_n^2}} = x_n + x_n - ax_n^2 = 2x_n - ax_n^2$$

(b) Let $a = 2.7153$, and $x_1 = 0.3$. Then

$x_2 = 0.35562$, $x_3 = 0.36785$, $x_4 = x_5 = 0.36828$,

Thus $\frac{1}{2.7153} \simeq 0.36828$.

16. (a) $f(x) = x^4 - 7x^2 - 8x - 3$, $f'(x) = 4x^3 - 14x - 8$, $f''(x) = 12x^2 - 14$.

$f'(2) = -4 < 0$ and $f'(3) = 58 > 0$; f' has a zero in $(2, 3)$.

$f''(x) = 12x^2 - 14 > 0$ on $(2, 3)$. Therefore, f' has exactly one zero in this interval.

(b) $x_{n+1} = x_n - \dfrac{4x_n^3 - 14x_n - 8}{12x_n^2 - 14} = \dfrac{4x_n^3 + 4}{6x_n^2 - 7}$; $x_3 \cong 2.1091$. Since $f''(x_3) > 0$, f has a local minimum at c.

17. (a) $f(x) = \sin x + \frac{1}{2}x^2 - 2x$, $f'(x) = \cos x + x - 2$, $f''(x) = -\sin x + 1$.

     $f'(2) = \cos 2 \cong -0.4161 < 0$ and $f'(3) = \cos 3 + 1 \cong 0.0100 > 0$; f' has a zero in $(2,3)$.

     $f''(x) = -\sin x + 1 > 0$ on $(2,3)$. Therefore, f' has exactly one zero in this interval.

   (b) $x_{n+1} = x_n - \dfrac{\cos x_n + x_n - 2}{1 - \sin x_n} = \dfrac{2 - x_n \sin x_n - \cos x_n}{1 - \sin x_n}$; $x_3 \cong 2.9883$. Since $f''(x_3) > 0$,

     f has a local minimum at c.

18. (a) $x_{n+1} = x_n - \dfrac{\sin x_n}{\cos x_n} = x_n - \tan x_n$, $x_1 = 3$; $x_4 \cong 3.14159$

   (b) $x_4 \cong 6.28319$

19. (a) $x_{n+1} = x_n - \dfrac{x_n + \tan x_n}{1 + \sec^2 x_n}$, $x_1 = 2\pi/3$; $r_1 \cong 2.029$

   (b) $x_1 = 5\pi/3$; $r_2 \cong 4.913$

REVIEW EXERCISES

1. f is differentiable on $(-1,1)$ and continuous on $[-1,1]$; $f(1) = f(-1) = 0$.

    $f'(x) = 3x^2 - 1$; $3c^2 - 1 = 0$ $\implies$ $c = \pm\dfrac{\sqrt{3}}{3}$.

2. f is differentiable on $(0, 2\pi)$ and continuous on $[0, 2\pi]$; $f(0) = f(2\pi) = 0$.

    $f'(x) = \cos x - \sin x$; $\cos c - \sin c = 0$ $\implies$ $c = \frac{1}{4}\pi, \; \frac{5}{4}\pi$.

3. f is differentiable on $(-2,3)$ and continuous on $[-2,3]$.

$$f'(c) = 3c^2 - 2 = \frac{f(3) - f(-2)}{5} = 5 \quad \implies \quad 3c^2 = 7 \quad \implies \quad c = \pm\sqrt{\frac{7}{3}}$$

4. f is differentiable on $(2,5)$ and continuous on $[2,5]$.

$$f'(c) = \frac{1}{2\sqrt{c-1}} = \frac{f(5) - f(2)}{3} = \frac{1}{3} \quad \implies \quad c = \frac{13}{4}.$$

5. f is differentiable on $(2,4)$ and continuous on $[2,4]$.

$$f'(c) = -\frac{2}{(c-1)^2} = \frac{f(4) - f(2)}{2} = -\frac{2}{3} \quad \implies \quad c = 1 + \sqrt{3}.$$

6. f is differentiable on $(0,16)$ and continuous on $[0,16]$.

$$f'(c) = \frac{3}{4c^{1/4}} = \frac{f(16) - f(0)}{16} = \frac{1}{2} \quad \implies \quad c = \frac{81}{16}.$$

7. $f'(x) = \dfrac{1 + x^{2/3}}{3x^{2/3}} \neq 0$ for all $x \in (-1,1)$. $f'(0)$ does not exist. Therefore f is not differentiable on $(-1,1)$.

8. $f'(x) = \dfrac{-3}{(x-2)^2} < 0$ for all $x \in (1,4)$; $\quad \dfrac{f(4) - f(1)}{3} = \dfrac{9}{6} > 0$. $f'(2)$ does not exist. Therefore f is not differentiable on $(1,4)$.

9. No. Reason: If such a function did exist, then, by the mean-value theorem, there is a number $c \in (1,4)$ such that $f'(c) = -\frac{4}{3} < -1$.

10. (a) $f'(x) = 3x^2 - 3 = 3(x^2 - 1) < 0$ on $(-1,1)$. Therefore, by Rolle's theorem, f can have at most one zero in $[-1,1]$.

  (b) Since f decreases on $[-1,1]$, we must have $f(-1) = -1 + 3 + k \geq 0$ and $f(1) = 1 - 3 + k \leq 0$. These conditions imply that $k \in [-2,2]$.

11. $f'(x) = 6x(x+1)$.

  f':

  f increases on $(-\infty, -1] \cup [0, \infty)$ and decreases on $[-1, 0]$.

  The critical points are $x = -1, 0$; $f(-1) = 2$ is a local max and $f(0) = 1$ is a local min.

12. $f'(x) = 4(x-1)(x^2 + x + 1)$.

  f':

  f increases on $[1, \infty)$ and decreases on $(-\infty, 1]$. The critical point is $x = 1$; $f(1) = 0$ is a local min.

13. $f' = (x+2)(x-1)^2(5x+4)$.

  f':

  f increases on $(-\infty, -2] \cup [-\frac{4}{5}, \infty)$ and decreases on $[-2, -\frac{4}{5}]$.

  The critical points are $x = -2, 1, -\frac{4}{5}$. $f(-2) = 0$ is a local max and $f(-\frac{4}{5}) \cong -8.981$ is a local min.

14. $f'(x) = 1 - \dfrac{8}{x^3} = \dfrac{x^3 - 8}{x^3}$.

  f':

  f increases on $(-\infty, 0) \cup [2, \infty)$ and decreases on $(0, 2)$. $x = 2$ is the only critical point; 0 is not a critical point. $f(2) = 3$ is a local min.

15. $f'(x) = \dfrac{1 - x^2}{(1 + x^2)^2}$.

  f':

  f increases on $[-1, 1]$ and decreases on $(-\infty, -1] \cup [1, \infty)$. The critical points are $x = -1, 1$. $f(-1) = -\frac{1}{2}$ is a local min and $f(1) = \frac{1}{2}$ is a local max.

16. $f'(x) = \sin x + \cos x.$

f':

f increases on $[0, \frac{3\pi}{4}] \cup [\frac{7\pi}{4}, 2\pi]$ and decreases on $[\frac{3\pi}{4}, \frac{7\pi}{4}]$. The critical points are $x = \frac{3\pi}{4}, \frac{7\pi}{4}$. $f(\frac{3\pi}{4}) = \sqrt{2}$ is a local max and $f(\frac{7\pi}{4}) = -\sqrt{2}$ is a local min.

17. $f'(x) = 3x^2 + 4x + 1 = (3x + 1)(x + 1);$ critical points: $x = -\frac{1}{3}, -1.$

f':

$f(-2) = -1$ endpoint and abs. min; $f(-1) = 1$ local max; $f(-\frac{1}{3}) = \frac{23}{27}$ local min; $f(1) = 5$ endpoint and abs. max.

18. $f'(x) = 4x^3 - 16x = 4x(x - 2)(x + 2);$ critical points: $x = 0, 2.$

f':

$f(-1) = -5$ endpoint min; $f(0) = 2$ local max; $f(2) = -14$ local and abs. min; $f(3) = 11$ endpoint and abs. max.

19. $f'(x) = 2x - \dfrac{8}{x^3} = \dfrac{2(x^4 - 4)}{x^3};$ critical point: $x = 4^{1/4} = \sqrt{2}.$

f':

$f(1) = 5$ endpoint max; $f(\sqrt{2}) = 4$ local and abs. min; $f(4) = \frac{65}{4}$ endpoint and abs. max.

20. $f'(x) = \cos x(1 - 2\sin x);$ critical points: $x = \frac{\pi}{6}, \frac{\pi}{2}, \frac{5\pi}{6}, \frac{3\pi}{2}.$

f':

$f(0) = 1$ endpoint min; $f(\frac{\pi}{6}) = 5/4$ local and abs. max; $f(\frac{\pi}{2}) = 1$ local min; $f(\frac{5\pi}{6}) = \frac{5}{4}$ local and abs. max; $f(\frac{3\pi}{2}) = -1$ local and abs. min; $f(2\pi) = 1$ endpoint max.

21. $f'(x) = \dfrac{-x}{2\sqrt{1 - x}} + \sqrt{1 - x} = \dfrac{2 - 3x}{2\sqrt{1 - x}};$ critical point: $x = \frac{2}{3}.$

$f(1) = 0$ endpoint min; $f(\frac{2}{3}) = \frac{2\sqrt{3}}{9}$ local and abs. max.

22. $f'(x) = \dfrac{(x - 2)2x - x^2}{(x - 2)^2} = \dfrac{x(x - 4)}{(x - 2)^2};$ critical point: $x = 4;$

$f(4) = 8$ local and abs. min.

23. $f(x) = \dfrac{3x(x - 3)}{(x - 4)(x + 3)}.$ Vertical asymptotes: $x = -3$ and $x = 4;$ horizontal asymptote: $y = 3.$

24. $f(x) = \dfrac{(x-2)(x+2)}{(x-2)(x-3)} = \dfrac{x+2}{x-3},\ x \neq 2.$ Vertical asymptote: $x = 3$; horizontal asymptote: $y = 1$.

25. $f(x) = x - \dfrac{x}{(x-1)(x^2+x+1)}.$ Vertical asymptote: $x = 1$; oblique asymptote: $y = x$.

26. $f'(x) = \dfrac{3}{5(x-1)^{2/5}},$ vertical tangent.

27. $f'(x) = x^{2/5} - 2x^{-3/5} = \dfrac{x-2}{x^{3/5}},$ vertical cusp.

28. $f'(x) = 2x^{-2/3} + 4x^{1/3} = \dfrac{2+4x}{x^{2/3}},$ vertical tangent.

29. $f(x) = 6 + 4x^3 - 3x^4,$ domain: $(-\infty, \infty)$

$f'(x) = 12x^2(1-x)$

critical pts. $x = 0,\ 1$

$f''(x) = 12x(2 - 3x)$

$f'(x) > 0$ on $(-\infty, 1)$,

$f'(x) < 0$ on $(1, \infty)$

$f''(x) > 0$ on $(0, 2/3)$

$f''(x) < 0$ on $(-\infty, 0) \cup (2/3, \infty)$

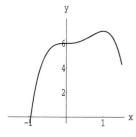

30. $f(x) = 3x^5 - 5x^3 + 1,$ domain: $(-\infty, \infty)$

$f'(x) = 15x^2(x-1)(x+1)$

critical pts. $x = -1,\ 0,\ 1$

$f''(x) = 30x\left(2x^2 - 1\right)$

$f'(x) > 0$ on $(-\infty, -1) \cup (1, \infty)$

$f'(x) < 0$ on $(-1, 1)$

$f''(x) > 0$ on $\left(-\frac{\sqrt{2}}{2}, 0\right) \cup \left(\frac{\sqrt{2}}{2}, \infty\right)$

$f''(x) < 0$ on $\left(-\infty, -\frac{\sqrt{2}}{2}\right) \cup \left(0, \frac{\sqrt{2}}{2}\right)$

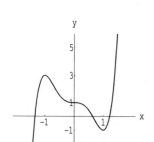

31. $f(x) = \dfrac{2x}{x^2 + 4},$ domain: $(-\infty, \infty)$

symmetric with respect to the origin.

$f'(x) = \dfrac{2\left(4 - x^2\right)}{\left(x^2 + 4\right)^2}$

critical pts. $x = 2,\ x = -2$

$$f''(x) = \frac{4x\left(x^2 - 12\right)}{\left(x^2 + 4\right)^3}$$

$f'(x) > 0$ on $(-2, 2)$,

$f'(x) < 0$ on $(-\infty, 2) \cup (2, \infty)$

$f''(x) > 0$ on $\left(-\sqrt{12}, 0\right) \cup \left(\sqrt{12}, \infty\right)$

$f''(x) < 0$ on $\left(-\infty, -\sqrt{12}\right) \cup \left(0, \sqrt{12}\right)$

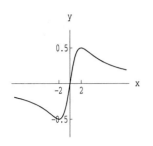

32. $f(x) = x^{2/3}(x - 10)$, domain: $(-\infty, \infty)$

$$f'(x) = \frac{5(x - 4)}{3x^{1/3}}$$

critical pts. $x = 4,\ x = 0$

$$f''(x) = \frac{10(x + 2)}{9x^{4/3}}$$

$f'(x) > 0$ on $(-\infty, 0) \cup (4, \infty)$,

$f'(x) < 0$ on $(0, 4)$

$f''(x) > 0$ on $(-2, 0) \cup (0, \infty)$

$f''(x) < 0$ on $(-\infty, -2)$

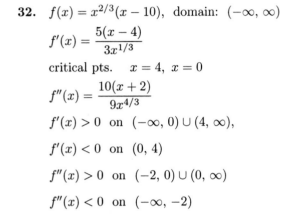

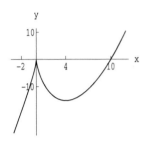

33. $f(x) = x\sqrt{4 - x}$, domain: $(-\infty, 4]$

$$f'(x) = \frac{8 - 3x}{2\sqrt{4 - x}}$$

$$f''(x) = \frac{3x - 16}{4(4 - x)^{3/2}}$$

$f'(x) > 0$ on $\left(-\infty, \frac{8}{3}\right]$,

$f'(x) < 0$ on $\left[\frac{8}{3}, 4\right)$

$f''(x) < 0$ on $(-\infty, 4)$

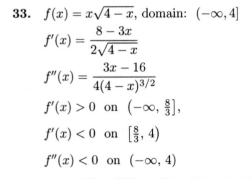

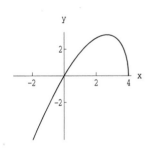

34. $f(x) = x^4 - 2x^2 + 3$, domain $(-\infty, \infty)$;

symmetric with respect to the y-axis

$f'(x) = 4x^3 - 4x = 4x(x - 1)(x + 1)$

critical pts. $x = -1,\ 0,\ 1$

$f''(x) = 12x^2 - 4 = 12(x - 1/\sqrt{3})(x + 1/\sqrt{3})$

$f'(x) > 0$ on $[-1, 0] \cup [1, \infty)$

$f'(x) < 0$ on $(-\infty, -1] \cup [0, 1]$

$f''(x) > 0$ on $(\infty, -1/\sqrt{3}) \cup (1/\sqrt{3}, \infty)$

$f''(x) < 0$ on $[-1/\sqrt{3}, 1/\sqrt{3}]$

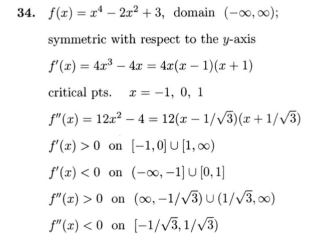

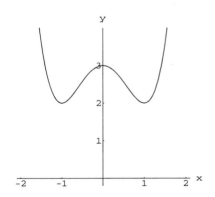

35. $f(x) = \sin x + \sqrt{3}\cos x,$ domain $[0, 2\pi]$

$f'(x) = \cos x - \sqrt{3}\sin x$

critical pts. $x = \frac{1}{6}\pi,\ x = \frac{7}{6}\pi$

$f''(x) = -\sin x - \sqrt{3}\cos x$

$f'(x) > 0$ on $[0, \pi/6) \cup (7\pi/6, 2\pi]$

$f'(x) < 0$ on $(\pi/6, 7\pi/6)$

$f''(x) > 0$ on $(2\pi/3, 5\pi/3)$

$f''(x) < 0$ on $[0, \pi/3) \cup (5\pi/3, 2\pi]$

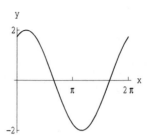

36. $f(x) = \sin^2 x - \cos x,\ x \in [0, 2\pi]$

$f'(x) = 2\sin x \cos x + \sin x$

critical pts. $x = \frac{2}{3}\pi,\ \pi,\ \frac{4}{3}\pi$

$f''(x) = 4\cos^2 x + \cos x - 2$

$f'(x) > 0$ on $\left(0, \frac{2}{3}\pi\right) \cup \left(\pi, \frac{4}{3}\pi\right),$

$f'(x) < 0$ on $\left(\frac{2}{3}\pi, \pi\right) \cup \left(\frac{4}{3}\pi, 2\pi\right)$

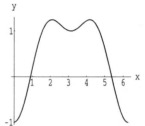

37.

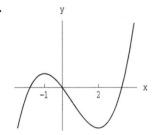

38. Let r be the radius of the sphere and x the side length of the cube. Then the sum of the surface areas equals constant implies

$$4\pi r^2 + 6x^2 = C \quad \Longrightarrow \quad x = \sqrt{\frac{C - 4\pi r^2}{6}}.$$

The sum of the volumes is:

$$V = \frac{4}{3}\pi r^3 + \left(\frac{C - 4\pi r^2}{6}\right)^{3/2}, \quad 0 \le r \le \sqrt{\frac{C}{4\pi}}.$$

Now

$$V' = 4\pi r^2 + \frac{3}{2}\left(\frac{C - 4\pi r^2}{6}\right)^{1/2}\left(-\frac{4}{3}\pi r\right)$$

$$= 4\pi r^2 - 2\pi r \left(\frac{C - 4\pi r^2}{6}\right)^{1/2}$$

Set $V'(r) = 0$:

$$2r = \left(\frac{C - 4\pi r^2}{6}\right)^{1/2} \implies 4r^2 = \frac{C - 4\pi r^2}{6} \implies r = \sqrt{\frac{C}{24 + 4\pi}}, \text{ and } x = 2\sqrt{\frac{C}{24 + 4\pi}}.$$

Without loss of generality, set $C = 1$. Then

$$V(0) = (1/6)^{3/2} \cong 0.07, \ V(1/\sqrt{24 + 4\pi}) \cong 0.055, \ V(1/\sqrt{4\pi}) \cong 0.09.$$

Thus, the sum of the volumes is a minimum when the side length of the cube equals the diameter of the sphere. The sum of the volumes is a maximum when the side length of the cube is 0.

39. Let x be the side length of the square base. Then, since the volume is 27, the height $y = 27/x^2$. The conditions: $x^2 \le 18$, $0 \le y \le 4$ imply $\frac{3\sqrt{3}}{2} \le x \le 3\sqrt{2}$. Thus, the surface area S of the box is given by

$$S(x) = 2x^2 + \frac{108}{x}, \quad \frac{3\sqrt{3}}{2} \le x \le 3\sqrt{2}$$

$$S'(x) = 4x - \frac{108}{x^2}; \quad S'(x) = 0 \implies 4x^3 = 27 \implies x = 3$$

Now,

$$S\left(\frac{3\sqrt{3}}{2}\right) = \frac{27}{2} + 24\sqrt{3} \cong 55.07,$$

$$S(3) = 54$$

$$S(3\sqrt{2}) = 36 + 18\sqrt{2} \cong 61.46$$

(a) The minimal surface area is 54 sq. ft. at $x = 3$.

(b) The maximal surface area is 61.46 sq. ft. (approx.) at $x = 3\sqrt{2}$.

40. An equation for the line with x-intercept a and y-intercept b is: $\dfrac{x}{a} + \dfrac{y}{b} = 1$

Since the line passes through $(1, 2)$, $\dfrac{1}{a} + \dfrac{2}{b} = 1 \implies b = \dfrac{2a}{a - 1}$

The area of the right triangle OAB is: $S = \frac{1}{2}ab = \dfrac{a^2}{a - 1}$

Now,

$$S'(a) = \frac{a^2 - 2a}{(a - 1)^2} \quad \text{and} \quad S'(a) = 0 \text{ at } a = 0, \ a = 2$$

The triangle OAB will have minimum area when $A = (2, 0)$ and $B = (0, 4)$.

41. Introduce a rectangular coordinate system so that the rectangle is in the first quadrant and the base and left side lie on the coordinate axes. Let x denote the length of the base, and y the height. Let $P = 2x + 2y$ be the given perimeter. If the rectangle is rotated about the y-axis, then the volume V is given by:

$$V(x) = \frac{\pi P}{2}x^2 - \pi x^3, \ 0 \le x \le P/2.$$

Now,

$$V'(x) = \pi Px - 3\pi x^2 \quad \text{and} \quad V'(x) = 0 \quad \implies \quad x = 0 \text{ or } x = \tfrac{1}{3}P.$$

At $x = 0, V = 0$; at $x = P/3$, $y = P/6$, $V = \pi P^3/54$; at $x = P/2$, $V = 0$. The dimensions that generate the cylinder of maximum volume are $x = P/3$, $y = P/6$.

42. Let x be the width of the page and y the height. Then $xy = 80$. The area available for print is:

$$A = (x - 2)(y - 2.5) = (x - 2)\left(\frac{80}{x} - 2.5\right)$$

$$= 85 - 2.5x - \frac{160}{x}, \quad 0 < x < \infty.$$

$$A'(x) = -2.5 + \frac{160}{x^2}; \qquad A'(x) = 0: \quad 2.5x^2 = 160, \quad x^2 = 64, \quad x = 8$$

Since $A''(x) < 0$, A has an absolute maximum at $x = 8$. The dimensions of the page that will maximize the printed area are: width 8, height 10.

43. $x'(t) = 1 - 2\sin t, \quad x''(t) = -2\cos t.$

The object is slowing down when $x'(t)x''(t) < 0 \implies t \in [0, \tfrac{1}{6}\pi) \cup (\tfrac{1}{2}\pi, \tfrac{5}{6}\pi) \cup (\tfrac{3}{2}\pi, 2\pi].$

44. $x(t) = (4t - 1)(t - 1)^2, \ t \geq 0; \quad x'(t) = v(t) = 2(4t - 1)(t - 1) + 4(t - 1)^2 = 6(2t^2 - 3t + 1)$
$= 6(2t - 1)(t - 1)$

   (a) The object is moving to the right: $v(t) > 0$ when $0 \leq t < \tfrac{1}{2}$ and when $t > 1$; moving to the left: $v(t) < 0$ when $\tfrac{1}{2} < t < 1$; the object changes direction at $t = \tfrac{1}{2}$ and $t = 1$.

   (b) Speed when moving left: $\nu = |v| = 6(3t - 2t^2 - 1)$ on $[\tfrac{1}{2}, 1]$. $d\nu/dt = 6(3 - 4t)$; $d\nu/dt = 0$ at $t = 3/4$. The maximum speed when moving left is $\nu(3/4) = 3/4$ units per unit time.

45. (a) $x(t) = \sqrt{t + 1}$
$$x'(t) = v(t) = \frac{1}{2\sqrt{t + 1}}$$
$$x''(t) = a(t) = -\frac{1}{4(t + 1)^{3/2}} = -2[v(t)]^3,$$
$$x'''(t) = a'(t) = \frac{3}{8(t + 1)^{5/2}}$$

   (b) Position: $x(17) = x(15 + 2) \cong x(15) + 2x'(15) = 4.25;$

   Velocity: $v(17) = v(15 + 2) \cong v(15) + 2v'(15) = \frac{15}{128} \cong 0.1172;$

   Acceleration: $a(17) = a(15 + 2) \cong a(15) + 2a'(15) \cong 0.0032.$

46. The height of the rocket at time t is given by: $y(t) = 128t - 16t^2$.

   (a) At maximum height, $v(t) = y'(t) = 128 - 32t = 0 \implies t = 4$. The rocket reaches its maximum height at 4 seconds; the maximum height is: $y(4) = 256$ feet.

   (b) $y(t) = 0 \implies 16t(8 - t) = 0 \implies t = 0, \ t = 8$; the rocket hits the ground at $t = 8$ seconds; $v(8) = -128$ ft/sec.

47. The height at time t is: $y(t) = -16t^2 + 8t + y_0$. $y(10) = 0 = -1600 + 80 + y_0$ $\implies$ $y_0 = 1520$ ft.

48. The height at time t is: $y(t) = -16t^2 + v_0 t$. $y(1) = 24 = -16 + v_0$ $\implies$ $v_0 = 40$. Therefore, $y(t) = -16t^2 + 40t$. The ball achieves maximum height at the instant $v(t) = -32t + 40 = 0$ $\implies$ $t = \frac{5}{4}$. The maximum height is $y(5/4) = -16(5/4)^2 + 40(5/4) = 25$.

49. Let x be the position of the boy and let y be the length of his shadow. Then
$$\frac{dx}{dt} = 168 \quad \text{and} \quad \frac{x+y}{12} = \frac{y}{5} \implies 7y = 5x \quad \text{and} \quad y = \frac{5}{7}x$$
Differentiating implicitly with respect to t, we get
$$7\frac{dy}{dt} = 5\frac{dx}{dt} \implies \frac{dy}{dt} = \left(\frac{5}{7}\right)(168) = 120$$
The shadow is lengthening at the rate of 120 ft/min.

50. $V = \frac{1}{3}\pi r^2 h$, $\quad V$ constant
$$0 = \frac{d}{dt}V = \frac{d}{dt}\left(\frac{1}{3}\pi r^2 h\right) = \frac{2\pi r h}{3}\frac{dr}{dt} + \frac{\pi r^2}{3}\frac{dh}{dt}$$
Solving for $\dfrac{dh}{dt}$ we get: $\dfrac{dh}{dt} = -\dfrac{2h}{r} \cdot \dfrac{dr}{dt}$.

At the instant $r = 4$ and $h = 15$, $\dfrac{dh}{dt} = -\dfrac{30}{4}(0.3) = -2.25$; the height is decreasing at the rate of 2.25 in/min.

51. Let x be the position of the locomotive at time t and let y be the position of the car. By the law of cosines the distance between the locomotive and the car is given by
$$s^2 = x^2 + y^2 - 2xy\cos 60° = x^2 + y^2 - xy.$$
Differentiating with respect to t, we get
$$2s\frac{ds}{dt} = 2x\frac{dx}{dt} + 2y\frac{dy}{dt} - x\frac{dy}{dt} - y\frac{dx}{dt}.$$
Setting $dx/dt = 60$, $dy/dt = -30$ and converting to feet, gives
$$2s\frac{ds}{dt} = 2x(60)(5280) - 2y(30)(5280) + x(30)(5280) - y(60)(5280) = 5280(150x - 120y).$$
When $x = y = 500$, $s = 500$ and
$$2(500)\frac{ds}{dt} = 5280(30)(500) \implies \frac{ds}{dt} = 15(5280).$$
The distance between the locomotive and the car is increasing at the rate of 15 miles per hour.

52. Let r be the radius of the circle. Then $dr/dt = 5$. The square has side length $r\sqrt{2}$ and area $A = 2r^2$. Differentiating with respect to t, we get
$$\frac{dA}{dt} = 4r\frac{dr}{dt} = 20r; \quad \text{and} \quad \left.\frac{dA}{dt}\right|_{r=10} = 200.$$

53. The cross-section of the water at time t is an isosceles triangle with base x and height y where x and y are related by $\dfrac{\frac{1}{2}x}{y} = \frac{1}{2}$ (similar triangles). Thus $x = y$.

The volume of water in the trough at time t is

$$V = 12(\tfrac{1}{2}xy) = 6y^2.$$

Differentiating with respect to t gives

$$\frac{dV}{dt} = 12y\frac{dy}{dt} \quad \Longrightarrow \quad \frac{dy}{dt} = \frac{1}{12y}\frac{dV}{dt}.$$

Since $dV/dt = -3$, $\quad \dfrac{dy}{dt} = -\dfrac{1}{4y}\quad$ and $\quad \dfrac{dy}{dt}\Big|_{t=1.5} = -\dfrac{1}{6}$. The water level is falling at the rate of $1/6$ feet per minute.

54. $f(3.8) \cong f(4) + f'(4)(-0.2) = 2 + 2(-0.2) = 1.6.$

55. $f(x) = \sqrt{x} + 1/\sqrt{x}, \quad f'(x) = \frac{1}{2}x^{-1/2} - \frac{1}{2}x^{-3/2}$

$f(4.2) \cong f(4) + f'(4)(0.2) = \frac{5}{2} + \frac{3}{16}(0.2) = 2.5375.$

56. $f(x) = x^{1/4}, \quad f'(x) = \frac{1}{4}x^{-3/4}$

$f(83) \cong f(81) + f'(81)(2) = 3 + \frac{1}{54} \cong 3.01852.$

57. $f(x) = \tan x, \quad f'(x) = \sec^2 x$

$f(43°) = f(\frac{1}{4}\pi - \frac{1}{90}\pi) \cong f(\pi/4) + f'(\pi/4)(-\pi/90) = 1 - 2(\pi/90) = 1 - (\pi/45) \cong 0.9302.$

58. $V = \frac{4}{3}\pi r^3, \quad dV = 4\pi r^2\, dr$. Calculate dV at $r = 10$ ft. and $dr = 0.05$ in. $\cong 0.00417$ ft.

$$dV\big|_{r=10,\, dr=0.00417} = 4\pi\,(10)^2\,(0.00417) \cong 5.240 \text{ cu. ft.} = 9055.025 \text{ cu. in.;}$$

$$= 9055.025/231 \cong 39.20 \text{ gal.}$$

59. $f(x) = x^3 - 10$

(a) $x_{n+1} = x_n - \dfrac{x_n^3 - 10}{3x_n^2} = \dfrac{2x_n^3 + 10}{3x_n^2}$

(b) $x_4 \cong 2.15443; \quad f(2.15443) \cong -0.00007$

60. $f(x) = x \sin x - \cos x$

(a) $x_{n+1} = \dfrac{x_n^2 \cos x_n + x_n \sin x_n + \cos x_n}{2 \sin x_n + x_n \cos x_n}$

(b) $x_4 \cong 0.86057; \quad f(0.86057) \cong 0.00049$

CHAPTER 5

SECTION 5.2

1. $L_f(P) = 0(\frac{1}{4}) + \frac{1}{2}(\frac{1}{4}) + 1(\frac{1}{2}) = \frac{5}{8}$, $U_f(P) = \frac{1}{2}(\frac{1}{4}) + 1(\frac{1}{4}) + 2(\frac{1}{2}) = \frac{11}{8}$

2. $L_f(P) = \frac{2}{3}(\frac{1}{3}) + \frac{1}{4}(\frac{5}{12}) + 0(\frac{1}{4}) + (-1)(1) = -\frac{97}{144}$,

$U_f(P) = 1(\frac{1}{3}) + \frac{2}{3}(\frac{5}{12}) + \frac{1}{4}(\frac{1}{4}) + 0(1) = \frac{97}{144}$

3. $L_f(P) = \frac{1}{4}(\frac{1}{2}) + \frac{1}{16}(\frac{1}{4}) + 0(\frac{1}{4}) = \frac{9}{64}$, $U_f(P) = 1(\frac{1}{2}) + \frac{1}{4}(\frac{1}{4}) + \frac{1}{16}(\frac{1}{4}) = \frac{37}{64}$

4. $L_f(P) = \frac{15}{16}(\frac{1}{4}) + \frac{3}{4}(\frac{1}{4}) + 0(\frac{1}{2}) = \frac{27}{64}$, $U_f(P) = 1(\frac{1}{4}) + \frac{15}{16}(\frac{1}{4}) + \frac{3}{4}(\frac{1}{2}) = \frac{55}{64}$

5. $L_f(P) = 1(\frac{1}{2}) + \frac{9}{8}(\frac{1}{2}) = \frac{17}{16}$, $U_f(P) = \frac{9}{8}(\frac{1}{2}) + 2(\frac{1}{2}) = \frac{25}{16}$

6. $L_f(P) = 0(\frac{1}{25}) + \frac{1}{5}(\frac{3}{25}) + \frac{2}{5}(\frac{5}{25}) + \frac{3}{5}(\frac{7}{25}) + \frac{4}{5}(\frac{9}{25}) = \frac{14}{25}$,

$U_f(P) = \frac{1}{5}(\frac{1}{25}) + \frac{2}{5}(\frac{3}{25}) + \frac{3}{5}(\frac{5}{25}) + \frac{4}{5}(\frac{7}{25}) + 1(\frac{9}{25}) = \frac{19}{25}$

7. $L_f(P) = \frac{1}{16}(\frac{3}{4}) + 0(\frac{1}{2}) + \frac{1}{16}(\frac{1}{4}) + \frac{1}{4}(\frac{1}{2}) = \frac{3}{16}$, $U_f(P) = 1(\frac{3}{4}) + \frac{1}{16}(\frac{1}{2}) + \frac{1}{4}(\frac{1}{4}) + 1(\frac{1}{2}) = \frac{43}{32}$

8. $L_f(P) = \frac{9}{16}(\frac{1}{4}) + \frac{1}{16}(\frac{1}{2}) + 0(\frac{1}{2}) + \frac{1}{16}(\frac{1}{4}) + \frac{1}{4}(\frac{1}{2}) = \frac{5}{16}$,

$U_f(P) = 1(\frac{1}{4}) + \frac{9}{16}(\frac{1}{2}) + \frac{1}{16}(\frac{1}{2}) + \frac{1}{4}(\frac{1}{4}) + 1(\frac{1}{2}) = \frac{9}{8}$

9. $L_f(P) = 0(\frac{\pi}{6}) + \frac{1}{2}(\frac{\pi}{3}) + 0(\frac{\pi}{2}) = \frac{\pi}{6}$, $U_f(P) = \frac{1}{2}(\frac{\pi}{6}) + 1(\frac{\pi}{3}) + 1(\frac{\pi}{2}) = \frac{11\pi}{12}$

10. $L_f(P) = \frac{1}{2}(\frac{\pi}{3}) + 0(\frac{\pi}{6}) + (-1)(\frac{\pi}{2}) = -\frac{\pi}{3}$, $U_f(P) = 1(\frac{\pi}{3}) + \frac{1}{2}(\frac{\pi}{6}) + 0(\frac{\pi}{2}) = \frac{5\pi}{12}$

11. (a) $L_f(P) \leq U_f(P)$ but $3 \nleq 2$.

(b) $L_f(P) \leq \displaystyle\int_{-1}^{1} f(x)\,dx \leq U_f(P)$ but $3 \nleq 2 \leq 6$.

(c) $L_f(P) \leq \displaystyle\int_{-1}^{1} f(x)\,dx \leq U_f(P)$ but $3 \leq 10 \nleq 6$.

12. (a) $L_f(P) = (x_0 + 3)(x_1 - x_0) + (x_1 + 3)(x_2 - x_1) + \cdots + (x_{n-1} + 3)(x_n - x_{n-1})$,

$U_f(P) = (x_1 + 3)(x_1 - x_0) + (x_2 + 3)(x_2 - x_1) + \cdots + (x_n + 3)(x_n - x_{n-1})$

(b) For each index i

$$x_{i-1} + 3 \leq \frac{1}{2}(x_{i-1} + x_i) + 3 \leq x_i + 3$$

Multiplying by $\Delta x_i = x_i - x_{i-1}$ gives

$$(x_{i-1} + 3)\Delta x_i \leq \frac{1}{2}(x_i^2 - x_{i-1}^2) + 3(x_i - x_{i-1}) \leq (x_i + 3)\Delta x_i.$$

Summing from $i = 1$ to $i = n$, we find that

$$L_f(P) \leq \frac{1}{2}(x_1^2 - x_0^2) + 3(x_1 - x_0) \cdots + \frac{1}{2}(x_n^2 - x_{n-1}{}^2) + 3(x_n - x_{n-1}) \leq U_f(P)$$

The middle sum collapses to

$$\frac{1}{2}\left(x_n{}^2 - x_0{}^2\right) + 3(x_n - x_0) = \frac{1}{2}(b^2 - a^2) + 3(b - a)$$

Thus

$$\int_a^b (x + 3)dx = \frac{1}{2}(b^2 - a^2) + 3(b - a)$$

13. (a) $L_f(P) = -3x_1(x_1 - x_0) - 3x_2(x_2 - x_1) - \cdots - 3x_n(x_n - x_{n-1})$,

   $U_f(P) = -3x_0(x_1 - x_0) - 3x_1(x_2 - x_1) - \cdots - 3x_{n-1}(x_n - x_{n-1})$

(b) For each index i

$$-3x_i \le -\frac{3}{2}\left(x_i + x_{i-1}\right) \le -3x_{i-1}.$$

Multiplying by $\Delta x_i = x_i - x_{i-1}$ gives

$$-3x_i\,\Delta x_i \le -\frac{3}{2}\left(x_i{}^2 - x_{i-1}^2\right) \le -3x_{i-1}\,\Delta x_i.$$

Summing from $i = 1$ to $i = n$, we find that

$$L_f(P) \le -\frac{3}{2}\left(x_1{}^2 - x_0{}^2\right) - \cdots - \frac{3}{2}\left(x_n{}^2 - x_{n-1}^2\right) \le U_f(P).$$

The middle sum collapses to

$$-\frac{3}{2}\left(x_n{}^2 - x_0{}^2\right) = -\frac{3}{2}(b^2 - a^2).$$

Thus

$$L_f(P) \le -\frac{3}{2}(b^2 - a^2) \le U_f(P) \quad \text{so that} \quad \int_a^b -3x\,dx = -\frac{3}{2}(b^2 - a^2).$$

14. (a) $L_f(P) = (1 + 2x_0)(x_1 - x_0) + (1 + 2x_1)(x_2 - x_1) + \cdots + (1 + 2x_{n-1})(x_n - x_{n-1})$,

   $U_f(P) = (1 + 2x_1)(x_1 - x_0) + (1 + 2x_2)(x_2 - x_1) + \cdots + (1 + 2x_n)(x_n - x_{n-1})$

(b) For each index i

$$1 + 2x_{i-1} \le 1 + (x_{i-1} + x_i) \le 1 + 2x_i$$

Multiplying by $\Delta x_i = x_i - x_{i-1}$ gives

$$(1 + 2x_{i-1})\,\Delta x_i \le (x_i - x_{i-1}) + \left(x_i{}^2 - x_{i-1}^2\right) \le (1 + 2x_i)\,\Delta x_i.$$

Proceeding as before, we get

$$\int_a^b (1 + 2x)\,dx = (b - a) + (b^2 - a^2)$$

15. $\displaystyle\int_{-1}^2 (x^2 + 2x - 3)\,dx$ **16.** $\displaystyle\int_0^3 (x^3 - 3x)\,dx$

17. $\displaystyle\int_0^{2\pi} t^2 \sin(2t + 1)\,dt$ **18.** $\displaystyle\int_1^4 \frac{\sqrt{t}}{t^2 + 1}\,dt$

19.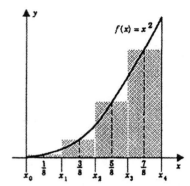

20.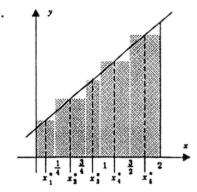

21. $\Delta x_1 = \Delta x_2 = \frac{1}{8}, \quad \Delta x_3 = \Delta x_4 = \Delta x_5 = \frac{1}{4}$

$m_1 = 0, \quad m_2 = \frac{1}{4}, \quad m_3 = \frac{1}{2}, \quad m_4 = 1, \quad m_5 = \frac{3}{2}$

$f(x_1^*) = \frac{1}{8}, \quad f(x_2^*) = \frac{3}{8}, \quad f(x_3^*) = \frac{3}{4}, \quad f(x_4^*) = \frac{5}{4}, \quad f(x_5^*) = \frac{3}{2}$

$M_1 = \frac{1}{4}, \quad M_2 = \frac{1}{2}, \quad M_3 = 1, \quad M_4 = \frac{3}{2}, \quad M_5 = 2$

(a) $L_f(P) = \frac{25}{32}$ (b) $S^*(P) = \frac{15}{16}$ (c) $U_f(P) = \frac{39}{32}$

22. $\displaystyle\int_0^1 2x \, dx = 1.$

23. $L_f(P) = x_0^3(x_1 - x_0) + x_1^3(x_2 - x_1) + \cdots + x_{n-1}^3(x_n - x_{n-1})$

$U_f(P) = x_1^3(x_1 - x_0) + x_2^3(x_2 - x_1) + \cdots + x_n^3(x_n - x_{n-1})$

For each index i

$$x_{i-1}^3 \leq \tfrac{1}{4}\left(x_i^3 + x_i^2 x_{i-1} + x_i x_{i-1}^2 + x_{i-1}^3\right) \leq x_i^3$$

and thus by the hint

$$x_{i-1}^3(x_i - x_{i-1}) \leq \tfrac{1}{4}\left(x_i^4 - x_{i-1}^4\right) \leq x_i^3(x_i - x_{i-1}).$$

Adding up these inequalities, we find that

$$L_f(P) \leq \tfrac{1}{4}\left(x_n^4 - x_0^4\right) \leq U_f(P).$$

Since $x_n = 1$ and $x_0 = 0$, the middle term is $\dfrac{1}{4}$: $\displaystyle\int_0^1 x^3 \, dx = \frac{1}{4}.$

24. (a) $L_f(P) = x_0^4(x_1 - x_0) + x_1^4(x_2 - x_1) + \cdots + x_{n-1}^4(x_n - x_{n-1}),$

   $U_f(P) = x_1^4(x_1 - x_0) + x_2^4(x_2 - x_1) + \cdots + x_n^4(x_n - x_{n-1})$

(b) For each index i

$$x_{i-1}^4 \leq \frac{x_i^4 + x_i^3 x_{i-1} + x_i^2 x_{i-1}^2 + x_i x_{i-1}^3 + x_{i-1}^4}{5} \leq x_i^4$$

Multiplying by $\Delta x_i = x_i - x_{i-1}$ gives

$$x_{i-1}^4 \, \Delta x_i \leq \frac{1}{5}\left(x_i^5 - x_{i-1}^5\right) \leq x_i^4 \, \Delta x_i.$$

Summing and collapsing the middle sum gives

$$L_f(P) \le \frac{1}{5}\left(x_n{}^5 - x_0{}^5\right) \le U_f(P),$$

Thus

$$\int_0^1 x^4\,dx = \frac{1}{5}(1^5 - 0^5) = \frac{1}{5}.$$

25. Necessarily holds: $L_g(P) \le \int_a^b g(x)\,dx < \int_a^b f(x)\,dx \le U_f(P)$.

26. Need not hold. Consider the partition $\{0,2,3\}$ on $[0,3]$ where $f(x) = x$ and $g(x) = 1$.
Then $\int_a^b f(x)\,dx = 4\frac{1}{2}$ and $\int_a^b g(x)\,dx = 3$, but $L_g(P) = 3$ and $L_f(P) = 2$.

27. Necessarily holds: $L_g(P) \le \int_a^b g(x)\,dx < \int_a^b f(x)\,dx$

28. Need not hold. Consider the partition $\{0,1,3\}$ on $[0,3]$ where $f(x) = 2$ and $g(x) = 3 - x$.
Then $\int_a^b f(x)\,dx = 6$ and $\int_a^b g(x)\,dx = 4\frac{1}{2}$, but $U_g(P) = 7$ and $U_f(P) = 6$.

29. Necessarily holds: $U_f(P) \ge \int_a^b f(x)\,dx > \int_a^b g(x)\,dx$

30. Need not hold. Use the same counter example as Exercise 30.

31. Let $P = \{x_0, x_1, x_2, \ldots, x_n\}$ be a regular partition of $[a, b]$ and let $\Delta x = (b - a)/n$.
Since f is increasing on $[a, b]$,

$$L_f(P) = f(x_0)\Delta x + f(x_1)\Delta x + \cdots + f(x_{n-1})\Delta x$$

and

$$U_f(P) = f(x_1)\Delta x + f(x_2)\Delta x + \cdots + f(x_n)\Delta x.$$

Now,

$$U_f(P) - L_f(P) = [f(x_n) - f(x_0)]\Delta x = [f(b) - f(a)]\Delta x.$$

32. Proceed as in Exercise 31.

33. (a) $f'(x) = \dfrac{x}{\sqrt{1 + x^2}} > 0$ for $x \in [0, 2]$. Thus, f is increasing on $[0, 2]$.

(b) Let $P = \{x_0, x_1, \ldots, x_n\}$ be a regular partition of $[0, 2]$ and let $\Delta x = 2/n$
By Exercise 30,

$$\int_0^2 f(x)\,dx - L_f(P) \le |f(2) - f(0)|\frac{2}{n} = \frac{2(\sqrt{5} - 1)}{n} \cong \frac{2.47}{n}$$

It now follows that $\int_0^2 f(x)\,dx - L_f(P) < 0.1$ if $n > 25$.

(c) $\displaystyle\int_0^2 f(x)\,dx \cong 2.96$

34. (a) $f'(x) = \dfrac{-2x}{(1 + x^2)^2} < 0$ on $(0, 1)$ $\Rightarrow$ f is decreasing.

(b) $U_f(P) - \int_0^1 f(x)\, dx \le |f(1) - f(0)|\Delta x = |\tfrac{1}{2} - 1|\tfrac{1}{n} = \dfrac{1}{2n}$.

so need $\dfrac{1}{2n} = 0.05$, or $n = 10$.

(c) Using $U_f(P)$ with $n = 10$, we have $\displaystyle\int_0^1 \frac{1}{1+x^2}\, dx \cong 0.78$

35. Let S be the set of positive integers for which the statement is true. Since $\dfrac{1(2)}{2} = 1, 1 \in S$. Assume that $k \in S$. Then

$$1 + 2 + \cdots + k + k + 1 = (1 + 2 + \cdots + k) + k + 1 = \frac{k(k+1)}{2} + k + 1$$

$$= \frac{(k+1)(k+2)}{2}$$

Thus, $k + 1 \in S$ and so S is the set of positive integers.

36. See Exercise 5 in section 1.8.

37. Let $f(x) = x$ and let $P = \{x_0, x_1, x_2, \ldots, x_n\}$ be a regular partition of $[0, b]$. Then $\Delta x = b/n$ and $x_i = \dfrac{ib}{n}, \; i = 0, 1, 2, \ldots, n$.

(a) Since f is increasing on $[0, b]$,

$$L_f(P) = \left[f(0) + f\left(\frac{b}{n}\right) + f\left(\frac{2b}{n}\right) + \cdots + f\left(\frac{(n-1)b}{n}\right)\right]\frac{b}{n}$$

$$= \left[0 + \frac{b}{n} + \frac{2b}{n} + \cdots + \frac{(n-1)b}{n}\right]\frac{b}{n}$$

$$= \frac{b^2}{n^2}\left[1 + 2 + \cdots + (n-1)\right]$$

(b)
$$U_f(P) = \left[f\left(\frac{b}{n}\right) + f\left(\frac{2b}{n}\right) + \cdots + f\left(\frac{(n-1)b}{n}\right) + f(b)\right]\frac{b}{n}$$

$$= \left[\frac{b}{n} + \frac{2b}{n} + \cdots + \frac{(n-1)b}{n} + b\right]\frac{b}{n}$$

$$= \frac{b^2}{n^2}\left[1 + 2 + \cdots + (n-1) + n\right]$$

(c) By Exercise 35,

$$L_f(P) = \frac{b^2}{n^2} \cdot \frac{(n-1)n}{2} = \frac{1}{2}b^2\left(\frac{n^2 - n}{n^2}\right) = \frac{1}{2}b^2\left(1 - \frac{1}{n}\right) = \frac{1}{2}b^2(1 - \|P\|)$$

$$L_f(P) = \frac{b^2}{n^2} \cdot \frac{n(n+1)}{2} = \frac{1}{2}b^2\left(\frac{n^2 + n}{n^2}\right) = \frac{1}{2}b^2\left(1 + \frac{1}{n}\right) = \frac{1}{2}b^2(1 + \|P\|)$$

(d) For any partition $P, L_f(P) \le \S^*(P) \le U_f(P)$. Since

$$\lim_{\|P\|\to 0} L_f(P) = \lim_{\|P\|\to 0} U_f(P) = \frac{1}{2}b^2,$$

$\displaystyle\lim_{\|P\|\to 0} S^*(P) = \frac{1}{2}b^2$ by the pinching theorem.

38. Let $f(x) = x^2$ and let $P = \{x_0, x_1, x_2, \ldots, x_n\}$ be a regular partition of $[0, b]$. Then $\Delta x = b/n$ and
$$x_i = \frac{ib}{n}, \quad i = 0, 1, 2, \ldots, n.$$

(a) Since f is increasing on $[0, b]$,
$$L_f(P) = \left[f(0) + f\left(\frac{b}{n}\right) + f\left(\frac{2b}{n}\right) + \cdots + f\left(\frac{(n-1)b}{n}\right) \right] \frac{b}{n}$$
$$= \left[0 + \frac{b^2}{n^2} + \frac{4b^2}{n^2} + \cdots + \frac{(n-1)^2 b^2}{n} \right] \frac{b}{n}$$
$$= \frac{b^3}{n^3} \left[1 + 2^2 + \cdots + (n-1)^2 \right]$$

(b)
$$U_f(P) = \left[f\left(\frac{b}{n}\right) + f\left(\frac{2b}{n}\right) + \cdots + f\left(\frac{(n-1)b}{n}\right) + f(b) \right] \frac{b}{n}$$
$$= \left[\frac{b^2}{n^2} + \frac{4b^2}{n^2} + \cdots + \frac{n^2 b^2}{n^2} \right] \frac{b}{n}$$
$$= \frac{b^3}{n^3} \left[1 + 2^2 + \cdots + n^2 \right]$$

(c) By Exercise 36,
$$L_f(P) = \frac{b^3}{n^3} \cdot \frac{(n-1)n(2n-1)}{6} = b^3 \left(\frac{2n^3 - 3n^2 + n}{6n^3} \right) = \frac{1}{6} b^3 = (2 - 3\|P\| + \|P\|^2)$$
$$U_f(P) = \frac{b^3}{n^3} \cdot \frac{n(n+1)(2n-1)}{6} = b^3 \left(\frac{2n^3 + 3n^2 + n}{6n^3} \right) = \frac{1}{6} b^3 = (2 + 3\|P\| + \|P\|^2)$$

(d) For any partition $P, L_f(P) \le \S^*(P) \le U_f(P)$. Since
$$\lim_{\|P\| \to 0} L_f(P) = \lim_{\|P\| \to 0} U_f(P) = \frac{1}{3} b^3,$$
$\displaystyle\lim_{\|P\| \to 0} S^*(P) = \frac{1}{3} b^3$ by the pinching theorem.

39. Choose each x_i^* so that $f(x_i^*) = m_i$. Then $S_i^*(P) = L_f(P)$.

Similarly, choosing each x_i^* so that $f(x_i^*) = M_i$ gives $S_i^*(P) = U_f(P)$.

Also, choosing each x_i^* so that $f(x_i^*) = \frac{1}{2}(m_i + M_i)$ (they exist by the intermediate value theorem) gives
$$S_i^*(P) = \frac{1}{2}(m_1 + M_1)\Delta x_1 + \cdots + \frac{1}{2}(m_n + M_n)\Delta x_n$$
$$= \frac{1}{2}[m_1 \Delta x_1 + \cdots + m_n \Delta x_n + M_1 \Delta x_1 + \cdots + M_n \Delta x_n]$$
$$= \frac{1}{2}[L_f(P) + U_f(P)].$$

40. (a) $L_f(P) = \dfrac{181}{25} \cong 7.24, \quad U_f(P) = \dfrac{221}{25} \cong 8.84$

(b) $\dfrac{1}{2}[L_f(P) + U_f(P)] = \dfrac{402}{50} \cong 8.04 \qquad$ (c) $S^*(P) \cong 7.98$

41. (a) $L_f(P) \cong 0.6105$, $U_f(P) \cong 0.7105$

  (b) $\dfrac{1}{2}[L_f(P) + U_f(P)] \cong 0.6605$ (c) $S^*(P) \cong 0.6684$

42. (a) $L_f(P) \cong 1.0224$, $U_f(P) \cong 1.1824$

  (b) $\dfrac{1}{2}[L_f(P) + U_f(P)] \cong 1.1024$ (c) $S^*(P) \cong 1.1074$

43. (a) $L_f(P) \cong 0.53138$, $U_f(P) \cong 0.73138$

  (b) $\dfrac{1}{2}[L_f(P) + U_f(P)] \cong 0.63138$ (c) $S^*(P) \cong 0.63926$

SECTION 5.3

1. (a) $\displaystyle\int_0^5 f(x)\,dx = \int_0^2 f(x)\,dx + \int_2^5 f(x)\,dx = 4 + 1 = 5$

  (b) $\displaystyle\int_1^2 f(x)\,dx = \int_0^2 f(x)\,dx - \int_0^1 f(x)\,dx = 4 - 6 = -2$

  (c) $\displaystyle\int_1^5 f(x)\,dx = \int_0^5 f(x)\,dx - \int_0^1 f(x)\,dx = 5 - 6 = -1$

  (d) 0   (e) $\displaystyle\int_2^0 f(x)\,dx = -\int_0^2 f(x)\,dx = -4$

  (f) $\displaystyle\int_5^1 f(x)\,dx = -\int_1^5 f(x)\,dx = 1$

2. (a) $\displaystyle\int_4^8 f(x)\,dx = \int_1^8 f(x)\,dx - \int_1^4 f(x)\,dx = 11 - 5 = 6$

  (b) $\displaystyle\int_4^3 f(x)\,dx = -\int_3^4 f(x)\,dx = -7$

  (c) $\displaystyle\int_1^3 f(x)\,dx = \int_1^4 f(x)\,dx - \int_3^4 f(x)\,dx = 5 - 7 = -2$

  (d) $\displaystyle\int_3^8 f(x)\,dx = \int_1^8 f(x)\,dx - \int_1^3 f(x)\,dx = 11 - (-2) = 13$

  (e) $\displaystyle\int_8^4 f(x)\,dx = -\int_4^8 f(x)\,dx = -6$

  (f) $\displaystyle\int_4^4 f(x)\,dx = 0$

3. With $P = \left\{1, \dfrac{3}{2}, 2\right\}$ and $f(x) = \dfrac{1}{x}$, we have

$$0.5 \le \frac{7}{12} = L_f(P) \le \int_1^2 \frac{dx}{x} \le U_f(P) = \frac{5}{6} < 1.$$

4. Using $P = \{0, \frac{1}{2}, 1\}$, we have $0.6 < 0.65 = L_f(P) \leq \displaystyle\int_0^1 \frac{1}{1+x^2}\,dx \leq U_f(P) = 0.9 < 1.$

5. (a) $F(0) = 0$　　　　(b) $F'(x) = x\sqrt{x+1}$　　　　(c) $F'(2) = 2\sqrt{3}$

   (d) $F(2) = \displaystyle\int_0^2 t\sqrt{t+1}\,dt$　　　　　　　　(e) $-F(x) = \displaystyle\int_x^0 t\sqrt{t+1}\,dt$

6. (a) $F(\pi) = \displaystyle\int_\pi^\pi t\sin t\,dt = 0$　　　　　　(b) By Theorem 5.3.5, $F'(x) = x\sin x.$

   (c) $F'(\frac{\pi}{2}) = \frac{\pi}{2}\sin\frac{\pi}{2} = \frac{\pi}{2}$　　　　　　(d) $F(2\pi) = \displaystyle\int_\pi^{2\pi} t\sin t\,dt$

   (e) $-F(x) = \displaystyle\int_x^\pi t\sin t\,dt.$

7. $F'(x) = \dfrac{1}{x^2+9};$　　(a) $\dfrac{1}{10}$　　(b) $\dfrac{1}{9}$　　(c) $\dfrac{4}{37}$　　(d) $\dfrac{-2x}{(x^2+9)^2}$

8. $F'(x) = -\sqrt{x^2+1}$　　(a) $-\sqrt{2}$　　(b) -1　　(c) $-\frac{1}{2}\sqrt{5}$　　(d) $\dfrac{-x}{\sqrt{x^2+1}}$

9. $F'(x) = -x\sqrt{x^2+1};$　　(a) $\sqrt{2}$　　(b) 0　　(c) $-\frac{1}{4}\sqrt{5}$　　(d) $-\left(\sqrt{x^2+1} + \dfrac{x^2}{\sqrt{x^2+1}}\right)$

10. $F'(x) = \sin\pi x$　　(a) 0　　(b) 0　　(c) 1　　(d) $\pi\cos\pi x$

11. $F'(x) = \cos\pi x;$　　(a) -1　　(b) 1　　(c) 0　　(d) $-\pi\sin\pi x$

12. $F'(x) = (x+1)^3$　　(a) 0　　(b) 1　　(c) $\dfrac{27}{8}$　　(d) $3(x+1)^2$

13. (a) Since $P_1 \subseteq P_2$, $U_f(P_2) \leq U_f(P_1)$　　but　　$5 \nleq 4.$

    (b) Since $P_1 \subseteq P_2$, $L_f(P_1) \leq L_f(P_2)$　　but　　$5 \nleq 4.$

14. (a) constant functions.　　　　　　(b) constant functions.

15. constant functions

16. We know this is true for $a < c < b$. Assume $a < b$. If $c = a$ or $c = b$, the equality becomes $\displaystyle\int_a^b f(x)\,dt = \int_a^b f(x)\,dt$, trivially true. If $c < a$, we get

$$\int_a^c f(t)\,dt + \int_c^b f(t)\,dt = -\int_c^a f(t)\,dt + \int_c^b f(t)\,dt = \int_a^b f(t)\,dt, \text{ as desired}$$

The other possible cases are proved in a similar manner.

17. $F'(x) = \dfrac{x-1}{1+x^2} = 0 \implies x = 1$ is a critical number.

$F''(x) = \dfrac{(1+x^2) - 2x(x-1)}{(1+x^2)^2}$, so $F''(1) = \dfrac{1}{2} > 0$ means $x = 1$ is a local minimum.

18. $F'(x) = \dfrac{x-4}{1+x^2} = 0 \implies x = 4$ is a critical number.

$F''(x) = \dfrac{(1+x^2) - 2x(x-4)}{(1+x^2)^2}$, so $F''(4) = \dfrac{1}{17} > 0$ means $x = 4$

is a local minimum.

19.   (a) $F'(x) = \dfrac{1}{x} > 0$ for $x > 0$.

  Thus, F is increasing on $(0, \infty)$;

  there are no critical numbers.

(b) $F''(x) = -\dfrac{1}{x^2} < 0$ for $x > 0$.

  The graph of F is concave down on $(0, \infty)$;

  there are no points of inflection.

(c)

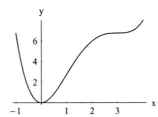

20.   (a) $F'(x) = x(x-3)^2$,

  F is increasing on $[0, \infty)$;

  F is decreasing on $(-\infty, 0]$;

  critical numbers $0, -3$.

(b) $F''(x) = (x-3)^2 + 2x(x-3) = 3(x-3)(x-1)$.

  The graph of F is concave up on $(-\infty, 1) \cup (3, \infty)$;

  The graph of F is concave down on $(1, 3)$;

  Inflection points at $x = 1$, $x = 3$.

(c)

21.   (a) F is differentiable, therefore continuous

(b) $F'(x) = f(x)$ f is differentiable; $F''(x) = f'(x)$

(c) $F'(1) = f(1) = 0$

(d) $F''(1) = f'(1) > 0$

(e) $f(1) = 0$ and f increasing $(f' > 0)$ implies $f < 0$ on $(0, 1)$ and $f > 0$ on $(1, \infty)$.

  Since $F' = f, F$ is decreasing on $(0, 1)$ and increasing on $(1, \infty)$;

  $F(0) = 0$ implies $F(1) < 0$.

22.   (a) G is differentiable, therefore continuous

(b) $G'(x) = g(x)$ and g is differentiable; $G''(x) = g'(x)$

(c) $G'(1) = g(1) = 0$

(d) $G''(x) = g'(x) < 0$ for $x < 1$

(e) $G'(x) = g(x) > 0$ for all $x \neq 0$.

  $G''(x) = g'(x) > 0$ for $x > 1$

23. (a)

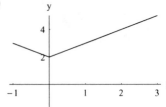

(b)

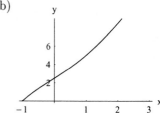

$$F(x) = \begin{cases} 2x - \frac{1}{2}x^2 + \frac{5}{2} & -1 \le x \le 0 \\ 2x + \frac{1}{2}x^2 + \frac{5}{2} & 0 < x \le 3 \end{cases}$$

(c) f is discontinuous at $x = 0$, but not differentiable; F is continuous but differentiable at $x = 0$.

24. (a)

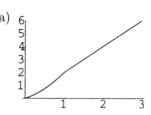

(b)

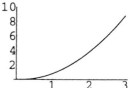

$$F(x) = \begin{cases} \dfrac{x^3}{3} + \dfrac{x^2}{2} & 0 \le x \le 1 \\ x^2 - \dfrac{1}{6} & 1 < x \le 3 \end{cases}$$

(c) f is continuous at $x = 1$, but not differentiable. F is continuous and differentiable at $x = 1$.

25. Let $u = x^3$. Then $F(u) = \displaystyle\int_1^u t \cos t\, dt$ and

$$\frac{dF}{dx} = \frac{dF}{du}\frac{du}{dx} = u \cos u\, (3x^2) = 3x^5 \cos x^3.$$

26. Let $u = \cos x$. $\dfrac{dF}{dx} = \dfrac{dF}{du}\dfrac{du}{dx} = \sqrt{1 - u^2}\,(-\sin x) = \sqrt{1 - \cos^2 x}\,(-\sin x) = -|\sin x|\,\sin x$

27. $F(x) = \displaystyle\int_{x^2}^1 (t - \sin^2 t)\, dt = -\int_1^{x^2} (t - \sin^2 t)\, dt$. Let $u = x^2$. Then

$$\frac{dF}{dx} = \frac{dF}{du}\frac{du}{dx} = -(u - \sin^2 u)(2x) = 2x\left[\sin^2(x^2) - x^2\right].$$

28. Let $u = \sqrt{x}$. $\dfrac{dF}{dx} = \dfrac{dF}{du}\dfrac{du}{dx} = \dfrac{u^2}{1 + u^4}\dfrac{1}{2\sqrt{x}} = \dfrac{x}{1 + x^2}\dfrac{1}{2\sqrt{x}}$

29. (a) $F(0) = 0$

(b) $F'(0) = 2 + \dfrac{\sin 2(0)}{1 + 0^2} = 2$

(c) $F''(0) = \dfrac{(1 + 0)^2 2 \cos 2(0) - \sin 2(0)(2)(0)}{(1 + 0)^2} = 2$

30. (a) $F(0) = 0$

   (b) Let $u = x^2$. Then $f(u) = 2\sqrt{u} + \int_0^u \dfrac{\sin 2t}{1 + t^2}\, dt.$

   $$\frac{dF}{dx} = \frac{dF}{du}\frac{du}{dx} = 2 + \frac{\sin 2u}{1 + u^2}(2x) = 2 + \frac{\sin 2x^2}{1 + x^4}(2x)$$

31. $f(x) = \dfrac{d}{dx}\left(\dfrac{2x}{4 + x^2}\right) = \dfrac{8 - 2x^2}{(4 + x^2)^2}$

   (a) $f(0) = \dfrac{1}{2}$ (b) $f(x) = 0$ at $x = -2,\ 2$

32. (a) $F(x) = \displaystyle\int_0^x t f(t)\, dt = \sin x - x\cos x.$

   $F'(x) = x f(x) = \cos x - \cos x + x\sin x = x\sin x \implies f(\frac{\pi}{2}) = \sin\frac{\pi}{2} = 1$

   (b) $f'(x) = \cos x$

33. By Theorem 5.3.5 and the mean-value theorem (Theorem 4.1.1)

   $$\frac{F(b) - F(a)}{b - a} = F'(c) \quad \text{for some } c \in (a,\, b)$$

   The result follows by observing that

   $$F(b) = \int_a^b f(t)\, dt, \quad F(a) = 0, \quad \text{and} \quad F'(c) = f(c).$$

34. Choose point $c \in (a, b)$ and set $F(x) = \displaystyle\int_c^x f(t)\, dt$. Since

   $$\int_c^x f(t)\, dt = \int_a^x f(t)\, dt - \int_a^c f(t)\, dt, \quad \text{(Exercise 16)}$$

   it follows that

   $$F'(x) = \frac{d}{dx}\left(\int_a^x f(t)\, dt - \int_a^c f(t)\, dt\right) = f(x)$$

   by Theorem 5.3.5.

35. (a) $F'(x) = f(x) = G'(x)$, on $[a, b]$. Therefore, by Theorem 4.2.4, F and G differ by a constant.

   (b) $F(x) = -\int_a^c f(t)\, dt + \int_a^x f(t)\, dt$ and $G(x) = -\int_a^d f(t)\, dt + \int_a^x f(t)\, dt$.

   Thus $F(x) - G(x) = -\int_a^c f(t)\, dt + \int_a^d f(t)\, dt = \int_c^d f(t)\, dt$, a constant

36. (a) $F'(x) = x\int_1^x f(u)\, du$ (b) $F'(1) = 0$

   (c) $F''(x) = x f(x) + \int_1^x f(u)\, du$ (d) $F''(1) = f(1)$

37. (a) $F'(x) = 0$ at $x = -1, 4$; F is increasing on $(-\infty, -1]$, $[4, \infty)$; F is decreasing on $[-1, 4]$

   (b) $F''(x) = 0$ at $x = \frac{3}{2}$;   the graph of F is concave up on $\left(\frac{3}{2}, \infty\right)$; concave down on $\left(-\infty, \frac{3}{2}\right)$

38. (a) $F'(x) = 2 - 3\cos x = 0$ $\Longrightarrow$ $\cos x = \frac{2}{3}$ $\Longrightarrow$ $x \cong 0.8411, 5.4421$;

   F is decreasing on $[0, 0.8411] \cup [5.4421, 2\pi]$; F is increasing on $[0.8411, 5.4421]$.

   (b) $F''(x) = \sin x = 0$ $\Longrightarrow$ $x = \pi$;

   the graph of F is concave up on $(0, \pi)$ and concave down on $(\pi, 2\pi)$.

39. (a) $F'(x) = 0$ at $x = 0, \frac{\pi}{2}, \pi, \frac{3\pi}{2}, 2\pi$

   F is increasing on $[\frac{\pi}{2}, \pi], [\frac{3\pi}{2}, 2\pi]$; F is decreasing on $[0, \frac{\pi}{2}], [\pi, \frac{3\pi}{2}]$

   (b) $F''(x) = 0$ at $x = \frac{\pi}{4}, \frac{3\pi}{4}, \frac{5\pi}{4}, \frac{7\pi}{4}$;

   the graph of F is concave up on $\left(\frac{\pi}{4}, \frac{3\pi}{4}\right), \left(\frac{5\pi}{4}, \frac{7\pi}{4}\right)$; concave down on $\left(0, \frac{\pi}{4}\right), \left(\frac{3\pi}{4}, \frac{5\pi}{4}\right), \left(\frac{7\pi}{4}, 2\pi\right)$

40. (a) $F'(x) = -(2 - x)^2 = 0$ at $x = 2; F'(x) < 0$ for all $x \neq 2$ $\Longrightarrow$ F is decreasing on $(-\infty, \infty)$.

   (b) $F''(x) = (2 - x) = 0$ at $x = 2$; the graph of F is concave up on $(-\infty, 2)$ and concave down on $(2, \infty)$.

SECTION 5.4

1. $\displaystyle\int_0^1 (2x - 3)\, dx = [x^2 - 3x]_0^1 = (-2) - (0) = -2$

2. $\displaystyle\int_0^1 (3x + 2)\, dx = \left[\frac{3x^2}{2} + 2x\right]_0^1 = \frac{7}{2}$

3. $\displaystyle\int_{-1}^0 5x^4\, dx = [x^5]_{-1}^0 = (0) - (-1) = 1$

4. $\displaystyle\int_1^2 (2x + x^2)\, dx = \left[x^2 + \frac{1}{3}x^3\right]_1^2 = \frac{16}{3}$

5. $\displaystyle\int_1^4 2\sqrt{x}\, dx = 2\int_1^4 x^{1/2}\, dx = 2\left[\frac{2}{3}x^{3/2}\right]_1^4 = \frac{4}{3}\left[x^{3/2}\right]_1^4 = \frac{4}{3}(8 - 1) = \frac{28}{3}$

6. $\displaystyle\int_0^4 \sqrt[3]{x}\, dx = \int_0^4 x^{\frac{1}{3}}\, dx = \left[\frac{3}{4}x^{4/3}\right]_0^4 = \frac{3}{4}4^{4/3} = 3\sqrt[3]{4}$

7. $\displaystyle\int_1^5 2\sqrt{x - 1}\, dx = \int_1^5 2(x - 1)^{1/2}\, dx = \left[\frac{4}{3}(x - 1)^{3/2}\right]_1^5 = \frac{4}{3}[4^{3/2} - 0] = \frac{32}{3}$

8. $\displaystyle\int_1^2 \left(\frac{3}{x^3} + 5x\right) dx = \left[-\frac{3}{2}x^{-2} + \frac{5}{2}x^2\right]_1^2 = \frac{69}{8}$

9. $\displaystyle\int_{-2}^0 (x+1)(x-2)\,dx = \int_{-2}^0 (x^2 - x - 2)\,dx = \left[\frac{x^3}{3} - \frac{x^2}{2} - 2x\right]_{-2}^0 = \left[0 - \left(\frac{-8}{3} - 2 + 4\right)\right] = \frac{2}{3}$

10. $\displaystyle\int_1^0 (t^3 + t^2)\,dt = \left[\frac{1}{4}t^4 + \frac{1}{3}t^3\right]_1^0 = -\frac{7}{12}$

11. $\displaystyle\int_1^2 \left(3t + \frac{4}{t^2}\right) dt = \int_1^2 (3t + 4t^{-2})\,dt = \left[\frac{3}{2}t^2 - 4t^{-1}\right]_1^2 = \left[(6-2) - \left(\frac{3}{2} - 4\right)\right] = \frac{13}{2}$

12. $\displaystyle\int_{-1}^{-1} 7x^6\,dx = 0$

13. $\displaystyle\int_0^1 (x^{3/2} - x^{1/2})\,dx = \left[\frac{2}{5}x^{5/2} - \frac{2}{3}x^{3/2}\right]_0^1 = \left[\left(\frac{2}{5} - \frac{2}{3}\right) - 0\right] = -\frac{4}{15}$

14. $\displaystyle\int_0^1 (x^{3/4} - 2x^{1/2})\,dx = \left[\frac{4}{7}x^{7/4} - \frac{4}{3}x^{3/2}\right]_0^1 = -\frac{16}{21}$

15. $\displaystyle\int_0^1 (x+1)^{17}\,dx = \left[\frac{1}{18}(x+1)^{18}\right]_0^1 = \frac{1}{18}(2^{18} - 1)$

16. $\displaystyle\int_0^a (a^2 x - x^3)\,dx = \left[\frac{a^2 x^2}{2} - \frac{x^4}{4}\right]_0^a = \frac{a^4}{4}$

17. $\displaystyle\int_0^a (\sqrt{a} - \sqrt{x})^2\,dx = \int_0^a (a - 2\sqrt{a}\,x^{1/2} + x)\,dx = \left[ax - \frac{4}{3}\sqrt{a}\,x^{3/2} + \frac{x^2}{2}\right]_0^a = a^2 - \frac{4}{3}a^2 + \frac{a^2}{2} = \frac{1}{6}a^2$

18. $\displaystyle\int_{-1}^1 (x-2)^2\,dx = \left[\frac{1}{3}(x-2)^3\right]_{-1}^1 = \frac{26}{3}$

19. $\displaystyle\int_1^2 \frac{6-t}{t^3}\,dt = \int_1^2 (6t^{-3} - t^{-2})\,dt = \left[-3t^{-2} + t^{-1}\right]_1^2 = \left[\frac{-3}{4} + \frac{1}{2}\right] - [-3 + 1] = \frac{7}{4}$

20. $\displaystyle\int_1^3 \left(x^2 - \frac{1}{x^2}\right) dx = \left[\frac{1}{3}x^3 + \frac{1}{x}\right]_1^3 = 8$

21. $\displaystyle\int_1^2 2x(x^2 + 1)\,dx = \int_1^2 (2x^3 + 2x)\,dx = \left[\frac{x^4}{2} + x^2\right]_1^2 = 12 - \frac{3}{2} = \frac{21}{2}$

22. $\displaystyle\int_0^1 3x^2(x^3 + 1)\,dx = \int_0^1 (3x^5 + 3x^2)\,dx = \left[\frac{1}{2}x^6 + x^3\right]_0^1 = \frac{3}{2}$

23. $\displaystyle\int_0^{\pi/2} \cos x\,dx = [\sin x]_0^{\pi/2} = 1$

24. $\displaystyle\int_0^\pi 3\sin x\,dx = [-3\cos x]_0^\pi = 6$

25. $\displaystyle\int_0^{\pi/4} 2\sec^2 x\,dx = 2\,[\tan x]_0^{\pi/4} = 2$

26. $\displaystyle\int_{\pi/6}^{\pi/3} \sec x\tan x\,dx = [\sec x]_{\pi/6}^{\pi/3} = 2 - \dfrac{2\sqrt{3}}{3}$

27. $\displaystyle\int_{\pi/6}^{\pi/4} \csc u\cot u\,dx = [-\csc u]_{\pi/6}^{\pi/4} = -\sqrt{2} - (-2) = 2 - \sqrt{2}$

28. $\displaystyle\int_{\pi/4}^{\pi/3} -\csc^2 u\,du = [\cot u]_{\pi/4}^{\pi/3} = \dfrac{\sqrt{3}}{3} - 1$

29. $\displaystyle\int_0^{2\pi} \sin x\,dx = [-\cos x]_0^{2\pi} = -1 - (-1) = 0$

30. $\displaystyle\int_0^\pi \dfrac{1}{2}\cos x\,dx = \left[\dfrac{1}{2}\sin x\right]_0^\pi = 0$

31. $\displaystyle\int_0^{\pi/3} \left(\dfrac{2}{\pi}x - 2\sec^2 x\right) dx = \left[\dfrac{1}{\pi}x^2 - 2\tan x\right]_0^{\pi/3} = \dfrac{\pi}{9} - 2\sqrt{3}$

32. $\displaystyle\int_{\pi/4}^{\pi/2} \csc x(\cot x - 3\csc x)\,dx = \int_{\pi/4}^{\pi/2} (\csc x\cot x - 3\csc^2 x)\,dx = [-\csc x + 3\cot x]_{\pi/4}^{\pi/2} = \sqrt{2} - 4$

33. $\displaystyle\int_0^3 \left[\dfrac{d}{dx}\left(\sqrt{4+x^2}\right)\right] dx = \left[\sqrt{4+x^2}\right]_0^3 = \sqrt{13} - 2$

34. $\displaystyle\int_0^{\pi/2} \left[\dfrac{d}{dx}\left(\sin^3 x\right)\right] = [\sin^3 x]_0^{\pi/2} = 1$

35. (a) $F(x) = \displaystyle\int_1^x (t+2)^2\,dt \qquad \Longrightarrow \qquad F'(x) = (x+2)^2$

(b) $\displaystyle\int_1^x (t+2)^2\,dt = \left[\dfrac{t^3}{3} + 2t^2 + 4t\right]_1^x = \dfrac{x^3}{3} + 2x^2 + 4x - 6\dfrac{1}{3}$

$\Longrightarrow \quad F'(x) = x^2 + 4x + 4 = (x+2)^2$

36. (a) $F(x) = \displaystyle\int_0^x (\cos t - \sin t)\,dt \qquad \Longrightarrow \qquad F'(x) = \cos x - \sin x$

(b) $\displaystyle\int_0^x (\cos t - \sin t)\,dt = [\sin t + \cos t]_0^x = \sin x - \cos x - 1$

$\Longrightarrow \quad F'(x) = \cos x - \sin x$

37. (a) $F(x) = \displaystyle\int_1^{2x+1} \tfrac{1}{2}\sec u\tan u\,du \quad \Rightarrow \quad F'(x) = \sec(2x+1)\tan(2x+1)$

(b) $\displaystyle\int_1^{2x+1} \dfrac{1}{2}\sec u\tan u\,du = \left[\dfrac{1}{2}\sec u\right]_1^{2x+1} = \dfrac{1}{2}\sec(2x+1) - \dfrac{1}{2}\sec 1$

$\Longrightarrow \quad F'(x) = \sec(2x+1)\tan(2x+1)$

38. (a) $F(x) = \int_{x^2}^{2} t(t-1)\, dt \quad \Rightarrow \quad F'(x) = -x^2(x^2-1)2x$

(b) $\int_{x^2}^{2} t(t-1)\, dt = \left[\dfrac{t^3}{3} - \dfrac{t^2}{2} \right]_{x^2}^{2} = \dfrac{2}{3} - \dfrac{x^6}{3} + \dfrac{x^4}{2}$

$\quad \Longrightarrow \quad F'(x) = -2x^5 + 2x^3 = -2x^3(x^2-1)$

39. (a) $F(x) = \int_{2}^{x} \dfrac{dt}{t}$ $\qquad\qquad\qquad$ (b) $F(x) = -3 + \int_{2}^{x} \dfrac{dt}{t}$

40. (a) $F(x) = \int_{3}^{x} \sqrt{1+t^2}\, dt$ $\qquad\qquad$ (b) $F(x) = 1 + \int_{3}^{x} \sqrt{1+t^2}\, dt$

41. $\text{Area} = \int_{0}^{4} (4x - x^2)\, dx = \left[2x^2 - \dfrac{x^3}{3} \right]_{0}^{4} = \dfrac{32}{3}$

42. $\text{Area} = \int_{1}^{9} (x\sqrt{x} + 1)\, dx = \int_{1}^{9} (x^{3/2} + 1)\, dx = \left[\dfrac{2}{5}x^{5/2} + x \right]_{1}^{9} = \dfrac{524}{5}$

43. $\text{Area} = \int_{-\pi/2}^{\pi/4} 2\cos x\, dx = 2\left[\sin x \right]_{-\pi/2}^{\pi/4} = \sqrt{2} + 2$

44. $\text{Area} = \int_{0}^{\pi/3} (\sec x \tan x)\, dx = \left[\sec x \right]_{0}^{\pi/3} = 2 - 1 = 1$

45. (a) $\int_{2}^{5} (x-3)\, dx = \left[\dfrac{x^2}{2} - 3x \right]_{2}^{5} = \dfrac{3}{2}$ $\qquad$ (b) $\int_{2}^{5} |x-3|\, dx = \int_{2}^{3} (3-x)\, dx + \int_{3}^{5} (x-3)\, dx$

$\qquad\qquad\qquad\qquad\qquad\qquad\qquad\qquad\qquad\qquad = \left[3x - \dfrac{x^2}{2} \right]_{2}^{3} + \left[\dfrac{x^2}{2} - 3x \right]_{3}^{5} = \dfrac{5}{2}$

46. (a) $\int_{-4}^{2} (2x+3)\, dx = \left[x^2 + 3x \right]_{-4}^{2} = 6$

(b) $\int_{-4}^{2} |2x+3|\, dx = \int_{-4}^{-3/2} (-2x-3)\, dx + \int_{-3/2}^{2} (2x+3)\, dx = \left[-x^2 - 3x \right]_{-4}^{-3/2} + \left[x^2 + 3x \right]_{-3/2}^{2} = \dfrac{37}{2}$

47. (a) $\int_{-2}^{2} (x^2-1)\, dx = \left[\dfrac{x^3}{3} - x \right]_{-2}^{2} = \dfrac{4}{3}$

(b) $\int_{-2}^{2} |x^2-1|\, dx = \int_{-2}^{-1} (x^2-1)\, dx + \int_{-1}^{1} (1-x^2)\, dx + \int_{1}^{2} (x^2-1)\, dx$

$\qquad\qquad\qquad = \left[\dfrac{x^3}{3} - x \right]_{-2}^{-1} + \left[x - \dfrac{x^3}{3} \right]_{-1}^{1} + \left[\dfrac{x^3}{3} - x \right]_{1}^{2} = 4$

48. (a) $\int_{-\pi/2}^{\pi} \cos x\, dx = \left[\sin x \right]_{-\pi/2}^{\pi} = 1$

(b) $\int_{-\pi/2}^{\pi} |\cos x|\, dx = \int_{-\pi/2}^{\pi/2} \cos x\, dx + \int_{\pi/2}^{\pi} -\cos x\, dx = \left[\sin x \right]_{-\pi/2}^{\pi/2} + \left[-\sin x \right]_{\pi/2}^{\pi} = 3$

49. valid

50. not valid; $\sec^2 x$ is not defined at $x = \frac{1}{2}\pi$, and $x = \frac{3}{2}\pi$

51. not valid; $1/x^3$ is not defined at $x = 0$

52. valid

53. (a) $x(t) = \displaystyle\int_0^t (10u - u^2)\, du = \left[5u^2 - \frac{u^3}{3}\right]_0^t = 5t^2 - \frac{t^3}{3}, \quad 0 \le t \le 10$

(b) $v'(t) = 10 - 2t$; v has an absolute maximum at $t = 5$. The object's position at $t = 5$ is
$$x(5) = \frac{250}{3}.$$

54. (a) We need $x(t)$ such that $x'(t) = 3\sin t + 4\cos t$ and $x(0) = 1$.

Then $x(t) = -3\cos t + 4\sin t + \mathrm{C}, \quad x(0) = -3 + \mathrm{C} = 1 \implies \mathrm{C} = 4$

$\implies x(t) = -3\cos t + 4\sin t + 4.$

(b) Maximum displacement when $v(t) = 0$: $3\sin t + 4\cos t = 0$

$\implies \tan t = -\frac{4}{3} \implies \sin t = \frac{4}{5}, \quad \cos t = -\frac{3}{5}$

So $\quad x_{max} = -3\left(\frac{-3}{5}\right) + 4\left(\frac{4}{5}\right) + 4 = 9$

55. $\displaystyle\int_0^4 f(x)\, dx = \int_0^1 (2x + 1)\, dx + \int_1^4 (4 - x)\, dx = \left[x^2 + x\right]_0^1 + \left[4x - \frac{x^2}{2}\right]_1^4 = \frac{13}{2}$

56. $\displaystyle\int_{-2}^4 f(x)\, dx = \int_{-2}^0 (2 + x^2)\, dx + \int_0^4 (\tfrac{1}{2}x + 2)\, dx = \left[2x + \frac{x^3}{3}\right]_{-2}^0 + \left[\frac{x^2}{4} + 2x\right]_0^4 = \frac{56}{3}$

57. $\displaystyle\int_{-\pi/2}^\pi f(x)\, dx = \int_{-\pi/2}^{\pi/3} (1 + 2\cos x\, dx + \int_{\pi/3}^\pi \left[\frac{3}{\pi}x + 1\right] dx = \left[x + 2\sin x\right]_{-\pi/2}^{\pi/3} + \left[\frac{3x^2}{2\pi} + x\right]_{\pi/3}^\pi$

$$= 2 + \sqrt{3} + \tfrac{17}{6}\pi$$

58. $\displaystyle\int_0^{3\pi/2} f(x)\, dx = \int_0^{\pi/2} 2\sin x\, dx + \int_{\pi/2}^{3\pi/2} (2 + \cos x)\, dx = \left[-2\cos x\right]_0^{\pi/2} + \left[2x + \sin x\right]_{\pi/2}^{3\pi/2} = 2\pi$

59. (a) f is continuous on $[-2, 2]$.

For $x \in [-2, 0], \quad g(x) = \displaystyle\int_{-2}^x (t + 2)dt = \left[\frac{1}{2}t^2 + 2t\right]_{-2}^x = \frac{1}{2}x^2 + 2x + 2.$

For $x \in [0, 1], \quad g(x) = \displaystyle\int_{-2}^0 (t + 2)\, dt + \int_0^x 2\, dt = 2 + [2t]_0^x = 2 + 2x.$

For $x \in [1, 2], \quad g(x) = \displaystyle\int_{-2}^0 (t + 2)\, dt + \int_0^1 2\, dt + \int_1^x (4 - 2t)\, dt = 2 + 2 + \left[4t - t^2\right]_1^x = 1 + 4x - x^2.$

Thus $\quad g(x) = \begin{cases} \frac{1}{2}x^2 + 2x + 2, & -2 \le x \le 0 \\ 2x + 2, & 0 \le x \le 1 \\ 1 + 4x - x^2, & 1 \le x \le 2 \end{cases}$

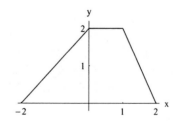

(b)

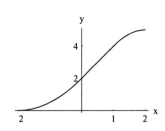

(c) f is continuous on $[-2, 2]$; f is differentiable on $(-2, 0)$, $(0, 1)$, and $(1, 2)$.

   g is differentiable on $(-2, 2)$.

60. (a) $g(x) = \int_{-1}^{x} f(x)\,dt = \int_{-1}^{x}(1 - t^2)\,dt = \left[t - \dfrac{t^3}{3}\right]_{-1}^{x} = 2x - \dfrac{x^3}{3} + \dfrac{5}{3}$, for $-1 \le x \le 1$

   $g(x) = \int_{-1}^{1}(1 - t^2)\,dt + \int_{1}^{x} 1\,dt = \dfrac{10}{3} + [t]_{1}^{x} = \dfrac{7}{3} + x$, for $1 < x < 3$

   $g(x) = \int_{-1}^{3} f(t)\,dt + \int_{3}^{x}(2t - 5)\,dt = \dfrac{16}{3} + [t^2 - 5t]_{3}^{x} = \dfrac{34}{3} + x^2 - 5x$, for $3 \le x \le 5$

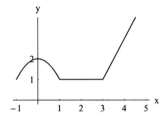

(b)

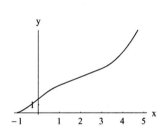

(c) f is continuous on $[-1, 1) \cup (1, 5]$, f is differentiable on $(-1, 1) \cup (1, 3) \cup (3, 5)$.

   g is differentiable on $(-1, 1) \cup (1, 5)$.

61. Follows from Theorem 5.3.2 since $f(x)$ is an antiderivative of $f'(x)$.

62. Let $F(x) = f^2(x)$.   Then $F'(x) = 2f(x)f'(x)$.

   Thus $\displaystyle\int_{a}^{b} f(t)f'(t)\,dt = \dfrac{1}{2}\int_{a}^{b} F'(t)\,dt = \dfrac{1}{2}[F(b) - F(a)] = \dfrac{1}{2}[f^2(b) - f^2(a)]$.

63. $\dfrac{d}{dx}\left[\displaystyle\int_{a}^{x} f(t)\,dt\right] = f(x)$; $\displaystyle\int_{a}^{x} \dfrac{d}{dt}[f(t)]\,dt = f(x) - f(a)$

64. $F(x) = \displaystyle\int_{0}^{x} xf(t)\,dt = x\int_{0}^{x} f(t)\,dt$; F is a product.

   $F'(x) = x\,f(x) + \displaystyle\int_{0}^{x} f(t)\,dt$

SECTION 5.5

1. $A = \displaystyle\int_0^1 (2 + x^3)\, dx = \left[2x + \dfrac{x^4}{4}\right]_0^1 = \dfrac{9}{4}$

2. $A = \displaystyle\int_0^2 (x + 2)^{-2}\, dx = \left[\dfrac{-1}{x + 2}\right]_0^2 = \dfrac{1}{4}$

3. $A = \displaystyle\int_3^8 \sqrt{x + 1}\, dx = \int_3^8 (x + 1)^{1/2}\, dx = \left[\dfrac{2}{3}(x + 1)^{3/2}\right]_3^8 = \dfrac{2}{3}[27 - 8] = \dfrac{38}{3}$

4. $A = \displaystyle\int_0^8 (3x^2 + x^3)\, dx = \left[x^3 + \dfrac{1}{4}x^4\right]_0^8 = 1536$

5. $A = \displaystyle\int_0^1 (2x^2 + 1)^2\, dx = \int_0^1 (4x^4 + 4x^2 + 1)\, dx = \left[\dfrac{4}{5}x^5 + \dfrac{4}{3}x^3 + x\right]_0^1 = \dfrac{47}{15}$

6. $A = \displaystyle\int_0^8 \dfrac{1}{2\sqrt{x + 1}}\, dx = \left[\sqrt{x + 1}\right]_0^8 = 2$

7. $A = \displaystyle\int_1^2 [0 - (x^2 - 4)]\, dx = \int_1^2 (4 - x^2)\, dx = \left[4x - \dfrac{x^3}{3}\right]_1^2 = \left[8 - \dfrac{8}{3}\right] - \left[4 - \dfrac{1}{3}\right] = \dfrac{5}{3}$

8. $A = \displaystyle\int_{\pi/6}^{\pi/3} \cos x\, dx = [\sin x]_{\pi/6}^{\pi/3} = \dfrac{\sqrt{3} - 1}{2}$

9. $A = \displaystyle\int_{\pi/3}^{\pi/2} \sin x\, dx = [-\cos x]_{\pi/3}^{\pi/2} = (0) - \left(-\dfrac{1}{2}\right) = \dfrac{1}{2}$

10. $A = -\displaystyle\int_{-2}^{-1} (x^3 + 1)\, dx = -\left[\dfrac{x^4}{4} + x\right]_{-2}^{-1} = \dfrac{11}{4}$

11.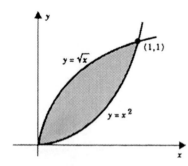

$A = \displaystyle\int_0^1 [x^{1/2} - x^2]\, dx$

$= \left[\tfrac{2}{3}x^{3/2} - \tfrac{1}{3}x^3\right]_0^1 = \tfrac{1}{3}$

12.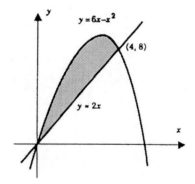

$A = \displaystyle\int_0^4 (6x - x^2 - 2x)\, dx = \dfrac{32}{3}$

13.

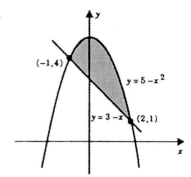

$$A = \int_{-1}^{2} [(5 - x^2) - (3 - x)] \, dx$$

$$= \int_{-1}^{2} (2 + x - x^2) \, dx$$

$$= \left[2x + \frac{x^2}{2} - \frac{x^3}{3} \right]_{-1}^{2}$$

$$= \left[4 + 2 - \tfrac{8}{3} \right] - \left[-2 + \tfrac{1}{2} + \tfrac{1}{3} \right] = \tfrac{9}{2}$$

14.

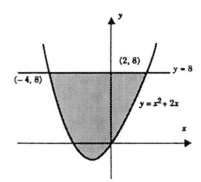

$$A = \int_{-4}^{2} (8 - x^2 - 2x) \, dx = 36$$

15.

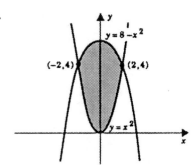

$$A = \int_{-2}^{2} [(8 - x^2) - (x^2)] \, dx$$

$$= \int_{-2}^{2} (8 - 2x^2) \, dx$$

$$= \left[8x - \tfrac{2}{3}x^3 \right]_{-2}^{2}$$

$$= \left[16 - \tfrac{16}{3} \right] - \left[-16 + \tfrac{16}{3} \right] = \tfrac{64}{3}$$

16.

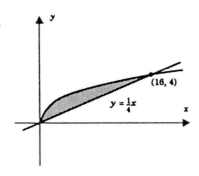

$$A = \int_{0}^{16} \left(\sqrt{x} - \tfrac{1}{4}x \right) \, dx = \frac{32}{3}$$

17.

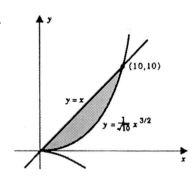

$$A = \int_0^{10} \left[x - \frac{1}{\sqrt{10}} x^{3/2} \right] dx$$

$$= \left[\frac{x^2}{2} - \frac{2\sqrt{10}}{50} x^{5/2} \right]_0^{10}$$

$$= 50 - \frac{2\sqrt{10}}{50}(10)^{5/2} = 10$$

18.

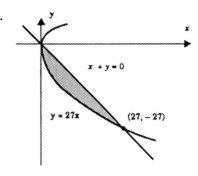

$$A = \int_0^{27} \left[-x - \left(-\sqrt{27x} \right) \right] dx = \frac{243}{2}$$

19.

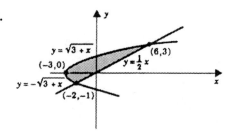

$$A = \int_{-3}^{-2} \left[(\sqrt{3+x}) - (-\sqrt{3+x}) \right] dx + \int_{-2}^{6} \left[(\sqrt{3+x}) - \left(\frac{1}{2}x \right) \right] dx$$

$$= \int_{-3}^{-2} 2(3+x)^{1/2} dx + \int_{-2}^{6} \left[(3+x)^{1/2} - \frac{1}{2}x \right] dx$$

$$= \left[\frac{4}{3}(3+x)^{3/2} \right]_{-3}^{-2} + \left[\frac{2}{3}(3+x)^{3/2} - \frac{x^2}{4} \right]_{-2}^{6} = \left[\frac{4}{3} - 0 \right] + \left[(18-9) - \left(\frac{2}{3} - 1 \right) \right] = \frac{32}{3}$$

20.

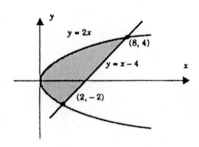

$$A = \int_0^2 \left[\sqrt{2x} - (-\sqrt{2x}) \right] dx + \int_2^8 \left[\sqrt{2x} - x + 4 \right] dx$$

$$= 18$$

21.

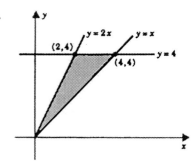

$$A = \int_0^2 [2x - x]\, dx + \int_2^4 [4 - x]\, dx$$

$$= \left[\tfrac{1}{2}x^2\right]_0^2 + \left[4x - \tfrac{1}{2}x^2\right]_2^4$$

$$= 2 + [8 - 6] = 4$$

22.

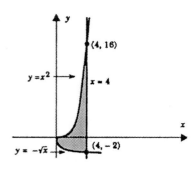

$$A = \int_0^4 \left(x^2 + \sqrt{x}\right)\, dx = \frac{80}{3}$$

23.

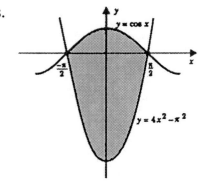

$$A = \int_{-\pi/2}^{\pi/2} [\cos x - (4x^2 - \pi^2)]\, dx$$

$$= \left[\sin x - \tfrac{4}{3}x^3 + \pi^2 x\right]_{-\pi/2}^{\pi/2}$$

$$= [1 - \tfrac{1}{6}\pi^3 + \tfrac{1}{2}\pi^3] - [-1 + \tfrac{1}{6}\pi^3 - \tfrac{1}{2}\pi^3]$$

$$= 2 + \tfrac{2}{3}\pi^3$$

24.

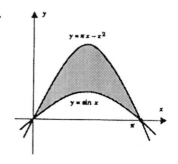

$$A = \int_0^\pi (\pi x - x^2 - \sin x)\, dx = \frac{\pi^3}{6} - 2$$

25.

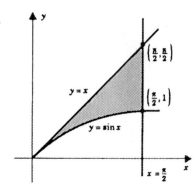

$$A = \int_0^{\pi/2} [x - \sin x]\, dx$$

$$= \left[\frac{x^2}{2} + \cos x \right]_0^{\pi/2}$$

$$= \frac{\pi^2}{8} - 1$$

26.

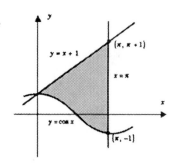

$$A = \int_0^\pi (x + 1 - \cos x)\, dx = \frac{\pi}{2}(\pi + 2)$$

27. (a) $\displaystyle\int_{-3}^4 (x^2 - x - 6)\, dx = \left[\frac{1}{3}x^3 - \frac{1}{2}x^2 - 6x \right]_{-3}^4 = -\frac{91}{6}$;

the area of the region bounded by the graph of f and the x-axis for $x \in [-3, -2] \cup [3, 4]$
minus the area of the region bounded the graph of f and the x-axis for $x \in [-2, 3]$.

(b) $\displaystyle A = \int_{-3}^{-2} (x^2 - x - 6)\, dx + \int_{-2}^3 (-x^2 + x + 6)\, dx + \int_3^4 (x^2 - x - 6)\, dx$

$$= \left[\tfrac{1}{3}x^3 - \tfrac{1}{2}x^2 - 6x \right]_{-3}^{-2} + \left[-\tfrac{1}{3}x^3 + \tfrac{1}{2}x^2 + 6x \right]_{-2}^3 + \left[\tfrac{1}{3}x^3 - \tfrac{1}{2}x^2 - 6x \right]_3^4 = \frac{17}{6} + \frac{125}{6} + \frac{17}{6} = \frac{53}{2}$$

(c) $\displaystyle A = -\int_{-2}^3 (x^2 - x - 6)\, dx = \frac{125}{6}$

28. (a) $\displaystyle\int_{-\pi/2}^{3\pi/4} 2\sin x\, dx = [-2\cos x]_{-\pi/2}^{3\pi/4} = \sqrt{2} = \text{area above} - \text{area below}$

(b) $\displaystyle A = \int_{-\pi/2}^0 -2\sin x\, dx + \int_0^{3\pi/4} 2\sin x\, dx = [2\cos x]_{-\pi/2}^0 + [-2\cos x]_0^{3\pi/4} = \sqrt{2} + 4$

(c) $\displaystyle A = \int_{-\pi/2}^0 -2\sin x\, dx = [2\cos x]_{-\pi/2}^0 = 2$

29. (a) $\displaystyle\int_{-2}^{2} (x^3 - x)\,dx = \left[\frac{1}{4}x^4 - \frac{1}{2}x^2\right]_{-2}^{2} = 0$

(b)

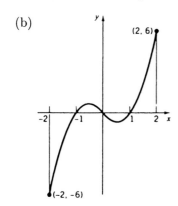

$$A = 2\left[-\int_{0}^{1} (x^3 - x)\,dx + \int_{1}^{2} (x^3 - x)\,dx\right]$$

$$= -2\left[\tfrac{1}{4}x^4 - \tfrac{1}{2}x^2\right]_0^1 + 2\left[\tfrac{1}{4}x^4 - \tfrac{1}{2}x^2\right]_1^2$$

$$= \frac{1}{2} + \frac{9}{2} = 5$$

30. (a) $\displaystyle\int_{-\pi}^{\pi} (\cos x + \sin x)\,dx = [\sin x - \cos x]_{-\pi}^{\pi} = 0$

(b)

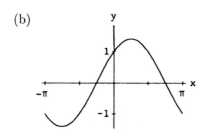

$$A = -\int_{-\pi}^{-\pi/4} f(x)\,dx + \int_{-\pi/4}^{3\pi/4} f(x)\,dx - \int_{3\pi/4}^{\pi} f(x)\,dx$$

$$= 4\sqrt{2}$$

31. (a) $\displaystyle\int_{-2}^{3} (x^3 - 4x + 2)\,dx = \left[\frac{1}{4}x^4 - 2x^2 + 2x\right]_{-2}^{3} = \frac{65}{4}$

(b)

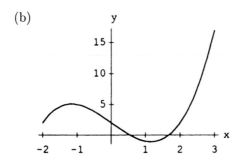

$$A \cong \int_{-2}^{0.54} (x^3 - 4x + 2)\,dx - \int_{0.54}^{1.68} (x^3 - 4x + 2)\,dx + \int_{1.68}^{3} (x^3 - 4x + 2)\,dx$$

$$= \left[\tfrac{1}{4}x^4 - 2x^2 + 2x\right]_{-2}^{0.54} - \left[\tfrac{1}{4}x^4 - 2x^2 + 2x\right]_{0.54}^{1.68} + \left[\tfrac{1}{4}x^4 - 2x^2 + 2x\right]_{1.68}^{3}$$

$$= 8.52 + .81 + 8.54 = 17.87$$

32. (a) $\int_{-\pi/2}^{\pi/2} (3x^2 - 2\cos x)\, dx = \left[x^3 - 2\sin x \right]_{-\pi/2}^{\pi/2} = \dfrac{\pi^3}{4} - 4$

(b)

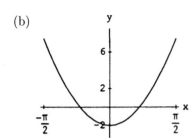

$A \cong -\int_{-0.71}^{0.71} (3x^2 - 2\cos x)\, dx + 2\int_{0.71}^{\pi/2} (3x^2 - 2\cos x)\, dx$

$\cong 7.53$

33.

$A = \int_0^1 (x^2 + 1)\, dx + \int_1^3 (3 - x)\, dx$

$= \left[\tfrac{1}{3} x^3 + x \right]_0^1 + \left[3x - \tfrac{1}{2} x^2 \right]_1^3$

$= \dfrac{4}{3} + 2 = \dfrac{10}{3}$

34.

$A = \int_0^1 3\sqrt{x}\, dx + \int_1^2 (4 - x^2)\, dx$

$= [2x^{3/2}]_0^1 + \left[4x - \dfrac{x^3}{3} \right]_1^2$

$= 2 + \dfrac{5}{3} = \dfrac{11}{3}$

35.

$y = \sin x$

$y = \cos x$

Area $= 2 - \sqrt{2}$

$A = \int_0^{\pi/4} \sin x\, dx + \int_{\pi/4}^{\pi/2} \cos x\, dx$

$= [-\cos x]_0^{\pi/4} + [\sin x]_{\pi/4}^{\pi/2}$

$= 2 - \sqrt{2}$

36.

$A = 2\int_0^{\pi/2} (1 + \cos x - 1)\, dx$

$= [2\sin x]_0^{\pi/2} = 2$

37.

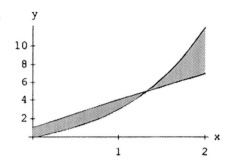

$$A \cong \int_0^{1.32} [3x + 1 - (x^3 + 2x)]\, dx + \int_{1.32}^2 [x^3 + 2x - (3x + 1)]\, dx$$

$$= \int_0^{1.32} (x + 1 - x^3)\, dx + \int_{1.32}^2 (x^3 - x - 1)\, dx$$

$$= \left[\tfrac{1}{2} x^2 + x - \tfrac{1}{4} x^4\right]_0^{1.32} + \left[\tfrac{1}{4} x^4 - \tfrac{1}{2} x^2 - x\right]_{1.32}^2 = 2.86$$

38.

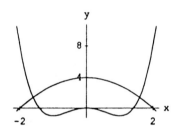

$$A = \int_{-\sqrt{\frac{1+\sqrt{17}}{2}}}^{\sqrt{\frac{1+\sqrt{17}}{2}}} (4 - x^2 - x^4 + 2x^2)\, dx$$

$$\cong 11.34$$

39. $h \cong 9.44892$

40. $h \cong 0.0355$

PROJECT 5.5

1. (a) $g(x) = \begin{cases} -1, & x = 0 \\ 0, & 0 < x \le 1 \\ 1, & 1 < x \le 2 \\ \vdots \\ 4, & 4 < x \le 5. \end{cases}$

$$\int_0^5 g(x)\, dx = \int_0^1 g(x)\, dx + \int_1^2 g(x)\, dx + \cdots + \int_4^5 g(x)\, dx = 0 + 1 + 2 + 3 + 4 = 10.$$

(b) $g(x) = \begin{cases} 0, & 0 \le x < 1 \\ 1, & 1 \le x < 2 \\ \vdots \\ 4, & 4 \le< x < 5 \\ 5 & x = 5. \end{cases}$

$$\int_0^5 g(x)\,dx = \int_0^1 g(x)\,dx + \int_1^2 g(x)\,dx + \cdots + \int_4^5 g(x)\,dx = 0 + 1 + 2 + 3 + 4 = 10.$$

(c) $g(x) = \begin{cases} 1 & x = 1, 2, \cdots 5 \\ 0 & \text{otherwise} \end{cases}$; $\int_0^5 g(x)\,dx = 0.$

2. (a) $\displaystyle\int_0^2 g(x)\,dx = \int_0^1 (2 - x)\,dx + \int_1^2 (2 + x)\,dx = \left[2x - \tfrac{1}{2}x^2\right]_0^1 + \left[2x + \tfrac{1}{2}x^2\right]_1^2 = 5.$

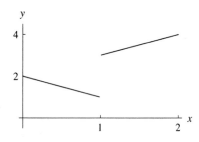

(b) $\displaystyle\int_0^5 g(x)\,dx = \int_0^2 x^2\,dx + \int_2^5 x\,dx = \left[\tfrac{1}{3}x^3\right]_0^2 + \left[\tfrac{1}{2}x^2\right]_2^5 = \tfrac{79}{6}.$

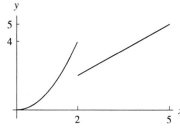

(c) $\displaystyle\int_0^{2\pi} g(x)\,dx = \int_0^{\pi/2} \cos x\,dx + \int_{\pi/2}^{\pi} \sin x\,dx + \int_{\pi}^{2\pi} \tfrac{1}{2}\,dx = \left[\sin x\right]_0^{\pi/2} - \left[\cos x\right]_{\pi/2}^{\pi} + \left[\tfrac{1}{2}x\right]_{\pi}^{2\pi}$

$= 2 + \tfrac{1}{2}\pi$

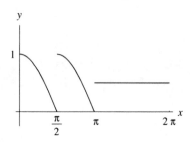

3.   (a)   For $0 \leq x < 1$, $G(x) = \displaystyle\int_0^x (2 - t)\, dt = \left[2t - \tfrac{1}{2}t^2\right]_0^x = 2x - \tfrac{1}{2}x^2$

For $1 \leq x \leq 2$, $G(x) = \displaystyle\int_0^1 (2 - t)\, dt + \int_1^x (2 + t)\, dt = \tfrac{3}{2} + \left[2t + \tfrac{1}{2}t^2\right]_1^x = 2x + \tfrac{1}{2}x^2 - 1$

Thus, $G(x) = \begin{cases} 2x - \tfrac{1}{2}x^2, & 0 \leq x < 1 \\ 2x + \tfrac{1}{2}x^2 - 1, & 1 \leq x \leq 2 \end{cases}$.

$$\lim_{h \to 0^-} \frac{G(1 + h) - G(1)}{h} = \lim_{h \to 0^-} \frac{2(1 + h) - \tfrac{1}{2}(1 + h)^2 - \tfrac{3}{2}}{h} = \lim_{h \to 0^-} \frac{h - \tfrac{1}{2}h^2}{h} = 1$$

$$\lim_{h \to 0^+} \frac{G(1 + h) - G(1)}{h} = \lim_{h \to 0^+} \frac{2(1 + h) + \tfrac{1}{2}(1 + h)^2 - 1 - \tfrac{3}{2}}{h} = \lim_{h \to 0^+} \frac{3h + \tfrac{1}{2}h^2}{h} = 3$$

Thus, G is not differentiable at $x = 1$.

(b) $G(x) = \begin{cases} \tfrac{1}{3}x^3, & 0 \leq x < 2 \\ \tfrac{1}{2}x^2 + \tfrac{2}{3}, & 2 \leq x \leq 5 \end{cases}$.

$$\lim_{h \to 0^-} \frac{G(2 + h) - G(2)}{h} = \lim_{h \to 0^-} \frac{\tfrac{1}{3}(2 + h)^3 - \tfrac{8}{3}}{h} = \lim_{h \to 0^-} \frac{4h + 2h^2 + \tfrac{1}{3}h^3}{h} = 4$$

$$\lim_{h \to 0^+} \frac{G(2 + h) - G(2)}{h} = \lim_{h \to 0^+} \frac{\tfrac{1}{2}(2 + h)^2 + \tfrac{2}{3} - \tfrac{8}{3}}{h} = \lim_{h \to 0^+} \frac{2h + \tfrac{1}{2}h^2}{h} = 2$$

Thus, G is not differentiable at $x = 2$.

(c) $G(x) = \begin{cases} \sin x, & 0 \leq x < \pi/2 \\ 1 - \cos x, & \pi/2 \leq x < \pi \\ 2 + \tfrac{1}{2}x - \pi/2, & \pi \leq x \leq 2\pi \end{cases}$.

G is not differentiable at $x = \pi/2,\ \pi$;

see the graph of G:

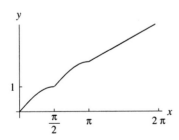

SECTION 5.6

1. $\displaystyle\int \frac{dx}{x^4} = \int x^{-4}\, dx = -\frac{1}{3}x^{-3} + C$

2. $\displaystyle\int (x - 1)^2\, dx = \int (x^2 - 2x + 1)\, dx = \frac{1}{3}x^3 - x^2 + x + C$

3. $\displaystyle\int (ax + b)\, dx = \frac{1}{2}\, ax^2 + bx + C$

4. $\displaystyle\int (ax^2 + b)\, dx = \frac{1}{3}ax^3 + bx + C$

5. $\displaystyle\int \frac{dx}{\sqrt{1+x}} = \int (1+x)^{-1/2}\, dx = 2(1+x)^{1/2} + C$

6. $\displaystyle\int \frac{x^3 + 1}{x^5}\, dx = \int x^{-2} + x^{-5}\, dx = -x^{-1} - \frac{1}{4}x^{-4} + C$

7. $\displaystyle\int \left(\frac{x^3 - 1}{x^2}\right) dx = \int (x - x^{-2})\, dx = \frac{1}{2}\, x^2 + x^{-1} + C$

8. $\displaystyle\int \left(\sqrt{x} - \frac{1}{\sqrt{x}}\right) dx = \int (x^{1/2} - x^{-1/2})\, dx = \frac{2}{3}x^{3/2} - 2x^{1/2} + C$

9. $\displaystyle\int (t - a)(t - b)\, dt = \int [t^2 - (a + b)t + ab]\, dt = \frac{1}{3}\, t^3 - \frac{a+b}{2}t^2 + abt + C$

10. $\displaystyle\int (t^2 - a)(t^2 - b)\, dt = \int \left(t^4 - (a + b)t^2 + ab\right)\, dt = \frac{1}{5}t^5 - \frac{1}{3}(a + b)t^3 + abt + C$

11. $\displaystyle\int \frac{(t^2 - a)(t^2 - b)}{\sqrt{t}}\, dt = \int [t^{7/2} - (a + b)t^{3/2} + abt^{-1/2}]\, dt$

$$= \tfrac{2}{9}t^{9/2} - \tfrac{2}{5}(a + b)t^{5/2} + 2abt^{1/2} + C$$

12. $\displaystyle\int (2 - \sqrt{x})(2 + \sqrt{x})\, dx = \int (4 - x)\, dx = 4x - \frac{1}{2}x^2 + C$

13. $\displaystyle\int g(x)g'(x)\, dx = \frac{1}{2}[g(x)]^2 + C$

14. $\displaystyle\int \sin x \cos x\, dx = \frac{1}{2}\sin^2 x + C$

15. $\displaystyle\int \tan x \sec^2 x\, dx = \int \sec x \frac{d}{dx}\, [\sec x]\, dx = \frac{1}{2}\sec^2 x + C$

$\displaystyle\int \tan x \sec^2 x\, dx = \int \tan x \frac{d}{dx}\, [\tan x]\, dx = \frac{1}{2}\tan^2 x + C$

16. $\displaystyle\int \frac{g'(x)}{[g(x)]^2}\, dx = -\frac{1}{g(x)} + C$

17. $\displaystyle\int \frac{4}{(4x + 1)^2}\, dx = \int 4(4x + 1)^{-2}\, dx = -(4x + 1)^{-1} + C$

18. $\displaystyle\int \frac{3x^2}{(x^3 + 1)^2}\, dx = -\frac{1}{x^3 + 1} + C,\quad$ (use Exercise 16)

19. $f(x) = \displaystyle\int f'(x)\,dx = \int (2x - 1)\,dx = x^2 - x + C.$

Since $f(3) = 4$, we get $4 = 9 - 3 + C$ so that $C = -2$ and
$$f(x) = x^2 - x - 2.$$

20. $f(x) = \displaystyle\int (3 - 4x)\,dx = 3x - 2x^2 + C, \quad f(1) = 6 \implies f(x) = -2x^2 + 3x + 5$

21. $f(x) = \displaystyle\int f'(x)\,dx = \int (ax + b)\,dx = \frac{1}{2}ax^2 + bx + C.$

Since $f(2) = 0$, we get $0 = 2a + 2b + C$ so that $C = -2a - 2b$ and
$$f(x) = \tfrac{1}{2}ax^2 + bx - 2a - 2b.$$

22. $f(x) = \displaystyle\int (ax^2 + bx + c)\,dx = \frac{1}{3}ax^3 + \frac{1}{2}bx^2 + cx + K,$

$f(0) = 0 \implies f(x) = \dfrac{a}{3}x^3 + \dfrac{b}{2}x^2 + cx$

23. $f(x) = \displaystyle\int f'(x)\,dx = \int \sin x\,dx = -\cos x + C.$

Since $f(0) = 2$, we get $2 = -1 + C$ so that $C = 3$ and
$$f(x) = 3 - \cos x.$$

24. $f(x) = \displaystyle\int \cos x\,dx = \sin x + C, \quad f(\pi) = 3 \implies f(x) = 3 + \sin x$

25. First,
$$f'(x) = \int f''(x)\,dx = \int (6x - 2)\,dx = 3x^2 - 2x + C.$$

Since $f'(0) = 1$, we get $1 = 0 + C$ so that $C = 1$ and
$$f'(x) = 3x^2 - 2x + 1.$$

Next,
$$f(x) = \int f'(x)\,dx = \int (3x^2 - 2x + 1)\,dx = x^3 - x^2 + x + K.$$

Since $f(0) = 2$, we get $2 = 0 + K$ so that $K = 2$ and
$$f(x) = x^3 - x^2 + x + 2.$$

26. $f'(x) = \displaystyle\int -12x^2\,dx = -4x^3 + C, \quad f'(0) = 1 \implies f'(x) = -4x^3 + 1$

$f(x) = \displaystyle\int (-4x^3 + 1)\,dx = -x^4 + x + K, \quad f(0) = 2 \implies f(x) = -x^4 + x + 2$

27. First,

$$f'(x) = \int f''(x)\,dx = \int (x^2 - x)\,dx = \frac{1}{3}x^3 - \frac{1}{2}x^2 + C.$$

Since $f'(1) = 0$, we get $0 = \frac{1}{3} - \frac{1}{2} + C$ so that $C = \frac{1}{6}$ and

$$f'(x) = \frac{1}{3}x^3 - \frac{1}{2}x^2 + \frac{1}{6}.$$

Next,

$$f(x) = \int f'(x)\,dx = \int \left(\frac{1}{3}x^3 - \frac{1}{2}x^2 + \frac{1}{6}\right)dx = \frac{1}{12}x^4 - \frac{1}{6}x^3 + \frac{1}{6}x + K.$$

Since $f(1) = 2$, we get $2 = \frac{1}{12} - \frac{1}{6} + \frac{1}{6} + K$ so that $K = \frac{23}{12}$ and

$$f(x) = \frac{x^4}{12} - \frac{x^3}{6} + \frac{x}{6} + \frac{23}{12} = \frac{1}{12}(x^4 - 2x^3 + 2x + 23).$$

28. $f'(x) = \int (1 - x)\,dx = x - \dfrac{x^2}{2} + C,\quad f'(2) = 1 \implies f'(x) = x - \dfrac{x^2}{2} + 1$

$f(x) = \int \left(x - \dfrac{x^2}{2} + 1\right)dx = \dfrac{x^2}{2} - \dfrac{x^3}{6} + x + K,\quad f(2) = 0 \implies f(x) = -\dfrac{x^3}{6} + \dfrac{x^2}{2} + x - \dfrac{8}{3}$

29. First,

$$f'(x) = \int f''(x)\,dx = \int \cos x\,dx = \sin x + C.$$

Since $f'(0) = 1$, we get $1 = 0 + C$ so that $C = 1$ and

$$f'(x) = \sin x + 1.$$

Next,

$$f(x) = \int f'(x)\,dx = \int (\sin x + 1)\,dx = -\cos x + x + K.$$

Since $f(0) = 2$, we get $2 = -1 + 0 + K$ so that $K = 3$ and

$$f(x) = -\cos x + x + 3.$$

30. $f'(x) = \int \sin x\,dx = -\cos x + C,\quad f'(0) = -2 \implies f'(x) = -\cos x - 1$

$f(x) = \int (-\cos x - 1)\,dx = -\sin x - x + K,\quad f(0) = 1 \implies f(x) = 1 - \sin x - x$

31. First,

$$f'(x) = \int f''(x)\,dx = \int (2x - 3)\,dx = x^2 - 3x + C.$$

Then,

$$f(x) = \int f'(x)\,dx = \int (x^2 - 3x + C)\,dx = \frac{1}{3}x^3 - \frac{3}{2}x^2 + Cx + K.$$

Since $f(2) = -1$, we get

(1) $$-1 = \tfrac{8}{3} - 6 + 2C + K;$$

and, from $f(0) = 3$, we conclude that

(2) $$3 = 0 + K.$$

Solving (1) and (2) simultaneously, we get $K = 3$ and $C = -\tfrac{1}{3}$ so that

$$f(x) = \tfrac{1}{3}x^3 - \tfrac{3}{2}x^2 - \tfrac{1}{3}x + 3.$$

32. $f'(x) = \displaystyle\int (5 - 4x)\,dx = 5x - 2x^2 + C,$

$f(x) = \displaystyle\int (5x - 2x^2 + C)\,dx = \frac{5}{2}x^2 - \frac{2}{3}x^3 + Cx + K$

$f(1) = \dfrac{5}{2} - \dfrac{2}{3} + C + K = 1, \quad f(0) = K = -2 \implies f(x) = -\dfrac{2}{3}x^3 + \dfrac{5}{2}x^2 + \dfrac{7}{6}x - 2$

33. $\dfrac{d}{dx}\left[\displaystyle\int f(x)\,dx\right] = f(x); \quad \displaystyle\int \dfrac{d}{dx}[f(x)]\,dx = f(x) + C$

34. $\displaystyle\int [f(x)g''(x) - g(x)f''(x)]\,dx = \int [f(x)g''(x) + f'(x)g'(x) - f'(x)g'(x) - g(x)f''(x)]\,dx$

$= \displaystyle\int \left(\dfrac{d}{dx}[f(x)g'(x)] - \dfrac{d}{dx}[f'(x)g(x)]\right)\,dx = f(x)g'(x) - g(x)f'(x) + C$

35. (a) $x(t) = \displaystyle\int v(t)\,dt = \int (6t^2 - 6)\,dt = 2t^3 - 6t + C.$

Since $x(0) = -2$, we get $-2 = 0 + C$ so that $C = -2$ and

$x(t) = 2t^3 - 6t - 2.$ Therefore $x(3) = 34.$

Three seconds later the object is 34 units to the right of the origin.

(b) $s = \displaystyle\int_0^3 |v(t)|\,dt = \int_0^3 |6t^2 - 6|\,dt = \int_0^1 (6 - 6t^2)\,dt + \int_1^3 (6t^2 - 6)\,dt$

$= [6t - 2t^3]_0^1 + [2t^3 - 6t]_1^3 = 4 + [36 - (-4)] = 44.$

The object traveled 44 units.

36. (a) $v(t) = \displaystyle\int a(t)\,dt = \int (t + 2)^3\,dt = \frac{1}{4}(t + 2)^4 + C,$

$v(0) = 3 \implies v(t) = \dfrac{1}{4}(t + 2)^4 - 1$

(b) $x(t) = \displaystyle\int \left[\frac{(t + 2)^4}{4} - 1\right]dt = \frac{(t + 2)^5}{20} - t + K, \quad x(0) = 0 \implies x(t) = \frac{(t + 2)^5}{20} - t - \frac{8}{5}$

37. (a) $v(t) = \displaystyle\int a(t)\,dt = \int (t + 1)^{-1/2}\,dt = 2(t + 1)^{1/2} + C.$

Since $v(0) = 1$, we get $1 = 2 + C$ so that $C = -1$ and

$$v(t) = 2(t + 1)^{1/2} - 1.$$

(b)   We know $v(t)$ by part (a). Therefore,

$$x(t) = \int v(t)\, dt = \int [2(t+1)^{1/2} - 1]\, dt = \frac{4}{3}(t+1)^{3/2} - t + C.$$

Since $x(0) = 0$, we get $0 = \frac{4}{3} - 0 + C$ so that

$C = -\frac{4}{3}$ and $x(t) = \frac{4}{3}(t+1)^{3/2} - t - \frac{4}{3}.$

38.   (a) $x(t) = \int t(1-t)\, dt = \frac{t^2}{2} - \frac{t^3}{3} + C,$ $x(0) = -2 \implies x(t) = \frac{t^2}{2} - \frac{t^3}{3} - 2$

$$x(10) = -\frac{856}{3}\ :\quad 285\frac{1}{3}\ \text{units to the left of the origin.}$$

(b) $s = \displaystyle\int_0^{10} |v(t)|\, dt = \int_0^1 t(1-t)\, dt + \int_1^{10} -t(1-t)\, dt = \left[\frac{t^2}{2} - \frac{t^3}{3}\right]_0^1 - \left[\frac{t^2}{2} - \frac{t^3}{3}\right]_1^{10}$

$$= \frac{851}{3} = 283\frac{2}{3}\ \text{units.}$$

39.   (a) $v_0 = 60$ mph $= 88$ feet per second. In general, $v(t) = at + v_0$. Here, in feet and seconds, $v(t) = -20t + 88$. Thus $v(t) = 0$ at $t = 4.4$ seconds.

(b)   In general, $x(t) = \frac{1}{2}at^2 + v_0 t + x_0$. Here we take $x_0 = 0$. In feet and seconds

$$x(t) = -10(4.4)^2 + 88(4.4) = 10(4.4)^2 = 193.6\ \text{ft.}$$

40.   Let acceleration $= a$.   Then $v(t) = \displaystyle\int a\, dt = at + v_0.$

$$x(t) = \int v(t)\, dt = \int (at + v_0)\, dt = \frac{1}{2}at^2 + v_0 t + x_0 = \frac{1}{2}\left[v(t) + 2v_0\right] t + x_0$$

41.
$$[v(t)]^2 = (at + v_0)^2 = a^2 t^2 + 2av_0 t + v_0{}^2$$

$$= v_0{}^2 + a(at^2 + 2v_0 t)$$

$$= v_0{}^2 + 2a(\tfrac{1}{2}at^2 + v_0 t)$$

(set $x(t) = \frac{1}{2}at^2 + v_0 t + x_0$)

$$= v_0{}^2 + 2a\left[x(t) - x_0\right]$$

42.   (a) $v(t) = at + v_0,$ and by Exercise 40 $x(t) = \dfrac{1}{2}\left[v(t) + 2v_0\right] t,$ so

$$a = \frac{v(t) - v_0}{t} = \frac{v(t) - v_0}{2x(t)}(v(t) + 2v_0) = \frac{58.7^2 + (58.7)(88) - 2(88)^2}{2(264)} = -13.02\ \text{ft/sec}^2$$

[Note   60 mph $= 88$ ft/sec,   40 mph $= 58\frac{2}{3}$ ft/sec.]

(b) $t = \dfrac{2x(t)}{v(t) + 2v_0} = \dfrac{2(264)}{58\frac{2}{3} + 176} = 2.24\ \text{sec}$

(c) We don't know $x(t)$, so we will use $t = \dfrac{v(t) - v_0}{a} = \dfrac{0 - 88}{-13.02} = 6.8\ \text{sec}$

(d) $x(t) = \dfrac{1}{2}\left[v(t) + 2v_0\right] t = (88)(6.8) = 598.4\ \text{ft}$

43. The car can accelerate to 60 mph (88 ft/sec) in 20 seconds thereby covering a distance of 880 ft. It can decelerate from 88 ft/sec to 0 ft/sec in 4 seconds thereby covering a distance of 176 ft. At full speed, 88 ft/sec, it must cover a distance of

$$\frac{5280}{2} - 880 - 176 = 1584 \text{ ft.}$$

This takes $\dfrac{1584}{88} = 18$ seconds. The run takes at least $20 + 18 + 4 = 42$ seconds.

44. $v(t) = \displaystyle\int \sin t\, dt = -\cos t + C, \quad v(0) = v_0 \Longrightarrow v(t) = -\cos t + v_0 + 1$

$x(t) = \displaystyle\int (-\cos t + v_0 + 1)\, dt = -\sin t + (v_0 + 1)t + K, \quad x(0) = x_0 \Longrightarrow x(t) = x_0 + (v_0 + 1)t - \sin t$

45. $v(t) = \displaystyle\int a(t)\, dt = \int (2A + 6Bt)\, dt = 2At + 3Bt^2 + C.$

Since $v(0) = v_0$, we have $v_0 = 0 + C$ so that $v(t) = 2At + 3Bt^2 + v_0$.

$$x(t) = \int v(t)\, dt = \int (2At + 3Bt^2 + v_0)\, dt = At^2 + Bt^3 + v_0 t + K.$$

Since $x(0) = x_0$, we have $x_0 = 0 + K$ so that $K = x_0$ and

$$x(t) = x_0 + v_0 t + At^2 + Bt^3.$$

46. $v(t) = \displaystyle\int \cos t\, dt = \sin t + C, \quad v(0) = v_0 \Longrightarrow v(t) = \sin t + v_0$

$x(t) = \displaystyle\int (\sin t + v_0)\, dt = -\cos t + v_0 t + K, \quad x(0) = x_0 \Longrightarrow x(t) = x_0 + 1 + v_0 t - \cos t$

47.

$$x'(t) = t^2 - 5, \qquad\qquad y'(t) = 3t,$$
$$x(t) = \tfrac{1}{3}t^3 - 5t + C. \qquad y(t) = \tfrac{3}{2}t^2 + K.$$

When $t = 2$, the particle is at $(4, 2)$. Thus, $x(2) = 4$ and $y(2) = 2$.

$$4 = \tfrac{8}{3} - 10 + C \implies C = \tfrac{34}{3}. \qquad\qquad 2 = 6 + K \implies K = -4.$$
$$x(t) = \tfrac{1}{3}t^3 - 5t + \tfrac{34}{3}, \qquad\qquad y(t) = \tfrac{3}{2}t^2 - 4.$$

Four seconds later the particle is at $(x(6), y(6)) = \left(\tfrac{160}{3}, 50\right)$.

48. $x(t) = \displaystyle\int (t - 2)\, dt = \frac{t^2}{2} - 2t + C, \quad x(4) = 3 \implies x(t) = \frac{t^2}{2} - 2t + 3$

$y(t) = \displaystyle\int \sqrt{t}\, dt = \frac{2}{3}t^{3/2} + K, \quad y(4) = 1 \implies y(t) = \frac{2}{3}t^{3/2} - \frac{13}{3}$

5 seconds later, $t = 9$, so position is $(x(9), y(9)) = \left(\tfrac{51}{2}, \tfrac{41}{3}\right)$.

49. Since $v(0) = 2$, we have $2 = A \cdot 0 + B$ so that $B = 2$. Therefore

$$x(t) = \int v(t)\, dt = \int (At + 2)\, dt = \frac{1}{2}At^2 + 2t + C.$$

Since $x(2) = x(0) - 1$, we have

$$2A + 4 + C = C - 1 \quad \text{so that} \quad A = -\tfrac{5}{2}.$$

50. $x(t) = \displaystyle\int (At^2 + 1)\, dt = \frac{1}{3}At^3 + t + C$

$x(1) - x(0) = (\frac{1}{3}A + 1 + C) - C = \dfrac{A}{3} + 1 = 0 \implies A = -3$

Distance traveled $= \displaystyle\int_0^{1/\sqrt{3}} (1 - 3t^2)\, dt + \int_{1/\sqrt{3}}^1 (3t^2 - 1)\, dt = \left[t - t^3\right]_0^{1/\sqrt{3}} + \left[t^3 - t\right]_{1/\sqrt{3}}^1 = \dfrac{4\sqrt{3}}{9}$

51. $x(t) = \displaystyle\int v(t)\, dt = \int \sin t\, dt = -\cos t + C$

Since $x(\pi/6) = 0$, we have $\quad 0 = -\dfrac{\sqrt{3}}{2} + C \quad$ so that $\quad C = \dfrac{\sqrt{3}}{2} \quad$ and $\quad x(t) = \frac{\sqrt{3}}{2} - \cos t$.

(a)   At $t = 11\pi/6$ sec.

(b) We want to find the smallest $t_0 > \pi/6$ for which $x(t_0) = 0$ and $v(t_0) > 0$. We get

$$t_0 = 13\pi/6 \text{ seconds.}$$

52. $x(t) = \displaystyle\int \cos t\, dt = \sin t + C, \quad x\left(\dfrac{\pi}{6}\right) = \sin\dfrac{\pi}{6} + C = 0 \implies x(t) = \sin t - \dfrac{1}{2}$

(a) $x(t) = 0$ at $t = \frac{5}{6}\pi$ sec.

(b) $x(t) = 0$ and $v(t) > 0 \implies t = \dfrac{13\pi}{6}$ sec.

53. The mean-value theorem. With obvious notation

$$\frac{x(1/12) - x(0)}{1/12} = \frac{4}{1/12} = 48.$$

By the mean-value theorem there exists some time t_0 at which

$$x'(t_0) = \frac{x(1/12) - x(0)}{1/12}.$$

54. (Taking the direction of motion as positive, speed and velocity are the same.) Let v be the speed of the motorcycle at time 0, the time when the brakes are applied. The distance between the motorcycle and the hay wagon t time units later is given by

$$d(t) = -\frac{1}{2}at^2 + (v_1 - v)t + s$$

[$v_1 t + s$ gives the position of the hay wagon, $\frac{1}{2}at^2 + vt$ gives the position of the motorcycle]. Collision can be avoided only if the quadratic

$$d(t) = -\frac{1}{2}at^2 + (v_1 - v)t + s$$

remains positive. This can be true only if the discriminant of the quadratic,

$$B^2 - 4AC = (v_1 - v)^2 + 2as = (v - v_1)^2 + 2as$$

remains negative. Observe that

$$(v - v_1)^2 + 2as < 0 \quad \text{iff} \quad v < v_1 + \sqrt{2|a|s}$$

55. $\dfrac{v'(t)}{[v(t)]^2} = 2 \implies -[v(t)]^{-1} = 2t - v_0^{-1}.$

$\implies [v(t)]^{-1} = v_0^{-1} - 2t \implies v(t) = \dfrac{1}{v_0^{-1} - 2t} = \dfrac{v_0}{1 - 2tv_0}.$

56. $ds\dfrac{d}{dx}\left(\displaystyle\int \dfrac{x^2 - x^3 + x^4}{\sqrt{x}}\, dx\right) = \dfrac{x^2 - x^3 + x^4}{\sqrt{x}}; \qquad ds\displaystyle\int \dfrac{d}{dx}\left(\dfrac{x^2 - x^3 + x^4}{\sqrt{x}}\right)\, dx = \dfrac{x^2 - x^3 + x^4}{\sqrt{x}} + C$

57. $\displaystyle\int (\cos x - 2\sin x)\, dx = \sin x + 2\cos x + C$ and so

$$\frac{d}{dx}\left(\int (\cos x - 2\sin x)\, dx\right) = \frac{d}{dx}[\sin x + 2\cos x] = \cos x - 2\sin x;$$

$\dfrac{d}{dx}[\cos x - 2\sin x] = -\sin x - 2\cos x$ and so

$$\int \frac{d}{dx}[\cos x - 2\sin x]\, dx = \int (-\sin x - 2\cos x)\, dx = \cos x - 2\sin x + C$$

58. $f(x) = x + 2\sqrt{x} - 6$

59. $f(x) = \sin x + 2\cos x + 1$

60. $f(x) = 3x + 2 - 2\cos x - 3\sin x$

61. $\frac{1}{12}x^4 - \frac{1}{2}x^3 + \frac{5}{2}x^2 + 4x - 3$

SECTION 5.7

1. $\left\{\begin{array}{l} u = 2 - 3x \\ du = -3\, dx \end{array}\right\};$ $\quad \displaystyle\int \dfrac{dx}{(2 - 3x)^2} = \int (2 - 3x)^{-2}\, dx = -\dfrac{1}{3}\int u^{-2}\, du = \dfrac{1}{3}u^{-1} + C$

$$= \tfrac{1}{3}(2 - 3x)^{-1} + C$$

2. $\left\{\begin{array}{l} u = 2x + 1 \\ du = 2\, dx \end{array}\right\};$ $\quad \displaystyle\int \dfrac{dx}{\sqrt{2x + 1}} = \dfrac{1}{2}\int \dfrac{du}{\sqrt{u}} = \sqrt{u} + C = \sqrt{2x + 1} + C$

3. $\left\{\begin{array}{l} u = 2x + 1 \\ du = 2\, dx \end{array}\right\};$ $\quad \displaystyle\int \sqrt{2x + 1}\, dx = \int (2x + 1)^{1/2}\, dx = \dfrac{1}{2}\int u^{1/2}\, du = \dfrac{1}{3}u^{3/2} + C$

$$= \frac{1}{3}(2x + 1)^{3/2} + C$$

4. $\left\{\begin{array}{l} u = ax + b \\ du = a\, dx \end{array}\right\};$ $\quad \displaystyle\int \sqrt{ax + b} = \dfrac{1}{a}\int \sqrt{u}\, du = \dfrac{2}{3a}u^{3/2} + C = \dfrac{2}{3a}(ax + b)^{3/2} + C$

5. $\left\{\begin{array}{l} u = ax + b \\ du = a\,dx \end{array}\right\}$; $\displaystyle \int (ax+b)^{3/4}\,dx = \frac{1}{a}\int u^{3/4}\,du = \frac{4}{7a}u^{7/4} + C$

$$= \frac{4}{7a}(ax+b)^{7/4} + C$$

6. $\left\{\begin{array}{l} u = ax^2 + b \\ du = 2ax\,dx \end{array}\right\}$; $\displaystyle \int 2ax(ax^2+b)^4\,dx = \int u^4\,du = \frac{1}{5}u^5 + C = \frac{1}{5}(ax^2+b)^5 + C$

7. $\left\{\begin{array}{l} u = 4t^2 + 9 \\ du = 8t\,dt \end{array}\right\}$; $\displaystyle \int \frac{t}{(4t^2+9)^2}\,dt = \frac{1}{8}\int \frac{du}{u^2} = -\frac{1}{8}u^{-1} + C = -\frac{1}{8}\left(4t^2+9\right)^{-1} + C$

8. $\left\{\begin{array}{l} u = t^2 + 1 \\ du = 2t\,dt \end{array}\right\}$; $\displaystyle \int \frac{3t}{(t^2-1)^2}\,dt = \frac{3}{2}\int \frac{du}{u^2} = -\frac{3}{2u} + C = \frac{-3}{2(t^2+1)} + C$

9. $\left\{\begin{array}{l} u = 1 + x^3 \\ du = 3x^2\,dx \end{array}\right\}$; $\displaystyle \int x^2(1+x^3)^{1/4}\,dx = \frac{1}{3}\int u^{1/4}\,du = \frac{4}{15}u^{5/4} + C = \frac{4}{15}(1+x^3)^{5/4} + C$

10. $\left\{\begin{array}{l} u = a + bx^n \\ du = nbx^{n-1}\,dx \end{array}\right\}$; $\displaystyle \int x^{n-1}\sqrt{a+bx^n}\,dx = \frac{1}{bn}\int \sqrt{u}\,du = \frac{2}{3bn}u^{3/2} + C = \frac{2}{3bn}\left(a+bx^n\right)^{3/2} + C$

11. $\left\{\begin{array}{l} u = 1 + s^2 \\ du = 2s\,ds \end{array}\right\}$; $\displaystyle \int \frac{s}{(1+s^2)^3}\,ds = \frac{1}{2}\int \frac{du}{u^3} = -\frac{1}{4}u^{-2} + C = -\frac{1}{4}(1+s^2)^{-2} + C$

12. $\left\{\begin{array}{l} u = 6 - 5s^2 \\ du = -10s\,ds \end{array}\right\}$; $\displaystyle \int \frac{2s}{\sqrt[3]{6-5s^2}}\,ds = -\frac{1}{5}\int u^{-1/3}\,du = -\frac{3}{10}u^{2/3} + C = \frac{-3}{10}\left(6-5s^2\right)^{2/3} + C$

13. $\left\{\begin{array}{l} u = x^2 + 1 \\ du = 2x\,dx \end{array}\right\}$; $\displaystyle \int \frac{x}{\sqrt{x^2+1}}\,dx = \int (x^2+1)^{-1/2}\,x\,dx = \frac{1}{2}\int u^{-1/2}\,du = u^{1/2} + C = \sqrt{x^2+1} + C$

14. $\left\{\begin{array}{l} u = 1 - x^3 \\ du = -3x^2\,dx \end{array}\right\}$; $\displaystyle \int \frac{x^2}{(1-x^3)^{2/3}} = -\frac{1}{3}\int \frac{du}{u^{2/3}} = -u^{1/3} + C = -(1-x^3)^{1/3} + C$

15. $\left\{\begin{array}{l} u = x^2 + 1 \\ du = 2x \end{array}\right\}$; $\displaystyle \int 5x\left(x^2+1\right)^{-3}\,dx = \frac{5}{2}\int u^{-3}\,du = -\frac{5}{4}u^{-2} + C = -\frac{5}{4}(x^2+1)^{-2} + C$

16. $\left\{\begin{array}{l} u = 1 - x^4 \\ du = -4x^3\,dx \end{array}\right\}$; $\displaystyle \int 2x^3(1-x^4)^{-1/4}\,dx = -\frac{1}{2}\int u^{-1/4}\,du = -\frac{2}{3}u^{3/4} + C = -\frac{2}{3}(1-x^4)^{3/4} + C$

17. $\left\{ \begin{array}{l} u = x^{1/4} + 1 \\ du = \frac{1}{4} x^{-3/4}\, dx \end{array} \right\}$; $\quad \displaystyle\int x^{-3/4} \left(x^{1/4} + 1 \right)^{-2} dx = 4 \int u^{-2}\, du = -4u^{-1} + C = -4(x^{1/4} + 1)^{-1} + C$

18. $\left\{ \begin{array}{l} u = x^2 + 3x + 1 \\ du = (2x + 3)\, dx \end{array} \right\}$; $\quad \displaystyle\int \frac{4x + 6}{\sqrt{x^2 + 3x + 1}}\, dx = 2 \int \frac{du}{\sqrt{u}} = 4\sqrt{u} + C = 4\sqrt{x^2 + 3x + 1} + C$

19. $\left\{ \begin{array}{l} u = 1 - a^4 x^4 \\ du = -4a^4 x^3\, dx \end{array} \right\}$;

$\displaystyle\int \frac{b^3 x^3}{\sqrt{1 - a^4 x^4}}\, dx = -\frac{b^3}{4a^4} \int u^{-1/2}\, du = -\frac{b^3}{2a^4} u^{1/2} + C = -\frac{b^3}{2a^4} \sqrt{1 - a^4 x^4} + C$

20. $\left\{ \begin{array}{l} u = a + bx^n \\ du = bnx^{n-1}\, dx \end{array} \right\}$;

$\displaystyle\int \frac{x^{n-1}}{\sqrt{a + bx^n}}\, dx = \frac{1}{bn} \int \frac{du}{\sqrt{u}} = \frac{2}{bn} \sqrt{u} + C = \frac{2}{bn} \sqrt{a + bx^n} + C$

21. $\left\{ \begin{array}{l} u = x^2 + 1 \\ du = 2x\, dx \end{array} \right\}$; $\quad \displaystyle\int x \left(x^2 + 1 \right)^3 dx = \frac{1}{2} \int_1^2 u^3\, du = \frac{1}{8} u^4 + C = \frac{1}{8} (x^2 + 1)^4 + C$

$\displaystyle\int_0^1 x(x^2 + 1)\, dx = \left[\frac{1}{8}(x^2 + 1)^4 \right]_0^1 = \frac{1}{8}[16 - 1] = \frac{15}{8}$

22. $\left\{ \begin{array}{ll} u = 4 + 2x^3 & \,\, x = -1 \,\Rightarrow\, u = 2 \\ du = 6x^2\, dx & \,\, x = 0 \,\Rightarrow\, u = 4 \end{array} \right\}$; $\quad \displaystyle\int_{-1}^0 3x^2 \left(4 + 2x^3 \right)^2 dx = \frac{1}{2} \int_2^4 u^2\, du = \left[\frac{1}{6} u^3 \right]_2^4 = \frac{28}{3}$

23. 0; the integrand is an odd function

24. $\left\{ \begin{array}{ll} u = r^2 + 16 & \,\, r = 0 \,\Rightarrow\, u = 16 \\ du = 2r\, dr & \,\, r = 3 \,\Rightarrow\, u = 25 \end{array} \right\}$; $\quad \displaystyle\int_0^3 \frac{r}{\sqrt{r^2 + 16}}\, dr = \frac{1}{2} \int_{16}^{25} \frac{du}{\sqrt{u}} = \left[\sqrt{u} \right]_{16}^{25} = 1$

25. $\left\{ \begin{array}{l} u = a^2 - y^2 \\ du = -2y\, dy \end{array} \right\}$; $\quad \displaystyle\int y\sqrt{a^2 - y^2}\, dy = -\frac{1}{2} \int u^{1/2}\, du = -\frac{1}{3} u^{3/2} + C = -\frac{1}{3}(a^2 - y^2)^{3/2} + C$

$\displaystyle\int_0^a y\sqrt{a^2 - y^2}\, dy = -\frac{1}{3} \left[(a^2 - y^2)^{3/2} \right]_0^a = \frac{1}{3}(a^2)^{3/2} = \frac{1}{3}|a|^3$

26. $\left\{\begin{array}{l} u = 1 - \dfrac{y^3}{a^3} \\[3mm] du = -\dfrac{3y^2}{a^3}\,dy \end{array}\right.\left|\begin{array}{l} y = -a \;\Rightarrow\; u = 2 \\[2mm] y = 0 \;\Rightarrow\; u = 1 \end{array}\right\}$

$$\int_{-a}^{0} y^2\left(1 - \frac{y^3}{a^3}\right)^{-2} dy = -\frac{a^3}{3}\int_{2}^{1} u^{-2}\,du = \frac{a^3}{3}\int_{1}^{2} u^{-2}\,du = -\frac{a^3}{3}\left[\frac{1}{u}\right]_{1}^{2} = \frac{a^3}{6}$$

27. $\left\{\begin{array}{l} u = 2x^2 + 1 \\[2mm] du = 4x\,dx \end{array}\right\}; \qquad \int x\sqrt{2x^2 + 1}\,dx = \frac{1}{4}\int u^{1/2}\,du\,\tfrac{1}{6}u^{3/2} + C = \tfrac{1}{6}(2x^2+1)^{3/2} + C$

$$\int_{0}^{2} x\sqrt{2x^2+1}\,dx = \left[\tfrac{1}{6}(2x^2+1)^{3/2}\right]_{0}^{2} = \frac{13}{3}$$

28. $\left\{\begin{array}{l} u = 2x^2 + 1 \\[2mm] du = 4x\,dx \end{array}\right.\left|\begin{array}{l} x = 0 \;\Rightarrow\; u = 1 \\[2mm] x = 2 \;\Rightarrow\; u = 9 \end{array}\right\}; \qquad \int_{0}^{2} \frac{x}{(2x^2+1)^2}\,dx = \frac{1}{4}\int_{1}^{9} u^{-2}\,du = \left[-\frac{1}{4u}\right]_{1}^{9} = \frac{2}{9}$

29. $\left\{\begin{array}{l} u = 1 + x^{-2} \\[2mm] du = -2x^{-3}\,dx \end{array}\right\}; \qquad \int x^{-3}(1 + x^{-2})^{-3}\,dx = -\frac{1}{2}\int u^{-3}\,du = \tfrac{1}{4}u^{-2} + C = \tfrac{1}{4}(1 + x^{-2})^{-2} + C$

$$\int_{0}^{2} x^{-3}(1 + x^{-2})^{-3}\,dx = \left[\frac{1}{4}(1 + x^{-2})^{-2}\right]_{1}^{2} = \frac{39}{400}$$

30. $\left\{\begin{array}{l} u = (x+2)(x+3) \\[2mm] du = (2x+5)dx \end{array}\right.\begin{array}{l} x = 0 \;\Rightarrow\; u = 6 \\[2mm] x = 1 \;\Rightarrow\; u = 12 \end{array}\right\};$

$$\int_{0}^{1} \frac{2x+5}{(x+2)^2(x+3)^2}\,dx = \int_{6}^{12} \frac{1}{u^2}\,du = \left[-\frac{1}{u}\right]_{6}^{12} = \frac{1}{12}$$

31. $\left\{\begin{array}{l} u = x + 1 \\[2mm] du = dx \end{array}\right\}; \qquad \int x\sqrt{x+1}\,dx = \int (u-1)\sqrt{u}\,du = \int (u^{3/2} - u^{1/2})\,du$

$$= \tfrac{2}{5}u^{5/2} - \tfrac{2}{3}u^{3/2} + C = \tfrac{2}{5}(x+1)^{5/2} - \tfrac{2}{3}(x+1)^{3/2} + C$$

32. $\left\{\begin{array}{l} u = x - 1 \\[2mm] du = dx \end{array}\right\}; \qquad \int 2x\sqrt{x-1}\,dx = \int 2(u+1)\sqrt{u}\,du = 2\int (u^{3/2} + u^{1/2})\,du$

$$= \frac{4}{5}u^{5/2} + \frac{4}{3}u^{3/2} + C = \frac{4}{5}(x-1)^{5/2} + \frac{4}{3}(x-1)^{3/2} + C$$

33. $\left\{\begin{array}{l} u = 2x - 1 \\[2mm] du = dx \end{array}\right\}; \qquad \int x\sqrt{2x-1}\,dx = \frac{1}{2}\int \frac{(u-1)}{2}\sqrt{u}\,du = \frac{1}{4}\int (u^{3/2} + u^{1/2})\,du$

$$= \frac{1}{10}u^{5/2} + \frac{1}{6}u^{3/2} + C = \frac{1}{10}(2x-1)^{5/2} + \frac{1}{6}(2x-1)^{3/2} + C$$

34. $\left\{ \begin{array}{l} u = 2t + 3 \\ du = 2\,dt \end{array} \right\}; \qquad \int t(2t+3)^8\,dt = \dfrac{1}{2}\int \dfrac{1}{2}(u-3)u^8\,du = \dfrac{1}{4}\int (u^9 - 3u^8)\,du$

$$= \dfrac{1}{40}u^{10} - \dfrac{1}{12}u^9 + C = \dfrac{1}{40}(2t+3)^{10} - \dfrac{1}{12}(2t+3)^9 + C$$

35. $\displaystyle\int \dfrac{1}{\sqrt{x\sqrt{x}+x}}\,dx = \int \dfrac{1}{\sqrt{x}\sqrt{\sqrt{x}+1}}\,dx$

$\left\{ \begin{array}{l} u = \sqrt{x} + 1 \\ du = dx/2\sqrt{x} \end{array} \right\}; \qquad \displaystyle\int \dfrac{1}{\sqrt{x}\sqrt{\sqrt{x}+1}}\,dx = 2\int u^{-1/2}\,du = 4\sqrt{u} + C = 4\sqrt{\sqrt{x}+1} + C$

36. $\left\{ \begin{array}{l|l} u = x^2 + 1 & x = -1 \Rightarrow u = 2 \\ du = 2x\,dx & x = 0 \;\;\Rightarrow u = 1 \end{array} \right\}; \qquad \displaystyle\int_{-1}^{0} x^3 \left(x^2+1\right)^6\,dx = \dfrac{1}{2}\int_{2}^{1}(u-1)u^6\,du$

$$= \dfrac{1}{2}\int_{2}^{1}(u^7 - u^6)\,du = \left[\dfrac{1}{16}u^8 - \dfrac{1}{14}u^7\right]_{2}^{1} = -\dfrac{255}{16} + \dfrac{127}{14} = -\dfrac{769}{112}$$

37. $\left\{ \begin{array}{l|l} u = x + 1 & x = 0 \Rightarrow u = 1 \\ du = dx & x = 1 \Rightarrow u = 2 \end{array} \right\};$

$$\int_{0}^{1} \dfrac{x+3}{\sqrt{x+1}}\,dx = \int_{1}^{2} \dfrac{u+2}{\sqrt{u}}\,du = \int_{1}^{2}(u^{1/2} + 2u^{-1/2})\,du = \left[\dfrac{2}{3}u^{3/2} + 4u^{1/2}\right]_{1}^{2} = \dfrac{16}{3}\sqrt{2} - \dfrac{14}{3}$$

38. Set $u = x - 1$. Then $du = dx$, $x = u + 1$, $x^2 = u^2 + 2u + 1$; $u(2) = 1$, $u(5) = 4$.

$$\int_{2}^{5} \dfrac{x^2}{\sqrt{x-1}}\,dx = \int_{1}^{4}(u^{3/2} + 2u^{1/2} + u^{-1/2})\,du = \left[\dfrac{2}{5}u^{5/2} + \dfrac{4}{3}u^{3/2} + 2u^{1/2}\right]_{1}^{4} = \dfrac{356}{15}$$

39. $\left\{ \begin{array}{l} u = x^2 + 1 \\ du = 2x\,dx \end{array} \right\}; \qquad \displaystyle\int x\sqrt{x^2+1}\,dx = \dfrac{1}{2}\int \sqrt{u}\,du = \dfrac{1}{3}u^{3/2} + C = \dfrac{1}{3}(x^2+1)^{3/2} + C.$

Also, $1 = \dfrac{1}{3}(0^2+1) + C \implies C = \dfrac{2}{3}.$ Thus $y = \dfrac{1}{3}(x^2+1)^{\frac{3}{2}} + \dfrac{2}{3}.$

40. $\left\{ \begin{array}{l} u = 1 + \sqrt{x} \\ du = \dfrac{1}{2\sqrt{x}}dx \end{array} \right\}; \qquad -\displaystyle\int \dfrac{1}{2\sqrt{x}(1+\sqrt{x})^2}\,dx = -\int u^{-2}\,du = \dfrac{1}{u} + C = \dfrac{1}{1+\sqrt{x}} + C.$

Also, $\dfrac{1}{3} = \dfrac{1}{1+\sqrt{4}} + C \implies C = 0$ Thus $y = \dfrac{1}{1+\sqrt{x}}$

41. $\displaystyle\int \cos(3x+1)\,dx = -\dfrac{1}{3}\sin(3x+1) + C$ **42.** $\displaystyle\int \sin 2\pi x\,dx = -\dfrac{1}{2\pi}\cos 2\pi x + C$

43. $\displaystyle\int \csc^2 \pi x\,dx = -\dfrac{1}{\pi}\cot \pi x + C$ **44.** $\displaystyle\int \sec 2x \tan 2x\,dx = \dfrac{1}{2}\sec 2x + C$

45. $\left\{\begin{array}{l} u = 3 - 2x \\ du = -2\,dx \end{array}\right\}$; $\displaystyle\int \sin\left(3 - 2x\right)dx = \int -\tfrac{1}{2}\sin u\,du = \tfrac{1}{2}\cos u + C = \tfrac{1}{2}\cos\left(3 - 2x\right) + C$

46. $\left\{\begin{array}{l} u = \sin x \\ du = \cos x\,dx \end{array}\right\}$; $\displaystyle\int \sin^2 x \cos x\,dx = \int u^2\,du = \frac{1}{3}u^3 + C = \frac{1}{3}\sin^3 x + C$

47. $\left\{\begin{array}{l} u = \cos x \\ du = -\sin x\,dx \end{array}\right\}$; $\displaystyle\int \cos^4 x \sin x\,dx = \int -u^4\,du = -\frac{1}{5}u^5 + C = -\frac{1}{5}\cos^5 x + C$

48. $\left\{\begin{array}{l} u = x^2 \\ du = 2x\,dx \end{array}\right\}$; $\displaystyle\int x \sec^2 x^2\,dx = \frac{1}{2}\int \sec^2 u\,du = \frac{1}{2}\tan u + C = \frac{1}{2}\tan x^2 + C$

49. $\left\{\begin{array}{l} u = x^{1/2} \\ du = \frac{1}{2}x^{-1/2}\,dx \end{array}\right\}$; $\displaystyle\int x^{-1/2}\sin x^{1/2}\,dx = \int 2\sin u\,du = -2\cos u + C = -2\cos x^{1/2} + C$

50. $\left\{\begin{array}{l} u = 1 - 2x \\ du = -2\,dx \end{array}\right\}$; $\displaystyle\int \csc(1 - 2x)\cot(1 - 2x)\,dx = -\frac{1}{2}\int \csc u \cot u\,du = \frac{1}{2}\csc u + C = \frac{1}{2}\csc(1 - 2x) +$
C

51. $\left\{\begin{array}{l} u = 1 + \sin x \\ du = \cos x\,dx \end{array}\right\}$; $\displaystyle\int \sqrt{1 + \sin x}\,\cos x\,dx = \int u^{1/2}\,du = \frac{2}{3}u^{3/2} + C = \tfrac{2}{3}(1 + \sin x)^{3/2} + C$

52. $\left\{\begin{array}{l} u = 1 + \cos x \\ du = -\sin x\,dx \end{array}\right\}$; $\displaystyle\int \frac{\sin x}{\sqrt{1 + \cos x}}\,dx = -\int \frac{du}{\sqrt{u}} = -2\sqrt{u} + C = -2\sqrt{1 + \cos x} + C$

53. $\left\{\begin{array}{l} u = \sin \pi x \\ du = \pi \cos \pi x\,dx \end{array}\right\}$; $\displaystyle\int \sin \pi x \cos \pi x\,dx = \frac{1}{\pi}\int u\,du = \frac{1}{2\pi}u^2 + C = \frac{1}{2\pi}\sin^2 \pi x + C$

54. $\left\{\begin{array}{l} u = \sin \pi x \\ du = \pi \cos \pi x\,dx \end{array}\right\}$; $\displaystyle\int \sin^2 \pi x \cos \pi x\,dx = \frac{1}{\pi}\int u^2\,du = \frac{1}{3\pi}u^3 + C = \frac{1}{3\pi}\sin^3 \pi x + C$

55. $\left\{\begin{array}{l} u = \cos \pi x \\ du = -\pi \cos \pi x\,dx \end{array}\right\}$; $\displaystyle\int \cos^2 \pi x \sin \pi x\,dx = -\frac{1}{\pi}\int u^2\,du = -\frac{1}{3\pi}u^3 + C = -\frac{1}{3\pi}\cos^3 \pi x + C$

56. $\displaystyle\int \left(1 + \tan^2 x\right)\sec^2 x\,dx = \int \sec^2 x\,dx + \int \tan^2 x \sec^2 x\,dx$

$\displaystyle = \tan x + \int u^2\,du = \tan x + \frac{1}{3}u^3 + C = \tan x + \frac{1}{3}\tan^3 x + C$

57. $\left\{\begin{array}{l} u = \sin x^2 \\ du = 2x \cos x^2\,dx \end{array}\right\}$; $\displaystyle\int x \sin^3 x^2 \cos x^2\,dx = \frac{1}{2}\int u^3\,du = \frac{1}{8}u^4 + C = \frac{1}{8}\sin^4 x^2 + C$

58. $\left\{ \begin{array}{l} u = \sin{(x^2 - \pi)} \\ du = 2x \cos{(x^2 - \pi)}\, dx \end{array} \right\};$

$$\int x \sin^4(x^2 - \pi) \cos{(x^2 - \pi)}\, dx = \frac{1}{2} \int u^4\, du = \frac{1}{10} u^5 + C = \frac{1}{10} \sin^5(x^2 - \pi) + C$$

59. $\left\{ \begin{array}{l} u = 1 + \tan{x} \\ du = \sec^2{x}\, dx \end{array} \right\}; \qquad \int \frac{\sec^2{x}}{\sqrt{1 + \tan{x}}}\, dx = \int u^{-1/2}\, du = 2u^{1/2} + C = 2(1 + \tan{x})^{1/2} + C$

60. $\left\{ \begin{array}{l} u = 2 + \cot{2x} \\ du = -2\csc^2{2x}\, dx \end{array} \right\}; \qquad \int \frac{\csc^2{2x}}{\sqrt{2 + \cot{2x}}}\, dx = -\frac{1}{2} \int u^{-1/2}\, du = -u^{1/2} + C = -\sqrt{2 + \cot{2x}} + C$

61. $\left\{ \begin{array}{l} u = 1/x \\ du = -1/x^2\, dx \end{array} \right\}; \qquad \int \frac{\cos{(1/x)}}{x^2}\, dx = -\int \cos{u}\, du = -\sin{u} + C = -\sin{(1/x)} + C.$

62. $\left\{ \begin{array}{l} u = 1/x \\ du = -1/x^2\, dx \end{array} \right\}; \qquad \int \frac{\sin{(1/x)}}{x^2}\, dx = -\int \sin{u}\, du = \cos{u} + C = \cos{(1/x)} + C.$

63. $\left\{ \begin{array}{l} u = \tan{(x^3 + \pi)} \\ du = 3x^2 \sec^2{(x^3 + \pi)}\, dx \end{array} \right\};$

$$\int x^2 \tan{(x^3 + \pi)} \sec^2(x^3 + \pi)\, dx = \frac{1}{3} \int u\, du = \frac{1}{6} u^2 + C = \frac{1}{6} \tan^2(x^3 + \pi) + C$$

64. $\int \left(x \sin^2{x} - x^2 \sin{x} \cos{x} \right)\, dx = \int x \sin{x}(\sin{x} - x \cos{x})\, dx$

$\left\{ \begin{array}{l} u = \sin{x} - x \cos{x} \\ du = x \sin{x}\, dx \end{array} \right\};$

$$\int x \sin{x}(\sin{x} - x \cos{x})\, dx = \int u\, du = \tfrac{1}{2} u^2 + C = \tfrac{1}{2} (\sin{x} - x \cos{x})^2 + C$$

65. $\left\{ \begin{array}{l} u = \sin{x} \\ du = \cos{x}\, dx \end{array} \left| \begin{array}{l} x = -\pi \;\Rightarrow\; u = 0 \\ x = \pi \;\Rightarrow\; u = 0 \end{array} \right. \right\}; \quad \int_{-\pi}^{\pi} \sin^4{x} \cos{x}\, dx = \int_{0}^{0} u^4\, du = 0.$

66. $\displaystyle \int_{-\pi/3}^{\pi/3} \sec{x} \tan{x}\, dx = [\sec{x}]_{-\pi/3}^{\pi/3} = 0$ \qquad **67.** $\displaystyle \int_{1/4}^{1/3} \sec^2{\pi x}\, dx = \frac{1}{\pi} [\tan{\pi x}]_{1/4}^{1/3} = \frac{1}{\pi}(\sqrt{3} - 1)$

68. $\displaystyle \int_{0}^{1} \cos^2\left(\frac{\pi}{2}x\right) \sin\left(\frac{\pi}{2}x\right)\, dx = \frac{-2}{3\pi} \left[\cos^3{\frac{\pi}{2}x}\right]_{0}^{1} = \frac{2}{3\pi}$

69. $\left\{ \begin{array}{l} u = \cos x \\ du = -\sin x\, dx \end{array} \middle| \begin{array}{l} x = 0 \;\Rightarrow\; u = 1 \\ x = \pi/2 \;\Rightarrow\; u = 0 \end{array} \right\};$

$$\int_0^{\pi/2} \sin x \cos^3 x\, dx = -\int_1^0 u^3\, du = \int_0^1 u^3\, du = \left[\frac{u^4}{4} \right]_0^1 = \frac{1}{4}.$$

70. $\int_0^{\pi} x \cos x^2\, dx = \frac{1}{2} \left[\sin x^2 \right]_0^{\pi} = \frac{1}{2} \sin \pi^2$

71. $\int \sin^2 x\, dx = \int \frac{1 - \cos 2x}{2}\, dx = \frac{1}{2} x - \frac{1}{4} \sin 2x + C$

72. $\int \cos^2 dx = \int \frac{1 + \cos 2x}{2}\, dx = \frac{1}{2} x + \frac{1}{4} \sin 2x + C$

73. $\int \cos^2 5x\, dx = \int \frac{1 + \cos 10x}{2}\, dx = \frac{1}{2} x + \frac{1}{20} \sin 10x + C$

74. $\int \sin^2 3x\, dx = \int \frac{1 - \cos 6x}{2}\, dx = \frac{1}{2} x - \frac{1}{12} \sin 6x + C$

75. $\int_0^{\pi/2} \cos^2 2x\, dx = \int_0^{\pi/2} \frac{1 + \cos 4x}{2}\, dx = \left[\frac{1}{2} x + \frac{1}{8} \sin 4x \right]_0^{\pi/2} = \frac{\pi}{4}$

76. $\int_0^{2\pi} \sin^2 x\, dx = \left[\frac{1}{2} x - \frac{1}{4} \sin 2x \right]_0^{2\pi} = \pi$

77. $A = \int_0^{\frac{\pi}{2}} [\cos x - (-\sin x)]\, dx = [\sin x - \cos x]_0^{\frac{\pi}{2}} = 2$

78. $A = \int_0^{1/4} (\cos \pi x - \sin \pi x)\, dx = \frac{1}{\pi} [\sin \pi x + \cos \pi x]_0^{1/4} = \frac{1}{\pi} (\sqrt{2} - 1)$

79. $A = \int_0^{1/4} (\cos^2 \pi x - \sin^2 \pi x)\, dx = \int_0^{1/4} \cos 2\pi x\, dx = \frac{1}{2\pi} [\sin 2\pi x]_0^{1/4} = \frac{1}{2\pi}$

80. $\int_0^{1/4} (\cos^2 \pi x + \sin^2 \pi x)\, dx = \int_0^{1/4} 1\, dx = [x]_0^{1/4} = \frac{1}{4}$

81. $A = \int_{1/6}^{1/4} (\csc^2 \pi x - \sec^2 \pi x)\, dx = \left[\frac{1}{\pi} (-\cot \pi x - \tan \pi x) \right]_{1/6}^{1/4}$

$$= \frac{1}{\pi} \left(-2 + \cot \frac{\pi}{6} + \tan \frac{\pi}{6} \right)$$

$$= \frac{1}{\pi} \left(-2 + \sqrt{3} + \frac{1}{\sqrt{3}} \right) = \frac{1}{3\pi} (4\sqrt{3} - 6)$$

82. (a) $\left\{ \begin{array}{l} u = \sin x \\ du = \cos x\, dx \end{array} \right\};$ $\quad \int \sin x \cos x\, dx = \int u\, du = \frac{1}{2} u^2 + C = \frac{1}{2} \sin^2 x + C$

(b) $\begin{cases} u = \cos x \\ du = -\sin x\, dx \end{cases}$; $\quad \int \sin x \cos x\, dx = -\int u\, du = -\tfrac{1}{2}u^2 + C' = -\tfrac{1}{2}\cos^2 x + C'$

(c) $C' = C + \tfrac{1}{2}$

83. (a) $\begin{cases} u = \sec x \\ du = \sec x \tan x\, dx \end{cases}$; $\quad \int \sec^2 x \tan x\, dx = \int u\, du = \tfrac{1}{2}u^2 + C = \tfrac{1}{2}\sec^2 x + C$

(b) $\begin{cases} u = \tan x \\ du = \sec^2 x\, dx \end{cases}$; $\quad \int \sec^2 x \tan x\, dx = \int u\, du = \tfrac{1}{2}u^2 + C' = \tfrac{1}{2}\tan^2 x + C'$

(c) $C' = \tfrac{1}{2} + C$

84. (a) Set $u = x - c$. Then $dx = du$; $u(a+c) = a$, $u(b+c) = b$.

$$\int_{a+c}^{b+c} f(x - c)\, dx = \int_a^b f(u)\, du = \int_a^b f(x)\, dx$$

(b) Set $u = x/c$. Then $du = (1/c)\, dx$; $u(ac) = a$, $u(bc) = b$.

$$\frac{1}{c}\int_{ac}^{bc} f(x/c)\, dx = \int_a^b f(u)\, du = \int_a^b f(x)\, dx$$

85. $A = 4\displaystyle\int_0^r \sqrt{r^2 - x^2}\, dx = 4\int_0^{\pi/2} \sqrt{r^2 - r^2 \sin^2 u}\,(r\cos u)\, du \quad (x = r \sin u)$

$$= 4\int_0^{\pi/2} r^2 \cos^2 u\, du = 4r^2\left[\frac{1}{2}u + \frac{1}{4}\sin 2u\right]_0^{\pi/2} = \pi r^2$$

86. $A = \dfrac{4b}{a}\displaystyle\int_0^a \sqrt{a^2 - x^2}\, dx = \dfrac{4b}{a}\left(\dfrac{\text{area of circle of radius } a}{4}\right) = \dfrac{4b}{a}\left(\dfrac{\pi a^2}{4}\right) = \pi ab$

SECTION 5.8

1. Yes; $\displaystyle\int_a^b [f(x) - g(x)]\, dx = \int_a^b f(x)\, dx - \int_a^b g(x)\, dx > 0$.

2. No; take, for example, the function $f(x) = x$ and $g(x) = 0$ on $\left[-\tfrac{1}{2}, 1\right]$.

3. Yes; otherwise we would have $f(x) \le g(x)$ for all $x \in [a, b]$ and it would follow that

$$\int_a^b f(x)\, dx \le \int_a^b g(x)\, dx.$$

4. No; take, for example, the function $f(x) = 0$ and $g(x) = -1$ on $[0, 1]$.

5. No; take $f(x) = 0$, $g(x) = -1$ on $[0, 1]$.

6. Yes; $\displaystyle\int_a^b |f(x)|\, dx \geq \int_a^b f(x)\, dx$ and we are assuming that $\displaystyle\int_a^b f(x)\, dx > \int_a^b g(x)\, dx$.

7. No;   take, for example, any odd function on an interval of the form $[-c, c]$.

8. Yes;   if $f(x) \neq 0$ for each $x \in [a, b]$, then by continuity either $f(x) > 0$ for all $x \in [a, b]$,

or $f(x) < 0$ for all $x \in [a, b]$. In either case

$$\int_a^b f(x)\, dx \neq 0$$

9. No; $\displaystyle\int_{-1}^1 x\, dx = 0$ but $\displaystyle\int_{-1}^1 |x|\, dx \neq 0$.

10. Yes; $\displaystyle\left| \int_a^b f(x)\, dx \right| = |0| = 0$ **11.**   Yes; $\displaystyle U_f(P) \geq \int_a^b f(x)\, dx = 0$.

12. No;   if $f(x) = 0$ for all $x \in [a, b]$, then

$$\int_a^b f(x)\, dx = 0, \quad \text{and} \quad U_f(P) = 0 \quad \text{for all} \quad P.$$

13. No; $\displaystyle L_f(P) \leq \int_a^b f(x)\, dx = 0$.

14. No;   take $f(x) = x$ on $[-1.1]$;

$$\int_{-1}^1 x\, dx = 0 \quad \text{but} \quad \int_{-1}^1 x^2\, dx = \frac{2}{3}$$

15. Yes; $\displaystyle\int_a^b [f(x) + 1]\, dx = \int_a^b f(x)\, dx + \int_a^b 1\, dx = 0 + b - a = b - a$.

16. $\displaystyle \frac{d}{dx}\left[\int_u^b f(t)\, dt \right] = \frac{d}{du}\left[\int_u^b f(t)\, dt \right]\frac{du}{dx} = \frac{d}{du}\left[-\int_b^u f(t)\, dt \right]\frac{du}{dx} = -f(u)\frac{du}{dx}.$

17. $\displaystyle \frac{d}{dx}\left[\int_0^{1+x^2} \frac{dt}{\sqrt{2t+5}} \right] = \frac{1}{\sqrt{2(1+x^2)+5}}\,\frac{d}{dx}\left(1+x^2\right) = \frac{2x}{\sqrt{2x^2+7}}$

18. $\displaystyle \frac{d}{dx}\left[\int_1^{x^2} \frac{dt}{t} \right] = \frac{1}{x^2}\,2x = \frac{2}{x}.$

19. $\dfrac{d}{dx}\left[\displaystyle\int_x^a f(t)\,dt\right] = \dfrac{d}{dx}\left[-\displaystyle\int_a^x f(t)\,dt\right] = -f(x)$

20. $\dfrac{d}{dx}\left[\displaystyle\int_0^{x^3}\dfrac{dt}{\sqrt{1+t^2}}\right] = \dfrac{3x^2}{\sqrt{1+x^6}}.$

21. $\dfrac{d}{dx}\left[\displaystyle\int_{x^2}^3\dfrac{\sin t}{t}\,dt\right] = -\dfrac{d}{dx}\left[\displaystyle\int_3^{x^2}\dfrac{\sin t}{t}\,dt\right] = -\dfrac{\sin(x^2)}{x^2}(2x) = -\dfrac{2\sin(x^2)}{x}$

22. $\dfrac{d}{dx}\left[\displaystyle\int_{\tan x}^4 \sin(t^2)\,dt\right] = -\sin\left(\tan^2 x\right)\sec^2 x.$

23. $\dfrac{d}{dx}\left[\displaystyle\int_1^{\sqrt{x}}\dfrac{t^2}{1+t^2}\,dt\right] = \dfrac{x}{1+x}\cdot\dfrac{1}{2\sqrt{x}} = \dfrac{\sqrt{x}}{2(1+x)}$

24. $\dfrac{d}{dx}\left[\displaystyle\int_u^v f(t)\,dt\right] = \dfrac{d}{dx}\left[\displaystyle\int_a^v f(t)\,dt - \displaystyle\int_a^u f(t)\,dt\right] = f(v)\dfrac{dv}{dx} - f(u)\dfrac{du}{dx}.$

25. $\dfrac{d}{dx}\left[\displaystyle\int_x^{x^2}\dfrac{dt}{t}\right] = \dfrac{1}{x^2}\dfrac{d}{dx}\left(x^2\right) - \dfrac{1}{x}\dfrac{d}{dx}\left(x\right) = \dfrac{2x}{x^2} - \dfrac{1}{x} = \dfrac{1}{x}$

26. $\dfrac{d}{dx}\left[\displaystyle\int_{\sqrt{x}}^{x^2+x}\dfrac{dt}{2+\sqrt{t}}\right] = \dfrac{1}{2+\sqrt{x^2+x}}(2x+1) - \dfrac{1}{2+\sqrt[4]{x}}\cdot\dfrac{1}{2\sqrt{x}}$

27. $\dfrac{d}{dx}\left[\displaystyle\int_{\tan x}^{2x} t\sqrt{1+t^2}\,dt\right] = 2x\sqrt{1+(2x)^2}\,(2) - \tan x\,\sqrt{1+\tan^2 x}\,\left(\sec^2 x\right)$

$$= 4x\sqrt{1+4x^2} - \tan x\,\sec^2 x\,|\sec x|$$

28. $\dfrac{d}{dx}\left[\displaystyle\int_{3x}^{1/x}\cos 2t\,dt\right] = \cos\left(\dfrac{2}{x}\right)\left(\dfrac{-1}{x^2}\right) - \cos 6x\,(3) = -\dfrac{\cos(2/x)}{x^2} - 3\cos 6x$

29. Set $h(x) = g(x) - f(x)$ and apply (5.8.2) to h.

30. Suppose $f(c) > 0$ for some $c \in (a,b)$. Then by Exercise 48, Section 2.4, there exists $\delta > 0$ such that $f(x) > 0$ for all $x \in (c-\delta, c+\delta)$. Also, we can choose δ such that $(c-\delta, c+\delta) \subset (a,b)$.

Then $\displaystyle\int_a^b |f(x)|\,dx \ge \int_{c-\delta}^{c+\delta} |f(x)|dx > 0$, a contradiction. The same holds if $f(c) < 0$ for some c.

Thus $f(x) = 0$ for all $x \in (a,b)$. Then since f is continuous on $[a,b]$, we must have

$f(a) = f(b) = 0$, so $f(x) = 0$ for all $x \in [a,b]$.

31. $\displaystyle H(x) = \int_{2x}^{x^3-4} \frac{x\,dt}{1+\sqrt{t}} = x \int_{2x}^{x^3-4} \frac{dt}{1+\sqrt{t}},$

$\displaystyle H'(x) = x \cdot \left[\frac{3x^2}{1+\sqrt{x^3-4}} - \frac{2}{1+\sqrt{2x}} \right] + 1 \cdot \int_{2x}^{x^3-4} \frac{dt}{1+\sqrt{t}},$

$\displaystyle H'(2) = 2\left[\frac{12}{3} - \frac{2}{3} \right] + \underbrace{\int_4^4 \frac{dt}{1+\sqrt{t}}}_{=\,0} = \frac{20}{3}$

32. $\displaystyle H(x) = \frac{1}{x} \int_3^x [2t - 3H'(t)]\,dt,$

$\displaystyle H'(x) = \frac{-1}{x^2} \int_3^x [2t - 3H'(t)]\,dt + \frac{1}{x}[2x - 3H'(t)],$

$\displaystyle H'(3) = \frac{-1}{3^2} \int_3^3 [2t - 3H'(t)]\,dt + \frac{1}{3}[2\cdot 3 - 3H'(3)]$

$\displaystyle = \frac{-1}{3^2} \cdot 0 + 2 - H'(3) \implies H'(3) = 1.$

33. (a) Let $u = -x$. Then $du = -dx$; and $u = 0$ when $x = 0$, $u = a$ when $x = -a$.

$$\int_{-a}^0 f(x)\,dx = -\int_a^0 f(-u)\,du = \int_0^a f(-u)\,du = \int_0^a f(-x)\,dx$$

(b) $\displaystyle \int_{-a}^a f(x)\,dx = \int_{-a}^0 f(x)\,dx + \int_0^a f(x)\,dx = -\int_a^0 f(u)\,du + \int_0^a f(x)\,dx$

$$\overset{\displaystyle \big\llcorner\, u=-x,\ du=-dx}{}$$

$$= \int_0^a f(u)\,du + \int_0^a f(x)\,dx = \int_0^a [f(x) + f(-x)]\,dx$$

34. (a) $\displaystyle \int_{-a}^a f(x)\,dx = \int_{-a}^0 f(x)\,dx + \int_0^a f(x)\,dx$

In first integral, use $u = -x, du = -dx, u(-a) = a, u(0) = 0, x = -u$, and note that

$f(x) = f(-u) = -f(u)$ since f is odd. Then

$$\int_{-a}^0 f(x)\,dx = -\int_a^0 f(-u)\,du = \int_0^a f(-u)\,du = -\int_0^a f(u)\,du$$

So $\displaystyle \int_{-a}^a f(x)\,dx = -\int_0^a f(u)\,du + \int_0^a f(x)\,dx = 0$

(b) As above, but now $f(x) = f(-u) = f(u)$ since f is even, so

$$\int_{-a}^{0} f(x)\,dx = -\int_{a}^{0} f(-u)\,du = \int_{0}^{a} f(-u)\,du = \int_{0}^{a} f(u)\,du,$$

hence $\displaystyle \int_{-a}^{a} f(x)\,dx = 2\int_{0}^{a} f(x)\,dx$

35. $\displaystyle \int_{-\pi/4}^{\pi/4} (x + \sin 2x)\,dx = 0$ since $f(x) = x + \sin 2x$ is an odd function.

36. $\dfrac{t^3}{1+t^2}$ is an odd function, so $\displaystyle \int_{-3}^{3} \dfrac{t^3}{1+t^2}\,dt = 0$

37. $\displaystyle \int_{-\pi/3}^{\pi/3} (1 + x^2 - \cos x)\,dx = 2\int_{0}^{\pi/3} (1 + x^2 - \cos x)\,dx$ since $f(x) = 1 + x^2 - \cos x$ is an even function.

$$2\int_{0}^{\pi/3} (1 + x^2 - \cos x)\,dx = 2\left[x + \frac{1}{3}x^3 - \sin x \right]_{0}^{\pi/3} = \frac{2}{3}\pi + \frac{2}{81}\pi^3 - \sqrt{3}$$

38. $2x$ and $\sin x$ are odd, and x^2 and $\cos 2x$ are even, so

$$\int_{-\pi/4}^{\pi/4} (x^2 - 2x + \sin x + \cos 2x)\,dx = 2\int_{0}^{\pi/4} (x^2 + \cos 2x)\,dx = 2\left[\frac{x^3}{3} + \frac{1}{2}\sin 2x \right]_{0}^{\pi/4} = \frac{\pi^3}{96} + 1$$

SECTION 5.9

1. A.V. $= \dfrac{1}{c}\displaystyle\int_{0}^{c} (mx + b)\,dx = \dfrac{1}{c}\left[\dfrac{m}{2}x^2 + bx \right]_{0}^{c} = \dfrac{mc}{2} + b;$ at $x = c/2$

2. A.V. $= \dfrac{1}{2}\displaystyle\int_{-1}^{1} x^2\,dx = \dfrac{1}{2}\left[\dfrac{x^3}{3} \right]_{-1}^{1} = \dfrac{1}{3};$ at $x = \pm\dfrac{\sqrt{3}}{3}.$

3. A.V. $= \frac{1}{2}\displaystyle\int_{-1}^{1} x^3\,dx = 0$ since the integrand is odd;   at $x = 0$

4. A.V. $= \dfrac{1}{3}\displaystyle\int_{1}^{4} x^{-2}\,dx = \dfrac{1}{3}\left[-\dfrac{1}{x} \right]_{1}^{4} = \dfrac{1}{3}\cdot\dfrac{3}{4} = \dfrac{1}{4};$ at $x = 2.$

5. A.V. $= \dfrac{1}{4}\displaystyle\int_{-2}^{2} |x|\,dx = \dfrac{1}{2}\displaystyle\int_{0}^{2} |x|\,dx = \dfrac{1}{2}\displaystyle\int_{0}^{2} x\,dx = \dfrac{1}{2}\left[\dfrac{x^2}{2} \right]_{0}^{2} = 1;$ at $x = \pm 1$

6. A.V. $= \dfrac{1}{16}\displaystyle\int_{-8}^{8} x^{1/3}\,dx = 0$ (odd function);   at; $x = 0$

7. A.V. $= \dfrac{1}{2}\displaystyle\int_{0}^{2} (2x - x^2)\,dx = \dfrac{1}{2}\left[x^2 - \dfrac{x^3}{3} \right]_{0}^{2} = \dfrac{2}{3};$ at $x = 1 \pm \dfrac{1}{3}\sqrt{3}$

8. A.V. $= \dfrac{1}{3} \displaystyle\int_0^3 (3-2x)\,dx = \dfrac{1}{3}\left[3x - x^2\right]_0^3 = 0;$ at $x = \dfrac{3}{2}.$

9. A.V. $= \dfrac{1}{9} \displaystyle\int_0^9 \sqrt{x}\,dx = \dfrac{1}{9}\left[\dfrac{2}{3}x^{3/2}\right]_0^9 = 2;$ at $x = 4$

10. A.V. $= \dfrac{1}{4} \displaystyle\int_{-2}^2 (4-x^2)\,dx = \dfrac{1}{4}\left[4x - \dfrac{x^3}{3}\right]_{-2}^2 = \dfrac{8}{3};$ at $x = \pm\dfrac{2\sqrt{3}}{3}.$

11. A.V. $= \dfrac{1}{2\pi} \displaystyle\int_0^{2\pi} \sin x\,dx = \dfrac{1}{2\pi}\left[-\cos x\right]_0^{2\pi} = 0;$ at $x = \pi$

12. A.V. $= \dfrac{1}{\pi} \displaystyle\int_0^{\pi} \cos x\,dx = \dfrac{1}{\pi}\left[\sin x\right]_0^{\pi} = 0;$ at $x = \dfrac{\pi}{2}.$

13. A.V. $= \dfrac{1}{b-a} \displaystyle\int_a^b x^n\,dx = \dfrac{1}{b-a}\left[\dfrac{x^{n+1}}{n+1}\right]_a^b = \dfrac{b^{n+1} - a^{n+1}}{(n+1)(b-a)}.$

14. (a) for constant f, $f(b)(b-a) = \displaystyle\int_a^b f(x)\,dx$

   (b) for increasing f, $f(b)(b-a) > \displaystyle\int_a^b f(x)\,dx$

   (c) for decreasing f, $f(b)(b-a) < \displaystyle\int_a^b f(x)\,dx$

15. Average of f' on $[a, b] = \dfrac{1}{b-a} \displaystyle\int_a^b f'(x)\,dx = \dfrac{1}{b-a}\left[f(x)\right]_a^b = \dfrac{f(b) - f(a)}{b-a}.$

16. (a) True, because $\displaystyle\int_a^b (f+g)\,dx = \int_a^b f\,dx + \int_a^b g\,dx.$

   (b) True, because $\displaystyle\int_a^b \alpha f\,dx = \alpha \int_a^b f\,dx.$

   (c) False;   take $f(x) = g(x) = x$ on $[0,1]:$ A.V.$(f\,g) = \dfrac{1}{3},$ $(A.V.(f))(A.V.(g)) = \dfrac{1}{2}\cdot\dfrac{1}{2} = \dfrac{1}{4}.$

   (d) False;   take $f(x) = x^2$ and $g(x) = x$ on $[0,1]:$

   $\qquad A.V.(f/g) = A.V.(x) = \dfrac{1}{2},$ $\dfrac{A.V.(f)}{A.V.(g)} = \dfrac{1/3}{1/2} = \dfrac{2}{3}.$

17. Distance from (x, y) to the origin: $\sqrt{x^2 + y^2}.$ Since $y = x^2,$ $D(x) = \sqrt{x^2 + x^4}.$

   On $\left[0, \sqrt{3}\right]$,   A.V.$= \dfrac{1}{\sqrt{3}} \displaystyle\int_0^{\sqrt{3}} x\sqrt{1 + x^2}\,dx = \dfrac{1}{\sqrt{3}}\left[\dfrac{1}{3}(1 + x^2)^{3/2}\right]_0^{\sqrt{3}} = \dfrac{1}{3\sqrt{3}}\,7 = \dfrac{7}{9}\sqrt{3}.$

18. Distance from (x, y) to the origin: $\sqrt{x^2 + y^2}$. Since $y = mx$, $\quad D(x) = \sqrt{x^2 + m^2 x^2} = |x|\sqrt{1 + m^2}$.

On $[0, 1]$,   A.V.$= \dfrac{1}{1 - 0}\displaystyle\int_0^1 x\sqrt{1 + m^2}\,dx = \left[\tfrac{1}{2}\sqrt{1 + m^2}\,x^2\right]_0^1 = \dfrac{\sqrt{1 + m^2}}{2}$

19. The distance the stone has fallen after t seconds is given by $s(t) = 16t^2$.

(a) The terminal velocity after x seconds is $s'(x) = 32x$. The average velocity

is $\dfrac{s(x) - s(0)}{x - 0} = 16x$. Thus the terminal velocity is twice the average velocity.

(b) For the first $\tfrac{1}{2}x$ seconds, aver. vel. $= \dfrac{s\left(\tfrac{1}{2}x\right) - s(0)}{\tfrac{1}{2}x - 0} = 8x$.

For the next $\tfrac{1}{2}x$ seconds, aver. vel. $= \dfrac{s(x) - s\left(\tfrac{1}{2}x\right)}{x - \tfrac{1}{2}x} = 24x$.

Thus, for the first $\tfrac{1}{2}x$ seconds the average velocity is one-third of the average velocity during the next $\tfrac{1}{2}x$ seconds.

20. Obvious since $\displaystyle\int_{-a}^{a} f(x)\,dx = 0$

21. Suppose $f(x) \neq 0$ for all x in (a, b). Then, since f is continuous, either

$$f(x) > 0 \text{ on } (a, b) \quad \text{or} \quad f(x) > 0 \text{ on } (a, b).$$

In either case, $\displaystyle\int_a^b f(x)\,dx \neq 0$.

22. $\dfrac{1}{(a + 2n) - a}\displaystyle\int_a^{a+2n} \sin \pi x\,dx = \dfrac{1}{2n}\left[-\dfrac{1}{\pi}\cos \pi x\right]_a^{a+2n} = -\dfrac{1}{2n\pi}(\cos(a\pi + 2n\pi) - \cos a\pi) = 0$

Similarly for the average value of $\cos \pi x$ on $[a, a + 2n]$.

23. (a) $v(t) - v(0) = \displaystyle\int_0^t a\,du; \quad v(0) = 0$. Thus $v(t) = at$.

$x(t) - x(0) = \displaystyle\int_0^t v(u)\,du; \quad x(0) = x_0$. Thus $x(t) = \displaystyle\int_0^t au\,du + x_0 = \dfrac{1}{2}at^2 + x_0$.

(b) $v_{avg} = \dfrac{1}{t_2 - t_1}\displaystyle\int_{t_1}^{t_2} at\,dt = \dfrac{1}{t_2 - t_1}\left[\dfrac{1}{2}at^2\right]_{t_1}^{t_2}$

$= \dfrac{at_2^2 - at_1^2}{2(t_2 - t_1)} = \dfrac{v(t_1) + v(t_2)}{2}$

24. Let c be the point that divides the rod into two pieces of equal mass:

$$\int_0^c kx\,dx = \int_c^L kx\,dx \implies \tfrac{1}{2}kc^2 = \tfrac{1}{2}kL^2 - \tfrac{1}{2}kc^2 \implies kc^2 = \tfrac{1}{2}kL^2 \implies c = \dfrac{\sqrt{2}}{2}L$$

25. (a) $M = \displaystyle\int_0^6 \frac{12}{\sqrt{x+1}}\, dx = 12 \int_0^6 \frac{1}{\sqrt{x+1}}\, dx = 24\left[\sqrt{x+1}\right]_0^6 = 24(\sqrt{7}-1)$

$$x_M M = \int_0^6 \frac{12x}{\sqrt{x+1}}\, dx = 12 \int_1^7 \left(u^{1/2} - u^{-1/2}\right) du$$

$$\llcorner\quad u = x+1,\ \ du = dx,\ \ x = u-1$$

$$= 12\left[\tfrac{2}{3} u^{3/2} - 2u^{1/2}\right]_1^7 = 16 + 32\sqrt{7};$$

$$x_M = \frac{16 + 32\sqrt{7}}{24(\sqrt{7}-1)} = \frac{4\sqrt{7}+2}{3\sqrt{7}-3}$$

(b) A.V.$= \dfrac{1}{6} \displaystyle\int_0^6 \frac{12}{\sqrt{x+1}}\, dx = \frac{1}{6}\left[24(\sqrt{7}-1)\right] = 4(\sqrt{7}-1)$

26. $\displaystyle\int_a^b (x - x_M)\,\lambda(x)\, dx = \int_a^b x\lambda(x)\, dx - x_M \int_a^b \lambda(x)\, dx = x_M\, M - x_M\, M = 0$

27. (a)
$$M = \int_0^L k\sqrt{x}\, dx = k\left[\frac{2}{3} x^{3/2}\right]_0^L = \frac{2}{3} kL^{3/2}$$

$$x_M M = \int_0^L x\left(k\sqrt{x}\,\right) dx = \int_0^L kx^{3/2}\, dx = \left[\frac{2}{5} kx^{5/2}\right]_0^L = \frac{2}{5} kL^{5/2}$$

$$x_M = \left(\tfrac{2}{5}kL^{5/2}\right) / \left(\tfrac{2}{3}kL^{3/2}\right) = \tfrac{3}{5}L$$

(b)
$$M = \int_0^L k\left(L-x\right)^2 dx = \left[-\frac{1}{3} k\left(L-x\right)^3\right]_0^L = \frac{1}{3} kL^3$$

$$x_M M = \int_0^L x\left[k\left(L-x\right)^2\right] dx = \int_0^L k\left(L^2 x - 2Lx^2 + x^3\right) dx$$

$$= k\left[\tfrac{1}{2}L^2 x^2 - \tfrac{2}{3}Lx^3 + \tfrac{1}{4}x^4\right]_0^L = \tfrac{1}{12}kL^4$$

$$x_M = \left(\tfrac{1}{12}kL^4\right) / \left(\tfrac{1}{3}kL^3\right) = \tfrac{1}{4}L$$

28. $x_M\, M = \displaystyle\int_a^b x\lambda(x)\, dx$

$$= \int_{x_0}^{x_1} x\lambda(x)\, dx + \int_{x_1}^{x_2} x\lambda(x)\, dx + \cdots + \int_{x_{n-1}}^{x_n} x\lambda(x)\, dx$$

$$= x_{M_1}\, M_1 + x_{M_2}\, M_2 + \cdots + x_{M_n}\, M_n$$

29. $\tfrac{1}{4}LM = \tfrac{1}{8}LM_1 + x_{M_2} M_2$

$$x_{M_2} = \frac{1}{M_2}\left(\frac{1}{4}LM - \frac{1}{8}LM_1\right) = \frac{L}{8M_2}(2M - M_1)$$

30. By Exercise 28, $x_M M = x_{M_1} M_1 + x_{M_2} M_2$, so $\dfrac{2}{3}LM = \dfrac{1}{4}LM_1 + \dfrac{7}{8}LM_2$.

Also, $M_1 + M_2 = M$. Solving gives: $M_1 = \dfrac{1}{3}M$, $M_2 = \dfrac{2}{3}M$.

31. Let $M = \int_a^{a+L} kx\,dx$, where a is the point of the first cut.

Thus $M = \left[\dfrac{kx^2}{2}\right]_a^{a+L} = \dfrac{k}{2}(2aL + L^2)$. Hence $a = \dfrac{2M - kL^2}{2kL}$, and $a + L = \dfrac{2M + kL^2}{2kL}$.

32. Yes. Suppose $g(x) < 0$ on $[a, b]$. Let m be the minimum value of f and M the maximum value of f on $[a, b]$. Then

$$m \le f(x) \le M \quad\text{and}\quad mg(x) \ge f(x)g(x) \ge Mg(x) \quad\text{since}\quad g(x) < 0.$$

The proof now proceeds exactly as in the proof of Theorem 5.9.3, only the inequalities are reversed.

33. If f is continuous on $[a, b]$, then, by Theorem 5.2.5, F satisfies the conditions of the mean-value theorem of differential calculus (Theorem 4.1.1). Therefore, by that theorem, there is at least one number c in (a, b) for which

$$F'(c) = \dfrac{F(b) - F(a)}{b - a}.$$

Then

$$\int_a^b f(x)\,dx = F(b) - F(a) = F'(c)(b - a) = f(c)(b - a).$$

34. $\left(\begin{array}{c}\text{min of } f \\ \text{on } [c - h, c + h]\end{array}\right) \le \left(\begin{array}{c}\text{average of } f \\ \text{on } [c - h, c + h]\end{array}\right) \le \left(\begin{array}{c}\text{max of } f \\ \text{on } [c - h, c + h]\end{array}\right).$

By continuity, as $h \to 0^+$

$\left(\begin{array}{c}\text{min of } f \\ \text{on } [c - h, c + h]\end{array}\right) \to f(c)$ and $\left(\begin{array}{c}\text{max of } f \\ \text{on } [c - h, c + h]\end{array}\right) \to f(c).$

By the pinching theorem the middle term must also tend to $f(c)$.

35. If f and g take on the same average value on every interval $[a, x]$, then

$$\dfrac{1}{x - a} \int_a^x f(t)\,dt = \dfrac{1}{x - a} \int_a^x g(t)\,dt.$$

Multiplication by $(x - a)$ gives

$$\int_a^x f(t)\,dt = \int_a^x g(t)\,dt.$$

Differentiation with respect to x gives $f(x) = g(x)$. This shows that, if the averages are everywhere the same, then the functions are everywhere the same.

36. Partition $[a, b]$ into n subintervals of equal length $\dfrac{b-a}{n}$,   where $P = \{x_0, \ldots, x_n\}$

and x_i^* is a point from $[x_{i-1}, x_i]$.   Then the average value of f on $[a, b]$ is:

$$\frac{1}{b-a}\int_a^b f(x)\,dx = \frac{1}{b-a}\lim_{\|P\|\to 0}\left[f(x_1^*)\left(\frac{b-a}{n}\right) + \cdots + f(x_n^*)\left(\frac{b-a}{n}\right)\right]$$

$$= \left(\frac{1}{b-a}\right)\lim_{n\to\infty}\frac{b-a}{n}[f(x_1^*) + \cdots + f(x_n^*)]$$

$$= \lim_{n\to\infty}\frac{1}{n}[f(x_1^*) + \cdots + f(x_n^*)],$$

which is the limit of arithmetic averages of values of f on $[a, b]$.

37. Let $P = \{a = x_0,\, x_1,\, x_2,\, \ldots,\, x_n = b\}$ be a partition of the interval $[a, b]$. Then

$$\int_a^b f(x)\,dx = \int_{x_0}^{x_1} f(x)\,dx + \int_{x_1}^{x_2} f(x)\,dx + \cdots + \int_{x_{n-1}}^{x_n} f(x)\,dx$$

By the mean-value theorem for integrals, there exists a number $x_i^* \in (x_{i-1}, x_i)$ such that

$$\int_{x_{i-1}}^{x_i} f(x)\,dx = f(x_i^*)(x_i - x_{i-1}) = f(x_i^*)\,\Delta x_i,\qquad i = 1,\, 2,\, \ldots,\, n$$

Thus

$$\int_a^b f(x)\,dx = f(x_1^*)\,\Delta x_1 + f(x_2^*)\,\Delta x_2 + \cdots + f(x_n^*)\,\Delta x_n$$

38. (a) A.V.$= \dfrac{1}{3}\displaystyle\int_{-1}^{2}(x^3 - x + 1)\,dx = \dfrac{1}{3}\left[\dfrac{x^4}{4} - \dfrac{x^2}{2} + x\right]_{-1}^{2} = \dfrac{7}{4}$

  (b) $x^3 - x + 1 = \dfrac{7}{4}$;   at $x \cong 1.263$

39. (a) A.V. $= \dfrac{1}{\pi}\displaystyle\int_0^{\pi}\sin x\,dx = \dfrac{1}{\pi}[-\cos x]_0^{\pi} = \dfrac{2}{\pi}$

  (b) $\sin x = \dfrac{2}{\pi} \implies x = 0.690$

  (c)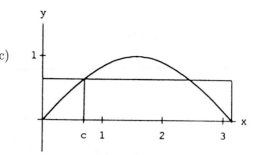

40. (a) A.V. $= \dfrac{12}{5\pi}\displaystyle\int_{-\pi/4}^{\pi/6} 2\cos 2x\,dx = \dfrac{12}{5\pi}[\sin 2x]_{-\pi/4}^{\pi/6} = \dfrac{12}{5\pi}\left[\dfrac{\sqrt{3}}{2} + 1\right] = \dfrac{6}{5\pi}(\sqrt{3} + 2) \cong 1.426$

  (b) $2\cos 2x = 1.426$ at $x \cong \pm 0.389$

41. (a) $f(x) = 0$ at $a \cong -3.4743$ and $b \cong 3.4743$.

(b)

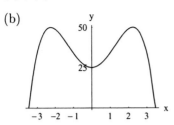

(c) $\dfrac{1}{b-a} \displaystyle\int_a^b (-x^4 + 10x^2 + 25)\, dx \cong 36.0984;$

solving $f(c) = 36.0984$ for c we get $c \cong \pm 2.9545, \ \pm 1.1274$

42. (a) $f(x) = 0$ at $a \cong -1.8364$ and $b \cong 1.8364$.

(b)

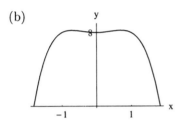

(c) $\dfrac{1}{b-a} \displaystyle\int_a^b (-x^4 + x^2 + 8)\, dx \cong 6.8496;$

solving $f(c) = 6.8496$ for c we get $c \cong \pm 1.2975$

REVIEW EXERCISES

1. $\displaystyle\int \frac{x^3 - 2x + 1}{\sqrt{x}}\, dx = \int (x^{5/2} - 2\sqrt{x} + x^{-1/2})\, dx = \frac{2}{7}x^{\frac{7}{2}} - \frac{4}{3}x^{\frac{3}{2}} + 2x^{\frac{1}{2}} + C$

2. $\displaystyle\int (x^{3/5} - 3x^{5/3})\, dx = \frac{5}{8}x^{\frac{8}{5}} - \frac{9}{8}x^{\frac{8}{3}} + C$

3. $\left\{ \begin{array}{l} u = 1 + t^3 \\ du = 3t^2\, dt \end{array} \right\};\qquad \displaystyle\int t^2(1+t^3)^{10}\, dt = \frac{1}{3}\int u^{10}\, du = \frac{1}{33}u^{11} + C = \frac{1}{33}(1+t^3)^{11} + C$

4. $\displaystyle\int (1 + 2\sqrt{x})^2\, dx = \int (1 + 4\sqrt{x} + 4x)\, dx = x + \frac{8}{3}x^{3/2} + 2x^2 + C$

5. $\left\{ \begin{array}{l} u = t^{2/3} - 1 \\ du = \dfrac{3}{3t^{1/3}}\, dt \end{array} \right\};\ \displaystyle\int \frac{(t^{2/3} - 1)^2}{t^{1/3}}\, dt = \frac{3}{2}\int u^2\, du = \frac{1}{2}u^3 + C = \frac{1}{2}(t^{2/3} - 1)^3 + C$

6. $\left\{ \begin{array}{l} u = x^2 - 2 \\ du = 2x\, dx \end{array} \right\};\qquad \displaystyle\int x\sqrt{x^2 - 2}\, dx = \frac{1}{2}\int u^{1/2}\, du = \frac{1}{3}u^{3/2} + C = \frac{1}{3}(x^2 - 2)^{3/2} + C$

7. Set $u = 2 - x$. Then $du = -dx$ and $x = 2 - u$.

$$\int x\sqrt{2-x}\,dx = -\int (2-u)u^{1/2}\,du = \int \left(u^{3/2} - 2u^{1/2}\right) du$$

$$= \tfrac{2}{5}u^{5/2} - \tfrac{4}{3}u^{3/2} + C = \tfrac{2}{5}(2-x)^{5/2} - \tfrac{4}{3}(2-x)^{3/2} + C$$

8. $\left\{ \begin{array}{c} u = 2 + 2x^3 \\ du = 6x^2\,dx \end{array} \right\}$; $\int x^2(2+2x^3)^4\,dx = \tfrac{1}{6}\int u^4\,du = \tfrac{1}{30}u^5 + C = \tfrac{1}{30}(2+2x^3)^5 + C$

9. $\left\{ \begin{array}{c} u = 1 + \sqrt{x} \\ du = \dfrac{1}{\sqrt{x}}dx \end{array} \right\}$; $\displaystyle\int \frac{(1+\sqrt{x})^5}{\sqrt{x}}\,dx = \int 2u^5\,du = \tfrac{1}{3}u^6 + C = \tfrac{1}{3}(1+\sqrt{x})^6 + C$

10. $\left\{ \begin{array}{c} u = 1/x \\ du = -dx/x^2 \end{array} \right\}$; $\displaystyle\int \frac{\sin(1/x)}{x^2}\,dx = -\int \sin u\,du = \cos(1/x) + C$

11. $\left\{ \begin{array}{c} u = 1 + \sin x \\ du = \cos x\,dx \end{array} \right\}$; $\displaystyle\int \frac{\cos x}{\sqrt{1 + \sin x}}\,dx = \int \frac{1}{\sqrt{u}}\,du = 2\sqrt{u} + C = 2\sqrt{1+\sin x} + C$

12. $\displaystyle\int (\sec\theta - \tan\theta)^2\,d\theta = \int \left(\sec^2\theta - 2\sec\theta\tan\theta + \tan^2\theta\right) d\theta$

$$= \int \left(2\sec^2\theta - 2\sec\theta\tan\theta - 1\right) d\theta = 2\tan\theta - 2\sec\theta - \theta + C$$

13. $\displaystyle\int (\tan 3\theta - \cot 3\theta)^2\,d\theta = \int (\tan^2 3\theta + \cot^2 3\theta - 2)\,d\theta = \int (\sec^2 3\theta + \csc^2 3\theta - 4)\,d\theta$

$$= \frac{1}{3}\tan 3\theta - \frac{1}{3}\cot 3\theta - 4\theta + C$$

14. $\left\{ \begin{array}{c} u = x^2 \\ du = 2x\,dx \end{array} \right\}$; $\displaystyle\int x\sin^3 x^2\cos x^2\,dx = \tfrac{1}{2}\int \sin^3 u\cos u\,du = \tfrac{1}{8}\sin^4 u + C = \tfrac{1}{8}\sin^4 (x^2)^2 + C$

15. $\displaystyle\int \frac{1}{1+\cos 2x}\,dx = \int \frac{1}{1+2\cos^2 x - 1}\,dx = \tfrac{1}{2}\int \frac{1}{\cos^2 x}\,dx = \tfrac{1}{2}\int \sec^2 x\,dx = \frac{1}{2}\tan x + C$

16. $\displaystyle\int \frac{1}{1-\sin 2x}\,dx = \int \frac{1}{1-\sin 2x}\frac{1+\sin 2x}{1+\sin 2x}\,dx = \int (\sec^2 2x + \sec 2x\tan 2x)\,dx$

$$= \tfrac{1}{2}\tan 2x + \tfrac{1}{2}\sec 2x + C$$

17. $\left\{ \begin{array}{c} u = \sec\pi x \\ du = \pi\sec\pi x\tan\pi x\,dx \end{array} \right\}$;

$$\int \sec^3\pi x\tan\pi x\,dx = \int \sec^2\pi x(\sec\pi x\tan\pi x)\,dx = \int \frac{1}{\pi}u^2\,du = \frac{1}{3\pi}u^3 + C = \frac{1}{3\pi}(\sec\pi x)^3 + C$$

18. $\left\{ \begin{array}{c} u = 1 + bx^2 \\ du = 2bx\,dx \end{array} \right\}$; $\displaystyle\int ax\sqrt{1+bx^2}\,dx = \frac{a}{2b}\int u^{1/2}\,du = \frac{a}{3b}u^{3/2} + C = \frac{a}{3b}(1+bx^2)^{3/2} + C$

19. Set $u = 1 + bx$. Then $du = b\,dx$ and $x = \frac{1}{b}(u-1)$.

$$\int ax\sqrt{1+bx}\,dx = \frac{a}{b^2}\int u^{1/2}(u-1)\,du = \frac{a}{b^2}\int\left(u^{3/2}-u^{1/2}\right)du$$

$$= \frac{a}{b^2}\left[\frac{2}{5}u^{5/2}-\frac{2}{3}u^{3/2}\right]+C = \frac{2a}{5b^2}(1+bx)^{5/2}-\frac{2a}{3b^2}(1+bx)^{3/2}+C$$

20. Set $u = 1 + bx$, $du = b\,dx$ and $x^2 = \frac{1}{b^2}\left(u^2-2u+1\right)$

$$\int ax^2\sqrt{1+bx}\,dx = \frac{a}{b^3}\int\left(u^2-2u+1\right)u^{1/2}\,du$$

$$= \frac{a}{b^3}\left(\frac{2}{7}u^{7/2}-\frac{4}{5}u^{5/2}+\frac{2}{3}u^{3/2}\right)+C$$

$$= \frac{2a}{7b^3}(1+bx)^{7/2}-\frac{4a}{5b^3}(1+bx)^{5/2}+\frac{2a}{3b^3}(1+bx)^{3/2}+C$$

21. $\left\{\begin{array}{l} u = 1 + g^2(x) \\ du = 2g(x)g'(x)dx \end{array}\right\}$; $\quad \int \frac{g(x)g'(x)}{\sqrt{1+g^2(x)}}\,dx = \frac{1}{2}\int\frac{1}{\sqrt{u}}\,du = \sqrt{u}+C = \sqrt{1+g^2(x)}+C$

22. Set $u = g(x)$. Then $du = g'(x)\,dx$; $\quad \int\frac{g'(x)}{g^3(x)}\,dx = \int u^{-3}\,du = -\frac{1}{2}g^{-2}(x)+C = \frac{1}{2g^2(x)}+C$

23. $\int_{-1}^{2}(x^2-2x+3)\,dx = \int_{-1}^{2}x^2\,dx - \int_{-1}^{2}2x\,dx + \int_{-1}^{2}3\,dx = \frac{1}{3}\left[x^3\right]_{-1}^{2}-\left[x^2\right]_{-1}^{2}+3\left[x\right]_{-1}^{2} = 9$

24. Set $u = 1 + x^2$. Then $du = 2x\,dx$, $u(0) = 1$, $u(1) = 2$.

$$\int_0^1\frac{x}{(x^2+1)^3}\,dx = \frac{1}{2}\int_1^2 u^{-3}\,du = -\frac{1}{4}\left[u^{-2}\right]_1^2 = \frac{3}{16}.$$

25. Set $u = \sin 2x$. Then $du = 2\cos 2x\,dx$, $\quad u(0) = 0$, $u(\pi/4) = 1$.

$$\int_0^{\pi/4}\sin^3 2x\cos 2x\,dx = \frac{1}{2}\int_0^1 u^3\,du = \frac{1}{8}\left[u^4\right]_0^1 = \frac{1}{8}$$

26. $\int_0^{\pi/8}\left(\tan^2 2x + \sec^2 2x\right)dx = \int_0^{\pi/8}\left(2\sec^2 2x - 1\right)dx = \left[\tan 2x - x\right]_0^{\pi/8} = 1 - \frac{\pi}{8}$

27. Set $u = x^3 + 3x - 6$. Then $du = 3\left(x^2+1\right)dx$, $u(0) = -6$, $u(2) = 8$.

$$\int_0^2(x^2+1)(x^3+3x-6)^{1/3}\,dx = \frac{1}{3}\int_{-6}^{8}u^{1/3}\,du = \frac{1}{4}\left[u^{4/3}\right]_{-6}^{8} = 4 - \frac{1}{4}(6)^{4/3}$$

28. Set $u = 1 + x^{1/3}$. Then $du = \frac{1}{3}x^{-2/3}\,dx$, $u(1) = 2$, $u(8) = 3$.

$$\int_1^8\frac{\left(1+x^{1/3}\right)^2}{x^{2/3}}\,dx = 3\int_2^3 u^2\,du = \left[u^3\right]_2^3 = 19$$

29. (a) $\displaystyle\int_2^3 f(x)\,dx = \int_0^3 f(x)\,dx - \int_0^2 f(x)\,dx = -2$

(b) $\displaystyle\int_3^5 f(x)\,dx = \int_0^3 f(x)\,dx + \int_3^5 f(x)\,dx - \int_0^2 f(x)\,dx = 6$;

(c) Mean-value theorem: there exists a $c \in (3,5)$ such that $f(c) = \dfrac{1}{2}\displaystyle\int_3^5 f(x)\,dx = 4$.

(d) If $f(x) \geq 0$ on $[2,3]$, then $\displaystyle\int_2^3 f \geq 0$. But $\displaystyle\int_2^3 f = -2 < 0$.

30. $\displaystyle\int_{-2}^8 g(x)\,dx = \int_{-2}^8 [f(x)+3]\,dx = \int_{-2}^8 f(x)\,dx + \int_{-2}^8 3\,dx = 4 + \Big[3x\Big]_{-2}^8 = 4 + 30 = 34$

31. $A = \displaystyle\int_{-2}^1 \left[(4-x^2)-(x+2)\right]\,dx = \int_{-2}^1 (-x^2 - x + 2)\,dx = \dfrac{9}{2}$

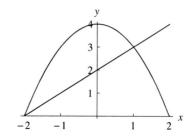

32. $A = \displaystyle\int_{-2}^3 \left[(4-x^2)-(-2-x)\right]\,dx = \int_{-2}^3 (6 + x - x^2)\,dx = \dfrac{125}{6}$

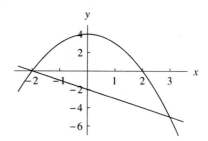

33. $A = \displaystyle\int_0^3 (3y - y^2)\,dy = \dfrac{9}{2}$

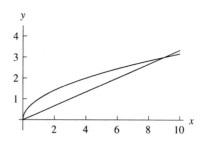

34. $A = \displaystyle\int_0^4 \sqrt{x}\,dx + \int_4^6 (6-x)\,dx$;

$A = \displaystyle\int_0^2 (6 - y - y^2)\,dy = \dfrac{22}{3}$

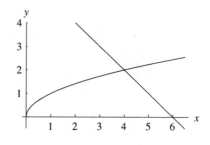

35. $A = \int_0^1 [x^3] \, dx + \int_1^2 (2 - x) \, dx = \dfrac{3}{4}$

$A = \int_0^1 (2 - y - y^{1/3}] \, dy = \dfrac{3}{4}$

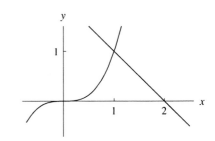

36. $A = \int_{-1}^2 \left[\frac{1}{4}(x^2 - x^4) - (-x - 1)\right] dx$

$= \int_{-1}^2 \left[1 + x + \frac{1}{4}x^2 - \frac{1}{4}x^4\right] dx = \dfrac{18}{5}$

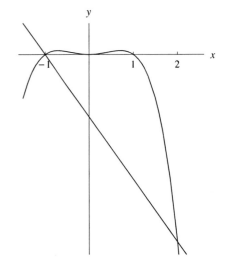

37. $\dfrac{d}{dx}\left(\int_0^x \dfrac{dt}{1 + t^2}\right) = \dfrac{1}{1 + x^2}$

38. $\dfrac{d}{dx}\int_0^{x^2} \dfrac{dt}{1 + t^2} = \dfrac{1}{1 + (x^2)^2}\,(2x) = \dfrac{2x}{1 + x^4}$

39. Fix a number a. Then

$$\dfrac{d}{dx}\left(\int_x^{x^2} \dfrac{dt}{1 + t^2}\right) = \dfrac{d}{dx}\left[\int_a^{x^2} \dfrac{dt}{1 + t^2} - \int_a^x \dfrac{dt}{1 + t^2}\right] = \dfrac{2x}{1 + x^4} - \dfrac{1}{1 + x^2}$$

40. $\dfrac{d}{dx}\left(\int_0^{\sin x} \dfrac{dt}{1 - t^2}\right) = \dfrac{1}{1 - \sin^2 x}\,\cos x = \dfrac{1}{\cos x} = \sec x$

41. $\dfrac{d}{dx}\left(\int_0^{\cos x} \dfrac{dt}{1 - t^2}\right) = \dfrac{1}{1 - \cos^2 x}\,(-\sin x) = -\dfrac{1}{\sin x} = -\csc x$

42. $f'(x) = x\sqrt{1 + x^2} \implies f(x) = \int x\sqrt{1 + x^2}\, dx = \frac{1}{3}(x^2 + 1)^{3/2} + C$. Since γ passes through $(0, 1)$, $1 = \frac{1}{3} + C$. Therefore, $C = \frac{2}{3}$ and $y = \frac{1}{3}(x^2 + 1)^{3/2} + \frac{2}{3}$.

43. (a) Yes, at $x = 0$.

(b) $F'(x) = \dfrac{1}{x^2 + 2x + 2} > 0 \Longrightarrow F$ increases on $(-\infty, \infty)$.

(c) $F''(x) = \dfrac{-2(x+1)}{x^2 + 2x + 2}$. The graph of F is concave up on $(-\infty, -1)$ and concave down on $(-1, \infty)$

(d)

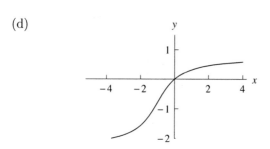

44. $\displaystyle\int_0^x t f(t)\, dt = x \sin x + \cos x - 1 \;\Rightarrow\; x f(x) = \dfrac{d}{dx} \int_0^x t f(t)\, dt = \dfrac{d}{dx}(x \sin x + \cos x - 1) = x \cos x$

(a) $\pi f(\pi) = -\pi;\quad f(\pi) = -1$

(b) For $x \neq 0$, $f(x) = \cos x$ and $f'(x) = \sin x$.

45. $f_{avg} = \dfrac{1}{4} \displaystyle\int_0^4 \dfrac{x}{\sqrt{x^2 + 9}}\, dx = \dfrac{1}{4}\left[\sqrt{x^2 + 9}\,\right]_0^4 = \dfrac{1}{2}$

46. $\displaystyle\int_0^\pi f(x)\, dx = \int_0^\pi (x + 2 \sin x)\, dx = \left[\dfrac{1}{2}x^2 - 2 \cos x\right]_0^\pi = 4 + \dfrac{\pi^2}{2};$

$f_{avg} = \dfrac{1}{\pi - 0} \displaystyle\int_0^\pi f(x)\, dx = \dfrac{1}{\pi}\left[4 + \dfrac{\pi^2}{2}\right] = \dfrac{4}{\pi} + \dfrac{\pi}{2}$

47. $f_{avg} = \dfrac{1}{2\pi} \displaystyle\int_a^{a+2\pi} \cos x\, dx = \Big[\sin x\Big]_a^{a+2\pi} = \sin(a + 2\pi) - \sin a = 0$

48. $\displaystyle\int_\alpha^\beta f(x)\, dx$

49. $\displaystyle\int_\alpha^\beta |f(x)|\, dx$

50. $\dfrac{1}{2}\left[\displaystyle\int_\alpha^\beta |f(x)|\, dx + \int_\alpha^\beta f(x)\, dx\right]$

51. $\dfrac{1}{2}\left[\displaystyle\int_\alpha^\beta |f(x)|\, dx - \int_\alpha^\beta f(x)\, dx\right]$

52. $\lambda(x) = k(2a - x)$, $k > 0$. for some positive number k.

$$M = \int_0^a k(2a - x)dx = k\left[2ax - \tfrac{1}{2}x^2\right]_0^a = \tfrac{3}{2}ka^2$$

$$x_M M = \int_0^a kx(2a - x)dx = k\left[ax^2 - \tfrac{1}{3}x^3\right]_0^a = \tfrac{2}{3}ka^3; \qquad x_M = \frac{\tfrac{2}{3}ka^3}{\tfrac{3}{2}ka^2} = \frac{4a}{9}.$$

53. $\lambda(x) = k\left(\tfrac{1}{4}a - x\right)$ for $0 \le x \le \tfrac{1}{4}a$ and $\lambda(x) = k(x - \tfrac{1}{4}a)$ for $\tfrac{1}{4}a \le x \le a$; $k > 0$.

$$M = \int_0^{a/4} k\left(\tfrac{1}{4}a - x\right) dx + \int_{a/4}^a k\left(x - \tfrac{1}{4}a\right) dx = \tfrac{5}{16}ka^2$$

$$x_M M = \int_0^{a/4} kx\left(\tfrac{1}{4}a - x\right) dx + \int_{a/4}^a kx\left(x - \tfrac{1}{4}a\right) dx = \frac{41}{192}ka^3; \qquad x_M = \frac{\tfrac{41}{192}ka^3}{\tfrac{5}{16}ka^2} = \frac{41a}{60}$$

SECTION 6.1

1.

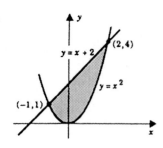

(a) $\int_{-1}^{2} [(x+2) - x^2]\, dx$

(b) $\int_{0}^{1} [(\sqrt{y}) - (-\sqrt{y})]\, dy + \int_{1}^{4} [(\sqrt{y}) - (y-2)]\, dy$

2.

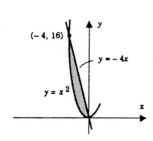

(a) $\int_{-4}^{0} [(-4x) - x^2]\, dx$

(b) $\int_{0}^{16} \left[\left(-\frac{1}{4}y\right) - (-\sqrt{y})\right]\, dy$

3.

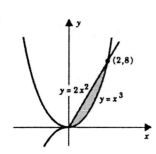

(a) $\int_{0}^{2} [(2x^2) - (x^3)]\, dx$

(b) $\int_{0}^{8} \left[(y^{1/3}) - \left(\frac{1}{2}y\right)^{1/2}\right]\, dy$

4.

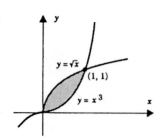

(a) $\int_{0}^{1} [\sqrt{x} - x^3]\, dx$

(b) $\int_{0}^{1} [y^{1/3} - y^2]\, dy$

5.

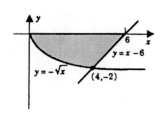

(a) $\int_{0}^{4} [(0) - (-\sqrt{x})]\, dx + \int_{4}^{6} [(0) - (x-6)]\, dx$

(b) $\int_{-2}^{0} [(y+6) - (y^2)]\, dy$

6.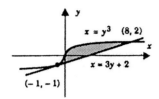

(a) $\displaystyle\int_{-1}^{8}\left[x^{1/3}-\left(\frac{x-2}{3}\right)\right]dx$

(b) $\displaystyle\int_{-1}^{2}\left[(3y+2)-y^{3}\right]dy$

7.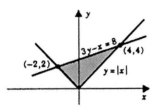

(a) $\displaystyle\int_{-2}^{0}\left[\left(\frac{8+x}{3}\right)-(-x)\right]dx+\int_{0}^{4}\left[\left(\frac{8+x}{3}\right)-(x)\right]dx$

(b) $\displaystyle\int_{0}^{2}\left[(y)-(-y)\right]dy+\int_{2}^{4}\left[(y)-(3y-8)\right]dy$

8.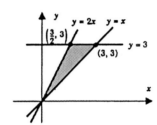

(a) $\displaystyle\int_{0}^{3/2}\left[2x-x\right]dx+\int_{3/2}^{3}\left[3-x\right]dx$

(b) $\displaystyle\int_{0}^{3}\left[y-\frac{1}{2}y\right]dy$

9.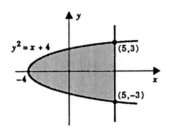

(a) $\displaystyle\int_{-4}^{5}\left[\left(\sqrt{4+x}\right)-\left(-\sqrt{4+x}\right)\right]dx$

(b) $\displaystyle\int_{-3}^{3}\left[(5)-\left(y^{2}-4\right)\right]dy$

10.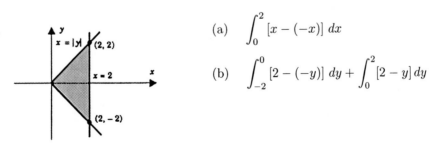

(a) $\displaystyle\int_{0}^{2}\left[x-(-x)\right]dx$

(b) $\displaystyle\int_{-2}^{0}\left[2-(-y)\right]dy+\int_{0}^{2}\left[2-y\right]dy$

11.

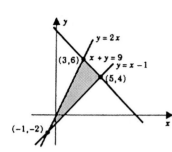

(a) $\displaystyle\int_{-1}^{3} [(2x) - (x-1)]\ dx + \int_{3}^{5} [(9-x) - (x-1)]\ dx$

(b) $\displaystyle\int_{-2}^{4} \left[(y+1) - \left(\tfrac{1}{2}y\right)\right] dy + \int_{4}^{6} \left[(9-y) - \left(\tfrac{1}{2}y\right)\right] dy$

12.

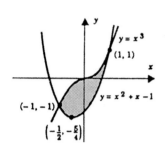

(a) $\displaystyle\int_{-1}^{1} \left[x^3 - (x^2 + x - 1)\right] dx$

(b) $\displaystyle\int_{-5/4}^{-1} \left[\left(-\tfrac{1}{2} + \tfrac{1}{2}\sqrt{4y+5}\right) - \left(-\tfrac{1}{2} - \tfrac{1}{2}\sqrt{4y+5}\right)\right] dy$

$\displaystyle + \int_{-1}^{1} \left[\left(-\tfrac{1}{2} + \tfrac{1}{2}\sqrt{4y+5}\right) - y^{1/3}\right] dy$

13.

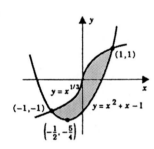

(a) $\displaystyle\int_{-1}^{1} \left[\left(x^{1/3}\right) - (x^2 + x - 1)\right] dx$

(b) $\displaystyle\int_{-5/4}^{-1} \left[\left(-\tfrac{1}{2} + \tfrac{1}{2}\sqrt{4y+5}\right) - \left(-\tfrac{1}{2} - \tfrac{1}{2}\sqrt{4y+5}\right)\right] dy$

$\displaystyle + \int_{-1}^{1} \left[\left(-\tfrac{1}{2} + \tfrac{1}{2}\sqrt{4y+5}\right) - (y^3)\right] dy$

14.

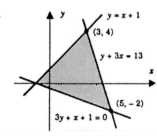

(a) $\displaystyle\int_{-1}^{3} \left[(x+1) - \left(\tfrac{-x-1}{3}\right)\right] dx + \int_{3}^{5} \left[(13-3x) - \left(\tfrac{-x-1}{3}\right)\right] dx$

(b) $\displaystyle\int_{-2}^{0} \left[\left(\tfrac{13-y}{3}\right) - (-3y-1)\right] dy + \int_{0}^{4} \left[\left(\tfrac{13-y}{3}\right) - (y-1)\right] dy$

15.

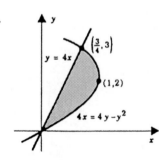

$\displaystyle A = \int_{0}^{3} \left[\left(\tfrac{4y - y^2}{4}\right) - \left(\tfrac{y}{4}\right)\right] dy$

$\displaystyle = \int_{0}^{3} \left(\tfrac{3}{4}y - \tfrac{1}{4}y^2\right) dy$

$\displaystyle = \left[\tfrac{3}{8}y^2 - \tfrac{1}{12}y^3\right]_{0}^{3} = \tfrac{9}{8}$

16.

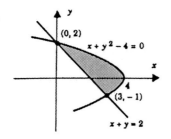

$$A = \int_{-1}^{2} \left[(4 - y^2) - (2 - y) \right] \, dy$$

$$= \int_{-1}^{2} (2 + y - y^2) \, dy$$

$$= \left[2y + \frac{y^2}{2} - \frac{y^3}{3} \right]_{-1}^{2} = \frac{9}{2}$$

17.

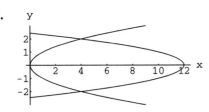

$$A = 2 \int_{0}^{2} \left[(12 - 2y^2) - (y^2) \right] \, dy$$

$$= 2 \int_{0}^{2} (12 - 3y^2) \, dy$$

$$= 2 \left[12y - y^3 \right]_{0}^{2} = 2 \,(16) = 32$$

18.

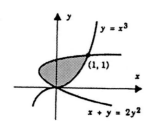

$$A = \int_{0}^{1} \left[y^{1/3} - (2y^2 - y) \right] \, dy$$

$$= \int_{0}^{1} (y^{1/3} + y - 2y^2) \, dy$$

$$= \left[\frac{3}{4} y^{4/3} + \frac{y^2}{2} - \frac{2}{3} y^3 \right]_{0}^{1} = \frac{7}{12}$$

19.

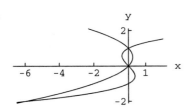

$$A = \int_{-2}^{0} \left[(y^3 - y) - (y - y^2) \right] \, dy + \int_{0}^{1} \left[(y - y^2) - (y^3 - y) \right] \, dy$$

$$= \int_{-2}^{0} (y^3 + y^2 - 2y) \, dy + \int_{0}^{1} (2y - y^2 - y^3) \, dy$$

$$= \left[\frac{1}{4} y^4 + \frac{1}{3} y^3 - y^2 \right]_{-2}^{0} + \left[y^2 - \frac{1}{3} y^3 - \frac{1}{4} y^4 \right]_{0}^{1} = \frac{8}{3} + \frac{5}{12} = \frac{37}{12}$$

20.

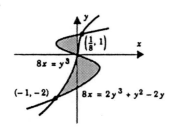

$$A = \int_{-2}^{0} \left[\frac{1}{8} (2y^3 + y^2 - 2y) - \frac{1}{8} y^3 \right] \, dy$$

$$+ \int_{0}^{1} \left[\frac{1}{8} y^3 - \frac{1}{8} (2y^3 + y^2 - 2y) \right] \, dy$$

$$= \frac{1}{8} \left[\frac{y^4}{4} + \frac{y^3}{3} - y^2 \right]_{-2}^{0} + \frac{1}{8} \left[-\frac{y^4}{4} - \frac{y^3}{3} + y^2 \right]_{0}^{1} = \frac{37}{96}$$

21.

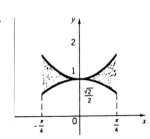

$$A = \int_{-\pi/4}^{\pi/4} \left[\sec^2 x - \cos x \right] dx$$

$$= 2 \int_0^{\pi/4} \left[\sec^2 x - \cos x \right] dx$$

$$= 2 \left[\tan x + \sin x \right]_0^{\pi/4} = 2 \left[1 + \sqrt{2}/2 \right] = 2 + \sqrt{2}$$

22.

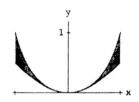

$$A = \int_{-\pi/4}^{\pi/4} (\tan^2 x - \sin^2 x)\, dx = \int_{-\pi/4}^{\pi/4} \left(\sec^2 x - \frac{3}{2} + \frac{\cos 2x}{2} \right) dx$$

$$= \left[\tan x - \frac{3}{2} x + \frac{\sin 2x}{4} \right]_{-\pi/4}^{\pi/4} = \frac{5}{2} - \frac{3\pi}{4}$$

23.

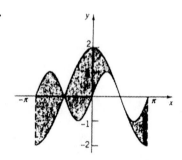

$$A = \int_{-\pi}^{-\pi/2} \left[\sin 2x - 2\cos x \right] dx + \int_{-\pi/2}^{\pi/2} \left[2\cos x - \sin 2x \right] dx$$

$$+ \int_{\pi/2}^{\pi} \left[\sin 2x - 2\cos x \right] dx$$

$$= \left[-\tfrac{1}{2} \cos 2x - 2\sin x \right]_{-\pi}^{-\pi/2} + \left[2\sin x + \tfrac{1}{2} \cos 2x \right]_{-\pi/2}^{\pi/2}$$

$$+ \left[-\tfrac{1}{2} \cos 2x - 2\sin x \right]_{\pi/2}^{\pi} = 8$$

24.

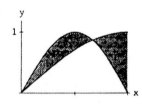

$$A = \int_0^{\pi/3} (\sin 2x - \sin x)\, dx + \int_{\pi/3}^{\pi/2} (\sin x - \sin 2x)\, dx$$

$$= \left[-\frac{\cos 2x}{2} + \cos x \right]_0^{\pi/3} + \left[-\cos x + \frac{\cos 2x}{2} \right]_{\pi/3}^{\pi/2} = \frac{1}{2}$$

25.

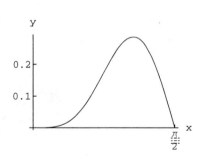

$$A = \int_0^{\pi/2} (\sin^4 x \cos x)\, dx$$

$$= \int_0^1 u^4\, du, \quad (u = \sin x)$$

$$= \left[\frac{u^5}{5} \right]_0^1 = \frac{1}{5}$$

26.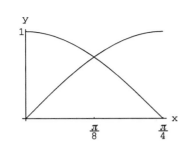

$$A = \int_0^{\pi/8} (\cos 2x - \sin 2x)\, dx + \int_{\pi/8}^{\pi/4} (\sin 2x - \cos 2x)\, dx$$

$$= \left[\frac{\sin 2x}{2} + \frac{\cos 2x}{2} \right]_0^{\pi/8} + \left[-\frac{\cos 2x}{2} - \frac{\sin 2x}{2} \right]_{\pi/8}^{\pi/4} = 2\sqrt{2} - 1$$

27. $\displaystyle A = \int_0^1 \left[3x - \frac{1}{3}x \right] dx + \int_1^3 \left[-x + 4 - \frac{1}{3}x \right] dx = \left[\frac{4}{3}x^2 \right]_0^1 + \left[-\frac{2}{3}x^2 + 4x \right]_1^3 = 4$

28. $\displaystyle A = \int_0^2 \left[(x+1) - \left(1 - \frac{x}{2}\right) \right] dx + \int_2^3 \left[(x+1) - (4x-8) \right] dx = \left[\frac{x^2}{4} \right]_0^2 + \left[-\frac{3x^2}{2} + 9x \right]_2^3 = \frac{5}{2}$

29. $\displaystyle A = \int_{-2}^1 [x - (-2)]\, dx + \int_1^5 [1 - (-2)]\, dx + \int_5^7 \left[-\frac{3}{2}x + \frac{17}{2} - (-2) \right] dx$

$$= \left[\tfrac{1}{2}x^2 + 2x \right]_{-2}^1 + \left[3x \right]_1^5 + \left[-\tfrac{3}{4}x^2 + \tfrac{21}{2}x \right]_5^7 = \frac{39}{2}$$

30.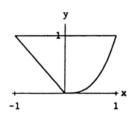

$$A = \int_0^1 \left[y^{1/3} - (-y) \right] dy$$

$$= \int_0^1 (y^{1/3} + y)\, dy$$

$$= \left[\frac{3}{4}y^{4/3} + \frac{y^2}{2} \right]_0^1 = \frac{5}{4}$$

31.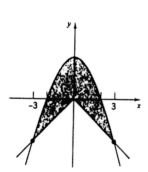

$$A = \int_{-3}^0 \left[6 - x^2 - x \right] dx + \int_0^3 \left[6 - x^2 - (-x) \right] dx$$

$$= \left[6x - \frac{1}{3}x^3 - \frac{1}{2}x^2 \right]_{-3}^0 + \left[6x - \frac{1}{3}x^3 + \frac{1}{2}x^2 \right]_0^3 = 27$$

32.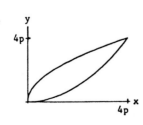

$$A = \int_0^{4p} \left(\sqrt{4px} - \frac{x^2}{4p} \right) dx$$

$$= \left[\sqrt{4p}\, \frac{2}{3}x^{3/2} - \frac{x^3}{12p} \right]_0^{4p} = \frac{16}{3}p^2$$

33.

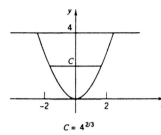

$C = 4^{2/3}$

$$\int_0^{\sqrt{c}} \left[c - x^2\right] dx = \frac{1}{2} \int_0^2 \left[4 - x^2\right] dx$$

$$\left[cx - \frac{1}{3}x^3\right]_0^{\sqrt{c}} = \frac{1}{2}\left[4x - \frac{1}{3}x^3\right]_0^2$$

$$\frac{2}{3}c^{3/2} = \frac{8}{3} \quad \text{and} \quad c = 4^{2/3}$$

34.   We want $\displaystyle\int_0^c \cos x\, dx = \frac{1}{2}\int_0^{\pi/2} \cos x\, dx \implies \left[\sin x\right]_0^c = \frac{1}{2}\left[\sin x\right]_0^{\pi/2} = \frac{1}{2}$

$$\implies \sin c = \frac{1}{2} \implies c = \frac{\pi}{6}$$

35. $\displaystyle A = \int_0^1 \sqrt{3}\, dx + \int_1^2 \sqrt{4 - x^2}\, dx;$

$$A = \int_0^{\sqrt{3}} \left(\sqrt{4 - y^2} - \frac{y}{\sqrt{3}}\right) dy$$

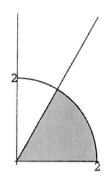

36. $\displaystyle A = \int_0^1 \left(\sqrt{4 - x^2} - \sqrt{3}\, x\right) dx$

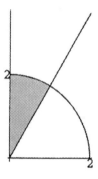

37. $\displaystyle A = \int_0^2 \left[\sqrt{4 - x^2} - \left(2 - \sqrt{4x - x^2}\right)\right] dx$

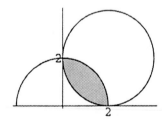

38. $\displaystyle A = \int_0^{\sqrt{3}} \left(\sqrt{4 - x^2} - \frac{x^2}{3}\right) dx$

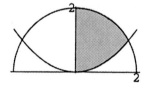

39.

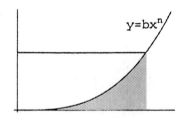

The area under the curve is $\displaystyle A_c = \int_0^a bx^n\, dx = \frac{ba^{n+1}}{n+1}.$

For the rectangle, $A_r = ba^{n+1}.$ Thus the ratio is $\dfrac{1}{n+1}.$

40.

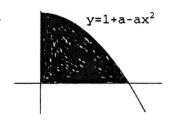

$y=1+a-ax^2$

(a) $A = \displaystyle\int_0^{\sqrt{\frac{1+a}{a}}} (1 + a - ax^2)\, dx$

$= \left[x + ax - \dfrac{ax^3}{3} \right]_0^{\sqrt{\frac{1+a}{a}}} = \dfrac{2(1+a)^{3/2}}{2a^{1/2}}.$

(b) $A' = \dfrac{2}{3} \left[\dfrac{3a(1+a)^{1/2} - (1+a)^{3/2}}{2a^{3/2}} \right] = 0$

$\implies \quad a = \dfrac{1}{2}.$

41.

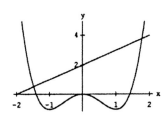

$A \cong \displaystyle\int_{-1.49}^{1.79} \left[x + 2 - (x^4 - 2x^2) \right]\, dx$

$= \left[\tfrac{1}{2}\, x^2 - 2x - \tfrac{1}{5}\, x^5 + \tfrac{2}{3}\, x^3 \right]_{-1.49}^{1.78} \cong 7.93$

42.

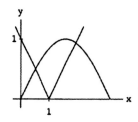

$A \cong 0.67$

43.

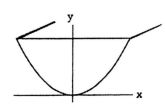

$V = 8 \cdot 12 \displaystyle\int_{-3}^{3} \left[4 - \dfrac{4}{9}\, x^2 \right]\, dx$

$= 96 \cdot 2 \displaystyle\int_0^3 \left[4 - \dfrac{4}{9}\, x^2 \right]\, dx$

$= 192 \left[4x - \tfrac{4}{27}\, x^3 \right]_0^3$

$= 1536 \text{ cu. in.} \cong 0.89 \text{ cu. ft.}$

44.

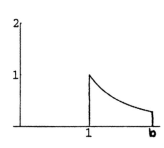

(a) $A = \displaystyle\int_1^b \dfrac{1}{x^2}\, dx = \left[-\dfrac{1}{x} \right]_1^b = 1 - \dfrac{1}{b}.$

(b) $1 - \dfrac{1}{b} \to 1 \text{ as } b \to \infty.$

45.

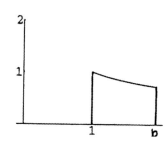

(a) $A = \int_1^b \frac{1}{\sqrt{x}}\, dx = \left[2\sqrt{x}\right]_1^b = 2\sqrt{b} - 2.$

(b) $2\sqrt{b} - 2 \to \infty$ as $b \to \infty.$

46.

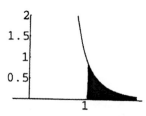

(a) $r > 1:$ $A = \int_1^b \frac{1}{x^r}\, dx = \left[\frac{x^{1-r}}{1-r}\right]_1^b = \frac{1}{r-1}\left[1 - b^{1-r}\right];$

$A \to \frac{1}{r-1}$ as $b \to \infty$

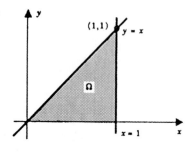

(b) $0 < r < 1:$ $A = \frac{b^{1-r}}{1-r} - \frac{1}{1-r};$ $A \to \infty$ as $b \to \infty$

SECTION 6.2

1.

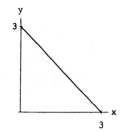

$V = \int_0^1 \pi \left[(x)^2 - (0)^2\right] dx = \pi \left[\frac{x^3}{3}\right]_0^1 = \frac{\pi}{3}$

2.

$V = \int_0^3 \pi \, (3-x)^2 \, dx = \left[-\frac{\pi(3-x)^3}{3}\right]_0^3 = 9\pi$

3.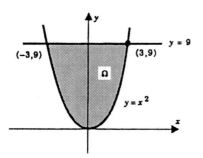

$$V = \int_{-3}^{3} \pi \left[(9)^2 - \left(x^2 \right)^2 \right] dx = 2 \int_{0}^{3} \pi \left(81 - x^4 \right) dx$$

$$= 2\pi \left[81x - \frac{x^5}{5} \right]_{0}^{3} = \frac{1944\pi}{5}$$

4.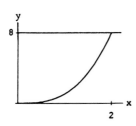

$$V = \int_{0}^{2} \pi \left[8^2 - x^6 \right] dx = \pi \left[64x - \frac{x^7}{7} \right]_{0}^{2} = \frac{768}{7} \pi$$

5.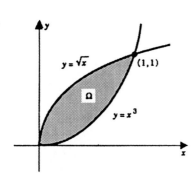

$$V = \int_{0}^{1} \pi \left[\left(\sqrt{x} \right)^2 - \left(x^3 \right)^2 \right] dx$$

$$= \int_{0}^{1} \pi \left(x - x^6 \right) dx$$

$$= \pi \left[\frac{1}{2}x^2 - \frac{1}{7}x^7 \right]_{0}^{1} = \frac{5\pi}{14}$$

6.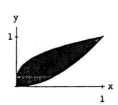

$$V = \int_{0}^{1} \pi \left[x^{2/3} - x^4 \right] dx = \pi \left[\frac{3}{5}x^{5/3} - \frac{x^5}{5} \right]_{0}^{1} = \frac{2\pi}{5}$$

7.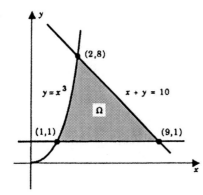

$$V = \int_{1}^{2} \pi \left[\left(x^3 \right)^2 - (1)^2 \right] dx + \int_{2}^{9} \pi \left[(10 - x)^2 - (1)^2 \right] dx$$

$$= \int_{1}^{2} \pi \left(x^6 - 1 \right) dx + \int_{2}^{9} \pi \left(99 - 20x + x^2 \right) dx$$

$$= \pi \left[\frac{1}{7}x^7 - x \right]_{1}^{2} + \pi \left[99x - 10x^2 + \frac{1}{3}x^3 \right]_{2}^{9} = \frac{3790\pi}{21}$$

8.

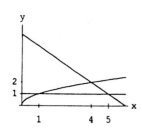

$$V = \int_1^4 \pi[x-1]\,dx + \int_4^5 \pi\left[(6-x)^2 - 1\right]\,dx$$

$$= \pi\left[\frac{x^2}{2} - x\right]_1^4 + \pi\left[-\frac{(6-x)^3}{3} - x\right]_4^5 = \frac{35}{6}\pi$$

9.

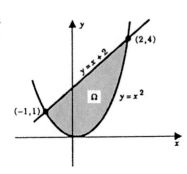

$$V = \int_{-1}^2 \pi\left[(x+2)^2 - \left(x^2\right)^2\right]\,dx$$

$$= \int_{-1}^2 \pi\left(x^2 + 4x + 4 - x^4\right)\,dx$$

$$= \pi\left[\tfrac{1}{3}x^3 + 2x^2 + 4x - \tfrac{1}{5}x^5\right]_{-1}^2 = \frac{72}{5}\pi$$

10.

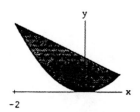

$$V = \int_{-2}^1 \pi\left[(2-x)^2 - x^4\right]\,dx = \pi\left[-\frac{(2-x)^3}{3} - \frac{x^5}{5}\right]_{-2}^1$$

$$= \frac{72}{5}\pi$$

11.

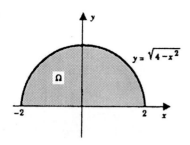

$$V = \int_{-2}^2 \pi\left[\sqrt{4-x^2}\,\right]^2\,dx = 2\int_0^2 \pi\left(4-x^2\right)\,dx$$

$$= 2\pi\left[4x - \frac{x^3}{3}\right]_0^2 = \frac{32}{3}\pi$$

12.

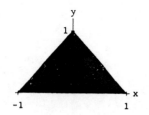

$$V = 2\,(\text{Volume of cone of radius 1, height 1}) = 2\cdot\frac{1}{3}\pi = \frac{2}{3}\pi$$

13.

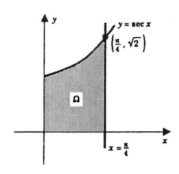

$$V = \int_0^{\pi/4} \pi \sec^2 x \, dx = \pi \left[\tan x\right]_0^{\pi/4} = \pi$$

14.

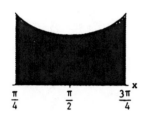

$$V = \int_{\pi/4}^{3\pi/4} \pi \left[\csc^2 x - 0\right] \, dx = \pi \left[-\cot x\right]_{\pi/4}^{3\pi/4} = 2\pi$$

15.

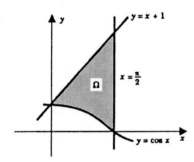

$$V = \int_0^{\pi/2} \pi \left[(x+1)^2 - (\cos x)^2\right] \, dx$$

$$= \int_0^{\pi/2} \pi \left[(x+1)^2 - \left(\frac{1}{2} + \frac{1}{2}\cos 2x\right)\right] \, dx$$

$$= \pi \left[\frac{1}{3}(x+1)^3 - \frac{1}{2}x - \frac{1}{4}\sin 2x\right]_0^{\pi/2}$$

$$= \frac{\pi^2}{24}\left(\pi^2 + 6\pi + 6\right)$$

16.

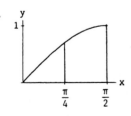

$$V = \int_{\pi/4}^{\pi/2} \pi \sin^2 x \, dx = \pi \left[\frac{x}{2} - \frac{1}{4}\sin 2x\right]_{\pi/4}^{\pi/2} = \frac{1}{8}\pi(\pi + 2)$$

17.

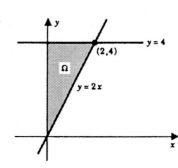

$$V = \int_0^4 \pi \left(\frac{y}{2}\right)^2 dy = \frac{\pi}{12} \left[y^3\right]_0^4 = \frac{16\pi}{3}$$

18.

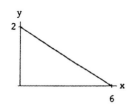

$$V = (\text{Volume of cone of radius 6, height 2})$$

$$= \frac{1}{3} \pi \, 6^2 \cdot 2 = 24\pi$$

19.

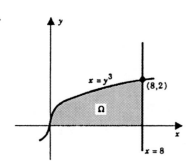

$$V = \int_0^2 \pi \left[(8)^2 - \left(y^3\right)^2\right] dy$$

$$= \int_0^2 \pi \left(64 - y^6\right) dy$$

$$= \pi \left[64y - \tfrac{1}{7}y^7\right]_0^2 = \frac{768}{7}\pi$$

20.

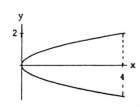

$$V = \int_{-2}^2 \pi \left[4^2 - y^4\right] dy = \pi \left[16y - \frac{y^5}{5}\right]_{-2}^2 = \frac{256}{5}\pi$$

21.

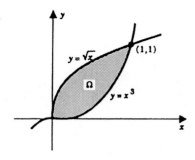

$$V = \int_0^1 \pi \left[\left(y^{1/3}\right)^2 - \left(y^2\right)^2\right] dy$$

$$= \int_0^1 \pi \left[y^{2/3} - y^4\right] dy$$

$$= \pi \left[\tfrac{3}{5}y^{5/3} - \tfrac{1}{5}y^5\right]_0^1 = \tfrac{2}{5}\pi$$

22.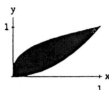

$$V = \int_0^1 \pi \left[(\sqrt{y})^2 - (y^3)^2 \right] dy = \int_0^1 \pi (y - y^6) \, dy$$

$$= \pi \left[\frac{y^2}{2} - \frac{y^7}{7} \right]_0^1 = \frac{5}{14} \pi$$

23.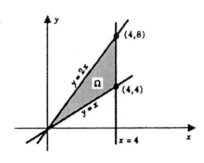

$$V = \int_0^4 \pi \left[y^2 - \left(\frac{y}{2} \right)^2 \right] dy + \int_4^8 \pi \left[4^2 - \left(\frac{y}{2} \right)^2 \right] dy$$

$$= \int_0^4 \pi \left[\frac{3}{4} y^2 \right] dy + \int_4^8 \pi \left[16 - \frac{1}{4} y^2 \right] dy$$

$$= \pi \left[\frac{1}{4} y^3 \right]_0^4 + \pi \left[16y - \frac{1}{12} y^3 \right]_4^8 = \frac{128}{3} \pi$$

24.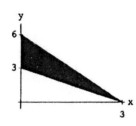

$$V = \int_0^3 \pi \left[\left(3 - \frac{y}{2} \right)^2 - (3 - y)^2 \right] dy + \int_3^6 \pi \left(3 - \frac{y}{2} \right)^2 dy$$

$$= \pi \left[-\frac{2}{3} \left(3 - \frac{y}{2} \right)^3 + \frac{(3 - y)^3}{3} \right]_0^3 + \pi \left[-\frac{2}{3} \left(3 - \frac{y}{2} \right)^3 \right]_3^6$$

$$= 9\pi$$

25.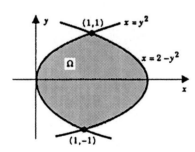

$$V = \int_{-1}^1 \pi \left[(2 - y^2)^2 - (y^2)^2 \right] dy$$

$$= 2 \int_0^1 \pi \left[4 - 4y^2 \right] dy = 2\pi \left[4y - \frac{4}{3} y^3 \right]_0^1 = \frac{16}{3} \pi$$

26.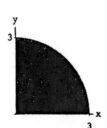

$$V = \int_0^3 \pi (9 - y^2) \, dy = \pi \left[9y - \frac{y^3}{3} \right]_0^3 = 18\pi \quad \text{(half sphere of radius 3.)}$$

27. (a) $V = \int_{-r}^{r} \left(2\sqrt{r^2 - x^2}\right)^2 dx = 8\int_0^r (r^2 - x^2)\, dx = 8\left[r^2 x - \frac{1}{3}x^3\right]_0^r = \frac{16}{3}r^3$

(b) $V = \int_{-r}^{r} \frac{\sqrt{3}}{4}\left(2\sqrt{r^2 - x^2}\right)^2 dx = 2\sqrt{3}\int_0^r (r^2 - x^2)\, dx = \frac{4\sqrt{3}}{3}r^3$

28. For each $x \in [-3, 3]$, the length of the base of the cross-section at x is $2y = \frac{4}{3}\sqrt{9 - x^2}$.

(a) The area of each triangle is $\frac{\sqrt{3}}{4}s^2$.

Thus $V = \int_{-3}^{3} \frac{\sqrt{3}}{4} \cdot \frac{16}{9}(9 - x^2)\, dx = \frac{4\sqrt{3}}{9}\int_{-3}^{3}(9 - x^2)\, dx$

$= \frac{4\sqrt{3}}{9}\left[9x - \frac{x^3}{3}\right]_{-3}^{3} = 16\sqrt{3}.$

(b) The area of each square is s^2

Thus $V = \int_{-3}^{3} \frac{16}{9}(9 - x^2)\, dx = \frac{16}{9}\int_{-3}^{3}(9 - x^2)\, dx$

$= \frac{16}{9}\left[9x - \frac{x^3}{3}\right]_{-3}^{3} = 64.$

29. (a) $V = \int_{-2}^{2}(4 - x^2)^2\, dx = 2\int_0^2 (16 - 8x^2 + x^4)\, dx = 2\left[16x - \frac{8}{3}x^3 + \frac{1}{5}x^5\right]_0^2 = \frac{512}{15}$

(b) $V = \int_{-2}^{2} \frac{\pi}{2}\left(\frac{4 - x^2}{2}\right)^2 dx = \frac{\pi}{4}\int_0^2 (4 - x^2)^2\, dx = \frac{\pi}{4}\left(\frac{256}{15}\right) = \frac{64}{15}\pi$

(c) $V = \int_{-2}^{2} \frac{\sqrt{3}}{4}(4 - x^2)^2\, dx = \frac{\sqrt{3}}{2}\int_0^2 (4 - x^2)^2\, dx = \frac{\sqrt{3}}{2}\left(\frac{256}{15}\right) = \frac{128}{15}\sqrt{3}$

30. (a) $V = \int_0^1 2\sqrt{x}h\, dx + \int_1^3 2 \cdot \frac{1}{\sqrt{2}}\sqrt{3 - x}\, h\, dx = \frac{4h}{3}\left[x^{3/2}\right]_0^1 + \left[\frac{-2\sqrt{2}h}{3}(3 - x)^{3/2}\right]_1^3 = 4h$

(b) $V = \int_0^1 \left(\frac{1}{2} \cdot 2\sqrt{x} \cdot \sqrt{3}\sqrt{x}\right) dx + \int_1^3 \left(\frac{1}{2} \cdot \sqrt{2}\sqrt{3 - x} \cdot \frac{\sqrt{3}}{\sqrt{2}} \cdot \sqrt{3 - x}\right) dx$

$= \sqrt{3}\int_0^1 x\, dx + \frac{\sqrt{3}}{2}\int_1^3 (3 - x)\, dx = \frac{3\sqrt{3}}{2}$

(c) $V = \int_0^1 \left(\frac{1}{2} \cdot 2\sqrt{x} \cdot \sqrt{x}\right) dx + \int_1^3 \left(\frac{1}{2} \cdot \sqrt{2}\sqrt{3 - x} \cdot \frac{\sqrt{2}}{2} \cdot \sqrt{3 - x}\right) dx$

$= \int_0^1 x\, dx + \frac{1}{2}\int_1^3 (3 - x)\, dx = \frac{3}{2}$

31. (a) $V = \int_0^4 \left[(\sqrt{y}\,) - (-\sqrt{y}\,) \right]^2 dy = \int_0^4 4y\, dy = \left[2y^2 \right]_0^4 = 32$

(b) $V = \int_0^4 \frac{\pi}{2} \left(\sqrt{y} \right)^2 dy = \frac{\pi}{2} \int_0^4 y\, dy = \frac{\pi}{2} \left[\frac{1}{2} y^2 \right]_0^4 = 4\pi$

(c) $V = \int_0^4 \frac{\sqrt{3}}{4} \left[(\sqrt{y}\,) - (-\sqrt{y}\,) \right]^2 dy = \sqrt{3} \int_0^4 y\, dy = 8\sqrt{3}$

32. (a) $V = \int_{-1}^1 (3 - 3y^2) h\, dy = h \left[3y - y^3 \right]_{-1}^1 = 4h$

(b) $V = \int_{-1}^1 \frac{1}{2} (3 - 3y^2) \frac{\sqrt{3}}{2} (3 - 3y^2)\, dy = \frac{9\sqrt{3}}{4} \int_{-1}^1 (1 - 2y^2 + y^4)\, dy$

$= \frac{9\sqrt{3}}{4} \left[y - \frac{2}{3} y^3 + \frac{y^5}{5} \right]_{-1}^1 = \frac{12}{5} \sqrt{3}$

(c) $V = \int_{-1}^1 \frac{1}{2} (3 - 3y^2) \frac{1}{2} (3 - 3y^2)\, dy = \frac{1}{\sqrt{3}} \cdot [\text{Volume in (b)}] = \frac{12}{5}$

33. (a) $V = \int_0^4 (4 - x)^2\, dx = \left[16x - 4x^2 + \frac{x^3}{3} \right]_0^4 = \frac{64}{3}.$

(b) $V = \frac{1}{4} \int_0^4 (4 - x)^2\, dx = \frac{1}{4} \left[16x - 4x^2 + \frac{x^3}{3} \right]_0^4 = \frac{16}{3}.$

34. (a) Area of each triangle $= y^2$,   thus $V = 2 \int_0^a \left(b^2 - \frac{b^2}{a^2} x^2 \right) dx = 2 \left[b^2 x - \frac{b^2}{3a^2} x^3 \right]_0^a = \frac{4}{3} ab^2.$

(b) Area of each square $= 4y^2$,   thus $V = 4 \cdot$ the answer to part (a) $= \frac{16}{3} ab^2.$

(c) Area of each triangle $= 2y$,   thus $V = 2 \int_0^a 2\sqrt{b^2 - \frac{b^2}{a^2} x^2}\, dx$

$= \frac{4b}{a} \int_0^a \sqrt{a^2 - x^2}\, dx = \frac{4b}{a} \left[\frac{x}{2} \sqrt{a^2 - x^2} + \frac{a^2}{2} \sin^{-1} \frac{x}{a} \right]_0^a = ab\pi.$

35. (a) $V = \int_0^{\pi/2} \sqrt{3} \sin x\, dx = -\sqrt{3} \left[\cos x \right]_0^{\pi/2} = \sqrt{3}$

(b) $V = \int_0^{\pi/2} 4 \sin x\, dx = -4 \left[\cos x \right]_0^{\pi/2} = 4$

36. (a) $V = \int_0^{\pi/4} \pi \left[\frac{1}{2} (\sec x - \tan x)^2 \right] dx = \frac{\pi}{4} \int_0^{\pi/4} \left(2\sec^2 x - 2\sec x \tan x - 1 \right) dx$

$= \frac{\pi}{4} \left[4 - 2\sqrt{2} - \frac{\pi}{4} \right]$

(b) $V = \int_0^{\pi/4} \pi \left(\sec x - \tan x \right)^2 dx = \int_0^{\pi/4} \left(2\sec^2 x - 2\sec x \tan x - 1 \right) dx = 4 - 2\sqrt{2} - \frac{\pi}{4}$

37. $V = \int_{-a}^{a} \pi \left(b\sqrt{1 - \dfrac{x^2}{a^2}} \right)^2 dx = \dfrac{2\pi b^2}{a^2} \int_{0}^{a} \pi \left(a^2 - x^2 \right) dx = \dfrac{2\pi b^2}{a^2} \pi \left[a^2 x - \dfrac{1}{3}x^3 \right]_{0}^{a}$

$$= \dfrac{2\pi b^2}{a^2} \left(\dfrac{2}{3}a^3 \right) = \dfrac{4}{3}\pi ab^2$$

38. $V = \int_{-b}^{b} \pi \left(\dfrac{a}{b} \sqrt{b^2 - y^2} \right)^2 dy = \dfrac{\pi a^2}{b^2} \int_{-b}^{b} (b^2 - y^2) \, dy$

$$= \dfrac{\pi a^2}{b^2} \left[b^2 y - \dfrac{y^3}{3} \right]_{-b}^{b} = \dfrac{4}{3}\pi a^2 b$$

39.

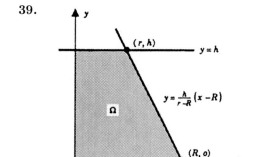

The specified frustum is generated by revolving the region Ω about the y-axis.

$$V = \int_{0}^{h} \pi \left[\dfrac{r - R}{h}y + R \right]^2 dy$$

$$= \pi \left[\dfrac{h}{3(r - R)} \left(\dfrac{r - R}{h}y + R \right)^3 \right]_{0}^{h}$$

$$= \dfrac{\pi h}{3(r - R)} \left(r^3 - R^3 \right) = \dfrac{\pi h}{3} \left(r^2 + rR + R^2 \right)$$

40. $V = 2 \int_{0}^{a/2} \pi \left(\sqrt{3}x \right)^2 dx = 6\pi \left[\dfrac{x^3}{3} \right]_{0}^{a/2} = \dfrac{1}{4}\pi a^3$ (twice the volume of cone of radius $\dfrac{\sqrt{3}}{2}a$, height $\dfrac{a}{2}$)

41. Capacity of basin $= \dfrac{1}{2} \left(\dfrac{4}{3}\pi r^3 \right) = \dfrac{2}{3}\pi r^3$.

(a) Volume of water $= \int_{r/2}^{r} \pi \left[\sqrt{r^2 - x^2} \right]^2 dx$

$$= \pi \int_{r/2}^{r} (r^2 - x^2) \, dx = \pi \left[r^2 x - \dfrac{1}{3}x^3 \right]_{r/2}^{r} = \dfrac{5}{24}\pi r^3.$$

The basin is $\left(\dfrac{5}{24}\pi r^3 \right) (100) / \left(\dfrac{2}{3}\pi r^3 \right) = 31\dfrac{1}{4}\%$ full.

(b) Volume of water $= \int_{2r/3}^{r} \pi \left[\sqrt{r^2 - x^2} \right]^2 dx = \pi \int_{2r/3}^{r} (r^2 - x^2) \, dx = \dfrac{8}{81}\pi r^3.$

The basin is $\left(\dfrac{8}{81}\pi r^3 \right) (100) / \left(\dfrac{2}{3}\pi r^3 \right) = 14\dfrac{22}{27}\%$ full.

42. $V = \int_{a}^{b} \pi \left(\sqrt{r^2 - x^2} \right)^2 dx = \pi \int_{a}^{b} (r^2 - x^2) \, dx = \pi \left[r^2 x - \dfrac{x^3}{3} \right]_{a}^{b} = \pi r^2 (b - a) - \dfrac{1}{3}\pi (b^3 - a^3).$

43. $V = \int_h^r \pi(r^2 - y^2)\, dy = \pi \left[r^2 y - \dfrac{y^3}{3} \right]_h^r = \dfrac{\pi}{3} \left[2r^3 - 3r^2 h + h^3 \right].$

44. Imagine the punchbowl upside down on the x, y-plane centered over the origin.

(a) $V = \pi \int_1^{12} (144 - y^2)\, dy = \pi \left[144y - \dfrac{y^3}{3} \right]_1^{12} = \pi(1584 - \dfrac{1727}{3})$ in^3 or about 13.7 gallons.

(b) $V = \pi \int_1^{10} (144 - y^2)\, dy = \pi \left[144y - \dfrac{y^3}{3} \right]_1^{10} = 963\pi$ in^3 or about 13.1 gallons.

45. (a)

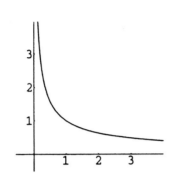

(b) $A(b) = \displaystyle\int_1^b x^{-\frac{2}{3}}\, dx = 3(b^{\frac{1}{3}} - 1)$

(c) $V(b) = \displaystyle\int_1^b \pi (x^{-\frac{2}{3}})^2\, dx = 3\pi(1 - b^{-\frac{1}{3}})$

(d) As $b \to \infty$, $A(b) \to \infty$ and $V(b) \to 3\pi$.

46. (a) $A(c) = \displaystyle\int_c^1 x^{-\frac{2}{3}}\, dx = \left[3x^{\frac{1}{3}} \right]_c^1 = 3(1 - c^{\frac{1}{3}}).$

(c) $V(c) = \pi \displaystyle\int_c^1 (x^{-\frac{2}{3}})^2\, dx = \pi \left[-3x^{-\frac{1}{3}} \right]_c^1 = 3\pi(c^{-\frac{1}{3}} - 1).$

(d) As $c \to 0$, $c^{\frac{1}{3}} \to 0$ and $c^{-\frac{1}{3}} \to \infty$. Thus $A(c) \to 3$, and $V(c) \to \infty$.

47. If the depth of the liquid in the container is h feet, then the volume of the liquid is:

$$V(h) = \int_0^h \pi \left(\sqrt{y+1} \right)^2 dy = \int_0^h \pi\, [y+1]\, dy.$$

Differentiation with respect to t gives

$$\frac{dV}{dt} = \frac{dV}{dh} \cdot \frac{dh}{dt} = \pi(h+1)\frac{dh}{dt}.$$

Now, since $\dfrac{dV}{dt} = 2$, it follows that $\dfrac{dh}{dt} = \dfrac{2}{\pi(h+1)}.$ Thus

$$\left. \frac{dh}{dt} \right|_{h=1} = \frac{2}{2\pi} = \frac{1}{\pi} \text{ ft/min} \quad \text{and} \quad \left. \frac{dh}{dt} \right|_{h=2} = \frac{2}{3\pi} \text{ ft/min}.$$

48. At time t the water in the container is at height h. The volume V of water is given by

$$V = \int_0^h \pi[f(y)]^2\, dy.$$

Differentiating with respect to t gives

$$\frac{dV}{dt} = \frac{dV}{dh}\frac{dh}{dt} = \pi[f(h)]^2\frac{dh}{dt}.$$

The surface area of the water at time t is given by $S = \pi[f(h)]^2$ and $dV/dt = kS = k\pi[f(h)]^2$, $k < 0$ the constant of proportionality. Thus, $dh/dt = k$

49.

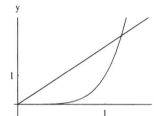

(b) The x-coordinates of the points of intersection are: $x = 0$,
$\quad x = 2^{1/4} \cong 1.1892$.

(c) $A \cong 2\displaystyle\int_0^{1.1892} \left(2x - x^5\right) dx \cong 2(0.9428) = 1.8856$

(d) $V = \displaystyle\int_0^{1.1892} \pi\left(4x^2 - x^{10}\right) dx \cong 5.1234$

50.

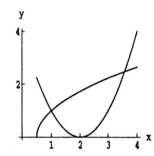

(b) x-coordinates of the points of intersection: $x = 1,\ 3.5747$

(c) $A \cong 3.1148$

(d) $V \cong 22.2025$

51. $V = \displaystyle\int_0^4 \pi\left[4 - \left(2 - \sqrt{x}\right)^2\right] dx = \int_0^4 \pi\left[4\sqrt{x} - x\right] dx = \pi\left[\frac{8}{3}x^{3/2} - \frac{1}{2}x^2\right]_0^4 = \frac{40\pi}{3}$

52. $V = \displaystyle\int_0^3 \pi\left([(x + 1) + 1]^2 - \left[(x - 1)^2 + 1\right]^2\right) dx = \int_0^3 \pi\left[(x + 2)^2 - (x^2 - 2x + 2)^2\right] dx$

$= \displaystyle\int_0^3 \pi(12x - 7x^2 + 4x^3 - x^4) dx = \pi\left[6x^2 - \frac{7}{3}x^3 + x^4 - \frac{x^5}{5}\right]_0^3 = \frac{117}{5}\pi$

53. $V = \displaystyle\int_0^\pi \pi\left[2\sin x - \sin^2 x\right] dx = \pi\left[-2\cos x\right]_0^\pi - \frac{\pi}{2}\int_0^\pi (1 - \cos 2x) dx$

$= 4\pi - \dfrac{\pi}{2}\left[x - \dfrac{1}{2}\sin 2x\right]_0^\pi = 4\pi - \dfrac{1}{2}\pi^2$

54. $V = \displaystyle\int_{\pi/4}^\pi \pi\left[(1 - \cos x)^2 - (1 - \sin x)^2\right] dx = \pi\int_{\pi/4}^\pi (2\sin x - 2\cos x + \cos 2x) dx$

$= \pi\left[-2\cos x - 2\sin x + \dfrac{1}{2}\sin 2x\right]_{\pi/4}^\pi = \dfrac{3}{2} + 2\sqrt{2}$

55.

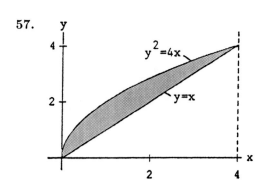

$$V = \int_0^5 \pi \left([3x - (-1)]^2 - [x^2 - 2x - (-1)]^2 \right) dx$$

$$= \pi \int_0^5 [3x + 1]^2 \, dx - \int_0^5 [x - 1]^4 \, dx$$

$$= \pi \left[\tfrac{1}{9}(3x + 1)^3 \right]_0^5 - \left[\frac{(x - 1)^5}{5} \right]_0^5$$

$$= 250\pi$$

56. (a) $V = \int_0^1 \pi \left[(2 + \sqrt{y})^2 - (2 + y^2)^2 \right] dy = \pi \int_0^1 \left(4y^{1/2} + y - 4y^2 - y^4 \right) dy = \frac{49}{30}\pi$

(b) $V = \int_0^1 \pi \left[(3 - y^2)^2 - (3 - \sqrt{y})^2 \right] dy = \pi \int_0^1 \left(6\sqrt{y} + y^4 - 6y^2 - y \right) dy = \frac{17}{10}\pi$

57.

(a) $V = \int_0^4 \pi \left[\left(\sqrt{4x} \right)^2 - x^2 \right] dx$

$$= \pi \int_0^4 \left[4x - x^2 \right] dx$$

$$= \pi \left[2x^2 - \tfrac{1}{3} x^3 \right]_0^4 = \frac{32\pi}{3}$$

(b) $V = \int_0^4 \pi \left[\left(\frac{1}{4} y^2 - 4 \right)^2 - (y - 4)^2 \right] dy$

$$= \pi \int_0^4 \left[\frac{1}{16} y^4 - 3y^2 + 8y \right] dy$$

$$= \pi \left[\tfrac{1}{80} y^5 - y^3 + 4y^2 \right]_0^4 = \frac{64\pi}{5}$$

58. (a) $V = \int_0^{16} \pi \left[\left(5 - \frac{y}{4} \right)^4 - (5 - \sqrt{y})^2 \right] dy = \pi \int_0^{16} \left(10\sqrt{y} - \frac{7}{2}y + \frac{y^2}{16} \right) dy$

$$= \pi \left[\frac{20}{3} y^{3/2} - \frac{7}{4} y^2 + \frac{y^3}{48} \right]_0^{16} = 64\pi$$

(b) $V = \int_0^{16} \pi \left[(\sqrt{y} + 1)^2 - \left(\frac{y}{4} + 1 \right)^2 \right] dy = \pi \int_0^{16} \left(\frac{y}{2} + 2\sqrt{y} - \frac{y^2}{16} \right) dy$

$$= \pi \left[\frac{y^2}{4} + \frac{4}{3} y^{3/2} - \frac{y^3}{48} \right]_0^{16} = 64\pi$$

59. (a) $V = \int_0^4 \pi \left(x^{3/2} \right)^2 dx = \pi \int_0^4 x^3 \, dx = \pi \left[\frac{1}{4} x^4 \right]_0^4 = 64\pi$

(b) $V = \int_0^8 \pi \left(4 - y^{2/3} \right)^2 dy = \pi \int_0^8 \left(16 - 8y^{2/3} + y^{4/3} \right) dy$

$= \pi \left[16y - \frac{24}{5} y^{5/3} + \frac{3}{7} y^{7/3} \right]_0^8 = \frac{1024}{35} \pi$

(c) $V = \int_0^4 \pi \left[(8)^2 - \left(8 - x^{3/2} \right)^2 \right] dx = \pi \int_0^4 \left(16x^{3/2} - x^3 \right) dx$

$= \pi \left[\frac{32}{5} x^{5/2} - \frac{1}{4} x^4 \right]_0^4 = \frac{704}{5} \pi$

(d) $V = \int_0^8 \pi \left[(4)^2 - \left(y^{2/3} \right)^2 \right] dy = \pi \int_0^8 \left(16 - y^{4/3} \right) dy = \pi \left[16y - \frac{3}{7} y^{7/3} \right]_0^8 = \frac{512}{7} \pi$

60. (a) $V = \int_0^8 \pi y^{4/3} \, dy = \pi \left[\frac{3}{7} y^{7/3} \right]_0^8 = \frac{384}{7} \pi$

(b) $V = \int_0^4 \pi (8 - x^{3/2})^2 \, dx = \pi \int_0^4 \left(64 - 16x^{3/2} + x^3 \right) dx = \pi \left[64x - \frac{32}{5} x^{5/2} + \frac{x^4}{4} \right]_0^4 = \frac{576}{5} \pi$

(c) $V = \int_0^8 \pi \left[4^2 - (4 - y^{2/3})^2 \right] dy = \pi \int_0^8 (8y^{2/3} - y^{4/3}) \, dy = \pi \left[\frac{24}{5} y^{5/3} - \frac{3}{7} y^{7/3} \right]_0^8 = \frac{3456}{35} \pi$

(d) $V = \int_0^4 \pi \left[8^2 - x^3 \right] dx = \pi \left[64x - \frac{x^4}{4} \right]_0^4 = 192\pi$

SECTION 6.3

1.

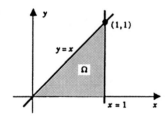

$V = \int_0^1 2\pi x \left[x - 0 \right] dx = 2\pi \int_0^1 x^2 \, dx$

$= 2\pi \left[\frac{1}{3} x^3 \right]_0^1 = \frac{2\pi}{3}$

2.

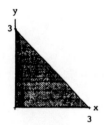

$V = \int_0^3 2\pi x (3 - x) \, dx = 2\pi \int_0^3 (3x - x^2) \, dx$

$= 2\pi \left[\frac{3x^2}{2} - \frac{x^3}{3} \right]_0^3 = 9\pi$

3.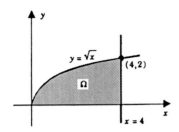

$$V = \int_0^4 2\pi x \left[\sqrt{x} - 0 \right] dx = 2\pi \int_0^4 x^{3/2}\, dx$$

$$= 2\pi \left[\tfrac{2}{5} x^{5/2} \right]_0^4 = \frac{128}{5}\pi$$

4.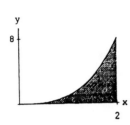

$$V = \int_0^2 2\pi x x^3\, dx = 2\pi \left[\frac{x^5}{5} \right]_0^2 = \frac{64}{5}\pi$$

5.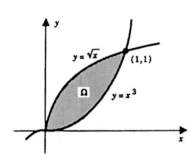

$$V = \int_0^1 2\pi x \left[\sqrt{x} - x^3 \right] dx$$

$$= 2\pi \int_0^1 \left(x^{3/2} - x^4 \right) dx$$

$$= 2\pi \left[\frac{2}{5} x^{5/2} - \frac{1}{5} x^5 \right]_0^1 = \frac{2\pi}{5}$$

6.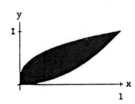

$$V = \int_0^1 2\pi x \left[x^{1/3} - x^2 \right] dx = 2\pi \int_0^1 \left(x^{4/3} - x^3 \right) dx$$

$$= 2\pi \left[\frac{3}{7} x^{7/3} - \frac{x^4}{4} \right]_0^1 = \frac{5\pi}{14}$$

7.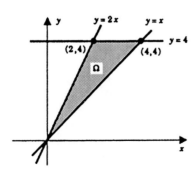

$$V = \int_0^2 2\pi x \left[2x - x \right] dx + \int_2^4 2\pi x \left[4 - x \right] dx$$

$$= 2\pi \int_0^2 x^2\, dx + 2\pi \int_2^4 \left(4x - x^2 \right) dx$$

$$= 2\pi \left[\tfrac{1}{3} x^3 \right]_0^2 + 2\pi \left[2x^2 - \tfrac{1}{3} x^3 \right]_2^4 = 16\pi$$

8.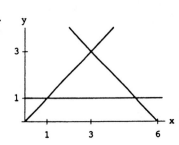

$$V = \int_1^3 2\pi x(x-1)\,dx + \int_3^5 2\pi x(6-x-1)\,dx$$

$$= 2\pi \int_1^3 (x^2-x)\,dx + 2\pi \int_3^5 (5x-x^2)\,dx$$

$$= 2\pi \left[\frac{x^3}{3} - \frac{x^2}{2}\right]_1^3 + 2\pi \left[\frac{5x^2}{2} - \frac{x^3}{3}\right]_3^5 = 24\pi$$

9.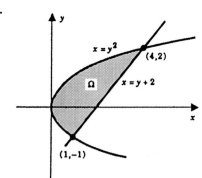

$$V = \int_0^1 2\pi x \left[(\sqrt{x}) - (-\sqrt{x})\right]\,dx + \int_1^4 2\pi x \left[(\sqrt{x}) - (x-2)\right]\,dx$$

$$= 4\pi \int_0^1 x^{3/2}\,dx + 2\pi \int_1^4 \left(x^{3/2} - x^2 + 2x\right)\,dx$$

$$= 4\pi \left[\tfrac{2}{5}x^{5/2}\right]_0^1 + 2\pi \left[\tfrac{2}{5}x^{5/2} - \tfrac{1}{3}x^3 + x^2\right]_1^4 = \tfrac{72}{5}\pi$$

10.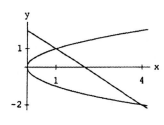

$$V = \int_0^1 2\pi x \cdot 2\sqrt{x}\,dx + \int_1^4 2\pi x(2-x+\sqrt{x})\,dx$$

$$= 4\pi \int_0^1 x^{3/2}\,dx + 2\pi \int_1^4 (2x - x^2 + x^{3/2})\,dx$$

$$= 4\pi \left[\frac{2}{5}x^{5/2}\right]_0^1 + 2\pi \left[x^2 - \frac{x^3}{3} + \frac{2}{5}x^{5/2}\right]_1^4 = \frac{72}{5}\pi$$

11.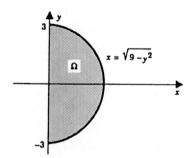

$$V = \int_0^3 2\pi x \left[\sqrt{9-x^2} - \left(-\sqrt{9-x^2}\right)\right]\,dx$$

$$= 4\pi \int_0^3 x\left(9-x^2\right)^{1/2}\,dx$$

$$= 4\pi \left[-\tfrac{1}{3}\left(9-x^2\right)^{3/2}\right]_0^3 = 36\pi$$

12.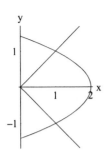

$$V = \int_0^1 2\pi x \cdot 2x \, dx + \int_1^2 2\pi x \cdot 2\sqrt{2-x} \, dx$$

$$= 4\pi \int_0^1 x^2 \, dx - 4\pi \int_1^0 (2-u)\sqrt{u} \, du \qquad (u = 2 - x)$$

$$= 4\pi \left[\frac{x^3}{3} \right]_0^1 - 4\pi \left[\frac{4}{3} u^{3/2} - \frac{2}{5} u^{5/2} \right]_1^0 = \frac{76}{15}\pi$$

13.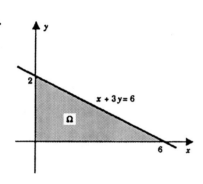

$$V = \int_0^2 2\pi y \, [6 - 3y] \, dy$$

$$= 6\pi \int_0^2 (2y - y^2) \, dy$$

$$= 6\pi \left[y^2 - \tfrac{1}{3} y^3 \right]_0^2 = 8\pi$$

14.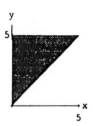

$$V = \int_0^5 2\pi y \cdot y \, dy = 2\pi \left[\frac{y^3}{3} \right]_0^5 = \frac{250}{3}\pi$$

15.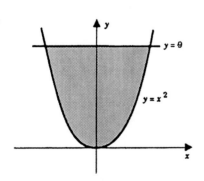

$$V = \int_0^9 2\pi y \, [(\sqrt{y}) - (-\sqrt{y})] \, dy$$

$$= 4\pi \int_0^9 y^{3/2} \, dy$$

$$= 4\pi \left[\tfrac{2}{5} y^{5/2} \right]_0^9 = \tfrac{1944}{5}\pi$$

16.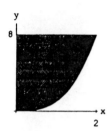

$$V = \int_0^8 2\pi y y^{1/3}\, dy = 2\pi \int_0^8 y^{4/3}\, dy$$

$$= 2\pi \left[\frac{3}{7}y^{7/3}\right]_0^8 = \frac{768}{7}\pi$$

17.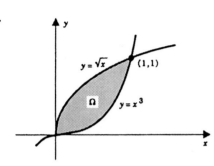

$$V = \int_0^1 2\pi y \left[y^{1/3} - y^2\right] dy$$

$$= 2\pi \int_0^1 \left(y^{4/3} - y^3\right) dy$$

$$= 2\pi \left[\frac{3}{7}y^{7/3} - \frac{1}{4}y^4\right]_0^1 = \frac{5}{14}\pi$$

18.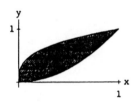

$$V = \int_0^1 2\pi y(\sqrt{y} - y^3)\, dy = 2\pi \int_0^1 (y^{3/2} - y^4)\, dy$$

$$= 2\pi \left[\frac{2}{5}y^{5/2} - \frac{y^5}{5}\right]_0^1 = \frac{2}{5}\pi$$

19.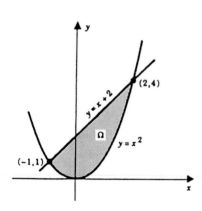

$$V = \int_0^1 2\pi y\left[(\sqrt{y}) - (-\sqrt{y})\right] dy + \int_1^4 2\pi y\left[(\sqrt{y}) - (y - 2)\right] dy$$

$$= 4\pi \int_0^1 y^{3/2}\, dy + 2\pi \int_1^4 \left(y^{3/2} - y^2 + 2y\right) dy$$

$$= 4\pi \left[\frac{2}{5}y^{5/2}\right]_0^1 + 2\pi \left[\frac{2}{5}y^{5/2} - \frac{1}{3}y^3 + y^2\right]_1^4 = \frac{72}{5}\pi$$

20.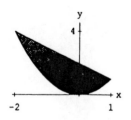

$$V = \int_0^1 2\pi y 2\sqrt{y}\, dy + \int_1^4 2\pi y(2 - y + \sqrt{y})\, dy$$

$$= 4\pi \int_0^1 y^{3/2}\, dy + 2\pi \int_1^4 (2y - y^2 + y^{3/2})\, dy$$

$$= 4\pi \left[\frac{2}{5}y^{5/2}\right]_0^1 + 2\pi \left[y^2 - \frac{y^3}{3} + \frac{2}{5}y^{5/2}\right]_1^4 = \frac{72}{5}\pi$$

21.

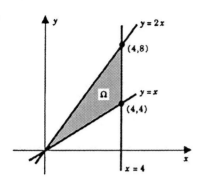

$$V = \int_0^4 2\pi y \left[y - \frac{y}{2} \right] dy + \int_4^8 2\pi y \left[4 - \frac{y}{2} \right] dy$$

$$= \pi \int_0^4 y^2 \, dy + \pi \int_4^8 \left(8y - y^2 \right) \, dy$$

$$= \pi \left[\tfrac{1}{3} y^3 \right]_0^4 + \pi \left[4y^2 - \tfrac{1}{3} y^3 \right]_4^8 = 64\pi$$

22.

$$V = \int_0^1 2\pi y(y) \, dy + \int_1^7 2\pi y \, dy + \int_7^8 2\pi y(8 - y) \, dy$$

$$= 2\pi \int_0^1 y^2 dy + 2\pi \int_1^7 y \, dy + 2\pi \int_7^8 \left(8y - y^2 \right) dy$$

$$= 2\pi \left[\frac{y^3}{3} \right]_0^1 + 2\pi \left[\frac{y^2}{2} \right]_1^7 + 2\pi \left[4y^2 - \frac{y^3}{3} \right]_7^8 = 56\pi$$

23.

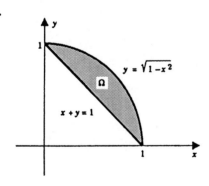

$$V = \int_0^1 2\pi y \left[\sqrt{1 - y^2} - (1 - y) \right] dy$$

$$= 2\pi \int_0^1 \left[y \left(1 - y^2 \right)^{1/2} - y + y^2 \right] dy$$

$$= 2\pi \left[-\frac{1}{3} \left(1 - y^2 \right)^{3/2} - \frac{1}{2} y^2 + \frac{1}{3} y^3 \right]_0^1 = \frac{\pi}{3}$$

24.

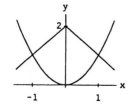

$$V = \int_0^1 2\pi y 2\sqrt{y} \, dy + \int_1^2 2\pi y 2(2 - y) \, dy$$

$$= 4\pi \int_0^1 y^{3/2} \, dy + 4\pi \int_1^2 \left(2y - y^2 \right) dy$$

$$= 4\pi \left[\frac{2}{5} y^{5/2} \right]_0^1 + 4\pi \left[y^2 - \frac{y^3}{3} \right]_1^2 = \frac{64}{15} \pi$$

25. (a) $V = \displaystyle\int_0^1 2\pi x \left[1 - \sqrt{x}\right] dx$

(b) $V = \displaystyle\int_0^1 \pi\, y^4 \, dy$

$= \pi \left[\tfrac{1}{5}\, y^5\right]_0^1 = \tfrac{1}{5}\pi$

26. (a) $V = \displaystyle\int_0^1 \pi \left[(2 - \sqrt{x})^2 - 1^2\right] dx$

(b) $V = \displaystyle\int_0^1 2\pi(2 - y)y^2 \, dy = 2\pi \int_0^1 (2y^2 - y^3)\, dy = 2\pi \left[\dfrac{2}{3}y^3 - \dfrac{y^4}{4}\right]_0^1 = \dfrac{5}{6}\pi$

27. (a) $V = \displaystyle\int_0^1 \pi \left(x - x^4\right) dx$

(b) $V = \displaystyle\int_0^1 2\pi y \left(\sqrt{y} - y^2\right) dy$

$= \pi \left[\tfrac{1}{2}\, x^2 - \tfrac{1}{5}\, x^5\right]_0^1 = \dfrac{3\pi}{10}$

28. (a) $V = \displaystyle\int_0^1 2\pi(x + 3)(\sqrt{x} - x^2)\, dx = 2\pi \int_0^1 \left(x^{3/2} + 3\sqrt{x} - 3x^2 - x^3\right) dx$

$= 2\pi \left[\dfrac{2}{5}x^{5/2} + 2x^{3/2} - x^3 - \dfrac{x^4}{4}\right]_0^1 = \dfrac{23}{10}\pi$

(b) $V = \displaystyle\int_0^1 \pi \left[(\sqrt{y} + 3)^2 - (y^2 + 3)^2\right] dy$

29. (a) $V = \displaystyle\int_0^1 2\pi x \cdot x^2 \, dx = 2\pi \int_0^1 x^3 \, dx$

(b) $V = \displaystyle\int_0^1 \pi(1 - y)\, dy$

$= 2\pi \left[\tfrac{1}{4}x^4\right]_0^1 = \dfrac{\pi}{2}$

30. (a) $V = \displaystyle\int_0^1 \pi \left[(x^2 + 1)^2 - 1\right] dx = \pi \int_0^1 (x^4 + 2x^2)\, dx = \dfrac{13}{15}\pi$

(b) $V = \displaystyle\int_0^1 2\pi(y + 1)(1 - \sqrt{y})\, dy$

31. $V = \displaystyle\int_0^a 2\pi x \left[2b\sqrt{1 - \dfrac{x^2}{a^2}}\,\right] dx = \dfrac{4\pi b}{a} \int_0^a x\left(a^2 - x^2\right)^{1/2} dx = \dfrac{4\pi b}{a} \left[-\dfrac{1}{3}\left(a^2 - x^2\right)^{3/2}\right]_0^a = \dfrac{4}{3}\pi a^2 b$

32. $V = \displaystyle\int_0^b 2\pi y\,\dfrac{2a}{b}\,\sqrt{b^2 - y^2}\, dy = \dfrac{4\pi a}{b} \int_0^b y(b^2 - y^2)^{1/2}\, dy$

$= \dfrac{4\pi a}{b} \left[-\dfrac{1}{3}(b^2 - y^2)^{3/2}\right]_0^b = \dfrac{4}{3}\pi ab^2$

33.

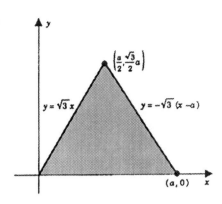

By the shell method

$$V = \int_0^{a/2} 2\pi x \left(\sqrt{3}\,x\right) dx + \int_{a/2}^a 2\pi x \left[\sqrt{3}\,(a-x)\right] dx$$

$$= 2\pi\sqrt{3} \int_0^{a/2} x^2\, dx + 2\pi\sqrt{3} \int_{a/2}^a (ax - x^2)\, dx$$

$$= 2\pi\sqrt{3} \left[\frac{1}{3}x^3\right]_0^{a/2} + 2\pi\sqrt{3} \left[\frac{a}{2}x^2 - \frac{1}{3}x^3\right]_{a/2}^a = \frac{\sqrt{3}}{4}a^3\pi$$

34. $V = \int_0^{\sqrt{r^2-a^2}} 2\pi x \left(\sqrt{r^2-x^2} - a\right) dx = 2\pi \int_0^{\sqrt{r^2-a^2}} \left[x(r^2-x^2)^{1/2} - ax\right] dx$

$$= 2\pi \left[-\frac{1}{3}(r^2-x^2)^{3/2} - \frac{a}{2}x^2\right]_0^{\sqrt{r^2-a^2}} = \frac{1}{3}\pi(2r^3 + a^3 - 3ar^2)$$

35. (a) $V = \int_0^8 2\pi y \left[4 - y^{2/3}\right] dy = 2\pi \int_0^8 \left(4y - y^{5/3}\right) dy = 2\pi \left[2y^2 - \frac{3}{8}y^{8/3}\right]_0^8 = 64\pi$

(b) $V = \int_0^4 2\pi (4-x) \left[x^{3/2}\right] dx = 2\pi \int_0^4 \left(4x^{3/2} - x^{5/2}\right) dx$

$$= 2\pi \left[\frac{8}{5}x^{5/2} - \frac{2}{7}x^{7/2}\right]_0^4 = \frac{1024}{35}\pi$$

(c) $V = \int_0^8 2\pi (8-y) \left[4 - y^{2/3}\right] dy = 2\pi \int_0^8 \left(32 - 4y - 8y^{2/3} + y^{5/3}\right) dy$

$$= 2\pi \left[32y - 2y^2 - \frac{24}{5}y^{5/3} + \frac{3}{8}y^{8/3}\right]_0^8 = \frac{704}{5}\pi$$

(d) $V = \int_0^4 2\pi x \left[x^{3/2}\right] dx = 2\pi \int_0^4 x^{5/2}\, dx = 2\pi \left[\frac{2}{7}x^{7/2}\right]_0^4 = \frac{512}{7}\pi$

36. (a) $V = \int_0^4 2\pi x(8 - x^{3/2})\, dx = 2\pi \int_0^4 (8x - x^{5/2})\, dx = 2\pi \left[4x^2 - \frac{2}{7}x^{7/2}\right]_0^4 = \frac{384}{7}\pi$

(b) $V = \int_0^8 2\pi(8-y)y^{2/3}\, dy = 2\pi \int_0^8 (8y^{2/3} - y^{5/3})\, dy = 2\pi \left[\frac{24}{5}y^{5/3} - \frac{3}{8}y^{8/3}\right]_0^8 = \frac{576}{5}\pi$

(c) $V = \int_0^4 2\pi(4-x)(8 - x^{3/2})\, dx = 2\pi \int_0^4 (32 - 8x - 4x^{3/2} + x^{5/2})\, dx$

$$= 2\pi \left[32x - 4x^2 - \frac{8}{5}x^{5/2} + \frac{2}{7}x^{7/2}\right]_0^4 = \frac{3456}{35}\pi$$

(d) $V = \int_0^8 2\pi y\, y^{2/3}\, dy = 2\pi \int_0^8 y^{5/3}\, dy = 2\pi \left[\frac{3}{8}y^{8/3}\right]_0^8 = 192\pi$

37. (a) $F'(x) = \sin x + x\cos x - \sin x = x\cos x = f(x).$

(b) $V = \int_0^{\pi/2} 2\pi x \cdot \cos x\, dx = 2\pi \left[x\sin x + \cos x\right]_0^{\pi/2} = \pi^2 - 2\pi$

38. (a)

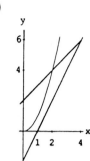

(b) By the shell method

$$V = \int_0^2 2\pi x(x^2 - 2x + 2)\, dx + \int_2^4 2\pi x\, (x + 2 - 2x + 2)\, dx$$

$$= 2\pi \int_0^2 (x^3 - 2x^2 + 2x)\, dx + 2\pi \int_2^4 (4x - x^2)\, dx$$

$$= 2\pi \left[\frac{x^4}{4} - \frac{2}{3}x^3 + x^2\right]_0^2 + 2\pi \left[2x^2 - \frac{x^3}{3}\right]_2^4 = 16\pi$$

39. (a) $V = \int_0^1 2\sqrt{3}\pi x^2\, dx + \int_1^2 2\pi x\sqrt{4 - x^2}\, dx$
(b) $V = \int_0^{\sqrt{3}} \pi \left[4 - \frac{4}{3}y^2\right] dy$

(c) $V = \int_0^{\sqrt{3}} \pi \left[4 - \frac{4}{3}y^2\right] dy = \pi \left[4y - \frac{4}{9}y^3\right]_0^{\sqrt{3}} = \frac{8\pi\sqrt{3}}{3}$

40. (a) $V = \int_0^1 \pi(\sqrt{3}x)^2\, dx + \int_1^2 \pi \left(\sqrt{4 - x^2}\right)^2 dx$

(b) $V = \int_0^{\sqrt{3}} 2\pi y \left[\sqrt{4 - y^2} - \frac{y}{\sqrt{3}}\right] dy$

(c) use (a): $V = 3\pi \int_0^1 x^2\, dx + \pi \int_1^2 (4 - x^2)\, dx = \pi \left[x^3\right]_0^1 + \pi \left[4x - \frac{x^3}{3}\right]_1^2 = \frac{8}{3}\pi$

41. (a) $V = \int_0^1 2\sqrt{3}\,\pi x(2 - x)\, dx + \int_1^2 2\pi(2 - x)\sqrt{4 - x^2}\, dx$

(b) $V = \int_0^{\sqrt{3}} \pi \left[\left(2 - \frac{y}{\sqrt{3}}\right)^2 - \left(2 - \sqrt{4 - y^2}\right)^2\right] dy$

42. (a) $V = \int_0^1 \pi \left[(\sqrt{3}\,x + 1)^2 - 1^2\right] dx + \int_1^2 \pi \left[\left(\sqrt{4 - x^2} + 1\right)^2 - 1^2\right] dx$

(b) $V = \int_0^{\sqrt{3}} 2\pi(y + 1)\left(\sqrt{4 - y^2} - \frac{y}{\sqrt{3}}\right) dy$

43. (a) $V = 2\int_{b-a}^{b+a} 2\pi x\sqrt{a^2 - (x - b)^2}\, dx$

(b) $V = \int_{-a}^a \pi \left[\left(b + \sqrt{a^2 - y^2}\right)^2 - \left(b - \sqrt{a^2 - y^2}\right)^2\right] dy$

44. $V = \int_{-a}^a 2\pi(a - x)2\sqrt{a^2 - x^2}\, dx = 4\pi a \int_{-a}^a \sqrt{a^2 - x^2}\, dx - 4\pi \int_{-a}^a x\sqrt{a^2 - x^2}\, dx$

$$= 4\pi a(\text{ Area of half circle }) - 4\pi \left[-\frac{1}{3}(a^2 - x^2)^{3/2}\right]_{-a}^a = 4\pi a \cdot \pi a^2 - 0 = 4\pi^2 a^3$$

45. $V = \int_0^r 2\pi x (h - \frac{h}{r} x) \, dx = 2\pi h \left[\frac{x^2}{2} - \frac{x^3}{3r} \right]_0^r = \frac{\pi r^2 h}{3}.$

46. $V = 2 \int_{\sqrt{r^2 - h^2/4}}^r 2\pi x \sqrt{r^2 - x^2} \, dx = -2\pi \int_{h^2/4}^0 u^{\frac{1}{2}} \, du = 2\pi \left[\frac{2}{3} u^{\frac{3}{2}} \right]_0^{h^2/4} = \frac{\pi h^3}{6}.$

$u = r^2 - x^2, \ du = -2x \, dx$

47. (a) $V = \int_a^r 2\pi x (r^2 - x^2) \, dx = 2\pi \int_a^r (r^2 x - x^3) \, dx = 2\pi \left[\frac{1}{2} r^2 x^2 - \frac{1}{4} x^4 \right]_a^r = \frac{1}{2} \pi \left(r^2 - a^2 \right)^2$

(b) $V = \int_0^{r^2 - a^2} \pi \left[\left(\sqrt{r^2 - y} \right)^2 - a^2 \right] dy = \pi \int_0^{r^2 - a^2} (r^2 - y - a^2) \, dy = \pi \left[(r^2 - a^2) y - \frac{1}{2} y^2 \right]_0^{r^2 - a^2}$

$= \frac{1}{2} \pi (r^2 - a^2)^2$

48. (a)

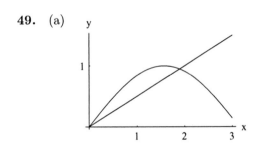

(b) $V = \int_0^1 \pi \sin^2 \pi x^2 \, dx \cong 1.1873$

(c) $V = \int_0^1 2\pi x \sin \pi x^2 \, dx = \left[-\cos \pi x^2 \right]_0^1 = 2$

49. (a)

(b) points of intersection: $x = 0, \ x \cong 1.8955$

(c) $A \cong \int_0^{1.8955} \left(\sin x - \frac{1}{2} x \right) dx \cong 0.4208$

(d) $V \cong \int_0^{1.8955} 2\pi \left(\sin x - \frac{1}{2} x \right) dx \cong 2.6226$

50. (a)

(b) first quadrant points of intersection: $x \cong 0.1939, \ x \cong 2.7093$

(c) $A \cong \int_{0.1939}^{2.7093} \left(\frac{3}{2} - \frac{1}{2} x - \frac{2}{(x+1)^2} \right) dx \cong 0.8114$

(d) $V \cong \int_{0.1939}^{2.7093} 2\pi x \left(\frac{3}{2} - \frac{1}{2} x - \frac{2}{(x+1)^2} \right) dx \cong 6.4873$

SECTION 6.4

1.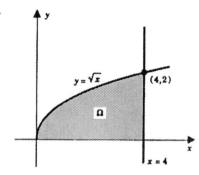

$$A = \int_0^4 \sqrt{x}\, dx = \frac{16}{3}$$

$$\overline{x}A = \int_0^4 x\sqrt{x}\, dx = \frac{64}{5}, \quad \overline{x} = \frac{12}{5}$$

$$\overline{y}A = \int_0^4 \frac{1}{2}\left(\sqrt{x}\right)^2 dx = 4, \quad \overline{y} = \frac{3}{4}$$

$$V_x = 2\pi\overline{y}A = 8\pi, \quad V_y = 2\pi\overline{x}A = \frac{128}{5}\pi$$

2.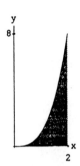

$$A = \int_0^2 x^3\, dx = 4$$

$$\overline{x}A = \int_0^2 x\,x^3\, dx = \frac{32}{5}, \quad \overline{x} = \frac{8}{5}$$

$$\overline{y}A = \int_0^2 \frac{1}{2}(x^3)^2\, dx = \frac{64}{7}, \quad \overline{y} = \frac{16}{7}$$

$$V_x = 2\pi\overline{y}A = \frac{128}{7}\pi, \quad V_y = 2\pi\overline{x}A = \frac{64}{5}\pi$$

3.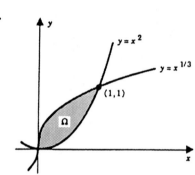

$$A = \int_0^1 \left(x^{1/3} - x^2\right) dx = \frac{5}{12}$$

$$\overline{x}A = \int_0^1 x\left(x^{1/3} - x^2\right) dx = \frac{5}{28}, \quad \overline{x} = \frac{3}{7}$$

$$\overline{y}A = \int_0^1 \frac{1}{2}\left[\left(x^{1/3}\right)^2 - \left(x^2\right)^2\right] dx = \frac{1}{5}, \quad \overline{y} = \frac{12}{25}$$

$$V_x = 2\pi\overline{y}A = \tfrac{2}{5}\pi, \quad V_y = 2\pi\overline{x}A = \tfrac{5}{14}\pi$$

4.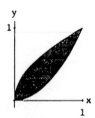

$$A = \int_0^1 (\sqrt{x} - x^3)\, dx = \frac{5}{12}$$

$$\overline{x}A = \int_0^1 x(\sqrt{x} - x^3)\, dx = \frac{1}{5}, \quad \overline{x} = \frac{12}{25}$$

$$\overline{y}A = \int_0^1 \frac{1}{2}(x - x^6)\, dx = \frac{5}{28}, \quad \overline{y} = \frac{3}{7}$$

$$V_x = 2\pi\overline{y}A = \frac{5}{14}\pi, \quad V_y = 2\pi\overline{x}A = \frac{2}{5}\pi$$

5.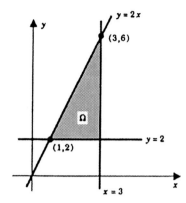

$$A = \int_1^3 (2x - 2) \, dx = 4$$

$$\overline{x}A = \int_1^3 x(2x - 2) \, dx = \frac{28}{3}, \quad \overline{x} = \frac{7}{3}$$

$$\overline{y}A = \int_1^3 \frac{1}{2}\left[(2x)^2 - (2)^2\right] dx = \frac{40}{3}, \quad \overline{y} = \frac{10}{3}$$

$$V_x = 2\pi\overline{y}A = \frac{80}{3}\pi, \quad V_y = 2\pi\overline{x}A = \frac{56}{3}\pi$$

6.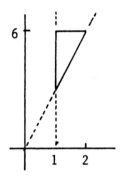

$$A = \frac{1}{2} \cdot 3 \cdot 1 = \frac{3}{2}$$

$$\overline{x}A = \int_1^2 x(6 - 3x) \, dx = 2, \overline{x} = \frac{4}{3}$$

$$\overline{y}A = \int_1^2 \frac{1}{2}\left(6^2 - (3x)^2\right) dx = \frac{15}{2}, \quad \overline{y} = 5$$

$$V_x = 2\pi\overline{y}A = 15\pi, \quad V_y = 2\pi\overline{x}A = 4\pi$$

7.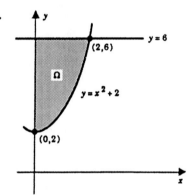

$$A = \int_0^2 \left[6 - \left(x^2 + 2x\right)\right] dx = \frac{16}{3}$$

$$\overline{x}A = \int_0^2 x\left[6 - \left(x^2 + 2\right)\right] dx = 4, \quad \overline{x} = \frac{3}{4}$$

$$\overline{y}A = \int_0^2 \frac{1}{2}\left[(6)^2 - \left(x^2 + 2\right)^2\right] dx = \frac{352}{15}, \quad \overline{y} = \frac{22}{5}$$

$$V_x = 2\pi\overline{y}A = \frac{704}{15}\pi, \quad V_y = 2\pi\overline{x}A = 8\pi$$

8.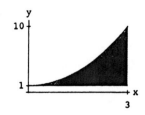

$$A = \int_0^3 \left[(x^2 + 1) - 1\right] dx = 9$$

$$\overline{x}A = \int_0^3 x\left[(x^2 + 1) - 1\right] dx = \frac{81}{4}, \quad \overline{x} = \frac{9}{4}$$

$$\overline{y}A = \int_0^3 \frac{1}{2}\left[(x^2 + 1)^2 - 1^2\right] dx = \frac{999}{30}, \quad \overline{y} = \frac{111}{30}$$

$$V_x = 2\pi\overline{y}A = \frac{333}{5}\pi, \quad V_y = 2\pi\overline{x}A = \frac{81}{2}\pi$$

9.

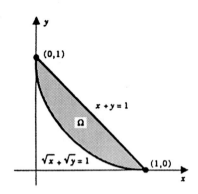

$$A = \int_0^1 \left[(1-x) - \left(1 - \sqrt{x}\right)^2 \right] dx = \frac{1}{3}$$

$$\overline{x}A = \int_0^1 x \left[(1-x) - \left(1 - \sqrt{x}\right)^2 \right] dx = \frac{2}{15}, \quad \overline{x} = \frac{2}{5}$$

$$\overline{y} = \frac{2}{5} \qquad \text{by symmetry}$$

$$V_x = 2\pi\overline{y}A = \frac{4}{15}\pi, \quad V_y = \frac{4}{15}\pi \qquad \text{by symmetry}$$

10.

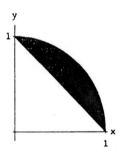

$$A = \frac{\pi}{4} - \frac{1}{2} = \frac{\pi - 2}{4}$$

$$\overline{x}A = \int_0^1 x \left(\sqrt{1 - x^2} + x - 1 \right) dx = \frac{1}{6}, \quad \overline{x} = \frac{2}{3(\pi - 2)}$$

$$\overline{y} = \frac{2}{3(\pi - 2)} \qquad \text{by symmetry}$$

$$V_x = 2\pi\overline{y}A = \frac{\pi}{3}, \quad V_y = 2\pi\overline{x}A = \frac{\pi}{3}$$

11.

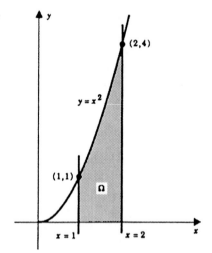

$$A = \int_1^2 x^2 \, dx = \frac{7}{3}$$

$$\overline{x}A = \int_1^2 x \left(x^2 \right) dx = \frac{15}{4}, \quad \overline{x} = \frac{45}{28}$$

$$\overline{y}A = \int_1^2 \frac{1}{2} \left(x^2 \right)^2 dx = \frac{31}{10}, \quad \overline{y} = \frac{93}{70}$$

$$V_x = 2\pi\overline{y}A = \frac{31}{5}\pi, \quad V_y = 2\pi\overline{x}A = \frac{15}{2}\pi$$

12.

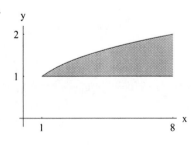

$$A = \int_1^8 (x^{1/3} - 1) \, dx = \frac{17}{4}$$

$$\overline{x}A = \int_1^8 x(x^{1/3} - 1) \, dx = \frac{321}{14}, \quad \overline{x} = \frac{642}{119}$$

$$\overline{y}A = \int_1^8 \frac{1}{2}(x^{2/3} - 1^2) \, dx = \frac{58}{10}, \quad \overline{y} = \frac{116}{85}$$

$$V_x = 2\pi\overline{y}A = \frac{58}{5}\pi, \quad V_y = 2\pi\overline{x}A = \frac{321}{7}\pi$$

13.

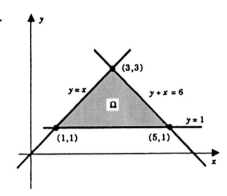

$$A = \tfrac{1}{2}bh = 4; \qquad \text{by symmetry,} \quad \overline{x} = 3$$

$$\overline{y}A = \int_1^3 y\left[(6-y) - y\right] dy = \frac{20}{3}, \quad \overline{y} = \frac{5}{3}$$

$$V_x = 2\pi\overline{y}A = \tfrac{40}{3}\pi, \quad V_y = 2\pi\overline{x}A = 24\pi$$

14.

$$A = \frac{9}{2}$$

$$\overline{x}A = \int_0^3 x(2x - x)\,dx = 9, \quad \overline{x} = 2$$

$$\overline{y}A = \int_0^3 \frac{1}{2}\left(4x^2 - x^2\right) dx = \frac{27}{2}, \quad \overline{y} = 3$$

$$V_x = 2\pi\overline{y}A = 27\pi, \quad V_y = 2\pi\overline{x}A = 18\pi$$

15. $\left(\frac{5}{2}, 5\right)$

16. $A = \int_{-1}^3 (4x - x^2 - 2x + 3)\,dx = \frac{32}{3}$

$$\overline{x}A = \int_{-1}^3 x(2x - x^2 + 3)\,dx = \frac{32}{3} \implies \overline{x} = 1$$

$$\overline{y}A = \int_{-1}^3 \frac{1}{2}\left[(4x - x^2)^2 - (2x - 3)^2\right] dx = \frac{32}{5} \implies \overline{y} = \frac{3}{5}$$

17. $\left(1, \frac{8}{5}\right)$

18. $A = \int_{-1}^3 (2x + 3 - x^2)\,dx = \frac{32}{3}$

$$\overline{x}A = \int_{-1}^3 x(2x + 3 - x^2)\,dx = \frac{32}{3} \implies \overline{x} = 1$$

$$\overline{y}A = \int_{-1}^3 \frac{1}{2}\left[(2x + 3)^2 - x^4\right] dx = \frac{544}{15} \implies \overline{y} = \frac{17}{5}$$

19. $\left(\frac{10}{3}, \frac{40}{21}\right)$

20. $A = \displaystyle\int_0^2 (x - x^2 + \sqrt{2x})\, dx = 2$

$\bar{x}A = \displaystyle\int_0^2 x(x - x^2 + \sqrt{2x})\, dx = \dfrac{28}{15} \implies \bar{x} = \dfrac{14}{15}$

$\bar{y}A = \displaystyle\int_0^2 \dfrac{1}{2}\left[(x - x^2)^2 - 2x\right] dx = -\dfrac{22}{15} \implies \bar{y} = -\dfrac{11}{15}$

21. $(2, 4)$

22. $A = \displaystyle\int_1^6 (6x - x^2 - 6 + x)\, dx; \quad \bar{x}A = \displaystyle\int_1^6 x(6x - x^2 - 6 + x)\, dx;$

$\bar{y}A = \displaystyle\int_1^6 \dfrac{1}{2}\left[(6x - x^2)^2 - (6 - x)^2\right] dx. \implies \bar{x} = \dfrac{7}{2}, \quad \bar{y} = 5$

23. $\left(-\dfrac{3}{5}, 0\right)$

24. $A = \displaystyle\int_0^a (\sqrt{a} - \sqrt{x})^2\, dx; \quad \bar{x}A = \displaystyle\int_0^a x(\sqrt{a} - \sqrt{x})^2\, dx;$

$\bar{y}A = \displaystyle\int_0^a \dfrac{1}{2}(\sqrt{a} - \sqrt{x})^4\, dx. \implies \bar{x} = \dfrac{a}{5}, \quad \bar{y} = \dfrac{a}{5}$

25.   (a) $(0, 0)$ by symmetry

(b) Ω_1 smaller quarter disc, Ω_2 the larger quarter disc

$A_1 = \dfrac{1}{16}\pi, \quad A_2 = \pi; \quad \bar{x}_1 = \bar{y}_1 = \dfrac{2}{3\pi}, \quad \bar{x}_2 = \bar{y}_2 = \dfrac{8}{3\pi} \quad$ (Example 1)

$\bar{x}A = \left(\dfrac{8}{3\pi}\right)(\pi) - \dfrac{2}{3\pi}\left(\dfrac{1}{16}\pi\right) = \dfrac{63}{24}, \quad A = \dfrac{15}{16}\pi$

$\bar{x} = \left(\dfrac{63}{24}\right) \Big/ \left(\dfrac{15\pi}{16}\right) = \dfrac{14}{5\pi}, \quad \bar{y} = \bar{x} = \dfrac{14}{5\pi} \quad$ (symmetry)

(c) $\bar{x} = 0, \quad \bar{y} = \dfrac{14}{5\pi}$

26. $A = \dfrac{1}{2}\pi ab; \quad \bar{x} = 0$ by symmetry.

$\bar{y}A = \displaystyle\int_{-a}^a \dfrac{1}{2}\left(\dfrac{b}{a}\sqrt{a^2 - x^2}\right)^2 dx = \dfrac{2}{3}ab^2 \implies \bar{y} = \dfrac{4b}{3\pi}.$

27.   Use theorem of Pappus. Centroid of rectangle is located

$$c + \sqrt{\left(\dfrac{a}{2}\right)^2 + \left(\dfrac{b}{2}\right)^2} \text{ units}$$

from line l. The area of the rectangle is ab. Thus,

$$\text{volume} = 2\pi \left[c + \sqrt{\left(\frac{a}{2}\right)^2 + \left(\frac{b}{2}\right)^2} \right] (ab) = \pi ab \left(2c + \sqrt{a^2 + b^2} \right).$$

28.

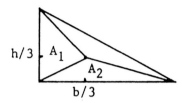

(a) $A_1 = \frac{1}{2}h \cdot \frac{b}{3} = \frac{1}{6}hb;$ $A_2 = \frac{1}{2}b \cdot \frac{h}{3} = \frac{1}{6}hb.$

$A_3 = \frac{1}{2}bh - (A_1 + A_2) = \frac{1}{2}bh - \frac{1}{3}bh = \frac{1}{6}bh.$

(b) Hypotenuse has equation $hx + by - bh = 0$, so distance from $\left(\frac{b}{3}, \frac{h}{3}\right)$ to hypotenuse is

$$d = \frac{\left| \frac{hb}{3} + \frac{bh}{3} - bh \right|}{\sqrt{h^2 + b^2}} = \frac{bh}{3\sqrt{h^2 + b^2}}$$

(c) $V = 2\pi d A = 2\pi \dfrac{bh}{3\sqrt{h^2 + b^2}} \cdot \dfrac{1}{2}bh = \dfrac{\pi b^2 h^2}{3\sqrt{h^2 + b^2}}$

29. (a) $\left(\frac{2}{3}a, \frac{1}{3}h\right)$ (b) $\left(\frac{2}{3}a + \frac{1}{3}b, \frac{1}{3}h\right)$ (c) $\left(\frac{1}{3}a + \frac{1}{3}b, \frac{1}{3}h\right)$

30. (a) $V = 2\pi \overline{y} A = 2\pi \dfrac{1}{3}h \dfrac{1}{2}bh = \dfrac{1}{3}\pi bh^2$ [using Exercise 29(c)]

(b) $V = 2\pi \overline{x} A = 2\pi \dfrac{1}{3}(a + b) \dfrac{1}{2}bh = \dfrac{1}{3}\pi(a + b)bh.$

31. (a) $V = \frac{2}{3}\pi R^3 \sin^3\theta + \frac{1}{3}\pi R^3 \sin^2\theta\cos\theta = \frac{1}{3}\pi R^3 \sin^2\theta\,(2\sin\theta + \cos\theta)$

(b) $\overline{x} = \dfrac{V}{2\pi A} = \dfrac{\frac{1}{3}\pi R^3 \sin^2\theta\,(2\sin\theta + \cos\theta)}{2\pi\left(\frac{1}{2}R^2\sin\theta\cos\theta + \frac{1}{4}\pi R^2\sin^2\theta\right)} = \dfrac{2R\sin\theta\,(2\sin\theta + \cos\theta)}{3\,(\pi\sin\theta + 2\cos\theta)}$

32. (a) $\overline{x} = 0, \overline{y} = 0$ (by symmetry).

(b) $\overline{x} = 0$ (by symmetry about y-axis), $\overline{y} = r + \dfrac{4r}{3\pi}$ by Example 6.

(c) $\overline{x} = 0$ (by symmetry about y-axis).

$$\overline{y} = \frac{1}{A} \int_{-r}^{r} \frac{1}{2}\left[\left(r + \sqrt{r^2 - x^2}\right)^2 - (-r)^2 \right] dx = \frac{1}{A}\left[r\int_{-r}^{r} \sqrt{r^2 - x^2}\,dx + \frac{1}{2}\int_{-r}^{r}(r^2 - x^2)\,dx \right]$$

$$= \frac{1}{\frac{\pi r^2}{2} + 4r^2}\left[r\frac{\pi r^2}{2} + \frac{2}{3}r^3 \right] = \frac{r}{3}\left(\frac{3\pi + 4}{\pi + 8} \right)$$

(d) as in (c): $\overline{x} = 0$,   (by symmetry about y-axis), $\overline{y} = -\frac{r}{3}\left(\frac{3\pi + 4}{\pi + 8}\right)$

(e) $\overline{x} = 0$, $\overline{y} = 0$ (by symmetry).

(f) $A = \pi r^2 + 4r^2$

$$\bar{x}A = \int_{-r}^{r} x \left(2r + \sqrt{r^2 - x^2}\right) dx + \int_{r}^{2r} x\, 2\sqrt{r^2 - (x - r)^2}\, dx = 0 + \bar{x}_{\Omega_2} A_{\Omega_2}$$

$$\implies \bar{x} = \bar{x}_{\Omega_2} \frac{A_{\Omega_2}}{A} = \left(r + \frac{4r}{3\pi}\right) \frac{\pi r^2 / 2}{\pi r^2 + 4r^2} = \frac{r}{6}\left(\frac{3\pi + 4}{\pi + 4}\right)$$

By symmetry about the line $y = x$, $\bar{y} = \bar{x} = \dfrac{r}{6}\left(\dfrac{3\pi + 4}{\pi + 4}\right)$

(g) $A = \dfrac{3}{2}\pi r^2 + 4r^2$

$$\bar{x}A = \bar{x}_{S \cup \Omega_1 \cup \Omega_2}\, A_{S \cup \Omega_1 \cup \Omega_2} \implies \bar{x} = \frac{r}{6}\left(\frac{3\pi + 4}{\pi + 4}\right) \frac{\pi r^2 + 4r^2}{\frac{3}{2}\pi r^2 + 4r^2} = \frac{r}{3}\left(\frac{3\pi + 4}{\pi + 8}\right)$$

$$\bar{y} = 0, \qquad \text{(by symmetry about x-axis).}$$

33.   An annular region; see Exercise 25(a).

34.   Extend the figure as indicated in the diagram below. Denote the area of Ω by A and the centroid by $(\bar{x}, \bar{y})$. The centroid of the parallelogram is the center $([b + 2a]/2, h/2)$ and the area of the parallelogram is $2A$.

The upper triangle has centroid $(b + \bar{x}, h + \bar{y})$ and, by congruence, area A.
The union of the configurations is a triangle with centroid $(2\bar{x}, 2\bar{y})$ and area $4A$. (The linear dimensions have been doubled.) Therefore, by Principle 2,

$$\bar{x}\,A + \left(\frac{b + 2a}{2}\right)2A + (b + \bar{x})A = (2\bar{x})4A$$

$$\bar{y}\,A + \left(\frac{h}{2}\right)2A + (b + \bar{y})AT = (2\bar{y})4A$$

Solve these equations for $\bar{x}$ and $\bar{y}$

and you will see that $\bar{x} = \frac{a+b}{3}, \quad \bar{y} = \frac{h}{3}.$

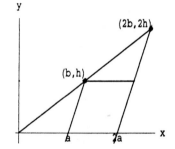

Now verify that these values of $\bar{x}$ and $\bar{y}$ satisfy the equations of the medians.

35.   (a) $A = \frac{1}{2}$ (b) $\left(\frac{16}{35}, \frac{16}{35}\right)$ (c) $V = \frac{16}{35}\pi$ (d) $V = \frac{16}{35}\pi$

36.   (a) $A = \frac{4}{3}$ (b) $\left(2, \frac{23}{5}\right)$ (c) $V = \frac{184}{15}\pi$ (d) $V = \frac{16}{3}\pi$

37.   (a) $A = \frac{250}{3}$ (b) $\left(-\frac{9}{8}, \frac{290}{21}\right) \cong (-1.125, 13.8095)$

38.   (a) $A = \frac{92}{9}$ (b) $\left(-\frac{6}{115}, \frac{127}{69}\right) \cong (-0.0522, 1.8406)$

PROJECT 6.4

1. Let $P = \{x_0, x_1, \ldots, x_n\}$ be a partition of $[a, b]$. P breaks up $[a, b]$ into n subintervals $[x_{i-1}, x_i]$. Choose x_i^* as the midpoint of $[x_{i-1}, x_i]$. By revolving the ith midpoint rectangle about x-axis, we obtain a solid cylinder of volume $V_i = \pi [f(x_i^*)]^2 \Delta x_i$ and centroid (center) on the x-axis at $x = x_i^*$. The union of all these cylinders has centroid at $x = \bar{x}_P$ where

$$x_p V_p = \pi x_1^* [f(x_1^*)]^2 \Delta x_1 + \cdots + \pi x_n^* [f(x_n^*)]^2 \Delta x_n$$

(Here V_P represents the union of the n cylinders.) As $\|P\| \to 0$, the union of the cylinders tends to the shape of S and the equation just derived tends to the given formula.

2. Let $P = \{x_0, x_1, \ldots, x_n\}$ be a partition of $[a, b]$. P breaks up $[a, b]$ into n subintervals $[x_{i-1}, x_i]$. Choose x_i^* as the midpoint of $[x_{i-1}, x_i]$. By revolving the ith midpoint rectangle about y-axis, we obtain a solid cylinder of volume $V_i = 2\pi x_i^* f(x_i^*) \Delta x_i$ and centroid (center) on the y-axis at $y = \frac{1}{2} f(x_i^*)$. The union of all these cylindrical shells has centroid at $y = \bar{y}_P$ where

$$\bar{y}_p V_p = \pi x_1^* [f(x_1^*)]^2 \Delta x_1 + \cdots + \pi x_n^* [f(x_n^*)]^2 \Delta x_n$$

(Here V_P represents volume of the union of the n cylindrical shells.) As $\|P\| \to 0$, the union of the cylinders tends to the shape of S and the equation just derived tends to given formula.

3. (a)

$$\bar{x}\left(\frac{1}{3}\pi r^2 h\right) = \int_0^h \pi x \left(\frac{r}{h}x\right)^2 dx = \frac{1}{4}\pi r^2 h^2$$

$$\bar{x} = \left(\frac{1}{4}\pi r^2 h^2\right) / \left(\frac{1}{3}\pi r^2 h\right) = \frac{3}{4}h.$$

The centroid of the cone lies on the axis of the cone at a distance $\frac{3}{4}h$ from the vertex.

(b) The ball is obtained by rotating $f(x) = \sqrt{r^2 - x^2}, x \in [-r, r]$, around the x-axis.

$$V_x = \frac{2}{3}\pi r^3; \quad \bar{x}V_x = \int_{-r}^r \pi x (r^2 - x^2)\, dx = \pi \left[r^2 \frac{x^2}{2} - \frac{x^4}{4}\right]_{-r}^r = 0$$

$\implies \bar{x} = 0$; no surprise here.

(c) $V_x = \displaystyle\int_0^a \pi \frac{b^2}{a^2}\left(a^2 - x^2\right) dx = \frac{2}{3}\pi a b^2, \quad \bar{x}V_x = \int_0^a \pi x \frac{b^2}{a^2}\left(a^2 - x^2\right) dx = \frac{1}{4}\pi a^2 b^2$

$\bar{x} = \left(\frac{1}{4}\pi a^2 b^2\right) / \left(\frac{2}{3}\pi a b^2\right) = \frac{3}{8}a;$ centroid $\left(\frac{3}{8}a, 0\right)$

(d) (i) $V_x = \displaystyle\int_0^1 \pi \left(\sqrt{x}\right)^2 dx = \frac{1}{2}\pi, \quad \bar{x}V_x = \int_0^1 \pi x \left(\sqrt{x}\right)^2 dx = \frac{1}{3}\pi$

$\bar{x} = \left(\frac{1}{3}\pi\right) / \left(\frac{1}{2}\pi\right) = \frac{2}{3};$ centroid $\left(\frac{2}{3}, 0\right)$

(ii) $V_y = \displaystyle\int_0^1 2\pi x \sqrt{x}\, dx = \frac{4}{5}\pi, \quad \bar{y}V_y = \int_0^1 \pi x \left(\sqrt{x}\right)^2 dx = \frac{1}{3}\pi$

$\bar{y} = \left(\frac{1}{3}\pi\right) / \left(\frac{4}{5}\pi\right) = \frac{5}{12};$ centroid $\left(0, \frac{5}{12}\right)$

(e) (i) $V_x = \displaystyle\int_0^2 \pi(4 - x^2)^2\, dx = \dfrac{256}{15}\pi$

$\overline{x}V_x = \displaystyle\int_0^2 \pi x(4 - x^2)^2\, dx = \pi\left[-\dfrac{(4-x^2)^3}{6}\right]_0^2 = \dfrac{32\pi}{3} \implies \overline{x} = \dfrac{5}{8}; \quad \overline{y} = 0$

(ii) $V_y = \displaystyle\int_0^2 2\pi x(4 - x^2)\, dx = 8\pi$

$\overline{y}V_y = \displaystyle\int_0^2 \pi x(4 - x^2)^2\, dx = \dfrac{32}{3}\pi \implies \overline{y} = \dfrac{4}{3}; \quad \overline{x} = 0$

SECTION 6.5

1. $W = \displaystyle\int_1^4 x\left(x^2 + 1\right)^2 dx = \dfrac{1}{6}\left[(x^2 + 1)^3\right]_1^4 = 817.5$ ft-lb

2. $W = \displaystyle\int_3^8 2x\sqrt{x+1}\, dx = \int_4^9 2(u-1)\sqrt{u}\, du = 2\left[\dfrac{2}{5}u^{5/2} - \dfrac{2}{3}u^{3/2}\right]_4^9 = \dfrac{2152}{15}$ ft-lb

3. $W = \displaystyle\int_1^3 x\sqrt{x^2 + 7}\, dx = \dfrac{1}{3}\left[(x^2 + 7)^{\frac{3}{2}}\right]_0^3 = \dfrac{1}{3}(64 - 7^{\frac{3}{2}})$ newton-meters

4. $W = \displaystyle\int_0^{\frac{\pi}{4}} (x^2 + \cos 2x)\, dx = \left[\dfrac{x^3}{3} + \dfrac{\sin 2x}{2}\right]_0^{\frac{\pi}{4}} = (\dfrac{\pi^3}{192} + \dfrac{1}{2})$ newton-meters

5. $W = \displaystyle\int_{\pi/6}^{\pi} (x + \sin 2x)\, dx = \left[\dfrac{1}{2}x^2 - \dfrac{1}{2}\cos 2x\right]_{\pi/6}^{\pi} = \dfrac{35}{72}\pi^2 - \dfrac{1}{4}$ newton-meters

6. $W = \displaystyle\int_0^{\pi/2} \dfrac{\cos 2x}{\sqrt{2 + \sin 2x}}\, dx = \left[(2 + \sin 2x)^{1/2}\right]_0^{\pi/2} = 0$ newton-meters

7. By Hooke's law, we have $600 = -k(-1)$. Therefore $k = 600$.

The work required to compress the spring to 5 inches is given by

$$W = \int_{10}^5 600(x - 10)\, dx = 600\left[\dfrac{1}{2}x^2 - 10x\right]_{10}^5$$

$$= 7500 \text{ in-lb, or } 625 \text{ ft-lb}$$

8. Work done by spring $= -5 = \displaystyle\int_1^3 -kx\, dx = -k\left[\dfrac{x^2}{2}\right]_1^3 = -4k \implies k = \dfrac{5}{4}$.

We want s such that $-6 = \displaystyle\int_0^s -\dfrac{5}{4}x\, dx = -\dfrac{5}{4}\dfrac{s^2}{2} \implies s = \dfrac{4\sqrt{3}}{\sqrt{5}}$ feet.

9. To counteract the restoring force of the spring we must apply a force $F(x) = kx$.
Since $F(4) = 200$, we see that $k = 50$ and therefore $F(x) = 50x$.

(a) $W = \displaystyle\int_0^1 50x\, dx = 25$ ft-lb   (b) $W = \displaystyle\int_0^{3/2} 50x\, dx = \dfrac{225}{4}$ ft-lb

10. (a) $W = \int_0^a -kx\, dx = -\frac{k}{2}a^2 \implies \int_0^{2a} -kx\, dx = -\frac{k}{2}(2a)^2 = 4\left(-\frac{k}{2}a^2\right) = 4W$

(b) $\int_0^{na} -kx\, dx = -\frac{k}{2}n^2 a^2 = n^2 W$

(c) $\int_a^{2a} -kx\, dx = -\frac{k}{2}(4a^2 - a^2) = 3\left(-\frac{k}{2}a^2\right) = 3W$

(d) $\int_a^{na} -kx\, dx = \int_0^{na} -kx\, dx - \int_0^a -kx\, dx = n^2 W - W = (n^2 - 1)W$

11. Let L be the natural length of the spring.

$$\int_{2-L}^{2.1-L} kx\, dx = \frac{1}{2}\int_{2.1-L}^{2.2-L} kx\, dx$$

$$\left[\tfrac{1}{2}kx^2\right]_{2-L}^{2.1-L} = \tfrac{1}{2}\left[\tfrac{1}{2}kx^2\right]_{2.1-L}^{2.2-L}$$

$$(2.1 - L)^2 - (2 - L)^2 = \tfrac{1}{2}\left[(2.2 - L)^2 - (2.1 - L)^2\right].$$

Solve this equation for L and you will find that $L = 1.95$.   Answer: 1.95 ft

12. (a) $W = \int_a^b \sigma s(x) A(x)\, dx = \int_0^6 62.5\, x\, 4\pi\, dx = 4{,}500\pi$ ft-lb

(b) $W = \int_0^6 62.5\,(x+5)\cdot 4\pi\, dx = 12{,}000\pi$ ft-lb.

13.

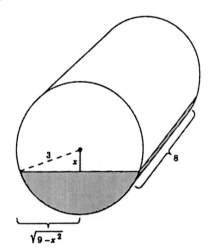

$\sqrt{9-x^2}$

(a) $W = \int_0^3 (x+3)\,(60)\,(8)\left(2\sqrt{9-x^2}\,\right)\, dx$

$= 960 \int_0^3 x\left(9-x^2\right)^{1/2}\, dx$

$+ 2880 \underbrace{\int_0^3 \sqrt{9-x^2}\, dx}_{\substack{\text{area of quarter}\\\text{circle of radius 3}}}$

$= 960\left[-\tfrac{1}{3}\left(9-x^2\right)^{3/2}\right]_0^3 + 2880\left[\tfrac{9}{4}\pi\right]$

$= (8640 + 6480\pi)$ ft-lb

(b) $W = \int_0^3 (x+7)(60)(8)\left(2\sqrt{9-x^2}\,\right)\, dx = 960\int_0^3 x\left(9-x^2\right)^{1/2}\, dx + 6720\int_0^3 \sqrt{9-x^2}\, dx$

$= (8640 + 15120\pi)$ ft-lb

14. (a) $W = \int_0^2 62.5(4-x)(6)\left(2\sqrt{4-(2-x)^2}\right) dx + \int_2^4 62.5(4-x)(6)\left(2\sqrt{4-(x-2)^2}\right) dx$

$750 \int_0^2 (4-x)\sqrt{4-(2-x)^2}\, dx = 750 \int_0^2 (2+u)\sqrt{4-u^2}\, du = 750\left[\frac{8}{3} + 2\pi\right]$

$(u = 2-x,\ du = -dx,\ 4-x = 2+(2-x);\ u(0) = 2,\ u(2) = 0)$

$750 \int_2^4 (4-x)\sqrt{4-(x-2)^2}\, dx = 750 \int_2^4 (2-u)\sqrt{4-u^2}\, du = 750\left[-\frac{8}{3} + 2\pi\right]$

$(u = x-2,\ du = dx,\ 4-x = 2-(x-2);\ u(2) = 0,\ u(4) = 2)$

$W = 750(4\pi) = 3000\pi$ ft-lb

(b) $W = \int_0^2 62.5(9-x)(6)\left(2\sqrt{4-(2-x)^2}\right) dx + \int_2^4 62.5(9-x)(6)\left(2\sqrt{4-(x-2)^2}\right) dx$

$ = 10,500\pi$ ft-lb

15. Set $w(x) =$ the weight of the chain from height x to the ground. (Weight measured in pounds, height measured in feet.) Then $w(x) = 1.5x$ and

$$W = \int_0^{50} 1.5x\, dx = 1.5\left[\frac{x^2}{2}\right]_0^{50} = 1875 \text{ ft-lb}$$

16. The lifting force also acts in the negative direction.

17.

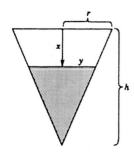

By similar triangles

$$\frac{h}{r} = \frac{h-x}{y} \quad \text{so that} \quad y = \frac{r}{h}(h-x).$$

Thus, the area of a cross section of the fluid at a depth of x feet is

$$\pi y^2 = \pi \frac{r^2}{h^2}(h-x)^2.$$

(a) $W = \int_0^{h/2} x\sigma\left[\pi\frac{r^2}{h^2}(h-x)^2\right] dx = \frac{\sigma\pi r^2}{h^2}\int_0^{h/2}\left(h^2 x - 2hx^2 + x^3\right) dx = \frac{11}{192}\sigma\pi r^2 h^2$ ft-lb

(b) $W = \int_0^{h/2} (x+k)\sigma\left[\pi\frac{r^2}{h^2}(h-x)^2\right] dx = \frac{11}{192}\pi r^2 h^2\sigma + \frac{7}{24}\pi r^2 hk\sigma$ ft-lb

18. $W = \int_0^h \sigma(h-x)\pi\frac{r^2}{h^2}(h-x)^2\, dx = \sigma\pi\frac{r^2}{h^2}\left[-\frac{(h-x)^4}{4}\right]_0^h = \frac{1}{4}\sigma\pi r^2 h^2$ ft-lb.

19. $y = \dfrac{3}{4}x^2,\ 0 \le x \le 4$

(a) $W = \displaystyle\int_0^{12} \sigma(12 - y)\pi x^2\,dy = \dfrac{4}{3}\pi\sigma \int_0^{12}(12y - y^2)\,dy$

$= \dfrac{4}{3}\pi\sigma \left[6y^2 - \dfrac{y^3}{3}\right]_0^{12} = \dfrac{4}{3}\pi\sigma(288) = 384\pi\sigma \text{ newton-meters.}$

(b) $W = \displaystyle\int_0^{12} \sigma(13 - y)\pi x^2\,dy = \dfrac{4}{3}\pi\sigma \int_0^{12}(13y - y^2)\,dy$

$= \dfrac{4}{3}\pi\sigma \left[\dfrac{13y^2}{2} - \dfrac{y^3}{3}\right]_0^{12} = \dfrac{4}{3}\pi\sigma(360) = 480\pi\sigma \text{ newton-meters.}$

20. $W = \displaystyle\int_{r_1}^{r_2} F\,dr = \int_{r_1}^{r_2} -\dfrac{GmM}{r^2}\,dr = \left[\dfrac{GmM}{r}\right]_{r_1}^{r_2} = GmM\left[\dfrac{1}{r_2} - \dfrac{1}{r_1}\right]$

21. $W = \displaystyle\int_0^{80}(80 - x)15\,dx = 15\left[80x - \dfrac{1}{2}x^2\right]_0^{80} = 48,000 \text{ ft-lb}$

22. (a) $W = wd$ ft-lb

(b) component of force along the inclined plane $= w\sin\theta$ lb

distance traveled $= \dfrac{d}{\sin\theta}$ ft; $W = (w\sin\theta)\dfrac{d}{\sin\theta} = wd$ ft-lb.

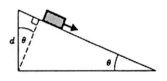

23. (a) $W = 200 \cdot 100 = 20,000$ ft-lb

(b) $W = \displaystyle\int_0^{100}\left[(100 - x)2 + 200\right]dx$

$= \displaystyle\int_0^{100}(400 - 2x)\,dx$

$= \left[400x - x^2\right]_0^{100} = 30,000 \text{ ft-lb}$

24. (a) Loses 50 pounds over 100 feet, so weight x feet from top is

$$W(x) = 200 - \dfrac{50}{100}(100 - x) = 150 + \dfrac{x}{2}$$

Thus, work $= \displaystyle\int_0^{100}\left(150 + \dfrac{x}{2}\right)dx = 17,500 \text{ ft-lb}$

(b) Just add the work done lifting the chain, namely $\displaystyle\int_0^{100} 2x\,dx = 10,000$ ft-lb.

Total work $= 27,500$ ft-lb

25. The bag is raised 8 feet and loses a total of 1 pound at a constant rate. Thus, the bag loses sand at the rate of 1/8 lb/ft. After the bag has been raised x feet it weighs $100 - \dfrac{x}{8}$ pounds.

$$W = \int_0^8\left(100 - \dfrac{x}{8}\right)dx = \left[100x - \dfrac{x^2}{16}\right]_0^8 = 796 \text{ ft-lb.}$$

26.   Weight at depth x is $40 - \dfrac{1}{20}(8.3)(40 - x) = 0.415x + 23.4$.

Thus $W = \displaystyle\int_0^{40} (0.415x + 23.4)\,dx = \left[0.415\dfrac{x^2}{2} + 23.4x\right]_0^{40} = 1268$ ft-lb.

27.   (a) $W = \displaystyle\int_0^l x\sigma\,dx = \dfrac{1}{2}\sigma l^2$ ft-lb

   (b) $W = \displaystyle\int_0^l (x + l)\,\sigma\,dx = \dfrac{3}{2}\sigma l^2$ ft-lb

28.   Work $= wh + \displaystyle\int_0^h \sigma x\,dx = \left(wh + \dfrac{1}{2}\sigma h^2\right)$ ft-lb.

29.   Thirty feet of cable and the steel beam weighing a total of: $800 + 30(6) = 980$ lbs. are raised 20 feet. The work required is: $(20)(980) = 19,600$ ft-lb.

Next, the remaining 20 feet of cable is raised a varying distance and wound onto the steel drum. Thus the total work is given by

$$W = 19,600 + \int_0^{20} 6x\,dx = 19,600 + 1,200 = 20,800 \text{ ft-lb.}$$

30.   Reaches height x at time $t = x/n$, and at that time weighs $w - 8.3pt = w - 8.3px/n$ pounds,

Therefore,   work $= \displaystyle\int_0^m \left(w - 8.3p\dfrac{x}{n}\right)\,dx = \left(wm - \dfrac{4.15pm^2}{n}\right)$ ft-lb.

31.   Let $\lambda(x)$ be the mass density of the chain at the point x units above the ground. Let g be the gravitational constant. the work done to pull the chain to the top of the building is given by:

$$W = \int_0^H (H - x)\,g\,\lambda(x)\,dx = Hg\int_0^H \lambda(x)\,dx - g\int_0^H x\,\lambda(x)\,dx$$

$$= HgM - g\,\overline{x}\,M = (H - \overline{x})\,gM$$

$$= \text{(weight of chain)} \times \text{(distance from the center of mass to top of building)}$$

32.   By the hint

$$W = \int_a^b F(x)\,dx = \int_a^b ma\,dx = \int_a^b mv\dfrac{dv}{dx}\,dx = \int_{v_a}^{v_b} mv\,dv = \left[\dfrac{1}{2}mv^2\right]_{v_a}^{v_b} = \dfrac{1}{2}\,mv_b^2 - \dfrac{1}{2}\,mv_a^2$$

33.   The height of the object at time t is $y(t) = -\frac{1}{2}gt^2 + h$;   time at impact is $t = \sqrt{2h/g}$;   velocity at time t is $v(t) = -gt$;   velocity at impact is $v\left(\sqrt{2h/g}\right) = -g\sqrt{2h/g} = -\sqrt{2gh}$.

34.   If the speed of the second object is 3 times the speed of the first, then the mass of the first object is 9 times the mass of the second.

35. Assume that $v_a = 0$ and $v_b = 95$ mph.

mass of ball: $\dfrac{5/16}{32} = \dfrac{1}{512}$;   speed in feet per second: $\dfrac{(95)(5280)}{3600}$

$$W = \frac{1}{2}\frac{1}{512}\left[\frac{(95)(5280)}{3600}\right]^2 \cong 18.96 \text{ ft-lbs.}$$

36. Mass of the vehicle: $\dfrac{2000}{32} = 62.5$;

speed in ft/sec: 30 mph $= 44$ ft/sec; 55 mph $\cong 80.67$ ft/sec.

$$W = \tfrac{1}{2}(62.5)(80.67)^2 - \tfrac{1}{2}(62.5)(44)^2 \cong 142{,}864 \text{ ft-lbs.}$$

37. Assume that $v_a = 0$ and $v_b = 17{,}000$ mph.

mass of satellite: $\dfrac{1000}{32} = 31.25$;   speed in feet per second: $\dfrac{(17{,}000)(5280)}{3600}$

$$W = \frac{1}{2}(31.25)\left[\frac{(17{,}000)(5280)}{3600}\right]^2 \cong 9.714 \times 10^9 \text{ ft-lbs.}$$

38. (a) Acceleration: $a = \dfrac{88}{15}$ feet/sec^2,   Force: $F = ma = \dfrac{w}{g}a = \dfrac{3000}{32}\cdot\dfrac{88}{15} = 550$ lbs

$v(t) = at$,   so $p = \dfrac{dW}{dt} = F\left(x(t)\right)v(t) = 550\cdot\dfrac{88}{15}t$

The engine must be able to sustain this until $t = 15$,

so need $p = 550(88)$ ft-lb/s $= 88$ horse power.

(b) Now there is a $\dfrac{4}{\sqrt{10016}}\cdot 3000$ component of gravity acting against the motion, so the total force

needed is $550 + \dfrac{4\cdot 3000}{\sqrt{10016}} \cong 670$ lbs, so the required power is:

$$p = 670(88) \text{ ft-lb/sec} = 107.2 \text{ horse power.}$$

39. (a) The work required to pump the water out of the tank is given by

$$W = \int_5^{10}(62.5)\pi\,5^2 x\,dx = 1562.5\pi\left[\frac{1}{2}x^2\right]_5^{10} \cong 184{,}078 \text{ ft-lb}$$

A $\frac{1}{2}$-horsepower pump can do 275 ft-lb of work per second. Therefore it will take

$\dfrac{184{,}078}{275} \cong 669$ seconds $\cong 11$ min, 10 sec, to empty the tank.

(b) The work required to pump the water to a point 5 feet above the top of the tank is given by

$$W = \int_5^{10}(62.5)\pi\,5^2(x+5)\,dx = \int_5^{10}(62.5)\pi\,5^2 x\,dx + \int_5^{10}(62.5)\pi\,5^3\,dx \cong 306796 \text{ ft-lb}$$

It will take a $\frac{1}{2}$-horsepower pump approximately $1{,}116$ sec, or 18 min, 36 sec, in this case.

40.   (a)
$$W = \int_0^8 60x16\pi \, dx + \int_0^4 60(x+8)\pi(16 - x^2) \, dx$$

$$= 960\pi \left[\frac{x^2}{2} \right]_0^8 + 60\pi \int_0^4 (128 + 16x - 8x^2 - x^3) \, dx$$

$$= 30,720\pi + 60\pi \left[128x + 8x^2 - \frac{8x^3}{3} - \frac{x^4}{4} \right]_0^4 = 30,720\pi + 24,320\pi$$

$$= 55,040\pi \cong 172,913 \text{ ft-lbs.}$$

(b)   Pump does $\frac{1}{2}(550) = 275$ ft-lbs/sec. Therefore, it will take $\dfrac{172,913}{275} \cong 629$ seconds, or 10.5 minutes.

41. $KE = \frac{1}{2}mv^2; \quad \dfrac{d\,KE}{dt} = mv\dfrac{dv}{dt} = mav = Fv = P.$

SECTION 6.6

1. $F = \displaystyle\int_0^6 (62.5) \cdot x \cdot 8 \, dx = 250 \left[x^2 \right]_0^6 = 9{,}000 \text{ lbs}$

2. $F = \displaystyle\int_1^7 62.5 \cdot x \cdot 6 \, dx = (62.5)(3) \left[x^2 \right]_1^7 = 9{,}000 \text{ lbs}$

3.   The width of the plate x meters below the surface is given by $w(x) = 60 + 2(20 - x) = 100 - 2x$ (see the figure). The force against the dam is

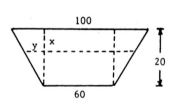

$$F = \int_0^{20} 9800x(100 - 2x) \, dx$$

$$= 9800 \int_0^{20} (100x - 2x^2) \, dx$$

$$= 9800 \left[50x^2 - \tfrac{2}{3}x^3 \right]_0^{20}$$

$$\cong 1.437 \times 10^8 \text{ newtons}$$

4. $F = \displaystyle\int_{15}^{20} 9800 \cdot x \cdot 5 \, dx = (4900)(5) \left[x^2 \right]_{15}^{20} = 4{,}287{,}500 \text{ N}$

5.   The width of the gate x meters below its top is given by $w(x) = 4 + \frac{2}{3}x$ (see the figure). The force of the water against the gate is

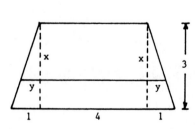

$$F = \int_0^3 9800(10 + x)\left(4 + \frac{2}{3}x \right) dx$$

$$= 9800 \int_0^3 \left[\frac{2}{3}x^2 + \frac{32}{3}x + 40 \right] dx$$

$$= 9800 \left[\tfrac{2}{9}x^3 + \tfrac{16}{3}x^2 + 40x \right]_0^3$$

$$\cong 1.7052 \times 10^6 \text{ Newtons}$$

6. (a) $F = \int_0^{75} 62.5x \cdot 1000 \, dx = (62.5)(1000)\left(\dfrac{75^2}{2}\right) = 175,781,250$ lbs.

(b) $F = \int_0^{50} 62.5x \cdot 1000 \, dx = (62.5)(1000)\left(\dfrac{50^2}{2}\right) = 78,125,000$ lbs.

7.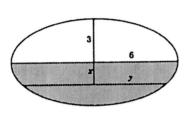

$$F = \int_0^3 (60)\,x\left[12\sqrt{1 - \dfrac{x^2}{9}}\,\right] dx$$

$$= 240\int_0^3 x\left(9 - x^2\right)^{1/2} dx$$

$$= 240\left[-\tfrac{1}{3}\left(9 - x^2\right)^{3/2}\right]_0^3 = 2160 \text{ lb}$$

ellipse: $\dfrac{x^2}{3^2} + \dfrac{y^2}{6^2} = 1$

8.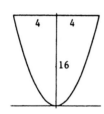

$$F = \int_0^{16} \sigma x w(x)\,dx = \int_0^{16} 70x2\sqrt{16-x}\,dx = 140\int_0^{16} x\sqrt{16-x}\,dx$$

$$= -140\int_{16}^0 (16-u)\sqrt{u}\,du = -40\left[\dfrac{32}{3}u^{3/2} - \dfrac{2}{5}u^{5/2}\right]_{16}^0$$

$$= \dfrac{114,688}{3} \text{ lb}$$

9.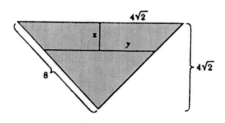

By similar triangles

$$\dfrac{4\sqrt{2}}{4\sqrt{2}} = \dfrac{y}{4\sqrt{2}-x} \quad\text{so}\quad y = 4\sqrt{2}-x.$$

$$F = \int_0^{4\sqrt{2}} (62.5)\,x\left[2\left(4\sqrt{2}-x\right)\right] dx$$

$$= 125\int_0^{4\sqrt{2}}\left(4\sqrt{2}\,x - x^2\right) dx = \dfrac{8000}{3}\sqrt{2} \text{ lb}$$

10.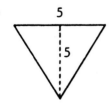

By similar triangles, $\dfrac{W(x)}{5} = \dfrac{5-x}{5} \implies W(x) = 5 - x.$

$$F = \int_0^5 62.5x(5-x)\,dx = 62.5\left[\dfrac{5x^2}{2} - \dfrac{x^3}{3}\right]_0^5 = \dfrac{15,625}{12} \text{ lb.}$$

11.

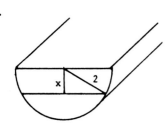

$$F = \int_0^2 (62.5) \cdot x \cdot 2\sqrt{4 - x^2}\, dx$$

$$= 125 \int_0^2 x\sqrt{4 - x^2}\, dx$$

$$= -\frac{125}{3}\left[\left(4 - x^2\right)^{3/2}\right]_0^2 = 333.33 \text{ lb}$$

12. $F = \int_0^4 62.5y\, 2\sqrt{4 - y}\, dy$

13.

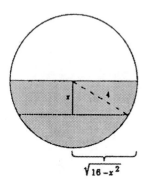

$$\sqrt{16 - x^2}$$

$$F = \int_0^4 60x\left(2\sqrt{16 - x^2}\right)\, dx$$

$$= 120 \int_0^4 x\left(16 - x^2\right)^{1/2}\, dx$$

$$= 120\left[-\frac{1}{3}\left(16 - x^2\right)^{3/2}\right]_0^4 = 2560 \text{ lb}$$

14. $F = \int_0^8 60x\, 2\sqrt{16 - (x - 4)^2}\, dx = 120 \int_{-4}^4 (u + 4)\sqrt{16 - u^2}\, du$

$$= 120 \int_{-4}^4 u\sqrt{16 - u^2}\, du + 120 \cdot 4 \int_{-4}^4 \sqrt{16 - u^2}\, du$$

$$= 0 + 120 \cdot 4\,[\text{Area of half circle}] = 480 \cdot \frac{1}{2} \cdot 16\pi = 3840\pi \text{ lb.}$$

15.   (a) The width of the plate is 10 feet and the depth of the plate ranges from 8 feet to 14 feet. Thus

$$F = \int_8^{14} 62.5x\,(10)\, dx = 41,250 \text{ lb.}$$

(b) The width of the plate is 6 feet and the depth of the plate ranges from 6 feet to 16 feet. Thus

$$F = \int_6^{16} 62.5x\,(6)\, dx = 41,250 \text{ lb.}$$

16. $W(x) = 2\pi r = 30\pi.$ $F = \int_0^{50} 60x 30\pi\, dx = 900\pi 50^2 = 2,250,000\pi \text{ lb.}$

17.   (a) Force on the sides:

$$F = \int_0^1 (9800)\, x\, 14\, dx + \int_0^2 (9800)(1 + x)\, 7(2 - x)\, dx$$

$$= 68,600\left[x^2\right]_0^1 + 68,600 \int_0^2 \left[2 + x - x^2\right]\, dx$$

$$= 68,600 + 68,600\left[2x + \frac{1}{2}x^2 - \frac{1}{3}x^3\right]_0^2$$

$$\cong 297,267 \text{ Newtons}$$

(b) Force at the shallow end:

$$F = \int_0^1 (9800) \cdot x \cdot 8\,dx = 39,200 \left[x^2\right]_0^1 = 39,200 \text{ Newtons}$$

Force at the deep end:

$$F = \int_0^3 (9800) \cdot x \cdot 8\,dx = 39,200 \left[x^2\right]_0^3 = 352,800 \text{ Newtons}$$

18. $\displaystyle F = \int_a^b \sigma x w(x)\,dx = \sigma \int_a^b x w(x)\,dx = \sigma \overline{x} A$

where A is the area of the submerged surface and $\overline{x}$ is the depth of the centroid.

19. From Exercise 18, $\quad F_1 = \sigma \overline{x}_1 A_1 = \sigma h_1 A_1, \ F_2 = \sigma h_2 A_2.$ Since $A_1 = A_2, \ F_2 = \dfrac{h_2}{h_1} F_1$

20. Following the argument given for a vertical plate, the approximate force on the ith strip is

$$\sigma x_i^* w(x_i) \sec \theta \Delta x_i$$

Therefore, the force F on the plate is given by

$$F = \int_a^b \sigma x w(x) \sec \theta\,dx$$

21. $\displaystyle F = \int_0^{14} (9800) \left(1 + \frac{1}{7}x\right) \cdot 8\frac{5\sqrt{2}}{7}\,dx = 392,000\frac{\sqrt{2}}{7} \left[x + \frac{1}{14}x^2\right]_0^{14} \cong 2.217 \times 10^6 \text{ Newtons}$

22. (a) $\displaystyle F = \int_0^{100} 62.5x(1000) \sec(\pi/6)\,dx$ $\qquad$ (b) $\displaystyle F = \int_0^{75} 62.5x(1000) \sec(\pi/6)\,dx$

$\qquad = \dfrac{125,000}{\sqrt{3}} \left[\dfrac{x^2}{2}\right]_0^{100} \cong 361,000,000 \text{ lbs}$ $\qquad = \dfrac{125,000}{\sqrt{3}} \left[\dfrac{x^2}{2}\right]_0^{75} \cong 203,000,000 \text{ lbs}$

REVIEW EXERCISES

1.

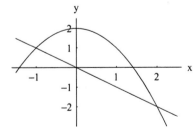

$$A = \int_{-1}^2 \left[2 - x^2 + x\right] dx$$

$$= \int_1^2 2\sqrt{2 - y}\,dy + \int_{-2}^1 \left[\sqrt{2 - y} + y\right] dy$$

$$A = \int_{-1}^2 \left[2 - x^2 + x\right] dx = \frac{9}{2}$$

2.

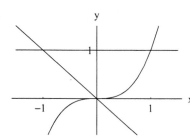

$$A = \int_0^1 \left[y^{1/3} + y \right] dy = \int_{-1}^0 (1 + x)\, dx + \int_0^1 (1 - x^3)\, dx$$

$$A = \int_0^1 [y^{1/3} + y]\, dy = 5/4$$

3.

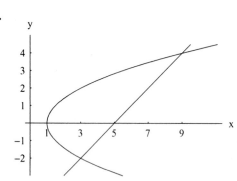

$$A = \int_{-2}^4 \left[4 + y - \frac{y^2}{2} \right] dy$$

$$= \int_1^3 2\sqrt{2(x-1)}\, dx + \int_3^9 \left[\sqrt{2(x-1)} - x + 5 \right] dx$$

$$A = \int_{-2}^4 \left[4 + y - \frac{y^2}{2} \right] dy = 18$$

4.

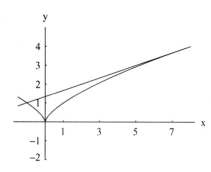

$$A = \int_{-1}^8 \left[\frac{x+4}{3} - x^{2/3} \right] dx$$

$$= \int_0^1 2y^{3/2}\, dy + \int_1^4 \left[y^{3/2} - 3y + 4 \right] dy$$

$$A = \int_{-1}^8 \left[\frac{x+4}{3} - x^{2/3} \right] dx = 27/10$$

5. Consecutive intersections of $\sin x$ and $\cos x$ occur at $x = \frac{\pi}{4}$, $x = \frac{5\pi}{4}$.

$$A = \int_{\pi/4}^{5\pi/4} (\sin x - \cos x)\, dx = \left[-\cos x - \sin x \right]_{\pi/4}^{5\pi/4} = 2\sqrt{2}$$

6. $A = \int_0^{\pi/4} \tan^2 x\, dx = \int_0^{\pi/4} (\sec^2 x - 1)\, dx = \left[\tan x - x \right]_0^{\pi/4} = 1 - \pi/4$

7. By symmetry, $A = 2 \int_0^1 (1 - x)\sqrt{x}\, dx = 2 \int_0^1 \left(x^{1/2} - x^{3/2} \right) dx = 2 \left[\frac{2}{3} x^{3/2} - \frac{2}{5} x^{5/2} \right]_0^1 = \frac{8}{15}$

8. $A = \int_0^a (a + x - 2\sqrt{ax})\, dx = \left[ax + \frac{1}{2} x^2 - \frac{4}{3} a^{1/2} x^{3/2} \right]_0^a = \frac{1}{6} a^2$

9. (a) $V = \int_{-r}^{r} A(x)\,dx = \int_{-r}^{r} \frac{1}{2}\pi \left(\sqrt{r^2 - x^2}\right)^2 dx = \int_{-r}^{r} \frac{\pi}{2}(r^2 - x^2)\,dx = \frac{\pi}{2}\left[r^2 x - \frac{1}{3}\pi x^3\right]_{-r}^{r} = \frac{2}{3}\pi r^3$

(b) $V = \int_{-r}^{r} A(x)\,dx = \int_{-r}^{r} \frac{1}{2}(2)\sqrt{r^2 - x^2}\sqrt{r^2 - x^2}\,dx = \int_{-r}^{r}(r^2 - x^2)\,dx = \left[r^2 x - \frac{1}{3}x^3\right]_{-r}^{r} = \frac{4}{3}r^3$

10. $V = \int_{0}^{a\sqrt{3}/2} \frac{4}{3}x^2\,dx = \frac{\sqrt{3}}{6}a^3$

11. $V = \int_{0}^{3} \frac{1}{2}\pi \left(\frac{3-x}{3}\right)^2 dx = \frac{1}{2}\pi \int_{0}^{3}\left(1 - \frac{2}{3}x + \frac{1}{9}x^2\right)dx = \frac{1}{2}\pi\left[x - \frac{1}{3}x^2 + \frac{1}{27}x^3\right]_{0}^{3} = \frac{1}{2}\pi$

12. $V = \int_{0}^{3} \frac{2}{\sqrt{3}}x\sqrt{9 - x^2}\,dx = -\frac{2}{3\sqrt{3}}\left[(9 - x^2)^{3/2}\right]_{0}^{3} = 2$

13. $V = \int_{0}^{2} \pi\left[(x/2)^2 - (x^2/4)^2\right]dx = \pi \int_{0}^{2}\left(\frac{x^2}{4} - \frac{x^4}{16}\right)dx = \pi\left[\frac{x^3}{12} - \frac{x^5}{80}\right]_{0}^{2} = \frac{4\pi}{15}$

14. $V = \int_{0}^{1} \pi[4y - 4y^2]\,dy = \frac{2\pi}{3}$ or $V = \int_{0}^{2} 2\pi x\left(\frac{x}{2} - \frac{x^2}{4}\right)dx = \frac{2\pi}{3}$

15. $V = \int_{0}^{1} \pi[(1)^2 - (x^3)^2]\,dx = \pi \int_{0}^{1}(1 - x^6)\,dx = \frac{6\pi}{7}$

16. $V = \int_{0}^{1} \pi\left(y^{1/3}\right)^2 dy = \pi \int_{0}^{2} y^{2/3}\,dy = \frac{3\pi}{5}$

17. $V = \int_{0}^{\pi/4} \pi\sec^2 x\,dx = \pi\left[\tan x\right]_{0}^{\pi/4} = \pi$

18. $V = \int_{-\pi/2}^{\pi/2} \pi\cos^2 x\,dx = \frac{\pi}{2}\int_{-\pi/2}^{\pi/2}[1 + \cos 2x]\,dx = \frac{\pi}{2}\left[x + \frac{1}{2}\sin 2x\right]_{-\pi/2}^{\pi/2} = \frac{\pi^2}{2}$

19. $V = \int_{0}^{\sqrt{\pi}} 2\pi x\sin x^2\,dx = \pi \int_{0}^{\sqrt{\pi}} 2x\sin x^2\,dx = \pi\left[-\cos x^2\right]_{0}^{\sqrt{\pi}} = 2\pi$

20. $V = \int_{0}^{\sqrt{\pi/2}} 2\pi x\cos x^2\,dx = \pi \int_{0}^{\sqrt{\pi/2}} 2x\cos x^2\,dx = \pi\left[\sin x^2\right]_{0}^{\sqrt{\pi/2}} = \pi$

21. By symmetry, $V = 2\int_{0}^{3} 2\pi x(3x - x^2)\,dx = 4\pi \int_{0}^{3}(3x^2 - x^3)\,dx = 4\pi\left[x^3 - \frac{1}{4}x^4\right]_{0}^{3} = 27\pi$

22. By symmetry, $V = 2\int_{0}^{3} 2\pi(4 - x)(3x - x^2)\,dx = 4\pi \int_{0}^{3}(12x - 7x^2 + x^3)\,dx$

$= 4\pi\left[6x^2 - \frac{7}{3}x^3 + \frac{1}{4}x^4\right]_{0}^{3} = 45\pi$

23. $V = \int_{0}^{3} \pi[(x + 1)^2 - (x - 1)^4]\,dx = \pi\left[\frac{1}{3}(x + 1)^3 - \frac{1}{5}(x - 1)^5\right]_{0}^{3} = \frac{72}{5}\pi$

24. $V = \int_0^5 2\pi x[3x - (x^2 - 2x)]\, dx = 2\pi \int_0^5 (5x^2 - x^3)\, dx = 2\pi \left[\frac{5}{3}x^3 - \frac{1}{4}x^4\right]_0^5 = \frac{625}{6}\pi$

25. With respect to x: $V = \int_0^4 \pi\left[(2^2 - (\sqrt{x})^2\right]\, dx = \pi \int_0^4 (4 - x)\, dx;$

with respect to y: $V = \int_0^2 2\pi y(y^2)\, dy = 2\pi \int_0^2 y^3\, dy = 8\pi$

26. With respect to x: $V = \int_0^4 \pi\left(2 - \sqrt{x}\right)^2 dx;$

with respect to y: $V = \int_0^2 2\pi(2 - y)y^2\, dy = 2\pi \int_0^2 (2y^2 - y^3)\, dy = \frac{8}{3}\pi$

27. With respect to x: $V = \int_0^4 2\pi(x + 1)\left(\sqrt{x} - \frac{1}{2}x\right) dx = 2\pi \int_0^4 \left(x^{3/2} + x^{1/2} - \frac{1}{2}x^2 - \frac{1}{2}x\right) dx = \frac{104}{15}\pi$

with respect to y: $V = \int_0^2 \pi\{[2y + 1]^2 - [y^2 + 1]^2\}\, dy$

28. With respect to x: $V = \int_0^4 2\pi x(\sqrt{x} - \frac{1}{2}x)\, dx$

with respect to y: $V = \int_0^2 \pi[(2y)^2 - (y^2)^2]\, dy = \pi \int_0^2 (4y^2 - y^4)\, dy = \frac{64}{15}\pi$

29. With respect to x: $V = \int_0^4 2\pi x(\frac{1}{2}x)\, dx = \pi \int_0^4 x^2\, dx = \frac{64}{3}\pi;$

with respect to y: $V = \int_0^2 \pi[4^2 - (2y)^2]\, dy$

30. With respect to x: $V = \int_0^4 \pi\left[\left(\frac{1}{2}x + 2\right)^2 - (2)^2\right] dx;$

with respect to y: $V = \int_0^2 2\pi(y + 2)(4 - 2y)\, dy = 2\pi \int_0^2 (8 - 2y^2)\, dy = \frac{64}{3}\pi$

31. Since the region is symmetric about the y-axis, $\overline{x} = 0$;

$A = \int_{-2}^2 (4 - x^2)\, dx = \frac{32}{3};$

$\overline{y}A = \int_{-2}^2 \frac{1}{2}\left(4 - x^2\right)^2 dx = \frac{1}{2}\int_{-2}^2 \left(16 - 8x^2 + x^4\right) dx = \frac{256}{15};\quad \overline{y} = \frac{8}{5}$

32. By symmetry, the centroid of the region is $(0, 0)$. The centroid of the first quadrant part is:

$A = \int_0^2 (4x - x^3)dx = \left[2x^2 - \frac{1}{4}x^4\right]_0^2 = 4$

$\overline{x}A = \int_0^2 x(4x - x^3)dx = \frac{64}{15} \implies \overline{x} = \frac{16}{15},\quad \overline{y}A = \int_0^2 \frac{1}{2}\left[(4x)^2 - (x^3)^2\right]dx = \frac{256}{21} \implies \overline{y} = \frac{64}{21}$

33. $A = \int_{-1}^{2} \left[(2x - x^2) - (x^2 - 4) \right] dx = \int_{-1}^{2} \left(2x - 2x^2 + 4 \right) dx = 9;$

$$\bar{x}A = \int_{-1}^{2} x \left(2x - 2x^2 + 4 \right) dx = \int_{-1}^{2} \left(2x^2 - 2x^3 + 4x \right) dx = \frac{9}{2};$$

$$\bar{y}A = \int_{-1}^{2} \frac{1}{2} \left[(2x - x^2) - (x^2 - 4) \right] dx = \frac{1}{2} \int_{-1}^{2} \left[12x^2 - 4x^3 - 16 \right] dx = -\frac{27}{2};$$

$$\bar{x} = \frac{1}{2}; \quad \bar{y} = -\frac{3}{2}$$

34. Since the region is symmetric about the y axis, $\bar{x} = 0$

$$A = \int_{-\frac{\pi}{2}}^{\frac{\pi}{2}} \cos x\, dx = 2$$

$$\bar{y}A = \int_{-\frac{\pi}{2}}^{\frac{\pi}{2}} \frac{1}{2} \cos^2 x\, dx = \frac{1}{4} \int_{-\frac{\pi}{2}}^{\frac{\pi}{2}} (\cos 2x + 1) dx = \frac{\pi}{4} \quad \Longrightarrow \quad \bar{y} = \frac{\pi}{8}$$

35. $A = \int_{0}^{1} \left[(2 - x^2) - x \right] dx = \int_{0}^{1} \left(2 - x^2 - x \right) dx = \frac{7}{6};$

$$\bar{x}A = \int_{0}^{1} x(2 - x^2 - x) dx = \int_{0}^{1} \left(2x - x^3 - x^2 \right) dx = \frac{5}{12}; \quad \bar{x} = \frac{5}{14}$$

$$\bar{y}A = \int_{0}^{1} \frac{1}{2} \left[(2 - x^2)^2 - x^2 \right] dx = \frac{1}{2} \int_{0}^{1} \left[4 - 5x^2 + x^4 \right] dx = \frac{19}{15}; \quad \bar{y} = \frac{38}{35}$$

around x-axis: $V = 2\pi \left(\dfrac{38}{35} \right) \left(\dfrac{7}{6} \right) = \dfrac{38\pi}{15}$

around y-axis: $V = 2\pi \left(\dfrac{5}{14} \right) \left(\dfrac{7}{6} \right) = \dfrac{5\pi}{6}$

36. Since the region is symmetric about the line $y = x$, $\bar{x} = \bar{y}$

$$A = \int_{0}^{1} (x^{\frac{1}{3}} - x^3) dx = \frac{1}{2}$$

$$\bar{x}A = \int_{0}^{1} x(x^{\frac{1}{3}} - x^3) dx = \frac{8}{35}$$

$$\bar{x} = \bar{y} = \frac{16}{35}$$

$$V_y = V_x = \int_{0}^{1} \pi(x^{2/3} - x^6) dx = \frac{16\pi}{35}$$

37. $W = \int_{0}^{3} F(x) dx = \int_{0}^{3} x\sqrt{7 + x^2}\, dx = \left[\frac{1}{3}(7 + x^2)^{3/2} \right]_{0}^{3} = \frac{1}{3}(64 - 7^{3/2})$ ft-lbs

38. $F(x) = -kx;$ $8000 = -k\left(\frac{-1}{2} \right)$ $\Longrightarrow$ $k = 16,000$

$$W = \int_{0}^{-3} (-16,000\, x)\, dx = 16,000 \int_{-3}^{0} x\, dx = 72,000 \text{ in-lbs} = 6,000 \text{ ft-lbs}$$

39. Assume the natural length is x_0. Then,

$$\int_9^{10} k(x - x_0)\, dx = \frac{3}{2} \int_8^9 k(x - x_0)\, dx$$

The solution of this equation is $x_0 = \frac{13}{2}$ inches.

40. (a)

$$W = \int_0^5 \sigma(10 - x)\pi \left(\frac{2}{5}x\right)^2 dx = \frac{4\pi\sigma}{25} \int_0^5 \left(10x^2 - x^3\right) dx$$

$$= \frac{4\pi\sigma}{25}\left[\frac{10}{3}x^3 - \frac{1}{4}x^4\right]_0^5 = \frac{125\pi\sigma}{3} \cong 8181.23 \text{ ft-lbs}$$

(b) $W = \int_0^5 \sigma(9 - x)\pi \left(\frac{2}{5}x\right)^2 dx = 35\pi\sigma \cong 6872.23$ ft-lbs

41. $W = \int_0^{25} 4(25 - x)\, dx = \left[100x - 2x^2\right]_0^{25} = 1250$ ft-lbs

42. $W = \int_0^{20} (5 + 60 - x)\, dx = \int_0^{20} (65 - x)\, dx = \left[65x - \frac{1}{2}x^2\right]_0^{20} = 1100$ ft-lbs.

43. $W = \int_0^{10} 60(20 - x)\pi(20x - x^2)\, dx = 60\pi \int_0^{10} (400x - 40x^2 + x^3)\, dx$

$$= 60\pi\left[200x^2 - \frac{40}{3}x^3 + \frac{1}{4}x^4\right]_0^{10} = 550,000\pi \text{ ft-lbs.}$$

44. (a) The force on the $1 \times 1/2$ side is: $F = \int_0^{1/2} 9800x\, dx = 1225$ newtons.

The force on the $1/2 \times 1/2$ side is: $F = \int_0^{1/2} 9800x(\frac{1}{2})\, dx = 612.5$ newtons.

(b) The force on the bottom is: $F = 9800 \times \frac{1}{2} \times \frac{1}{2} \times 1 = 2450$ newtons.

45. (a) $F = \int_0^{50} 9800x(100 - 2x)\, dx = 9800 \int_0^{50} (100x - 2x^2)\, dx$

$$= 9800\left[50x^2 - \frac{2}{3}x^3\right]_0^{50} = \frac{1225}{3} \times 10^6 \text{ newtons.}$$

(b) $F = \int_{10}^{50} 9800x(100 - 2x)\, dx = \frac{10976}{3} \times 10^5 \text{newtons.}$

CHAPTER 7

SECTION 7.1

1. Suppose $f(x_1) = f(x_2)$ $x_1 \neq x_2$. Then

    $5x_1 + 3 = 5x_2 + 3 \implies x_1 = x_2$;

    f is one-to-one

    $$f(y) = x$$
    $$5y + 3 = x$$
    $$5y = x - 3$$
    $$y = \tfrac{1}{5}(x - 3)$$
    $$f^{-1}(x) = \tfrac{1}{5}(x - 3)$$
    $$\operatorname{dom} f^{-1} = (-\infty, \infty)$$

2. $f^{-1}(x) = \dfrac{1}{3}(x - 5)$

    $\operatorname{dom} f^{-1} = (-\infty, \infty)$

3. f is not one-to-one; e.g. $f(1) = f(-1)$

4. $f^{-1}(x) = x^{1/5}$; $\operatorname{dom} f^{-1} = (-\infty, \infty)$

5. $f'(x) = 5x^4 \geq 0$ on $(-\infty, \infty)$ and

    $f'(x) = 0$ only at $x = 0$; f is increasing.

    Therefore, f is one-to-one.

    $$f(y) = x$$
    $$y^5 + 1 = x$$
    $$y^5 = x - 1$$
    $$y = (x - 1)^{1/5}$$
    $$f^{-1}(x) = (x - 1)^{1/5}$$
    $$\operatorname{dom} f^{-1} = (-\infty, \infty)$$

6. not one-to-one;   e.g. $f(0) = f(3)$

7. $f'(x) = 9x^2 \geq 0$ on $(-\infty, \infty)$ and

    $f'(x) = 0$ only at $x = 0$; f is increasing.

    Therefore, f is one-to-one.

    $$f(y) = x$$
    $$1 + 3y^3 = x$$
    $$y^3 = \tfrac{1}{3}(x - 1)$$
    $$y = \left[\tfrac{1}{3}(x - 1)\right]^{1/3}$$
    $$f^{-1}(x) = \left[\tfrac{1}{3}(x - 1)\right]^{1/3}$$
    $$\operatorname{dom} f^{-1} = (-\infty, \infty)$$

8. $f^{-1}(x) = (x + 1)^{1/3}$

    $\operatorname{dom} f^{-1} = (-\infty, \infty)$

9. $f'(x) = 3(1-x)^2 \geq 0$ on $(-\infty, \infty)$ and

$f'(x) = 0$ only at $x = 1$; f is increasing.

Therefore, f is one-to-one.

$$f(y) = x$$
$$(1-y)^3 = x$$
$$1-y = x^{1/3}$$
$$y = 1 - x^{1/3}$$
$$f^{-1}(x) = 1 - x^{1/3}$$
$$\text{dom } f^{-1} = (-\infty, \infty)$$

10. not one-to-one;   e.g. $f(0) = f(2)$.

11. $f'(x) = 3(x+1)^2 \geq 0$ on $(-\infty, \infty)$ and

$f'(x) = 0$ only at $x = -1$; f is increasing.

Therefore, f is one-to-one.

$$f(y) = x$$
$$(y+1)^3 + 2 = x$$
$$(y+1)^3 = x - 2$$
$$y+1 = (x-2)^{1/3}$$
$$y = (x-2)^{1/3} - 1$$
$$f^{-1}(x) = (x-2)^{1/3} - 1$$
$$\text{dom } f^{-1} = (-\infty, \infty)$$

12. $f^{-1}(x) = \dfrac{1}{4}(x^{1/3} + 1)$

$\text{dom } f^{-1} = (-\infty, \infty)$

13. $f'(x) = \dfrac{3}{5x^{2/5}} > 0$ for all $x \neq 0$;

f is increasing on $(-\infty, \infty)$

$$f(y) = x$$
$$y^{3/5} = x$$
$$y = x^{5/3}$$
$$f^{-1}(x) = x^{5/3}$$
$$\text{dom } f^{-1} = (-\infty, \infty)$$

14. $f^{-1}(x) = (1-x)^3 + 2$

$\text{dom } f^{-1} = (-\infty, \infty)$

15. $f'(x) = -3(2-3x)^2 \leq 0$ for all x and

$f'(x) = 0$ only at $x = 2/3$; f is decreasing

$$f(y) = x$$
$$(2-3y)^3 = x$$
$$2 - 3y = x^{1/3}$$
$$3y = 2 - x^{1/3}$$
$$y = \tfrac{1}{3}(2 - x^{1/3})$$
$$f^{-1}(x) = \tfrac{1}{3}(2 - x^{1/3})$$
$$\text{dom } f^{-1} = (-\infty, \infty)$$

16. not one-to-one;   e.g. $f(1) = f(-1)$

17. $f'(x) = \cos x \geq 0$ on $[-\pi/2, \pi/2]$; and $f'(x) = 0$ only at $x = -\pi/2, \pi/2$. Therefore f is increasing on $[-\pi/2, \pi/2]$ and f has an inverse. The inverse is denoted by $\arcsin x$; this function will be studied in Section 7.7. The domain of $\arcsin x$ is $[-1, 1] =$ range of $\sin x$ on $[-\pi/2, \pi/2]$.

18. f is not one-to-one on $(-\pi/2, \pi/2)$. For example, $f(-\pi/4) = f(\pi/4) = \dfrac{\sqrt{2}}{2}$.

19. $f'(x) = -\dfrac{1}{x^2} < 0$ for all $x \neq 0$;
   f is decreasing on $(-\infty, 0) \cup (0, \infty)$

$$f(y) = x \quad \Longrightarrow \quad \frac{1}{y} = x \quad \Longrightarrow \quad y = \frac{1}{x}$$

$f^{-1}(x) = \dfrac{1}{x}$ dom f^{-1}: $x \neq 0$

20. $f^{-1}(x) = 1 - \dfrac{1}{x}$
   dom f^{-1}: $x \neq 0$

21. f is not one-to-one;   e.g. $f\left(\frac{1}{2}\right) = f(2)$

22. not one-to-one;   e.g. $f(1) = f(2)$

23. $f'(x) = -\dfrac{3x^2}{(x^3 + 1)^2} \leq 0$ for all $x \neq -1$;
   f is decreasing on $(-\infty, -1) \cup (-1, \infty)$

$$f(y) = x$$
$$\frac{1}{y^3 + 1} = x$$
$$y^3 + 1 = \frac{1}{x}$$
$$y^3 = \frac{1}{x} - 1$$
$$y = \left(\frac{1}{x} - 1\right)^{1/3}$$
$$f^{-1}(x) = \left(\frac{1}{x} - 1\right)^{1/3}$$

dom f^{-1}: $x \neq 0$

24. $f^{-1}(x) = \dfrac{x}{1 + x}$
   dom f^{-1}: $x \neq -1$

25. $f'(x) = \dfrac{-1}{(x + 1)^2} < 0$ for all $x \neq -1$;
   f is decreasing on $(-\infty, -1) \cup (-1, \infty)$

$$f(y) = x$$
$$\frac{y + 2}{y + 1} = x$$
$$y + 2 = xy + x$$
$$y(1 - x) = x - 2$$
$$y = \frac{x - 2}{1 - x}$$
$$f^{-1}(x) = \frac{x - 2}{1 - x}$$

dom f^{-1}: $x \neq 1$

26. not one-to-one;   e.g. $f(1) = f(-3)$

27. They are equal.

28.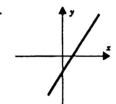

29.

30.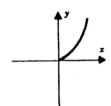

31.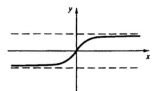

32. (a) Suppose f and g are one-to-one, and that $f(g(x_1)) = f(g(x_2))$. Then since f is one-to-one, $g(x_1) = g(x_2)$, and since g is one-to-one this implies $x_1 = x_2$.

(b) Since $g^{-1}(f^{-1}(f(g(x)))) = g^{-1}(g(x)) = x$, we have $(f \circ g)^{-1} = g^{-1} \circ f^{-1}$.

33. (a) $f'(x) = x^2 + 2x + k$: f will be increasing on $(-\infty, \infty)$ if f' does not change sign. This will occur if the discriminant of f', namely $4 - 4k$ is non-positive.

$$4 - 4k \le 0 \implies k \ge 1$$

(b) $g'(x) = 3x^2 + 2kx + 1$: g will be increasing on $(-\infty, \infty)$ if g' does not change sign. This will occur if the discriminant of g', namely $4k^2 - 12$ is non-positive.

$$4k^2 - 12 \le 0 \implies k^2 \le 3 \implies -\sqrt{3} \le k \le \sqrt{3}$$

34. (a) $\left(f^{-1}\right)'(5) = \dfrac{1}{f'(2)} = -\dfrac{4}{3}$

(b) $g' = \dfrac{-1}{\left(f^{-1}\right)^2} \cdot \left(f^{-1}\right)'$; $\quad g'(3) = \dfrac{-1}{2^2} \cdot \dfrac{1}{2/3} = -\dfrac{3}{8}$

35. $f'(x) = 3x^2 \ge 0$ on $I = (-\infty, \infty)$ and $f'(x) = 0$ only at $x = 0$; f is increasing on I and so it has an inverse.

$$f(2) = 9 \text{ and } f'(2) = 12; \quad (f^{-1})'(9) = \frac{1}{f'(2)} = \frac{1}{12}$$

36. $f'(x) = -2 - 3x^2$; $\quad (f^{-1})'(4) = \dfrac{1}{f'(f^{-1}(4))} = \dfrac{1}{f'(-1)} = -\dfrac{1}{5}$

37. $f'(x) = 1 + \dfrac{1}{\sqrt{x}} > 0$ on $I = (0, \infty)$; f is increasing on I and so it has an inverse.

$$f(4) = 8 \text{ and } f'(4) = 1 + \frac{1}{2} = \frac{3}{2}; \quad (f^{-1})'(8) = \frac{1}{f'(4)} = \frac{1}{3/2} = \frac{2}{3}$$

38. $f'(x) = 1 + \cos x; \quad (f^{-1})'(0) = \dfrac{1}{f'(f^{-1}(-1/2))} = \dfrac{1}{f'(-\pi/6)} = \dfrac{1}{2}$

39. $f'(x) = 2 - \sin x > 0$ on $I = (-\infty, \infty)$; f is increasing on I and so it has an inverse.

$$f(\pi/2) = \pi \text{ and } f'(\pi/2) = 1; \quad (f^{-1})'(\pi) = \frac{1}{f'(\pi/2)} = 1$$

40. $f'(x) = \dfrac{-4}{(x-1)^2}; \quad (f^{-1})'(3) = \dfrac{1}{f'(f^{-1}(3))} = \dfrac{1}{f'(3)} = \dfrac{1}{1/2} = 2$

41. $f'(x) = \sec^2 x > 0$ on $I = (-\pi/2, \pi/2)$; f is increasing on I and so it has an inverse

$$f(\pi/3) = \sqrt{3} \text{ and } f'(\pi/3) = 4; \quad (f^{-1})'(\sqrt{3}) = \frac{1}{f'(\pi/3)} = \frac{1}{4}$$

42. $f'(x) = 5x^4 + 6x^2 + 2; \quad (f^{-1})'(-5) = \dfrac{1}{f'(f^{-1}(-5))} = \dfrac{1}{f'(-1)} = \dfrac{1}{13}$

43. $f'(x) = 3 + \dfrac{3}{x^4} > 0$ on $I = (0, \infty)$; f is increasing on I and so it has an inverse.

$f(1) = 2$ and $f'(1) = 6; \quad f^{-1'}(2) = \dfrac{1}{f'(1)} = \dfrac{1}{6}.$

44. $f'(x) = 1 - \sin x \geq 0$ on $I = [0, \pi]$, with $f(x) = 0$ for only one value on I and so it has an inverse.

$f(\pi) = -1$ and $f'(\pi) = 1; \quad f^{-1'}(-1) = \dfrac{1}{f'(\pi)} = 1.$

45. Let $x \in \text{dom}\,(f^{-1})$ and let $f(z) = x$. Then

$$(f^{-1})'(x) = \frac{1}{f'(z)} = \frac{1}{f(z)} = \frac{1}{x}$$

46. $(f^{-1})'(x) = \dfrac{1}{f'(f^{-1}(x))} = \dfrac{1}{1 + [f(f^{-1}(x))]^2} = \dfrac{1}{1 + x^2}$

47. Let $x \in \text{dom}\,(f^{-1})$ and let $f(z) = x$. Then

$$(f^{-1})'(x) = \frac{1}{f'(z)} = \frac{1}{\sqrt{1 - [f(z)]^2}} = \frac{1}{\sqrt{1 - x^2}}$$

48. (a) The figure indicates that f is one-to-one.   (b)

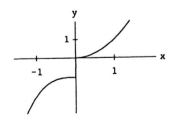

$$f^{-1}(x) = \begin{cases} (x+1)^{1/3}, & x < 0 \\ \sqrt{x} \quad , & x \geq 0 \end{cases}$$

49. (a)
$$f'(x) = \frac{(cx + d)a - (ax + b)c}{(cx + d)^2} = \frac{ad - bc}{(cx + d)^2}, \quad x \neq -d/c$$

Thus, $f'(x) \neq 0$ iff $ad - bc \neq 0$.

(b)
$$\frac{at + b}{ct + d} = x$$

$$at + b = ctx + dx$$

$$(a - cx)t = dx - b$$

$$t = \frac{dx - b}{a - cx}; \quad f^{-1}(x) = \frac{dx - b}{a - cx}$$

50. $f = f^{-1} \implies \dfrac{ax + b}{cx + d} = \dfrac{dx - b}{a - cx}$

$\implies a^2x + ab - acx^2 = cdx^2 + d^2x - bd$

$\implies a = -d$ as long as either b or $c \neq 0$.

If $b = c = 0$, then $a = \pm d$.

51. (a) $f'(x) = \sqrt{1 + x^2} > 0$, so f is always increasing, hence one-to-one.

(b) $(f^{-1})'(0) = \dfrac{1}{f'(f^{-1}(0))} = \dfrac{1}{f'(2)} = \dfrac{1}{\sqrt{5}}$.

52. (a) $f'(x) = \sqrt{16 + (2x)^4}\,(2) = 8\sqrt{1 + x^4} > 0$ for all x.

(b) Since $f(1/2) = 0$, we have
$$(f^{-1})'(0) = \frac{1}{f'(1/2)} = \frac{1}{2\sqrt{17}} = \frac{\sqrt{17}}{34}$$

53. (a) $g'(x) = \dfrac{1}{f'[g(x)]}; \qquad g''(x) = -\dfrac{1}{(f'[g(x)])^2}f''[g(x)]g'(x) = -\dfrac{f''[g(x)]}{(f'[g(x)])^3}$

(b) If f' is increasing, then the graphs of f and g have opposite concavity;
if f' is decreasing then the graphs of f and g have the same concavity.

54. (a) No. If p is a polynomial of even degree, then $\displaystyle\lim_{x \to \pm\infty} p(x) = \infty$ or $\displaystyle\lim_{x \to \pm\infty} p(x) = -\infty$.

(b) Yes, for instance $P(x) = x^3$ has an inverse. $P(x) = x^3 - x$ does not have an inverse.

55. Let $f(x) = \sin x$ and let $y = f^{-1}(x)$. Then

$$\sin y = x$$

$$\cos y \frac{dy}{dx} = 1$$

$$\frac{dy}{dx} = \frac{1}{\cos y} \quad (y \neq \pm\pi/2)$$

$$= \frac{1}{\sqrt{1 - \sin^2 y}} = \frac{1}{\sqrt{1 - x^2}} \quad (x \neq \pm 1)$$

56. Let $y = f^{-1}(x)$. Then $\tan y = x$, so $\sec^2 y \dfrac{dy}{dx} = 1$.

Thus $\dfrac{dy}{dx} = \dfrac{1}{\sec^2 y} = \cos^2 y = \dfrac{1}{1 + x^2}$

57. $f^{-1}(x) = \dfrac{x^2 - 8x + 25}{9}, \quad x \geq 4$

58. $f^{-1}(x) = \dfrac{5x}{3 - 2x}$

59. $f^{-1}(x) = 16 - 12x + 6x^2 - x^3$

60. $f^{-1}(x) = \dfrac{1 - x}{1 + x} = f(x)$

61. $f'(x) = 3x^2 + 3 > 0$ for all x;
f is increasing on $(-\infty, \infty)$

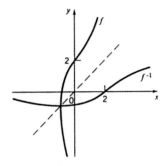

62. $f'(x) = \dfrac{3}{5} x^{-2/5} > 0 \ (x \neq 0)$
f is increasing on $(-\infty, \infty)$

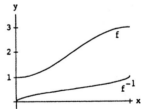

63. $f'(x) = 8 \cos 2x > 0, \ x \in (-\pi/4, \pi/4)$;
f is increasing on $[-\pi/4, \pi/4]$

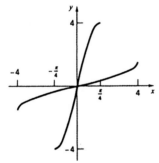

64. $f'(x) = 3 \sin 3x > 0, \ x \in (0, \pi/3)$
f is increasing on $[0, \pi/3]$

SECTION 7.2

1. $\ln 20 = \ln 2 + \ln 10 \cong 2.99$

2. $\ln 16 = \ln 2^4 = 4 \ln 2 \cong 2.78$

3. $\ln 1.6 = \ln \frac{16}{10} = 2 \ln 4 - \ln 10 \cong 0.48$

4. $\ln 3^4 = 4 \ln 3 \cong 4.39$

5. $\ln 0.1 = \ln \frac{1}{10} = \ln 1 - \ln 10 \cong -2.30$

6. $\ln 2.5 = \ln \frac{10}{4} = \ln 10 - \ln 4 \cong 0.92$

7. $\ln 7.2 = \ln \frac{72}{10} = \ln 8 + \ln 9 - \ln 10 \cong 1.98$

8. $\ln \sqrt{630} = \frac{1}{2} \ln(9 \cdot 7 \cdot 10) = \frac{1}{2}(\ln 9 + \ln 7 + \ln 10) \cong 3.22$

9. $\ln\sqrt{2} = \frac{1}{2}\ln 2 \cong 0.35$

10. $\ln 0.4 = \ln\frac{4}{10} = \ln 4 - \ln 10 \cong -0.92$

11. For any positive integer k: $\displaystyle\int_{k}^{2k}\frac{1}{x}\,dx = [\ln x]_{k}^{2k} = \ln 2k - \ln k = \ln\left(\frac{2k}{k}\right) = \ln 2.$

12. Fix a positive integer m. For any positive integer k:

$$\int_{k}^{km}\frac{1}{x}\,dx = [\ln x]_{k}^{km} = \ln km - \ln k = \ln\left(\frac{km}{k}\right) = \ln m.$$

13. $\frac{1}{2}[L_f(P) + U_f(P)] = \frac{1}{2}\left[\frac{763}{1980} + \frac{1691}{3960}\right] \cong 0.406$ **14.** $\ln 2.5 \cong \frac{1}{2}[L_f(P) + U_f(P)] \cong 0.921$

15. (a) $\ln 5.2 \cong \ln 5 + \frac{1}{5}(0.2) \cong 1.65$

(b) $\ln 4.8 \cong \ln 5 - \frac{1}{5}(0.2) \cong 1.57$ (b)

(c) $\ln 5.5 \cong \ln 5 + \frac{1}{5}(0.5) \cong 1.71$

16. (a) $\ln 10.3 \cong \ln 10 + \frac{1}{10}(0.3) \cong 2.33$

(b) $\ln 9.6 \cong \ln 10 + \frac{1}{10}(-0.4) \cong 2.26$

(c) $\ln 11 \cong \ln 10 + \frac{1}{10}(1) \cong 2.40$

17. $x = e^2$ **18.** $x = \dfrac{1}{e}$ **19.** $2 - \ln x = 0$ or $\ln x = 0$. Thus $x = e^2$ or $x = 1$.

20. $\ln x^{1/2} - \ln(2x - 1) = 0 \Longrightarrow \ln\dfrac{\sqrt{x}}{2x-1} = 0 \Longrightarrow \dfrac{\sqrt{x}}{2x-1} = 1 \Longrightarrow x = 1$

21.
$$\ln[(2x + 1)(x + 2)] = 2\ln(x + 2)$$
$$\ln[(2x + 1)(x + 2)] = \ln[(x + 2)^2]$$
$$(2x + 1)(x + 2) = (x + 2)^2$$
$$x^2 + x - 2 = 0$$
$$(x + 2)(x - 1) = 0$$
$$x = -2, 1$$

We disregard the solution $x = -2$ since it does not satisfy the initial equation.
Thus, the only solution is $x = 1$.

22. $2\ln(x + 2) - \dfrac{1}{2}\ln x^4 = \ln\dfrac{(x + 2)^2}{x^2} = 1 = \ln e \Longrightarrow \dfrac{(x + 2)^2}{x^2} = e \Longrightarrow x = \dfrac{-2}{(1 \pm \sqrt{e})}$

23. See Exercises 3.1, (3.1.6).

$$\lim_{x\to 1}\frac{\ln x}{x - 1} = \frac{d}{dx}(\ln x)\bigg|_{x=1} = \frac{1}{x}\bigg|_{x=1} = 1$$

24. Let $n > 2$ be a positive integer. Let P be the partition $\{1, 2, \cdots, n\}$. Let $f(x) = 1/x$.
Then the lower sum

$$\frac{1}{2} + \frac{1}{3} + \cdots + \frac{1}{n} < \ln n.$$

Therefore, $k = n$.

25. Continuing Exercise 24, the upper sum

$$\frac{1}{2} + \frac{1}{3} + \cdots + \frac{1}{n-1} > \ln n.$$

Therefore, $k = n - 1$.

26. (a) $\ln 1 - g(1) = 1 > 0,$ $\ln 2 - g(2) = \ln 2 - 2 < 0,$ so by the intermediate-value theorem
$\ln r - g(r) = 0$ for some $r \in [1, 2].$

(b) 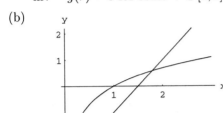 $r \cong 1.7915$

27. (a) Let $G(x) = \ln x - \sin x$. Then $G(2) = \ln 2 - \sin 2 \cong -0.22 < 0$ and $G(3) = \ln 3 - \sin 3 \cong 0.96 > 0.$
Thus, G has at least one zero on $[2, 3]$ which implies that there is at least one number $r \in [2, 3]$ such
that $\sin r = \ln r.$

(b) 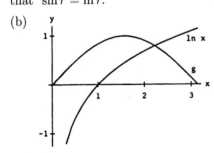 $r \cong 2.2191$

28. (a) $\ln 1 - \dfrac{1}{1^2} = -1 < 0,$ $\ln 2 - \dfrac{1}{2^2} \cong 0.69 - \dfrac{1}{4} > 0,$ so by the intermediate-value theorem

$\ln r - \dfrac{1}{r^2} = 0$ for some $r \in [1, 2].$

(b) 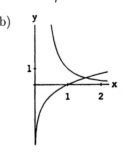 $r \cong 1.5316$

29. $L = 1$ **30.** $L = 0$

SECTION 7.3

1. $\text{dom}(f) = (0, \infty)$, $f'(x) = \dfrac{1}{4x}(4) = \dfrac{1}{x}$

2. $\text{dom}(f) = \left(-\dfrac{1}{2}, \infty\right)$, $f'(x) = \dfrac{2}{2x+1}$

3. $\text{dom}(f) = (-1, \infty)$, $f'(x) = \dfrac{1}{x^3+1}\dfrac{d}{dx}(x^3+1) = \dfrac{3x^2}{x^3+1}$

4. $\text{dom}(f) = (-1, \infty)$, $f'(x) = \dfrac{3}{x+1}$

5. $\text{dom}(f) = (-\infty, \infty)$, $f(x) = \dfrac{1}{2}\ln(1+x^2)$ so $f'(x) = \dfrac{1}{2}\left[\dfrac{1}{1+x^2}(2x)\right] = \dfrac{x}{1+x^2}$

6. $\text{dom}(f) = (0, \infty)$, $f'(x) = \dfrac{3(\ln x)^2}{x}$

7. $\text{dom}(f) = \{x \mid x \neq \pm 1\}$, $f'(x) = \dfrac{1}{x^4-1}\dfrac{d}{dx}(x^4-1) = \dfrac{4x^3}{x^4-1}$

8. $\text{dom}(f) = (1, \infty)$, $f'(x) = \dfrac{1}{x\ln x}$

9. $\text{dom}(f) = \left(-\dfrac{1}{2}, \infty\right)$,

$$f'(x) = (2x+1)^2\dfrac{d}{dx}[\ln(2x+1)] + 2(2x+1)(2)\ln(2x+1)$$

$$= (2x+1)^2\dfrac{2}{2x+1} + 4(2x+1)\ln(2x+1)$$

$$= 2(2x+1) + 4(2x+1)\ln(2x+1) = 2(2x+1)[1 + 2\ln(2x+1)]$$

10. $\text{dom}(f) = (-\infty, -2)\cup(-2, 1)\cup(1, \infty)$, $f'(x) = \dfrac{1}{x+2} - \dfrac{3x^2}{x^3-1}$ (rewrite $f(x)$ as $\ln|x+2| - \ln|x^3-1|$)

11. $\text{dom}(f) = (0, 1)\cup(1, \infty)$, $f(x) = (\ln x)^{-1}$ so $f'(x) = -(\ln x)^{-2}\dfrac{d}{dx}(\ln x) = -\dfrac{1}{x(\ln x)^2}$

12. $\text{dom}(f) = (-\infty, \infty)$, $f'(x) = \dfrac{x}{2(x^2+1)}$ (rewrite $f(x)$ as $\dfrac{1}{4}\ln(x^2+1)$.)

13. $\text{dom}(f) = (0, \infty)$, $f'(x) = \cos(\ln x)\left(\dfrac{1}{x}\right) = \dfrac{\cos(\ln x)}{x}$

14. $\text{dom}(f) = (0, \infty)$, $f'(x) = -\dfrac{\sin(\ln x)}{x}$

15. $\displaystyle\int \dfrac{dx}{x+1} = \ln|x+1| + C$

16. $\displaystyle \int \frac{dx}{3-x} = -\int \frac{dx}{x-3} = -\ln|x-3| + C$

17. $\left\{ \begin{array}{l} u = 3 - x^2 \\ du = -2x\,dx \end{array} \right\}$; $\displaystyle \int \frac{x}{3-x^2}\,dx = -\frac{1}{2}\int \frac{du}{u} = -\frac{1}{2}\ln|u| + C = -\frac{1}{2}\ln|3-x^2| + C$

18. $\displaystyle \int \frac{x+1}{x^2}\,dx = \int \left(\frac{1}{x} + \frac{1}{x^2} \right) dx = \ln|x| - \frac{1}{x} + C$

19. $\left\{ \begin{array}{l} u = 3x \\ du = 3dx \end{array} \right\}$; $\displaystyle \int \tan 3x\,dx = \frac{1}{3}\int \tan u\,du = \frac{1}{3}\ln|\sec u| + C = \frac{1}{3}\ln|\sec 3x| + C$

20. $\displaystyle \int \sec \frac{\pi}{2}\,x\,dx = \frac{2}{\pi}\ln\left|\sec \frac{\pi}{2}x + \tan \frac{\pi}{2}x\right| + C$

21. $\left\{ \begin{array}{l} u = x^2 \\ du = 2x\,dx \end{array} \right\}$; $\displaystyle \int x\sec x^2\,dx = \frac{1}{2}\int \sec u\,du = \frac{1}{2}\ln|\sec u + \tan u| + C$

$$= \tfrac{1}{2}\ln|\sec x^2 + \tan x^2| + C$$

22. $\left\{ \begin{array}{l} u = 2 + \cot x \\ du = -\csc^2 x\,dx \end{array} \right\}$; $\displaystyle \int \frac{\csc^2 x}{2+\cot x}\,dx = -\int \frac{du}{u} = -\ln|u| + C = -\ln|2+\cot x| + C.$

23. $\left\{ \begin{array}{l} u = 3 - x^2 \\ du = -2x\,dx \end{array} \right\}$; $\displaystyle \int \frac{x}{(3-x^2)^2}\,dx = -\frac{1}{2}\int \frac{du}{u^2} = \frac{1}{2u} + C = \frac{1}{2(3-x^2)} + C$

24. $\left\{ \begin{array}{l} u = \ln(x+a) \\ du = \dfrac{1}{x+a}\,dx \end{array} \right\}$; $\displaystyle \int \frac{\ln(x+a)}{x+a}\,dx = \int u\,du = \frac{u^2}{2} + C = \frac{1}{2}[\ln(x+a)]^2 + C$

25. $\left\{ \begin{array}{l} u = 2 + \cos x \\ du = -\sin x\,dx \end{array} \right\}$; $\displaystyle \int \frac{\sin x}{2+\cos x}\,dx = -\int \frac{1}{u}\,du = -\ln|u| + C = -\ln|2+\cos x| + C$

26. $\left\{ \begin{array}{l} u = 4 - \tan 2x \\ du = -2\sec^2 2x\,dx \end{array} \right\}$; $\displaystyle \int \frac{\sec^2 2x}{4-\tan 2x}\,dx = -\frac{1}{2}\int \frac{du}{u} = -\frac{1}{2}\ln|u| + C = -\frac{1}{2}\ln|4-\tan 2x| + C$

27. $\left\{ u = \ln x,\ du = \dfrac{dx}{x} \right\}$; $\displaystyle \int \frac{dx}{x\ln x} = \int \frac{du}{u} = \ln u + C = \ln|\ln x| + C$

28. $\left\{ \begin{array}{l} u = 2x^3 - 1 \\ du = 6x^2\,dx \end{array} \right\}$; $\displaystyle \int \frac{x^2}{2x^3-1}\,dx = \frac{1}{6}\int \frac{du}{u} = \frac{1}{6}\ln|u| + C = \frac{1}{6}\ln|2x^3-1| + C$

29. $\left\{ u = \ln x,\ du = \dfrac{dx}{x} \right\}$; $\displaystyle \int \frac{dx}{x(\ln x)^2} = \int \frac{du}{u^2} = -\frac{1}{u} + C = -\frac{1}{\ln x} + C$

30. $\left\{\begin{array}{l} u = 1 + \sec 2x \\ du = 2\sec 2x \tan 2x\, dx \end{array}\right\};$

$$\int \frac{\sec 2x \tan 2x}{1 + \sec 2x}\, dx = \frac{1}{2}\int \frac{1}{1+u}\, du = \frac{1}{2}\ln|1+u| + C = \frac{1}{2}\ln|1 + \sec 2x| + C$$

31. $\left\{\begin{array}{l} u = \sin x + \cos x \\ du = (\cos x - \sin x)\, dx \end{array}\right\};$

$$\int \frac{\sin x - \cos x}{\sin x + \cos x}\, dx = -\int \frac{1}{u}\, du = -\ln|u| + C = -\ln|\sin x + \cos x| + C$$

32. $\left\{\begin{array}{l} u = \sqrt{x} \\ du = \dfrac{1}{2\sqrt{x}}\, dx \end{array}\right\};$ $\quad \int \dfrac{1}{\sqrt{x}(1 + \sqrt{x})}\, dx = 2\int \dfrac{du}{1+u} = 2\ln|1+u| + C = 2\ln|1 + \sqrt{x}| + C$

33. $\left\{ u = 1 + x\sqrt{x}\,,\; du = \dfrac{3}{2}x^{1/2}\, dx \right\};$ $\quad \int \dfrac{\sqrt{x}}{1 + x\sqrt{x}}\, dx = \dfrac{2}{3}\int \dfrac{du}{u} = \dfrac{2}{3}\ln|u| + C$

$$= \frac{2}{3}\ln\left|1 + x\sqrt{x}\,\right| + C$$

34. $\left\{ u = \ln x,\; du = \dfrac{1}{x}dx \right\};$ $\quad \int \dfrac{\tan(\ln x)}{x}\, dx = \int \tan u\, du = \ln|\sec u| + C = \ln|\sec(\ln x)| + C$

35. $\displaystyle\int (1 + \sec x)^2\, dx = \int \left(1 + 2\sec x + \sec^2 x\right) dx = x + 2\ln|\sec x + \tan x| + \tan x + C$

36. $\displaystyle\int (3 - \csc x)^2\, dx = \int \left(9 - 6\csc x + \csc^2 x\right) dx = 9x - 6\ln|\csc x - \cot x| - \cot x + C$

37. $\displaystyle\int_1^e \frac{dx}{x} = [\ln x]_1^e = \ln e - \ln 1 = 1 - 0 = 1$

38. $\displaystyle\int_1^{e^2} \frac{dx}{x} = [\,\ln|x|\,]_1^{e^2} = \ln e^2 - \ln 1 = 2$

39. $\displaystyle\int_e^{e^2} \frac{dx}{x} = [\,\ln x\,]_e^{e^2} = \ln e^2 - \ln e = 2 - 1 = 1$

40. $\displaystyle\int_0^1 \left(\frac{1}{x+1} - \frac{1}{x+2}\right) dx = \left[\ln\left|\frac{x+1}{x+2}\right|\right]_0^1 = \ln\frac{2}{3} - \ln\frac{1}{2} = \ln\frac{4}{3}$

41. $\displaystyle\int_4^5 \frac{x}{x^2 - 1}\, dx = \left[\frac{1}{2}\ln|x^2 - 1|\right]_4^5 = \frac{1}{2}(\ln 24 - \ln 15) = \frac{1}{2}\ln\frac{8}{5}$

42. $\displaystyle\int_{1/4}^{1/3} \tan \pi x\, dx = \frac{1}{\pi}\left[\ln|\sec \pi x|\right]_{1/4}^{1/3} = \frac{1}{\pi}\left(\ln 2 - \ln\sqrt{2}\right) = \frac{\ln 2}{2\pi}.$

43. $\left\{\begin{array}{l} u = 1 + \sin x \;\big|\; x = \pi/6 \implies u = 3/2 \\ du = \cos x\, dx \;\big|\; x = \pi/2 \implies u = 2 \end{array}\right\};$ $\quad \displaystyle\int_{\pi/6}^{\pi/2} \frac{\cos x}{1 + \sin x}\, dx = \int_{3/2}^2 \frac{du}{u} = [\ln u]_{3/2}^2 = \ln\frac{4}{3}$

44. $\displaystyle\int_{\pi/4}^{\pi/2} (1 + \csc x)^2 \, dx = \int_{\pi/4}^{\pi/2} (1 + 2\csc x + \csc^2 x) \, dx = [x + 2\ln|\csc x - \cot x| - \cot x]_{\pi/4}^{\pi/2}$

$$= \frac{\pi}{4} + 1 - 2\ln(\sqrt{2} - 1)$$

45. $\displaystyle\int_{\pi/4}^{\pi/2} \cot x \, dx = [\,\ln|\sin x|\,]_{\pi/4}^{\pi/2} = \ln 1 - \ln\frac{\sqrt{2}}{2} = \ln\sqrt{2} = \frac{1}{2}\ln 2$

46. $\displaystyle\left\{ u = \ln x, \, du = \frac{dx}{x} \right\}; \quad \int_1^e \frac{\ln x}{x} \, dx = \left[\frac{(\ln x)^2}{2} \right]_1^e = \frac{1}{2}$

47. The integrand $\dfrac{1}{x - 2}$ is not defined at $x = 2$.

48. Let $f(x) = \ln x$. Then $f'(x) = \dfrac{1}{x}$ and $f'(1) = 1$.

By the definition of the derivative at $x = 1$, we have

$$f'(1) = \lim_{h \to 0} \frac{\ln(1 + h) - \ln(1)}{h} = \lim_{h \to 0} \frac{\ln(1 + h)}{h} = \lim_{x \to 0} \frac{\ln(1 + x)}{x} = 1$$

49. $\ln|g(x)| = 2\ln(x^2 + 1) + 5\ln|x - 1| + 3\ln x$

$$\frac{g'(x)}{g(x)} = 2\left(\frac{2x}{x^2 + 1} \right) + \frac{5}{x - 1} + \frac{3}{x}$$

$$g'(x) = (x^2 + 1)^2 (x - 1)^5 x^3 \left(\frac{4x}{x^2 + 1} + \frac{5}{x - 1} + \frac{3}{x} \right); \quad g'(1) = 0$$

50. $\ln|g(x)| = \ln|x| + \ln|x + a| + \ln|x + b| + \ln|x + c|$

$$\frac{g'(x)}{g(x)} = \frac{1}{x} + \frac{1}{x + a} + \frac{1}{x + b} + \frac{1}{x + c}$$

$$g'(x) = x(x + a)(x + b)(x + c)\left(\frac{1}{x} + \frac{1}{x + a} + \frac{1}{x + b} + \frac{1}{x + c} \right); \quad g'(-b) = -b(a - b)(c - b)$$

51. $\ln|g(x)| = 4\ln|x| + \ln|x - 1| - \ln|x + 2| - \ln(x^2 + 1)$

$$\frac{g'(x)}{g(x)} = \frac{4}{x} + \frac{1}{x - 1} - \frac{1}{x + 2} - \frac{2x}{x^2 + 1}$$

$$g'(x) = \frac{x^4(x - 1)}{(x + 2)(x^2 + 1)} \left(\frac{4}{x} + \frac{1}{x - 1} - \frac{1}{x + 2} - \frac{2x}{x^2 + 1} \right); \quad g'(0) = 0$$

52. $\ln|g(x)| = \frac{1}{2}\left(\ln|x - 1| + \ln|x - 2| - \ln|x - 3| - \ln|x - 4| \right)$

$$\frac{g'(x)}{g(x)} = \frac{1}{2}\left(\frac{1}{x - 1} + \frac{1}{x - 2} - \frac{1}{x - 3} - \frac{1}{x - 4} \right)$$

$$g'(x) = \frac{1}{2}\left(\sqrt{\frac{(x - 1)(x - 2)}{(x - 3)(x - 4)}} \right)^2 \left(\frac{1}{x - 1} + \frac{1}{x - 2} - \frac{1}{x - 3} - \frac{1}{x - 4} \right); \quad g'(2) = 0$$

53.

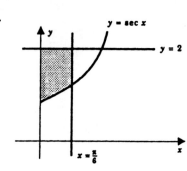

$$A = \int_0^{\pi/6} (2 - \sec x)\, dx$$

$$= [2x - \ln|\sec x + \tan x|]_0^{\pi/6}$$

$$= \frac{\pi}{3} - \ln\left|\frac{2}{\sqrt{3}} + \frac{1}{\sqrt{3}}\right| = \frac{\pi}{3} - \frac{1}{2}\ln 3$$

54.

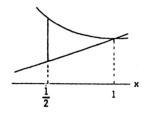

$$A = \int_{1/2}^1 \left(\csc \frac{\pi}{2} x - x\right) dx$$

$$= \left[\frac{2}{\pi}\ln\left|\csc \frac{\pi}{2} x - \cot \frac{\pi}{2} x\right| - \frac{x^2}{2}\right]_{1/2}^1$$

$$= -\frac{2}{\pi}\ln(\sqrt{2} - 1) - \frac{3}{8} = \frac{2}{\pi}\ln(1 + \sqrt{2}) - \frac{3}{8}$$

55.

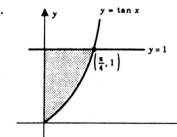

$$A = \int_0^{\pi/4} (1 - \tan x)\, dx$$

$$= [x - \ln|\sec x|]_0^{\pi/4}$$

$$= \frac{\pi}{4} - \ln \sqrt{2} = \frac{\pi}{4} - \frac{1}{2}\ln 2$$

56.

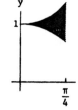

$$A = \int_0^{\pi/4} (\sec x - \cos x)\, dx$$

$$= [\ln|\sec x + \tan x| - \sin x]_0^{\pi/4}$$

$$= \ln(1 + \sqrt{2}) - \frac{\sqrt{2}}{2}$$

57. $A = \int_1^4 \left[\frac{5 - x}{4} - \frac{1}{x}\right] dx = \left[\frac{5}{4}x - \frac{1}{8}x^2 - \ln x\right]_1^4 = \frac{15}{8} - \ln 4$

58. $A = \int_1^2 \left(3 - x - \frac{2}{x}\right) dx = \left[3x - \frac{x^2}{2} - 2\ln x\right]_1^2 = \frac{3}{2} - 2\ln 2$

59. $V = \int_0^8 \pi \left(\frac{1}{\sqrt{1+x}}\right)^2 dx = \pi \int_0^8 \frac{1}{1+x}\, dx = \pi\left[\ln|1+x|\right]_0^8 = \pi \ln 9$

60. By shells: $V = \int_0^3 2\pi x \cdot \frac{3}{1+x^2}\, dx = \left[3\pi \ln(1 + x^2)\right]_0^3 = 3\pi \ln 10$

61.

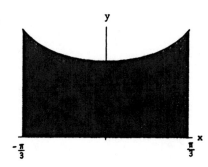

$$V = \int_{-\pi/3}^{\pi/3} \pi \left(\sqrt{\sec x} \right)^2 dx$$

$$= 2\pi \int_0^{\pi/3} \sec x \, dx$$

$$= 2\pi \left[\ln |\sec x + \tan x| \right]_0^{\pi/3} = 2\pi \ln \left(2 + \sqrt{3} \right)$$

62.

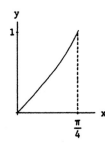

$$V = \int_0^{\pi/4} \pi \tan^2 x \, dx$$

$$= \pi \int_0^{\pi/4} \left(\sec^2 x - 1 \right) dx$$

$$= \pi \left[\tan x - x \right]_0^{\pi/4}$$

$$= \pi \left(1 - \frac{\pi}{4} \right)$$

63. $v(t) = \int a(t) \, dt = \int -(t+1)^{-2} \, dt = \dfrac{1}{t+1} + C.$

Since $v(0) = 1$, we get $1 = 1 + C$ so that $C = 0$. Then

$$s = \int_0^4 |v(t)| \, dt = \int_0^4 \frac{dt}{t+1} = \left[\ln (t+1) \right]_0^4 = \ln 5.$$

The particle traveled $\ln 5$ ft.

64. $v(t) = \int a(t) \, dt = \int -(t+1)^{-2} \, dt = \dfrac{1}{t+1} + C, \quad v(0) = 2 \implies v(t) = \dfrac{1}{t+1} + 1$

Then $\quad s = \int_0^4 v(t) \, dt = \int_0^4 \left(\dfrac{1}{1+t} + 1 \right) dt = \left[\ln(t+1) + t \right]_0^4 = 4 + \ln 5 \ $ ft

65.

$$\frac{d}{dx} (\ln x) = \frac{1}{x}$$

$$\frac{d^2}{dx^2} (\ln x) = -\frac{1}{x^2}$$

$$\frac{d^3}{dx^3} (\ln x) = \frac{2}{x^3}$$

$$\frac{d^4}{dx^4} (\ln x) = -\frac{2 \cdot 3}{x^4}$$

$$\vdots$$

$$\frac{d^n}{dx^n} (\ln x) = (-1)^{n-1} \frac{(n-1)!}{x^n}$$

66.

$$\frac{d}{dx} (\ln(1-x)) = \frac{-1}{1-x}$$

$$\frac{d^2}{dx^2} (\ln(1-x)) = \frac{-1}{(1-x)^2}$$

$$\frac{d^3}{dx^3} (\ln(1-x)) = \frac{-2}{(1-x)^3}$$

$$\frac{d^4}{dx^4} (\ln(1-x)) = \frac{-2 \cdot 3}{(1-x)^4}$$

$$\vdots$$

$$\frac{d^n}{dx^n} [\ln(1-x)] = -\frac{(n-1)!}{(1-x)^n}$$

67. $\displaystyle\int \csc x\,dx = \int \frac{\csc x(\csc x - \cot x)}{\csc x - \cot x}\,dx = \int \frac{\csc^2 - \csc x\,\cot x}{\csc x - \cot x}\,dx$

$\left\{\begin{array}{l} u = \csc x - \cot x \\[4pt] du = (-\csc x\,\cot x + \csc^2 x)\,dx \end{array}\right\}; \quad \displaystyle\int \csc x\,dx = \int \frac{du}{u} = \ln|u| + C$

$$= \ln|\csc x - \cot x| + C$$

68. (a) If $g(x) = g_1(x)g_2(x)$, $(7.3.7)$ gives $g'(x) = g(x)\left[\dfrac{g_1'(x)}{g_1(x)} + \dfrac{g_2'(x)}{g_2(x)}\right]$

$\Longrightarrow g'(x) = g_1(x)g_2(x)\left[\dfrac{g_1'(x)}{g_1(x)} + \dfrac{g_2'(x)}{g_2(x)}\right] = g_1'(x)\,g_2(x) + g_1(x)\,g_2'(x).$

(b) If $g(x) = \dfrac{g_1(x)}{g_2(x)} = g_1(x)\left(\dfrac{1}{g_2(x)}\right)$,   then

$\Longrightarrow g'(x) = g_1(x)\left(\dfrac{1}{g_2(x)}\right)\left(\dfrac{g_1'(x)}{g_1(x)} + \dfrac{\frac{d}{dx}(1/g_2(x))}{1/g_2(x)}\right) = \dfrac{g_1'(x)g_2(x) - g_1(x)g_2'(x)}{[g_2(x)]^2}$

69. $f(x) = \ln(4 - x), \quad x < 4$

$f'(x) = \dfrac{1}{x - 4}$

$f''(x) = \dfrac{-1}{(x - 4)^2}$

(i)   domain $(-\infty, 4)$

(ii)   decreases throughout

(iii)   no extreme values

(iv)   concave down throughout: no pts of inflection

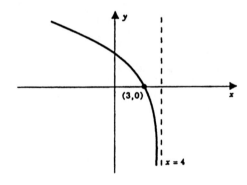

70. $f'(x) = 1 - \dfrac{1}{x}$

$f''(x) = \dfrac{1}{x^2}$

(i) domain $(0, \infty)$

(ii) decreases on $(0, 1]$,   increases on $[1, \infty)$

(iii) $f(1) = 1$ local and absolute min

(iv) concave up on $(0, \infty)$; no pts of inflection

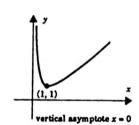

71. $f(x) = x^2 \ln x, \ x > 0$

$f'(x) = 2x \ln x + x$

$f''(x) = 2 \ln x + 3$

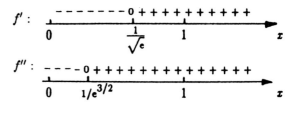

(i) domain $(0, \infty)$

(ii) decreases on $(0, 1/\sqrt{e}]$,   increases on $[1/\sqrt{e}, \infty)$

(iii) $f(1/\sqrt{e}) = -1/2e$ local and absolute min

(iv) concave down on $(0, 1/e^{3/2})$,

   concave up on $(1/e^{3/2}, \infty)$;

   pt of inflection at $(1/e^{3/2}, -3/2e^3)$

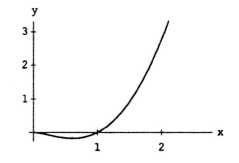

72. $f'(x) = -\dfrac{2x}{4 - x^2}$

$f''(x) = -\dfrac{2(4 + x^2)}{(4 - x^2)^2}$

(i) domain $(-2, 2)$

(ii) increases on $(-2, 0]$,   decreases on $[0, 2)$

(iii) $f(0) = \ln 4$ local and absolute max

(iv) concave down on $(-2, 2)$;

   no pts of inflection

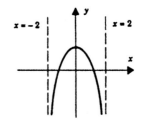

73. $f(x) = \ln \left[\dfrac{x}{1 + x^2} \right], \ x > 0$

$f'(x) = \dfrac{1 - x^2}{x + x^3}$

$f''(x) = \dfrac{x^4 - 4x^2 - 1}{(x + x^3)^2}$

(i) domain $(0, \infty)$

(ii) increases on $(0, 1]$,   decreases on $[1, \infty)$

(iii) $f(1) = $ local and absolute max

(iv) concave down on $(0, 2.0582)$,

   concave up on $(2.0582, \infty)$;

   pt of inflection at $(2.0582, -0.9338)$ (approx.)

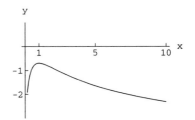

74. $f(x) = \ln\left[\dfrac{x^3}{x-1}\right]$

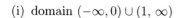

$f'(x) = \dfrac{2x-3}{x^2-x}$

$f''(x) = \dfrac{-(2x^2-6x+3)}{x^2(x-1)^2}$

(i) domain $(-\infty, 0) \cup (1, \infty)$

(ii) increases on $\left[\frac{3}{2}, \infty\right)$, decreases on $(-\infty, 0) \cup \left(1, \frac{3}{2}\right]$

(iii) $f(3/2) \cong 1.91$ local and absolute min

(iv) concave down on $(-\infty, 0) \cup (2.366, \infty))$, concave up on $(1, 2.366)$;

pt of inflection at $(2.366, 2.272)$ (approx.)

75. Average slope $= \dfrac{1}{b-a}\displaystyle\int_a^b \dfrac{1}{x}\,dx = \dfrac{1}{(b-a)}\ln\dfrac{b}{a}$

76. (a) $f'(x) = \dfrac{1}{2x}(2) = \dfrac{1}{x}$; $\quad g'(x) = \dfrac{1}{3x}(3) = \dfrac{1}{x}$ (b) $F'(x) = \dfrac{1}{kx}(k) = \dfrac{1}{x}$

(c) $F(x) = \ln kx = \ln k + \ln x$, $\quad$ so $\quad F'(x) = 0 + \dfrac{d}{dx}(\ln x) = \dfrac{1}{x}$.

77.

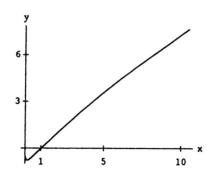

x-intercept: 1; abs min at $x = 1/e^2$;
abs max at $x = 10$

78.

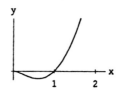

x-intercept at $x=1$; abs min at $x \cong 0.7165$;
abs max at $x = 2$

79.

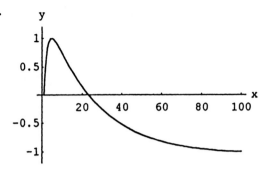

x-intercepts: $1, 23.1407$; abs min at $x = 100$;

abs max at $x = 4.8105$

80.

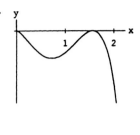

x-intercept at $x = \pi/2$; abs max at $x = \pi/2$;

local min at $x \cong 0.7269$; abs min at $x = 2$;

81. (a) $v(t) - v(0) = \displaystyle\int_0^t a(u)\,du, \quad 0 \le t \le 3$

$$v(t) = \int_0^t \left[4 - 2(u+1) + \frac{3}{u+1}\right] du + 2$$

$$= \left[2u - u^2 + 3\ln|u+1|\right]_0^t = 2 + 2t - t^2 + 3\ln(t+1)$$

(b)

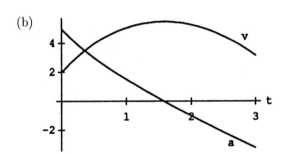

(c) max velocity at $t = 1.5811$; min velocity at $t = 0$

82. (a) $v(t) = \displaystyle\int a(t)\,dt = \int \left[2\cos 2(t+1) + \frac{2}{t+1}\right] dt = \sin 2(t+1) + 2\ln(t+1) + C$

$v(0) = 2 \implies v(t) = \sin 2(t+1) + 2\ln(t+1) + 2 - \sin 2$

(b)

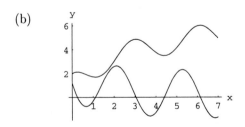

(c) max at $t \cong 6.1389$; min at $t \cong 1.1092$

83. (b) x-coordinates of the points of intersection: $x = 1,\ 3.3028$

(c) $A \cong 2.34042$

84. (b) x-coordinates of the points of intersection: $x_1 \cong -0.6180$, $x_2 \cong 1.6180$, $x_3 \cong 2.6180$

(c) $A = \displaystyle\int_{x_2}^{x_3} [f(x) - g(x)]\, dx \cong 0.2549$

85. (a) $f(x) = \dfrac{\ln x}{x^2}$, $\qquad f'(x) = \dfrac{1 - 2\ln x}{x^3}$, $\qquad f''(x) = \dfrac{-5 + 6\ln x}{x^4}$

(b) $f(1) = 0$, $\qquad f'(e^{1/2}) = 0$, $\qquad f''(e^{5/6}) = 0$

(c) $f(x) > 0$ on $(1, \infty)$ $\qquad f'(x) > 0$ on $(0, e^{1/2})$ $\qquad f''(x) > 0$ on $(e^{5/6}, \infty)$

$\quad\ f(x) < 0$ on $(0, 1)$ $\qquad f'(x) < 0$ on $(e^{1/2}, \infty)$ $\qquad f''(x) > 0$ on $(0, e^{5/6})$

(d) $f(e^{1/2})$ local and absolute maximum

86. (a) $f(x) = \dfrac{1 + 2\ln x}{2\sqrt{\ln x}}$, $\qquad f'(x) = \dfrac{2\ln x - 1}{4x(\ln x)^{3/2}}$, $\qquad f''(x) = \dfrac{3 - 4(\ln x)^2}{8x^2(\ln x)^{5/2}}$

(b) $f(x) \neq 0$, $\qquad f'(e^{1/2}) = 0$, $\qquad f''(e^{\sqrt{3}/2}) = 0$

(c) $f(x) > 0$ on $(1, \infty)$ $\qquad f'(x) > 0$ on $(e^{1/2}, \infty)$ $\qquad f''(x) > 0$ on $(1, e^{\sqrt{3}/2})$

$\qquad\qquad\qquad\qquad\qquad\qquad f'(x) < 0$ on $(1, e^{1/2})$ $\qquad f''(x) < 0$ on $(e^{\sqrt{3}/2}, \infty)$

(d) $f(e^{1/2})$ local and absolute maximum

SECTION 7.4

1. $\dfrac{dy}{dx} = e^{-2x}\dfrac{d}{dx}(-2x) = -2e^{-2x}$
$\qquad\qquad\qquad\qquad$ **2.** $\dfrac{dy}{dx} = 3e^{2x+1} \cdot 2 = 6e^{2x+1}$

3. $\dfrac{dy}{dx} = e^{x^2-1}\dfrac{d}{dx}(x^2 - 1) = 2xe^{x^2-1}$
$\qquad\qquad$ **4.** $\dfrac{dy}{dx} = 2e^{-4x}(-4) = -8e^{-4x}$

5. $\dfrac{dy}{dx} = e^x\dfrac{d}{dx}(\ln x) + \ln x\dfrac{d}{dx}(e^x) = e^x\left(\dfrac{1}{x} + \ln x\right)$

6. $\dfrac{dy}{dx} = 2xe^x + x^2e^x$

7. $\dfrac{dy}{dx} = x^{-1}\dfrac{d}{dx}(e^{-x}) + e^{-x}\dfrac{d}{dx}(x^{-1}) = -x^{-1}e^{-x} - e^{-x}x^{-2} = -(x^{-1} + x^{-2})e^{-x}$

8. $\dfrac{dy}{dx} = e^{\sqrt{x}+1}\left(\dfrac{1}{2\sqrt{x}}\right)$

9. $\dfrac{dy}{dx} = \dfrac{1}{2}(e^x - e^{-x})$
$\qquad\qquad\qquad\qquad$ **10.** $\dfrac{dy}{dx} = \dfrac{1}{2}(e^x - e^{-x}(-1)) = \dfrac{1}{2}(e^x + e^{-x})$

11. $\dfrac{dy}{dx} = e^{\sqrt{x}}\dfrac{d}{dx}(\ln\sqrt{x}\,) + \ln\sqrt{x}\,\dfrac{d}{dx}(e^{\sqrt{x}}) = e^{\sqrt{x}}\left(\dfrac{1}{\sqrt{x}} \cdot \dfrac{1}{2\sqrt{x}}\right) + \ln\sqrt{x}\dfrac{e^{\sqrt{x}}}{2\sqrt{x}} = \dfrac{1}{2}e^{\sqrt{x}}\left(\dfrac{1}{x} + \dfrac{\ln\sqrt{x}}{\sqrt{x}}\right)$

12. $\dfrac{dy}{dx} = 3(3 - 2e^{-x})^2(-2e^{-x}(-1)) = 6e^{-x}(3 - 2e^{-x})^2$

13. $\dfrac{dy}{dx} = 2(e^{x^2} + 1)\dfrac{d}{dx}(e^{x^2} + 1) = 2(e^{x^2} + 1)e^{x^2}\dfrac{d}{dx}(x^2) = 4xe^{x^2}(e^{x^2} + 1)$

14. $\dfrac{dy}{dx} = 2(e^{2x} - e^{-2x}) \cdot (2e^{2x} + 2e^{-2x}) = 4(e^{4x} - e^{-4x})$

15. $\dfrac{dy}{dx} = (x^2 - 2x + 2)\dfrac{d}{dx}(e^x) + e^x(2x - 2) = x^2 e^x$

16. $\dfrac{dy}{dx} = 2xe^x + x^2 e^x - e^{x^2} - xe^{x^2}\,2x = (x^2 + 2x)e^x - (2x^2 + 1)e^{x^2}$

17. $\dfrac{dy}{dx} = \dfrac{(e^x + 1)\,e^x - (e^x - 1)\,e^x}{(e^x + 1)^2} = \dfrac{2e^x}{(e^x + 1)^2}$

18. $\dfrac{dy}{dx} = \dfrac{2e^{2x}(e^{2x} + 1) - (e^{2x} - 1)2e^{2x}}{(e^{2x} + 1)^2} = \dfrac{4e^{2x}}{(e^{2x} + 1)^2}$

19. $y = e^{4\ln x} = (e^{\ln x})^4 = x^4$ so $\dfrac{dy}{dx} = 4x^3$.   **20.** $y = \ln e^{3x} = 3x \implies \dfrac{dy}{dx} = 3$

21. $f'(x) = \cos(e^{2x})\,e^{2x} \cdot 2 = 2e^{2x}\cos(e^{2x})$ **22.** $f'(x) = e^{\sin 2x} \cdot 2\cos 2x$

23. $f'(x) = e^{-2x}(-\sin x) + e^{-2x}(-2)\cos x = -e^{-2x}(2\cos x + \sin x)$

24. $f'(x) = -\dfrac{1}{\cos e^{2x}} \cdot \sin e^{2x}(2e^{2x}) = -2e^{2x}\tan(e^{2x})$

25. $\displaystyle\int e^{2x}\,dx = \dfrac{1}{2}e^{2x} + C$ **26.** $\displaystyle\int e^{-2x}\,dx = -\dfrac{1}{2}e^{-2x} + C$

27. $\displaystyle\int e^{kx}\,dx = \dfrac{1}{k}e^{kx} + C$ **28.** $\displaystyle\int e^{ax+b}\,dx = \dfrac{1}{a}e^{ax+b} + C$

29. $\left\{u = x^2,\quad du = 2x\,dx\right\};\quad \displaystyle\int xe^{x^2}\,dx = \dfrac{1}{2}\int e^u\,du = \dfrac{1}{2}e^u + C = \dfrac{1}{2}e^{x^2} + C$

30. $\displaystyle\int xe^{-x^2}\,dx = -\dfrac{1}{2}\int e^u\,du = -\dfrac{1}{2}e^u + C = -\dfrac{1}{2}e^{-x^2} + C$

31. $\left\{u = \dfrac{1}{x},\quad du = -\dfrac{1}{x^2}\,dx\right\};\quad \displaystyle\int \dfrac{e^{1/x}}{x^2}\,dx = -\int e^u\,du = -e^u + C = -e^{1/x} + C$

32. $\displaystyle\int \dfrac{e^{2\sqrt{x}}}{\sqrt{x}}\,dx = e^{2\sqrt{x}} + C$ **33.** $\displaystyle\int \ln e^x\,dx = \int x\,dx = \dfrac{1}{2}x^2 + C$

34. $\displaystyle\int e^{\ln x}\,dx = \int x\,dx = \dfrac{x^2}{2} + C$ **35.** $\displaystyle\int \dfrac{4}{\sqrt{e^x}}\,dx = \int 4e^{-x/2}\,dx = -8e^{-x/2} + C$

36. $\displaystyle\int \frac{e^x}{e^x+1}\,dx = \int \frac{du}{u} = \ln|u| + C = \ln(e^x+1) + C$

37. $\left\{\begin{array}{l} u = e^x + 1 \\ du = e^x\,dx \end{array}\right\};\qquad \displaystyle\int \frac{e^x}{\sqrt{e^x+1}}\,dx = \int \frac{du}{\sqrt{u}} = \int u^{-1/2}\,du = 2u^{1/2} + C = 2\sqrt{e^x+1} + C$

38. $\left\{\begin{array}{l} u = e^{ax^2} + 1 \\ du = 2axe^{ax^2}\,dx \end{array}\right\};\qquad \displaystyle\int \frac{xe^{ax^2}}{e^{ax^2}+1}\,dx = \frac{1}{2a}\int \frac{du}{u} = \frac{1}{2a}\ln|u| + C = \frac{1}{2a}\ln(e^{ax^2}+1) + C$

39. $\left\{\begin{array}{l} u = 2e^{2x} + 3 \\ du = 4e^{2x}\,dx \end{array}\right\};\qquad \displaystyle\int \frac{e^{2x}}{2e^{2x}+3}\,dx = \frac{1}{4}\int \frac{du}{u} = \frac{1}{4}\ln u + C = \frac{1}{4}\ln(2e^{2x}+3) + C$

40. $\left\{\begin{array}{l} u = e^{-2x} \\ du = -2e^{-2x}\,dx \end{array}\right\};\qquad \displaystyle\int \frac{\sin(e^{-2x})}{e^{2x}}\,dx = -\frac{1}{2}\int \sin u\,du = \frac{1}{2}\cos u + C = \frac{1}{2}\cos\left(e^{-2x}\right) + C$

41. $\{\,u = \sin x,\quad du = \cos x\,dx\,\};\qquad \displaystyle\int \cos x\, e^{\sin x}\,dx = \int e^u\,du = e^u + C = e^{\sin x} + C$

42. $\{\,u = e^{-x},\quad du = -e^{-x}\,dx\,\};$

$\displaystyle\int e^{-x}\left[1 + \cos\left(e^{-x}\right)\right]dx = -\int(1+\cos u)\,du = -u - \sin u + C = -e^{-x} - \sin\left(e^{-x}\right) + C$

43. $\displaystyle\int_0^1 e^x\,dx = \left[\,e^x\,\right]_0^1 = e - 1$

44. $\displaystyle\int_0^1 e^{-kx}\,dx = -\frac{1}{k}[e^{-kx}]_0^1 = \frac{1}{k}(1 - e^{-k})$

45. $\displaystyle\int_0^{\ln \pi} e^{-6x}\,dx = \left[-\frac{1}{6}e^{-6x}\right]_0^{\ln \pi} = -\frac{1}{6}e^{-6\ln \pi} + \frac{1}{6}e^0 = \frac{1}{6}\left(1 - \pi^{-6}\right)$

46. $\displaystyle\int_0^1 xe^{-x^2}\,dx = -\frac{1}{2}[e^{-x^2}]_0^1 = \frac{1}{2}\left(1 - \frac{1}{e}\right)$

47. $\displaystyle\int_0^1 \frac{e^x+1}{e^x}\,dx = \int_0^1 (1 + e^{-x})\,dx = [x - e^{-x}]_0^1 = \left(1 - e^{-1}\right) - (0 - 1) = 2 - \frac{1}{e}$

48. $\displaystyle\int_0^1 \frac{4 - e^x}{e^x}\,dx = \int_0^1 (4e^{-x} - 1)\,dx = [-4e^{-x} - x]_0^1 = 3 - 4e^{-1}$

49. $\displaystyle\int_0^{\ln 2} \frac{e^x}{e^x+1}\,dx = [\,\ln(e^x+1)\,]_0^{\ln 2} = \ln(e^{\ln 2}+1) - \ln(e^0+1) = \ln 3 - \ln 2 = \ln\frac{3}{2}$

50. $\displaystyle\int_0^1 \frac{e^x}{4 - e^x}\,dx = [\,-\ln|e^x - 4|\,]_0^1 = \ln\left(\frac{3}{4 - e}\right)$

51. $\displaystyle\int_0^1 x(e^{x^2} + 2)\, dx = \int_0^1 (xe^{x^2} + 2x)\, dx = \left[\frac{1}{2}e^{x^2} + x^2\right]_0^1 = \left(\frac{1}{2}e + 1\right) - \left(\frac{1}{2} + 0\right) = \frac{1}{2}(e + 1)$

52. $\displaystyle\int_0^{\ln \frac{\pi}{4}} e^x \sec e^x\, dx = \left[\ln|\sec e^x + \tan e^x|\right]_0^{\ln \frac{\pi}{4}} = \ln\left(\frac{\sqrt{2} + 1}{2}\right).$

53. (a) $f(x) = e^{ax},\ f'(x) = ae^{ax},\ f''(x) = a^2 e^{ax},\ \ldots,\ f^{(n)}(x) = a^n e^{ax}$

    (b) $f(x) = e^{-ax},\ f'(x) = -ae^{-ax},\ f''(x) = a^2 e^{-ax},\ \ldots,\ f^{(n)}(x) = (-1)^n a^n e^{-ax}$

54. (a) $x'(t) = kAe^{kt} - kBe^{-kt}$

$$x'(t) = 0 \implies kAe^{kt} - kBe^{-kt} = 0 \implies e^{2kt} = \frac{B}{A} \implies t = \frac{\ln B - \ln A}{2k}$$

    The particle is closest to the origin at time $\ t = \dfrac{\ln B - \ln A}{2k}.$

    (b) $x''(t) = k^2 Ae^{kt} + k^2 Be^{-kt} = k^2 x(t);\ k^2\ $ is the constant of proportionality.

55. $A = 2xe^{-x^2}$

$$A' = 2x(-2xe^{-x^2}) + 2e^{-x^2} = 2e^{-x^2}(1 - 2x^2) = 0 \implies x = \pm\sqrt{\frac{1}{2}}\ \text{and}\ y = \frac{1}{\sqrt{e}}.$$

    Put the vertices at $\ \left(\pm\dfrac{1}{\sqrt{2}}, \dfrac{1}{\sqrt{e}}\right).$

56. $y = e^x \implies x = \ln y.$ The area of the rectangle is given by: $A = y \ln y.$

$$\frac{dA}{dt} = (1 + \ln y)\frac{dy}{dt}.$$

    At $y = 3,\ \dfrac{dy}{dt} = \dfrac{1}{2}.$ Thus $\dfrac{dA}{dt} = \dfrac{1}{2}(1 + \ln 3)$ square units per minute.

57. $f(x) = e^{-x^2}$

    $f'(x) = -2xe^{-x^2}$

    $f''(x) = (4x^2 - 2)e^{-x^2}$

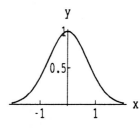

    (a)   symmetric with respect to the y-axis

          $f(-x) = f(x)$

    (b)   increases on $(-\infty, 0]$,   decreases on $[0, \infty)$

    (c) $f(0) = 1$ local and absolute max

    (d)   concave up on $\left(-\infty, -1/\sqrt{2}\right) \cup \left(1/\sqrt{2}, \infty\right),$ concave down on $\left(-1/\sqrt{2}, 1/\sqrt{2}\right)$

58. (a) $V = \displaystyle\int_0^1 \pi\, (e^x)^2\, dx = \pi \int_0^1 e^{2x}\, dx = \pi\left[\frac{1}{2}e^{2x}\right]_0^1 = \frac{1}{2}\pi\left[e^2 - 1\right]$

    (b) $V = \displaystyle\int_0^1 2\pi x\, e^x\, dx = 2\pi \int_0^1 x\, e^x\, dx.$

59. (a) $V = \int_0^1 2\pi\, x e^{-x^2}\, dx = \pi\left[-e^{-x^2}\right]_0^1 = \pi\left[1 - \tfrac{1}{e}\right]$

   (b) $V = \int_0^1 \pi\left[e^{-x^2}\right]^2 dx = \pi\int_0^1 e^{-2x^2}\, dx$

60.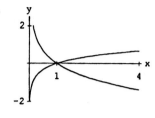

$$A = \int_{-\ln 4}^0 \left(4 - e^{-y}\right) dy + \int_0^{\frac{1}{2}\ln 4} \left(4 - e^{2y}\right) dy$$

$$= \left[4y + e^{-y}\right]_{-\ln 4}^0 + \left[4y - \frac{1}{2}e^{2y}\right]_0^{\frac{1}{2}\ln 4}$$

$$= 12\ln 2 - \frac{9}{2}$$

61.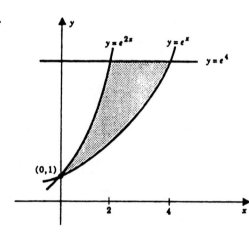

$$A = \int_0^2 \left(e^{2x} - e^x\right) dx + \int_2^4 \left(e^4 - e^x\right) dx$$

$$= \left[\tfrac{1}{2}e^{2x} - e^x\right]_0^2 + \left[e^4 x - e^x\right]_2^4$$

$$= \left(\tfrac{1}{2}e^4 - e^2 - \tfrac{1}{2} + 1\right) + \left(4e^4 - e^4 - 2e^4 + e^2\right)$$

$$= \tfrac{1}{2}\left(3e^4 + 1\right)$$

62.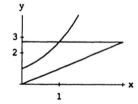

$$A = \text{triangle} \; - \; \text{upper left corner}$$

$$= \frac{1}{2}e^2 - \int_0^1 \left(e - e^x\right) dx$$

$$= \frac{1}{2}e^2 - \left[ex - e^x\right]_0^1$$

$$= \frac{1}{2}e^2 - 1$$

63.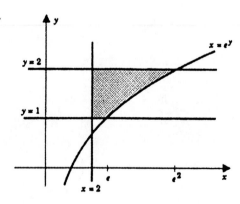

$$A = \int_1^2 \left(e^y - 2\right) dy$$

$$= \left[e^y - 2y\right]_1^2 = e^2 - e - 2$$

64. $f'(x) = -xe^x$

$f''(x) = -e^x - xe^x = -e^x(1+x)$

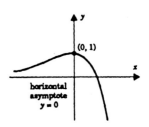

  (i) domain $(-\infty, \infty)$

  (ii) increases on $(-\infty, 0]$,   decreases on $[0, \infty)$

  (iii) $f(0) = 1$ local and absolute max

  (iv) concave up on $(-\infty, -1)$,   concave down on $(-1, \infty)$

       pt of inflection $(-1, 2/e)$

65. $f(x) = e^{(1/x)^2}$

$f'(x) = \dfrac{-2}{x^3} e^{(1/x)^2}$

$f''(x) = \dfrac{6x^2 + 4}{x^6} e^{(1/x)^2}$

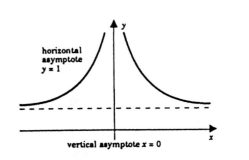

  (i) domain $(-\infty, 0) \cup (0, \infty)$

  (ii) increases on $(-\infty, 0)$,   decreases on $(0, \infty)$

  (iii) no extreme values

  (iv) concave up on $(-\infty, 0)$ and on $(0, \infty)$

66. $f'(x) = (2x - x^2)e^{-x}$

$f''(x) = e^{-x}(2 - 4x + x^2)$

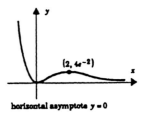

  (i) domain $(-\infty, \infty)$

  (ii) decreases on $(-\infty, 0]$, and on $[2, \infty)$,   increases on $[0, 2]$

  (iii) $f(0) = 0$ local and absolute min,

       $f(2) = 4e^{-2}$ local max

  (iv) concave up on $(-\infty, 2 - \sqrt{2})$ and $(2 + \sqrt{2}, \infty)$,

       concave down on $(2 - \sqrt{2}, 2 + \sqrt{2})$;

       points of inflection at $x = 2 \pm \sqrt{2}$.

67. $f(x) = x^2 \ln x$

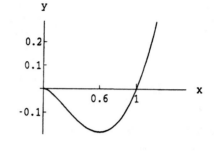

$f'(x) = 2x \ln x + x$

$f''(x) = 3 + 2 \ln x$

(i) domain $(0, \infty)$

(ii) decreases on $(0, e^{-1/2})$,   increases on $(e^{-1/2}, \infty)$

(iii) $f(e^{-1/2}) = -1/2e$ is a local and absolute min

(iv) concave down on $(0, e^{-3/2})$ and concave up

on $(e^{-3/2}, \infty)$; $(e^{-3/2}, -3/2e^3)$ is a point of inflection

68. $f(x) = (x - x^2)e^{-x}$

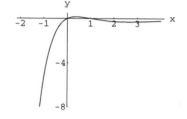

$f'(x) = (x^2 - 3x + 1)e^{-x}$

$f''(x) = -(x^2 - 5x + 4)e^{-x}$

(i) domain $(-\infty, \infty)$

(ii) decreases on (r_1, r_2), where $r_1 = \dfrac{3 - \sqrt{5}}{2}$, $r_2 = \dfrac{3 + \sqrt{5}}{2}$
increases on $(-\infty, r_1) \cup (r_2, \infty)$

(iii) $f(r_1))$ is a local and absolute max; $f(r_2)$ is a local min

(iv) concave down on $(-\infty, 1) \cup (4, \infty)$; concave up on $(1, 4)$
pts of inflection at $(1, 0)$ and $(4, -12e^{-4})$

69. $\displaystyle\int_0^{x_n} e^x \, dx = [e^x]_0^{x_n} = e^{x_n} - 1; \quad e^{x_n} - 1 = n \quad \Longrightarrow \quad x_n = \ln(n + 1).$

70. (a) $f(x) = x^k \ln x$, $f'(x) = x^{k-1} + kx^{k-1} \ln x = x^{k-1}(1 + k \ln x)$.

$f'(x) = 0:$ $1 + k \ln x = 0 \implies \ln x = -1/k \implies x = e^{-1/k}.$

$f'(x) < 0$ on $(0, e^{-1/k})$ and $f'(x) > 0$ on $(e^{-1/k}, \infty)$; f has a local and absolute

minimum at $x = e^{-1/k}$.

(b) $f(x) = x^k e^{-x}$, $f'(x) = -x^k e^{-x} + kx^{k-1}e^{-x} = x^{k-1}e^{-x}(k - x)$.

$f'(x) = 0:$ $k - x = 0 \implies x = k.$

$f'(x) > 0$ on $(0, k)$ and $f'(x) < 0$ on (k, ∞); f has a local and absolute maximum

at $x = e^{-1/k}$.

71. (a) For $y = e^{ax}$ we have $dy/dx = ae^{ax}$. Therefore the line tangent to the curve $y = e^{ax}$ at an arbitrary point (x_0, e^{ax_0}) has equation

$$y - e^{ax_0} = ae^{ax_0}(x - x_0).$$

The line passes through the origin iff $e^{ax_0} = (ae^{ax_0})x_0$ iff $x_0 = 1/a$. The point of tangency is $(1/a, e)$. This is point B. By symmetry, point A is $(-1/a, e)$.

(b) The tangent line at B has equation $y = aex$.   By symmetry

$$A_{\mathrm{I}} = 2\int_0^{1/a}(e^{ax} - aex)\,dx = 2\left[\frac{1}{a}e^{ax} - \frac{1}{2}aex^2\right]_0^{1/a} = \frac{1}{a}(e - 2).$$

(c) The normal at B has equation

$$y - e = -\frac{1}{ae}\left(x - \frac{1}{a}\right).$$

This can be written

$$y = -\frac{1}{ae}x + \frac{a^2e^2 + 1}{a^2e}.$$

Therefore

$$A_{\mathrm{II}} = 2\int_0^{1/a}\left(-\frac{1}{ae}x + \frac{a^2e^2 + 1}{a^2e} - e^{ax}\right)dx = \frac{1 + 2a^2e}{a^3e}.$$

72. By induction. True for $n = 0 : e^x > 1$ for $x > 0$.

Assume true for n. Then

$$e^x = 1 + \int_0^x e^t\,dt > 1 + \int_0^x\left(1 + t + \frac{t^2}{2!} + \cdots + \frac{t^n}{n!}\right)dt$$

$$= 1 + \left[t + \frac{t^2}{2} + \frac{t^3}{3!} + \cdots + \frac{t^{n+1}}{(n+1)!}\right]_0^x$$

$$= 1 + x + \frac{x^2}{2!} + \frac{x^3}{3!} + \cdots + \frac{x^{n+1}}{(n+1)!}$$

So the result is true for $n + 1$

73. For $x > (n+1)!$

$$e^x > 1 + x + \cdots + \frac{x^{n+1}}{(n+1)!} > \frac{x^{n+1}}{(n+1)!} = x^n\left[\frac{x}{(n+1)!}\right] > x^n.$$

74. (a)   **(b)** Intersect at $x \cong \pm 0.7531$

(c) Area $\cong 0.98$

75.   (a)

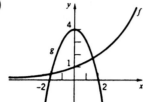

(b) $x_1 \cong -1.9646, x_2 \cong 1.0580$

(c) $A \cong \displaystyle\int_{-1.9646}^{1.0580} \left[4 - x^2 - e^x\right] dx = \left[4x - \frac{1}{3}x^3 - e^x\right]_{-1.9646}^{1.0580} \cong 6.4240$

76.

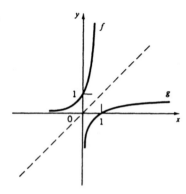

$f\left(g(x)\right) = e^{2\ln\sqrt{x}} = e^{2(1/2)\ln x} = e^{\ln x} = x$

77.

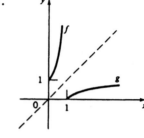

$f\left(g(x)\right) = e^{\left(\sqrt{\ln x}\right)^2} = e^{\ln x} = x$

78.

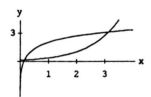

$f\left(g(x)\right) = e^{2 + \ln x - 2} = e^{\ln x} = x$

79.   (a) $f(x) = \sin\left(e^x\right);$ $f(x) = 0$ $\implies$ $e^x = n\pi$ $\implies$ $x = \ln n\pi, \; n = 1, 2, \cdots.$

(b)

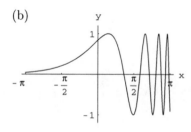

80. (a) $f(x) = e^{\sin x} - 1$; $f(x) = 0$ $\implies$ $\sin x = 0$ $\implies$ $x = n\pi$, n any integer.

(b)

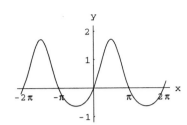

81. (a)

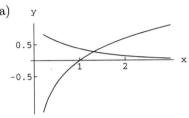

(b) $x \cong 1.3098$

(c) $f'(1.3098) \cong -0.26987$; $g'(1.3098) \cong 0.76348$

(d) the tangent lines are not perpendicular

82. (a)

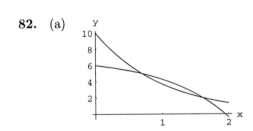

(b) The x-coordinates of the points of intersection are: $x = \ln 2$ and $x = \ln 5$.

(c) $A = \displaystyle\int_{\ln 2}^{\ln 5} \left[7 - e^x - 10 e^{-x}\right]\, dx \cong 0.4140$

83. (a) $\displaystyle\int \frac{1}{1 - e^x}\, dx = x - \ln|e^x - 1| + C$ 　　　　(c) $\displaystyle\int \frac{e^{\tan x}}{\cos^2 x}\, dx = e^{\tan x} + C$

(b) $\displaystyle\int e^{-x}\left(\frac{1 - e^x}{e^x}\right)^4 dx = -\frac{1}{5}\, e^{-5x} + e^{-4x} - 2\,e^{-3x} + 2\,e^{-2x} - e^{-x} + C$

PROJECT 7.4

Step 1. $\ln\left(1 + \dfrac{1}{n}\right) = \displaystyle\int_1^{1 + \frac{1}{n}} \frac{dt}{t} \le \int_1^{1 + \frac{1}{n}} 1\, dt = \dfrac{1}{n}$

(since $\dfrac{1}{t} \le 1$ throughout the interval of integration) $= 1$.

$\ln\left(1 + \dfrac{1}{n}\right) = \displaystyle\int_1^{1 + \frac{1}{n}} \frac{dt}{t} \ge \int_1^{1 + \frac{1}{n}} \frac{dt}{1 + \frac{1}{n}} = \dfrac{1}{1 + \frac{1}{n}}\, \dfrac{1}{n} = \dfrac{1}{n + 1}.$

(since $\dfrac{1}{t} \ge \dfrac{1}{1 + \frac{1}{n}}$ throughout the interval of integration)

Step 2. From Step 1, we get

$$1 + \frac{1}{n} \le e^{1/n} \implies \left(1 + \frac{1}{n}\right)^n \le e \quad \text{and} \quad e^{1/n+1} \le 1 + \frac{1}{n} \implies e \le \left(1 + \frac{1}{n}\right)^{n+1}$$

Combining these two inequalities, we have

$$\left(1 + \frac{1}{n}\right)^n \le e \le \left(1 + \frac{1}{n}\right)^{n+1}$$

SECTION 7.5

1. $\log_2 64 = \log_2 \left(2^6\right) = 6$

2. $\log_2 \dfrac{1}{64} = \log_2 2^{-6} = -6$

3. $\log_{64}\left(1/2\right) = \dfrac{\ln\left(1/2\right)}{\ln 64} = \dfrac{-\ln 2}{6\ln 2} = -\dfrac{1}{6}$

4. $\log_{10} 0.01 = \log_{10} 10^{-2} = -2$

5. $\log_5 1 = \log_5 \left(5^0\right) = 0$

6. $\log_5 0.2 = \log_5 5^{-1} = -1$

7. $\log_5\left(125\right) = \log_5\left(5^3\right) = 3$

8. $\log_2 4^3 = \log_2 2^6 = 6$

9. $\log_p xy = \dfrac{\ln xy}{\ln p} = \dfrac{\ln x + \ln y}{\ln p} = \dfrac{\ln x}{\ln p} + \dfrac{\ln y}{\ln p} = \log_p x + \log_p y$

10. $\log_p \dfrac{1}{x} = \dfrac{\ln\frac{1}{x}}{\ln p} = -\dfrac{\ln x}{\ln p} = -\log_p x.$

11. $\log_p x^y = \dfrac{\ln x^y}{\ln p} = y\dfrac{\ln x}{\ln p} = y\log_p x$

12. $\log_p \dfrac{x}{y} = \dfrac{\ln \frac{x}{y}}{\ln p} = \dfrac{\ln x - \ln y}{\ln p} = \log_p x - \log_p y$

13. $10^x = e^x \implies \left(e^{\ln 10^x}\right) = e^x \implies e^{x\ln 10} = e^x$
    $\implies x\ln 10 = x \implies x(\ln 10 - 1) = 0 \implies$ Thus, $x = 0.$

14. $\log_5 x = 0.04 \implies x = 5^{0.04}$

15. $\log_x 10 = \log_4 100 \implies \dfrac{\ln 10}{\ln x} = \dfrac{\ln 100}{\ln 4} \implies \dfrac{\ln 10}{\ln x} = \dfrac{2\ln 10}{2\ln 2}$
    $\implies \ln x = \ln 2$ Thus, $x = 2.$

16. $\log_x 2 = \log_3 x \implies \dfrac{\ln 2}{\ln x} = \dfrac{\ln x}{\ln 3} \implies \ln x = \pm\sqrt{(\ln 2)(\ln 3)} \implies x = e^{\pm\sqrt{(\ln 2)(\ln 3)}}$

17. The logarithm function is increasing. Thus,

$$e^{t_1} < a < e^{t_2} \implies t_1 = \ln e^{t_1} < \ln a < \ln e^{t_2} = t_2.$$

18. Since the exponential function is increasing, $e^{\ln x_1} < e^b < e^{\ln x_2}$,   so $x_1 < e^b < x_2$

19. $f'(x) = 3^{2x}(\ln 3)(2) = 2(\ln 3)3^{2x}$ $\qquad\qquad$ **20.** $g'(x) = 4^{3x^2}(\ln 4)\, 6x$

21. $f'(x) = 2^{5x}(\ln 2)(5)3^{\ln x} + 2^{5x}3^{\ln x}(\ln 3)\dfrac{1}{x} = 2^{5x}3^{\ln x}\left(5\ln 2 + \dfrac{\ln 3}{x}\right)$

22. $F'(x) = 5^{-2x^2+x}(\ln 5)(-4x+1)$

23. $g'(x) = \frac{1}{2}\left(\log_3 x\right)^{-1/2}\left(\dfrac{1}{\ln 3}\right)\dfrac{1}{x} = \dfrac{1}{2(\ln 3)x\sqrt{\log_3 x}}$

24. $h'(x) = 7^{\sin x^2}(\ln 7)(\cos x^2)\, 2x$

25. $f'(x) = \sec^2\left(\log_5 x\right)(\ln 5)\dfrac{1}{x} = \dfrac{\sec^2\left(\log_5 x\right)}{x\ln 5}$

26. $g'(x) = \dfrac{1}{\ln 10}\dfrac{\left(\frac{1}{x}\right)x^2 - \ln x(2x)}{x^4} = \dfrac{x - 2x\ln x}{x^4\ln 10} = \dfrac{1 - 2\ln x}{x^3\ln 10}$.

27. $F'(x) = -\sin\left(2^x + 2^{-x}\right)\left[2^x\ln 2 - 2^{-x}\ln 2\right] = \ln 2\left(2^{-x} - 2^x\right)\sin\left(2^x + 2^{-x}\right)$

28. $h'(x) = a^{-x}\ln a(-1)\cos bx + a^{-x}(-\sin bx)b = -(\ln a)a^{-x}\cos bx - ba^{-x}\sin bx$

29. $\displaystyle\int 3^x\, dx = \dfrac{3^x}{\ln 3} + C$ $\qquad\qquad$ **30.** $\displaystyle\int 2^{-x}\, dx = -\dfrac{2^{-x}}{\ln 2} + C$

31. $\displaystyle\int (x^3 + 3^{-x})\, dx = \dfrac{1}{4}x^4 - \dfrac{3^{-x}}{\ln 3} + C$

32. $\displaystyle\int x10^{-x^2}\, dx = -\dfrac{1}{2}\int 10^u\, du = -\dfrac{10^u}{2\ln 10} + C = -\dfrac{10^{-x^2}}{2\ln 10} + C$

33. $\displaystyle\int \dfrac{dx}{x\ln 5} = \dfrac{1}{\ln 5}\int \dfrac{dx}{x} = \dfrac{\ln|x|}{\ln 5} + C = \log_5|x| + C$

34. $\displaystyle\int \dfrac{\log_5 x}{x}\, dx = \dfrac{1}{\ln 5}\int \dfrac{\ln x}{x}\, dx = \dfrac{1}{\ln 5}\dfrac{1}{2}(\ln x)^2 + C = \dfrac{(\ln x)^2}{2\ln 5} + C$

35. $\qquad\displaystyle\int \dfrac{\log_2 x^3}{x}\, dx = \dfrac{1}{\ln 2}\int \dfrac{\ln x^3}{x}\, dx = \dfrac{3}{\ln 2}\int \dfrac{\ln x}{x}\, dx$

$\qquad\qquad\qquad\quad = \dfrac{3}{\ln 2}\left[\dfrac{1}{2}(\ln x)^2\right] + C = \dfrac{3}{\ln 4}(\ln x)^2 + C$

36. Write $c = b^{\log_b c}$ Then $\log_a c = \log_a\left(b^{\log_b c}\right) = (\log_b c)(\log_a b)$.

37. $f'(x) = \dfrac{1}{x\ln 3}$ so $f'(e) = \dfrac{1}{e\ln 3}$

38. $f(x) = x \log_3 x;$ $f'(x) = \log_3 x + x \cdot \dfrac{1}{x \ln 3} = \dfrac{\ln x + 1}{\ln 3};$ $f'(e) = \dfrac{2}{\ln 3}$

39. $f'(x) = \dfrac{1}{x \ln x}$ so $f'(e) = \dfrac{1}{e \ln e} = \dfrac{1}{e}$

40. $f(x) = \log_3 (\log_2 x) = \dfrac{\ln \left(\frac{\ln x}{\ln 2} \right)}{\ln 3} = \dfrac{\ln(\ln x) - \ln(\ln 2)}{\ln 3};$ $f'(x) = \dfrac{1}{\ln 3} \cdot \dfrac{1}{\ln x} \cdot \dfrac{1}{x} \implies f'(e) = \dfrac{1}{e \ln 3}$

41. $\quad f(x) = p^x$

$\ln f(x) = x \ln p$

$\dfrac{f'(x)}{f(x)} = \ln p$

$f'(x) = f(x) \ln p$

$f'(x) = p^x \ln p$

42. $\quad f(x) = p^{g(x)}$

$\ln f(x) = g(x) \ln p$

$\dfrac{f'(x)}{f(x)} = g'(x) \ln p$

$f'(x) = f(x) g'(x) \ln p = p^{g(x)} g'(x) \ln p$

43. $\quad y = (x + 1)^x$

$\ln y = x \ln (x + 1)$

$\dfrac{1}{y} \dfrac{dy}{dx} = \dfrac{x}{x + 1} + \ln (x + 1)$

$\dfrac{dy}{dx} = (x + 1)^x \left[\dfrac{x}{x + 1} + \ln (x + 1) \right]$

44. $\quad y = (\ln x)^x$

$\ln y = x \ln(\ln x)$

$\dfrac{1}{y} \dfrac{dy}{dx} = \ln(\ln x) + x \dfrac{1}{\ln x} \cdot \dfrac{1}{x}$

$\dfrac{dy}{dx} = (\ln x)^x \left[\ln(\ln x) + \dfrac{1}{\ln x} \right]$

45. $\quad y = (\ln x)^{\ln x}$

$\ln y = \ln x \left[\ln (\ln x) \right]$

$\dfrac{1}{y} \dfrac{dy}{dx} = \ln x \left[\dfrac{1}{x \ln x} \right] + \dfrac{1}{x} \left[\ln (\ln x) \right]$

$\dfrac{dy}{dx} = (\ln x)^{\ln x} \left[\dfrac{1 + \ln (\ln x)}{x} \right]$

46. $\quad y = \left(\dfrac{1}{x} \right)^x$

$\ln y = x \ln \dfrac{1}{x} = -x \ln x$

$\dfrac{1}{y} \dfrac{dy}{dx} = -\ln x - x \cdot \dfrac{1}{x} = -\ln x - 1$

$\dfrac{dy}{dx} = -\left(\dfrac{1}{x} \right)^x [1 + \ln x]$

47. $\quad y = x^{\sin x}$

$\ln y = (\sin x)(\ln x)$

$\dfrac{1}{y} \dfrac{dy}{dx} = (\cos x)(\ln x) + \sin x \left(\dfrac{1}{x} \right)$

$\dfrac{dy}{dx} = x^{\sin x} \left[(\cos x)(\ln x) + \dfrac{\sin x}{x} \right]$

48. $\quad y = (\cos x)^{x^2 + 1}$

$\ln y = (x^2 + 1) \ln(\cos x)$

$\dfrac{1}{y} \dfrac{dy}{dx} = 2x \ln(\cos x) + (x^2 + 1) \left(-\dfrac{\sin x}{\cos x} \right)$

$\dfrac{dy}{dx} = (\cos x)^{x^2 + 1} \left[2x \ln(\cos x) - (x^2 + 1) \tan x \right]$

49. $y = (\sin x)^{\cos x}$

$\ln y = (\cos x)(\ln[\sin x])$

$\dfrac{1}{y}\dfrac{dy}{dx} = (-\sin x)(\ln[\sin x]) + (\cos x)\left(\dfrac{1}{\sin x}\right)(\cos x)$

$\dfrac{dy}{dx} = (\sin x)^{\cos x}\left[\dfrac{\cos^2 x}{\sin x} - (\sin x)(\ln[\sin x])\right]$

50. $y = x^{x^2}$

$\ln y = x^2 \ln x$

$\dfrac{1}{y}\dfrac{dy}{dx} = 2x\ln x + x^2 \cdot \dfrac{1}{x}$

$\dfrac{dy}{dx} = x^{x^2+1}\left(2\ln x + 1\right)$

51. $y = x^{2^x}$

$\ln y = 2^x \ln x$

$\dfrac{1}{y}\dfrac{dy}{dx} = 2^x \ln 2 \ln x + 2^x\left(\dfrac{1}{x}\right)$

$\dfrac{dy}{dx} = x^{2^x}\left[2^x \ln 2 \ln x + \dfrac{2^x}{x}\right]$

52. $y = (\tan x)^{\sec x}$

$\ln y = \sec x \ln(\tan x)$

$\dfrac{1}{y}\dfrac{dy}{dx} = \sec x \tan x \ln(\tan x) + \sec x \cdot \dfrac{\sec^2 x}{\tan x}$

$\dfrac{dy}{dx} = (\tan x)^{\sec x}\left[\sec x \tan x \ln(\tan x)\right.$

$\left. + \sec^3 x \cot x\right]$

53. From the definition of the derivative, the derivative of $f(x) = \ln x$ at $x = 1$ is

$$f'(1) = \lim_{h \to 0} \frac{\ln(1+h) - \ln 1}{h}.$$

Since $f'(1) = 1$, we have

$$\frac{\ln(1+h) - \ln 1}{h} = \frac{1}{h}\ln(1+h) = \ln(1+h)^{1/h} \to 1 \text{ as } h \to 0.$$

Set $x = 1/h$. Then $h \to 0 \implies x \to \infty$ and

$$\ln\left(1 + \frac{1}{x}\right)^x \to 1 \text{ as } x \to \infty$$

Therefore

$$\left(1 + \frac{1}{x}\right)^x = e^{\ln(1+1/x)^x} \to e^1 = e \text{ as } x \to \infty.$$

54.

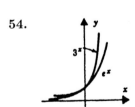

55.

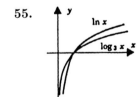

56.

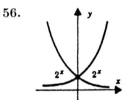

57.

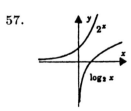

58.

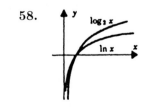

59. $\displaystyle\int_1^2 2^{-x}\, dx = \left[-\frac{2^{-x}}{\ln 2}\right]_1^2 = \frac{1}{4\ln 2}$

60. $\displaystyle\int_0^1 4^x\, dx = \left[\frac{4^x}{\ln 4}\right]_0^1 = \frac{3}{\ln 4}$

61. $\displaystyle\int_1^4 \frac{dx}{x\ln 2} = [\log_2 x]_1^4 = \log_2 4 - 0 = 2$

62. $\displaystyle\int_0^2 p^{x/2}\, dx = \left[\frac{2p^{x/2}}{\ln p}\right]_0^2 = \frac{2(p-1)}{\ln p}$

63. $\displaystyle\int_0^1 x 10^{1+x^2}\, dx = \left[\frac{1}{2\ln 10}10^{1+x^2}\right]_0^1 = \frac{1}{2\ln 10}(100-10) = \frac{45}{\ln 10}$

64. $\displaystyle\int_0^1 \frac{5p^{\sqrt{x+1}}}{\sqrt{x+1}}\, dx = \left[\frac{10p^{\sqrt{x+1}}}{\ln p}\right]_0^1 = \frac{10}{\ln p}\left(p^{\sqrt 2}-p\right)$

65. $\displaystyle\int_0^1 \left(2^x + x^2\right)\, dx = \left[\frac{2^x}{\ln 2} + \frac{x^3}{3}\right]_0^1 = \frac{1}{3} + \frac{1}{\ln 2}$

66. $7^{1/\ln 7} \cong 2.71828.$ $7^{1/\ln 7} = \left(e^{\ln 7}\right)^{1/\ln 7} = e^1 \cong 2.71828.$

67. approx   16.99999; $5^{\ln 17/\ln 5} = \left(e^{\ln 5}\right)^{\ln 17/\ln 5} = e^{\ln 17} = 17$

68. approx 54.59815; $16^{1/\ln 2} = \left(e^{\ln 16}\right)^{1/\ln 2} = e^{\ln 16/\ln 2} = e^{4\ln 2/\ln 2} = e^4 \cong 54.59815$

69. (b) the x-coordinates of the points of intersection are: $x_1 \cong -1.198$, $x_2 = 3$ and $x_3 \cong 3.408$

(c) for the interval $[-1.198, 3]$, $A \cong 5.5376$; for the interval $[3, 3.408]$, $A \cong 0.01373$

70. (b) the x-coordinates of the points of intersection are: $x_1 \cong -0.7667$, $x_2 = 2$ and $x_3 = 4$

(c) The area of the region bounded by the two graphs is: $A = \displaystyle\int_2^4 \left(2^{-x} - x^{-2}\right)\, dx \cong 0.0205$

SECTION 7.6

1. We begin with $A(t) = A_0 e^{rt}$

and take $A_0 = \$500$ and $t = 10$. The interest earned is given by

$$A(10) - A_0 = 500\left(e^{10r} - 1\right).$$

Thus, (a) $500\left(e^{0.6} - 1\right) \cong \$411.06$ (b) $500\left(e^{0.8} - 1\right) \cong \$612.77$

(c) $500\left(e - 1\right) \cong \$859.14.$

2. We want $A(t) = A_0 e^{rt} = 2A_0,$ so $e^{rt} = 2 \implies rt = \ln 2 \implies t = \dfrac{\ln 2}{r}$

(a) $t = \dfrac{\ln 2}{0.06} \cong 11.55$ years.   (b) $t = \dfrac{\ln 2}{0.08} \cong 8.66$ years.   (c) $t = \dfrac{\ln 2}{0.1} \cong 6.93$ years.

3. In general

$$A(t) = A_0 e^{rt}.$$

We set

$$3A_0 = A_0 e^{20r}$$

and solve for r:

$$3 = e^{20r}, \quad \ln 3 = 20r, \quad r = \frac{\ln 3}{20} \cong 5\frac{1}{2}\%.$$

4. We want $A(10) = A_0 e^{10r} = 2A_0$, so $r = \frac{\ln 2}{10} \cong 6.9\%$

5. $P(t) = \frac{9}{2} e^{\frac{t}{20} \ln (4/3)} = \frac{9}{2} e^{\ln (4/3)^{t/20}} = \frac{9}{2} \left(\frac{4}{3}\right)^{t/20}.$

6. $P(t) = P_0 e^{kt}. \quad P(4) = P_0 e^{k4} = 3P_0 \implies k = \frac{\ln 3}{4}$

   (a) $P(12) = P_0 e^{\frac{\ln 3}{4} 12} = P_0 e^{3 \ln 3} = 27P_0 = 1 \implies P_0 \cong 0.037$ square inches

   (b) $P(t) = P_0 e^{\frac{\ln 3}{4} t} = 2P_0 \implies \frac{\ln 3}{4} t = \ln 2 \implies t = \frac{4 \ln 2}{\ln 3} \cong 2.52$ hours.

7. (a) $P(t) = 10,000 e^{t \ln 2} = 10,000(2)^t$

   (b) $P(26) = 10,000(2)^{26}, \quad P(52) = 10,000(2)^{52}$

8. $qC = Ce^{kp} \implies q = e^{kp} \implies p = \frac{1}{k} \ln q$

9. (a) $P(10) = P(0)e^{0.035(10)t} = P(0)e^{0.35t}.$ Thus it increases by $e^{0.35}$.

   (b) $2P(0) = P(0)e^{15k} \implies k = \frac{\ln 2}{15}.$

10. Let $P(t)$ be the world population at time t years, and let 1990 correspond to $t = 0$.
    Then $P(t) = 249\, e^{kt}; \quad P(10) = 249\, e^{10k} = 281 \implies e^{10k} = \frac{281}{249} \implies k \cong 0.0121.$
    Thus $P(t) = 249\, e^{0.0121\, t}.$

    According to this model, the population in 1980 was $P(-10) = 249\, e^{-0.121} \cong 221$ million.

11. Using the data from Ex. 10, the growth constant $k \cong 0.0121$. Therefore

    $P(20) \cong 249\, e^{20k} \cong 249\, e^{0.242} \cong 317.1$ million.

    $P(11) \cong 249\, e^{11k} \cong 249\, e^{0.1331} \cong 284.4$ million.

12. $Pe^{kt} = 2P \implies kt = \ln 2.$

Since $k = \dfrac{1}{10} \ln\left(\dfrac{281}{249}\right),$ we get $t = \dfrac{10 \ln 2}{\ln\left(\dfrac{281}{249}\right)} \cong 57.3$ years.

13. $4.5e^{0.0143t} = 30 \implies 0.0143t = \ln\dfrac{30}{4.5} \implies t \simeq 115.7$ years.

Thus maximum population will be reached in 2095.

14. $\dfrac{ds}{dx} = -\dfrac{s}{V} \implies s(x) = Ce^{-x/V} = s_0 e^{-x/V}.$ We want $s(x) = \frac{1}{2}s_0,$ so $e^{-x/V} = \frac{1}{2},$

hence $x = -V \ln\frac{1}{2} = V \ln 2;$ $V \ln 2 = 10{,}000 \ln 2 \cong 6931$ gallons

15.
$$V'(t) = kV(t)$$
$$V'(t) - ktV(t) = 0$$
$$e^{-kt}V'(t) - ke^{-kt}V(t) = 0$$
$$\frac{d}{dt}\left[e^{-kt}V(t)\right] = 0$$
$$e^{-kt}V(t) = C$$
$$V(t) = Ce^{kt}.$$

Since $V(0) = C = 200,$ $V(t) = 200\, e^{kt}.$
Since $V(5) = 160,$ $200\, e^{5k} = 160,$ $e^{5k} = \frac{4}{5},$ $e^k = \left(\frac{4}{5}\right)^{1/5}$
and therefore $V(t) = 200\left(\frac{4}{5}\right)^{t/5}$ liters.

16. $A(t) = A_0 e^{kt}.$ $A(5) = A_0 e^{5k} = \dfrac{2}{3}A_0 \implies k = \dfrac{\ln(2/3)}{5} \implies A(t) = A_0 e^{\frac{\ln(2/3)}{5} t}$

$A(t) = \dfrac{1}{2}A_0 \implies \dfrac{\ln(2/3)}{5} t = \ln\dfrac{1}{2} \implies t = \dfrac{5\ln(1/2)}{\ln(2/3)} \cong 8.55$ years.

17. Take two years ago as time $t = 0.$ In general

$(*)$ $A(t) = A_0 e^{kt}.$

We are given that

$$A_0 = 5 \quad \text{and} \quad A(2) = 4.$$
Thus,

$$4 = 5e^{2k} \quad \text{so that} \quad \tfrac{4}{5} = e^{2k} \quad \text{or} \quad e^k = \left(\tfrac{4}{5}\right)^{1/2}.$$
We can write
$$A(t) = 5\left(\tfrac{4}{5}\right)^{t/2}$$
and compute $A(5)$ as follows:
$$A(5) = 5\left(\tfrac{4}{5}\right)^{5/2} = 5e^{\frac{5}{2}\ln(4/5)} \cong 5e^{-0.56} \cong 2.86.$$

About 2.86 gm will remain 3 years from now.

18. Let $t = 0$ correspond to a year ago. Then $A(t) = 4e^{kt}$, and $A(1) = 3 \Longrightarrow 4e^k = 3 \Longrightarrow k = \ln(3/4)$. Therefore, $A(t) = 4e^{\ln(3/4)t} = 4\left(\frac{3}{4}\right)^t$. Ten years ago, $t = -9$; $A(-9) = 4\left(\frac{3}{4}\right)^{-9} \cong 53.27$ grams.

19. A fundamental property of radioactive decay is that the percentage of substance that decays during any year is constant:

$$100\left[\frac{A(t) - A(t+1)}{A(t)}\right] = 100\left[\frac{A_0e^{kt} - A_0e^{k(t+1)}}{A_0e^{kt}}\right] = 100(1 - e^k)$$

If the half-life is n years, then

$$\tfrac{1}{2}A_0 = A_0e^{kn} \quad \text{so that} \quad e^k = \left(\tfrac{1}{2}\right)^{1/n}.$$

Thus, $100\left[1 - \left(\tfrac{1}{2}\right)^{1/n}\right]$% of the material decays during any one year.

20. $A(t) = ne^{kt}$. $A(5) = ne^{5k} = m \Longrightarrow 5k = \ln(m/n) \Longrightarrow k = \frac{1}{5}\ln(m/n)$ and $A(t) = ne^{\frac{\ln(m/n)}{5}t}$.

$$A(10) = ne^{2\ln(m/n)} = n\left(\frac{m}{n}\right)^2 = \frac{m^2}{n} \quad \text{grams.}$$

21. (a) $A(1620) = A_0e^{1620k} = \frac{1}{2}A_0 \Longrightarrow k = \frac{\ln\frac{1}{2}}{1620} \simeq -0.00043$.

Thus $A(500) = A_0e^{500k} = 0.807A_0$. Hence 80.7% will remain.

(b) $0.25A_0 = A_0e^{kt} \Longrightarrow t = 3240$ years.

22. $Ae^{5.3k} = \frac{1}{2}A \Longrightarrow k \simeq -0.1308$.

(a) $A(8) = A_0e^{8k} \simeq 0.351A_0$. Thus 35.1% will remain.

(b) $100 = A_0e^{3k} \Longrightarrow A_0 \simeq 148$ grams.

23. (a) $x_1(t) = 10^6t, \quad x_2(t) = e^t - 1$

(b) $\dfrac{d}{dt}[x_1(t) - x_2(t)] = \dfrac{d}{dt}[10^6t - (e^t - 1)] = 10^6 - e^t$

This derivative is zero at $t = 6\ln 10 \cong 13.8$. After that the derivative is negative.

(c) $x_2(15) < e^{15} = (e^3)^5 \cong 20^5 = 2^5(10^5) = 3.2(10^6) < 15(10^6) = x_1(15)$

$x_2(18) = e^{18} - 1 = (e^3)^6 - 1 \cong 20^6 - 1 = 64(10^6) - 1 > 18(10^6) = x_1(18)$

$x_2(18) - x_1(18) \cong 64(10^6) - 1 - 18(10^6) \cong 46(10^6)$

(d) If by time t_1 EXP has passed LIN, then $t_1 > 6\ln 10$. For all $t \geq t_1$ the speed of EXP is greater than the speed of LIN:

$$\text{for} \quad t \geq t_1 > 6\ln 10, \quad v_2(t) = e^t > 10^6 = v_1(t).$$

24. (a) $x_1(t) = t$, $x_3(t) = 10^6 \ln(t+1)$

(b) $\dfrac{d}{dt}[x_3(t) - x_1(t)] = \dfrac{d}{dt}\left[10^6 \ln(t+1) - t\right] = \dfrac{10^6}{t+1} - 1$

This derivative is 0 at $t = 10^6 - 1$. After that the derivative is negative.

(c) $x_1\left(10^7 - 1\right) = 10^7 - 1 < 7(\ln 10)10^6 = 10^6 \ln 10^7 = x_3(10^7 - 1)$

$x_3(10^8 - 1) = 10^6 \ln 10^8 = (10^6)8\ln 10 < (10^6)24 < 10^8 - 1 = x_1(10^8 - 1)$

(d) If by time t_1 LIN had passed LOG, then $t_1 > 10^6 - 1$. For all $t \ge t_1$ the speed of LIN is greater than the speed of LOG:

$$\text{for}\quad t \ge t_1 > 10^6 - 1, \quad v_1(t) = t > \frac{10^6}{t+1} = v_3(t).$$

25. Let $p(h)$ denote the pressure at altitude h. The equation $\dfrac{dp}{dh} = kp$ gives

$(*)$ $p(h) = p_0 e^{kh}$

where p_0 is the pressure at altitude zero (sea level).

Since $p_0 = 15$ and $p(10000) = 10$,

$$10 = 15e^{10000k}, \quad \tfrac{2}{3} = e^{10000k}, \quad \tfrac{1}{10000}\ln\tfrac{2}{3} = k.$$

Thus, $(*)$ can be written

$$p(h) = 15\left(\tfrac{2}{3}\right)^{h/10000}.$$

(a) $p(5000) = 15\left(\tfrac{2}{3}\right)^{1/2} \cong 12.25 \text{ lb/in.}^2$.

(b) $p(15000) = 15\left(\tfrac{2}{3}\right)^{3/2} \cong 8.16 \text{ lb/in.}^2$.

26. $P = 20{,}000\, e^{-(0.06)(4)} \cong \$15{,}732.56$.

27. From Exercise 26, we have $6000 = 10{,}000e^{-8r}$. Thus

$$e^{-8r} = \frac{6000}{10{,}000} = \frac{3}{5} \;\Rightarrow\; -8r = \ln(3/5) \quad\text{and}\quad r \cong 0.064 \;\text{ or }\; r = 6.4\%$$

28. (a) $P = 50{,}000\, e^{-(0.04)(20)} \cong \$22{,}466.45$

(b) $P = 50{,}000\, e^{-(0.06)(20)} \cong \$15{,}059.71$

(c) $P = 50{,}000\, e^{-(0.08)(20)} \cong \$10{,}094.83$

29. The future value of $\$25{,}000$ at an interest rate r, t years from now is given by $Q(t) = 25{,}000\, e^{rt}$.

Thus

(a) For $r = 0.05$: $P(3) = 25{,}000\, e^{(0.05)3} \cong \$29{,}045.86$.

(b) For $r = 0.08$: $P(3) = 25{,}000\, e^{(0.08)3} \cong \$31{,}781.23$.

(c) For $r = 0.12$: $P(3) = 25{,}000\, e^{(0.12)3} \cong \$35{,}833.24$.

30. $\dfrac{dv}{dt} = -kv \Longrightarrow v = ce^{-kt}$, $\quad v(0) = ce^0 = c$, $\quad$ so c is velocity when power is shut off.

31. By Exercise 30

$(*)$ $\qquad\qquad\qquad\qquad v(t) = Ce^{-kt}$, $\quad t$ in seconds.

We use the initial conditions

$$v(0) = C = 4 \text{ mph } = \tfrac{1}{900} \text{ mi/sec} \quad \text{and} \quad v(60) = 2 = \tfrac{1}{1800} \text{ mi/sec}$$

to determine e^{-k}:

$$\tfrac{1}{1800} = \tfrac{1}{900} e^{-60k}, \qquad e^{60k} = 2, \qquad e^k = 2^{1/60}.$$

Thus, $(*)$ can be written

$$v(t) = \tfrac{1}{900} 2^{-t/60}.$$

The distance traveled by the boat is

$$s = \int_0^{60} \frac{1}{900} 2^{-t/60}\, dt = \frac{1}{900} \left[\frac{-60}{\ln 2} 2^{-t/60} \right]_0^{60} = \frac{1}{30 \ln 2} \text{ mi } = \frac{176}{\ln 2} \text{ ft} \quad \text{(about 254 ft)}.$$

32. Since the amount $A(t)$ of raw sugar present after t hours decreases at a rate proportional to A, we have

$$A(t) = A_0 e^{kt}.$$

We are given $\quad A_0 = 1000 \quad$ and $\quad A(10) = 800$. $\quad$ Thus,

$$800 = 1000 e^{10k}, \quad \tfrac{4}{5} = e^{10k}, \quad e^k = \left(\tfrac{4}{5}\right)^{1/10}$$

so that

$$A(t) = 1000 \left(\tfrac{4}{5}\right)^{t/10}.$$

Now,

$$A(20) = 1000 \left(\frac{4}{5}\right)^{20/10} = 640;$$

after 10 more hours of inversion there will remain 640 pounds.

33. Let $A(t)$ denote the amount of ^{14}C remaining t years after the organism dies. Then $A(t) = A(0)e^{kt}$ for some constant k. Since the half-life of ^{14}C is 5700 years, we have

$$\frac{1}{2} = e^{5700k} \Rightarrow k = -\frac{\ln 2}{5700} \cong 0.000122 \text{ and } A(t) = A(0)e^{-0.000122t}$$

If 25% of the original amount of ^{14}C remains after t years, then

$$0.25 A(0) = A(0)e^{-0.000122t} \Rightarrow t = \frac{\ln 0.25}{-0.000122} \cong 11,400 \text{ (years)}$$

34. $A(t) = A_0 e^{-\frac{\ln 2}{5700} t}$; $\quad A(2000) = A_0 e^{-\frac{\ln 2}{5700} \cdot 2000} \cong 0.78 A_0$; $\quad 78\%$ remains

35.
$$f'(t) = tf(t)$$

$$f'(t) - tf(t) = 0$$

$$e^{-t^2/2} f'(t) - te^{-t^2/2} f(t) = 0$$

$$\frac{d}{dt}[e^{-t^2/2} f(t)] = 0$$

$$e^{-t^2/2} f(t) = C$$

$$f(t) = Ce^{t^2/2}$$

36.
$$f'(t) = \sin t\, f(t)$$

$$f'(t) - \sin t\, f(t) = 0$$

$$e^{\cos t} f'(t) - \sin t\, e^{\cos t} f(t) = 0$$

$$\frac{d}{dt}\left[e^{\cos t} f(t)\right] = 0$$

$$e^{\cos t} f(t) = C$$

$$f(t) = Ce^{-\cos t}$$

37.
$$f'(t) = \cos t\, f(t)$$

$$f'(t) - \cos t\, f(t) = 0$$

$$e^{-\sin t} f'(t) - \cos t\, e^{-\sin t} f(t) = 0$$

$$\frac{d}{dt}\left[e^{-\sin t} f(t)\right] = 0$$

$$e^{-\sin t} f(t) = C$$

$$f(t) = Ce^{\sin t}$$

38. Write the equation as $f'(t) - g(t)f(t) = 0$; and set $h(t) = -\int g(t)\, dt$. Then

$$f'(t) - g(t)f(t) = 0$$

$$e^{h(t)} f'(t) - g(t)e^{h(t)} f(t) = 0$$

$$[e^{h(t)} f(t)]' = 0$$

$$e^{h(t)} f(t) = C$$

$$f(t) = Ce^{-h(t)} = Ce^{\int g(t)\, dt}$$

SECTION 7.7

1. (a) 0   (b) $-\pi/3$

2. (a) $\pi/3$ (b) $\pi/3$

3. (a) $2\pi/3$ (b) $3\pi/4$

4. (a) $-2/\sqrt{3}$ (b) -2

5. (a) $1/2$ (b) $\pi/4$

6. (a) $-\pi/6$ (b) $-\pi/4$

7. (a) does not exist   **8.** (a) $4/5$ (b) $5/3$ **9.** (a) $\sqrt{3}/2$ (b) $-7/25$

     (b) does not exist

10. (a) arc cosine; domain: $[-1, 1]$, range: $[0, \pi]$

     (b) arc cotangent: domain: $(-\infty, \infty)$, range: $[0, \pi]$

11. $\dfrac{dy}{dx} = \dfrac{1}{1 + (x+1)^2} = \dfrac{1}{x^2 + 2x + 2}$ **12.** $\dfrac{dy}{dx} = \dfrac{1}{1 + (\sqrt{x})^2} \cdot \dfrac{1}{2\sqrt{x}} = \dfrac{1}{2\sqrt{x}(1+x)}$

13. $f'(x) = \dfrac{1}{|2x^2|\sqrt{(2x^2)^2 - 1}} \dfrac{d}{dx}\left(2x^2\right) = \dfrac{2}{x\sqrt{4x^4 - 1}}$

14. $f'(x) = e^x \arcsin x + e^x \dfrac{1}{\sqrt{1 - x^2}} = e^x \left[\arcsin x + \dfrac{1}{\sqrt{1 - x^2}}\right]$

15. $f'(x) = \arcsin 2x + x \dfrac{1}{\sqrt{1 - (2x)^2}} \dfrac{d}{dx}\left(2x\right) = \arcsin 2x + \dfrac{2x}{\sqrt{1 - 4x^2}}$

16. $f'(x) = e^{\arctan x} \cdot \dfrac{1}{1 + x^2}$ **17.** $\dfrac{du}{dx} = 2\left(\arcsin x\right)\dfrac{d}{dx}\left(\arcsin x\right) = \dfrac{2\arcsin x}{\sqrt{1 - x^2}}$

18. $\dfrac{dy}{dx} = \dfrac{1}{1 + (e^x)^2} \cdot e^x = \dfrac{e^x}{1 + e^{2x}}$

19. $\dfrac{dy}{dx} = \dfrac{x\left(\dfrac{1}{1 + x^2}\right) - (1)\arctan x}{x^2} = \dfrac{x - \left(1 + x^2\right)\arctan x}{x^2\left(1 + x^2\right)}$

20. $\dfrac{dy}{dx} = \dfrac{1}{|\sqrt{x^2 + 2}|\sqrt{\left(\sqrt{x^2 + 2}\right)^2 - 1}} \cdot \dfrac{x}{\sqrt{x^2 + 2}} = \dfrac{x}{(x^2 + 2)\sqrt{x^2 + 1}}$

21. $f'(x) = \dfrac{1}{2}\left(\arctan 2x\right)^{-1/2}\dfrac{d}{dx}\left(\arctan 2x\right) = \dfrac{1}{2}\left(\arctan 2x\right)^{-1/2}\dfrac{2}{1 + (2x)^2} = \dfrac{1}{\left(1 + 4x^2\right)\sqrt{\arctan 2x}}$

22. $f'(x) = \dfrac{1}{\arctan x} \cdot \dfrac{1}{1 + x^2} = \dfrac{1}{\left(1 + x^2\right)\arctan x}$

23. $\dfrac{dy}{dx} = \dfrac{1}{1 + (\ln x)^2}\dfrac{d}{dx}\left(\ln x\right) = \dfrac{1}{x[1 + (\ln x)^2]}$

24. $g'(x) = \dfrac{-\sin x}{|\cos x + 2|\sqrt{(\cos x + 2)^2 - 1}}$

25. $\dfrac{d\theta}{dr} = \dfrac{1}{\sqrt{1 - \left(\sqrt{1 - r^2}\,\right)^2}}\dfrac{d}{dr}\left(\sqrt{1 - r^2}\right) = \dfrac{1}{\sqrt{r^2}} \cdot \dfrac{-r}{\sqrt{1 - r^2}} = -\dfrac{r}{|r|\sqrt{1 - r^2}}$

26. $\dfrac{d\theta}{dr} = \dfrac{1}{\sqrt{1 - [r/(r+1)]^2}} \cdot \dfrac{1}{(r+1)^2} = \dfrac{1}{(r+1)\sqrt{2r+1}}$

27. $g'(x) = 2x \arcsec\left(\dfrac{1}{x}\right) + x^2 \cdot \dfrac{1}{\left|\dfrac{1}{x}\right|\sqrt{\dfrac{1}{x^2} - 1}} \cdot \left(-\dfrac{1}{x^2}\right) = 2x \sec^{-1}\left(\dfrac{1}{x}\right) - \dfrac{x^2}{\sqrt{1 - x^2}}$

28. $\dfrac{d\theta}{dr} = \dfrac{1}{1 + [1/(1 + r^2)]^2} \cdot \dfrac{-2r}{(1 + r^2)^2} = \dfrac{-2r}{r^4 + 2r^2 + 2}$

29. $\dfrac{dy}{dx} = \cos\left[\arcsec(\ln x)\right] \cdot \dfrac{1}{|\ln x|\sqrt{(\ln x)^2 - 1}} \cdot \dfrac{1}{x} = \dfrac{\cos\left[\arcsec(\ln x)\right]}{x\,|\ln x|\sqrt{(\ln x)^2 - 1}}$

30. $f'(x) = e^{\arcsec x} \cdot \dfrac{1}{|x|\sqrt{x^2 - 1}} = \dfrac{e^{\arcsec x}}{|x|\sqrt{x^2 - 1}}$

31. $f'(x) = \dfrac{-x}{\sqrt{c^2 - x^2}} + \dfrac{c}{\sqrt{1 - (x/c)^2}} \cdot \left(\dfrac{1}{c}\right) = \dfrac{c - x}{\sqrt{c^2 - x^2}} = \sqrt{\dfrac{c - x}{c + x}}$

32. $\dfrac{dy}{dx} = \dfrac{\sqrt{c^2 - x^2}\,(1) - x\left(\dfrac{-x}{\sqrt{c^2 - x^2}}\right)}{\left(\sqrt{c^2 - x^2}\,\right)^2} - \dfrac{1}{\sqrt{1 - (x/c)^2}}\left(\dfrac{1}{c}\right)$

$\qquad = \dfrac{c^2}{(c^2 - x^2)^{3/2}} - \dfrac{1}{(c^2 - x^2)^{1/2}} = \dfrac{x^2}{(c^2 - x^2)^{3/2}}$

33. (a) $\sin(\arcsin x) = x$ (b) $\cos(\arcsin x) = \sqrt{1 - x^2}$ (c) $\tan(\arcsin x) = \dfrac{x}{\sqrt{1 - x^2}}$

(d) $\cot(\arcsin x) = \dfrac{\sqrt{1 - x^2}}{x}$ (e) $\sec(\arcsin x) = \dfrac{1}{\sqrt{1 - x^2}}$ (f) $\csc(\arcsin x) = \dfrac{1}{x}$

34. (a) $\tan(\arctan x) = x$ (b) $\cot(\arctan x) = \dfrac{1}{x}$ (c) $\sin(\arctan x) = \dfrac{x}{\sqrt{1 + x^2}}$

(d) $\cos(\arctan x) = \dfrac{1}{\sqrt{1 + x^2}}$ (e) $\sec(\arctan x) = \sqrt{1 + x^2}$ (f) $\csc(\arctan x) = \dfrac{\sqrt{1 + x^2}}{x}$

35. $\left\{\begin{array}{l} au = x + b \\ a\,du = dx \end{array}\right\}$; $\displaystyle\int \frac{dx}{\sqrt{a^2 - (x+b)^2}} = \int \frac{a\,du}{\sqrt{a^2 - a^2 u^2}} = \int \frac{du}{\sqrt{1 - u^2}}$

$$= \arcsin u + C = \sin^{-1}\left(\frac{x+b}{a}\right) + C$$

36. $\left\{\begin{array}{l} au = x + b \\ a\,du = dx \end{array}\right\}$; $\displaystyle\int \frac{dx}{a^2 + (x+b)^2} = \frac{1}{a^2}\int \frac{a\,du}{1 + u^2} = \frac{1}{a}\arctan u + C = \frac{1}{a}\arctan\left(\frac{x+b}{a}\right) + C$

37. $\left\{\begin{array}{l} au = x + b \\ a\,du = dx \end{array}\right\}$; $\displaystyle\int \frac{dx}{(x+b)\sqrt{(x+b)^2 - a^2}} = \int \frac{a\,du}{au\sqrt{a^2 u^2 - a^2}} = \frac{1}{a}\int \frac{du}{u\sqrt{u^2 - 1}}$

$$= \frac{1}{a}\operatorname{arcsec}|u| + C = \frac{1}{a}\operatorname{arcsec}\left(\frac{|x+b|}{a}\right) + C$$

38.   (a) $\sec(\operatorname{arcsec} x) = \sec(\arccos 1/x) = \dfrac{1}{1/x} = x;$ $\csc(\operatorname{arccsc} x) = \csc(\arcsin 1/x) = \dfrac{1}{1/x} = x$

     (b)   arc secant: range: $[0, \pi/2) \cup (\pi/2, \pi)$ (c)   arc cosecant: range: $[-\pi/2, 0) \cup (0, \pi/2]$

39. $\displaystyle\int_0^1 \frac{dx}{1 + x^2} = [\arctan x]_0^1 = \frac{\pi}{4}$ **40.** $\displaystyle\int_{-1}^1 \frac{dx}{1 + x^2} = [\arctan x]_{-1}^1 = \frac{\pi}{4} - \left(-\frac{\pi}{4}\right) = \frac{\pi}{2}$

41. $\displaystyle\int_0^{1/\sqrt{2}} \frac{dx}{\sqrt{1 - x^2}} = [\arcsin x]_0^{1/\sqrt{2}} = \frac{\pi}{4}$ **42.** $\displaystyle\int_0^1 \frac{dx}{\sqrt{4 - x^2}} = \left[\arcsin \frac{x}{2}\right]_0^1 = \frac{\pi}{6}$

43. $\displaystyle\int_0^5 \frac{dx}{25 + x^2} = \left[\frac{1}{5}\arctan \frac{x}{5}\right]_0^5 = \frac{\pi}{20}$

44. $\displaystyle\int_5^8 \frac{dx}{x\sqrt{x^2 - 16}} = \frac{1}{4}\left[\operatorname{arcsec} \frac{x}{4}\right]_5^8 = \frac{1}{4}\left(\operatorname{arcsec} 2 - \operatorname{arcsec} \frac{5}{4}\right) = \frac{\pi}{12} - \frac{1}{4}\operatorname{arcsec} \frac{5}{4}$

45. $\left\{\begin{array}{l|l} 3u = 2x & x = 0 \implies u = 0 \\ 3\,du = 2\,dx & x = 3/2 \implies u = 1 \end{array}\right\}$; $\displaystyle\int_0^{3/2} \frac{dx}{9 + 4x^2} = \frac{1}{6}\int_0^1 \frac{du}{1 + u^2} = \frac{1}{6}[\arctan u]_0^1 = \frac{\pi}{24}$

46. $\displaystyle\int_2^5 \frac{dx}{9 + (x-2)^2} = \frac{1}{3}\left[\arctan\left(\frac{x-2}{3}\right)\right]_2^5 = \frac{1}{3}\left(\frac{\pi}{4} - 0\right) = \frac{\pi}{12}$

47. $\left\{\begin{array}{l|l} u = 4x & x = 3/2 \implies u = 6 \\ du = 4\,dx & x = 3 \implies u = 12 \end{array}\right\}$;

$$\int_{3/2}^3 \frac{dx}{x\sqrt{16x^2 - 9}} = \int_6^{12} \frac{du/4}{(u/4)\sqrt{u^2 - 9}} = \frac{1}{3}\left[\operatorname{arcsec}\left(\frac{|u|}{3}\right)\right]_6^{12} = \frac{1}{3}\operatorname{arcsec} 4 - \frac{\pi}{9}$$

48. $\displaystyle\int_4^6 \frac{dx}{(x-3)\sqrt{x^2 - 6x + 8}} = \int_4^6 \frac{dx}{(x-3)\sqrt{(x-3)^2 - 1}} = [\operatorname{arcsec}(x-3)]_4^6 = \operatorname{arcsec} 3$

49. $\displaystyle\int_{-3}^{-2} \frac{dx}{\sqrt{4 - (x + 3)^2}} = \left[\arcsin\left(\frac{x + 3}{2}\right)\right]_{-3}^{-2} = \frac{\pi}{6}$

50. $\displaystyle\int_{\ln 2}^{\ln 3} \frac{e^{-x}}{\sqrt{1 - e^{-2x}}}\, dx = \left[-\arcsin e^{-x}\right]_{\ln 2}^{\ln 3} = \arcsin\left(\frac{1}{2}\right) - \arcsin\left(\frac{1}{3}\right) = \frac{\pi}{6} - \arcsin\left(\frac{1}{3}\right)$

51. $\left\{ \begin{array}{l} u = e^x \\ du = e^x\, dx \end{array} \middle| \begin{array}{l} x = 0 \implies u = 1 \\ x = \ln 2 \implies u = 2 \end{array} \right\}$;

$\displaystyle\int_0^{\ln 2} \frac{e^x}{1 + e^{2x}}\, dx = \int_1^2 \frac{du}{1 + u^2} = [\arctan u]_1^2 = \arctan 2 - \frac{\pi}{4} \cong 0.322$

52. $\displaystyle\int_0^{1/2} \frac{dx}{\sqrt{3 - 4x^2}} = \frac{1}{\sqrt{3}} \int_0^{1/2} \frac{dx}{\sqrt{1 - \frac{4}{3}x^2}} = \frac{1}{2}\left[\arcsin\frac{2x}{\sqrt{3}}\right]_0^{1/2} = \frac{1}{2}\arcsin\left(\frac{\sqrt{3}}{3}\right)$

53. $\left\{ \begin{array}{l} u = x^2 \\ du = 2x\, dx \end{array} \right\}$; $\displaystyle\int \frac{x}{\sqrt{1 - x^4}}\, dx = \frac{1}{2} \int \frac{du}{\sqrt{1 - u^2}} = \frac{1}{2}\arcsin u + C = \frac{1}{2}\arcsin x^2 + C$

54. $\displaystyle\int \frac{\sec^2 x}{\sqrt{9 - \tan^2 x}}\, dx = \int \frac{du}{\sqrt{9 - u^2}} = \arcsin\left(\frac{u}{3}\right) + C = \arcsin\left(\frac{\tan x}{3}\right) + C$

55. $\left\{ \begin{array}{l} u = x^2 \\ du = 2x\, dx \end{array} \right\}$; $\displaystyle\int \frac{x}{1 + x^4}\, dx = \frac{1}{2} \int \frac{du}{1 + u^2} = \frac{1}{2}\arctan u + C = \frac{1}{2}\arctan x^2 + C$

56. $\displaystyle\int \frac{dx}{\sqrt{4x - x^2}} = \int \frac{dx}{\sqrt{4 - (x - 2)^2}} = \arcsin\left(\frac{x - 2}{2}\right) + C$

57. $\left\{ \begin{array}{l} u = \tan x \\ du = \sec^2 x\, dx \end{array} \right\}$; $\displaystyle\int \frac{\sec^2 x}{9 + \tan^2 x}\, dx = \int \frac{du}{9 + u^2} = \frac{1}{3}\arctan\left(\frac{u}{3}\right) + C = \frac{1}{3}\arctan\left(\frac{\tan x}{3}\right) + C$

58. $\displaystyle\int \frac{\cos x}{3 + \sin^2 x}\, dx = \int \frac{du}{3 + u^2} = \frac{1}{\sqrt{3}}\arctan\left(\frac{u}{\sqrt{3}}\right) + C = \frac{1}{\sqrt{3}}\arctan\left(\frac{\sin x}{\sqrt{3}}\right) + C$

59. $\left\{ \begin{array}{l} u = \arcsin x \\ du = \dfrac{1}{\sqrt{1 - x^2}}\, dx \end{array} \right\}$; $\displaystyle\int \frac{\arcsin x}{\sqrt{1 - x^2}}\, dx = \int u\, du = \frac{1}{2}u^2 + C = \frac{1}{2}(\arcsin x)^2 + C$

60. $\displaystyle\int \frac{\arctan x}{1 + x^2}\, dx = \int u\, du = \frac{u^2}{2} + C = \frac{1}{2}(\arctan x)^2 + C$

61. $\left\{ \begin{array}{l} u = \ln x \\ du = \dfrac{1}{x}\, dx \end{array} \right\}$; $\displaystyle\int \frac{dx}{x\sqrt{1 - (\ln x)^2}} = \int \frac{du}{\sqrt{1 - u^2}} = \arcsin u + C = \arcsin(\ln x) + C$

62. $\displaystyle\int \frac{1}{x} \cdot \frac{1}{1 + (\ln x)^2}\, dx = \int \frac{du}{1 + u^2} = \arctan u + C = \arctan(\ln x) + C$

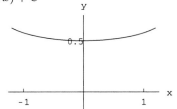

63. $\displaystyle A = \int_{-1}^{1} \frac{1}{\sqrt{4 - x^2}}\, dx = 2\int_{0}^{1} \frac{1}{\sqrt{4 - x^2}}\, dx$

$\displaystyle = 2\left[\arcsin\left(\frac{x}{2}\right)\right]_0^1 = \frac{\pi}{3}$

64. $\displaystyle A = \int_{-3}^{3} \frac{3}{9 + x^2}\, dx = 3\left[\frac{1}{3}\arctan\left(\frac{x}{3}\right)\right]_{-3}^{3} = \frac{\pi}{4} + \frac{\pi}{4} = \frac{\pi}{2}$

65. $\displaystyle \frac{8}{x^2 + 4} = \frac{1}{4}x^2 \quad\Rightarrow\quad x = \pm 2$

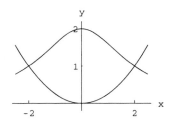

$\displaystyle A = \int_{-2}^{2}\left(\frac{8}{x^2 + 4} - \frac{1}{4}x^2\right) dx = 2\int_{0}^{2}\left(\frac{8}{x^2 + 4} - \frac{1}{4}x^2\right) dx$

$\displaystyle = 2\left[8 \cdot \frac{1}{2}\arctan\left(\frac{x}{2}\right) - \frac{1}{12}x^3\right]_0^2 = 2\pi - \frac{4}{3}$

66. $\displaystyle V = \int_{0}^{2} \pi \frac{1}{4 + x^2}\, dx = \frac{\pi}{2}\left[\arctan(x/2)\right]_0^2 = \frac{\pi^2}{8}$

67. $\displaystyle V = \int_{0}^{2} 2\pi x \frac{1}{\sqrt{4 + x^2}}\, dx = \pi \int_{0}^{2} \frac{2x}{\sqrt{4 + x^2}}\, dx = 2\pi\left[\sqrt{4 + x^2}\right]_0^2 = 4\pi\left(\sqrt{2} - 1\right)$

68. $\displaystyle V = \int_{2\sqrt{3}}^{6} 2\pi x \cdot \frac{1}{x^2\sqrt{x^2 - 9}}\, dx = 2\pi \int_{2\sqrt{3}}^{6} \frac{dx}{x\sqrt{x^2 - 9}} = 2\pi\left[\frac{1}{3}\operatorname{arcsec}\left(\frac{x}{3}\right)\right]_{2\sqrt{3}}^{6} = \frac{\pi^2}{9}$

69. Let x be the distance between the motorist and the point on the road where the line determined by the sign intersects the road. Then, from the given figure,

$$\theta = \arctan\left(\frac{s + k}{x}\right) - \arctan\frac{s}{x}, \quad 0 < x < \infty$$

and

$$\frac{d\theta}{dx} = \frac{1}{1 + \frac{(s + k)^2}{x^2}}\left(-\frac{s + k}{x^2}\right) - \frac{1}{1 + \frac{s^2}{x^2}}\left(-\frac{s}{x^2}\right)$$

$$= \frac{-(s + k)}{x^2 + (s + k)^2} + \frac{s}{x^2 + s^2} = \frac{s^2 k + sk^2 - kx^2}{[x^2 + (s + k)^2][x^2 + s^2]}$$

Setting $d\theta/dx = 0$ we get $x = \sqrt{s^2 + sk}$. Since θ is essentially 0 when x is close to 0 and when x is "large," we can conclude that θ is a maximum when $x = \sqrt{s^2 + sk}$.

70. $\displaystyle y = \arctan\frac{x}{30}; \quad \frac{dy}{dt} = \frac{1}{1 + (x/30)^2} \cdot \frac{1}{30} \cdot \frac{dx}{dt}$

If $\displaystyle \frac{dy}{dt} = 6$ and $x = 50$ then $\displaystyle \frac{dy}{dt} = \frac{30}{900 + (50)^2} \cdot 6 = \frac{9}{170}$ rad/sec

71. (b) $\displaystyle\int_{-a}^{a} \sqrt{a^2 - x^2}\, dx = \left[\frac{x}{2}\sqrt{a^2 - x^2} + \frac{a^2}{2}\arcsin\left(\frac{x}{a}\right)\right]_{-a}^{a} = \frac{\pi a^2}{2}$

The graph of $f(x) = \sqrt{a^2 - x^2}$ on the interval $[-a, a]$ is the upper half of the circle of radius a centered at the origin. Thus, the integral gives the area of the semi-circle: $A = \frac{1}{2}\pi a^2$.

72. (a) $\displaystyle f'(x) = \frac{1}{1 + \left(\frac{a+x}{1-ax}\right)^2}\cdot\frac{1+a^2}{(1-ax)^2} = \frac{1+a^2}{(1+a^2)(1+x^2)} = \frac{1}{1+x^2}$

(b) $\displaystyle\lim_{x\to(1/a)^-} f(x) = \pi/2; \qquad \lim_{x\to(1/a)^+} f(x) = -\pi/2$

(c) Let $g(x) = \arctan x$. For $x = 0 < \dfrac{1}{a}$, $g(0) = 0$, $f(0) = \arctan(a)$, so

$$f(x) = g(x) + \arctan a \quad \text{for} \quad x < \frac{1}{a}, \quad \text{i.e.,} \quad C_1 = \arctan a$$

For $x > \dfrac{1}{a}$, note that $\displaystyle\lim_{x\to\infty}\arctan x = \frac{\pi}{2}$, $\displaystyle\lim_{x\to\infty}\arctan\left(\frac{a+x}{1-ax}\right) = \arctan(-1/a)$ so

$$f(x) = g(x) + \arctan(-1/a) - \frac{\pi}{2} \quad \text{for} \quad x > \frac{1}{a}, \quad \text{i.e.,} \quad C_2 = \arctan(-1/a) - \frac{\pi}{2}$$

73. Set $y = \operatorname{arccot} x$. Then $\cot y = x$ and, by the hint, $\tan\left(\frac{1}{2}\pi - y\right) = x$. Therefore

$$\frac{1}{2}\pi - y = \arctan x, \qquad \arctan x + y = \frac{1}{2}\pi, \qquad \arctan x + \operatorname{arccot} x = \frac{1}{2}\pi.$$

74. Set $y = \operatorname{arccsc} x$. Then $\csc y = x$ and $\sec\left(\frac{1}{2}\pi - y\right) = x$ $\left[\sec\left(\frac{1}{2}\pi - \theta\right)\right] = \csc\theta\right]$. Therefore

$$\frac{1}{2}\pi - y = \operatorname{arcsec} x, \qquad \operatorname{arcsec} x + y = \frac{1}{2}\pi, \qquad \operatorname{arcsec} x + \operatorname{arccsc} x = \frac{1}{2}\pi.$$

75. The integrand is undefined for $x \geq 1$.

76. Numerical work suggests limit $\cong 1$. One way to see this is to note that the limit is the derivative of $f(x) = \arcsin x$ at $x = 0$ and this derivative is 1:

$$f'(x) = \frac{1}{\sqrt{1-x^2}}, \quad f'(0) = 1$$

77. $\displaystyle I = \int_0^{0.5}\frac{1}{\sqrt{1-x^2}}\, dx \cong \frac{1}{10}\left[f(0.05) + f(0.15) + f(0.25) + f(0.35) + f(0.45)\right] \cong 0.523;$

and $\sin(0.523) \cong 0.499$. Explanation: $I = \arcsin(0.5)$ and $\sin\left[\arcsin(0.5)\right] = 0.5$.

78. (a) 1.5698, 1.5704, 1.5706, 1.5707   (b) $1.5708 \cong \dfrac{\pi}{2}$

    (c) $1.570796 \cong \dfrac{\pi}{2}$

PROJECT 7.7

1. (a) 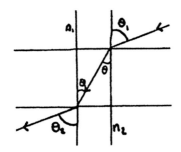 $$n_1 \sin \theta_1 = n \sin \theta = n_2 \sin \theta_2$$

2. (a) Think of n and θ as functions of altitude y.   Then

$$n \sin \theta = C$$

Differentiation with respect to y gives

$$n \cos \theta \frac{d\theta}{dy} + \frac{dn}{dy} \sin \theta = 0, \quad \cot \theta \frac{d\theta}{dy} + \frac{1}{n}\frac{dn}{dy} = 0$$

and so when $\alpha = \dfrac{\pi}{2} - \theta$,

$$\frac{1}{n}\frac{dn}{dy} = -\cot \theta \frac{d\theta}{dy} = -\frac{dy}{dx}\left(-\frac{d\alpha}{dy}\right) = \frac{d\alpha}{dx}$$

Now, $\alpha = \arctan\left(\dfrac{dy}{dx}\right)$ and $\dfrac{d\alpha}{dx} = \dfrac{\dfrac{d^2y}{d^2x}}{1+\left(\dfrac{dy}{dx}\right)^2}$.

(b) $1 + \left(\dfrac{dy}{dx}\right)^2 = 1 + \tan^2 \alpha = 1 + \cot^2 \theta = \csc^2 \theta = \dfrac{n^2}{C^2} = (\text{a constant}) \cdot [n(y)]^2$.

(c) $n(y) = \dfrac{k}{|y+b|}$, with b, k constants, $k > 0$.

SECTION 7.8

1. $\dfrac{dy}{dx} = \cosh x^2 \dfrac{d}{dx}\left(x^2\right) = 2x \cosh x^2$ **2.** $\dfrac{dy}{dx} = \sinh(x+a)$

3. $\dfrac{dy}{dx} = \dfrac{1}{2}\left(\cosh ax\right)^{-1/2}\left(a \sinh ax\right) = \dfrac{a \sinh ax}{2\sqrt{\cosh ax}}$

4. $\dfrac{dy}{dx} = a\cosh^2 ax + a\sinh^2 ax = a\left(\cosh^2 ax + \sinh^2 ax\right)$

5. $\dfrac{dy}{dx} = \dfrac{\left(\cosh x - 1\right)\left(\cosh x\right) - \sinh x\left(\sinh x\right)}{\left(\cosh x - 1\right)^2} = \dfrac{1}{1 - \cosh x}$

6. $\dfrac{dy}{dx} = \dfrac{x\cosh x - \sinh x}{x^2}$

7. $\dfrac{dy}{dx} = ab\cosh bx - ab\sinh ax = ab\,(\cosh bx - \sinh ax)$

8. $\dfrac{dy}{dx} = e^x(\cosh x + \sinh x) + e^x(\sinh x + \cosh x) = 2e^x(\cosh x + \sinh x)$

9. $\dfrac{dy}{dx} = \dfrac{1}{\sinh ax}\,(a\cosh ax) = a\coth ax$

10. $\dfrac{dy}{dx} = \dfrac{1}{1-\cosh ax}(-\sinh ax)a = \dfrac{-a\sinh ax}{1-\cosh ax}$

11. $\dfrac{dy}{dx} = \cosh(e^{2x})e^{2x}(2) = 2e^{2x}\cosh(e^{2x})$

12. $\dfrac{dy}{dx} = \sinh(\ln x^3)\cdot\dfrac{1}{x^3}\cdot 3x^2 = \dfrac{3\sinh(\ln x^3)}{x}$

13. $\dfrac{dy}{dx} = -e^{-x}\cosh 2x + 2e^{-x}\sinh 2x$

14. $\dfrac{dy}{dx} = \dfrac{1}{1+\sinh^2 x}\cdot\cosh x = \dfrac{1}{\cosh x}$

15. $\dfrac{dy}{dx} = \dfrac{1}{\cosh x}\,(\sinh x) = \tanh x$

16. $\dfrac{dy}{dx} = \dfrac{1}{\sinh x}\cdot\cosh x = \coth x$

17. $\ln y = x\ln\sinh x;\quad \dfrac{1}{y}\dfrac{dy}{dx} = \ln\sinh x + x\,\dfrac{\cosh x}{\sinh x}\;$ and $\;\dfrac{dy}{dx} = (\sinh x)^x\,[\ln\sinh x + x\coth x]$

18. $y = x^{\cosh x} \implies \ln y = \cosh x\ln x \implies \dfrac{1}{y}\dfrac{dy}{dx} = \sinh x\ln x + \cosh x\dfrac{1}{x}$

$\implies \dfrac{dy}{dx} = x^{\cosh x}\left(\sinh x\ln x + \dfrac{\cosh x}{x}\right)$

19. $\cosh^2 t - \sinh^2 t = \left(\dfrac{e^t+e^{-t}}{2}\right)^2 - \left(\dfrac{e^t-e^{-t}}{2}\right)^2$

$= \dfrac{1}{4}\left\{\left(e^{2t}+2+e^{-2t}\right)-\left(e^{2t}-2+e^{-2t}\right)\right\} = \dfrac{4}{4} = 1$

20. $\sinh t\cosh s + \cosh t\sinh s = \dfrac{1}{2}(e^t-e^{-t})\dfrac{1}{2}(e^s+e^{-s}) + \dfrac{1}{2}(e^t+e^{-t})\dfrac{1}{2}(e^s-e^{-s})$

$= \dfrac{1}{2}\left(e^{s+t}-e^{-(s+t)}\right) = \sinh(s+t)$

21. $\cosh t\cosh s + \sinh t\sinh s = \left(\dfrac{e^t+e^{-t}}{2}\right)\left(\dfrac{e^s+e^{-s}}{2}\right) + \left(\dfrac{e^t-e^{-t}}{2}\right)\left(\dfrac{e^s-e^{-s}}{2}\right)$

$= \dfrac{1}{4}\{2e^{t+s}+2e^{-(t+s)}\} = \dfrac{e^{t+s}+e^{-(t+s)}}{2} = \cosh(t+s)$

22. Follows from Exercise 20, with $s = t$.

23. Set $s = t$ in $\cosh(t+s) = \cosh t\cosh s + \sinh t\sinh s$ to get $\cosh(2t) = \cosh^2 t + \sinh^2 t$.
Then use Exercise 19 to obtain the other two identities.

24. $\cosh(-t) = \dfrac{1}{2}\left(e^{-t}+e^{-(-t)}\right) = \dfrac{1}{2}\left(e^{-t}+e^t\right) = \cosh t$

25. $\sinh(-t) = \dfrac{e^{(-t)}-e^{-(-t)}}{2} = -\dfrac{e^t-e^{-t}}{2} = -\sinh t$

26. $y = 5\cosh x + 4\sinh x = \dfrac{5}{2}(e^x + e^{-x}) + \dfrac{4}{2}(e^x - e^{-x}) = \dfrac{9}{2}e^x + \dfrac{1}{2}e^{-x}$

$\dfrac{dy}{dx} = \dfrac{9}{2}e^x - \dfrac{1}{2}e^{-x} = \dfrac{e^{-x}}{2}(9e^{2x} - 1); \quad \dfrac{dy}{dx} = 0 \implies e^{2x} = \dfrac{1}{9} \implies x = -\ln 3.$

$\dfrac{d^2y}{dx^2} = \dfrac{9}{2}e^x + \dfrac{1}{2}e^{-x} > 0 \quad \text{for all } x, \quad \text{so abs min occurs at } x = -\ln 3.$

At $x = -\ln 3, \quad y = \frac{9}{2}(\frac{1}{3}) + \frac{1}{2}(3) = 3.$

27.

$y = -5\cosh x + 4\sinh x = -\dfrac{5}{2}(e^x + e^{-x}) + \dfrac{4}{2}(e^x - e^{-x}) = -\dfrac{1}{2}e^x - \dfrac{9}{2}e^{-x}$

$\dfrac{dy}{dx} = -\dfrac{1}{2}e^x + \dfrac{9}{2}e^{-x} = \dfrac{e^{-x}}{2}(9 - e^{2x}); \quad \dfrac{dy}{dx} = 0 \implies e^x = 3 \text{ or } x = \ln 3$

$\dfrac{d^2y}{dx^2} = -\dfrac{1}{2}e^x - \dfrac{9}{2}e^{-x} < 0 \quad \text{for all } x \quad \text{so abs max occurs at } x = \ln 3.$

The abs max is $y = -\frac{1}{2}e^{\ln 3} - \frac{9}{2}e^{-\ln 3} = -\frac{1}{2}(3) - \frac{9}{2}\left(\frac{1}{3}\right) = -3.$

28. $y = 4\cosh x + 5\sinh x = \dfrac{4}{2}(e^x + e^{-x}) + \dfrac{5}{2}(e^x - e^{-x}) = \dfrac{9}{2}e^x - \dfrac{1}{2}e^{-x}$

$\dfrac{dy}{dx} = \dfrac{9}{2}e^x + \dfrac{1}{2}e^{-x} > 0 \quad \text{always increasing, so no extreme values.}$

29. $[\cosh x + \sinh x]^n = \left[\dfrac{e^x + e^{-x}}{2} + \dfrac{e^x - e^{-x}}{2}\right]^n$

$= [e^x]^n = e^{nx} = \dfrac{e^{nx} + e^{-nx}}{2} + \dfrac{e^{nx} - e^{-nx}}{2} = \cosh nx + \sinh nx$

30. $y = A\cosh cx + B\sinh cx, \quad y' = Ac\sinh cx + Bc\cosh cx, \quad y'' = Ac^2\cosh cx + Bc^2\sinh cx$

$\implies y'' = c^2 y$

31.

$y = A\cosh cx + B\sinh cx; \qquad\qquad y(0) = 2 \implies 2 = A.$

$y' = Ac\sinh cx + Bc\cosh cx; \qquad\qquad y'(0) = 1 \implies 1 = Bc.$

$y'' = Ac^2\cosh cx + Bc^2\sinh cx = c^2 y; \qquad y'' - 9y = 0 \implies (c^2 - 9)y = 0.$

Thus, $c = 3, \quad B = \frac{1}{3}, \quad \text{and} \quad A = 2.$

32. From Exercise 30, $y'' = c^2 y, \quad \text{so } c = \frac{1}{2}.$

$1 = y(0) = A\cosh 0 + B\sinh 0 = A \implies A = 1$

$2 = y'(0) = Ac\sinh 0 + Bc\cosh 0 = Bc \implies B = \dfrac{2}{c} = 4$

33. $\dfrac{1}{a}\sinh ax + C$ **34.** $\dfrac{1}{a}\cosh ax + C$ **35.** $\dfrac{1}{3a}\sinh^3 ax + C$

36. $\dfrac{1}{3a}\cosh^3 ax + C$ **37.** $\dfrac{1}{a}\ln(\cosh ax) + C$ **38.** $\dfrac{1}{a}\ln|\sinh ax| + C$

39. $-\dfrac{1}{a \cosh ax} + C$

40. $\displaystyle\int \sinh^2 x\, dx = \int \frac{1}{4}(e^{2x} - 2 + e^{-2x})\, dx = \frac{1}{4}\left(\frac{1}{2}e^{2x} - 2x - \frac{1}{2}e^{-2x}\right) + C$

$\qquad\qquad = \dfrac{1}{4}\sinh 2x - \dfrac{1}{2}x + C = \dfrac{1}{2}\sinh x \cosh x - \dfrac{1}{2}x + C$

41. From the identity $\cosh 2t = 2\cosh^2 t - 1$ (Exercise 23), we get

$\cosh^2 t = \frac{1}{2}(1 + \cosh 2t)$. Thus,

$$\int \cosh^2 x\, dx = \frac{1}{2}\int (1 + \cosh 2x)\, dx$$

$$= \frac{1}{2}\left(x + \frac{1}{2}\sinh 2x\right) + C$$

$$= \frac{1}{2}(x + \sinh x \cosh x) + C$$

42. $\displaystyle\int \sinh 2x\, e^{\cosh 2x}\, dx = \frac{1}{2}\int e^u\, du = \frac{1}{2}e^u + C = \frac{1}{2}e^{\cosh 2x} + C$

43. $\left\{\begin{array}{l} u = \sqrt{x} \\ du = dx/2\sqrt{x} \end{array}\right\}; \quad \displaystyle\int \frac{\sinh \sqrt{x}}{\sqrt{x}}\, dx = 2\int \sinh u\, du = 2\cosh u + C = 2\cosh \sqrt{x} + C$

44. $\displaystyle\int \frac{\sinh x}{1 + \cosh x}\, dx = \int \frac{du}{1 + u} = \ln|1 + u| + C = \ln(1 + \cosh x) + C$

45. $A.V. = \dfrac{1}{1 - (-1)}\displaystyle\int_{-1}^{1} \cosh x\, dx = \dfrac{1}{2}[\sinh x]_{-1}^{1} = \dfrac{e^2 - 1}{2e} \cong 1.175$

46. $A.V. = \dfrac{1}{4}\displaystyle\int_{0}^{4} \sinh 2x\, dx = \dfrac{1}{8}[\cosh 2x]_{0}^{4} = \dfrac{1}{8}\left[\dfrac{e^8 + e^{-8}}{2} - 1\right] \cong 186.185$

47. $A = \displaystyle\int_{0}^{\ln 10} \sinh x\, dx = \left[\cosh x\right]_{0}^{\ln 10} = \dfrac{e^{\ln 10} + e^{-\ln 10}}{2} - 1 = \dfrac{81}{20}$

48. $A = \displaystyle\int_{-\ln 5}^{\ln 5} \cosh 2x\, dx = \frac{1}{2}[\sinh 2x]_{-\ln 5}^{\ln 5} = \dfrac{312}{25}$

49. $V = \displaystyle\int_{0}^{1} \pi\left(\cosh^2 x - \sinh^2 x\right) dx = \int_{0}^{1} \pi\, dx = \pi$

50. $V = \int_0^{\ln 5} \pi [\sinh x]^2\, dx = \pi \int_0^{\ln 5} \sinh^2 x\, dx$

$$= \tfrac{1}{4}\pi \int_0^{\ln 5} \left(e^{2x} - 2 + e^{-2x} \right)\, dx$$

$$= \tfrac{1}{4}\pi \left[\tfrac{1}{2} e^{2x} - 2x - \tfrac{1}{2} e^{-2x} \right]_0^{\ln 5} = \tfrac{1}{4}\pi \left[\sinh(2\ln 5) - 2\ln 5 \right]$$

51. $V = \int_{-\ln 5}^{\ln 5} \pi [\cosh 2x]^2\, dx = 2\pi \int_0^{\ln 5} \cosh^2 2x\, dx$

$$= \tfrac{1}{2}\pi \int_0^{\ln 5} \left(e^{4x} + 2 + e^{-4x} \right)\, dx$$

$$= \tfrac{1}{2}\pi \left[\tfrac{1}{4} e^{4x} + 2x + \tfrac{1}{4} e^{-4x} \right]_0^{\ln 5} = \pi \left[\tfrac{1}{4} \sinh(4\ln 5) + \ln 5 \right]$$

52. (a) $\displaystyle \lim_{x\to\infty} \frac{\sinh x}{e^x} = \lim_{x\to\infty} \frac{e^x - e^{-x}}{2e^x} = \lim_{x\to\infty} \left(\frac{1}{2} - \frac{e^{-2x}}{2} \right) = \frac{1}{2}$

(b) $\displaystyle \lim_{x\to\infty} \frac{\cosh x}{e^{ax}} = \lim_{x\to\infty} \frac{e^x + e^{-x}}{2e^{ax}} = \lim_{x\to\infty} \frac{1}{2} \left(e^{x-ax} + e^{-x-ax} \right)$

For $0 < a < 1,$ limit $= \infty.$ For $a > 1,$ limit $= 0.$

53. (a) $(0.69315, 1.25)$ (b) $A \cong 0.38629$ **54.** (a) $(\pm 1.06128, 0.6180)$ (b) $A \cong 1.388$

SECTION 7.9

1. $\dfrac{dy}{dx} = 2\tanh x\, \text{sech}^2 x$

2. $\dfrac{dy}{dx} = 2\tanh 3x\, \text{sech}^2 3x \cdot 3 = 6\tanh 3x\, \text{sech}\, 3x$

3. $\dfrac{dy}{dx} = \dfrac{1}{\tanh x}\, \text{sech}^2 x = \text{sech}\, x\, \text{csch}\, x$

4. $\dfrac{dy}{dx} = \text{sech}^2(\ln x) \cdot \dfrac{1}{x}$

5. $\dfrac{dy}{dx} = \cosh\left(\arctan e^{2x} \right) \dfrac{d}{dx}\left(\arctan e^{2x} \right) = \dfrac{2e^{2x}\cosh\left(\arctan e^{2x} \right)}{1 + e^{4x}}$

6. $\dfrac{dy}{dx} = -\text{sech}\,(3x^2 + 1)\tanh(3x^2 + 1)(6x) = -6x\,\text{sech}\,(3x^2 + 1)\tanh(3x^2 + 1)$

7. $\dfrac{dy}{dx} = -\text{csch}^2\left(\sqrt{x^2 + 1}\, \right) \dfrac{d}{dx}\left(\sqrt{x^2 + 1}\, \right) = -\dfrac{x}{\sqrt{x^2 + 1}}\, \text{csch}^2\left(\sqrt{x^2 + 1}\, \right)$

8. $\dfrac{dy}{dx} = \dfrac{1}{\text{sech}\, x} \cdot (-\text{sech}\, x)(\tanh x) = -\tanh x$

9. $\dfrac{dy}{dx} = \dfrac{(1 + \cosh x)(-\operatorname{sech} x \tanh x) - \operatorname{sech} x (\sinh x)}{(1 + \cosh x)^2}$

$= \dfrac{-\operatorname{sech} x (\tanh x + \cosh x \tanh x + \sinh x)}{(1 + \cosh x)^2} = \dfrac{-\operatorname{sech} x (\tanh x + 2 \sinh x)}{(1 + \cosh x)^2}$

10. $\dfrac{dy}{dx} = \dfrac{\sinh x(1 + \operatorname{sech} x) - \cosh x(-\operatorname{sech} x) \tanh x}{(1 + \operatorname{sech} x)^2} = \dfrac{\sinh x + 2 \tanh x}{(1 + \operatorname{sech} x)^2}$

11. $\dfrac{d}{dx}(\coth x) = \dfrac{d}{dx}\left[\dfrac{\cosh x}{\sinh x}\right] = \dfrac{\sinh x (\sinh x) - \cosh x (\cosh x)}{\sinh^2 x}$

$= -\dfrac{\cosh^2 x - \sinh^2 x}{\sinh^2 x} = \dfrac{-1}{\sinh^2 x} = -\operatorname{csch}^2 x$

12. $\dfrac{d}{dx}(\operatorname{sech} x) = \dfrac{d}{dx}\left(\dfrac{1}{\cosh x}\right) = \dfrac{-1}{(\cosh x)^2} \cdot \sinh x = -\dfrac{1}{\cosh x} \cdot \dfrac{\sinh x}{\cosh x} = -\operatorname{sech} x \tanh x$

13. $\dfrac{d}{dx}(\operatorname{csch} x) = \dfrac{d}{dx}\left[\dfrac{1}{\sinh x}\right] = -\dfrac{\cosh x}{\sinh^2 x} = -\operatorname{csch} x \coth x$

14. $\tanh(t + s) = \dfrac{\sinh(t + s)}{\cosh(t + s)} = \dfrac{\sinh t \cosh s + \cosh t \sinh s}{\cosh t \cosh s + \sinh t \sinh s} = \dfrac{\tanh t + \tanh s}{1 + \tanh t \tanh s}$

15. **(a)** By the hint $\operatorname{sech}^2 x_0 = \dfrac{9}{25}$. Take $\operatorname{sech} x_0 = \dfrac{3}{5}$ since $\operatorname{sech} x = \dfrac{1}{\cosh x} > 0$ for all x.

(b) $\cosh x_0 = \dfrac{1}{\operatorname{sech} x_0} = \dfrac{5}{3}$ **(c)** $\sinh x_0 = \cosh x_0 \tanh x_0 = \left(\dfrac{5}{3}\right)\left(\dfrac{4}{5}\right) = \dfrac{4}{3}$

(d) $\coth x_0 = \dfrac{\cosh x_0}{\sinh x_0} = \dfrac{5/3}{4/3} = \dfrac{5}{4}$ **(e)** $\operatorname{csch} x_0 = \dfrac{1}{\sinh x_0} = \dfrac{3}{4}$

16. $\operatorname{sech}^2 t_0 = 1 - \tanh^2 t_0 = 1 - \dfrac{25}{144} = \dfrac{119}{144} \implies \operatorname{sech} t_0 = \dfrac{\sqrt{119}}{12}; \quad \cosh t_0 = \dfrac{1}{\operatorname{sech} t_0} = \dfrac{12}{\sqrt{119}};$

$\sinh t_0 = \cosh t_0 \tanh t_0 = \dfrac{12}{\sqrt{119}} \cdot \dfrac{-5}{12} = \dfrac{-5}{\sqrt{119}}; \quad \coth t_0 = \dfrac{1}{\tanh t_0} = \dfrac{-12}{5};$

$\operatorname{csch} t_0 = \dfrac{1}{\sinh t_0} = -\dfrac{\sqrt{119}}{5}.$

17. If $x \leq 0$, the result is obvious. Suppose then that $x > 0$. Since $x^2 \geq 1$, we have $x \geq 1$. Consequently

$$x - 1 = \sqrt{x - 1}\,\sqrt{x - 1} \leq \sqrt{x - 1}\,\sqrt{x + 1} = \sqrt{x^2 - 1}$$

and therefore $x - \sqrt{x^2 - 1} \leq 1.$

18. We will show that

$$\tanh\left[\dfrac{1}{2}\ln\left(\dfrac{1 + x}{1 - x}\right)\right] = x \quad \text{for all} \quad x \in [-1, 1].$$

First we observe that

$$\tanh s = \frac{e^s - e^{-s}}{e^s + e^{-s}} \quad \text{and therefore} \quad \tanh(\ln t) = \frac{t - 1/t}{t + 1/t} = \frac{t^2 - 1}{t^2 + 1}$$

It follows that

$$\tanh\left[\frac{1}{2}\ln\left(\frac{1+x}{1-x}\right)\right] = \tanh\left(\ln\sqrt{\frac{1+x}{1-x}}\right) = \frac{\frac{1+x}{1-x} - 1}{\frac{1+x}{1-x} + 1} = \frac{2x}{2} = x.$$

19. By Theorem 7.9.2,

$$\frac{d}{dx}\left(\sinh^{-1} x\right) = \frac{d}{dx}\left[\ln\left(x + \sqrt{x^2 + 1}\,\right)\right] = \frac{1}{x + \sqrt{x^2+1}}\left(1 + \frac{x}{\sqrt{x^2+1}}\right) = \frac{1}{\sqrt{x^2+1}}.$$

20. $\dfrac{d}{dx}(\cosh^{-1} x) = \dfrac{d}{dx}\left[\ln(x + \sqrt{x^2 - 1})\right] = \dfrac{1}{x + \sqrt{x^2-1}}\left(1 + \dfrac{x}{\sqrt{x^2-1}}\right) = \dfrac{1}{\sqrt{x^2-1}}.$

21. By Theorem 7.9.2

$$\frac{d}{dx}\left(\operatorname{arctan} x\right) = \frac{d}{dx}\left[\frac{1}{2}\ln\left(\frac{1+x}{1-x}\right)\right] = \frac{1}{2}\frac{1}{\left(\frac{1+x}{1-x}\right)}\left(\frac{(1-x)\,(1) - (1+x)\,(-1)}{(1-x)^2}\right)$$

$$= \frac{1}{\left(\frac{1+x}{1-x}\right)(1-x)^2} = \frac{1}{1-x^2}.$$

22. $y = \operatorname{sech}^{-1} x \implies \operatorname{sech} y = x \implies \cosh y = \dfrac{1}{x} \implies y = \cosh^{-1}\left(\dfrac{1}{x}\right),$

so $\quad \dfrac{dy}{dx} = \dfrac{1}{\sqrt{(1/x)^2 - 1}} \cdot \left(\dfrac{-1}{x^2}\right) = \dfrac{-1}{x\sqrt{1-x^2}}.$

23. Let $y = \operatorname{csch}^{-1} x$. Then $\operatorname{csch} y = x$ and $\sinh y = \dfrac{1}{x}$.

$$\sinh y = \frac{1}{x}$$

$$\cosh y \frac{dy}{dx} = -\frac{1}{x^2}$$

$$\frac{dy}{dx} = -\frac{1}{x^2 \cosh y} = -\frac{1}{x^2\sqrt{1 + (1/x)^2}} = -\frac{1}{|x|\sqrt{1 + x^2}}$$

24. $y = \coth^{-1} x \implies \coth y = x \implies \tanh y = \dfrac{1}{x} \implies y = \tanh^{-1}\left(\dfrac{1}{x}\right),$ so

$$\frac{dy}{dx} = \frac{1}{1 - (1/x)^2} \cdot \left(-\frac{1}{x^2}\right) = \frac{-1}{x^2 - 1} = \frac{1}{1 - x^2}$$

25.

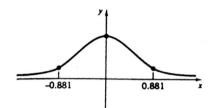

(a) $\dfrac{dy}{dx} = -\operatorname{sech} x \tanh x = -\dfrac{\sinh x}{\cosh^2 x}$

$\dfrac{dy}{dx} = 0$ at $x = 0$;

$\dfrac{dy}{dx} > 0$ if $x < 0$; $\quad \dfrac{dy}{dx} < 0$ if $x > 0$

f is increasing on $(-\infty, 0]$ and decreasing on $[0, \infty)$; $\ f(0) = 1$ is the absolute maximum of f.

(b) $\dfrac{d^2y}{dx^2} = -\dfrac{\cosh^2 x - 2\sinh^2 x}{\cosh^3 x} = \dfrac{\sinh^2 x - 1}{\cosh^3 x}$

$\dfrac{d^2y}{dx^2} = 0 \ \Rightarrow \ \sinh x = \pm 1$

$\sinh x = 1 \Rightarrow \dfrac{e^x - e^{-x}}{2} = 1 \ \Rightarrow \ e^{2x} - 2e^x - 1 = 0 \ \Rightarrow \ x = \ln(1 + \sqrt{2}) \cong 0.881$;

$\sinh x = -1 \Rightarrow \dfrac{e^x - e^{-x}}{2} = -1 \ \Rightarrow \ e^{2x} + 2e^x - 1 = 0 \ \Rightarrow \ x = -\ln(1 + \sqrt{2}) = -0.881$

(c) The graph of f is concave up on $(-\infty, -0.881) \cup (0.881, \infty)$ and concave down on $(-0.881, 0.881)$;

points of inflection at $x = \pm 0.881$

26. (a)

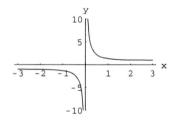

(b)

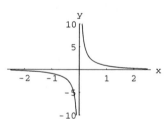

27. $y = \sinh x; \quad \dfrac{dy}{dx} = \cosh x; \quad \dfrac{d^2y}{dx^2} = \sinh x.$

$\dfrac{d^2y}{dx^2} = 0 \ \Rightarrow \ \sinh x = 0 \ \Rightarrow \ x = 0.$

$y = \sinh^{-1} x = \ln\left(x + \sqrt{x^2 + 1}\right); \quad \dfrac{dy}{dx} = \dfrac{1}{\sqrt{x^2 + 1}}; \quad \dfrac{d^2y}{dx^2} = -\dfrac{x}{(x^2 + 1)^{3/2}}.$

$\dfrac{d^2y}{dx^2} = 0 \ \Rightarrow \ -\dfrac{x}{(x^2 + 1)^{3/2}} = 0 \ \Rightarrow \ x = 0.$

It is easy to verify that $(0,0)$ is a point of inflection for both graphs.

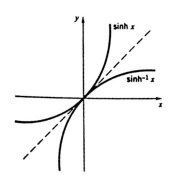

28. (a)

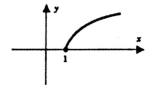

(b)

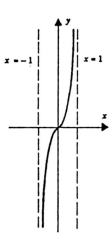

29. (a) $\tan \phi = \sinh x$

$$\phi = \arctan(\sinh x)$$

$$\frac{d\phi}{dx} = \frac{\cosh x}{1 + \sinh^2 x}$$

$$= \frac{\cosh x}{\cosh^2 x} = \frac{1}{\cosh x} = \operatorname{sech} x$$

(b) $\sinh x = \tan \phi$

$$x = \sinh^{-1}(\tan \phi)$$

$$= \ln \left(\tan \phi + \sqrt{\tan^2 \phi + 1} \right)$$

$$= \ln(\tan \phi + \sec \phi)$$

$$= \ln(\sec \phi + \tan \phi)$$

(c) $x = \ln(\sec \phi + \tan \phi)$

$$\frac{dx}{d\phi} = \frac{\sec \phi \tan \phi + \sec^2 \phi}{\tan \phi + \sec \phi} = \sec \phi$$

30. $V = \displaystyle\int_{-1}^{1} \pi \operatorname{sech}^2 x \, dx = [\pi \tanh x]_{-1}^{1} = \pi \left(\dfrac{e - e^{-1}}{e + e^{-1}} - \dfrac{e^{-1} - e}{e^{-1} + e} \right) = 2\pi \left(\dfrac{e^2 - 1}{e^2 + 1} \right)$

31. $\displaystyle\int \tanh x \, dx = \int \frac{\sinh x}{\cosh x} \, dx$

$$\left\{ \begin{array}{l} u = \cosh x \\ du = \sinh x \, dx \end{array} \right\} ; \quad \int \frac{\sinh x}{\cosh x} \, dx = \int \frac{1}{u} \, du = \ln|u| + C = \ln \cosh x + C$$

32. $\displaystyle\int \coth x \, dx = \int \frac{\cosh x}{\sinh x} \, dx = \int \frac{du}{u} = \ln|u| + C = \ln|\sinh x| + C$

33. $\displaystyle\int \operatorname{sech} x \, dx = \int \frac{1}{\cosh x} \, dx = \int \frac{2}{e^x + e^{-x}} \, dx = \int \frac{2e^x}{e^{2x} + 1} \, dx$

$$\left\{ \begin{array}{l} u = e^x \\ du = e^x \, dx \end{array} \right\} ; \quad \int \frac{2e^x}{e^{2x} + 1} \, dx = 2 \int \frac{1}{u^2 + 1} \, du = 2 \arctan u + C = 2 \arctan(e^x) + C$$

34. $\displaystyle\int \operatorname{csch} x \, dx = \int \frac{2}{e^x - e^{-x}} \, dx = 2 \int \frac{e^x}{e^{2x} - 1} \, dx = -2 \int \frac{du}{1 - u^2} \quad (u = e^x)$

$$= \left\{ \begin{array}{l} -2 \tanh^{-1} e^x + C, \, e^x < 1 \\ -2 \coth^{-1} e^x + C, \, e^x > 1. \end{array} \right.$$

35. $\left\{ \begin{array}{l} u = \operatorname{sech} x \\ du = - \operatorname{sech} s \tanh x \, dx \end{array} \right\} ; \quad \displaystyle\int \operatorname{sech}^3 x \tanh x \, dx = - \int u^2 \, du = -\frac{1}{3} u^3 + C$

$$= -\tfrac{1}{3} \operatorname{sech}^3 x + C$$

36. $\displaystyle\int x \operatorname{sech}^2 x^2 \, dx = \frac{1}{2} \int \operatorname{sech}^2 u \, du = \frac{1}{2} \tanh u + C = \frac{1}{2} \tanh x^2 + C$

37. $\left\{ \begin{array}{l} u = \ln(\cosh x) \\ du = \tanh x \, dx \end{array} \right\} ; \quad \displaystyle\int \tanh x \ln(\cosh x) \, dx = \int u \, du = \frac{1}{2} u^2 + C = \frac{1}{2} [\ln(\cosh x)]^2 + C$

38. $\displaystyle\int \frac{1 + \tanh x}{\cosh^2 x} \, dx = \int \left(\operatorname{sech}^2 x + \frac{\sinh x}{\cosh^3 x} \right) dx = \tanh x - \frac{1}{2 \cosh^2 x} + C = \tanh x - \frac{1}{2} \operatorname{sech}^2 x + C$

39. $\left\{ \begin{array}{l} u = 1 + \tanh x \\ du = \operatorname{sech}^2 x \, dx \end{array} \right\} ; \quad \displaystyle\int \frac{\operatorname{sech}^2 x}{1 + \tanh x} \, dx = \int \frac{1}{u} \, du = \ln|u| + C = \ln|1 + \tanh x| + C$

40. $\displaystyle\int \tanh^5 x \operatorname{sech}^2 x \, dx = \frac{1}{6} \tanh^6 x + C$

41. $\left\{ \begin{array}{l} x = a \sinh u \\ dx = a \cosh u \, du \end{array} \right\} ; \quad \displaystyle\int \frac{dx}{\sqrt{a^2 + x^2}} \, dx = \int \frac{a \cosh u}{\sqrt{a^2 + a^2 \sinh^2 u}} \, du$

$$= \int du = u + C = \sinh^{-1}\left(\frac{x}{a}\right) + C$$

42. $\displaystyle\int \frac{1}{\sqrt{x^2 - a^2}} \, dx = \frac{1}{a} \int \frac{1}{\sqrt{(x/a)^2 - 1}} \, dx = \int \frac{du}{\sqrt{u^2 - 1}} = \cosh^{-1}(u) + C = \cosh^{-1}\left(\frac{x}{a}\right) + C$

43. Suppose $|x| < a$.

$$\begin{cases} x = a \tanh u \\ dx = a \operatorname{sech}^2 u \, du \end{cases}; \quad \int \frac{dx}{a^2 - x^2} \, dx = \int \frac{a \operatorname{sech}^2 u}{a^2 - a^2 \tanh^2 u} \, du$$

$$= \frac{1}{a} \int du = \frac{u}{a} + C = \frac{1}{a} \tanh^{-1}\left(\frac{x}{a}\right) + C$$

The other case is done in the same way.

44. (a) $v(0) = \sqrt{\dfrac{mg}{k}} \tanh 0 = 0$

$$v'(t) = \sqrt{\frac{mg}{k}} \operatorname{sech}^2\left(\sqrt{\frac{gk}{m}}\, t\right)\left(\sqrt{\frac{gk}{m}}\right) = g \operatorname{sech}^2\left(\sqrt{\frac{gk}{m}}\, t\right)$$

$$mg - kv^2 = mg - k\frac{mg}{k}\tanh^2\left(\sqrt{\frac{gk}{m}}\, t\right) = mg\left[1 - \tanh^2\left(\sqrt{\frac{gk}{m}}\, t\right)\right]$$

$$= mg \operatorname{sech}^2\left(\sqrt{\frac{gk}{m}}\, t\right) = m\frac{dv}{dt}$$

(b) $\displaystyle\lim_{t\to\infty} v(t) = \lim_{t\to\infty} \sqrt{\frac{mg}{k}} \tanh\left(\sqrt{\frac{gk}{m}}\, t\right) = \sqrt{\frac{mg}{k}}.$

REVIEW EXERCISES

1. $f'(x) = \frac{1}{3}x^{-2/3} > 0$ except at $x = 0$; f is increasing, it is one-to-one; $f^{-1}(x) = (x - 2)^3$.

2. f is not one-to-one; $f(3) = f(-2) = 0$

3. Suppose $f(x_1) = f(x_2)$. Then

$$\frac{x_1 + 1}{x_1 - 1} = \frac{x_2 + 1}{x_2 - 1}$$
$$x_1 x_2 + x_2 - x_1 - 1 = x_1 x_2 + x_1 - x_2 - 1$$
$$2x_2 = 2x_1$$
$$x_1 = x_2$$

Thus f is one-to-one. $f^{-1}(x) = \dfrac{x + 1}{x - 1}.$

4. $f'(x) = 6(2x + 1)^2 > 0$ except at $x = -\frac{1}{2}$; f is increasing, it is one-to-one; $f^{-1}(x) = \dfrac{x^{1/3} - 1}{2}.$

5. $f'(x) = -\dfrac{1}{x^2}e^{\frac{1}{x}} < 0$ except at $x = 0$; f is decreasing, it is one-to-one; $f^{-1}(x) = \frac{1}{\ln x}.$

6. $f(x)$ is not one-to-one; $f(\pi/2) = f(3\pi/2) = 0$, .

7. f is not one-to-one. Reason: $f'(x) = 1 + \ln x \Longrightarrow f$ decreases on $(0, 1/e]$ and increases on $[1/e, \infty)$. There exist horizontal lines that intersect the graph in two points.

8. Suppose $f(x_1) = f(x_2)$. Then

$$\frac{2x_1 + 1}{3 - 2x_1} = \frac{2x_2 + 1}{3 - 2x_2}$$

$$-4x_1x_2 + 6x_1 - 2x_2 + 3 = -4x_1x_2 + 6x_2 - 2x_1 + 3$$

$$8x_1 = 8x_2$$

$$x_1 = x_2$$

Thus f is one-to-one. $\quad f^{-1}(x) = \dfrac{3x - 1}{2x + 2}$.

9. $f'(x) = \dfrac{-e^x}{(1 + e^x)^2} < 0$ for all x; f is one to one and has an inverse function.

Since $f(0) = \frac{1}{2}$, $\quad (f^{-1})'\left(\frac{1}{2}\right) = \dfrac{1}{f'(0)} = -4$

10. $f'(x) = 3 + \dfrac{3}{x^4} > 0$ for all $x > 0$; f is one-to-one and has an inverse function.

$f(1) = 2$; therefore $(f^{-1})'(2) = \dfrac{1}{f'(1)} = \dfrac{1}{6}$

11. $f'(x) = \sqrt{4 + x^2} > 0 \implies f$ has an inverse.

Since $f(0) = 0$, $\quad (f^{-1})'(0) = \dfrac{1}{f'(0)} = \dfrac{1}{2}$

12. $f' = 1 - \sin x \geq 0$ for all x; f has an inverse function.

Since $f(\pi) = -1$; $(f^{-1})'(-1) = \dfrac{1}{f'(\pi)} = 1$

13. $f'(x) = 3\left(\ln x^2\right)^2 \dfrac{1}{x^2}(2x) = \dfrac{6\left(\ln x^2\right)^2}{x} = \dfrac{24\left(\ln x\right)^2}{x}$

14. $y' = 2(\cos e^{3x})(e^{3x})(3) = 6e^{3x}\cos e^{3x}$

15. $g'(x) = \dfrac{\left(1 + e^{2x}\right)e^x - e^x\left(2e^{2x}\right)}{\left(1 + e^{2x}\right)^2} = \dfrac{e^x(1 - e^{2x})}{(1 + e^{2x})^2}$

16. $\ln f(x) = \sinh x \ln(x^2 + 1)$; $\quad \dfrac{f'(x)}{f(x)} = \sinh x \dfrac{2x}{x^2 + 1} + \ln(x^2 + 1)\cosh x$;

$f'(x) = (x^2 + 1)^{\sinh x}\left[\cosh x \ln(x^2 + 1) + \dfrac{2x \sinh x}{x^2 + 1}\right]$

17. $\dfrac{dy}{dx} = \dfrac{1}{x^3 + 3^x}\left(3x^2 + 3^x \ln 3\right) = \dfrac{3x^2 + 3^x \ln 3}{x^3 + 3^x}$

18. $g'(x) = \dfrac{\sinh x}{1 + \cosh^2 x}$

19. $\ln f(x) = \dfrac{1}{x}\ln\cosh x = \dfrac{\ln\cosh x}{x}$

$$\frac{f'(x)}{f(x)} = \frac{x\,\dfrac{\sinh x}{\cosh x} - \ln\cosh x}{x^2} = \frac{\sinh x}{x\cosh x} - \frac{\ln\cosh x}{x^2}.$$

Therefore $f'(x) = (\cosh x)^{1/x}\left[\dfrac{\sinh x}{x\cosh x} - \dfrac{\ln\cosh x}{x^2}\right]$

20. $f'(x) = 2x^3\dfrac{2x}{\sqrt{1-x^4}} + 6x^2\arcsin(x^2) = \dfrac{4x^4}{\sqrt{1-x^4}} + 6x^2\arcsin(x^2)$

21. $f(x) = \dfrac{1}{\ln 3}\ln\dfrac{1+x}{1-x};\qquad f'(x) = \dfrac{1}{\ln 3}\left(\dfrac{1-x}{1+x}\right)\left(\dfrac{2}{(1-x)^2}\right) = \dfrac{2}{\ln 3(1-x^2)}$

22. $f'(x) = \dfrac{1}{\sqrt{x^2+4}\sqrt{x^2+4-1}}\cdot\dfrac{1}{2}\dfrac{2x}{\sqrt{x^2+4}} == \dfrac{x}{(x^2+4)\sqrt{x^2+3}}$

23. Substitution: Set $u = e^x$, $du = e^x\,dx$.

$$\int\frac{e^x}{\sqrt{1-e^{2x}}}\,dx = \int\frac{1}{\sqrt{1-u^2}}\,dx = \arcsin u + C = \arcsin e^x + C$$

24. Substitution: let $u = \ln x$. Then $du = \dfrac{1}{x}\,dx$, $u(1) = 0$, $u(e) = 1$.

$$\int_1^e\frac{\sqrt{\ln x}}{x}\,dx = \int_0^1 u^{1/2}\,du = \tfrac{2}{3}\left[u^{3/2}\right]_0^1 = \tfrac{2}{3}$$

25. Substitution: let $u = \sin x$, $du = \cos x\,dx$

$$\int\frac{\cos x}{4+\sin^2 x}\,dx = \int\frac{1}{4+u^2}\,du = \frac{1}{2}\arctan\left(\frac{u}{2}\right) + C = \frac{1}{2}\arctan\left(\frac{\sin x}{2}\right) + C$$

26. Substitution: let $u = \ln\cos x$, $du = -\tan x\,dx$

$$\int\tan x\ln\cos x\,dx = -\int u\,du = -\frac{1}{2}u^2 + C = -\frac{1}{2}\ln^2\cos x + C$$

27. Substitution: let $u = \sqrt{x}$, $du = \dfrac{1}{2}\dfrac{1}{\sqrt{x}}\,dx$

$$\int\frac{\sec\sqrt{x}}{\sqrt{x}}\,dx = 2\int\sec u\,du = 2\ln|\sec u + \tan u| + C = 2\ln|\sec\sqrt{x} + \tan\sqrt{x}| + C$$

28. Substitution: let $u = x^2$, $du = 2x\,dx$

$$\int\frac{1}{x\sqrt{x^4-9}}\,dx = \frac{1}{2}\int\frac{1}{u\sqrt{u^2-3^2}}\,du = \frac{1}{6}\text{arcsec}\,\frac{x^2}{3} + C$$

29. Substitution: let $u = \ln x$, $du = \dfrac{1}{x}\,dx$

$$\int\frac{5^{\ln x}}{x}\,dx = \frac{5^{\ln x}}{\ln 5} + C$$

30. Substitution: let $u = x^3$, $du = 3x^2\, dx$

$$\int_0^2 x^2 e^{x^3}\, dx = \frac{1}{3} e^{x^3}\big|_0^2 = \frac{1}{3}(e^8 - 1)$$

31. Substitution: $u = x^{4/3} + 1$, $du = \frac{4}{3}x^{1/3}\, dx$, $u(1) = 2$, $u(8) = 17$

$$\int_1^8 \frac{x^{1/3}}{x^{4/3} + 1}\, dx = \frac{3}{4}\int_2^{17} \frac{1}{u}\, du = \frac{3}{4}[\ln u]_2^{17} = \frac{3}{4}(\ln 17 - \ln 2) = \frac{3}{4}\ln\left(\frac{17}{2}\right)$$

32. Substitution: let $u = \sec x$, $du = \sec x \tan x\, dx$

$$\int \frac{\sec x \tan x}{1 + \sec^2 x}\, dx = \int \frac{1}{1 + u^2}\, du = \arctan u + C = \arctan \sec x + C$$

33. Substitution: $u = 2^x$, $du = 2^x \ln 2\, dx$

$$\int 2^x \sinh 2^x\, dx = \frac{1}{\ln 2}\int \sinh u\, du = \frac{1}{\ln 2}\cosh u + C = \frac{1}{\ln 2}\cosh 2^x + C$$

34. Substitution: let $u = e^{2x}$, $du = 2e^{2x}\, dx$

$$\int \frac{e^x}{e^x + e^{-x}}\, dx = \int \frac{e^{2x}}{1 + e^{2x}}\, dx = \frac{1}{2}\ln(1 + e^{2x}) + C$$

35. $\displaystyle\int_2^5 \frac{1}{x^2 - 4x + 13}\, dx = \int_2^5 \frac{1}{(x-2)^2 + 3^2}\, dx = \frac{1}{3}\left[\arctan\frac{x-2}{3}\right]_2^5 = \frac{\pi}{12}$

36. Substitution: let $u = x - 1$, $du = dx$

$$\int \frac{1}{\sqrt{15 + 2x - x^2}}\, dx = \int \frac{1}{\sqrt{4^2 - (x-1)^2}}\, dx = \arcsin\frac{x-1}{4} + C$$

37. $\displaystyle\int_0^2 \operatorname{sech}^2\left(\frac{x}{2}\right) dx = 2\left[\tanh\frac{x}{2}\right]_0^2 = 2\tanh 1 = 2\frac{e^2 - 1}{e^2 + 1}$

38. $\displaystyle\int \tanh^2 2x\, dx = x - \frac{1}{2}\tanh 2x + C$

39. $A = \displaystyle\int_0^1 \frac{x}{1 + x^2}\, dx = \frac{1}{2}\left[\ln(x^2 + 1)\right]_0^1 = \frac{\ln 2}{2}$

40. $A = \displaystyle\int_0^1 \frac{1}{1 + x^2}\, dx = [\arctan x]_0^1 = \frac{\pi}{4}$

41. $A = \displaystyle\int_0^{1/2} \frac{1}{\sqrt{1 - x^2}}\, dx = [\arcsin x]_0^{1/2} = \frac{\pi}{6}$

42. $A = \displaystyle\int_0^{1/2} \frac{x}{\sqrt{1 - x^2}}\, dx = -\left[\sqrt{1 - x^2}\right]_0^{1/2} = \frac{2 - \sqrt{3}}{2}$

43. Fix $x > 0$. Then, by the mean-value theorem,

$$\ln(1+x) - \ln 1 = \frac{1}{1+c}(x-0) = \frac{x}{1+c} \quad \text{for some } c \in (0, x)$$

Since $\dfrac{x}{1+x} < \dfrac{x}{1+c} < x,$ it follows that

$$\frac{x}{1+x} < \ln(1+x) < x$$

Now fix $x \in (-1, 0)$. Then,

$$\ln 1 - \ln(1+x) = \frac{1}{1+c}(0-x) = \frac{-x}{1+c} \quad \text{for some } c \in (x, 0)$$

Since $-x < \dfrac{-x}{1+c} < \dfrac{-x}{1+x},$ it follows that

$$\frac{x}{1+x} < \ln(1+x) < x$$

Thus $\dfrac{x}{1+x} < \ln(1+x) < x$ for all $x > -1$ (the result is obvious if $x = 0$).

(b) The result follows from part (a) and the pinching theorem (Theorem 2.5.1)

44. Using the hint,

$$\ln \frac{n+1}{m} = \int_m^{n+1} \frac{dx}{x} = \int_m^{m+1} \frac{dx}{x} + \cdots + \int_n^{n+1} \frac{dx}{x} < \frac{1}{m} + \cdots + \frac{1}{n}$$

$$\ln \frac{n}{m-1} = \int_{m-1}^{n} \frac{dx}{x} = \int_{m-1}^{m} \frac{dx}{x} + \cdots + \int_{n-1}^{n} \frac{dx}{x} > \frac{1}{m} + \cdots + \frac{1}{n}.$$

45. Assume $a > 0$: $A = \displaystyle\int_a^{2a} \frac{a^2}{x} \, dx = \left[a^2 \ln x\right]_a^{2a} = a^2(\ln 2a - \ln a) = a^2 \ln 2$

46. $A = \displaystyle\int_{-1/2}^{1/2} \sec \tfrac{1}{2}\pi x \, dx = \frac{2}{\pi}\left[\ln\left|\sec \tfrac{1}{2}\pi x + \tan \tfrac{1}{2}\pi x\right|\right]_{-1/2}^{1/2} = \frac{2}{\pi}\ln \frac{\sqrt{2}+1}{\sqrt{2}-1}$

47. (a) $V = \displaystyle\int_0^{\sqrt{3}} \pi \frac{1}{1+x^2} \, dx = \pi[\arctan x]_0^{\sqrt{3}} = \frac{\pi^2}{3}$

(b) $V = \displaystyle\int_0^{\sqrt{3}} 2\pi x(1+x^2)^{-1/2} \, dx = 2\pi[\sqrt{1+x^2}]_0^{\sqrt{3}} = 2\pi$

48. (a) $V = \displaystyle\int_0^{1/2} \frac{\pi}{\sqrt{1+x^2}} \, dx = \pi \ln \frac{1+\sqrt{5}}{2}$

(b) $V = \displaystyle\int_0^{1/2} 2\pi x \frac{1}{(1+x^2)^{1/4}} \, dx = \frac{4\pi}{3}[(1+x^2)^{3/4}]_0^{1/2} = \frac{4\pi}{3}\left[\left(\frac{5}{4}\right)^{3/4} - 1\right]$

49. $f(x) = \dfrac{\ln x}{x}$, domain: $(0, \infty)$

$f'(x) = \dfrac{1 - \ln x}{x^2}$

critical pt. $x = e$

$f''(x) = \dfrac{2\ln x - 3}{x^3}$

$f'(x) > 0$ on $(0, e)$,

$f'(x) < 0$ on (e, ∞)

$f''(x) > 0$ on $\left(e^{3/2}, \infty\right)$

$f''(x) < 0$ on $\left(0, e^{3/2}\right)$

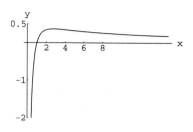

50. $f(x) = x^2 e^{-x^2}$, domain: $(-\infty, \infty)$

$f'(x) = 2\left(x - x^3\right)e^{-x^2} = 2x(1-x)(1+x)e^{-x^2}$

critical pts.. $x = 0$, $x = 1$, $x = -1$

$f''(x) = 2(2x^4 - 5x^2 + 1)e^{-x^2}$

$f'(x) > 0$ on $(-\infty, -1) \cup (0, 1)$,

$f'(x) < 0$ on $(-1, 0) \cup (1, \infty)$

$f''(x) > 0$ on $\left(-\infty, -\frac{1}{2}\sqrt{5 + \sqrt{17}}\right) \cup \left(-\frac{1}{2}\sqrt{5 - \sqrt{17}}, \frac{1}{2}\sqrt{5 - \sqrt{17}}\right) \cup \left(\frac{1}{2}\sqrt{5 + \sqrt{17}}, \infty\right)$

$f''(x) < 0$ on $\left(-\frac{1}{2}\sqrt{5 + \sqrt{17}}, -\frac{1}{2}\sqrt{5 - \sqrt{17}}\right) \cup \left(\frac{1}{2}\sqrt{5 - \sqrt{17}}, \frac{1}{2}\sqrt{5 + \sqrt{17}}\right)$

51. $\displaystyle\int_0^1 \dfrac{b}{\sqrt{1 - b^2 x^2}}\, dx$: substitution: $u = bx$, $du = b\, dx$, $u(0) = 0$, $u(1) = b$

$\displaystyle\int_0^1 \dfrac{b}{\sqrt{1 - b^2 x^2}}\, dx = \int_0^b \dfrac{1}{\sqrt{1 - u^2}}\, du = [\arcsin u]_0^b = \arcsin b.$

$\displaystyle\int_0^a \dfrac{1}{\sqrt{1 - x^2}}\, dx = \arcsin a; \Longrightarrow a = b$

52. $\displaystyle\int_0^1 \dfrac{a}{1 + a^2 x^2}\, dx = \arctan a = \int_0^a \dfrac{1}{1 + x^2}\, dx$

53. Let 6 a.m. correspond to $t = 0$; measure time in hours.

$$P(t) = 20e^{kt}; \quad P(2) = 40 = 20e^{2k} \implies k = \frac{\ln 2}{2} \implies P(t) = 20e^{\frac{t}{2}\ln 2} = 20\,(2)^{t/2}$$

(a) $P(6) = 20(2)^3 = 20(8) = 160$ grams

(b) $20(2)^{t/2} = 200 \implies \dfrac{t}{2}\ln 2 = \ln 10 \implies t = \dfrac{2\ln 10}{\ln 2} \cong 6.64$ hours

54. $A(t) = A(0)e^{kt}$. $A(1) = 0.8A(0)$ $\implies$ $e^k = 0.8$ $\implies$ $k = \ln 0.8$ $\implies$ $T = -\dfrac{\ln 2}{\ln 0.8} \cong 3.1$

The half-life of the substance is approximately 3.1 years.

55. (a) $A(t) = 100e^{kt}$. From the relation $k = \dfrac{-\ln 2}{T}$, where T is the half-life, $A(t) = 100e^{-\frac{t}{140}\ln 2}$

(b) $100e^{-\frac{t}{140}\ln 2} = 75$ $\implies$ $\dfrac{-t}{140}\ln 2 = \ln 0.75$ $\implies$ $t = \dfrac{-140\ln 0.75}{\ln 2} \cong 58.11$ days

56. $P_{US} = 227e^{kt}$; $P_{US}(10) = 249 = 227e^{10k}$ $\implies$ $k = \dfrac{\ln(249/227)}{10} \cong 0.00925$

$P_M = 62e^{mt}$; $P_M(10) = 79 = 62e^{10m}$ $\implies$ $m = \dfrac{\ln(79/62)}{10} \cong 0.02423$

To find when the populations will be equal, solve $227e^{kt} = 62e^{mt}$ for t:

$$227e^{kt} = 62e^{mt} \implies e^{(m-k)t} = \frac{227}{62} \implies t = \frac{\ln(227/62)}{m-k} \cong 86.64$$

If the populations continue to grow at the given rates, the two populations will be equal in the year 2067.

57. $P(t) = P_0 e^{kt}$. Since the doubling time is 2 years, $k = \frac{1}{2}\ln 2$ and $P(t) = P_0 e^{\frac{t}{2}\ln 2} = P_0(2)^{t/2}$.

(a) $40,000 = P(4) = P_0(2)^2$ $\implies$ $4P_0 = 25,000 \implies P_0 = 6250$.

(b) $P(t) = 6250(2)^{t/2} = 40,000$ $\implies$ $(2)^{t/2} = 6.4$ $\implies$ $t = \dfrac{2\ln 6.4}{\ln 2} \cong 5.36$; it will take

approximately 5.36 years for the population to reach 40,000.

58. $\dfrac{1}{p} + \dfrac{1}{q} = 1 \implies q = \dfrac{p}{p-1}$. Assume that $a \leq b$. See the figure: clearly $ab \leq$ area $\Omega_1 +$ area Ω_2.

$\text{area } \Omega_1 = \displaystyle\int_0^a x^{p-1}dx = \left[\dfrac{x^p}{p}\right]_0^a = \dfrac{a^p}{p}$.

$y = x^{p-1} \implies x = y^{1/(p-1)}$;

$\text{area } \Omega_2 = \displaystyle\int_0^b y^{1/(p-1)}dy = \left[\dfrac{y^{1+1/(p-1)}}{1+1/(p-1)}\right]_0^b = \dfrac{b^{1+1/(p-1)}}{1+1/(p-1)} = \dfrac{b^q}{q}$

The same argument applies if $b < a$.

CHAPTER 8

SECTION 8.1

1. $\displaystyle\int e^{2-x}dx = -e^{2-x} + C$

2. $\displaystyle\int \cos \frac{2}{3}x\, dx = \frac{3}{2}\sin \frac{2}{3}x + C$

3. $\displaystyle\int_0^1 \sin \pi x\, dx = \left[-\frac{1}{\pi}\cos \pi x\right]_0^1 = \frac{2}{\pi}$

4. $\displaystyle\int_0^t \sec \pi x \tan \pi x\, dx = \frac{1}{\pi}\left[\sec \pi x\right]_0^t = \frac{1}{\pi}(\sec \pi t - 1)$

5. $\displaystyle\int \sec^2 (1-x)\, dx = -\tan (1-x) + C$

6. $\displaystyle\int \frac{dx}{5^x} = \int 5^{-x}\, dx = \frac{-1}{\ln 5}5^{-x} + C = -\frac{1}{5^x \ln 5} + C$

7. $\displaystyle\int_{\pi/6}^{\pi/3} \cot x\, dx = \left[\ln (\sin x)\right]_{\pi/6}^{\pi/3} = \ln \frac{\sqrt{3}}{2} - \ln \frac{1}{2} = \frac{1}{2}\ln 3$

8. $\displaystyle\int_0^1 \frac{x^3}{1+x^4}\, dx = \frac{1}{4}\left[\ln(1+x^4)\right]_0^1 = \frac{1}{4}\ln 2$

9. $\displaystyle\left\{\begin{array}{l} u = 1-x^2 \\ du = -2x\, dx \end{array}\right\};\qquad \int \frac{x\, dx}{\sqrt{1-x^2}} = -\frac{1}{2}\int u^{-1/2}\, du = -u^{1/2} + C = -\sqrt{1-x^2} + C$

10. $\displaystyle\int_{-\pi/4}^{\pi/4} \frac{dx}{\cos^2 x} = \int_{-\pi/4}^{\pi/4} \sec^2 x\, dx = \left[\tan x\right]_{-\pi/4}^{\pi/4} = 2$

11. $\displaystyle\int_{-\pi/4}^{\pi/4} \frac{\sin x}{\cos^2 x}\, dx = \int_{-\pi/4}^{\pi/4} \sec x \tan x\, dx = \left[\sec x\right]_{-\pi/4}^{\pi/4} = 0$

12. $\displaystyle\int \frac{e^{\sqrt{x}}}{\sqrt{x}}\, dx = 2e^{\sqrt{x}} + C$

13. $\displaystyle\left\{\begin{array}{l|ll} u = 1/x & x = 1 & \Rightarrow \quad u = 1 \\ du = dx/x^2 & x = 2 & \Rightarrow \quad u = 1/2 \end{array}\right\};$

$\displaystyle\int_1^2 \frac{e^{1/x}}{x^2}\, dx = \int_1^{1/2} -e^u\, du = \left[-e^u\right]_1^{1/2} = e - \sqrt{e}$

14. $\displaystyle\int \frac{x^3}{\sqrt{1-x^4}}\, dx = -\frac{1}{4}\int \frac{du}{\sqrt{u}} = -\frac{1}{2}\sqrt{u} + C = -\frac{1}{2}\sqrt{1-x^4} + C$

15. $\displaystyle\int_0^c \frac{dx}{x^2 + c^2} = \left[\frac{1}{c}\arctan \left(\frac{x}{c}\right)\right]_0^c = \frac{\pi}{4c}$

16. $\displaystyle\int a^x e^x\, dx = \int (ae)^x\, dx = \frac{(ae)^x}{\ln(ae)} + C = \frac{a^x e^x}{1 + \ln a} + C$

17. $\left\{ \begin{array}{l} u = 3\tan\theta + 1 \\ du = 3\sec^2\theta\,d\theta \end{array} \right\};$

$$\int \frac{\sec^2\theta}{\sqrt{3\tan\theta + 1}}\,d\theta = \frac{1}{3}\int u^{-1/2}\,du = \frac{2}{3}u^{1/2} + C = \frac{2}{3}\sqrt{3\tan\theta + 1} + C$$

18. $\displaystyle\int \frac{\sin\phi}{3 - 2\cos\phi}\,d\phi = \frac{1}{2}\int \frac{du}{u} = \frac{1}{2}\ln|u| + C = \frac{1}{2}\ln(3 - 2\cos\phi) + C$

19. $\displaystyle\int \frac{e^x}{ae^x - b}\,dx = \frac{1}{a}\ln|ae^x - b| + C$

20. $\displaystyle\int \frac{dx}{x^2 - 4x + 13} = \int \frac{dx}{(x-2)^2 + 9} = \frac{1}{3}\arctan\left(\frac{x-2}{3}\right) + C$

21. $\left\{ \begin{array}{l} u = x + 1 \\ du = dx \end{array} \right\};$

$$\int \frac{x}{(x+1)^2 + 4}\,dx = \int \frac{u - 1}{u^2 + 4}\,du = \int \frac{u}{u^2 + 4}\,du - \int \frac{du}{u^2 + 4}$$

$$= \frac{1}{2}\ln|u^2 + 4| - \frac{1}{2}\arctan\frac{u}{2} + C$$

$$= \frac{1}{2}\ln|(x+1)^2 + 4| - \frac{1}{2}\arctan\left(\frac{x+1}{2}\right) + C$$

22. $\displaystyle\int \frac{\ln x}{x}\,dx = \frac{1}{2}(\ln x)^2 + C$

23. $\left\{ \begin{array}{l} u = x^2 \\ du = 2x\,dx \end{array} \right\};$ $\displaystyle\int \frac{x}{\sqrt{1 - x^4}}\,dx = \frac{1}{2}\int \frac{du}{\sqrt{1 - u^2}} = \frac{1}{2}\arcsin u + C = \tfrac{1}{2}\arcsin(x^2) + C$

24. $\displaystyle\int \frac{e^x}{1 + e^{2x}}\,dx = \int \frac{du}{1 + u^2} = \arctan u + C = \arctan e^x + C$

25. $\left\{ \begin{array}{l} u = x + 3 \\ du = dx \end{array} \right\}; \displaystyle\int \frac{dx}{x^2 + 6x + 10} = \int \frac{dx}{(x+3)^2 + 1} = \int \frac{du}{u^2 + 1} = \arctan u + C = \arctan(x+3) + C$

26. $\displaystyle\int e^x \tan e^x\,dx = \int \tan u\,du = \ln|\sec u| + C = \ln|\sec e^x| + C$

27. $\displaystyle\int x \sin x^2\,dx = -\frac{1}{2}\cos x^2 + C$

28. $\displaystyle\int \frac{x + 1}{\sqrt{1 - x^2}}\,dx = \int \frac{x}{\sqrt{1 - x^2}}\,dx + \int \frac{dx}{\sqrt{1 - x^2}} = -\sqrt{1 - x^2} + \arcsin x + C$

29. $\displaystyle\int \tan^2 x\,dx = \int (\sec^2 x - 1)\,dx = \tan x - x + C$

30. $\displaystyle\int \cosh 2x \sinh^3 2x \, dx = \frac{1}{8} \sinh^4 2x + C$

31. $\left\{ \begin{array}{l|ll} u = \ln x & x = 1 & \Rightarrow \quad u = 0 \\ du = dx/x & x = e & \Rightarrow \quad u = 1 \end{array} \right\};$

$$\int_1^e \frac{\ln x^3}{x} \, dx = \int_1^e \frac{3 \ln x}{x} \, dx = 3 \int_0^1 u \, du = 3 \left[\frac{u^2}{2}\right]_0^1 = \frac{3}{2}$$

32. $\displaystyle\int_0^{\pi/4} \frac{\arctan x}{1 + x^2} \, dx = \frac{1}{2}\left[(\arctan x)^2\right]_0^{\pi/4} = \frac{1}{2}$

33. $\left\{ \begin{array}{l} u = \arcsin x \\ du = \dfrac{dx}{\sqrt{1-x^2}} \end{array} \right\}; \quad \displaystyle\int \frac{\arcsin x}{\sqrt{1-x^2}} \, dx = \int u \, du = \frac{1}{2} u^2 + C = \frac{1}{2}(\arcsin x)^2 + C$

34. $\displaystyle\int e^x \cosh(2 - e^x) \, dx = -\int \cosh u \, du = -\sinh u + C = -\sinh(2 - e^x) + C$

35. $\left\{ \begin{array}{l} u = \ln x \\ du = dx/x \end{array} \right\}; \quad \displaystyle\int \frac{1}{x \ln x} \, dx = \int \frac{1}{u} \, du = \ln|u| + C = \ln|\ln x| + C$

36. $\displaystyle\int_{-1}^1 \frac{x^2}{x^2+1} \, dx = \int_{-1}^1 \frac{x^2 + 1 - 1}{x^2+1} \, dx = \int_{-1}^1 \left(1 - \frac{1}{x^2+1}\right) dx = \left[x - \arctan x\right]_{-1}^1 = 2 - \frac{\pi}{2}$

37. $\left\{ \begin{array}{l|ll} u = \cos x & x = 0 & \Rightarrow \quad u = 1 \\ du = -\sin x \, dx & x = \pi/4 & \Rightarrow \quad u = \sqrt{2}/2 \end{array} \right\};$

$$\int_0^{\pi/4} \frac{1 + \sin x}{\cos^2 x} \, dx = \int_0^{\pi/4} \sec^2 x \, dx + \int_0^{\pi/4} \frac{\sin x}{\cos^2 x} \, dx$$

$$= \left[\tan x\right]_0^{\pi/4} - \int_1^{\sqrt{2}/2} \frac{du}{u^2} = 1 + \left[\frac{1}{u}\right]_1^{\sqrt{2}/2} = \sqrt{2}$$

38. $\displaystyle\int_0^{1/2} \frac{1 + x}{\sqrt{1 - x^2}} \, dx = \left[\arcsin x - \sqrt{1 - x^2}\right]_0^{1/2} = \frac{\pi}{6} + 1 - \frac{\sqrt{3}}{2}$

39. (formula 99) $\displaystyle\int \sqrt{x^2 - 4} \, dx = \frac{x}{2}\sqrt{x^2 - 4} - 2 \ln|x + \sqrt{x^2 - 4}| + C$

40. (formula 87) $\displaystyle\int \sqrt{4 - x^2} \, dx = \frac{x}{2}\sqrt{4 - x^2} + 2 \arcsin \frac{x}{2} + C$

41. (formula 18) $\displaystyle\int \cos^3 2t \, dt = \frac{1}{2}\left[\sin 2t - \frac{1}{3}\sin^3 2t\right] + C$

42. (formula 38) $\displaystyle\int \sec^4 t \, dt = \frac{1}{3}\sec^2 t \sin t + \frac{2}{3}\int \sec^2 t \, dt = \frac{1}{3}\sec^2 t \sin t + \frac{2}{3}\tan t + C$

43. (formula 108) $\displaystyle\int \frac{1}{x(2x + 3)} \, dx = \frac{1}{3}\ln\left|\frac{x}{2x + 3}\right| + C$

44. (formula 106) $\displaystyle\int \frac{x}{2+3x}\,dx = \frac{1}{9}(2+3x-2\ln|2+3x|)+C$

45. (formula 81) $\displaystyle\int \frac{\sqrt{x^2+9}}{x^2}\,dx = -\frac{\sqrt{x^2+9}}{x} + \ln|x+\sqrt{x^2+9}|+C$

46. (formula 103) $\displaystyle\int \frac{1}{x^2\sqrt{x^2-2}}\,dx = \frac{\sqrt{x^2-2}}{2x}+C$

47. (formula 11) $\displaystyle\int x^3 \ln x\,dx = x^4\left(\frac{\ln x}{4}-\frac{1}{16}\right)+C$

48. (formulas 23,24) $\displaystyle\int x^3 \sin x\,dx = -x^3\cos x + 3\int x^2\cos x\,dx$

$= -x^3\cos x + 3x^2\sin x - 6\int x\sin x\,dx = -x^3\cos x + 3x^2\sin x + 6x\cos x - 6\sin x + C$

49.
$$\int_0^\pi \sqrt{1+\cos x}\,dx = \int_0^\pi \sqrt{2\cos^2\left(\frac{x}{2}\right)}\,dx$$

$$= \sqrt{2}\int_0^\pi \cos\left(\frac{x}{2}\right)dx \qquad \left[\cos\left(\frac{x}{2}\right)\geq 0 \ \text{ on } \ [0,\pi]\right]$$

$$= 2\sqrt{2}\left[\sin\left(\frac{x}{2}\right)\right]_0^\pi = 2\sqrt{2}$$

50. (a) Clear since $du = \sec^2 x\,dx$.   (b) Clear since $du = \sec x\tan x\,dx$

     (c) Since $\sec^2 x = 1+\tan^2 x$, $\frac{1}{2}\tan^2 x + C_1 = \frac{1}{2}\sec^2 x + C_2$ with $C_2 = C_1 - \frac{1}{2}$

51. (a) $\displaystyle\int_0^\pi \sin^2 nx\,dx = \int_0^\pi \left[\frac{1}{2}-\frac{\cos 2nx}{2}\right]dx = \left[\frac{x}{2}-\frac{\sin 2nx}{4n}\right]_0^\pi = \frac{\pi}{2}$

     (b) $\displaystyle\int_0^\pi \sin nx\cos nx\,dx = \frac{1}{2}\int_0^\pi \sin 2nx\,dx = -\left[\frac{\cos 2nx}{4n}\right]_0^\pi = 0$

     (c) $\displaystyle\int_0^{\pi/n} \sin nx\cos nx\,dx = \frac{1}{2}\int_0^{\pi/n} \sin 2nx\,dx = -\left[\frac{\cos 2nx}{4n}\right]_0^{\pi/n} = 0$

52. (a) $\displaystyle\int \sin^3 x\,dx = \int (1-\cos^2 x)\sin x\,dx = -\cos x + \frac{1}{3}\cos^3 x + C$

     (b) $\displaystyle\int \sin^5 x\,dx = \int (1-\cos^2 x)^2\sin x\,dx = \int (1-2\cos^2 x + \cos^4 x)\sin x\,dx$

$$= -\cos x + \frac{2}{3}\cos^3 x - \frac{1}{5}\cos^5 x + C$$

     (c) Write $\sin^{2k+1} x$ as $\sin^{2k} x\sin x = (1-\cos^2 x)^k \sin x$,   expand $(1-\cos^2 x)^k$ and then integrate.

53. (a) $\displaystyle\int \tan^3 x \, dx = \int \tan^2 x \tan x \, dx = \int (\sec^2 x - 1) \tan x \, dx$

$$= \int \sec^2 x \tan x \, dx - \int \tan x \, dx$$

$$= \int u \, du - \int \tan x \, dx \quad (u = \tan x, \ \ du = \sec^2 x \, dx)$$

$$= \frac{1}{2} u^2 - \ln|\sec x| + C = \frac{1}{2} \tan^2 x - \ln|\sec x| + C$$

(b) $\displaystyle\int \tan^5 x \, dx = \int \tan^3 x \tan^2 x \, dx = \int \tan^3 x \, (\sec^2 x - 1) \, dx$

$$= \int \tan^3 x \sec^2 x \, dx - \int \tan^3 x \, dx$$

$$= \int u^3 \, du - \int \tan^3 x \, dx \quad (u = \tan x \ \ du = \sec^2 x \, dx)$$

$$= \frac{1}{4} u^4 - \frac{1}{2} \tan^2 x + \ln|\sec x| + C$$

$$= \frac{1}{4} \tan^4 x - \frac{1}{2} \tan^2 x + \ln|\sec x| + C$$

(c) $\displaystyle\int \tan^7 x \, dx = \int \tan^5 x \sec^2 x \, dx - \int \tan^5 x \, dx$

$$= \frac{1}{6} \tan^6 x - \frac{1}{4} \tan^4 x + \frac{1}{2} \tan^2 x - \ln|\sec x| + C$$

(d) $\displaystyle\int \tan^{2k+1} x \, dx = \int \tan^{2k-1} x \tan^2 x \, dx = \int \tan^{2k-1} x \sec^2 x \, dx - \int \tan^{2k-1} x \, dx$

$$= \frac{1}{2k} \tan^{2k} x - \int \tan^{2k-1} x \, dx$$

54. (a)

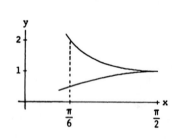

(b) $\displaystyle A = \int_{\pi/6}^{\pi/2} (\csc x - \sin x) \, dx$

$$= [\ln|\csc x - \cot x| + \cos x]_{\pi/6}^{\pi/2}$$

$$= -\ln(2 - \sqrt{3}) - \frac{\sqrt{3}}{2}$$

(c) $\displaystyle V = \int_{\pi/6}^{\pi/2} \pi(\csc^2 x - \sin^2 x) \, dx$

$$= \pi\left[-\cot x - \frac{x}{2} + \frac{1}{4} \sin 2x \right]_{\pi/6}^{\pi/2}$$

$$= \frac{7\pi\sqrt{3}}{8} - \frac{\pi^2}{6}$$

55. (a)

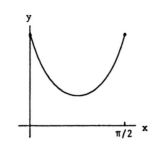

(b) $\sin x + \cos x = \sqrt{2}\,[\sin x\,\cos(\pi/4) + \cos x\,\sin(\pi/4)]$

$$= \sqrt{2}\,\sin\left(x + \frac{\pi}{4}\right);$$

$$A = \sqrt{2}, \quad B = \pi/4$$

(c) $\text{Area} = \displaystyle\int_{0}^{\pi/2} \frac{1}{\sin x + \cos x}\,dx = \frac{1}{\sqrt{2}}\int_{0}^{\pi/2} \frac{1}{\sin\left(x + \frac{\pi}{4}\right)}\,dx$

$$\left\{\begin{array}{l} u = x + \pi/4 \\ du = dx \end{array} \;\middle|\; \begin{array}{ll} x = 0 & \Rightarrow \quad u = \pi/4 \\ x = \pi/2 & \Rightarrow \quad u = 3\pi/4 \end{array}\right\};$$

$$\frac{1}{\sqrt{2}}\int_{0}^{\pi/2} \frac{1}{\sin\left(x + \frac{\pi}{4}\right)}\,dx = \frac{\sqrt{2}}{2}\int_{\pi/4}^{3\pi/4} \frac{1}{\sin u}\,du$$

$$= \frac{\sqrt{2}}{2}\int_{\pi/4}^{3\pi/4} \csc u\,du$$

$$= \frac{\sqrt{2}}{2}\Big[\,\ln|\csc u - \cot u|\,\Big]_{\pi/4}^{3\pi/4}$$

$$= \frac{\sqrt{2}}{2}\,\ln\left[\frac{\sqrt{2}+1}{\sqrt{2}-1}\right]$$

56. (a)

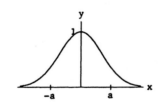

(b) $V = \displaystyle\int_{0}^{a} 2\pi x e^{-x^2}\,dx$

$$= \left[-\pi e^{-x^2}\right]_{0}^{a} = \pi\left(1 - e^{-a^2}\right)$$

(c) $\pi\left(1 - e^{-a^2}\right) = 2$

$$a = \sqrt{-\ln(1 - 2/\pi)} \cong 1.0061$$

57. (a)

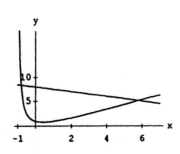

(b) $x_1 \cong -0.80, \quad x_2 \cong 5.80$

(c)
$$\text{Area} \cong \int_{-0.80}^{5.80} \left[8 - \frac{1}{2}x - \frac{x^2+1}{x+1} \right] dx = \int_{-0.80}^{5.80} \left[8 - \frac{1}{2}x - x + 1 - \frac{2}{x+1} \right] dx$$

$$= \int_{-0.80}^{5.80} \left[9 - \frac{3}{2}x - \frac{2}{x+1} \right] dx$$

$$= \left[9x - \frac{3}{4}x^2 - 2\ln|x+1| \right]_{-0.80}^{5.80} \cong 27.6$$

58. (a)

(b) Let $u = 1 - x$, $du = -dx$, $u(0) = 1$, $u(1) = 0$

$$A = 2 \int_0^1 x\sqrt{1-x}\, dx$$

$$A = -2 \int_1^0 (1-u)\sqrt{u}\, du$$

$$= -2 \left[\frac{2}{3}u^{3/2} - \frac{2}{5}u^{5/2} \right]_1^0 = \frac{8}{15}$$

SECTION 8.2

1.
$$\boxed{\begin{array}{ll} u = x, & dv = e^{-x}\, dx \\ du = dx, & v = -e^{-x} \end{array}}$$
$$\int xe^{-x}\, dx = -xe^{-x} - \int -e^{-x}\, dx = -xe^{-x} - e^{-x} + C$$

2.
$$\int_0^2 x2^x\, dx = \left[\frac{x2^x}{\ln 2} \right]_0^2 - \int_0^2 \frac{2^x}{\ln 2}\, dx$$

$$\boxed{\begin{array}{ll} u = x, & dv = 2^x\, dx \\ du = dx, & v = \dfrac{2^x}{\ln 2} \end{array}}$$
$$= \frac{8}{\ln 2} - \left[\frac{2^x}{(\ln 2)^2} \right]_0^2$$

$$= \frac{8}{\ln 2} - \frac{3}{(\ln 2)^2}$$

3. $\left\{ \begin{array}{l} t = -x^3 \\ dt = -3x^2\, dx \end{array} \right\}$; $\int x^2 e^{-x^3}\, dx = -\frac{1}{3}\int e^t\, dt = -\frac{1}{3}e^t + C = -\frac{1}{3}e^{-x^3} + C$

4. $\int x\ln x^2\, dx = \frac{1}{2}\int \ln u\, du = \frac{1}{2}u\ln u - \frac{1}{2}u + C = \frac{1}{2}x^2\ln x^2 - \frac{x^2}{2} + C = x^2\ln x - \frac{x^2}{2} + C$

5.

$$\int x^2 e^{-x}\,dx = -x^2 e^{-x} - \int -2x e^{-x}\,dx = -x^2 e^{-x} + 2\int x e^{-x}\,dx$$

$$\boxed{\begin{array}{ll} u = x^2, & dv = e^{-x}\,dx \\ du = 2x\,dx, & v = -e^{-x} \end{array}}$$

$$= -x^2 e^{-x} + 2\left[-x e^{-x} - \int -e^{-x}\,dx\right]$$

$$\boxed{\begin{array}{ll} u = x, & dv = e^{-x}\,dx \\ du = dx, & v = -e^{-x} \end{array}}$$

$$= -x^2 e^{-x} + 2\left(-x e^{-x} - e^{-x}\right) + C$$

$$= -e^{-x}\left(x^2 + 2x + 2\right) + C$$

$$\int_0^1 x^2 e^{-x}\,dx = \left[-e^{-x}\left(x^2 + 2x + 2\right)\right]_0^1 = 2 - 5e^{-1}$$

6.

$$\int x^3 e^{-x^2}\,dx = -\frac{x^2}{2} e^{-x^2} - \int -\frac{1}{2}\cdot 2x e^{-x^2}\,dx$$

$$\boxed{\begin{array}{ll} u = x^2, & dv = x e^{-x^2}\,dx \\ du = 2x\,dx, & v = -\dfrac{1}{2} e^{-x^2} \end{array}}$$

$$= -\frac{x^2 e^{-x^2}}{2} + \int x e^{-x^2}\,dx$$

$$= -\frac{x^2 e^{-x^2}}{2} - \frac{1}{2} e^{-x^2} + C$$

7.

$$\int x^2 \left(1 - x\right)^{-1/2}\,dx = -2x^2 \left(1 - x\right)^{1/2} + 4\int x\left(1 - x\right)^{1/2}\,dx$$

$$\boxed{\begin{array}{ll} u = x^2, & dv = (1 - x)^{-1/2}\,dx \\ du = 2x\,dx, & v = -2(1 - x)^{1/2} \end{array}}$$

$$= -2x^2 \left(1 - x\right)^{1/2} + 4\left[-\frac{2x}{3}\left(1 - x\right)^{3/2} + \int \frac{2}{3}\left(1 - x\right)^{3/2}\,dx\right]$$

$$\boxed{\begin{array}{ll} u = x, & dv = (1 - x)^{1/2}\,dx \\ du = dx, & v = -\frac{2}{3}(1 - x)^{3/2} \end{array}}$$

$$= -2x^2 \left(1 - x\right)^{1/2} - \frac{8x}{3}\left(1 - x\right)^{3/2} - \frac{16}{15}\left(1 - x\right)^{5/2} + C$$

Or, use the substitution $t = 1 - x$ (no integration by parts needed) to obtain:

$$-2\left(1 - x\right)^{1/2} + \tfrac{4}{3}\left(1 - x\right)^{3/2} - \tfrac{2}{5}\left(1 - x\right)^{5/2} + C.$$

8. $\displaystyle \int \frac{dx}{x(\ln x)^3} = \int \frac{du}{u^3} = -\frac{1}{2u^2} + C = \frac{-1}{2(\ln x)^2} + C$

9.

$$\int x \ln \sqrt{x}\,dx = \frac{1}{2}\int x \ln x\,dx = \frac{1}{2}\left[\frac{1}{2}x^2 \ln x - \frac{1}{2}\int x\,dx\right]$$

$$\boxed{\begin{array}{ll} u = \ln x, & dv = x\,dx \\ du = \dfrac{dx}{x}, & v = \dfrac{1}{2}x^2 \end{array}}$$

$$= \tfrac{1}{4}x^2 \ln x - \tfrac{1}{8}x^2 + C$$

$$\int_1^{e^2} x \ln \sqrt{x}\,dx = \left[\tfrac{1}{4}x^2 \ln x - \tfrac{1}{8}x^2\right]_1^{e^2} = \tfrac{3}{8}e^4 + \tfrac{1}{8}$$

10.
$$\left\{ \begin{array}{l} u = x + 1 \\ du = dx \end{array} \right. \left| \begin{array}{l} u(0) = 1 \\ u(3) = 4 \end{array} \right\} \quad \int_0^3 x\sqrt{x+1}\, dx = \int_1^4 (u-1)\sqrt{u}\, du = \left[\frac{2}{5} u^{5/2} - \frac{2}{3} u^{3/2} \right]_1^4 = \frac{116}{15}$$

11.
$$\int \frac{\ln(x+1)}{\sqrt{x+1}}\, dx = 2\sqrt{x+1}\ln(x+1) - \int \frac{2\, dx}{\sqrt{x+1}}$$

$$\boxed{\begin{array}{ll} u = \ln(x+1), & dv = \dfrac{dx}{\sqrt{x+1}} \\[2mm] du = \dfrac{dx}{x+1}, & v = 2\sqrt{x+1} \end{array}} \quad = 2\sqrt{x+1}\ln(x+1) - 4\sqrt{x+1} + C$$

12.
$$\int x^2(e^x - 1)\, dx = \int (x^2 e^x - x^2)\, dx = \int x^2 e^x\, dx - \frac{x^3}{3} = x^2 e^x - \int 2x e^x\, dx - \frac{x^3}{3}$$

$$= x^2 e^x - 2x e^x + \int 2 e^x\, dx - \frac{x^3}{3} = e^x(x^2 - 2x + 2) - \frac{x^3}{3} + C$$

13.
$$\int (\ln x)^2\, dx = x(\ln x)^2 - 2\int \ln x\, dx$$

$$\boxed{\begin{array}{ll} u = (\ln x)^2, & dv = dx \\[2mm] du = \dfrac{2\ln x}{x}\, dx, & v = x \end{array}}$$

$$= x(\ln x)^2 - 2\left[x\ln x - \int dx \right]$$

$$\boxed{\begin{array}{ll} u = \ln x, & dv = dx \\[2mm] du = \dfrac{dx}{x}, & v = x \end{array}}$$

$$= x(\ln x)^2 - 2x\ln x + 2x + C$$

14.
$$\left\{ \begin{array}{l} u = x + 5 \\ du = dx \end{array} \right\} \quad \int x(x+5)^{-14}\, dx = \int (u-5)u^{-14}\, du = \int (u^{-13} - 5u^{-14})\, du$$

$$= -\frac{1}{12} u^{-12} + \frac{5}{13} u^{-13} + C$$

$$= -\frac{1}{12}(x+5)^{-12} + \frac{5}{13}(x+5)^{-13} + C$$

15.
$$\int x^3 \, 3^x \, dx = \frac{x^3 \, 3^x}{\ln 3} - \frac{3}{\ln 3} \int x^2 \, 3^x \, dx$$

$$\boxed{\begin{array}{ll} u = x^3, & dv = 3^x \, dx \\ du = 3x^2 \, dx, & v = \dfrac{3^x}{\ln 3} \end{array}}$$

$$= \frac{x^3 \, 3^x}{\ln 3} - \frac{3}{\ln 3} \left[\frac{x^2 \, 3^x}{\ln 3} - \frac{2}{\ln 3} \int x \, 3^x \, dx \right]$$

$$\boxed{\begin{array}{ll} u = x^2, & dv = 3^x \, dx \\ du = 2x \, dx & v = \dfrac{3^x}{\ln 3} \end{array}}$$

$$= \frac{x^3 \, 3^x}{\ln 3} - \frac{3x^2 \, 3^x}{(\ln 3)^2} + \frac{6}{(\ln 3)^2} \int x \, 3^x \, dx$$

$$= \frac{x^3 \, 3^x}{\ln 3} - \frac{3x^2 \, 3^x}{(\ln 3)^2} + \frac{6}{(\ln 3)^2} \left[\frac{x 3^x}{\ln 3} - \frac{1}{\ln 3} \int 3^x \, dx \right]$$

$$\boxed{\begin{array}{ll} u = x, & dv = 3^x \, dx \\ du = dx, & v = \dfrac{3^x}{\ln 3} \end{array}}$$

$$= 3^x \left[\frac{x^3}{\ln 3} - \frac{3x^2}{(\ln 3)^2} + \frac{6x}{(\ln 3)^3} - \frac{6}{(\ln 3)^4} \right] + C$$

16.
$$\boxed{\begin{array}{ll} u = \ln x, & dv = \sqrt{x} \, dx \\ du = \dfrac{1}{x} \, dx, & v = \dfrac{2}{3} x^{3/2} \end{array}} \qquad \int \sqrt{x} \ln x \, dx = \frac{2}{3} x^{3/2} \ln x - \int \frac{2}{3} x^{3/2} \cdot \frac{1}{x} \, dx = \frac{2}{3} x^{3/2} \ln x - \frac{4}{9} x^{3/2} + C$$

17.
$$\int x \, (x+5)^{14} \, dx = \frac{x}{15} \, (x+5)^{15} - \frac{1}{15} \int (x+5)^{15} \, dx$$

$$\boxed{\begin{array}{ll} u = x, & dv = (x+5)^{14} \, dx \\ du = dx, & v = \frac{1}{15} \, (x+5)^{15} \end{array}} \qquad = \tfrac{1}{15} x \, (x+5)^{15} - \tfrac{1}{240} \, (x+5)^{16} + C$$

Or, use the substitution $t = x + 5$ (integration by parts not needed) to obtain:

$$\tfrac{1}{16} \, (x+5)^{16} - \tfrac{1}{3} \, (x+5)^{15} + C.$$

18.
$$\int (2^x + x^2)^2 \, dx = \int (2^{2x} + 2x^2 2^x + x^4) \, dx = \frac{2^{2x}}{2 \ln 2} + \frac{x^5}{5} + \int 2x^2 2^x \, dx$$

$$= \frac{4^x}{\ln 4} + \frac{x^5}{5} + \frac{2x^2 2^x}{\ln 2} - \int \frac{4x 2^x}{\ln 2} \, dx$$

$$= \frac{4^x}{\ln 4} + \frac{x^5}{5} + \frac{2x^2 2^x}{\ln 2} - \frac{4x 2^x}{(\ln 2)^2} + \int \frac{4 \cdot 2^x}{(\ln 2)^2} \, dx$$

$$= \frac{4^x}{\ln 4} + \frac{x^5}{5} + 2^x \left[\frac{2x^2}{\ln 2} - \frac{4x}{(\ln 2)^2} + \frac{4}{(\ln 2)^3} \right] + C$$

19.

$$\int x \cos \pi x \, dx = \frac{1}{\pi} x \sin x - \frac{1}{\pi} \int \sin \pi x \, dx$$

$$\boxed{\begin{array}{ll} u = x, & dv = \cos \pi x \, dx \\ du = dx, & v = \frac{1}{\pi} \sin \pi x \end{array}} \qquad = \frac{1}{\pi} x \sin \pi x + \frac{1}{\pi^2} \cos \pi x + C$$

$$\int_0^{1/2} x \cos \pi x \, dx = \left[\frac{1}{\pi} x \sin \pi x + \frac{1}{\pi^2} \cos \pi x \right]_0^{1/2} = \frac{1}{2\pi} - \frac{1}{\pi^2}$$

20. $\displaystyle \int_0^{\pi/2} x^2 \sin x \, dx = \left[-x^2 \cos x \right]_0^{\pi/2} + \int_0^{\pi/2} 2x \cos x \, dx = \left[-x^2 \cos x \right]_0^{\pi/2} + \left[2x \sin x \right]_0^{\pi/2} - \int_0^{\pi/2} 2 \sin x \, dx$

$$= \left[-x^2 \cos x + 2x \sin x + 2 \cos x \right]_0^{\pi/2} = \pi - 2$$

21.

$$\int x^2 (x+1)^9 \, dx = \frac{x^2}{10} (x+1)^{10} - \frac{1}{5} \int x (x+1)^{10} \, dx$$

$$\boxed{\begin{array}{ll} u = x^2, & dv = (x+1)^9 \, dx \\ du = 2x \, dx, & v = \frac{1}{10} (x+1)^{10} \end{array}} \qquad = \frac{x^2}{10} (x+1)^{10} - \frac{1}{5} \left[\frac{x}{11} (x+1)^{11} - \frac{1}{11} \int (x+1)^{11} \, dx \right]$$

$$\boxed{\begin{array}{ll} u = x, & dv = (x+1)^{10} \, dx \\ du = dx, & v = \frac{1}{11} (x+1)^{11} \end{array}} \qquad = \frac{x^2}{10} (x+1)^{10} - \frac{x}{55} (x+1)^{11} + \frac{1}{660} (x+1)^{12} + C$$

22.

$$\int x^2 (2x-1)^{-7} \, dx = -\frac{x^2}{12} (2x-1)^{-6} + \int \frac{x}{6} (2x-1)^{-6} \, dx$$

$$\boxed{\begin{array}{ll} u = x^2 & dv = (2x-1)^{-7} \, dx \\ du = 2x\,dx & v = \dfrac{(2x-1)^{-6}}{-12} \end{array}} \qquad = -\frac{x^2}{12} (2x-1)^{-6} - \frac{x}{60} (2x-1)^{-5} + \int \frac{1}{60} (2x-1)^{-5} \, dx$$

$$= -\frac{x^2}{12} (2x-1)^{-6} - \frac{x}{60} (2x-1)^{-5} - \frac{1}{480} (2x-1)^{-4} + C$$

23.

$$\int e^x \sin x \, dx = -e^x \cos x + \int e^x \cos x \, dx$$

$$\boxed{\begin{array}{ll} u = e^x, & dv = \sin x \, dx \\ du = e^x dx, & v = -\cos x \end{array}}$$

$$= -e^x \cos x + e^x \sin x - \int e^x \sin x \, dx$$

$$\boxed{\begin{array}{ll} u = e^x, & dv = \cos x \, dx \\ du = e^x dx, & v = \sin x \end{array}}$$

Adding $\int e^x \sin x \, dx$ to both sides, we get

$$2 \int e^x \sin x \, dx = -e^x \cos x + e^x \sin x$$

so that

$$\int e^x \sin x \, dx = \frac{1}{2} e^x (\sin x - \cos x) + C.$$

24.
$$\int (e^x + 2x)^2 \, dx = \int (e^{2x} + 4xe^x + 4x^2) \, dx = \frac{1}{2} e^{2x} + \frac{4}{3} x^3 + \int 4xe^x \, dx$$

$$= \frac{1}{2} e^{2x} + \frac{4}{3} x^3 + 4xe^x - \int 4e^x \, dx = \frac{1}{2} e^{2x} + \frac{4}{3} x^3 + 4xe^x - 4e^x + C$$

25.
$$\int \ln \left(1 + x^2 \right) \, dx = x \ln \left(1 + x^2 \right) - 2 \int \frac{x^2}{1 + x^2} \, dx$$

$$\boxed{\begin{array}{ll} u = \ln(1 + x^2), & dv = dx \\ du = \dfrac{2x}{1 + x^2} \, dx, & v = x \end{array}} \quad = x \ln \left(1 + x^2 \right) - 2 \int \frac{x^2 + 1 - 1}{1 + x^2} \, dx$$

$$= x \ln \left(1 + x^2 \right) - 2 \int \left(1 - \frac{1}{1 + x^2} \right) \, dx$$

$$= x \ln \left(1 + x^2 \right) - 2x + 2 \arctan x + C$$

$$\int_0^1 \ln(1 + x^2) \, dx = \left[x \ln(1 + x^2) - 2x + 2 \arctan x \right]_0^1 = \ln 2 - 2 + \frac{\pi}{2}$$

26.
$$\int x \ln(x + 1) \, dx = \int (x + 1) \ln(x + 1) \, dx - \int \ln(x + 1) \, dx$$

$$\text{(from Example 3)} = \frac{1}{2} (x + 1)^2 \ln(x + 1) - \frac{1}{4} (x + 1)^2 - (x + 1) \ln(x + 1) + (x + 1) + C_1$$

$$= \frac{1}{2} (x^2 - 1) \ln(x + 1) - \frac{1}{4} x^2 + \frac{1}{2} x + C$$

27.
$$\int x^n \ln x \, dx = \frac{x^{n+1} \ln x}{n + 1} - \frac{1}{n + 1} \int x^n \, dx$$

$$\boxed{\begin{array}{ll} u = \ln x, & dv = x^n \, dx \\ du = \dfrac{dx}{x}, & v = \dfrac{x^{n+1}}{n + 1} \end{array}} \quad = \frac{x^{n+1} \ln x}{n + 1} - \frac{x^{n+1}}{(n + 1)^2} + C$$

28.
$$\int e^{3x} \cos 2x \, dx = \frac{1}{2} e^{3x} \sin 2x - \int \frac{3}{2} e^{3x} \sin 2x \, dx$$

$$\boxed{\begin{array}{ll} u = e^{3x} & dv = \cos 2x \, dx \\ du = 3e^{3x} \, dx & v = \dfrac{1}{2} \sin 2x \end{array}} \quad = \frac{1}{2} e^{3x} \sin 2x + \frac{3}{4} e^{3x} \cos 2x - \int \frac{9}{4} e^{3x} \cos 2x \, dx$$

$$\implies \frac{13}{4} \int e^{3x} \cos 2x \, dx = \frac{1}{2} e^{3x} \sin 2x + \frac{3}{4} e^{3x} \cos 2x$$

$$\implies \int e^{3x} \cos 2x \, dx = \frac{2}{13} e^{3x} \sin 2x + \frac{3}{13} e^{3x} \cos 2x + C$$

29.

$$\{t = x^2, \quad dt = 2x \, dx\}\,; \qquad \int x^3 \sin x^2 \, dx = \frac{1}{2} \int t \sin t \, dt$$

$$\boxed{\begin{aligned} u &= t, & dv &= \sin t \, dt \\ du &= dt, & v &= -\cos t \end{aligned}}$$

$$= \frac{1}{2}\left[-t \cos t + \int \cos t \, dt \right]$$

$$= \tfrac{1}{2}\left(-t \cos t + \sin t \right) + C$$

$$= -\tfrac{1}{2}x^2 \cos x^2 + \tfrac{1}{2} \sin x^2 + C$$

30.

$$\int x^3 \sin x \, dx = -x^3 \cos x + \int 3x^2 \cos x \, dx$$

$$\boxed{\begin{aligned} u &= x^3 & dv &= \sin x \, dx \\ du &= 3x^2 \, dx & v &= -\cos x \end{aligned}}$$

$$= -x^3 \cos x + 3x^2 \sin x - \int 6x \sin x \, dx$$

$$= -x^3 \cos x + 3x^2 \sin x + 6x \cos x - \int 6 \cos x \, dx$$

$$= -x^3 \cos x + 3x^2 \sin x + 6x \cos x - 6 \sin x + C$$

31.

$$\left\{\begin{array}{l|l} u = 2x & x = 0 \implies u = 0 \\ du = 2\, dx & x = 1/4 \implies u = 1/2 \end{array}\right\};$$

$$\int_0^{1/4} \arcsin 2x \, dx = \frac{1}{2} \int_0^{1/2} \arcsin u \, du$$

$$= \frac{1}{2}\left[u \arcsin u + \sqrt{1 - u^2} \right]_0^{1/2} \qquad [\text{by}(8.2.5)]$$

$$= \frac{1}{2}\left[\frac{\pi}{12} + \frac{\sqrt{3}}{2} - 1 \right] = \frac{\pi}{24} + \frac{\sqrt{3} - 2}{4}$$

32.

$$\left\{\begin{aligned} u &= \arcsin 2x \\ du &= \frac{2}{\sqrt{1 - 4x^2}}\, dx \end{aligned}\right\} \quad \int \frac{\arcsin 2x}{\sqrt{1 - 4x^2}}\, dx = \frac{1}{2}\int u \, du = \frac{1}{4}u^2 + C = \frac{1}{4}\left(\arcsin 2x \right)^2 + C$$

33.

$$\left\{\begin{array}{l|l} u = x^2 & x = 0 \implies u = 0 \\ du = 2x \, dx & x = 1 \implies u = 1 \end{array}\right\};$$

$$\int_0^1 x \arctan (x)^2 \, dx = \frac{1}{2}\int_0^1 \arctan u \, du$$

$$= \frac{1}{2}\left[u \arctan u - \tfrac{1}{2}\left(1 + u^2 \right) \right]_0^1 \qquad [\text{by } 8.2.6]$$

$$= \frac{\pi}{8} - \frac{1}{4}\ln 2$$

34.

$$\int \cos \sqrt{x} \, dx = \int \sqrt{x}\frac{\cos \sqrt{x}}{\sqrt{x}}\, dx$$

$$\boxed{\begin{aligned} ux &= \sqrt{x} & dv &= \frac{\cos \sqrt{x}}{\sqrt{x}}\, dx \\ du &= \frac{1}{2\sqrt{x}}\, dx & v &= 2 \sin \sqrt{x} \end{aligned}}$$

$$= 2\sqrt{x} \sin \sqrt{x} - \int \frac{\sin \sqrt{x}}{\sqrt{x}}\, dx$$

$$= 2\sqrt{x} \sin \sqrt{x} + 2 \cos \sqrt{x} + C$$

35.
$$\int x^2 \cosh 2x \, dx = \tfrac{1}{2}x^2 \sinh 2x - \int x \sinh 2x \, dx$$

$$\boxed{\begin{array}{ll} u = x^2, & dv = \cosh 2x \, dx \\ du = 2x \, dx, & v = \tfrac{1}{2}\sinh 2x \end{array}}$$
$$= \tfrac{1}{2}x^2 \sinh 2x - \tfrac{1}{2}x \cosh 2x + \tfrac{1}{2}\int \cosh 2x \, dx$$

$$\boxed{\begin{array}{ll} u = x, & dv = \sinh 2x \, dx \\ du = dx, & v = \tfrac{1}{2}\cosh 2x \end{array}}$$
$$= \tfrac{1}{2}x^2 \sinh 2x - \tfrac{1}{2}x \cosh 2x + \tfrac{1}{4}\sinh 2x + C$$

36. $\displaystyle\int_{-1}^{1} x \sinh(2x^2) \, dx = \frac{1}{4}\left[\cosh(2x^2)\right]_{-1}^{1} = 0$

37. Let $u = \ln x$, $du = \tfrac{1}{x}\, dx$. Then
$$\int \frac{1}{x}\arcsin(\ln x)\, dx = \int \arcsin u \, du = u \arcsin u + \sqrt{(1 - u^2)} + C$$
$$= (\ln x)\arcsin(\ln x) + \sqrt{1 - (\ln x)^2} + C$$

38.

$$\boxed{\begin{array}{ll} u = \cos(\ln x) & dv = dx \\ du = -\dfrac{\sin(\ln x)}{x}\, dx & v = x \end{array}}$$
$$\int \cos(\ln x)\, dx = x\cos(\ln x) + \int \sin(\ln x)\, dx$$

$$\boxed{\begin{array}{ll} u = \sin(\ln x) & dv = dx \\ du = \dfrac{\cos(\ln x)}{x}\, dx & v = x \end{array}}$$
$$= x\cos(\ln x) + x\sin(\ln x) - \int \cos(\ln x)\, dx$$

$$\Longrightarrow \int \cos(\ln x)\, dx = \frac{1}{2}x\left[\cos(\ln x) + \sin(\ln x)\right] + C$$

39.
$$\int \sin(\ln x)\, dx = x\sin(\ln x) - \int \cos(\ln x)\, dx$$

$$\boxed{\begin{array}{ll} u = \sin(\ln x), & dv = dx \\ du = \cos(\ln x)\dfrac{1}{x}\, dx, & v = x \end{array}}$$
$$= x\sin(\ln x) - x\cos(\ln x) - \int \sin(\ln x)\, dx$$

$$\boxed{\begin{array}{ll} u = \cos(\ln x), & dv = dx \\ du = -\sin(\ln x)\dfrac{1}{x}\, dx, & v = x \end{array}}$$

Adding $\displaystyle\int \sin(\ln x)\, dx$ to both sides, we get

$$2\int \sin(\ln x)\, dx = x\sin(\ln x) - x\cos(\ln x)$$

so that

$$\int \sin(\ln x)\, dx = \tfrac{1}{2}\left[x\sin(\ln x) - x\cos(\ln x)\right] + C$$

40.

$$\begin{array}{|ll|}
\hline
u = (\ln x)^2 & dv = x^2 dx \\
du = 2\dfrac{\ln x}{x}\, dx & v = \dfrac{x^3}{3} \\
\hline
\end{array}$$

$$\int_1^{2e} x^2 (\ln x)^2\, dx = \left[\frac{x^3}{3}(\ln x)^2 \right]_1^{2e} - \int_1^{2e} \frac{2x^2}{3}\ln x\, dx$$

$$\begin{array}{|ll|}
\hline
u = \ln x & dv = \dfrac{2x^2}{3}\, dx \\
du = \dfrac{1}{x}\, dx & v = \dfrac{2x^3}{9} \\
\hline
\end{array}$$

$$= \left[\frac{x^3}{3}(\ln x)^2 - \frac{2x^3}{9}\ln x \right]_1^{2e} + \int_1^{2e} \frac{2x^2}{9}\, dx$$

$$= \left[\frac{x^3}{3}(\ln x)^2 - \frac{2x^3}{9}\ln x + \frac{2x^3}{27} \right]_1^{2e}$$

$$= \frac{8e^3}{3}\left[(\ln 2e)^2 - \frac{2}{3}\ln 2e + \frac{2}{9} \right] - \frac{2}{27}$$

41.

$$\begin{array}{|ll|}
\hline
u = \ln x & dv = dx \\
du = \dfrac{1}{x}\, dx & v = x \\
\hline
\end{array}$$

$$\int \ln x\, dx = x\ln x - \int dx = x\ln x - x + C$$

42.

$$\begin{array}{|ll|}
\hline
u = \arctan x & dv = dx \\
du = \dfrac{1}{1+x^2}\, dx & v = x \\
\hline
\end{array}$$

$$\int \arctan x\, dx = x\arctan x - \int \frac{x}{1+x^2}\, dx$$

$$= x\arctan x - \frac{1}{2}\ln(1+x^2) + C$$

43.

$$\begin{array}{|ll|}
\hline
u = \ln x & dv = x^k dx \\
du = \dfrac{1}{x}\, dx & v = \dfrac{1}{k+1}x^{k+1} \\
\hline
\end{array}$$

$$\int x^k \ln x\, dx = \frac{1}{k+1}x^{k+1}\ln x - \int \frac{1}{k+1}x^k\, dx$$

$$= \frac{1}{k+1}x^{k+1}\ln x - \frac{x^{k+1}}{(k+1)^2} + C$$

44.

$$\begin{array}{|ll|}
\hline
u = e^{ax} & dv = \cos bx\, dx \\
du = ae^{ax}\, dx & v = \dfrac{1}{b}\sin bx \\
\hline
\end{array}$$

$$\int e^{ax}\cos bx\, dx = \frac{1}{b}e^{ax}\sin bx - \frac{a}{b}\int e^{ax}\sin bx\, dx$$

$$\begin{array}{|ll|}
\hline
u = e^{ax} & dv = \sin bx\, dx \\
du = ae^{ax}\, dx & v = -\dfrac{1}{b}\cos bx \\
\hline
\end{array}$$

$$= \frac{1}{b}e^{ax}\sin bx + \frac{a}{b^2}e^{ax}\cos bx - \frac{a^2}{b^2}\int e^{ax}\cos bx\, dx.$$

$$\implies \int e^{ax}\cos bx\, dx = \frac{e^{ax}}{a^2+b^2}(a\cos bx + b\sin bx) + C$$

45.

$$\begin{array}{|ll|}
\hline
u = e^{ax} & dv = \sin bx\, dx \\
du = ae^{ax}\, dx & v = -\dfrac{1}{b}\cos bx \\
\hline
\end{array}$$

$$\int e^{ax}\sin bx\, dx = -\frac{1}{b}e^{ax}\cos bx + \frac{a}{b}\int e^{ax}\cos bx\, dx$$

$$\boxed{\begin{aligned} u &= e^{ax} & dv &= \cos bx\, dx \\ du &= ae^{ax}\, dx & v &= \frac{1}{b}\sin bx \end{aligned}} = -\frac{1}{b}e^{ax}\cos bx + \frac{a}{b^2}e^{ax}\sin bx - \frac{a^2}{b^2}\int e^{ax}\sin bx\, dx.$$

$$\implies \quad \int e^{ax}\sin bx\, dx = \frac{e^{ax}}{a^2+b^2}(a\sin bx - b\cos bx) + C$$

46. The integrals cancel each other if you integrate by parts.

$$\int e^{ax}\cosh ax\, dx = \frac{e^{2ax}}{4a} + \frac{x}{2} + C$$

47. $\displaystyle\int_0^{\pi} x\sin x\, dx = \big[-x\cos x + \sin x\big]_0^{\pi} = \pi$

48. $\displaystyle\int_0^{\pi} x\cos(x/2)\, dx = \big[2x\sin(x/2) + 4\cos(x/2)\big]_0^{\pi} = 2\pi - 4$

49. $A = \displaystyle\int_0^{1/2} \arcsin x\, dx = \big[x\arcsin x + \sqrt{1-x^2}\big]_0^{1/2} = \dfrac{\pi}{12} + \dfrac{\sqrt{3}-2}{2}$

50.

$$A = \int_0^2 xe^{-2x}\, dx = \left[-\frac{xe^{-2x}}{2}\right]_0^2 + \int_0^2 \frac{e^{-2x}}{2}\, dx$$

$$= \left[-\frac{xe^{-2x}}{2} - \frac{e^{-2x}}{4}\right]_0^2 = \frac{1}{4} - \frac{5}{4}e^{-4}$$

51. (a) $A = \displaystyle\int_1^e \ln x\, dx = [x\ln x - x]_1^e = 1$

(b) $\bar{x}A = \displaystyle\int_1^e x\ln x\, dx = \left[\frac{1}{2}x^2\ln x - \frac{1}{4}x^2\right]_1^e = \frac{1}{4}\left(e^2+1\right), \quad \bar{x} = \frac{1}{4}\left(e^2+1\right)$

$\bar{y}A = \displaystyle\int_1^e \frac{1}{2}(\ln x)^2\, dx = \frac{1}{2}\left[x(\ln x)^2 - 2x\ln x + 2x\right]_1^e = \frac{1}{2}e - 1, \quad \bar{y} = \frac{1}{2}e - 1$

(c) $V_x = 2\pi\bar{y}A = \pi(e-2), \quad V_y = 2\pi\bar{x}A = \frac{1}{2}\pi\left(e^2+1\right)$

52. (a) $A = \displaystyle\int_1^{2e} \frac{\ln x}{x}\, dx = \left[\frac{1}{2}(\ln x)^2\right]_1^{2e} = \frac{(\ln 2e)^2}{2} = \frac{(1+\ln 2)^2}{2}$

$$V = \int_1^{2e} \pi\left(\frac{\ln x}{x}\right)^2 dx$$

(b) $\boxed{\begin{aligned} u &= (\ln x)^2, & dv &= \frac{1}{x^2}\, dx \\ du &= \frac{2\ln x}{x}, dx & v &= -\frac{1}{x} \end{aligned}}$ $\displaystyle = \pi\left[-\frac{(\ln x)^2}{x}\right]_1^{2e} - \pi\int_1^{2e} -\frac{2\ln x}{x^2}\, dx$

$$= \pi\left[-\frac{(\ln 2e)^2}{2e}\right] + 2\pi\left[-\frac{\ln x}{x}\right]_1^{2e} - 2\pi\int_1^{2e} -\frac{1}{x^2}\, dx$$

$$= \pi\left[-\frac{(\ln 2e)^2}{2e} - \frac{\ln 2e}{e} - \frac{1}{e} + 2\right]$$

53. $\bar{x} = \dfrac{1}{e-1}, \qquad \bar{y} = \dfrac{1}{4}(e+1)$

54. $\bar{x} = \dfrac{e-2}{e-1}, \quad \bar{y} = \dfrac{1}{4}\left(\dfrac{e+1}{e}\right)$

55. $\bar{x} = \frac{1}{2}\pi, \qquad \bar{y} = \frac{1}{8}\pi$

56. $\bar{x} = \dfrac{\pi}{2} - 1, \quad \bar{y} = \dfrac{\pi}{8}$

57. (a) $M = \displaystyle\int_0^1 e^{kx}\,dx = \dfrac{1}{k}\left(e^k - 1\right)$

(b) $x_M M = \displaystyle\int_0^1 xe^{kx}\,dx = \dfrac{(k-1)\,e^k + 1}{k^2}, \qquad x_M = \dfrac{(k-1)\,e^k + 1}{k\,(e^k - 1)}$

58. (a) $M = \displaystyle\int_2^3 \ln x\,dx = \left[x\ln x - x\right]_2^3 = 3\ln 3 - 2\ln 2 - 1$

(b) $x_M M = \displaystyle\int_2^3 x\ln x\,dx = \left[\dfrac{x^2}{2}\ln x - \dfrac{x^2}{4}\right]_2^3 = \dfrac{9}{2}\ln 3 - 2\ln 2 - \dfrac{5}{4}$

$\Longrightarrow x_M = \dfrac{18\ln 3 - 8\ln 2 - 5}{4(3\ln 3 - 2\ln 2 - 1)}$

59. $V_y = \displaystyle\int_0^1 2\pi x\cos\dfrac{1}{2}\pi x\,dx = \left[4x\sin\dfrac{1}{2}\pi x + \dfrac{8}{\pi}\cos\dfrac{1}{2}\pi x\right]_0^1 = 4 - \dfrac{8}{\pi}$

60. $V_y = \displaystyle\int_0^\pi 2\pi x^2\sin x\,dx = 2\pi\left[-x^2\cos x + 2x\sin x + 2\cos x\right]_0^\pi = 2\pi(\pi^2 - 4)$

61. $V_y = \displaystyle\int_0^1 2\pi x^2 e^x\,dx = 2\pi\,(e - 2) \qquad$ (see Example 6)

62. $V_y = \displaystyle\int_0^{\pi/2} 2\pi x^2\cos x\,dx = 2\pi\left[x^2\sin x + 2x\cos x - 2\sin x\right]_0^{\pi/2} = \dfrac{\pi}{2}(\pi^2 - 8)$

63.

$V_x = \displaystyle\int_0^1 \pi e^{2x}\,dx = \pi\left[\dfrac{1}{2}e^{2x}\right]_0^1 = \dfrac{1}{2}\pi\left(e^2 - 1\right)$

$\bar{x}V_x = \displaystyle\int_0^1 \pi x e^{2x}\,dx = \pi\left[\dfrac{1}{2}xe^{2x} - \dfrac{1}{4}e^{2x}\right]_0^1 = \dfrac{1}{4}\pi\left(e^2 + 1\right);$

$\bar{x} = \dfrac{e^2 + 1}{2\left(e^2 - 1\right)}$

64.

$\bar{x}V_x = \displaystyle\int_0^{\pi/2} \pi x\sin^2 x\,dx = \dfrac{1}{8}\pi\left[2x^2 - 2x\sin 2x - \cos 2x\right]_0^{\pi/2} = \dfrac{1}{16}\pi(\pi^2 + 4)$

$V_x = \displaystyle\int_0^{\pi/2} \pi\sin^2 x\,dx = \pi\left[\dfrac{1}{2}x - \dfrac{1}{4}\sin 2x\right]_0^{\pi/2} = \dfrac{1}{4}\pi^2;$

$\bar{x} = \dfrac{\pi^2 + 4}{4\pi}$

65. $A = \int_0^1 \cosh x \, dx = [\sinh x]_0^1 = \sinh 1 = \dfrac{e - e^{-1}}{2} = \dfrac{e^2 - 1}{2e}$

$\overline{x}A = \int_0^1 x \cosh x \, dx = [x \sinh x - \cosh x]_0^1 = \sinh 1 - \cosh 1 + 1 = \dfrac{2\,(e - 1)}{2e}$

$\overline{y}A = \int_0^1 \dfrac{1}{2} \cosh^2 x \, dx = \dfrac{1}{4} [\sinh x \cosh x + x]_0^1 = \dfrac{1}{4}(\sinh 1 \cosh 1 + 1) = \dfrac{e^4 + 4e^2 - 1}{16e^2}$

Therefore $\overline{x} = \dfrac{2}{e + 1}$ and $\overline{y} = \dfrac{e^4 + 4e^2 - 1}{8e\,(e^2 - 1)}$.

66. (a) $\overline{x}V_x = \int_0^1 \pi x \cosh^2 x \, dx = \dfrac{\pi}{8} \left[2x^2 + 2x \sinh 2x - \cosh 2x \right]_0^1 = \dfrac{\pi}{8}(3 + 2 \sinh 2 - \cosh 2)$

$V_x = \int_0^1 \pi \cosh^2 x \, dx = \pi \left[\dfrac{x}{2} + \dfrac{1}{4} \sinh 2x \right]_0^1 = \dfrac{\pi}{4}(2 + \sinh 2)$

$\overline{x} = \dfrac{3 + 2 \sinh 2 - \cosh 2}{2(2 + \sinh 2)}$

(b) $\overline{y}V_y = \int_0^1 2\pi x \cosh^2 x \, dx = \dfrac{\pi}{8}(1 + 2 \sinh 2 - \cosh 2)$

$V_y = \int_0^1 2\pi x \cosh x \, dx = 2\pi \, [x \sinh x - \cosh x]_0^1 = 2\pi(\sinh 1 - \cosh 1 + 1)$

$\overline{y} = \dfrac{1 + 2 \sinh 2 - \cosh 2}{16(\sinh 1 - \cosh 1 + 1)}$

67.
$$\boxed{\begin{array}{ll} u = x^2, & dv = e^{-x}\,dx \\ du = 2x\,dx, & v = -e^{-x} \end{array}} \qquad \int x^n e^{ax}\,dx = \dfrac{x^n e^{ax}}{a} - \dfrac{n}{a} \int x^{n-1} e^{ax}\,dx$$

68.
$$\int (\ln x)^n \, dx = x(\ln x)^n - n \int \dfrac{x(\ln x)^{n-1}}{x}\,dx$$

$$\boxed{\begin{array}{ll} u = (\ln x)^n & dv = dx \\ du = \dfrac{n(\ln x)^{n-1}}{x}\,dx & v = x \end{array}} \qquad = x(\ln x)^n - n \int (\ln x)^{n-1}\,dx$$

69.
$$\int x^3 e^{2x}\,dx = \dfrac{1}{2}x^3 e^{2x} - \dfrac{3}{2} \int x^2 e^{2x}\,dx = \dfrac{1}{2}x^3 e^{2x} - \dfrac{3}{2} \left[\dfrac{1}{2}x^2 e^{2x} - \int x e^{2x}\,dx \right]$$

$$= \dfrac{1}{2}x^3 e^{2x} - \dfrac{3}{4}x^2 e^{2x} - \dfrac{3}{4}x e^{2x} - \dfrac{3}{4} \int e^{2x}\,dx$$

$$= \dfrac{1}{2}x^3 e^{2x} - \dfrac{3}{4}x^2 e^{2x} + \dfrac{3}{4}x e^{2x} - \dfrac{3}{8} e^{2x} + C$$

70.
$$\int x^2 e^{-x}\,dx = -x^2 e^{-x} + 2 \int x e^{-x}\,dx$$

$$= -x^2 e^{-x} + 2 \left[-x e^{-x} + \int e^{-x}\,dx \right]$$

$$= -x^2 e^{-x} - 2x e^{-x} - 2e^{-x} + C$$

71.

$$\int (\ln x)^3 \, dx = x(\ln x)^3 - 3 \int (\ln x)^2 \, dx = x(\ln x)^3 - 3x(\ln x)^2 + 6 \int \ln x \, dx$$

$$= x(\ln x)^3 - 3x(\ln x)^2 + 6x \ln x - 6 \int dx$$

$$= x(\ln x)^3 - 3x(\ln x)^2 + 6x \ln x - 6x + C$$

72.

$$\int (\ln x)^4 \, dx = x(\ln x)^4 - 4 \int (\ln x)^3 \, dx$$

$$= x(\ln x)^4 - 4 \left[x(\ln x)^3 - 3 \int (\ln x)^2 \, dx \right]$$

$$= x(\ln x)^4 - 4x(\ln x)^3 + 12 \int (\ln x)^2 \, dx$$

$$= x(\ln x)^4 - 4x(\ln x)^3 + 12 \left[x(\ln x)^2 - 2 \int \ln x \, dx \right]$$

$$= x(\ln x)^4 - 4x(\ln x)^3 + 12x(\ln x)^2 - 24x \ln x + 24x + C$$

73. (a) Differentiating, $x^3 e^x = Ax^3 e^x + 3Ax^2 e^x + 2Bxe^x + Bx^2 e^x + Ce^x + Cxe^x + De^x$

$\implies$ $A = 1$, $B = -3$, $C = 6$, and $D = -6$

(b)

$$\int x^3 e^x \, dx = x^3 e^x - 3 \int x^2 e^x \, dx$$

$$= x^3 e^x - 3x^2 e^x + 6 \int xe^x \, dx$$

$$= x^3 e^x - 3x^2 e^x + 6xe^x - 6 \int e^x \, dx$$

$$= x^3 e^x - 3x^2 e^x + 6xe^x - 6e^x + C$$

74. Use induction on k. For $k = 0$, $P(x)$ is a constant and the result follows immediately. Now suppose the statement is true for $k = n$ and let $P(x)$ have degree $n + 1$. Then $P(x) = ax^{n+1} + Q(x)$ for $Q(x)$ of degree n.

$$\int P(x)e^x \, dx = a \int x^{n+1} e^x \, dx + \int Q(x)e^x \, dx$$

$$= ax^{n+1} e^x - a(n+1) \int x^n e^x \, dx + \int Q(x)e^x \, dx$$

$$= \left[ax^{n+1} - a(n+1)(x^n - n(n-1)x^{n-1} + \cdots \pm a \right] e^x + \left[Q(x) - Q'(x) + \cdots \pm Q^{(n)}(x) \right] e^x + C$$

$$= \left[ax^{n+1} + Q(x) - [a(n+1)x^n + Q'(x)] + \cdots \pm Q^{(n)}(x) \right] e^x + C$$

$$= \left[P(x) - P'(x) + \cdots \pm P^{(n)}(x) \right] e^x + C$$

75. (a) Set $\int (x^2 - 3x + 1) e^x \, dx = Ax^2 e^x + Bxe^x + Ce^x$ and differentiate both sides of the equation.

$$x^2 e^x - 3xe^x + e^x = Ax^2 e^x + 2Axe^x + Bxe^x + Be^x + Ce^x$$

$$= Ax^2 e^x + (2A + B)xe^x + (B + C)e^x$$

Equating the coefficients, we find that $A = 1$, $B = -5$, $C = 6$

Thus $\int (x^2 - 3x + 1) e^x \, dx = x^2 e^x - 5xe^x + 6e^x + K$

(b) Set $\int (x^3 - 2x) e^x \, dx = Ax^3 e^x + Bx^2 e^x + Cxe^x + De^x$ and differentiate both sides of the equation.

$$x^3 e^x - 2xe^x = Ax^3 e^x + 3Ax^2 e^x + Bx^2 e^x + 2Bxe^x + Cxe^x + ce^x + De^x$$

$$= Ax^3 e^x + (3A + B)x^2 e^x + (2B + C)xe^x + (C + D)e^x$$

Equating the coefficients, we find that $A = 1$, $B = -3$, $C = 4$, $D = -4$

Thus $\int (x^3 - 2x) e^x \, dx = x^3 e^x - 3x^2 e^x + 4xe^x - 4e^x + K$

76. Set $u = f^{(-1)}(x)$ and $dv = dx$.

77. Let $u = f(x)$, $dv = g''(x) \, dx$. Then $du = f'(x) \, dx$, $v = g'(x)$, and

$$\int_a^b f(x)g''(x) \, dx = [f(x)g'(x)]_a^b - \int_a^b f'(x)g'(x) \, dx = -\int_a^b f'(x)g'(x) \, dx \quad \text{since } f(a) = f(b) = 0$$

Now let $u = f'(x)$, $dv = g'(x) \, dx$. Then $du = f''(x) \, dx$, $v = g(x)$, and

$$-\int_a^b f'(x)g'(x) \, dx = [-f'(x)g(x)]_a^b + \int f''(x)g(x) \, dx = \int g(x)f''(x) \, dx \quad \text{since } g(a) = g(b) = 0$$

Therefore, if f and g have continuous second derivatives, and if $f(a) = g(a) = f(b) = g(b) = 0$, then

$$\int_a^b f(x)g''(x) \, dx = \int_a^b g(x)f''(x) \, dx$$

78. (a)

$$f(b) - f(a) = \int_a^b f'(x) \, dx = [f'(x)(x - b)]_a^b - \int_a^b f''(x)(x - b) \, dx \ (a)$$

$u = f'(x)$	$dv = dx$
$du = f''(x) \, dx$	$v = (x - b)$

$$= f'(a)(b - a) - \int_a^b f''(x)(x - b) \, dx$$

(b)

$$f(b) - f(a) = f'(a)(b - a) - \left[f''(x)\frac{(x - b)^2}{2} \right]_a^b + \int_a^b \frac{f'''(x)}{2}(x - b)^2 \, dx$$

$u = f''(x)$	$dv = (x - b) \, dx$
$du = f'''(x) \, dx$	$v = \dfrac{(x - b)^2}{2}$

$$= f'(a)(b - a) + \frac{f''(a)}{2}(b - a)^2 + \int_a^b \frac{f'''(x)}{2}(x - b)^2 \, dx$$

79. (a) From Exercise 47, $A = \int_0^\pi x \sin x \, dx = \pi$

(b) $f(x) = x \sin x < 0$ on $(\pi, 2\pi)$. Therefore

$$A = -\int_\pi^{2\pi} x \sin x \, dx = \big[-x \cos x + \sin x \big]_\pi^{2\pi} = 3\pi$$

(c) $A = \int_{2\pi}^{3\pi} x \sin x \, dx = \big[-x \cos x + \sin x \big]_{2\pi}^{3\pi} = 5\pi$

(d) $A = (2n+1)\pi, \quad n = 1, 2, 3, \ldots$

80. (a) $\int_{\pi/2}^{3\pi/2} x \cos x \, dx = \big[x \sin x + \cos x \big]_{\pi/2}^{3\pi/2} = -2\pi; \quad A = 2\pi$

(b) $A = 4\pi$ (c) $A = 6\pi$ (d) $A = 2n\pi$

81. (a) area of R: $\int_0^\pi (1 - \sin x) \, dx = \big[x + \cos x \big]_0^\pi = \pi - 2$

(b) $V = \int_0^\pi 2\pi x \, (1 - \sin x) \, dx = 2\pi \big[\tfrac{1}{2} x^2 + x \cos x - \sin x \big]_0^\pi = \pi^3 - 2\pi^2$

(c) The region is symmetric about the line $x = \tfrac{1}{2}\pi$. Therefore $\overline{x} = \tfrac{1}{2}\pi$.

$$\overline{y}A = \tfrac{1}{2} \int_0^\pi (1 - \sin x)^2 \, dx$$

$$= \tfrac{1}{2} \int_0^\pi \left(1 - 2\sin x + \sin^2 x \right) dx$$

$$= \tfrac{1}{2} \int_0^\pi \left(\tfrac{3}{2} - 2\sin x - \tfrac{1}{2} \cos 2x \right) dx$$

$$= \tfrac{1}{2} \big[\tfrac{3}{2} x + 2\cos x - \tfrac{1}{4} \sin 2x \big]_0^\pi = \tfrac{3}{4}\pi - 2$$

Therefore, $\overline{y} = \dfrac{\tfrac{3}{4}\pi - 2}{\pi - 2} \cong 0.31202$

82. (a) $A = \int_0^{10} x e^{-x} \, dx \cong 0.9995$

$$\overline{x} = \frac{\int_0^{10} x^2 e^{-x} \, dx}{A} \cong 1.9955 \qquad \overline{y} = \frac{\tfrac{1}{2} \int_0^{10} \left[x e^{-x} \right]^2 dx}{A} \cong 0.1251$$

(b) around x-axis: $V \cong 0.7854$;   around y-axis: $V \cong 12.5316$

PROJECT 8.2

1. $\displaystyle \int_0^{2\pi} \sin^2 nx \, dx = \int_0^{2\pi} \left(\frac{1}{2} - \frac{1}{2} \cos 2nx \right) dx = \left[\frac{1}{2} x - \frac{1}{4n} \sin 2nx \right]_0^{2\pi} = \pi.$

$\displaystyle \int_0^{2\pi} \cos^2 nx \, dx = \int_0^{2\pi} \left(\frac{1}{2} + \frac{1}{2} \cos 2nx \right) dx = \left[\frac{1}{2} x + \frac{1}{4n} \sin 2nx \right]_0^{2\pi} = \pi$

2. $\displaystyle\int_0^{2\pi} \cos\left[(m+n)x\right] dx = \left[\frac{1}{m+n} \sin\left[(m+n)x\right]\right]_0^{2\pi} = 0.$

$\displaystyle 0 = \int_0^{2\pi} \cos\left[(m+n)x\right] dx = \int_0^{2\pi} \cos(mx+nx)\, dx = \int_0^{2\pi} \cos mx \cos nx\, dx - \int_0^{2\pi} \sin mx \sin nx\, dx$

Thus, $\displaystyle\int_0^{2\pi} \cos mx \cos nx\, dx = \int_0^{2\pi} \sin mx \sin nx\, dx$

SECTION 8.3

1. $\displaystyle\int \sin^3 x\, dx = \int (1 - \cos^2 x) \sin x\, dx = \frac{1}{3} \cos^3 x - \cos x + C$

2. $\displaystyle\int_0^{\pi/8} \cos^2 4x\, dx = \left[\frac{x}{2} + \frac{1}{16} \sin 8x\right]_0^{\pi/8} = \frac{\pi}{16}$

3. $\displaystyle\int_0^{\pi/6} \sin^2 3x\, dx = \int_0^{\pi/6} \frac{1 - \cos 6x}{2}\, dx = \left[\frac{1}{2}x - \frac{1}{12} \sin 6x\right]_0^{\pi/6} = \frac{\pi}{12}$

4. $\displaystyle\int \cos^3 x\, dx = \int (1 - \sin^2 x) \cos x\, dx = \sin x - \frac{1}{3} \sin^3 x + C$

5.
$$\int \cos^4 x \sin^3 x\, dx = \int \cos^4 x\,(1 - \cos^2 x) \sin x\, dx$$
$$= \int (\cos^4 x - \cos^6 x) \sin x\, dx$$
$$= -\tfrac{1}{5} \cos^5 x + \tfrac{1}{7} \cos^7 x + C$$

6. $\displaystyle\int \sin^3 x \cos^2 x\, dx = \int \cos^2(1 - \cos^2 x) \sin x\, dx = -\frac{1}{3} \cos^3 x + \frac{1}{5} \cos^5 x + C$

7.
$$\int \sin^3 x \cos^3 x\, dx = \int \sin^3 x\,(1 - \sin^2 x) \cos x\, dx = \int (\sin^3 x - \sin^5 x) \cos x\, dx$$
$$= \tfrac{1}{4} \sin^4 x - \tfrac{1}{6} \sin^6 x + C$$

8.
$$\int \sin^2 x \cos^4 x\, dx = \int (\sin x \cos x)^2 \cos^2 x\, dx = \int \frac{1}{4} \sin^2 2x \left(\frac{1}{2} + \frac{1}{2} \cos 2x\right) dx$$
$$= \frac{1}{8} \int \sin^2 2x\, dx + \frac{1}{8} \int \sin^2 2x \cos 2x\, dx$$
$$= \frac{1}{8} \left(\frac{1}{2}x - \frac{1}{8} \sin 4x\right) + \frac{1}{48} \sin^3 2x + C$$

9. $\displaystyle\int \sec^2 \pi x\, dx = \frac{1}{\pi} \tan \pi x + C$

10. $\displaystyle\int \csc^2 2x\, dx = -\frac{1}{2} \cot 2x + C$

11.
$$\int \tan^3 x\, dx = \int (\sec^2 x - 1) \tan x\, dx$$
$$= \int \tan x \sec^2 x\, dx - \int \tan x\, dx$$
$$= \tfrac{1}{2} \tan^2 x + \ln |\cos x| + C$$

12. $\displaystyle\int \cot^3 x\, dx = \int \cot x (\csc^2 x - 1)\, dx = -\frac{1}{2}\cot^2 x - \ln|\sin x| + C$

13.
$$\int \sin^4 x\, dx = \int \left(\frac{1 - \cos 2x}{2}\right)^2 dx$$
$$= \frac{1}{4} \int \left(1 - 2\cos 2x + \cos^2 2x\right) dx$$
$$= \frac{1}{4} \int \left(1 - 2\cos 2x + \frac{1 + \cos 4x}{2}\right) dx$$
$$= \int \left(\frac{3}{8} - \frac{1}{2}\cos 2x + \frac{1}{8}\cos 4x\right) dx$$
$$= \tfrac{3}{8}x - \tfrac{1}{4}\sin 2x + \tfrac{1}{32}\sin 4x + C$$
$$\int_0^\pi \sin^4 x\, dx = \left[\tfrac{3}{8}x - \tfrac{1}{4}\sin 2x + \tfrac{1}{32}\sin 4x\right]_0^\pi = \tfrac{3}{8}\pi$$

14.
$$\int \cos^3 x \cos 2x\, dx = \int (1 - \sin^2 x)(1 - 2\sin^2 x)\cos x\, dx = \int (1 - 3\sin^2 x + 2\sin^4 x)\cos x\, dx$$
$$= \sin x - \sin^3 x + \frac{2}{5}\sin^5 x + C$$

15.
$$\int \sin 2x \cos 3x\, dx = \int \frac{1}{2}\left[\sin(-x) + \sin 5x\right] dx$$
$$= \int \frac{1}{2}(-\sin x + \sin 5x)\, dx$$
$$= \tfrac{1}{2}\cos x - \tfrac{1}{10}\cos 5x + C$$

16.
$$\int_0^{\pi/2} \cos 2x \sin 3x\, dx = \int_0^{\pi/2} \frac{1}{2}\left[\sin(3x - 2x) + \sin(3x + 2x)\right] dx = \frac{1}{2}\int_0^{\pi/2}(\sin x + \sin 5x)\, dx$$
$$= \left[-\frac{1}{2}\cos x - \frac{1}{10}\cos 5x\right]_0^{\pi/2} = \frac{3}{5}$$

17. $\displaystyle\int \tan^2 x \sec^2 x\, dx = \frac{1}{3}\tan^3 x + C$

18. $\displaystyle\int \cot^2 x \csc^2 x\, dx = -\frac{1}{3}\cot^3 x + C$

19.
$$\int \sin^2 x \sin 2x\, dx = \int \sin^2 x\, (2\sin x \cos x)\, dx$$
$$= 2\int \sin^3 x \cos x\, dx$$
$$= \tfrac{1}{2}\sin^4 x + C$$

20.

$$\int_0^{\pi/2} \cos^4 x \, dx = \left[\frac{1}{4} \cos^3 x \sin x\right]_0^{\pi/2} + \frac{3}{4} \int_0^{\pi/2} \cos^2 x \, dx$$

$$= \left[\frac{1}{4} \cos^3 x \sin x + \frac{3}{8} x + \frac{3}{16} \sin 2x\right]_0^{\pi/2} = \frac{3\pi}{16}$$

21.

$$\int \sin^6 x \, dx = \int \left(\frac{1 - \cos 2x}{2}\right)^3 dx$$

$$= \frac{1}{8} \int (1 - 3\cos 2x + 3\cos^2 2x - \cos^3 2x) \, dx$$

$$= \frac{1}{8} \int \left[1 - 3\cos 2x + 3\left(\frac{1 + \cos 4x}{2}\right) - \cos 2x \left(1 - \sin^2 2x\right)\right] dx$$

$$= \frac{1}{8} \int \left(\frac{5}{2} - 4\cos 2x + \frac{3}{2} \cos 4x + \sin^2 2x \cos 2x\right) dx$$

$$= \tfrac{5}{16} x - \tfrac{1}{4} \sin 2x + \tfrac{3}{64} \sin 4x + \tfrac{1}{48} \sin^3 2x + C$$

22.

$$\int \cos^5 x \sin^5 x \, dx = \int \cos^5 x \sin^4 x \sin x \, dx = \int \cos^5 x (1 - \cos^2 x)^2 \sin x \, dx$$

$$= \int \cos^5 x (1 - 2\cos^2 x + \cos^4 x) \sin x \, dx$$

$$= -\frac{\cos^6 x}{6} + \frac{1}{4} \cos^8 x - \frac{1}{10} \cos^{10} x + C$$

Equivalently, $\int \cos^5 x \sin^4 x \sin x \, dx$ gives $-\tfrac{1}{6} \cos^x + \tfrac{1}{4} \cos^8 x - \tfrac{1}{10} \cos^{10} x + C$.

23. $\int_{\pi/6}^{\pi/2} \cot^2 x \, dx = \int_{\pi/6}^{\pi/2} (\csc^2 x - 1) \, dx = [-\cot x - x]_{\pi/6}^{\pi/2} = \sqrt{3} - \dfrac{\pi}{3}$

24.

$$\int \tan^4 x \, dx = \int \tan^2 x (\sec^2 x - 1) \, dx = \int \tan^2 x \sec^2 x \, dx - \int \tan^2 x \, dx$$

$$= \frac{1}{3} \tan^3 x - \int (\sec^2 x - 1) \, dx = \frac{1}{3} \tan^3 x - \tan x + x + C$$

25.

$$\int \cot^3 x \csc^3 x \, dx = \int (\csc^2 x - 1) \csc^3 x \cot x \, dx$$

$$= \int (\csc^4 x - \csc^2 x) \csc x \cot x \, dx$$

$$= -\frac{1}{5} \csc^5 x + \frac{1}{3} \csc^3 x + C$$

26.

$$\int \tan^3 x \sec^3 x \, dx = \int (\sec^2 x - 1) \sec^2 x \sec x \tan x \, dx$$

$$= \frac{1}{5} \sec^5 x - \frac{1}{3} \sec^3 x + C$$

27.
$$\int \sin 5x \sin 2x \, dx = \int \frac{1}{2}(\cos 3x - \cos 7x) \, dx$$
$$= \tfrac{1}{6} \sin 3x - \tfrac{1}{14} \sin 7x + C$$

28. $\displaystyle\int \sec^4 3x \, dx = \int (1 + \tan^2 3x) \sec^2 3x \, dx = \frac{1}{9} \tan^3 3x + \frac{1}{3} \tan 3x + C$

29.
$$\int \sin^{5/2} x \cos^3 x \, dx = \int \sin^{5/2} x \left(1 - \sin^2 x\right) \cos x \, dx$$
$$= \int \sin^{5/2} x \cos x \, dx - \int \sin^{9/2} x \cos x \, dx$$
$$= \tfrac{2}{7} \sin^{7/2} x - \tfrac{2}{11} \sin^{11/2} x + C$$

30. $\displaystyle\int \frac{\sin^3 x}{\cos x} \, dx = \int \frac{(1 - \cos^2 x) \sin x}{\cos x} \, dx = \int \left(\tan x - \frac{\sin 2x}{2}\right) dx = \ln |\sec x| + \tfrac{1}{4} \cos 2x + C$

31.
$$\int \tan^5 3x \, dx = \int \tan^3 3x \, (\sec^2 3x - 1) \, dx$$
$$= \int \tan^3 3x \sec^2 3x \, dx - \int \tan^3 3x \, dx$$
$$= \int \tan^3 3x \sec^2 3x \, dx - \int (\tan 3x \sec^2 3x - \tan 3x) \, dx$$
$$= \tfrac{1}{12} \tan^4 3x - \tfrac{1}{6} \tan^2 3x + \tfrac{1}{3} \ln |\sec 3x| + C$$

32.
$$\int \cot^5 2x \, dx = \int (\cot^3 2x \csc^2 2x - \cot^3 2x) \, dx$$
$$= \int (\cot^3 2x \csc^2 2x - \cot 2x \csc^2 2x + \cot 2x) \, dx$$
$$= -\frac{1}{8} \cot^4 2x + \frac{1}{4} \cot^2 2x + \frac{1}{2} \ln |\sin 2x| + C$$

33.
$$\int_{-1/6}^{1/3} \sin^4 3\pi x \cos^3 3\pi x \, dx = \int_{-1/6}^{1/3} \sin^4 3\pi x \cos^2 3\pi x \cos 3\pi x \, dx$$
$$= \int_{-1/6}^{1/3} \sin^4 3\pi x \left(1 - \sin^2 3\pi x \cos 3\pi x\right) dx$$
$$= \int_{-1}^{0} u^4(1 - u^2) \frac{1}{3\pi} \, du \quad [u = \sin 3\pi x, \quad du = 3\pi \cos 3\pi x \, dx]$$
$$= \frac{1}{3\pi} \left[\tfrac{1}{5} u^5 - \tfrac{1}{7} u^7\right]_{-1}^{0} = \frac{2}{105\pi}$$

34.

$$\int_0^{1/2} \cos \pi x \cos \frac{\pi}{2} x \, dx = \frac{1}{2} \int_0^{1/2} \left[\cos \left(\pi x - \frac{\pi}{2} x \right) + \cos \left(\pi x + \frac{\pi}{2} x \right) \right] dx$$

$$= \frac{1}{2} \int_0^{1/2} \left(\cos \frac{\pi}{2} x + \cos \frac{3\pi}{2} x \right) dx = \frac{1}{2} \left[\frac{2}{\pi} \sin \frac{\pi}{2} x + \frac{2}{3\pi} \sin \frac{3\pi}{2} x \right]_0^{1/2}$$

$$= \frac{2\sqrt{2}}{3\pi}$$

35.

$$\int_0^{\pi/4} \cos 4x \sin 2x \, dx = \int_0^{\pi/4} \frac{1}{2} (\sin 6x - \sin 2x) \, dx$$

$$= \left[-\tfrac{1}{12} \cos 6x + \tfrac{1}{4} \cos 2x \right]_0^{\pi/4} = -\tfrac{1}{6}$$

36.

$$\int (\sin 3x - \sin x)^2 \, dx = \int (\sin^2 3x - 2 \sin 3x \sin x + \sin^2 x) \, dx$$

$$= \int \left(\frac{1}{2} - \frac{1}{2} \cos 6x - \cos 2x + \cos 4x + \frac{1}{2} - \frac{1}{2} \cos 2x \right) dx$$

$$= x - \frac{1}{12} \sin 6x + \frac{1}{4} \sin 4x - \frac{3}{4} \sin 2x + C$$

37.

$$\int \tan^4 x \sec^4 x \, dx = \int \tan^4 x \left(\tan^2 x + 1 \right) \sec^2 x \, dx$$

$$= \int (\tan^6 x + \tan^4 x) \sec^2 x \, dx$$

$$= \frac{1}{7} \tan^7 x + \frac{1}{5} \tan^5 x + C$$

38. $\displaystyle \int \cot^4 x \csc^4 x \, dx = \int \cot^4 x (\cot^2 x + 1) \csc^2 x \, dx = -\frac{1}{7} \cot^7 x - \frac{1}{5} \cot^5 x + C$

39. $\displaystyle \int \sin(x/2) \cos 2x \, dx = \int \frac{1}{2} \left(\sin \left(\frac{5}{2} x \right) - \sin \left(\frac{3}{2} x \right) \right) dx = \frac{1}{3} \cos \left(\frac{3}{2} x \right) - \frac{1}{5} \cos \left(\frac{5}{2} x \right) + C$

40. $\displaystyle \int_0^{2\pi} \sin^2 ax \, dx = \int_0^{2\pi} \left(\frac{1}{2} - \frac{1}{2} \cos 2ax \right) dx = \left[\frac{x}{2} - \frac{1}{4a} \sin 2ax \right]_0^{2\pi} = \pi - \frac{\sin 4\pi a}{4a}$

41. $\left\{ \begin{array}{l} u = \tan x \\ du = \sec^2 x \, dx \end{array} \middle| \begin{array}{l} x = 0 \ \Rightarrow \ u = 0 \\ x = \pi/4 \ \Rightarrow \ u = 1 \end{array} \right\}; \quad \int_0^{\pi/4} \tan^3 x \sec^2 x \, dx = \int_0^1 u^3 \, du = \left[\frac{1}{4} u^4 \right]_0^1 = \frac{1}{4}$

42. $\displaystyle \int_{\pi/4}^{\pi/2} \csc^3 x \cot x \, dx = \int_{\pi/4}^{\pi/2} \csc^2 x \csc x \cot x \, dx = \left[-\frac{1}{3} \csc^3 x \right]_{\pi/4}^{\pi/2} = \frac{2\sqrt{2} - 1}{3}$

43. $\displaystyle\int_0^{\pi/6} \tan^2 2x\, dx = \int_0^{\pi/6} \left(\sec^2 2x - 1\right)\, dx = \left[\tfrac{1}{2}\tan 2x - x\right]_0^{\pi/6} = \dfrac{\sqrt{3}}{2} - \dfrac{\pi}{6}$

44. $\displaystyle\int_0^{\pi/3} \tan x \sec^{3/2} x\, dx = \int_0^{\pi/3} \sec^{1/2} x \sec x \tan x\, dx = \left[\tfrac{2}{3}\sec^{3/2} x\right]_0^{\pi/3} = \dfrac{4\sqrt{2}-2}{3}$

45. $A = \displaystyle\int_0^{\pi} \sin^2 x\, dx = \int_0^{\pi} \tfrac{1}{2}(1-\cos 2x)\, dx = \tfrac{1}{2}\left[x - \tfrac{1}{2}\sin 2x\right]_0^{\pi} = \dfrac{\pi}{2}$

46. $V = \displaystyle\int_{-\pi/2}^{\pi/2} \pi \cos^2 x\, dx = \pi \left[\dfrac{x}{2} + \dfrac{\sin 2x}{4}\right]_{-\pi/2}^{\pi/2} = \dfrac{\pi^2}{2}$

47.
$$V = \int_0^{\pi} \pi\left(\sin^2 x\right)^2 dx = \pi \int_0^{\pi} \sin^4 x\, dx = \pi \int_0^{\pi} \left(\tfrac{1}{2}(1-\cos 2x)\right)^2 dx$$

$$= \frac{\pi}{4}\int_0^{\pi}\left(1 - 2\cos 2x + \cos^2 2x\right) dx$$

$$= \frac{\pi}{4}\left[x - \sin 2x\right]_0^{\pi} + \frac{\pi}{8}\int_0^{\pi}\left(1 + \cos 4x\right) dx$$

$$= \frac{\pi^2}{4} + \frac{\pi}{8}\left[x + \tfrac{1}{4}\sin 4x\right]_0^{\pi} = \frac{3\pi^2}{8}$$

48. $V = \displaystyle\int_0^{\pi/4} \pi(\cos^2 x - \sin^2 x)\, dx = \pi \int_0^{\pi/4} \cos 2x\, dx = \dfrac{\pi}{2}\left[\sin 2x\right]_0^{\pi/4} = \dfrac{\pi}{2}$

49. $V = \displaystyle\int_0^{\pi/4} \pi\left[1^2 - \tan^2 x\right] dx = \pi \int_0^{\pi/4}\left[2 - \sec^2 x\right] dx = \pi\left[2x - \tan x\right]_0^{\pi/4} = \dfrac{\pi^2}{2} - \pi$

50. $V = \displaystyle\int_0^{\pi/4} \pi \tan^4 x\, dx = \pi \left[\dfrac{1}{3}\tan^3 x - \tan x + x\right]_0^{\pi/4} = \pi\left(\dfrac{\pi}{4} - \dfrac{2}{3}\right)$

51.
$$V = \int_0^{\pi/4} \pi\left[(\tan x + 1)^2 - 1^2\right] dx = \pi \int_0^{\pi/4}\left[\tan^2 x + 2\tan x\right] dx$$

$$= \pi \int_0^{\pi/4}\left(\sec^2 x + 2\tan x - 1\right) dx$$

$$= \pi\left[\tan x + 2\ln|\sec x| - x\right]_0^{\pi/4} = \pi\left[\ln 2 + 1 - \dfrac{\pi}{4}\right]$$

52. $V = \displaystyle\int_0^{\pi/4} \pi \sec^4 x\, dx = \pi \left[\dfrac{1}{3}\tan^3 x + \tan x\right]_0^{\pi/4} = \dfrac{4\pi}{3}$

53. (a) $\displaystyle\int \sin^n x\, dx = \int \sin^{n-1} x \sin x\, dx;\qquad \left\{\begin{array}{ll} u = \sin^{n-1} x & dv = \sin x\, dx \\ du = (n-1)\sin^{n-2} x\, dx & v = -\cos x \end{array}\right\}$

$$\int \sin^n x\, dx = -\sin^{n-1} x \cos x + (n-1)\int \sin^{n-2} x \cos^2 x\, dx$$

$$= -\sin^{n-1} x \cos x + (n-1)\int \sin^{n-2} x \left(1 - \sin^2 x\right) dx$$

Therefore,

$$n \int \sin^n x \, dx = -\sin^{n-1} x \cos x + (n-1) \int \sin^{n-2} x \, dx$$

and

$$\int \sin^n x \, dx = -\frac{1}{n} \sin^{n-1} x \cos x + \frac{n-1}{n} \int \sin^{n-2} x \, dx.$$

(b) $\displaystyle \int_0^{\pi/2} \sin^n x \, dx = \left[-\frac{1}{n} \sin^{n-1} x \cos x \right]_0^{\pi/2} + \frac{n-1}{n} \int_0^{\pi/2} \sin^{n-2} x \, dx = \frac{n-1}{n} \int_0^{\pi/2} \sin^{n-2} x \, dx$

(c) n even:

$$\int_0^{\pi/2} \sin^n x \, dx = \frac{n-1}{n} \int_0^{\pi/2} \sin^{n-2} x \, dx = \frac{n-1}{n} \frac{n-3}{n-2} \int_0^{\pi/2} \sin^{n-4} x \, dx$$

and so on,

$$\int_0^{\pi/2} \sin^n x \, dx = \frac{n-1}{n} \frac{n-3}{n-2} \cdots \frac{3}{4} \frac{1}{2} \int_0^{\pi/2} 1 \, dx = \frac{n-1}{n} \frac{n-3}{n-2} \cdots \frac{3}{4} \frac{1}{2} \frac{\pi}{2}.$$

n odd:

$$\int_0^{\pi/2} \sin^n x \, dx = \frac{n-1}{n} \frac{n-3}{n-2} \cdots \frac{4}{5} \frac{2}{3} \int_0^{\pi/2} \sin x \, dx$$

$$= \frac{n-1}{n} \frac{n-3}{n-2} \cdots \frac{4}{5} \frac{2}{3} \left[-\cos x \right]_0^{\pi/2} = \frac{n-1}{n} \frac{n-3}{n-2} \cdots \frac{4}{5} \frac{2}{3}$$

54. $\displaystyle \int_0^{\pi/2} \cos^n x \, dx = \int_0^{\pi/2} \sin^n \left(\frac{\pi}{2} - x \right) dx = -\int_{\pi/2}^0 \sin^n u \, du = \int_0^{\pi/2} \sin^n u \, du$

55. (a) $\displaystyle \int_0^{\pi/2} \sin^7 x \, dx = \frac{6 \cdot 4 \cdot 2}{7 \cdot 5 \cdot 3} = \frac{16}{35}$ (b) $\displaystyle \int_0^{\pi/2} \cos^6 x \, dx = \left(\frac{5 \cdot 3 \cdot 1}{6 \cdot 4 \cdot 2} \right) \frac{\pi}{2} = \frac{5\pi}{32}$

56. (a) $\displaystyle \int_0^\pi \pi (x + \sin 2x)^2 \, dx = \frac{\pi^4}{3} - \frac{\pi^2}{2} \cong 27.5349$

(b)

$$\int_0^\pi \pi (x + \sin 2x)^2 \, dx = \pi \int_0^\pi \left(x^2 + 2x \sin 2x + \sin^2 2x \right) dx$$

$$= \pi \left[\tfrac{1}{3} x^3 - x \cos 2x + \tfrac{1}{2} \sin 2x + \tfrac{1}{2} x - \tfrac{1}{8} \sin 4x \right]_0^\pi$$

$$= \frac{\pi^4}{3} - \frac{\pi^2}{2}$$

57. (a) $V \cong 4.9348$

(b) $\displaystyle V = \int_0^{\sqrt{\pi}} 2\pi x \sin^2 (x^2) \, dx = \pi \int_0^\pi \sin^2 u \, du = \tfrac{1}{2} \pi \int_0^\pi (1 - \cos 2u) \, du$

$$u = x^2 \underline{}\!\!\uparrow$$

$$= \tfrac{1}{2} \pi \left[u - \tfrac{1}{2} \sin 2u \right]_0^\pi$$

$$= \tfrac{1}{2} \pi^2 \cong 4.9348$$

58. (a) The x-coordinates of the points of intersection are: $x_1 = 1.7918$, $x_2 = 4.4914$;

$$A = \int_{x_1}^{x_2} \left[\sin(x/2) - 1 - \cos x\right] dx \cong 1.751$$

(b) $V = \int_{x_1}^{x_2} \pi \left[\sin^2(x/2) - (1 + \cos x)^2\right] dx \cong 6.173$

SECTION 8.4

1. $\left\{ \begin{array}{l} x = a \sin u \\ dx = a \cos u\, du \end{array} \right\}$; $\displaystyle\int \frac{dx}{\sqrt{a^2 - x^2}} = \int \frac{a \cos u\, du}{a \cos u}$

$$= \int du = u + C = \arcsin\left(\frac{x}{a}\right) + C$$

2. $\left\{ \begin{array}{l} u = x^2 - 4 \\ du = 2x\, dx \end{array} \right\}$; $\displaystyle\int_{5/2}^{4} \frac{x}{\sqrt{x^2 - 4}}\, dx = \frac{1}{2}\int_{9/4}^{12} u^{-1/2}\, du = \left[u^{1/2}\right]_{9/4}^{12} = \sqrt{12} - \frac{3}{2} = \frac{4\sqrt{3} - 3}{2}$

3. $\left\{ \begin{array}{l} x = \sec u \\ dx = \sec u \tan u\, du \end{array} \right\}$; $\displaystyle\int \sqrt{x^2 - 1}\, dx = \int \tan^2 u \sec u\, du$

$$= \int (\sec^3 u - \sec u)\, du$$

$$= \frac{1}{2}\sec u \tan u - \frac{1}{2}\ln|\sec u + \tan u| + C$$

Example 8, Section 8.3

$$= \frac{1}{2}x\sqrt{x^2 - 1} - \frac{1}{2}\ln|x + \sqrt{x^2 - 1}| + C$$

4. $\displaystyle\int \frac{x}{\sqrt{4 - x^2}}\, dx = -\int \frac{-2x}{2\sqrt{4 - x^2}}\, dx = -\int \frac{du}{2\sqrt{u}} = -\sqrt{u} + C = -\sqrt{4 - x^2} + C$

5. $\left\{ \begin{array}{l} x = 2 \sin u \\ dx = 2 \cos u\, du \end{array} \right\}$; $\displaystyle\int \frac{x^2}{\sqrt{4 - x^2}}\, dx = \int \frac{4\sin^2 u}{2\cos u} 2\cos u\, du$

$$= 2\int (1 - \cos 2u)\, du$$

$$= 2u - \sin 2u + C$$

$$= 2u - 2\sin u \cos u + C$$

$$= 2\arcsin\left(\frac{x}{2}\right) - \frac{1}{2}x\sqrt{4 - x^2} + C$$

6. $\left\{ \begin{array}{l} x = 2 \sec u \\ dx = 2 \sec u \tan u\, du \end{array} \right\}$; $\displaystyle\int \frac{x^2}{\sqrt{x^2 - 4}}\, dx = \int \frac{8\sec^3 u \tan u}{\sqrt{4(\sec^2 u - 1)}}\, du$

$$= 4\int \sec^3 u\, du$$

$$= 2\sec u \tan u + 2\ln|\sec u + \tan u| + C_1$$

$$= \frac{1}{2}x\sqrt{x^2 - 4} + 2\ln\left|\frac{x + \sqrt{x^2 - 4}}{2}\right| + C_1$$

$$= \frac{1}{2}x\sqrt{x^2 - 4} + 2\ln|x + \sqrt{x^2 - 4}| + C \quad (C = C_1 - 2\ln 2)$$

7. $\left\{ \begin{array}{l} u = 1 - x^2 \\ du = -2x\,dx \end{array} \right\}$; $\displaystyle \int \frac{x}{(1-x^2)^{3/2}}\,dx = -\frac{1}{2}\int \frac{du}{u^{3/2}} = u^{-1/2} + C = \frac{1}{\sqrt{1-x^2}} + C$

8. $\left\{ \begin{array}{l} x = 2\tan u \\ dx = 2\sec^2 u\,du \end{array} \right\}$; $\displaystyle \int \frac{x^2}{\sqrt{4+x^2}}\,dx = \int \frac{8\tan^2 u \sec^2 u}{\sqrt{4(1+\tan^2 u)}}\,du = 4\int \tan^2 u \sec u\,du$

$$= 4\int (\sec^3 u - \sec u)\,du$$

$$= 4\left(\frac{1}{2}\sec u \tan u - \frac{1}{2}\ln|\sec u + \tan u| \right) + C$$

$$= \frac{1}{2}x\sqrt{x^2+4} - 2\ln\left(x + \sqrt{x^2+4} \right) + C$$

(absorbing $2\ln 2$ into C)

9. $\left\{ \begin{array}{l} x = \sin u \\ dx = \cos u\,du \end{array} \right\}$; $\displaystyle \int \frac{x^2}{(1-x^2)^{3/2}}\,dx = \int \frac{\sin^2 u}{\cos^3 u}\cos u\,du = \int \tan^2 u\,du$

$$= \int (\sec^2 u - 1)\,du = \tan u - u + C$$

$$= \frac{x}{\sqrt{1-x^2}} - \arcsin x + C$$

$$\int_0^{1/2} \frac{x^2}{(1-x^2)^{3/2}}\,dx = \left[\frac{x}{\sqrt{1-x^2}} - \arcsin x \right]_0^{1/2} = \frac{2\sqrt{3}-\pi}{6}$$

10. $\left\{ \begin{array}{l} u = a^2 + x^2 \\ dx = 2x\,dx \end{array} \right\}$; $\displaystyle \int \frac{x}{a^2+x^2}\,dx = \frac{1}{2}\int \frac{du}{u} = \frac{1}{2}\ln|u| + C$

$$= \frac{1}{2}\ln(a^2+x^2) + C$$

11. $\left\{ \begin{array}{l} u = 4 - x^2 \\ du = -2x\,dx \end{array} \right\}$; $\displaystyle \int x\sqrt{4-x^2}\,dx = -\frac{1}{2}\int u^{1/2}\,du = -\frac{1}{3}u^{3/2} + C$

$$= -\frac{1}{3}(4-x^2)^{3/2} + C$$

12. $\left\{ \begin{array}{l} x = 4\sin u \\ dx = 4\cos u\,du \end{array} \right\}$; $\displaystyle \int \frac{x^3}{\sqrt{16-x^2}}\,dx = \frac{256\sin^3 u \cos u}{\sqrt{16-16\sin^2 u}}\,du = 64\int \sin^3 u\,du$

$$= 64\int (1-\cos^2 u)\sin u\,du = 64\left(-\cos u + \frac{\cos^3 u}{3} \right) + C$$

$$= \frac{1}{3}\left(\sqrt{16-x^2} \right)^3 - 16\sqrt{16-x^2} + C$$

$$\int_0^2 \frac{x^3}{\sqrt{16-x^2}}\,dx = \left[\frac{1}{3}(16-x^2)^{3/2} + 64\sqrt{16-x^2} \right]_0^2 = \frac{128}{3} - 12\sqrt{12}$$

13. $\begin{cases} x = 5\sin u \mid x = 0 \implies u = 0 \\ dx = 5\cos u\, du \mid x = 5 \implies u = \pi/2 \end{cases}$;

$$\int_0^5 x^2\sqrt{25 - x^2}\, dx = \int_0^{\pi/2} (5\sin u)^2 (5\cos u)^2\, du$$

$$= 625 \int_0^{\pi/2} \left(\sin^2 u - \sin^4 u\right)\, du$$

$$= 625 \left[\frac{1}{2}\cdot\frac{\pi}{2} - \frac{3\cdot 1}{4\cdot 2}\cdot\frac{\pi}{2}\right] = \frac{625\pi}{16} \qquad \text{[see Exercise 53, Section 8.3]}$$

14. $\begin{cases} x = \sin u \\ dx = \cos u\, du \end{cases}$; $\displaystyle \int \frac{\sqrt{1 - x^2}}{x^4}\, dx = \int \frac{\cos^2 u}{\sin^4 u}\, du = \int \cot^2 u \csc^2 u\, du$

$$= -\frac{\cot^3 u}{3} + C = -\frac{(1 - x^2)^{3/2}}{3x^3} + C$$

15. $\begin{cases} x = \sqrt{8}\tan u \\ dx = \sqrt{8}\sec^2 u\, du \end{cases}$; $\displaystyle \int \frac{x^2}{(x^2 + 8)^{3/2}}\, dx = \int \frac{8\tan^2 u}{(8\sec^2 u)^{3/2}}\sqrt{8}\sec^2 u\, du$

$$= \int \frac{\tan^2 u}{\sec u}\, du = \int \frac{\sec^2 u - 1}{\sec u}\, du$$

$$= \int (\sec u - \cos u)\, du$$

$$= \ln|\sec u + \tan u| - \sin u + C$$

$$= \ln\left(\frac{\sqrt{x^2 + 8} + x}{\sqrt{8}}\right) - \frac{x}{\sqrt{x^2 + 8}} + C$$

$$(\text{absorb} - \ln\sqrt{8} \text{ in } C)$$

$$= \ln\left(\sqrt{x^2 + 8} + x\right) - \frac{x}{\sqrt{x^2 + 8}} + C$$

16. $\displaystyle \int_0^a \sqrt{a^2 - x^2}\, dx = \frac{1}{4}(\text{Area of circle of radius } a) = \frac{\pi a^2}{4}$

17. $\begin{cases} x = a\sin u \\ dx = a\cos u\, du \end{cases}$; $\displaystyle \int \frac{dx}{x\sqrt{a^2 - x^2}} = \int \frac{a\cos u\, du}{a\sin u\,(a\cos u)} = \frac{1}{a}\int \csc u\, du$

$$= \frac{1}{a}\ln|\csc u - \cot u| + C$$

$$= \frac{1}{a}\ln\left|\frac{a - \sqrt{a^2 - x^2}}{x}\right| + C$$

18. $\begin{cases} x = \sec u \\ dx = \sec u \tan u\, du \end{cases}$; $\displaystyle \int \frac{\sqrt{x^2 - 1}}{x}\, dx = \int \frac{\tan u}{\sec u}\sec u \tan u\, du$

$$= \int \tan^2 u\, du = \int (\sec^2 u - 1)\, du$$

$$= \tan u - u + C = \sqrt{x^2 - 1} - \arctan\sqrt{x^2 - 1} + C$$

19. $\left\{ \begin{array}{l} x = 3\tan u \\ dx = 3\sec^2 u\,du \end{array} \right\};$ $\displaystyle\int \frac{x^3}{\sqrt{9+x^2}}\,dx = \int \frac{27\tan^3 u}{3\sec u}\cdot 3\sec^2 u\,du$

$$= 27\int \tan^3 u\,\sec u\,du$$

$$= 27\int \left(\sec^2 u - 1\right)\sec u\,\tan u\,du$$

$$= 27\left[\tfrac{1}{3}\sec^3 u - \sec u\right] + C$$

$$= \tfrac{1}{3}\left(9+x^2\right)^{3/2} - 9\left(9+x^2\right)^{1/2} + C$$

$$\int_0^3 \frac{x^3}{\sqrt{9+x^2}}\,dx = \left[\tfrac{1}{3}\left(9+x^2\right)^{3/2} - 9\left(9+x^2\right)^{1/2}\right]_0^3 = 18 - 9\sqrt{2}$$

20. $\left\{ \begin{array}{l} x = a\sin u \\ dx = a\cos u\,du \end{array} \right\};$ $\displaystyle\int \frac{dx}{x^2\sqrt{a^2-x^2}} = \int \frac{a\cos u}{a^2\sin^2 u\,a\cos u}\,du$

$$= \frac{1}{a^2}\int \csc^2 u\,du = -\frac{1}{a^2}\cot u + C$$

$$= -\frac{\sqrt{a^2-x^2}}{a^2 x} + C$$

21. $\left\{ \begin{array}{l} x = a\tan u \\ dx = a\sec^2 u\,du \end{array} \right\};$ $\displaystyle\int \frac{dx}{x^2\sqrt{a^2+x^2}} = \int \frac{a\sec^2 u\,du}{a^2\tan^2 u\,(a\sec u)}$

$$= \frac{1}{a^2}\int \frac{\sec u}{\tan^2 u}\,du$$

$$= \frac{1}{a^2}\int \cot u\,\csc u\,du$$

$$= -\frac{1}{a^2}\cos u + C = -\frac{1}{a^2 x}\sqrt{a^2+x^2} + C$$

22. $\left\{ \begin{array}{l} x = \sqrt{2}\tan u \\ dx = \sqrt{2}\sec^2 u\,du \end{array} \right\};$ $\displaystyle\int \frac{dx}{(x^2+2)^{3/2}} = \int \frac{\sqrt{2}\sec^2 u\,du}{(2\tan^2 u + 2)^{3/2}}$

$$= \int \frac{\sqrt{2}\sec^2 u}{2\sqrt{2}\sec^3 u}\,du$$

$$= \frac{1}{2}\int \cos u\,du = \frac{1}{2}\sin u + C$$

$$= \frac{1}{2}\frac{x}{\sqrt{x^2+2}} + C$$

23. $\left\{ \begin{array}{l} x = \sqrt{5}\sin u \\ dx = \sqrt{5}\cos u\,du \end{array} \right\};$ $\displaystyle\int \frac{dx}{(5-x^2)^{3/2}} = \int \frac{\sqrt{5}\cos u\,du}{(5\cos^2 u)^{3/2}}$

$$= \frac{1}{5}\int \sec^2 u\,du$$

$$= \frac{1}{5}\tan u + C = \frac{x}{5\sqrt{5-x^2}} + C$$

$$\int_0^1 \frac{dx}{(5-x^2)^{3/2}} = \left[\frac{x}{5\sqrt{5-x^2}}\right]_0^1 = \frac{1}{10}$$

24. $\left\{ \begin{array}{l} e^x = 2\tan u \\ e^x\,dx = 2\sec^2 u\,du \end{array} \right\}$; $\displaystyle\int \frac{dx}{e^x\sqrt{4+e^{2x}}} = \int \frac{e^x}{e^{2x}\sqrt{4+e^{2x}}}\,dx = \int \frac{2\sec^2 u}{4\tan^2 u \cdot 2\sec u}\,du$

$$= \frac{1}{4}\int \frac{\cos u}{\sin^2 u}\,du = -\frac{1}{4}\cdot\frac{1}{\sin u} + C$$

$$= -\frac{1}{4}\frac{\sqrt{4+e^{2x}}}{e^x} + C$$

25. $\left\{ \begin{array}{l} x = a\sec u \\ dx = a\sec u\tan u\,du \end{array} \right\}$; $\displaystyle\int \frac{dx}{x^2\sqrt{x^2-a^2}} = \int \frac{a\sec u\tan u\,du}{a^2\sec^2 u\,(a\tan u)}$

$$= \frac{1}{a^2}\int \cos u\,du$$

$$= \frac{1}{a^2}\sin u + C$$

$$= \frac{1}{a^2 x}\sqrt{x^2-a^2} + C$$

26. $\left\{ \begin{array}{l} u = e^x \\ du = e^x\,dx \end{array} \right\}$; $\displaystyle\int \frac{e^x}{\sqrt{9-e^{2x}}}\,dx = \int \frac{du}{\sqrt{9-u^2}}$

$$= \arcsin\left(\frac{u}{3}\right) + C = \arcsin\left(\frac{e^x}{3}\right) + C$$

27. $\left\{ \begin{array}{l} e^x = 3\sec u \\ e^x\,dx = 3\sec u\tan u\,du \end{array} \right\}$; $\displaystyle\int \frac{dx}{e^x\sqrt{e^{2x}-9}} = \int \frac{\tan u\,du}{3\sec u\,(3\tan u)}$

$$= \frac{1}{9}\int \cos u\,du$$

$$= \frac{1}{9}\sin u + C$$

$$= \frac{1}{9}e^{-x}\sqrt{e^{2x}-9} + C$$

28. $\left\{ \begin{array}{l} x-1 = 2\sec u \\ dx = 2\sec u\tan u\,du \end{array} \right\}$; $\displaystyle\int \frac{dx}{\sqrt{x^2-2x-3}} = \int \frac{dx}{\sqrt{(x-1)^2-4}}$

$$= \int \frac{2\sec u\tan u}{2\tan u}\,du$$

$$= \int \sec u\,du = \ln|\sec u + \tan u| + C$$

$$= \ln\left|x-1 + \sqrt{(x-1)^2-4}\right| + C$$

29. $(x^2-4x+4)^{\frac{3}{2}} = \left\{ \begin{array}{ll} (x-2)^3, & x > 2 \\ (2-x)^3, & x < 2 \end{array} \right.$

$$\int \frac{dx}{(x^2-4x+4)^{3/2}} = \left\{ \begin{array}{ll} -\dfrac{1}{2(x-2)^2} + C, & x > 2 \\[2mm] \dfrac{1}{2(2-x)^2} + C, & x < 2 \end{array} \right.$$

30. $\left\{\begin{array}{l} x - 3 = 3\sin u \\ dx = 3\cos u\,du \end{array}\right\}$; $\displaystyle\int \frac{x}{\sqrt{6x - x^2}}\,dx = \int \frac{x}{\sqrt{9 - (x-3)^2}}\,dx = \int \frac{3 + 3\sin u}{3\cos u}\cdot 3\cos u\,du$

$$= 3\int (1 + \sin u)\,du = 3u - 3\cos u + C$$

$$= 3\arcsin\left(\frac{x-3}{3}\right) - \sqrt{6x - x^2} + C$$

31. $\left\{\begin{array}{l} x - 3 = \sin u \\ dx = \cos u\,du \end{array}\right\}$; $\displaystyle\int x\sqrt{6x - x^2 - 8}\,dx = \int x\sqrt{1 - (x-3)^2}\,dx$

$$= \int (3 + \sin u)(\cos u)\cos u\,du$$

$$= \int (3\cos^2 u + \cos^2 u \sin u)\,du$$

$$= \int \left[3\left(\frac{1 + \cos 2u}{2}\right) + \cos^2 u \sin u\right]\,du$$

$$= \frac{3u}{2} + \frac{3}{4}\sin 2u - \frac{1}{3}\cos^3 u + C$$

$$= \frac{3}{2}\arcsin(x - 3) + \frac{3}{2}(x-3)\sqrt{6x - x^2 - 8} - \frac{1}{3}(6x - x^2 - 8)^{3/2} + C$$

32. $\left\{\begin{array}{l} x + 2 = 3\tan u \\ dx = 3\sec^2 u\,du \end{array}\right\}$; $\displaystyle\int \frac{x + 2}{\sqrt{x^2 + 4x + 13}}\,dx = \int \frac{x + 2}{\sqrt{(x+2)^2 + 9}}\,dx$

$$= \int \frac{3\tan u}{3\sec u}\cdot 3\sec^2 u\,du = 3\int \sec u \tan u\,du$$

$$= 3\sec u + C = \sqrt{x^2 + 4x + 13} + C$$

33. $\left\{\begin{array}{l} x + 1 = 2\tan u \\ dx = 2\sec^2 u\,du \end{array}\right\}$; $\displaystyle\int \frac{x}{(x^2 + 2x + 5)^2}\,dx = \int \frac{x}{[(x+1)^2 + 4]^2}\,dx$

$$= \int \frac{2\tan u - 1}{(4\sec^2 u)^2}\,2\sec^2 u\,du$$

$$= \frac{1}{8}\int \frac{2\tan u - 1}{\sec^2 u}\,du$$

$$= \frac{1}{8}\int (2\sin u \cos u - \cos^2 u)\,du$$

$$= \frac{1}{8}\int \left(2\sin u \cos u - \frac{1 + \cos 2u}{2}\right)\,du$$

$$= \frac{1}{8}\left(\sin^2 u - \frac{u}{2} - \frac{\sin 2u}{4}\right) + C$$

$$= \frac{1}{8}\left[\left(\frac{x+1}{\sqrt{x^2+2x+5}}\right)^2 - \frac{1}{2}\arctan\left(\frac{x+1}{2}\right) - \frac{1}{4}(2)\overbrace{\left(\frac{x+1}{\sqrt{x^2+2x+5}}\right)\left(\frac{2}{\sqrt{x^2+2x+5}}\right)}^{\sin 2u = 2\sin u\cos u}\right] + C$$

$$= \frac{x^2+x}{8(x^2+2x+5)} - \frac{1}{16}\arctan\left(\frac{x+1}{2}\right) + C$$

34. $\left\{\begin{array}{l} x-1 = 2\sec u \\ dx = 2\sec u\tan u\, du \end{array}\right\};\quad \displaystyle\int \frac{x}{\sqrt{x^2-2x-3}}\,dx = \int \frac{x}{\sqrt{(x-1)^2-4}}\,dx$

$$= \int \frac{1+2\sec u}{2\tan u}\cdot 2\sec u\tan u\, du$$

$$= \int (\sec u + 2\sec^2 u)\, du$$

$$= \ln|\sec u + \tan u| + 2\tan u + C$$

$$= \ln\left|x-1+\sqrt{x^2-2x-3}\right| + \sqrt{x^2-2x-3} + C$$

35.

$$\boxed{\begin{array}{ll} u = \sec^{-1}x & dv = dx \\ du = \dfrac{1}{x\sqrt{x^2-1}}\,dx & v = x \end{array}}\quad \int \sec^{-1}x\, dx = x\sec^{-1}x - \int \frac{1}{\sqrt{x^2-1}}\,dx$$

$$\boxed{x = \sec u \quad dx = \sec u\tan u\, du}\quad = x\sec^{-1}x - \int \sec u\, du$$

$$= x\sec^{-1}x - \ln\left[\sec u + \tan u\right] + C$$

$$= x\sec^{-1}x - \ln\left[x + \sqrt{x^2-1}\right] + C$$

36. (a) Set $u = \sqrt{a^2-x^2}$. Then $u^2 = a^2 - x^2$ and $2u\, du = -2x\, dx$.

$$\int \frac{\sqrt{a^2-x^2}}{x}\,dx = \int \frac{\sqrt{a^2-x^2}}{x^2}\,x\, dx = -\int \frac{u}{a^2-u^2}\,du$$

$$= \int \frac{u^2}{u^2-a^2}\,du$$

$$= \int \left(1 - \frac{a^2}{a^2-u^2}\right)du$$

$$= u + \frac{a}{2}\ln\left|\frac{a-u}{a+u}\right| + C \quad \text{Formula 96}$$

$$= \sqrt{a^2-x^2} + \frac{a}{2}\ln\left|\frac{a-\sqrt{a^2-x^2}}{a+\sqrt{a^2-x^2}}\right| + C$$

(b) Set $x = a \sin u$. Then $dx = a \cos u \, du$ and $\sqrt{a^2 - x^2} = a \cos u$.

$$\int \frac{\sqrt{a^2 - x^2}}{x} \, dx = \int \frac{a \cos u}{a \sin u} \, a \cos u \, du$$

$$= a \int \frac{\cos^2 u}{\sin u} \, du = a \int \frac{1 - \sin^2 u}{\sin u} \, du$$

$$= a \int (\csc u - \sin u) \, du$$

$$= a \ln |\csc u - \cot u| + a \cos u + C$$

$$= \sqrt{a^2 - x^2} + a \ln \left| \frac{a}{x} - \frac{\sqrt{a^2 - x^2}}{x} \right| + C$$

$$= \sqrt{a^2 - x^2} + a \ln \left| \frac{a - \sqrt{a^2 - x^2}}{x} \right| + C$$

(c) $\dfrac{a - \sqrt{a^2 - x^2}}{a + \sqrt{a^2 - x^2}} = \left(\dfrac{a - \sqrt{a^2 - x^2}}{x} \right)^2$

37. Let $x = a \tan u$. Then $dx = a \sec^2 u \, du$, $\sqrt{x^2 + a^2} = a \sec u$, and

$$\int \frac{dx}{(x^2 + a^2)^n} = \int \frac{a \sec^2 u}{a^{2n} \sec^{2n} u} \, du = \frac{1}{a^{2n-1}} \int \cos^{2n-2} u \, du$$

38.
$$\int \frac{1}{(x^2 + 1)^2} \, dx = \int \cos^2 u \, du = \frac{1}{2} \int (1 + \cos 2u) \, du$$

$$= \frac{1}{2} \left[u + \frac{1}{2} \sin 2u \right] + C$$

$$= \frac{1}{2} [u + \sin u \, \cos u] + C$$

$$= \frac{1}{2} \left[\arctan x + \frac{x}{x^2 + 1} \right] + C$$

39.
$$\int \frac{1}{(x^2 + 1)^3} \, dx = \int \cos^4 u \, du$$

$$= \frac{1}{4} \cos^3 u \, \sin u + \frac{3}{8} \cos u \, \sin u + \frac{3}{8} u + C \quad \text{[by the reduction formula (8.3.2)]}$$

$$= \frac{3}{8} \arctan x + \frac{3x}{8 (x^2 + 1)} + \frac{x}{4 (x^2 + 1)^2} + C$$

40.

$$\boxed{\begin{array}{ll} u = \arctan x & dv = x\,dx \\ du = \dfrac{1}{1+x^2}\,dx & v = \dfrac{x^2}{2} \end{array}}$$

$$\int x \arctan x \, dx = \frac{x^2}{2} \arctan x - \frac{1}{2} \int \frac{x^2}{1+x^2}\,dx$$

$$\boxed{x = \tan u \quad dx = \sec^2 u \, du} \qquad = \frac{x^2}{2} \arctan x - \frac{1}{2} \int \frac{\tan^2 u}{1 + \tan^2 u} \sec^2 u \, du$$

$$= \frac{x^2}{2} \arctan x - \frac{1}{2} \int \tan^2 u \, du$$

$$= \frac{x^2}{2} \arctan x - \frac{1}{2} \tan u + \frac{1}{2} u + C$$

$$= \frac{x^2}{2} \arctan x - \frac{1}{2} x + \frac{1}{2} \arctan x + C$$

$$= \arctan x \left(\frac{x^2}{2} + \frac{1}{2} \right) - \frac{1}{2} x + C$$

41.

$$\boxed{\begin{array}{ll} u = \arcsin x & dv = x\,dx \\ du = \dfrac{1}{\sqrt{1-x^2}}\,dx & v = \dfrac{x^2}{2} \end{array}}$$

$$\int x \arcsin x \, dx = \frac{x^2}{2} \arcsin x - \frac{1}{2} \int \frac{x^2}{\sqrt{1-x^2}}\,dx$$

$$\boxed{x = \sin u \quad dx = \cos u \, du} \qquad = \frac{x^2}{2} \arcsin x - \frac{1}{2} \int \sin^2 u \, du$$

$$= \frac{x^2}{2} \arcsin x - \frac{1}{2} \left[\frac{1}{2} u - \frac{1}{4} \sin 2u \right] + C$$

$$= \frac{x^2}{2} \arcsin x - \frac{1}{4} \arcsin x + \frac{1}{8}(2x\sqrt{1-x^2}) + C$$

$$= \arcsin x \left(\frac{x^2}{2} - \frac{1}{4} \right) + \frac{1}{4} x\sqrt{1-x^2} + C$$

42.

$$\boxed{x = 3 \sec u \quad dx = 3 \sec u \tan u \, du} \qquad \int_3^5 \frac{\sqrt{x^2-9}}{x}\,dx = 3 \int_0^b \tan^2 u \, du \quad (\text{where } b = \sec^{-1} \frac{5}{3})$$

$$= 3 \left[\tan u - u \right]_0^b$$

$$= 4 - 3 \sec^{-1} \frac{5}{3}$$

43.
$$V = \int_0^1 \pi \left(\frac{1}{1+x^2} \right)^2 dx = \pi \int_0^1 \frac{1}{(1+x^2)^2} dx$$

$$= \pi \int_0^{\pi/4} \cos^2 u \, du \qquad [x = \tan u, \text{ see Ex.45}]$$

$$= \frac{\pi}{2} \int_0^{\pi/4} (1 + \cos 2u) \, du$$

$$= \frac{\pi}{2} \left[u + \tfrac{1}{2} \sin 2u \right]_0^{\pi/4} = \frac{\pi^2}{8} + \frac{\pi}{4}$$

44.
$$A = 2 \int_0^{\sqrt{r^2 - h^2}} \left(\sqrt{r^2 - x^2} - h \right) dx = 2 \left[\frac{1}{2} x \sqrt{r^2 - x^2} + \frac{1}{2} r^2 \arcsin \left(\frac{x}{r} \right) - hx \right]_0^{\sqrt{r^2 - h^2}}$$

$$= r^2 \arcsin \left(\frac{\sqrt{r^2 - h^2}}{r} \right) - h\sqrt{r^2 - h^2}$$

45. We need only consider angles θ between 0 and π. Assume first that $0 \le \theta \le \frac{\pi}{2}$.
The area of the triangle is: $\frac{1}{2} r^2 \sin \theta \cos \theta$.

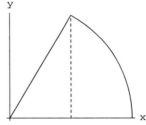

The area of the other region is given by:

$$\int_{r\cos\theta}^r \sqrt{r^2 - x^2} \, dx = \left[\frac{x}{2} \sqrt{r^2 - x^2} + \frac{r^2}{2} \arcsin \frac{x}{r} \right]_{r\cos\theta}^r$$

$$= \frac{\pi r^2}{4} - \frac{r^2}{2} \sin\theta \cos\theta - \frac{r^2}{2} \arcsin(\cos\theta)$$

$$= \frac{r^2\theta}{2} - \frac{r^2}{2} \sin\theta \cos\theta$$

Thus, the area of the sector is $A = \frac{1}{2} r^2 \theta$. If $\frac{\pi}{2} < \theta \le \pi$, then
$$A = \tfrac{1}{2} \pi r^2 - \tfrac{1}{2} r^2 (\pi - \theta) = \tfrac{1}{2} r^2 \theta.$$

46. $A = 4 \int_0^b \frac{a}{b} \sqrt{b^2 + y^2} dy = \frac{4a}{b} \left[\frac{y}{2} \sqrt{b^2 + y^2} + \frac{b^2}{2} \ln \left(y + \sqrt{b^2 + y^2} \right) \right]_0^b = 2ab \left[\sqrt{2} + \ln \left(1 + \sqrt{2} \right) \right]$

47. $A = 2 \int_3^5 4\sqrt{\frac{x^2}{9} - 1} dx = 24 \int_1^{\frac{5}{3}} \sqrt{u^2 - 1} du \quad (\text{where } u = \frac{x}{3})$

$$= 24 \left[\frac{u}{2} \sqrt{u^2 - 1} - \frac{1}{2} \ln |u + \sqrt{u^2 - 1}| \right]_1^{\frac{5}{3}} = \frac{80}{3} - 12 \ln 3.$$

48. $V = 2 \int_{b-a}^{b+a} 2\pi x \sqrt{a^2 - (x - b)^2} dx = 4\pi \int_{-a}^a (u + b) \sqrt{a^2 - u^2} \, du = 2\pi^2 a^2 b.$

49. $M = \displaystyle\int_0^a \frac{dx}{\sqrt{x^2 + a^2}} = \left[\ln\left(x + \sqrt{x^2 + a^2}\right)\right]_0^a = \ln\left(1 + \sqrt{2}\right)$

$\qquad x_M M = \displaystyle\int_0^a \frac{x}{\sqrt{x^2 + a^2}}\, dx = \left[\sqrt{x^2 + a^2}\right]_0^a = (\sqrt{2} - 1)a \qquad x_M = \dfrac{(\sqrt{2} - 1)a}{\ln\left(1 + \sqrt{2}\right)}$

50. $M = \displaystyle\int_0^a (x^2 + a^2)^{-3/2}\, dx = \left[\frac{1}{a^2}\sin u\right]_0^{\pi/4} = \dfrac{\sqrt{2}}{2a^2}$

$\qquad x_M M = \displaystyle\int_0^a \frac{x}{(x^2 + a^2)^{3/2}}\, dx = \left[-(x^2 + a^2)^{-1/2}\right]_0^a = \dfrac{2 - \sqrt{2}}{2a} \implies x_M = (\sqrt{2} - 1)a$

51.

$A = \displaystyle\int_a^{\sqrt{2}a} \sqrt{x^2 - a^2}\, dx = \left[\frac{1}{2}x\sqrt{x^2 - a^2} - \frac{1}{2}a^2 \ln\left|x + \sqrt{x^2 - a^2}\right|\right]_a^{\sqrt{2}a}$

$\qquad\qquad = \frac{1}{2}a^2\left[\sqrt{2} - \ln\left(\sqrt{2} + 1\right)\right]$

$\overline{x}A = \displaystyle\int_a^{\sqrt{2}a} x\sqrt{x^2 - a^2}\, dx = \frac{1}{3}a^3, \qquad \overline{y}A = \displaystyle\int_a^{\sqrt{2}a} \left[\frac{1}{2}(x^2 - a^2)\right] dx = \frac{1}{6}a^3(2 - \sqrt{2})$

$\overline{x} = \dfrac{2a}{3\left[\sqrt{2} - \ln\left(\sqrt{2} + 1\right)\right]}, \qquad \overline{y} = \dfrac{(2 - \sqrt{2})a}{3\left[\sqrt{2} - \ln\left(\sqrt{2} + 1\right)\right]}$

52. Using $\overline{y}$ and A found in Exercise 51,

$V_x = 2\pi\overline{y}A = 2\pi \dfrac{(2 - \sqrt{2})a}{3\left[\sqrt{2} - \ln(\sqrt{2} + 1)\right]} \cdot \frac{1}{2}a^2\left[\sqrt{2} - \ln(\sqrt{2} + 1)\right] = \frac{1}{3}\pi a^3(2 - \sqrt{2})$

$\overline{x}V_x = \displaystyle\int_a^{\sqrt{2}a} \pi x(x^2 - a^2)\, dx = \frac{1}{4}\pi a^4 \implies \overline{x} = \dfrac{3a}{4(2 - \sqrt{2})} = \frac{3}{8}a(2 + \sqrt{2})$

53. $V_y = 2\pi\overline{R}A = \dfrac{2}{3}\pi a^3 \qquad \overline{y}V_y = \displaystyle\int_a^{\sqrt{2}a} \pi x(x^2 - a^2)\, dx = \frac{1}{4}\pi a^4, \quad \overline{y} = \frac{3}{8}a$

54. Let $x = a\tan u$. Then $dx = a\sec^2 u\, du$ and $\sqrt{x^2 - a^2} = a\sec u$.

$\qquad \displaystyle\int \frac{1}{\sqrt{a^2 + x^2}}\, dx = \int \sec u\, du = \ln|\sec u + \tan u| + C$

$\qquad\qquad\qquad\qquad\qquad = \ln\left|\dfrac{\sqrt{a^2 + x^2}}{a} + \dfrac{x}{a}\right| + C$

$\qquad\qquad\qquad\qquad\qquad = \ln\left|x + \sqrt{a^2 + x^2}\right| + K, \quad K = C - \ln a$

55. Let $x = a\sec u$. Then $dx = a\sec u\tan u\, du$ and $\sqrt{x^2 - a^2} = a\tan u$.

$\qquad \displaystyle\int \frac{1}{\sqrt{x^2 - a^2}}\, dx = \int \sec u\, du = \ln|\sec u + \tan u| + C$

$\qquad\qquad\qquad\qquad\qquad = \ln\left|\dfrac{x}{a} + \dfrac{\sqrt{x^2 - a^2}}{a}\right| + C$

$\qquad\qquad\qquad\qquad\qquad = \ln\left|x + \sqrt{x^2 - a^2}\right| + K, \quad K = C - \ln a$

56. (a)

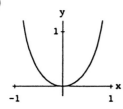

(b) $A = \displaystyle\int_0^{1/2} \dfrac{x^2}{\sqrt{1-x^2}}\,dx$

$= \displaystyle\int_0^{\pi/6} \sin^2 u\,du = \left[\dfrac{u}{2} - \dfrac{\sin 2u}{4}\right]_0^{\pi/6} = \dfrac{2\pi - 3\sqrt{3}}{24}$

(c) $V = \displaystyle\int_0^{1/2} \dfrac{\pi x^4}{1 - x^2}\,dx = \int_0^{\pi/6} \dfrac{\pi \sin^4 u}{\cos^2 u}\cos u\,du$

$= \pi \displaystyle\int_0^{\pi/6} \left(\sec u - \cos u - \sin^2 u\cos u\right)du$

$= \pi\left[\ln|\sec u + \tan u| - \sin u - \dfrac{1}{3}\sin^3 u\right]_0^{\pi/6} = \pi\left(\ln\sqrt{3} - \dfrac{13}{24}\right)$

57. (a)

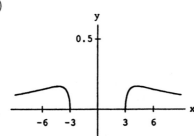

(b) $A = \displaystyle\int_3^6 \dfrac{\sqrt{x^2 - 9}}{x^2}\,dx$

$= \displaystyle\int_0^{\pi/3} \dfrac{3\tan u}{9\sec^2 u}\cdot 3\sec u\tan u\,du \quad [x = 3\sec u]$

$= \displaystyle\int_0^{\pi/3} \left(\sec u - \cos u\right)du$

$= \left[\ln|\sec u + \tan u| - \sin u\right]_0^{\pi/3}$

$= \ln(2 + \sqrt{3}) - \dfrac{\sqrt{3}}{2}$

(c) $\overline{x}A = \displaystyle\int_3^6 x\cdot\dfrac{\sqrt{x^2-9}}{x^2}\,dx = \int_0^{\pi/3}\left(3\sec^2 u - 3\right)du = [3\tan u - 3u]_0^{\pi/3} = 3\sqrt{3} - \pi$

Thus, $\overline{x} = \dfrac{2\left(3\sqrt{3} - \pi\right)}{2\ln\left(2 + \sqrt{3}\right) - \sqrt{3}}$.

$\overline{y}A = \displaystyle\int \dfrac{1}{2}\left(\dfrac{\sqrt{x^2-9}}{x^2}\right)^2 dx = \dfrac{1}{2}\int_3^6\left(\dfrac{1}{x^2} - \dfrac{9}{x^4}\right)dx = \dfrac{1}{2}\left[-\dfrac{1}{x} + \dfrac{3}{x^3}\right]_3^6 = \dfrac{5}{144}$

Thus, $\overline{y} = \dfrac{5}{72\left[2\ln\left(2 + \sqrt{3}\right) - \sqrt{3}\right]}$.

SECTION 8.5

1. $\dfrac{1}{x^2 + 7x + 6} = \dfrac{1}{(x+1)(x+6)} = \dfrac{A}{x+1} + \dfrac{B}{x+6}$

$1 = A(x+6) + B(x+1)$

$x = -6: \quad 1 = -5B \quad\Longrightarrow\quad B = -1/5$

$x = -1: \quad 1 = 5A \quad\Longrightarrow\quad A = 1/5$

$\dfrac{1}{x^2 + 7x + 6} = \dfrac{1/5}{x+1} - \dfrac{1/5}{x+6}$

2. $\dfrac{x^2}{(x-1)(x^2+4x+5)} = \dfrac{A}{x-1} + \dfrac{Bx+C}{x^2+4x+5}$

$x^2 = A(x^2+4x+5) + (Bx+C)(x-1)$

$x = 1: \quad 1 = 10A \implies A = 1/10$

$x = 0: \quad 0 = 5A - C \implies C = 5A = 1/2$

Coefficient of x^2 is $\quad A + B = 1 \implies B = 1 - A = 9/10$

$R(x) = \dfrac{\frac{1}{10}}{x-1} + \dfrac{\frac{9}{10}x + \frac{1}{2}}{x^2+4x+5}$

3. $\dfrac{x}{x^4-1} = \dfrac{x}{(x^2+1)(x+1)(x-1)} = \dfrac{Ax+B}{x^2+1} + \dfrac{C}{x+1} + \dfrac{D}{x-1}$

$x = (Ax+B)(x^2-1) + C(x-1)(x^2+1) + D(x+1)(x^2+1)$

$x = 1: \quad 1 = 4D \implies D = 1/4$

$x = -1: \quad -1 = -4C \implies C = 1/4$

$x = 0: \quad -B - C + D = 0 \implies B = 0$

$x = 2: \quad 6A + 5C + 15D = 2 \implies A = -1/2$

$\dfrac{x}{x^4-1} = \dfrac{1/4}{x-1} + \dfrac{1/4}{x+1} - \dfrac{x/2}{x^2+1}$

4. $\dfrac{x^4}{(x-1)^3} = \dfrac{[(x-1)+1]^4}{(x-1)^3} = \dfrac{(x-1)^4 + 4(x-1)^3 + 6(x-1)^2 + 4(x-1) + 1}{(x-1)^3}$

$\qquad = x + 3 + \dfrac{6}{x-1} + \dfrac{4}{(x-1)^2} + \dfrac{1}{(x-1)^3}$

5. $\dfrac{x^2-3x-1}{x^3+x^2-2x} = \dfrac{x^2-3x-1}{x(x+2)(x+1)} = \dfrac{A}{x} + \dfrac{B}{x+2} + \dfrac{C}{x-1}$

$x^2 - 3x - 1 = A(x+2)(x-1) + Bx(x-1) + Cx(x+2)$

$x = 0: \quad -1 = -2A \implies A = 1/2$

$x = -2: \quad 9 = 6B \implies B = 3/2$

$x = 1: \quad -3 = 3C \implies C = -1$

$\dfrac{x^2-3x-1}{x^3+x^2-2x} = \dfrac{1/2}{x} + \dfrac{3/2}{x+2} - \dfrac{1}{x-1}$

6. $\dfrac{x^3+x^2+x+2}{x^4+3x^2+2} = \dfrac{x^3+x^2+x+2}{(x^2+1)(x^2+2)} = \dfrac{Ax+B}{x^2+1} + \dfrac{Cx+D}{x^2+2}$

$x^3 + x^2 + x + 2 = (Ax+B)(x^2+2) + (Cx+D)(x^2+1)$

$\qquad = (A+C)x^3 + (B+D)x^2 + (2A+C)x + 2B + D$

$A + C = 1 = 2A + 1 \implies A = 0, \; C = 1; \quad B + D = 1, \quad 2B + D = 2 \implies B = 1, \; D = 0$

$R(x) = \dfrac{1}{x^2+1} + \dfrac{x}{x^2+2}$

7. $\dfrac{2x^2+1}{x^3-6x^2+11x-6} = \dfrac{2x^2+1}{(x-1)(x-2)(x-3)} = \dfrac{A}{x-1} + \dfrac{B}{x-2} + \dfrac{C}{x-3}$

$2x^2 + 1 = A(x-2)(x-3) + B(x-1)(x-3) + C(x-1)(x-2)$

$x = 1$: $3 = 2A$ $\implies$ $A = 3/2$

$x = 2$: $9 = -B$ $\implies$ $B = -9$

$x = 3$: $19 = 2C \implies C = 19/2$

$$\frac{2x^2 + 1}{x^3 - 6x^2 + 11x - 6} = \frac{3/2}{x - 1} - \frac{9}{x - 2} + \frac{19/2}{x - 3}$$

8. $\dfrac{1}{x(x^2 + 1)^2} = \dfrac{A}{x} + \dfrac{Bx + C}{x^2 + 1} + \dfrac{Dx + E}{(x^2 + 1)^2}$

$1 = A(x^2 + 1)^2 + (Bx + C)x(x^2 + 1) + (Dx + E)x$

Constant term: $A = 1$, x^4 term: $0 = A + B \implies B = -1$

x^3 term: $0 = C$, x^2 term: $0 = 2A + B + D \implies D = -1$

x term: $0 = E$

$R(x) = \dfrac{1}{x} - \dfrac{x}{x^2 + 1} - \dfrac{x}{(x^2 + 1)^2}$

9. $\dfrac{7}{(x - 2)(x + 5)} = \dfrac{A}{x - 2} + \dfrac{B}{x + 5}$

$7 = A(x + 5) + B(x - 2)$

$x = -5$: $7 = -7B$ $\implies$ $B = -1$

$x = 2$: $7 = 7A$ $\implies$ $A = 1$

$\displaystyle\int \frac{7}{(x - 2)(x + 5)}\, dx = \int \left(\frac{1}{x - 2} - \frac{1}{x + 5} \right) = \ln|x - 2| - \ln|x + 5| + C = \ln\left| \frac{x - 2}{x + 5} \right| + C$

10. $\displaystyle\int \frac{x}{(x + 1)(x + 2)(x + 3)}\, dx = \int \left(\frac{-1/2}{x + 1} + \frac{2}{x + 2} - \frac{3/2}{x + 3} \right) dx$

$$= -\frac{1}{2} \ln|x + 1| + 2 \ln|x + 2| - \frac{3}{2} \ln|x + 3| + C$$

11.   We carry out the division until the numerator has degree smaller than the denominator:

$$\frac{2x^4 - 4x^3 + 4x^2 + 3}{x^3 - x^2} = 2x - 2 + \frac{2x^2 + 3}{x^2(x - 1)}$$

$\dfrac{2x^2 + 3}{x^2(x - 1)} = \dfrac{A}{x} + \dfrac{B}{x^2} + \dfrac{C}{x - 1}$ $\implies$ $2x^2 + 3 = Ax(x - 1) + B(x - 1) + Cx^2$

$x = 0$: $3 = -B$ $\implies$ $B = -3$

$x = 1$: $5 = C$ $\implies$ $C = 5$

$x = -1$: $5 = 2A - 2B + C$ $\implies$ $A = -3$

$$\int \left(2x - 2 + \frac{2x^2 + 3}{x^2(x - 1)} \right) dx = x^2 - 2x + \int \left(-\frac{3}{x} - \frac{3}{x^2} + \frac{5}{x - 1} \right) dx$$

$$= x^2 - 2x - 3 \ln|x| + \frac{3}{x} + 5 \ln|x - 1| + C$$

12. $\displaystyle\int \frac{x^2 + 1}{x(x^2 - 1)}\, dx = \int \left(\frac{-1}{x} + \frac{1}{x - 1} + \frac{1}{x + 1} \right) dx = -\ln|x| + \ln|x - 1| + \ln|x + 1| + C = \ln\left| \frac{x^2 - 1}{x} \right| + C$

13. We carry out the division until the numerator has degree smaller than the denominator:

$$\frac{x^5}{(x-2)^2} = \frac{x^5}{x^2-4x+4} = x^3 + 4x^2 + 12x + 32 + \frac{80x-128}{(x-2)^2}.$$

Then,

$$\frac{80x-128}{(x-2)^2} = \frac{80x-160+32}{(x-2)^2} = \frac{80}{x-2} + \frac{32}{(x-2)^2}$$

$$\int \frac{x^5}{(x-2)^2}\, dx = \int \left(x^3 + 4x^2 + 12x + 32 + \frac{80}{x-2} + \frac{32}{(x-2)^2} \right) dx$$

$$= \frac{1}{4}x^4 + \frac{4}{3}x^3 + 6x^2 + 32x + 80\ln|x-2| - \frac{32}{x-2} + C.$$

14. $\displaystyle \int \frac{x^5}{x-2}\, dx = \int \left(x^4 + 2x^3 + 4x^2 + 8x + 16 + \frac{32}{x-2} \right) dx$

$$= \frac{1}{5}x^5 + \frac{1}{2}x^4 + \frac{4}{3}x^3 + 4x^2 + 16x + 32\ln|x-2| + C$$

15. $\displaystyle \frac{x+3}{x^2-3x+2} = \frac{A}{x-1} + \frac{B}{x-2}$

$$x + 3 = A(x-2) + B(x-1)$$

$x = 1: \quad 4 = -A \quad \Longrightarrow \quad A = -4$

$x = 2: \quad 5 = B \quad \Longrightarrow \quad B = 5$

$$\int \frac{x+3}{x^2-3x+2}\, dx = \int \left(\frac{-4}{x-1} + \frac{5}{x-2} \right) dx = -4\ln|x-1| + 5\ln|x-2| + C$$

16. $\displaystyle \int \frac{x^2+3}{x^2-3x+2}\, dx = \int \left(1 + \frac{7}{x-2} - \frac{4}{x-1} \right) dx = x - 7\ln|x-2| - 4\ln|x-1| + C$

17. $\displaystyle \int \frac{dx}{(x-1)^3} = \int (x-1)^{-3}\, dx = -\frac{1}{2}(x-1)^{-2} + C = -\frac{1}{2(x-1)^2} + C$

18. $\displaystyle \int \frac{dx}{x^2+2x+2} = \int \frac{dx}{(x+1)^2+1} = \arctan(x+1) + C$

19. $\displaystyle \frac{x^2}{(x-1)^2(x+1)} = \frac{A}{x-1} + \frac{B}{(x-1)^2} + \frac{C}{x+1}$

$$x^2 = A(x-1)(x+1) + B(x+1) + C(x-1)^2$$

$x = 1: \quad 1 = 2B \quad \Longrightarrow \quad B = 1/2$

$x = -1: \quad 1 = 4C \quad \Longrightarrow \quad C = 1/4$

$$x = 0: \quad 0 = -A + B + C \quad \Longrightarrow \quad A = 3/4$$

$$\int \frac{x^2}{(x-1)^2(x+1)} \, dx = \int \left(\frac{3/4}{x-1} + \frac{1/2}{(x-1)^2} + \frac{1/4}{x+1} \right) dx$$

$$= \frac{3}{4} \ln|x-1| - \frac{1}{2(x-1)} + \frac{1}{4} \ln|x+1| + C$$

20. $\displaystyle \int \frac{2x-1}{(x+1)^2(x-2)^2} \, dx = \int \left[\frac{-1/3}{(x+1)^2} + \frac{1/3}{(x-2)^2} \right] dx = \frac{1}{3} \left(\frac{1}{x+1} - \frac{1}{x-2} \right) + C$

21. $\quad x^4 - 16 = (x^2 - 4)(x^2 + 4) = (x-2)(x+2)(x^2+4)$

$$\frac{1}{x^4 - 16} = \frac{A}{x-2} + \frac{B}{x+2} + \frac{Cx+D}{x^2+4}$$

$$1 = A(x+2)(x^2+4) + B(x-2)(x^2+4) + (Cx+D)(x^2-4)$$

$$x = 2: \quad 1 = 32A \quad \Longrightarrow \quad A = 1/32$$

$$x = -2: \quad 1 = -32B \quad \Longrightarrow \quad B = -1/32$$

$$x = 0: \quad 1 = 8A - 8B - 4D \quad \Longrightarrow \quad D = -1/8$$

$$x = 1: \quad 1 = 15A - 5B - 3C - 3D \quad \Longrightarrow \quad C = 0$$

$$\int \frac{dx}{x^4 - 16} = \int \left(\frac{1/32}{x-2} - \frac{1/32}{x+2} - \frac{1/8}{x^2+4} \right) dx$$

$$= \frac{1}{32} \ln|x-2| - \frac{1}{32} \ln|x+2| - \frac{1}{8} \left(\frac{1}{2} \arctan \frac{x}{2} \right) + C$$

$$= \frac{1}{32} \ln \left| \frac{x-2}{x+2} \right| - \frac{1}{16} \arctan \frac{x}{2} + C$$

22. Divide the denominator into the numerator: $\dfrac{3x^5 - 3x^2 + x}{x^3 - 1} = 3x^2 + \dfrac{x}{x^3 - 1}$

$$\int \left(3x^2 + \frac{x}{x^3 - 1} \right) dx = x^3 + \frac{1}{3} \int \left(\frac{1}{x-1} + \frac{1-x}{x^2+x+1} \right) dx$$

$$= x^3 + \frac{1}{3} \int \frac{1}{x-1} \, dx - \frac{1}{6} \int \frac{2x+1}{x^2+x+1} \, dx + \frac{1}{2} \int \frac{1}{x^2+x+1} \, dx$$

$$= x^3 + \frac{1}{3} \ln|x-1| - \frac{1}{6} \ln|x^2+x+1| + \frac{1}{2} \int \frac{1}{(x+\frac{1}{2})^2 + \frac{3}{4}} \, dx$$

$$= x^3 + \frac{1}{6} \ln \left(\frac{x^2 - 2x + 1}{x^2 + x + 1} \right) + \frac{1}{\sqrt{3}} \arctan \left[\frac{2}{\sqrt{3}} \left(x + \frac{1}{2} \right) \right] + C$$

23. $\dfrac{x^3 + 4x^2 - 4x - 1}{(x^2 + 1)^2} = \dfrac{Ax + B}{x^2 + 1} + \dfrac{Cx + D}{(x^2 + 1)^2}$

$x^3 + 4x^2 - 4x - 1 = (Ax + B)(x^2 + 1) + (Cx + D)$

$x = 0$: $-1 = B + D$ $\implies$ $D = -B - 1$

$\implies$ $B = 4, \ D = -5$

$x = 1$: $0 = 2A + 2B + C + D$ $\implies$ $6 = 4B + 2D$

$x = -1$: $6 = -2A + 2B - C + D$ $6 = -2A + 8 - C - 5$

$\implies A = 1,$

$C = -5$

$x = 2$: $15 = 10A + 5B + 2C + D$ $15 = 10A + 20 + 2C - 5$

$$\int \frac{x^3 + 4x^2 - 4x - 1}{(x^2 + 1)^2}\, dx = \int \left(\frac{x}{x^2 + 1} + \frac{4}{x^2 + 1} - \frac{5x}{(x^2 + 1)^2} - \frac{5}{(x^2 + 1)^2} \right) dx$$

(*)

$$= \frac{1}{2} \ln(x^2 + 1) + 4\arctan x + \frac{5}{2(x^2 + 1)} - 5 \int \frac{dx}{(x^2 + 1)^2}$$

For this last integral we set

$$\left\{ \begin{array}{c} x = \tan u \\ dx = \sec^2 u\, du \end{array} \right\} ; \int \frac{dx}{(x^2 + 1)^2} = \int \frac{\sec^2 u\, du}{(1 + \tan^2 u)^2} = \int \cos^2 u\, du$$

$$= \frac{1}{2} \int (1 + \cos 2u)\, du$$

$$= \tfrac{1}{2}\left(u + \tfrac{1}{2}\sin 2u\right) + C = \tfrac{1}{2}(u + \sin u \cos u) + C$$

$$= \frac{1}{2}\left(\arctan x + \frac{x}{1 + x^2} \right) + C.$$

Substituting this result in (*) and rearranging the terms, we get

$$\int \frac{x^3 + 4x^2 - 4x + 1}{(x^2 + 1)^2}\, dx = \frac{1}{2} \ln(x^2 + 1) + \frac{3}{2}\arctan x + \frac{5(1 - x)}{2(1 + x^2)} + C.$$

24. $$\int \frac{dx}{(x^2 + 16)^2} = \frac{1}{64} \int \cos^2 u\, du = \frac{1}{128}\left[u + \frac{\sin 2u}{2} \right] + C$$

$$= \frac{1}{128} \arctan\left(\frac{x}{4}\right) + \frac{1}{32} \cdot \frac{x}{(x^2 + 16)} + C$$

25.
$$\frac{1}{x^4 + 4} = \frac{Ax + B}{x^2 + 2x + 2} + \frac{Cx + D}{x^2 - 2x + 2} \qquad \text{(using the hint)}$$

$$1 = (Ax + B)(x^2 - 2x + 2) + (Cx + D)(x^2 + 2x + 2)$$

$$
\left.
\begin{array}{rl}
x = 0: & 1 = 2B + 2D \\
x = 1: & 1 = A + B + 5C + 5D \\
x = -1: & 1 = -5A + 5B - C + D \\
x = 2: & 1 = 4A + 2B + 20C + 10D
\end{array}
\right\}
\implies
\begin{array}{rl}
A = & 1/8 \\
B = & 1/4 \\
C = & -1/8 \\
D = & 1/4
\end{array}
$$

$$\int \frac{dx}{x^4 + 4} = \frac{1}{8} \int \frac{x + 2}{x^2 + 2x + 2}\, dx - \frac{1}{8} \int \frac{x - 2}{x^2 - 2x + 2}\, dx$$

$$= \frac{1}{8} \int \frac{x + 1}{x^2 + 2x + 2}\, dx + \frac{1}{8} \int \frac{dx}{(x + 1)^2 + 1} - \frac{1}{8} \int \frac{x - 1}{x^2 - 2x + 2}\, dx + \frac{1}{8} \int \frac{dx}{(x - 1)^2 + 1}$$

$$= \frac{1}{16} \ln(x^2 + 2x + 2) + \frac{1}{8} \arctan(x + 1) - \frac{1}{16} \ln(x^2 - 2x + 2) + \frac{1}{8} \arctan(x - 1) + C$$

$$= \frac{1}{16} \ln \left(\frac{x^2 + 2x + 2}{x^2 - 2x + 2} \right) + \frac{1}{8} \arctan(x + 1) + \frac{1}{8} \arctan(x - 1) + C$$

26. $\displaystyle \int \frac{dx}{x^4 + 16} = \frac{\sqrt{2}}{64} \int \left(\frac{2x + 4\sqrt{2}}{x^2 + 2\sqrt{2}x + 4} - \frac{2x - 4\sqrt{2}}{x^2 - 2\sqrt{2}x + 4} \right) dx$

$$= \frac{\sqrt{2}}{64} \int \left(\frac{2x + 2\sqrt{2}}{x^2 + 2\sqrt{2}x + 4} + \frac{2\sqrt{2}}{x^2 + 2\sqrt{2}x + 4} \right) dx$$

$$- \frac{\sqrt{2}}{64} \int \left(\frac{2x - 2\sqrt{2}}{x^2 - 2\sqrt{2}x + 4} - \frac{2\sqrt{2}}{x^2 - 2\sqrt{2}x + 4} \right) dx$$

$$= \frac{\sqrt{2}}{64} \ln|x^2 + 2\sqrt{2}x + 4| + \frac{1}{16} \cdot \frac{1}{\sqrt{2}} \arctan \left(\frac{x + \sqrt{2}}{\sqrt{2}} \right)$$

$$- \frac{\sqrt{2}}{64} \ln|x^2 - 2\sqrt{2}x + 4| + \frac{1}{16} \cdot \frac{1}{\sqrt{2}} \arctan \left(\frac{x - \sqrt{2}}{\sqrt{2}} \right) + C$$

$$= \frac{\sqrt{2}}{64} \ln \left(\frac{x^2 + 2\sqrt{2}x + 4}{x^2 - 2\sqrt{2}x + 4} \right) + \frac{\sqrt{2}}{32} \arctan \left(\frac{x + \sqrt{2}}{\sqrt{2}} \right) + \frac{\sqrt{2}}{32} \arctan \left(\frac{x - \sqrt{2}}{\sqrt{2}} \right) + C$$

27.
$$\frac{x - 3}{x^3 + x^2} = \frac{x - 3}{x^2(x + 1)} = \frac{A}{x} + \frac{B}{x^2} + \frac{C}{x + 1}$$

$$x - 3 = Ax(x + 1) + B(x + 1) + Cx^2$$

$$x = 0: \quad -3 = B$$

$$x = -1: \quad -4 = C$$

$$x = 1: \quad -2 = 2A + 2B + C \quad \Longrightarrow \quad A = 4$$

$$\int \frac{x-3}{x^3 + x^2}\, dx = 4\int \frac{1}{x}\, dx - 3\int \frac{1}{x^2}\, dx - 4\int \frac{1}{x+1}\, dx$$

$$= 4\ln|x| + \frac{3}{x} - 4\ln|x+1| + C = \frac{3}{x} + 4\ln\left|\frac{x}{x+1}\right| + C$$

28. $\displaystyle\int \frac{1}{(x-1)(x^2+1)^2}\, dx$

$$= \frac{1}{4}\int \left[\frac{1}{x-1} - \frac{x+1}{x^2+1} - \frac{2x+2}{(x^2+1)^2}\right] dx$$

$$= \frac{1}{4}\int \left[\frac{1}{x-1} - \frac{x}{x^2+1} - \frac{1}{x^2+1} - \frac{2x}{(x^2+1)^2} - \frac{2}{(x^2+1)^2}\right] dx$$

$$= \frac{1}{4}\ln|x-1| - \frac{1}{8}\ln(x^2+1) - \frac{1}{4}\arctan x + \frac{1}{4}\cdot\frac{1}{(x^2+1)} - \frac{1}{4}\left(\arctan x + \frac{x}{x^2+1}\right) + C$$

$$= \frac{1}{8}\ln\left(\frac{x^2 - 2x + 1}{x^2+1}\right) - \frac{1}{2}\arctan x - \frac{x-1}{4(x^2+1)} + C$$

29. $\displaystyle\frac{x+1}{x^3 + x^2 - 6x} = \frac{x+1}{x(x-2)(x+3)} = \frac{A}{x} + \frac{B}{x-2} + \frac{C}{x+3}$

$$x + 1 = A(x-2)(x+3) + Bx(x+3) + Cx(x-2)$$

$$x = 0: \quad 1 = -6A \quad \Longrightarrow \quad A = -1/6$$

$$x = 2: \quad 3 = 10B \quad \Longrightarrow \quad B = 3/10$$

$$x = -3: \quad -2 = 15C \quad \Longrightarrow \quad C = -2/15$$

$$\int \frac{x+1}{x^3 + x^2 - 6x}\, dx = -\tfrac{1}{6}\int \tfrac{1}{x}\, dx + \tfrac{3}{10}\int \tfrac{1}{x-2}\, dx - \tfrac{2}{15}\int \tfrac{1}{x+3}\, dx$$

$$= -\tfrac{1}{6}\ln|x| + \tfrac{3}{10}\ln|x-2| - \tfrac{2}{15}\ln|x+3| + C$$

30. $\displaystyle\int \frac{x^3 + x^2 + x + 3}{(x^2+1)(x^2+3)}\, dx = \int \left(\frac{1}{x^2+1} + \frac{x}{x^2+3}\right) dx = \arctan x + \frac{1}{2}\ln(x^2+3) + C$

31. $\displaystyle\int_0^2 \frac{x}{x^2 + 5x + 6}\, dx = \int_0^2 \frac{x}{(x+2)(x+3)}\, dx = \int_0^2 \left(\frac{3}{x+3} - \frac{2}{x+2}\right) dx$

$$= [3\ln|x+3| - 2\ln|x+2|]_0^2 = \ln\left(\frac{125}{108}\right)$$

32. $\displaystyle\int_1^3 \frac{1}{x^3 + x}\, dx = \int_1^3 \left(\frac{1}{x} - \frac{x}{x^2+1}\right) dx = \left[\ln|x| - \frac{1}{2}\ln(x^2+1)\right]_1^3 = \ln\left(\frac{3}{\sqrt{5}}\right)$

33.

$$\int_1^3 \frac{x^2 - 4x + 3}{x^3 + 2x^2 + x}\, dx = \int_1^3 \frac{x^2 - 4x + 3}{x(x+1)^2}\, dx = \int_1^3 \left(\frac{3}{x} - \frac{2}{x+1} - \frac{8}{(x+1)^2}\right) dx$$

$$= \left[3\ln|x| - 2\ln|x+1| + \frac{8}{x+1}\right]_1^3 = \ln\left(\frac{27}{4}\right) - 2$$

34. $\displaystyle \int_0^2 \frac{x^3}{(x^2+2)^2}\,dx = \int_0^2 \left(\frac{x}{x^2+2} - \frac{2x}{(x^2+2)^2} \right)\,dx = \left[\frac{1}{2}\ln(x^2+2) + \frac{1}{x^2+2} \right]_0^2 = \ln\sqrt{3} - \frac{1}{3}$

35. $\displaystyle \int \frac{\cos\theta}{\sin^2\theta - 2\sin\theta - 8}\,d\theta = \frac{1}{6}\int \frac{\cos\theta}{\sin\theta - 4}\,d\theta - \frac{1}{6}\int \frac{\cos\theta}{\sin\theta + 2}\,d\theta$

$$= \frac{1}{6}\ln|\sin\theta - 4| - \frac{1}{6}\ln|\sin\theta + 2| + C$$

$$= \frac{1}{6}\ln\left| \frac{\sin\theta - 4}{\sin\theta + 2} \right| + C$$

36. $\displaystyle \int \frac{e^t}{e^{2t} + 5e^t + 6}\,dt = \int \frac{e^t}{e^t + 2}\,dt - \int \frac{e^t}{e^t + 3}\,dt$

$$= \ln|e^t + 2| - \ln|e^t + 3| + C$$

$$= \ln\left| \frac{e^t + 2}{e^t + 3} \right| + C$$

37. $\displaystyle \int \frac{1}{t[(\ln t)^2 - 4]}\,dt = \frac{1}{4}\int \frac{1}{t(\ln t - 2)}\,dt - \frac{1}{4}\int \frac{1}{t(\ln t + 2)}\,dt$

$$= \frac{1}{4}\ln|\ln t - 2| - \frac{1}{4}\ln|\ln t + 2| + C$$

$$= \frac{1}{4}\ln\left| \frac{\ln t - 2}{\ln t + 2} \right| + C$$

38. $\displaystyle \int \frac{\sec^2\theta}{\tan^3\theta - \tan^2\theta}\,d\theta = \int \frac{\sec^2\theta}{\tan\theta - 1}\,d\theta - \int \frac{\sec^2\theta}{\tan\theta}\,d\theta - \int \frac{\sec^2\theta}{\tan^2\theta}\,d\theta$

$$= \ln|\tan\theta - 1| - \ln|\tan\theta| + \cot\theta + C$$

$$= \ln\left| \frac{\tan\theta - 1}{\tan\theta} \right| + \cot\theta + C$$

39. First we carry out the division: $\displaystyle \frac{u}{a+bu} = \frac{1}{b} - \frac{a/b}{a+bu}$.

$$\int \frac{u}{a+bu}\,du = \int \left(\frac{1}{b} - \frac{a/b}{a+bu} \right)\,du$$

$$= \frac{1}{b}u - \frac{a}{b^2}\ln|a+bu| + K$$

$$= \frac{1}{b^2}\left(a + bu - a\ln|a+bu| \right) + C, \quad C = K - \frac{a}{b^2}$$

40.

$$\int \frac{1}{u(a+bu)}\,du = \frac{1}{a}\int \frac{1}{u}\,du - \frac{1}{a}\int \frac{b}{a+bu}\,du$$

$$= \frac{1}{a}\left(\ln|u| - \ln|a+bu|\right) + C$$

$$= \frac{1}{a}\ln\left|\frac{u}{a+bu}\right| + C$$

41.

$$\int \frac{1}{u^2(a+bu)}\,du = -\frac{b}{a^2}\int \frac{1}{u}\,du + \frac{1}{a}\int \frac{1}{u^2}\,du + \frac{b}{a^2}\int \frac{b}{a+bu}\,du$$

$$= \frac{b}{a^2}\left(\ln|a+bu| - \ln|u|\right) - \frac{1}{au} + C$$

$$= \frac{b}{a^2}\ln\left|\frac{u}{a+bu}\right| - \frac{1}{au} + C$$

42.

$$\int \frac{1}{u(a+bu)^2}\,du = \frac{1}{a^2}\int \frac{1}{u}\,du - \frac{1}{a^2}\int \frac{b}{a+bu}\,du - \frac{b}{ab}\int \frac{b}{(a+bu)^2}\,du$$

$$= \frac{1}{a^2}\left(\ln|u| - \ln|a+bu|\right) + \frac{b}{ab(a+bu)} + C$$

$$= \frac{1}{a(a+bu)} - \frac{1}{a^2}\ln\left|\frac{a+bu}{u}\right| + C$$

43.

$$\int \frac{1}{a^2-u^2}\,du = \frac{1}{2a}\int \frac{1}{a+u}\,du + \frac{1}{2a}\int \frac{1}{a-u}\,du$$

$$= \frac{1}{2a}\left(\ln\left|\frac{a+u}{a-u}\right|\right) + C$$

44.

$$\int \frac{u}{a^2-u^2}\,du = -\frac{1}{2}\int \frac{1}{v}\,dv, \quad \text{where} \quad v = a^2 - u^2$$

$$= -\frac{1}{2}\ln v + C$$

$$= -\frac{1}{2}\ln|a^2-u^2| + C$$

45.

$$\int \frac{u^2}{a^2-u^2}\,du = \int \frac{a^2}{a^2-u^2}\,du - \int du$$

$$= a^2\left(\frac{1}{2a}\right)\left(\ln\left|\frac{a+u}{a-u}\right|\right) - u + C \quad \text{by Exercise 44}$$

$$= -u + \frac{a}{2}\left(\ln\left|\frac{a+u}{a-u}\right|\right) + C$$

46. If $ad = bc$, then $\dfrac{a}{b} = \dfrac{c}{d}$ and $\dfrac{1}{(a+bu)(c+du)} = \dfrac{1}{bd(a/b+u)^2}$.

Therefore

$$\int \frac{1}{(a+bu)(c+du)}\, du = \frac{1}{bd}\int \frac{1}{(a/b+u)^2}\, du$$

$$= -\frac{1}{bd(a/b+u)} + C = -\frac{1}{d(a+bu)} + C.$$

47. $\displaystyle\int \frac{1}{(a+bu)(c+du)}\, du = -\frac{1}{ad-bc}\int \frac{b}{a+bu}\, du + \frac{1}{ad-bc}\int \frac{d}{c+du}\, du$

$$= \frac{1}{ad-bc}\left(\ln\left|\frac{c+du}{a+bu}\right|\right) + C$$

48. Note that $\qquad y = \dfrac{1}{x^2-1} = \dfrac{1}{2}\left[\dfrac{1}{x-1} - \dfrac{1}{x+1}\right]$

and thus $\qquad \dfrac{d^0 y}{dx^0} = \left(\dfrac{1}{2}\right)(-1)^0\, 0!\left[\dfrac{1}{(x-1)^{0+1}} - \dfrac{1}{(x+1)^{0+1}}\right].$

The rest is a routine induction.

49. (a) $V = \displaystyle\int_0^{3/2} \pi\left(\frac{1}{\sqrt{4-x^2}}\right)^2 dx = \pi\int_0^{3/2}\frac{1}{4-x^2}\, dx = \frac{\pi}{4}\ln\left|\frac{2+x}{2-x}\right|\Big|_0^{3/2} = \frac{1}{4}\pi\ln 7$

(b) $V = \displaystyle\int_0^{3/2} 2\pi x\frac{1}{\sqrt{4-x^2}}\, dx = \pi\int_0^{3/2}\frac{2x}{\sqrt{4-x^2}}\, dx = -2\pi\left[\sqrt{4-x^2}\right]_0^{3/2} = \pi\left[4-\sqrt{7}\right]$

50.

$$\int x^3 \arctan x\, dx = \frac{x^4}{4}\arctan x - \frac{1}{4}\int \frac{x^4}{x^2+1}\, dx$$

$$\boxed{\begin{array}{ll} u = \arctan x & dv = x^3\, dx \\[4pt] du = \dfrac{1}{x^2+1}\, dx & v = \dfrac{x^4}{4} \end{array}} \qquad = \frac{x^4}{4}\arctan x - \frac{1}{4}\int\left(x^2 - 1 + \frac{1}{x^2+1}\right) dx$$

51.

$$A = \int_0^1 \frac{dx}{x^2+1} = [\arctan x]_0^1 = \frac{1}{4}\pi$$

$$\bar{x}A = \int_0^1 \frac{x}{x^2+1}\, dx = \frac{1}{2}\left[\ln(x^2+1)\right]_0^1 = \frac{1}{2}\ln 2$$

$$\bar{y}A = \int_0^1 \frac{dx}{2(x^2+1)^2} = \frac{1}{2}\left[\arctan x + \frac{x}{x^2+1}\right]_0^1 = \frac{1}{8}(\pi+2)$$

$$\bar{x} = \frac{2\ln 2}{\pi}; \quad \bar{y} = \frac{\pi+2}{2\pi}$$

52. $\overline{x}V_x = \overline{y}V_y = \displaystyle\int_0^1 \frac{\pi x}{(x^2+1)^2}\,dx = \frac{1}{2}\left[-\frac{1}{x^2+1}\right]_0^1 = \frac{1}{4}\pi$

(a) $V_x = \displaystyle\int_0^1 \frac{\pi}{(x^2+1)^2}\,dx = \pi\left[\arctan x + \frac{x}{x^2+1}\right]_0^1 = \frac{1}{4}\pi(\pi+2);\quad \overline{x} = \frac{1}{\pi+2}$

(b) $V_y = \displaystyle\int_0^1 \frac{2\pi x}{x^2+1}\,dx = \left[\ln(x^2+1)\right]_0^1 = \pi\ln 2;\quad \overline{y} = \frac{1}{4\ln 2}$

53.   (a) $\dfrac{6x^4+11x^3-2x^2-5x-2}{x^2(x+1)^3} = \dfrac{1}{x} - \dfrac{2}{x^2} + \dfrac{5}{x+1} - \dfrac{4}{(x+1)^3}$

(b) $-\dfrac{x^3+20x^2+4x+93}{(x^2+4)(x^2-9)} = \dfrac{1}{x^2+4} + \dfrac{3}{x+3} - \dfrac{4}{x-3}$

(c) $\dfrac{x^2+7x+12}{x(x^2+2x+4)} = \dfrac{2x-1}{x^2+2x+4} - \dfrac{3}{x}$

54. $\dfrac{2x^6 - 13x^5 + 23x^4 - 15x^3 + 40x^2 - 24x + 9}{x^5 - 6x^4 + 9x^3} = 2x - 1 + \dfrac{3}{x} - \dfrac{2}{x^2} + \dfrac{1}{x^3} + \dfrac{2}{(x-3)^2} - \dfrac{4}{x-3}$

55. $\dfrac{x^8 + 2x^7 + 7x^6 + 23x^5 + 10x^4 + 95x^3 - 19x^2 + 133x - 52}{x^6 + 2x^5 + 5x^4 + 16x^3 - 8x^2 + 32x - 48}$

$= x^2 + 2 + \dfrac{2}{x-1} + \dfrac{1}{x+3} + \dfrac{x}{(x^2+4)^2} + \dfrac{3}{x^2+4}$

56. $n=0: \displaystyle\int \frac{1}{x^2+2x}\,dx = \frac{1}{2}\ln\left|\frac{x}{x+2}\right| + C;\quad n=1: \int \frac{1}{x^2+2x+1}\,dx = \frac{-1}{x+1} + C$

$n=2: \displaystyle\int \frac{1}{x^2+2x+2}\,dx = \arctan(x+1) + C$

57.   (a)

(b) $A = \displaystyle\int_0^4 \frac{x}{x^2+5x+6}\,dx$

$= \displaystyle\int_0^4 \frac{x}{(x+2)(x+3)}\,dx$

$= \displaystyle\int_0^4 \left(\frac{3}{x+3} - \frac{2}{x+2}\right)dx$

$= \left[3\ln|x+3| - 2\ln|x+2|\right]_0^4 = 3\ln 7 - 5\ln 3$

58.   (a) $V_y = \displaystyle\int_0^4 \frac{2\pi x^2}{x^2+5x+6}\,dx = 2\pi\int_0^4 \left(1 + \frac{4}{x+2} - \frac{9}{x+3}\right)dx$

$= 2\pi\left[x + 4\ln|x+2| - 9\ln|x+3|\right]_0^4 = 2\pi(4 + 13\ln 3 - 9\ln 7)$

(b) $\displaystyle \overline{y} V_y = \int_0^4 \frac{\pi x^3}{(x^2 + 5x + 6)^2}\, dx = \pi \int_0^4 \left[\frac{28}{x+2} - \frac{8}{(x+2)^2} - \frac{27}{x+3} - \frac{27}{(x+3)^2} \right] dx$

$\displaystyle = \pi \left[28 \ln|x+2| + \frac{8}{x+2} - 27 \ln|x+3| + \frac{27}{x+3} \right]_0^4$

$\displaystyle = \pi \left[\frac{4}{3} + \frac{27}{7} - 13 + 55 \ln 3 - 27 \ln 7 \right]$

$\displaystyle \overline{y} = \frac{-\frac{164}{21} + 55 \ln 3 - 27 \ln 7}{2(4 + 13 \ln 3 - 9 \ln 7)}$

59. (a)

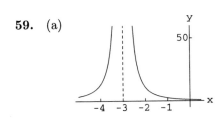

(b) $\displaystyle A = \int_{-2}^9 \frac{9-x}{(x+3)^2}\, dx$

$\displaystyle = \int_{-2}^9 \left(\frac{12}{(x+3)^2} - \frac{1}{x+3} \right) dx$

$\displaystyle = \left[-\ln|x+3| - \frac{12}{x+3} \right]_{-2}^9 = 11 - \ln 12$

60. (a) $\displaystyle V_x = \int_{-2}^9 \pi \frac{(9-x)^2}{(x+3)^4}\, dx = \pi \int_{-2}^9 \left[\frac{1}{(x+3)^2} - \frac{24}{(x+3)^3} + \frac{144}{(x+3)^4} \right] dx$

$\displaystyle = \pi \left[\frac{-1}{x+3} + \frac{12}{(x+3)^2} - \frac{48}{(x+3)^3} \right]_{-2}^9 = \frac{1331}{36} \pi$

(b) $\displaystyle \overline{x} V_x = \int_{-2}^9 \pi x \frac{(9-x)^2}{(x+3)^4}\, dx = \pi \int_{-2}^9 \left[\frac{1}{x+3} - \frac{27}{(x+3)^2} + \frac{216}{(x+3)^2} - \frac{432}{(x+3)^4} \right] dx$

$\displaystyle = \pi \left[\ln|x+3| + \frac{27}{x+3} - \frac{108}{(x+3)^2} + \frac{144}{(x+3)^3} \right]_{-2}^9 = \pi \left(\ln 12 - \frac{737}{12} \right)$

$\displaystyle \overline{x} = \frac{36}{1331} \left(\ln 12 - \frac{737}{12} \right)$

SECTION 8.6

1. $\displaystyle \left\{ \begin{array}{l} x = u^2 \\ dx = 2u\, du \end{array} \right\} ; \quad \int \frac{dx}{1 - \sqrt{x}} = \int \frac{2u\, du}{1 - u} = 2 \int \left(\frac{1}{1-u} - 1 \right) du$

$\displaystyle = -2(\ln|1-u| + u) + C = -2(\sqrt{x} + \ln|1 - \sqrt{x}|) + C$

2.
$$\left\{ \begin{array}{l} u = \sqrt{x} \\ du = \frac{1}{2\sqrt{x}}\,dx \end{array} \right\};$$

$$\int \frac{\sqrt{x}}{1+x}\,dx = \int \frac{2u^2}{1+u^2}\,du = \int \left(2 - \frac{2}{1+u^2}\right) du$$

$$= 2u - 2\arctan u + C$$

$$= 2\sqrt{x} - 2\arctan \sqrt{x} + C$$

3.
$$\left\{ \begin{array}{l} u^2 = 1 + e^x \\ 2u\,du = e^x\,dx \end{array} \right\};$$

$$\int \sqrt{1+e^x}\,dx = \int u \cdot \frac{2u\,du}{u^2 - 1} = 2\int \left(1 + \frac{1}{u^2-1}\right) du$$

$$= 2\int \left(1 + \frac{1}{2}\left[\frac{1}{u-1} - \frac{1}{u+1}\right]\right) du$$

$$= 2u + \ln|u-1| - \ln|u+1| + C$$

$$= 2\sqrt{1+e^x} + \ln\left[\frac{\sqrt{1+e^x}-1}{\sqrt{1+e^x}+1}\right] + C$$

$$= 2\sqrt{1+e^x} + \ln\left[\frac{\left(\sqrt{1+e^x}-1\right)^2}{e^x}\right] + C$$

$$= 2\sqrt{1+e^x} + 2\ln\left(\sqrt{1+e^x}-1\right) - x + C$$

4.
$$\left\{ \begin{array}{l} u^3 = x \\ 3u^2\,du = dx \end{array} \right\};$$

$$\int \frac{dx}{x(x^{1/3}-1)} = \int \frac{3\,du}{u(u-1)} = 3\int \left(\frac{1}{u-1} - \frac{1}{u}\right) du$$

$$= 3\ln|u-1| - 3\ln|u| + C = 3\ln|u-1| - \ln|u^3| + C$$

$$= 3\ln|x^{1/3}-1| - \ln|x| + C$$

5. (a)
$$\left\{ \begin{array}{l} u^2 = 1 + x \\ 2u\,du = dx \end{array} \right\};$$

$$\int x\sqrt{1+x}\,dx = \int (u^2-1)(u)2u\,du$$

$$= \int (2u^4 - 2u^2)\,du$$

$$= \tfrac{2}{5}u^5 - \tfrac{2}{3}u^3 + C$$

$$= \tfrac{2}{5}(x+1)^{5/2} - \tfrac{2}{3}(x+1)^{3/2} + C$$

(b)
$$\left\{ \begin{array}{l} u = 1 + x \\ du = dx \end{array} \right\};$$

$$\int x\sqrt{1+x}\,dx = \int (u-1)\sqrt{u}\,du = \int \left(u^{3/2} - u^{1/2}\right) du$$

$$= \tfrac{2}{5}u^{5/2} - \tfrac{2}{3}u^{3/2} + C$$

$$= \tfrac{2}{5}(1+x)^{5/2} - \tfrac{2}{3}(1+x)^{3/2} + C$$

6. (a)
$$\left\{ \begin{array}{l} u^2 = 1 + x \\ 2u\,du = dx \end{array} \right\};$$

$$\int x^2\sqrt{1+x}\,dx = \int (u^2-1)^2 u \cdot 2u\,du$$

$$= \int (2u^6 - 4u^4 + 2u^2)\,du$$

$$= \frac{2}{7}u^7 - \frac{4}{5}u^5 + \frac{2}{3}u^3 + C$$

$$= \frac{2}{7}(1+x)^{7/2} - \frac{4}{5}(1+x)^{5/2} + \frac{2}{3}(1+x)^{3/2} + C$$

(b) $\begin{Bmatrix} u = 1 + x \\ du = dx \end{Bmatrix}$; $\quad \int x^2 \sqrt{1+x}\, dx = \int (u-1)^2 \sqrt{u}\, du = \int (u^2 - 2u + 1)\sqrt{u}\, du$

$$= \frac{2}{7}u^{7/2} - \frac{4}{5}u^{5/2} + \frac{2}{3}u^{3/2} + C$$

$$= \frac{2}{7}(1+x)^{7/2} - \frac{4}{5}(1+x)^{5/2} + \frac{2}{3}(1+x)^{3/2} + C$$

7. $\begin{Bmatrix} u^2 = x - 1 \\ 2u\, du = dx \end{Bmatrix}$; $\quad \int (x+2)\sqrt{x-1}\, dx = \int (u+3)(u)2u\, du$

$$= \int (2u^4 + 6u^2)\, du$$

$$= \tfrac{2}{5}u^5 + 2u^3 + C$$

$$= \tfrac{2}{5}(x-1)^{5/2} + 2(x-1)^{3/2} + C$$

8. $\begin{Bmatrix} u = x + 2 \\ du = dx \end{Bmatrix}$; $\quad \int (x-1)\sqrt{x+2}\, dx = \int (u-3)\sqrt{u}\, du$

$$= \frac{2}{5}u^{5/2} - 2u^{3/2} + C$$

$$= \frac{2}{5}(x+2)^{5/2} - 2(x+2)^{3/2} + C$$

9. $\begin{Bmatrix} u^2 = 1 + x^2 \\ 2u\, du = 2x\, dx \end{Bmatrix}$; $\quad \int \frac{x^3}{(1+x^2)^3}\, dx = \int \frac{x^2}{(1+x^2)^3} x\, dx = \int \frac{u^2 - 1}{u^6} u\, du$

$$= \int (u^{-3} - u^{-5})\, du = \frac{1}{2}u^{-2} + \frac{1}{4}u^{-4} + C$$

$$= \frac{1}{4(1+x^2)^2} - \frac{1}{2(1+x^2)} + C$$

$$= -\frac{1 + 2x^2}{4(1+x^2)^2} + C$$

10. $\begin{Bmatrix} u = 1 + x \\ du = dx \end{Bmatrix}$; $\quad \int x(1+x)^{1/3}\, dx = \int (u-1)u^{1/3}\, du$

$$= \frac{3}{7}u^{7/3} - \frac{3}{4}u^{4/3} + C$$

$$= \frac{3}{7}(1+x)^{7/3} - \frac{3}{4}(1+x)^{4/3} + C$$

11. $\begin{Bmatrix} u^2 = x \\ 2u\, du = dx \end{Bmatrix}$; $\quad \int \frac{\sqrt{x}}{\sqrt{x}-1}\, dx = \int \left(\frac{u}{u-1}\right)2u\, du = 2\int \left(u + 1 + \frac{1}{u-1}\right)du$

$$= u^2 + 2u + 2\ln|u-1| + C$$

$$= x + 2\sqrt{x} + 2\ln|\sqrt{x}-1| + C$$

12.

$$\left\{ \begin{array}{l} u = x + 1 \\ du = dx \end{array} \right\}; \qquad \int \frac{x}{\sqrt{1+x}}\, dx = \int \frac{u-1}{\sqrt{u}}\, du$$

$$= \frac{2}{3} u^{3/2} - 2u^{1/2} + C$$

$$= \frac{2}{3}(x+1)^{3/2} - 2(x+1)^{1/2} + C$$

13.

$$\left\{ \begin{array}{l} u^2 = x - 1 \\ 2u\, du = dx \end{array} \right\}; \qquad \int \frac{\sqrt{x-1}+1}{\sqrt{x-1}-1}\, dx = \int \frac{u+1}{u-1} 2u\, du = \int \left(2u + 4 + \frac{4}{u-1} \right) du$$

$$= u^2 + 4u + 4\ln|u-1| + C$$

$$= x - 1 + 4\sqrt{x-1} + 4\ln|\sqrt{x-1}-1| + C$$

(absorb -1 in C)

$$= x + 4\sqrt{x-1} + 4\ln|\sqrt{x-1}-1| + C$$

14.

$$\left\{ \begin{array}{l} u = e^x \\ du = e^x\, dx \end{array} \right\}; \qquad \int \frac{1-e^x}{1+e^x}\, dx = \int \frac{1-u}{u(1+u)}\, du$$

$$= \int \left(\frac{1}{u} - \frac{2}{1+u} \right) du$$

$$= \ln|u| - 2\ln|1+u| + C$$

$$= x - 2\ln(1+e^x) + C$$

15.

$$\left\{ \begin{array}{l} u^2 = 1 + e^x \\ 2u\, du = e^x\, dx \end{array} \right\}; \qquad \int \frac{dx}{\sqrt{1+e^x}} = \int \left(\frac{1}{u} \right) \frac{2u\, du}{u^2-1} = \int \left[\frac{1}{u-1} - \frac{1}{u+1} \right] du$$

$$= \ln|u-1| - \ln|u+1| + C$$

$$= \ln \left[\frac{\sqrt{1+e^x}-1}{\sqrt{1+e^x}+1} \right] + C$$

$$= \ln \left[\frac{(\sqrt{1+e^x}-1)^2}{e^x} \right] + C$$

$$= 2\ln(\sqrt{1+e^x}-1) - x + C$$

16.

$$\left\{ \begin{array}{l} u = e^{-x} \\ du = -e^{-x}\, dx \end{array} \right\}; \qquad \int \frac{dx}{1+e^{-x}} = -\int \frac{du}{u(1+u)}$$

$$= -\int \left(\frac{1}{u} - \frac{1}{1+u} \right) du$$

$$= -\ln|u| + \ln|1+u| + C$$

$$= \ln \left| \frac{1+u}{u} \right| + C$$

$$= \ln \left| \frac{1+e^x}{e^{-x}} \right| + C = \ln(1+e^x) + C$$

17. $\left\{\begin{array}{l} u^2 = x + 4 \\ 2u\,du = dx \end{array}\right\}$; $\displaystyle\int \frac{x}{\sqrt{x+4}}\,dx = \int \frac{u^2 - 4}{u}2u\,du = \int (2u^2 - 8)\,du$

$$= \tfrac{2}{3}u^3 - 8u + C$$

$$= \tfrac{2}{3}(x+4)^{3/2} - 8(x+4)^{1/2} + C$$

$$= \tfrac{2}{3}(x-8)\sqrt{x+4} + C$$

18. $\left\{\begin{array}{l} u^2 = x - 2 \\ 2u\,du = dx \end{array}\right\}$; $\displaystyle\int \frac{x+1}{x\sqrt{x-2}}\,dx = 2\int \frac{u^2 + 3}{u^2 + 2}\,du = 2\int \left(1 + \frac{1}{u^2 + 2}\right)\,du$

$$= 2u + \sqrt{2}\arctan\left(\frac{u}{\sqrt{2}}\right) + C$$

$$= 2\sqrt{x-2} + \sqrt{2}\arctan\left(\sqrt{\frac{x-2}{2}}\right) + C$$

19. $\left\{\begin{array}{l} u^2 = 4x + 1 \\ 2u\,du = 4\,dx \end{array}\right\}$; $\displaystyle\int 2x^2(4x+1)^{-5/2}\,dx = \int 2\left(\frac{u^2 - 1}{4}\right)^2 (u^{-5})\frac{u}{2}\,du$

$$= \frac{1}{16}\int (1 - 2u^{-2} + u^{-4})\,du = \frac{1}{16}u + \frac{1}{8}u^{-1} - \frac{1}{48}u^{-3} + C$$

$$= \tfrac{1}{16}(4x+1)^{1/2} + \tfrac{1}{8}(4x+1)^{-1/2} - \tfrac{1}{48}(4x+1)^{-3/2} + C$$

20. $\left\{\begin{array}{l} u = x - 1 \\ du = dx \end{array}\right\}$; $\displaystyle\int x^2\sqrt{x-1}\,dx = \int (u+1)^2\sqrt{u}\,du$

$$= \frac{2}{7}u^{7/2} + \frac{4}{5}u^{5/2} + \frac{2}{3}u^{3/2} + C$$

$$= \frac{2}{7}(x-1)^{7/2} + \frac{4}{5}(x-1)^{5/2} + \frac{2}{3}(x-1)^{3/2} + C$$

21. $\left\{\begin{array}{l} u^2 = ax + b \\ 2u\,du = a\,dx \end{array}\right\}$; $\displaystyle\int \frac{x}{(ax+b)^{3/2}}\,dx = \int \frac{\frac{u^2 - b}{a}}{u^3}\frac{2u}{a}\,du$

$$= \frac{2}{a^2}\int (1 - bu^{-2})\,du$$

$$= \frac{2}{a^2}(u + bu^{-1}) + C = \frac{2u^2 + 2b}{a^2 u} + C$$

$$= \frac{4b + 2ax}{a^2\sqrt{ax+b}} + C$$

22. $\left\{\begin{array}{l} u = ax + b \\ du = a\,dx \end{array}\right\}$ $\displaystyle\int \frac{x}{\sqrt{ax+b}}\,dx = \frac{1}{a^2}\int \frac{u - b}{\sqrt{u}}\,du$

$$= \frac{1}{a^2}\left[\frac{2}{3}u^{3/2} - 2bu^{1/2}\right] + C$$

$$= \frac{2}{3a^2}(ax - 2b)\sqrt{ax+b} + C$$

23.

$$\begin{cases} u = \tan(x/2), & dx = \dfrac{2}{1+u^2}\,du \\[2mm] \sin x = \dfrac{2u}{1+u^2}, & \cos x = \dfrac{1-u^2}{1+u^2} \end{cases};$$

$$\int \frac{1}{1+\cos x - \sin x}\,dx = \int \frac{1}{1+\dfrac{1-u^2}{1+u^2} - \dfrac{2u}{1+u^2}} \cdot \frac{2}{1+u^2}\,du$$

$$= \int \frac{1}{1-u}\,du$$

$$= -\ln|1-u| + C = -\ln\left|1 - \tan\left(\frac{x}{2}\right)\right| + C$$

24.

$$\begin{cases} u = \tan(x/2), & dx = \dfrac{2}{1+u^2}\,du \\[2mm] \cos x = \dfrac{1-u^2}{1+u^2} \end{cases};$$

$$\int \frac{1}{2+\cos x}\,dx = \int \frac{1}{2+\frac{1-u^2}{1+u^2}} \cdot \frac{2}{1+u^2}\,du = \int \frac{2}{2+2u^2+1-u^2}\,du = \int \frac{2}{3+u^2}\,du$$

$$= \frac{2}{\sqrt{3}} \arctan\left(\frac{u}{\sqrt{3}}\right) + C = \frac{2}{\sqrt{3}} \arctan\left(\frac{\tan(x/2)}{\sqrt{3}}\right) + C$$

25.

$$\begin{cases} u = \tan(x/2), & dx = \dfrac{2}{1+u^2}\,du \\[2mm] \sin x = \dfrac{2u}{1+u^2} \end{cases};$$

$$\int \frac{1}{2+\sin x}\,dx = \int \frac{1}{2+\dfrac{2u}{1+u^2}} \cdot \frac{2}{1+u^2}\,du$$

$$= \int \frac{1}{u^2+u+1}\,du = \int \frac{1}{\left(u+\frac{1}{2}\right)^2 + \left(\frac{\sqrt{3}}{2}\right)^2}\,du$$

$$= \frac{2}{\sqrt{3}} \arctan\left(\frac{u+\frac{1}{2}}{\frac{\sqrt{3}}{2}}\right) + C$$

$$= \frac{2}{\sqrt{3}} \arctan\left[\frac{1}{\sqrt{3}}\left(2\tan(x/2)+1\right)\right] + C$$

26.

$$\begin{cases} u = \tan(x/2), & dx = \dfrac{2}{1+u^2}\,du \\[2mm] \sin x = \dfrac{2u}{1+u^2} \end{cases};$$

$$\int \frac{\sin x}{1 + \sin^2 x}\, dx = \int \frac{\frac{2u}{1+u^2}}{1 + \frac{4u^2}{(1+u^2)^2}} \cdot \frac{2}{1+u^2}\, du$$

$$\left\{ \begin{aligned} v &= u^2 \\ dv &= 2u\, du \end{aligned} \right\} = \int \frac{4u}{1 + 6u^2 + u^4}\, du$$

$$= \int \frac{2}{1 + 6v + v^2}\, dv = \int \frac{2}{(v+3)^2 - 8}\, dv$$

$$= \frac{2}{\sqrt{8}} \ln \left| \frac{v + 3 - \sqrt{8}}{\sqrt{(v+3)^2 - 8}} \right| + C$$

$$= \frac{2}{\sqrt{8}} \ln \left| \frac{\tan^2(x/2) + 3 - \sqrt{8}}{\sqrt{[\tan^2(x/2) + 3]^2 - 8}} \right| + C$$

27.

$$\left\{ \begin{aligned} u &= \tan(x/2), & dx &= \frac{2}{1+u^2}\, du \\ \sin x &= \frac{2u}{1+u^2}, & \tan x &= \frac{2u}{1-u^2} \end{aligned} \right\};$$

$$\int \frac{1}{\sin x + \tan x}\, dx = \int \frac{1}{\frac{2u}{1+u^2} + \frac{2u}{1-u^2}} \cdot \frac{2}{1+u^2}\, du$$

$$= \int \frac{1 - u^2}{2u}\, du = \frac{1}{2} \int \left(\frac{1}{u} - u \right) du$$

$$= \tfrac{1}{2} \left(\ln |u| - \tfrac{1}{2} u^2 \right) + C = \tfrac{1}{2} \ln |\tan(x/2)| - \tfrac{1}{4} [\tan(x/2)]^2 + C$$

28.

$$\int \frac{1}{1 + \sin x + \cos x}\, dx = \int \frac{2}{1 + \frac{2u}{1+u^2} + \frac{1-u^2}{1+u^2}} \cdot \frac{2}{1+u^2}\, du$$

$$= \int \frac{2}{2 + 2u}\, du = \int \frac{du}{1 + u}$$

$$= \int \ln |1 + u| + C = \ln \left| 1 + \tan \left(\frac{x}{2} \right) \right| + C$$

29.

$$\left\{ \begin{aligned} u &= \tan(x/2), & dx &= \frac{2}{1+u^2}\, du \\ \sin x &= \frac{2u}{1+u^2}, & \cos x &= \frac{1-u^2}{1+u^2} \end{aligned} \right\};$$

$$\int \frac{1 - \cos x}{1 + \sin x}\, dx = \int \frac{1 - \dfrac{1 - u^2}{1 + u^2}}{1 + \dfrac{2u}{1 + u^2}} \cdot \frac{2}{1 + u^2}\, du$$

$$= \int \frac{4u^2}{(1 + u^2)(u + 1)^2}\, du$$

$$= \int \left[\frac{2u}{1 + u^2} - \frac{2}{u + 1} + \frac{2}{(u + 1)^2} \right] du$$

$$= \ln(u^2 + 1) - 2\ln|u + 1| - \frac{2}{u + 1} + C$$

$$= \ln\left[\frac{u^2 + 1}{(u + 1)^2} - \frac{2}{u + 1} \right] + C$$

$$= \ln\left[\frac{\tan^2(x/2) + 1}{(\tan(x/2) + 1)^2} \right] - \frac{2}{\tan(x/2) + 1} + C = \ln\left| \frac{1}{1 + \sin x} \right| - \frac{2}{\tan(x/2) + 1} + C$$

30.

$$\int \frac{1}{5 + 3\sin x}\, dx = \int \frac{1}{5 + \frac{6u}{1 + u^2}} \cdot \frac{2}{1 + u^2}\, du$$

$$= \int \frac{2}{5 + 5u^2 + 6u}\, du = \frac{2}{5} \int \frac{du}{(u + \frac{3}{5})^2 + \frac{16}{25}}$$

$$= \frac{2}{5} \cdot \frac{5}{4} \arctan\left(\frac{u + 3/5}{4/5} \right) + C$$

$$= \frac{1}{2} \arctan\left[\frac{5}{4}\left(\tan\left(\frac{x}{2} \right) + \frac{3}{5} \right) \right] + C$$

31. $\left\{ \begin{array}{l} u^2 = x \\ 2u\, du = dx \end{array} \right\}$; $\displaystyle \int \frac{x^{3/2}}{x + 1}\, dx = \int \frac{u^3}{u^2 + 1}\, 2u\, du$

$$= \int \frac{2u^4}{u^2 + 1}\, du = \int \left[2u^2 - 2 + \frac{2}{u^2 + 1} \right] du$$

$$= \frac{2}{3}u^3 - 2u + 2\arctan u + C = \frac{2}{3}x^{3/2} - 2x^{1/2} + 2\arctan x^{1/2} + C$$

$$\int_0^4 \frac{x^{3/2}}{x + 1}\, dx = \left[\frac{2}{3}x^{3/2} - 2x^{1/2} + 2\arctan x^{1/2} \right]_0^4 = \frac{4}{3} + 2\arctan 2$$

32. $\left\{ \begin{array}{l} u^2 = x \\ 2u\, du = dx \end{array} \right\}$ $\displaystyle \int_0^4 \frac{1}{1 + \sqrt{x}}\, dx = \int_0^2 \frac{2u}{1 + u}\, du$

$$= \int_0^2 \left(2 - \frac{2}{1 + u} \right) du = [2u - 2\ln|1 + u|]_0^2$$

$$= 4 - 2\ln 3$$

33.
$$\int_0^{\pi/2} \frac{\sin 2x}{2 + \cos x}\, dx = \int_0^{\pi/2} \frac{2 \sin x \cos x}{2 + \cos x}\, dx$$
$$= \int_0^1 \frac{2u}{2 + u}\, du \qquad [u = \cos x, \quad du = -\sin x\, dx]$$
$$= \int_0^1 \left(2 - \frac{4}{2 + u} \right) du$$
$$= [2u - 4\ln|2 + u|]_0^1 = 2 + 4\ln\left(\tfrac{2}{3}\right)$$

34.
$$\left\{ \begin{array}{l} u = \tan(x/2), \quad dx = \dfrac{2}{1 + u^2}\, du \\[2mm] \sin x = \dfrac{2u}{1 + u^2} \end{array} \right\};$$

$$\int_0^{\pi/2} \frac{1}{1 + \sin x}\, dx = \int_0^1 \frac{1}{1 + \frac{2u}{1+u^2}} \cdot \frac{2}{1 + u^2}\, du$$
$$= \int_0^1 \frac{2}{1 + 2u + u^2}\, du = \int_0^1 \frac{2}{(u + 1)^2}\, du$$
$$= -\left[\frac{2}{u + 1} \right]_0^1 = 1$$

35.
$$\left\{ \begin{array}{ll} u = \tan(x/2), & dx = \dfrac{2}{1 + u^2}\, du \\[2mm] \sin x = \dfrac{2u}{1 + u^2}, & \cos x = \dfrac{1 - u^2}{1 + u^2} \end{array} \right\};$$

$$\int \frac{1}{\sin x - \cos x - 1}\, dx = \int \frac{1}{\frac{2u}{1+u^2} - \frac{1 - u^2}{1 + u^2} - 1} \cdot \frac{2}{1 + u^2}\, du$$
$$= \int \frac{1}{u - 1}\, du$$
$$= \ln|u - 1| + C = \ln|\tan(x/2) - 1| + C$$
$$\int_0^{\pi/3} \frac{1}{\sin x - \cos x - 1}\, dx = [\ln|\tan(x/2) - 1|]_0^{\pi/3} = \ln\left(\frac{\sqrt{3} - 1}{\sqrt{3}} \right)$$

36.
$$\left\{ \begin{array}{l} u^2 = x \\[1mm] 2u\, du = dx \end{array} \right\} \quad \int_0^1 \frac{\sqrt{x}}{1 + \sqrt{x}}\, dx = \int_0^1 \frac{2u^2}{1 + u}\, du$$
$$= \int_0^1 \left(2u - 2 + \frac{2}{1 + u} \right) du$$
$$= [u^2 - 2u + 2\ln|1 + u|]_0^1 = 2\ln 2 - 1$$

37.
$$\left\{ \begin{array}{ll} u = \tan(x/2), & dx = \dfrac{2}{1 + u^2}\, du \\[2mm] \sin x = \dfrac{2u}{1 + u^2}, & \cos x = \dfrac{1 - u^2}{1 + u^2} \end{array} \right\};$$

$$\int \sec x \, dx = \int \frac{1}{\cos x} \, dx = \int \frac{1}{\frac{1-u^2}{1+u^2}} \cdot \frac{2}{1+u^2} \, du$$

$$= 2 \int \frac{1}{1-u^2} \, du = 2 \int \left[\frac{1/2}{1-u} + \frac{1/2}{1+u} \right] du$$

$$= \int \left[\frac{1}{1-u} + \frac{1}{1+u} \right] du = - \ln|1-u| + \ln|1+u| + C$$

$$= \ln \left| \frac{1 + \tan(x/2)}{1 - \tan(x/2)} \right| + C$$

38. (a)

$$\left\{ \begin{array}{l} u = \sin x \\ du = \cos x \, dx \end{array} \right\} \quad \int \sec x \, dx = \int \frac{du}{1-u^2} = \frac{1}{2} \int \left(\frac{1}{1-u} + \frac{1}{1+u} \right) du$$

$$= -\frac{1}{2} \ln|1-u| + \frac{1}{2} \ln|1+u| + C$$

$$= \ln \sqrt{\frac{1 + \sin x}{1 - \sin x}} + C$$

(b) $\sqrt{\dfrac{1 + \sin x}{1 - \sin x}} = \sqrt{\dfrac{(1 + \sin x)^2}{1 - \sin^2 x}} = \left| \dfrac{1 + \sin x}{\cos x} \right| = |\sec x + \tan x|$

39.

$$\int \csc x \, dx = \int \frac{\sin x}{\sin^2 x} \, dx = \int \frac{\sin x}{1 - \cos^2 x} \, dx$$

$$= - \int \frac{1}{1-u^2} \, du \qquad [u = \cos x, \quad du = - \sin x \, dx]$$

$$= \frac{1}{2} \int \left[\frac{1}{u-1} - \frac{1}{u+1} \right] du$$

$$= \frac{1}{2} \left[\ln|u-1| - \ln|u+1| \right] + C = \ln \sqrt{\frac{1 - \cos x}{1 + \cos x}} + C$$

40. $\quad \cosh\left(\dfrac{x}{2}\right) = \dfrac{1}{\operatorname{sech}\left(\frac{x}{2}\right)} = \dfrac{1}{\sqrt{1 - \tanh^2(\frac{x}{2})}} = \dfrac{1}{\sqrt{1 - u^2}}$

$\sinh\left(\dfrac{x}{2}\right) = \sqrt{\cosh^2\left(\dfrac{x}{2}\right) - 1} = \dfrac{u}{\sqrt{1 - u^2}}$

$\sinh x = 2 \sinh\left(\dfrac{x}{2}\right) \cosh\left(\dfrac{x}{2}\right) = \dfrac{2u}{1 - u^2}, \quad \cosh x = \cosh^2\left(\dfrac{x}{2}\right) + \sinh^2\left(\dfrac{x}{2}\right) = \dfrac{1 + u^2}{1 - u^2}$

$du = \dfrac{1}{2} \operatorname{sech}^2\left(\dfrac{x}{2}\right) dx \implies dx = 2 \cosh^2\left(\dfrac{x}{2}\right) du = \dfrac{2}{1 - u^2} \, du$

41.

$$\left\{ \begin{array}{l} u = \tanh(x/2), \qquad dx = \dfrac{2}{1-u^2} \, du \\[2mm] \cosh x = \dfrac{1+u^2}{1-u^2}, \qquad \operatorname{sech} x = \dfrac{1-u^2}{1+u^2} \end{array} \right\} ;$$

$$\int \text{sech}\, x\, dx = \int \frac{1-u^2}{1+u^2} \cdot \frac{2}{1-u^2}\, du$$

$$= \int \frac{2}{1+u^2}\, du = 2\arctan u + C = 2\arctan\left(\tanh(x/2)\right) + C$$

42. $\displaystyle \int \frac{1}{1+\cosh x}\, dx = \int \frac{1}{1+\frac{1+u^2}{1-u^2}} \cdot \frac{2}{1-u^2}\, du = \int du = u + C = \tanh\left(\frac{x}{2}\right) + C$

43.
$$\left\{ \begin{array}{c} u = \tanh(x/2), \quad dx = \dfrac{2}{1-u^2}\, du \\[2mm] \sinh x = \dfrac{2u}{1-u^2}, \quad \cosh x = \dfrac{1+u^2}{1-u^2} \end{array} \right\};$$

$$\int \frac{1}{\sinh x + \cosh x}\, dx = \int \frac{1}{\dfrac{2u}{1-u^2} + \dfrac{1+u^2}{1-u^2}} \cdot \frac{2}{1-u^2}\, du$$

$$= \int \frac{2}{(1+u)^2}\, du = \frac{-2}{u+1} + C = \frac{-2}{\tanh(x/2)+1} + C$$

44.
$$\int \frac{1-e^x}{1+e^x}\, dx = \int \frac{1-\cosh x - \sinh x}{1+\cosh x + \sinh x}\, dx$$

$$= \int \frac{1 - \dfrac{1+u^2}{1-u^2} - \dfrac{2u}{1-u^2}}{1 + \dfrac{1+u^2}{1-u^2} + \dfrac{2u}{1-u^2}} \cdot \frac{2}{1-u^2}\, du$$

$$= \int -\frac{2u}{1-u^2}\, du = \ln|1-u^2| + C = \ln\left|1 - \tanh^2\left(\frac{x}{2}\right)\right| + C$$

SECTION 8.7

1. (a) $\quad L_{12} = \frac{12}{12}[0 + 1 + 4 + 9 + 16 + 25 + 36 + 49 + 64 + 81 + 100 + 121] = 506$

(b) $\quad R_{12} = \frac{12}{12}[1 + 4 + 9 + 16 + 25 + 36 + 49 + 64 + 81 + 100 + 121 + 144] = 650$

(c) $\quad M_6 = \frac{12}{6}[1 + 9 + 25 + 49 + 81 + 121] = 572$

(d) $\quad T_{12} = \frac{12}{24}[0 + 2(1 + 4 + 9 + 16 + 25 + 36 + 49 + 64 + 81 + 100 + 121) + 144] = 578$

(e) $\quad S_6 = \frac{12}{36}[0 + 144 + 2(4 + 16 + 36 + 64 + 100) + 4(1 + 9 + 25 + 49 + 81 + 121)] = 576$

$$\int_0^{12} x^2\, dx = \left[\frac{1}{3}x^3\right]_0^{12} = 576$$

2. (a) 0.500000 $\qquad\qquad$ (b) 0.500000 $\qquad\qquad$ (c) 0.500000

$$\int_0^1 \sin^2 \pi x\, dx = \frac{1}{\pi}\left[\frac{\pi x}{2} - \frac{\sin 2\pi x}{4}\right]_0^1 = \frac{1}{2}$$

3. (a) $L_6 = \dfrac{3}{6}\left[\dfrac{1}{1+0} + \dfrac{1}{1+1/8} + \dfrac{1}{1+1} + \dfrac{1}{1+27/8} + \dfrac{1}{1+8} + \dfrac{1}{1+125/8}\right]$

$\qquad\qquad = \frac{1}{2}\left[1 + \frac{8}{9} + \frac{1}{2} + \frac{8}{35} + \frac{1}{9} + \frac{8}{133}\right] \cong 1.394$

(b) $R_6 = \dfrac{3}{6}\left[\dfrac{1}{1+1/8} + \dfrac{1}{1+1} + \dfrac{1}{1+27/8} + \dfrac{1}{1+8} + \dfrac{1}{1+125/8} + \dfrac{1}{1+27}\right]$

$\qquad\qquad = \frac{1}{2}\left[\frac{8}{9} + \frac{1}{2} + \frac{8}{35} + \frac{1}{9} + \frac{8}{133} + \frac{1}{28}\right] \cong 0.9122$

(c) $M_3 = \dfrac{3}{3}\left[\dfrac{1}{1+1/8} + \dfrac{1}{1+27/8} + \dfrac{1}{1+125/8}\right] = \dfrac{8}{9} + \dfrac{8}{35} + \dfrac{8}{133} \cong 1.1776$

(d) $T_6 = \frac{3}{12}\left[1 + 2\left(\frac{8}{9} + \frac{1}{2} + \frac{8}{35} + \frac{1}{9} + \frac{8}{133}\right) + \frac{1}{28}\right] \cong 1.1533$

(e) $S_3 = \frac{3}{18}\left\{1 + \frac{1}{28} + 2\left[\frac{1}{2} + \frac{1}{9}\right] + 4\left[\frac{8}{9} + \frac{8}{35} + \frac{8}{133}\right]\right\} \cong 1.1614$

4. (a) 0.4229 (b) 0.4339

5. (a) $\dfrac{1}{4}\pi \cong T_4 = \dfrac{1}{8}\left[1 + 2\left(\dfrac{1}{1+1/16} + \dfrac{1}{1+1/4} + \dfrac{1}{1+9/16}\right) + \dfrac{1}{1+1}\right]$

$\qquad\qquad\qquad = \frac{1}{8}\left[1 + 2\left(\frac{16}{17} + \frac{4}{5} + \frac{16}{25}\right) + \frac{1}{2}\right] \cong 0.7828$

$\qquad\qquad \pi \cong 4(0.7828) = 3.1312$

(b) $\frac{1}{4}\pi \cong S_4 = \frac{1}{24}\left[1 + \frac{1}{2} + 2\left(\frac{16}{17} + \frac{4}{5} + \frac{16}{25}\right) + 4\left(\frac{64}{65} + \frac{64}{73} + \frac{64}{89} + \frac{64}{113}\right)\right] \cong 0.7854$

$\qquad\qquad \pi \cong 4(0.7854) = 3.1416$

6. (a) 0.8511 (b) 0.8542

7. (a) $M_4 = \dfrac{2}{4}\left[\cos\left(\frac{-3}{4}\right)^2 + \cos\left(\frac{-1}{4}\right)^2 + \cos\left(\frac{1}{4}\right)^2 + \cos\left(\frac{3}{4}\right)^2\right] \cong 1.8440$

(b) $T_8 = \dfrac{2}{16}\left[\cos(-1)^2 + 2\cos\left(\frac{-3}{4}\right)^2 + 2\cos\left(\frac{-1}{2}\right)^2 + 2\cos\left(\frac{-1}{4}\right)^2 + \right.$

$\qquad\qquad\qquad \left. 2\cos(0)^2 + 2\cos\left(\frac{1}{4}\right)^2 + 2\cos\left(\frac{1}{2}\right)^2 + 2\cos\left(\frac{3}{4}\right)^2 + \cos(1)^2\right] \cong 1.7915$

(c) $S_4 = \dfrac{2}{24}\left\{\cos(-1)^2 + \cos(1)^2 + 2\left[\cos\left(\frac{-1}{2}\right)^2 + \cos(0)^2 + \cos\left(\frac{1}{2}\right)^2\right] + \right.$

$\qquad\qquad\qquad \left. 4\left[\cos\left(\frac{-3}{4}\right)^2 + \cos\left(\frac{-1}{4}\right)^2 + \cos\left(\frac{1}{4}\right)^2 + \cos\left(\frac{3}{4}\right)^2\right]\right\} \cong 1.8090$

8. (a) 3.0543 (b) 3.0615 (c) 3.0591

9. (a) $T_{10} = \dfrac{2}{20}\left[e^{-0^2} + 2e^{-(1/5)^2} + 2e^{-(2/5)^2} + 2e^{-(3/5)^2} + 2e^{-(4/5)^2} + 2e^{-1^2} + 2e^{-(6/5)^2}\right.$

$\qquad\qquad\qquad \left. + 2e^{-(7/5)^2} + 2e^{-(8/5)^2} + 2e^{-(9/5)^2} + e^{-2^2}\right] \cong 0.8818$

(b) $\quad S_5 = \dfrac{2}{30}\left\{ e^{-0^2} + e^{-2^2} + 2\left[e^{-(2/5)^2} + e^{-(4/5)^2} + e^{-(6/5)^2} + e^{-(8/5)^2}\right]\right.$

$\left. + 4\left[e^{-(1/5)^2} + e^{-(3/5)^2} + e^{-1^2} + e^{-(7/5)^2} + e^{-(9/5)^2}\right]\right\} \cong 0.8821$

10. (a) $\quad 1.9133$ (b) $\quad 1.9271$ (c) $\quad 1.9225$

11. Such a curve passes through the three points

$$(a_1, b_1), \quad (a_2, b_2), \quad (a_3, b_3)$$

iff

$$b_1 = a_1{}^2 A + a_1 B + C, \quad b_2 = a_2{}^2 A + a_2 B + C, \quad b_3 = a_3{}^2 A + a_3 B + C,$$

which happens iff

$$A = \frac{b_1(a_2 - a_3) - b_2(a_1 - a_3) + b_3(a_1 - a_2)}{(a_1 - a_3)(a_1 - a_2)(a_2 - a_3)},$$

$$B = -\frac{b_1(a_2{}^2 - a_3{}^2) - b_2(a_1{}^2 - a_3{}^2) + b_3(a_1{}^2 - a_2{}^2)}{(a_1 - a_3)(a_1 - a_2)(a_2 - a_3)},$$

$$C = \frac{a_1{}^2(a_2 b_3 - a_3 b_2) - a_2{}^2(a_1 b_3 - a_3 b_1) + a_3{}^2(a_1 b_2 - a_2 b_1)}{(a_1 - a_3)(a_1 - a_2)(a_2 - a_3)}.$$

12. $\quad \dfrac{b-a}{6}\left[g(a) + 4g\left(\dfrac{a+b}{2}\right) + g(b)\right]$

$$= \frac{b-a}{6}\left\{ (Aa^2 + Ba + C) + 4\left[A\left(\frac{a+b}{2}\right)^2 + B\left(\frac{a+b}{2}\right) + C\right] + (Ab^2 + Bb + C)\right\}$$

$$= \frac{b-a}{6}\left\{ A(b^2 + a^2) + B(b + a) + 2C + A(a^2 + 2ab + b^2) + 2B(a + b) + 4C\right\}$$

$$= \frac{b-a}{6}\left\{ 2A(b^2 + ab + a^2) + 3B(b + a) + 6C\right\}$$

$$= \frac{1}{3} A(b^3 - a^3) + \frac{1}{2} B(b^2 - a^2) + C(b - a)$$

$$\int_a^b A x^2\, dx + \int_a^b Bx\, dx + \int_a^b C\, dx = \int_a^b g(x)\, dx$$

13. (a) $\quad \left|\dfrac{(b-a)^3}{12n^2} f''(c)\right| = \dfrac{27}{12n^2}\dfrac{1}{4c^{3/2}} \leq \dfrac{9}{16n^2} < 0.01 \implies n^2 > \left(\dfrac{15}{2}\right)^2 \implies n \geq 8$

(b) $\quad \left|\dfrac{(b-a)^5}{2880n^4} f^{(4)}(c)\right| = \dfrac{243}{2880n^4}\dfrac{15}{16c^{7/2}} \leq \dfrac{81}{1024n^4} < 0.01 \implies n \geq 2$

14. (a) $\quad \left|\dfrac{(b-a)^3}{12n^2} f''(c)\right| = \dfrac{8}{12n^2}20c^3 \leq \dfrac{8 \cdot 20 \cdot 27}{12n^2} < 0.01 \implies n \geq 190$

(b) $\left| \dfrac{(b-a)^5}{2880n^4} f^{(4)}(c) \right| = \dfrac{32}{2880n^4} \cdot 120c \leq \dfrac{32 \cdot 120 \cdot 3}{2880n^4} < 0.01 \implies n \geq 5$

15. (a) $\left| \dfrac{(b-a)^3}{12n^2} f''(c) \right| = \dfrac{27}{12n^2} \dfrac{1}{4c^{3/2}} \leq \dfrac{9}{16n^2} < 0.00001 \implies n > 75\sqrt{10} \implies n \geq 238$

(b) $\left| \dfrac{(b-a)^5}{2880n^4} f^{(4)}(c) \right| = \dfrac{243}{2880n^4} \dfrac{15}{16c^{7/2}} \leq \dfrac{81}{1024n^4} < 0.00001 \implies n \geq 10$

16. (a) $\dfrac{8 \cdot 20 \cdot 27}{12n^2} < 0.00001 \implies n \geq 6000$

(b) $\dfrac{32 \cdot 120 \cdot 3}{2880n^4} < 0.00001 \implies n \geq 26$

17. (a) $\left| \dfrac{(b-a)^3}{12n^2} f''(c) \right| = \dfrac{\pi^3}{12n^2} \sin c \leq \dfrac{\pi^3}{12n^2} < 0.001 \implies n > 5\pi\sqrt{\dfrac{10\pi}{3}} \implies n \geq 51$

(b) $\left| \dfrac{(b-a)^5}{2880n^4} f^{(4)}(c) \right| = \dfrac{\pi^5}{2880n^4} \sin c \leq \dfrac{\pi^5}{2880n^4} < 0.001 \implies n \geq 4$

18. (a) $\left| \dfrac{(b-a)^3}{12n^2} f''(c) \right| = \dfrac{\pi^3}{12n^2} \cos c \leq \dfrac{\pi^3}{12n^2} < 0.001 \implies n \geq 51$

(b) $\left| \dfrac{(b-a)^5}{2880n^4} f^{(4)}(c) \right| = \dfrac{\pi^5}{2880n^4} \cos c \leq \dfrac{\pi^5}{2880n^4} < 0.001 \implies n \geq 4$

19. (a) $\left| \dfrac{(b-a)^3}{12n^2} f''(c) \right| = \dfrac{8}{12n^2} e^c \leq \dfrac{8}{12n^2} e^3 < 0.01 \implies n > 10e\sqrt{\dfrac{2e}{3}} \implies n \geq 37$

(b) $\left| \dfrac{(b-a)^5}{2880n^4} f^{(4)}(c) \right| = \dfrac{32}{2880n^4} e^c \leq \dfrac{1}{90n^4} e^3 < 0.01 \implies n \geq 3$

20. (a) $\left| \dfrac{(b-a)^3}{12n^2} f''(c) \right| = \dfrac{(e-1)^3}{12n^2} \cdot \dfrac{1}{c^2} \leq \dfrac{(e-1)^3}{12n^2} < 0.01 \implies n \geq 7$

(b) $\left| \dfrac{(b-a)^5}{2880n^4} f^{(4)}(c) \right| = \dfrac{(e-1)^5}{2880n^4} \cdot \dfrac{6}{c^4} \leq \dfrac{6(e-1)^5}{2880n^4} < 0.01 \implies n \geq 2$

21. (a) $\left| \dfrac{(b-a)^3}{12n^2} f''(c) \right| = \left| \dfrac{8}{12n^2} 2e^{-c^2}(2c^2 - 1) \right| \leq \dfrac{8}{3n^2} e^{-3/2} < 0.0001$

$\implies n > 100\sqrt{\dfrac{8}{3} e^{-3/2}} \implies n \geq 78$

(b) $\left| \dfrac{(b-a)^5}{2880n^4} f^{(4)}(c) \right| = \left| \dfrac{32}{2880n^4} 4e^{-c^2} \left(4c^4 - 12c^2 + 3 \right) \right| \leq \dfrac{32}{2880n^4} 12 < 0.0001$

$\implies \quad n > 10 \left[\dfrac{32 \cdot 12}{2880} \right]^{1/4} \quad \implies \quad n \geq 7$

22. (a) $\left| \dfrac{(b-a)^3}{12n^2} f''(c) \right| = \dfrac{8}{12n^2} \cdot e^c \leq \dfrac{8}{12n^2} \cdot e^2 < 0.00001 \implies n \geq 702$

(b) $\left| \dfrac{(b-a)^5}{2880n^4} f^{(4)}(c) \right| = \dfrac{32}{2880n^4} \cdot e^c \leq \dfrac{32}{2880n^4} \cdot e^2 < 0.00001 \implies n \geq 10$

23. $f^{(4)}(x) = 0$ for all x;   therefore by (8.7.3) the theoretical error is zero

24. If f is linear, $f''(x) = 0$ for all x, so the theoretical error is zero

25. (a) $\left| T_2 - \displaystyle\int_0^1 x^2 \, dx \right| = \dfrac{3}{8} - \dfrac{1}{3} = \dfrac{1}{24} = E_2^T$

(b) $\left| S_1 - \displaystyle\int_0^1 x^4 \, dx \right| = \dfrac{5}{24} - \dfrac{1}{5} = \dfrac{1}{120} = E_1^S$

26. Since $m_i \leq f(x_{i-1}) \leq M_i, \quad m_i \leq f(x_i) \leq M_i, \quad$ and $\quad m_i \leq f\left(\tfrac{x_{i-1}+x_i}{2} \right) \leq M_i, \quad$ we get

$$m_i \leq \dfrac{1}{2} \left[f(x_{i-1}) + f(x_i) \right] \leq M_i \quad \text{and} \quad m_i \leq \dfrac{1}{6} \left[f(x_{i-1}) + 4f\left(\dfrac{x_{i-1} + x_i}{2} \right) + f(x_i) \right] \leq M_i$$

So by the intermediate value theorem, we can find $a_i, \ b_i \in [x_{i-1}, x_i] \quad$ such that

$$f(a_i) = \dfrac{1}{2} \left[f(x_{i-1}) + f(x_i) \right], \quad f(b_i) = \dfrac{1}{6} \left[f(x_{i-1}) + 4f\left(\dfrac{x_{i-1} + x_i}{2} \right) + f(x_i) \right]$$

So using a_i or b_i as x_i^*, we can write T_n and S_n as Riemann sums.

27. (a) Let f be twice differentiable on $[a, b]$ with $f(x) > 0$ and $f''(x) > 0$, and let $P = \{x_0, x_1, x_2, \ldots, x_n\}$ be a regular partition of $[a, b]$. Figure A shows a typical subinterval with the approximating trapezoid ABCD. Since the area under the curve is less than the area of the trapezoid, we can conclude that

$$\int_a^b f(x) \, dx \leq T_n.$$

Figure A

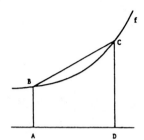

Figure B

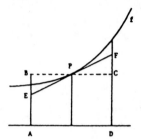

Now consider Figure B. Since the triangles EBP and PFC are congruent, the area of the rectangle ABCD equals the area of the trapezoid AEFD, and since the area under the curve is greater than the area of AEFD it follows that

$$M_n \leq \int_a^b f(x)\, dx.$$

(b) A similar argument in the case $f(x) > 0,\ f''(x) < 0$ gives

$$T_n \leq \int_a^b f(x)\, dx \leq M_n.$$

28.
$$\frac{1}{3}T_n + \frac{2}{3}M_n$$

$$= \frac{1}{3}\frac{b-a}{2n}\left[f(x_0) + 2f(x_1) + \cdots + 2f(x_{n-1}) + f(x_n)\right] + \frac{2}{3}\cdot\frac{b-a}{n}\left[f\left(\frac{x_0 + x_1}{2}\right) + \cdots + f\left(\frac{x_{n-1} + x_n}{2}\right)\right]$$

$$= \frac{b-a}{6n}\left\{f(x_0) + 2f(x_1) + \cdots + 2f(x_{n-1}) + f(x_n) + 4\left[f\left(\frac{x_0 + x_1}{2}\right) + \cdots + f\left(\frac{x_{n-1} + x_n}{2}\right)\right]\right\}$$

$$= S_n$$

29.   (a) $\int_0^{10} (x + \cos x)\, dx \cong 49.4578$

(b) $\int_{-4}^7 \left(x^5 - 5x^4 + x^3 - 3x^2 - x + 4\right)\, dx \cong 1280.56$

30.   (a) $\int_{-4}^3 \frac{x^2}{x^2 + 4}\, dx \cong 2.8201$ (b) $\int_0^{\pi/6} (x + \tan x)\, dx \cong 0.2809$

31. $\left|E_{20}^S\right| \leq \dfrac{4^5}{2880(20)^4}\max\left|f^{(4)}(t)\right| \leq \dfrac{1024}{2880(20)^4}(0.1806) \leq 4.01 \times 10^{-7}$

32. $\left|E_{30}^T\right| \leq 0.00036$

REVIEW EXERCISES

1. Substitution: let $u = \sin x$, $du = \cos x\, dx$

$$\int \frac{\cos x}{4 + \sin^2 x}\, dx = \int \frac{1}{4 + u^2}\, du = \frac{1}{2} \arctan\left(\frac{u}{2}\right) + C = \frac{1}{2} \arctan\left(\frac{\sin x}{2}\right) + C$$

2. $\dfrac{x^2}{1 + x^2} = 1 - \dfrac{1}{x^2}$

$$\int_0^{\pi/4} \frac{x^2}{1 + x^2}\, dx = \int_0^{\pi/4} \left(1 - \frac{1}{1 + x^2}\right) dx = \Big[x - \arctan x\Big]_0^{\pi/4} = \frac{\pi}{4} - 1$$

3. Integration by parts: $\begin{Bmatrix} u = x & dv = \sinh x\, dx \\ du = dx & v = \cosh x \end{Bmatrix}$

$$\int 2x \sinh x\, dx = 2\int x \sinh x\, dx = 2\left[x \cosh x - \int \cosh x\, dx\right] = 2x \cosh x - 2 \sinh x + C$$

4. $\displaystyle\int (\tan x + \cot x)^2 dx = \int (\tan^2 x + \cot^2 x + 2)\, dx = \int (\sec^2 x + \csc^2 x)\, dx = \tan x - \cot x + C$

5. Partial fractions: $\dfrac{x - 3}{x^2(x + 1)} = \dfrac{4}{x} - \dfrac{3}{x^2} - \dfrac{4}{x + 1}$

$$\int \frac{x - 3}{x^2(x + 1)}\, dx = \int \left(\frac{4}{x} - \frac{3}{x^2} - \frac{4}{x + 1}\right) dx = 4\ln|x| + \frac{3}{x} - 4\ln|x + 1| + C$$

6. Integration by parts: $\begin{Bmatrix} u = \arctan x & dv = x\, dx \\ du = \dfrac{1}{1 + x^2}\, dx & v = \frac{1}{2}x^2 \end{Bmatrix}$

$$\int x \arctan x\, dx = \frac{1}{2}\left[x^2 \arctan - \int \frac{x^2}{1 + x^2}\, dx\right]$$

$$= \frac{1}{2}\left[x^2 \arctan x - \int \left(1 - \frac{1}{1 + x^2}\right) dx\right] = \frac{1}{2}(x^2 + 1) \arctan x - \frac{1}{2}x + C$$

7. $\displaystyle\int \sin 2x \cos x\, dx = \int (2\sin x \cos x) \cos x\, dx = 2 \int \cos^2 x \sin x\, dx$

Substitution: $u = \cos x$, $du = -\sin x\, dx$

$$2 \int \cos^2 x \sin x\, dx = -2 \int u^2\, du = -\tfrac{2}{3} u^3 + C = -\tfrac{2}{3} \cos^3 x + C$$

8. Integration by parts: $\begin{Bmatrix} u = x & dv = e^{-3x}\, dx \\ du = dx & v = -\frac{1}{3}e^{-3x} \end{Bmatrix}$

$$\int 3xe^{-3x}\, dx = 3 \int xe^{-3x}\, dx = 3\left[-\frac{1}{3} xe^{-3x} + \frac{1}{3} \int e^{-3x}\, dx\right]$$

$$= 3\left[-\frac{1}{3} xe^{-3x} - \frac{1}{9} e^{-3x}\right] + C$$

$$= -xe^{-3x} - \tfrac{1}{3}e^{-3x} + C$$

9. $\int \ln \sqrt{x+1}\, dx = \int \ln(x+1)^{1/2}\, dx = \frac{1}{2}\int \ln(x+1)\, dx$

Integration by parts: $\quad \begin{Bmatrix} u = \ln(x+1) & dv = dx \\ du = \dfrac{1}{x+1}\, dx & v = x+1 \end{Bmatrix}$

$$\frac{1}{2}\int \ln(x+1)\, dx = \frac{1}{2}\left[(x+1)\ln(x+1) - \int dx\right] = \frac{1}{2}\left[(x+1)\ln(x+1) - x\right] + C$$

$$\int_0^3 \ln \sqrt{x+1}\, dx = \frac{1}{2}\left[(x+1)\ln(x+1) - x\right]\Big|_0^3 = 4\ln 2 - \frac{3}{2}$$

Note: This answer can also be written: $\frac{1}{2}\left[(x+1)\ln(x+1) - (x+1)\right] + C$;

set $w = x+1$, $dw = dx$, and integrate $\int \ln w\, dw$ by parts.

10. Partial fractions: $\quad \dfrac{2}{x(1+x^2)} = \dfrac{2}{x} - \dfrac{2x}{x^2+1}$

$$\int \frac{2}{x(1+x^2)}\, dx = \int \frac{2}{x}\, dx - \int \frac{2x}{x^2+1}\, dx = 2\ln|x| - \ln|x^2+1| + C = \ln \frac{x^2}{x^2+1} + C$$

11. $\int \dfrac{\sin^3 x}{\cos x}\, dx = \int \dfrac{(1-\cos^2 x)\sin x}{\cos x}\, dx = \int \tan x\, dx - \int \cos x \sin x\, dx = \ln|\sec x| - \dfrac{1}{2}\sin^2 x + C$

12. Substitution: $u = \sin x$, $du = \cos x\, dx$

$$\int \frac{\cos x}{\sin^3 x}\, dx = \int u^{-3}\, du = -\frac{1}{2}u^{-2} + C = -\frac{1}{2}\csc^2 x + C$$

13. $\int_0^1 e^{-x}\cosh x\, dx = \int_0^1 e^{-x}\left[\dfrac{e^x + e^{-x}}{2}\right]dx = \dfrac{1}{2}\int_0^1 \left[1 + e^{-2x}\right]dx$

$$= \frac{1}{2}\left[x - \frac{e^{-2x}}{2}\right]_0^1 = \frac{1}{2}\left[1 - \frac{e^{-2}}{2} + \frac{1}{2}\right] = \frac{3}{4} - \frac{e^{-2}}{2}$$

14. Trigonometric substitution: set $x = 3\tan u$, $dx = 3\sec^2 u\, du$.

$$\int \frac{x^2+3}{\sqrt{x^2+9}}\, dx = \int \frac{3\tan^2 u + 3}{3\sec u}\, 3\sec^2 u\, du$$

$$= 3\int \sec^3 u\, du = 3\left[\frac{1}{2}\sec u \tan u + \frac{1}{2}\ln|\sec u + \tan u|\right] + C$$

15. $\int \dfrac{dx}{e^x - 4e^{-x}} = \int \dfrac{e^x}{e^{2x} - 4}\, dx$;   substitution: $u = e^x$, $du = e^x\, dx$

$$\int \frac{dx}{e^x - 4e^{-x}} = \int \frac{1}{u^2 - 4}\, du = \int \left(\frac{1/4}{u-2} - \frac{1/4}{u+2}\right)du = \frac{1}{4}\left[\ln|u-2| - \ln|u+2|\right] + C$$

$$= \frac{1}{4}\left[\ln|e^x - 2| - \ln|e^x + 2|\right] + C = \frac{1}{4}\ln\left|\frac{e^x - 2}{e^x + 2}\right| + C$$

16. Partial fractions: $\dfrac{1}{x^3 - 1} = \dfrac{1}{(x-1)(x^2 + x + 1)} = \dfrac{\frac{1}{3}}{x-1} + \dfrac{-\frac{1}{3}x - \frac{2}{3}}{x^2 + x + 1}$

$$\int \frac{1}{x^3 - 1}\, dx = \int \frac{\frac{1}{3}}{x-1}\, dx + \int \frac{-\frac{1}{3}x - \frac{2}{3}}{x^2 + x + 1}\, dx$$

$$= \frac{1}{3}\ln|x-1| - \frac{1}{3}\int \frac{x+2}{x^2 + x + 1}\, dx$$

$$= \frac{1}{3}\ln|x-1| - \frac{1}{6}\int \frac{2x + 1 - 1 + 4}{x^2 + x + 1}\, dx$$

$$= \frac{1}{3}\ln|x-1| - \frac{1}{6}\ln|x^2 + x + 1| - \frac{1}{2}\int \frac{1}{(x+\frac{1}{2})^2 + (\frac{\sqrt{3}}{2})^2}\, dx$$

$$= \frac{1}{3}\ln|x-1| - \frac{1}{6}\ln|x^2 + x + 1| - \frac{\sqrt{3}}{3}\arctan\frac{2x + 1}{\sqrt{3}} + C$$

17.
Integration by parts: $\begin{cases} u = x & dv = 2^x\, dx \\ du = dx & v = 2^x/\ln 2 \end{cases}$

$$\int x2^x\, dx = \frac{1}{\ln 2}x2^x - \frac{1}{\ln 2}\int 2^x\, dx = \frac{1}{\ln 2}\left(x2^x - \frac{1}{\ln 2}2^x\right) + C$$

18. $\displaystyle\int \ln(x\sqrt{x})\,dx = \int \ln x^{3/2}\, dx = \frac{3}{2}\int \ln x\, dx = \frac{3}{2}[x\ln x - x] + C$

19. Trigonometric substitution: set $x = a\sin u$, $dx = a\cos u\, du$

$$\int \frac{\sqrt{a^2 - x^2}}{x^2}\,dx = \int \frac{a\cos u}{a^2\sin^2 u}a\cos u\, du = \int \frac{\cos^2 u}{\sin^2 u}\, du = \int \frac{1 - \sin^2 u}{\sin^2 u}\, du$$

$$= \int \csc^2 u\, du - \int du = -\cot u - u + c$$

$$= -\frac{\sqrt{a^2 - x^2}}{x} - \arcsin\left(\frac{x}{a}\right) + C$$

20. substitution: $u = x^3$, $du = 3x^2\, dx$; $u(0) = 0$, $u(2) = 8$

$$\int_0^2 x^2 e^{x^3}\, dx = \frac{1}{3}\int_0^8 e^u\, du = \frac{1}{3}\Big[e^u\Big]_0^8 = \frac{1}{3}(e^8 - 1)$$

21. $\displaystyle\int x^3 e^{x^2}\, dx = \int x^2\, xe^{x^2}\, dx$; integration by parts: $\begin{cases} u = x^2 & dv = xe^{x^2}\, dx \\ du = 2x\, dx & v = \frac{1}{2}e^{x^2} \end{cases}$

$$\int x^3 e^{x^2}\, dx = \frac{1}{2}\left[x^2 e^{x^2} - \int 2xe^{x^2}\, dx^2\right] = \frac{1}{2}\left[x^2 e^{x^2} - e^{x^2}\right] + C$$

22. $\displaystyle\int \sin 2x \sin 3x\, dx = \frac{1}{2}\int [\cos x - \cos 5x]\, dx = \frac{1}{2}\sin x - \frac{1}{10}\sin 5x + C$

23. $\displaystyle\int \frac{\sin^5 x}{\cos^7 x}\, dx = \int \tan^5 \sec^2\, dx = \frac{1}{6}\tan^6 x + C$ (substitution: $u = \tan x$)

24. $\int \left(\dfrac{\sqrt{4+x^2}}{x} - \dfrac{x}{\sqrt{4+x^2}} \right) dx = \int \dfrac{4}{x\sqrt{4+x^2}}\, dx$

Trigonometric substitution: set $x = 2\tan u,\ dx = 2\sec^2 u\, du$

$$\int \frac{4}{x\sqrt{4+x^2}}\, dx = 2\int \frac{\sec u}{\tan u}\, du = 2\int \csc u\, du$$

$$= 2\ln|\csc u - \cot u| + C = 2\ln\left| \frac{\sqrt{4+x^2}-2}{x} \right| + C$$

25.

$$\int_{\pi/6}^{\pi/3} \frac{\sin x}{\sin 2x}\, dx = \int_{\pi/6}^{\pi/3} \frac{\sin x}{2\sin x \cos x}\, dx = \frac{1}{2}\int_{\pi/6}^{\pi/3} \sec x\, dx$$

$$= \frac{1}{2}\Big[\ln|\sec x + \tan x| \Big]_{\pi/6}^{\pi/3} = \frac{1}{2}\ln(2+\sqrt{3}) - \frac{1}{4}\ln 3$$

26.

$$\int \frac{x+3}{\sqrt{x^2+2x-8}}\, dx = \frac{1}{2}\int \frac{2x+6}{\sqrt{x^2+2x-8}}\, dx = \frac{1}{2}\int \frac{2x+2+4}{\sqrt{x^2+2x-8}}\, dx$$

$$= \frac{1}{2}\int \frac{2x+2}{\sqrt{x^2+2x-8}}\, dx + 2\int \frac{1}{\sqrt{(x+1)^2-9}}\, dx$$

$$= \sqrt{x^2+2x-8} + 2\int \frac{1}{\sqrt{(x+1)^2-9}}\, dx$$

Use the trigonometric substitution: $x+1 = 3\sec u,\ dx = 3\sec u\tan u\, du,$ on the remaining integral.

$$\int \frac{1}{\sqrt{(x+1)^2-9}}\, dx = \int \sec u\, du$$

$$= \ln|\sec u + \tan u| = \ln\left| \frac{x+1+\sqrt{x^2+2x-8}}{3} \right|$$

Thus $\displaystyle \int \frac{x+3}{\sqrt{x^2+2x-8}}\, dx = \sqrt{x^2+2x-8} + \ln\left| \frac{x+1+\sqrt{x^2+2x-8}}{3} \right| + C$

27. $\displaystyle \int \frac{x^2+x}{\sqrt{1-x^2}}\, dx = \int \frac{x^2}{\sqrt{1-x^2}}\, dx + \int \frac{x}{\sqrt{1-x^2}}\, dx$

$$\int \frac{x}{\sqrt{1-x^2}}\, dx = -\sqrt{1-x^2} + C_1,\quad \text{substitution: set } u = 1-x^2,\ du = -2x\, dx$$

For the first integral, use the trigonometric substitution: $x = \sin u,\ dx = \cos u\, du$

$$\int \frac{x^2}{\sqrt{1-x^2}}\, dx = \int \frac{\sin^2 u}{\cos u}\cos u\, du = \int \sin^2 u\, du = \int \frac{1-\cos 2u}{2}\, du$$

$$= \tfrac{1}{2}u - \tfrac{1}{4}\sin 2u + C_2 = \tfrac{1}{2}u - \tfrac{1}{2}\sin u \cos u + C_2$$

$$= \frac{1}{2}\arcsin x - \frac{1}{2}x\sqrt{1-x^2} + C_2$$

Therefore,

$$\int \frac{x^2 + x}{\sqrt{1 - x^2}} dx = \frac{1}{2} \arcsin x - \frac{1}{2}x\sqrt{1 - x^2} - \sqrt{1 - x^2} + C$$

28.
$$\int x \tan^2 2x \, dx = \int x \left(\sec^2 2x - 1\right) dx = -\int x \, dx + \int x \sec^2 2x \, dx$$

$$= -\frac{1}{2}x^2 + \frac{1}{2}x \tan 2x - \frac{1}{4} \ln|\sec 2x| + C \quad \text{(integration by parts)}$$

29. Since
$$\frac{\cos^4 x}{\sin^2 x} = \frac{(1 - \sin^2 x)^2}{\sin^2 x} = \frac{1 - 2\sin^2 x + \sin^4 x}{\sin^2 x} = \csc^x - 2 + \sin x$$

$$\int \frac{\cos^4 x}{\sin^2 x} dx = \int (\csc^x - 2 + \sin x) dx = -\cot x - \frac{3}{2}x - \frac{1}{4} \sin 2x + C$$

30.
$$\int x \ln \sqrt{x^2 + 1} dx = \frac{1}{2} \int \ln \sqrt{x^2 + 1} d(x^2) = \frac{1}{2} \left[x^2 \ln \sqrt{x^2 + 1} - \int \frac{x^3}{x^2 + 1} dx \right]$$

$$= \frac{1}{2} \left[x^2 \ln \sqrt{x^2 + 1} - \int x dx + \int \frac{x}{x^2 + 1} dx \right]$$

$$= \frac{1}{2} \left[x^2 \ln \sqrt{x^2 + 1} - \frac{1}{2}x^2 + \frac{1}{2} \ln(x^2 + 1) \right] + C$$

Hence

$$\int_0^3 x \ln \sqrt{x^2 + 1} dx = \frac{1}{2} \{x^2 \ln \sqrt{x^2 + 1} - \frac{1}{2}x^2 + \frac{1}{2} \ln(x^2 + 1)\}|_0^3 = 5 \ln \sqrt{10} - \frac{9}{4}$$

31.
$$\int (\sin 2x + \cos 2x)^2 dx = \int (1 + 2\sin 2x \cos 2x) \, dx = x + \int \sin 4x \, dx = x - \frac{1}{4} \cos 4x + C$$

32.
$$\int \sqrt{\cos x} \sin^3 x dx = -\int \sqrt{\cos x}(1 - \cos^2 x) d(\cos x) = -\int (\sqrt{\cos x} - \cos^{5/2} x) d(\cos x)$$

$$= -\frac{2}{3} \cos^{3/2} x + \frac{2}{7} \cos^{7/2} x + C$$

33.
$$\frac{5x + 1}{(x + 2)(x^2 - 2x + 1)} = \frac{-1}{x + 2} + \frac{1}{x - 1} + \frac{2}{(x - 1)^2}$$

$$\int \frac{5x + 1}{(x + 2)(x^2 - 2x + 1)} dx = \int \left\{ \frac{-1}{x + 2} + \frac{1}{x - 1} + \frac{2}{(x - 1)^2} \right\} dx$$

$$= -\ln|x + 2| + \ln|x - 1| - \frac{2}{x - 1} + C$$

34.
$$\int \tan^{3/2} x \sec^4 x dx = \int \tan^{3/2}(1 + \tan^2 x) d(\tan x) = \int \{\tan^{3/2} x + \tan^{7/2} x\} d(\tan x)$$

$$= \frac{2}{5} \tan^{5/2} x + \frac{2}{9} \tan^{9/2} x + C$$

35.
$$\int \frac{1}{\sqrt{x + 1} - \sqrt{x}} dx = \int \frac{\sqrt{x + 1} + \sqrt{x}}{(\sqrt{x + 1} - \sqrt{x})(\sqrt{x + 1} + \sqrt{x})} dx$$

$$= \int (\sqrt{x + 1} + \sqrt{x}) dx = \frac{2}{3}(x + 1)^{3/2} + \frac{2}{3}x^{3/2} + C$$

36. $\displaystyle\int \cos \pi x \cos \frac{1}{2}\pi x\, dx = \frac{1}{2}\int \left\{ \cos \frac{3}{2}\pi x + \cos \frac{1}{2}\pi x \right\} dx = \frac{1}{3\pi}\sin \frac{3\pi}{2}x + \frac{1}{\pi}\sin \frac{\pi}{2}x + C$

Hence

$$\int_0^{1/2} \cos \pi x \cos \frac{1}{2}\pi x\, dx = \left[\frac{1}{3\pi}\sin \frac{3\pi}{2}x + \frac{1}{\pi}\sin \frac{\pi}{2}x \right]\Big|_0^{1/2} = \frac{\sqrt{2}}{6\pi} + \frac{\sqrt{2}}{2\pi}$$

37. integration by parts: $\quad \begin{cases} u = x^2 & dv = \cos 2x\, dx \\ du = 2x\, dx & v = \frac{1}{2}\sin 2x \end{cases}$

$$\int x^2 \cos 2x\, dx = \frac{1}{2}x^2 \sin 2x - \int x \sin 2x\, dx$$

integration by parts again: $\quad \begin{cases} u = x & dv = \sin 2x\, dx \\ du = dx & v = -\frac{1}{2}\cos 2x \end{cases}$

$$\int x^2 \cos 2x\, dx = \frac{1}{2}x^2 \sin 2x - \left[-\frac{1}{2}x \cos 2x + \frac{1}{2}\int \cos 2x\, dx \right] = \frac{1}{2}x^2 \sin 2x + \frac{1}{2}x \cos 2x - \frac{1}{4}\sin 2x + C$$

38. See Exercise 45, Section 8.2.

$$\int e^{2x}\sin 4x\, dx = \frac{e^{2x}(2\sin 4x - 4\cos 4x)}{2^2 + 4^2} = \frac{e^{2x}(2\sin 4x - 4\cos 4x)}{20} + C$$

39. $\displaystyle \frac{1 - \sin 2x}{1 + \sin 2x} = \frac{1 - \sin 2x}{1 + \sin 2x}\frac{1 - \sin 2x}{1 - \sin 2x} = \frac{1 - 2\sin 2x + \sin^2 2x}{1 - \sin^2 2x} = \frac{1 - 2\sin 2x + \sin^2 2x}{\cos^2 2x}$

$$= \sec^2 2x - \frac{2\sin 2x}{\cos^2 2x} + \tan^2 2x = 2\sec^2 2x - 1 - \frac{2\sin 2x}{\cos^2 2x}$$

Therefore

$$\int \frac{1 - \sin 2x}{1 + \sin 2x}\, dx = \tan 2x - x - \frac{1}{\cos 2x} + C = \tan 2x - x - \sec 2x + C$$

40. partial fraction decomposition: $\quad \displaystyle \frac{5x + 3}{(x - 1)(x^2 + 2x + 5)} = \frac{1}{x - 1} + \frac{-x + 2}{x^2 + 2x + 5},$

$$\int \frac{5x + 3}{(x - 1)(x^2 + 2x + 5)}\, dx = \int \left\{ \frac{1}{x - 1} + \frac{-x + 2}{x^2 + 2x + 5} \right\} dx$$

$$= \ln|x - 1| + \int \frac{-x + 2}{x^2 + 2x + 5}\, dx$$

$$= \ln|x - 1| - \int \frac{-x - 1 + 3}{x^2 + 2x + 5}\, dx$$

$$= \ln|x - 1| - \frac{1}{2}\int \frac{2x + 2}{x^2 + 2x + 5} + 3\int \frac{1}{(x + 1)^2 + 5}\, dx$$

$$= \ln|x - 1| - \frac{1}{2}\ln(x^2 + 2x + 5) + \frac{3}{2}\arctan \frac{x + 1}{2} + C$$

41. (a)   integration by parts: $\begin{cases} u = x^n & dv = \cos ax\, dx \\ du = nx^{n-1}\, dx & v = \frac{1}{a}\sin ax \end{cases}$

$$\int x^n \cos ax\, dx = \frac{x^n \sin ax}{a} - \frac{n}{a}\int x^{n-1}\sin ax\, dx$$

(b)   integration by parts: $\begin{cases} u = x^n & dv = \sin ax\, dx \\ du = nx^{n-1}\, dx & v = -\frac{1}{a}\cos ax \end{cases}$

$$\int x^n \sin ax\, dx = -\frac{x^n \cos ax}{a} + \frac{n}{a}\int x^{n-1}\cos ax\, dx$$

42. (a) Using 41,

$$\int x^2 \cos 3x\, dx = \frac{x^2 \sin 3x}{3} - \frac{2}{3}\int x \sin 3x\, dx$$

$$= \frac{x^2 \sin 3x}{3} - \frac{2}{3}\left\{ -\frac{x\cos x}{3} + \frac{1}{3}\int \cos 3x\, dx \right\}$$

$$= \frac{x^2 \sin 3x}{3} + \frac{2}{9}x\cos 3x - \frac{2}{27}\sin 3x + C$$

(b) Similarly,

$$\int x^3 \sin 4x\, dx = -\frac{x^3 \cos 4x}{4} + \frac{3}{4}\int x^2 \cos 4x\, dx$$

$$= -\frac{x^3 \cos 4x}{4} + \frac{3}{4}\left\{ \frac{x^2 \sin 4x}{4} - \frac{1}{2}\int x \sin 4x\, dx \right\}$$

$$= -\frac{x^3 \cos 4x}{4} + \frac{3}{16}x^2 \sin 4x - \frac{3}{8}\int x \sin 4x\, dx$$

$$= -\frac{x^3 \cos 4x}{4} + \frac{3}{16}x^2 \sin 4x - \frac{3}{8}\left\{ -\frac{x\cos 4x}{4} + \frac{1}{4}\int \cos 4x\, dx \right\}$$

$$= -\frac{x^3 \cos 4x}{4} + \frac{3}{16}x^2 \sin 4x + \frac{3}{32}x\cos 4x - \frac{3}{128}\sin 4x + C$$

43. (a)   integration by parts: $\begin{cases} u = (\ln x)^n & dv = x^m\, dx \\ du = n(\ln x)^{n-1}\frac{1}{x}\, dx & v = \frac{1}{m+1}x^{m+1} \end{cases}$

$$\int x^m (\ln x)^n\, dx = \frac{x^{m+1}(\ln x)^n}{m+1} - \frac{n}{m+1}\int x^m (\ln x)^{n-1}\, dx$$

(b) From (a),

$$\int x^4 (\ln x)^3\, dx = \frac{x^5(\ln x)^3}{5} - \frac{3}{5}\int x^4 (\ln x)^2\, dx$$

$$= \frac{x^5(\ln x)^3}{5} - \frac{3}{5}\left\{ \frac{x^5(\ln x)^2}{5} - \frac{2}{5}\int x^4 \ln x\, dx \right\}$$

$$= \frac{1}{5}x^5(\ln x)^3 - \frac{3}{25}x^5(\ln x)^2 + \frac{6}{25}\left\{ \frac{x^5 \ln x}{5} - \frac{1}{5}\int x^4\, dx \right\}$$

$$= \frac{1}{5}x^5(\ln x)^3 - \frac{3}{25}x^5(\ln x)^2 + \frac{6}{125}x^5 \ln x - \frac{6}{125}x^5\, dx + C$$

44. Using integration by parts,

$$\int x^2 \arctan x \, dx = \frac{1}{3}x^3 \arctan x - \frac{1}{3}\int \frac{x^3}{1+x^2}dx$$

$$= \frac{1}{3}x^3 \arctan x - \frac{1}{3}\int (x - \frac{x}{1+x^2})dx$$

$$= \frac{1}{3}x^3 \arctan x - \frac{1}{3}\left(\frac{1}{2}x^2 - \frac{1}{2}\ln(1+x^2)\right)$$

$$= \frac{1}{3}x^3 \arctan x - \frac{1}{6}x^2 + \frac{1}{6}\ln(1+x^2) + C$$

Therefore, $A = \displaystyle\int_0^1 x^2 \arctan x \, dx = \frac{\pi}{12} - \frac{1}{6} - \frac{\ln 2}{6}$

45.

$$A = \int_0^{\frac{1}{2}}(1-x^2)^{-1/2}\,dx = \Big[\arcsin x\Big]_0^{1/2} = \frac{\pi}{6}$$

$$\overline{x}A = \int_0^{1/2} x(1-x^2)^{-1/2}\,dx = -\Big[\sqrt{1-x^2}\Big]_0^{1/2} = 1 - \frac{\sqrt{3}}{2} \quad\Longrightarrow\quad \overline{x} = \frac{6}{\pi}\left(1 - \frac{\sqrt{3}}{2}\right)$$

$$\overline{y}A = \int_0^{1/2}\frac{1}{2}(1-x^2)^{-1}\,dx = \frac{1}{4}\int_0^{1/2}\left[\frac{1}{1-x} + \frac{1}{1+x}\right]\,dx = \frac{1}{4}\left[\ln\left(\frac{1+x}{1-x}\right)\right]_0^{1/2} = \frac{1}{4}\ln 3$$

Therefore, $\quad \overline{y} = \dfrac{3\ln 3}{2\pi}$

46. **(a)**

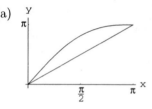

(b) $A = \displaystyle\int_0^\pi (x + \sin x - x)\,dx = \int_0^\pi \sin x \, dx = -\Big[\cos x\Big]_0^\pi = 2$

(c) $\overline{x}A = \displaystyle\int_0^\pi x(x + \sin x - x)\,dx = \int_0^\pi x \sin x \, dx = \Big[-x\cos x + \sin x\Big]_0^\pi = \pi$

$$\text{integration by parts} \nearrow$$

Therefore $\quad \overline{x} = \frac{\pi}{2}$

$$\overline{y}A = \int_0^\pi \frac{1}{2}\Big[(x + \sin x)^2 - x^2\Big]dx = \frac{1}{2}\int_0^\pi \Big[2x\sin x + \sin^2 x\Big]\,dx$$

$$= \int_0^\pi x\sin x \, dx + \frac{1}{4}\int_0^\pi (1 + \cos 2x)\,dx$$

$$= \pi + \frac{1}{4}\Big[x + \frac{1}{2}\sin 2x\Big]_0^\pi = \frac{5\pi}{4}$$

Therefore, $\quad \overline{y} = \dfrac{5\pi}{8}$

47. Use Pappus's theorem and Problem 45.

(a) about the x-axis: $V = 2\pi \left(\frac{3\ln 3}{2\pi}\right) \frac{\pi}{6} = \frac{\pi \ln 3}{2}$

(b) about the y-axis: $V = 2\pi \left[\frac{6}{\pi}\left(1 - \frac{\sqrt{3}}{2}\right)\right] \frac{\pi}{6} = 2\pi\left(1 - \frac{\sqrt{3}}{2}\right)$

48. $V = \int_1^e 2\pi x \ln 2x \, dx = \left[\pi x^2 \ln(2x) - \frac{\pi}{2}x^2\right]_1^e = \frac{\pi}{2}\left[e^2(1 + 2\ln 2) - 2\ln 2 + 1\right].$

$\uparrow$
$\quad$ integration by parts

49. Let $f = \sqrt{x^3 + x}$.

(a) $\int_0^2 \sqrt{x^3 + x}\,dx = \frac{2}{4}\left[f\left(\frac{0 + 1/2}{2}\right) + f\left(\frac{1/2 + 1}{2}\right) + f\left(\frac{1 + 3/2}{2}\right) + f\left(\frac{3/2 + 2}{2}\right)\right]$

$= \frac{1}{2}[f(1/4) + f(3/4) + f(5/4) + f(7/4)] \approx 3.0270$

(b) $\int_0^2 \sqrt{x^3 + x}\,dx$

$= \frac{2}{16}[f(0) + 2f(1/4) + 2f(1/2) + 2f(3/4) + 2f(1) + 2f(5/4) + 2f(3/2) + 2f(7/4) + f(2)]$

$= \frac{1}{8}[2f(1/4) + 2f(1/2) + 2f(3/4) + 2f(1) + 2f(5/4) + 2f(3/2) + 2f(7/4) + f(2)] \approx 3.0120$

(c) $\int_0^2 \sqrt{x^3 + x}\,dx$

$= \frac{2}{24}\{f(0) + 2[f(1/2) + 2f(1) + f(3/2)] + 4[f(1/4) + f(3/4) + f(5/4) + f(7/4)] + f(2)]\}$

$= \frac{1}{12}\{2[f(1/2) + 2f(1) + f(3/2)] + 4[f(1/4) + f(3/4) + f(5/4) + f(7/4)] + f(2)]\} \approx 3.0170$

50. Let $f = \sqrt{1 + 3x}$.

(a) $\int_0^2 \sqrt{1 + 3x}\,dx$

$= \frac{2}{16}[f(0) + 2f(1/4) + 2f(1/2) + 2f(3/4) + 2f(1) + 2f(5/4) + 2f(3/2) + 2f(7/4) + f(2)]$

$= \frac{1}{8}[1 + 2f(1/4) + 2f(1/2) + 2f(3/4) + 2f(1) + 2f(5/4) + 2f(3/2) + 2f(7/4) + f(2)] \approx 3.8886$

(b) $\int_0^2 \sqrt{1 + 3x}\,dx$

$= \frac{2}{24}\{f(0) + 2[f(1/2) + 2f(1) + f(3/2)] + 4[f(1/4) + f(3/4) + f(5/4) + f(7/4)] + f(2)]\}$

$= \frac{1}{12}\{1 + 2[f(1/2) + 2f(1) + f(3/2)] + 4[f(1/4) + f(3/4) + f(5/4) + f(7/4)] + f(2)]\} \approx 3.8932$

(c) $\int_0^2 \sqrt{1 + 3x}\,dx = \frac{2}{9}(1 + 3x)^{3/2}\big|_0^2 = \frac{2}{9}[7^{3/2} - 1] \approx 3.8934$

51. (a) $|E_n^T| \le \dfrac{(b-a)^3}{12n^2} M.$

With $a = 0$, $b = 2$ and

$$|f''(x)| = \frac{9}{4}(1 + 3x)^{-3/2} \le \frac{9}{4} = M \quad \text{on } [0, 2],$$

we have

$$|E_n^T| \le \frac{2^3}{12n^2} \frac{9}{4} = \frac{3}{2n^2}.$$

Solving $\dfrac{3}{2n^2} \le 0.0001$ for n gives $n \ge 123$.

(b) $|E_n^S| \le \dfrac{(b-a)^5}{2880n^4} M.$

With $a = 0, b = 2$ and

$$|f^{(4)}(x)| = \frac{1215}{16}(1 + 3x)^{-7/2} \le \frac{1215}{16} = M,$$

we have

$$|E_n^S| \le= \frac{2^5}{2880n^4} \frac{1215}{16} = \frac{27}{32n^4}.$$

Solving $\dfrac{27}{32n^4} \le 0.0001$ for n gives $n \ge 11$.

52. Let $f(x) = \dfrac{e^{-x}}{x}.$

(a) $\displaystyle\int_1^3 \frac{e^{-x}}{x}\, dx$

$$= \frac{2}{16}[f(1) + 2f(1.25) + 2f(1.5) + 2f(1.75) + 2f(2) + 2f(2.25) + 2f(2.5) + 2f(2.75) + f(3)]$$

$$\approx 0.2100$$

(b) $\displaystyle\int_1^3 \frac{e^{-x}}{x}\, dx$

$$= \frac{2}{24}\{f(0) + 2[f(1/2) + 2f(1) + f(3/2)] + 4[f(1/4) + f(3/4) + f(5/4) + f(7/4)] + f(2)]\}$$

$$\approx 0.2064$$

CHAPTER 9

SECTION 9.1

1. $y_1'(x) = \frac{1}{2} e^{x/2};$ $2y_1' - y_1 = 2\left(\frac{1}{2}\right)e^{x/2} - e^{x/2} = 0;$ y_1 is a solution.

$y_2'(x) = 2x + e^{x/2};$ $2y_2' - y_2 = 2\left(2x + e^{x/2}\right) - \left(x^2 + 2e^{x/2}\right) = 4x - x^2 \neq 0;$

y_2 is not a solution.

2. $y_1' + xy_1 = -xe^{-x^2/2} + xe^{-x^2/2} = 0;$ not a solution

$y_2' + xy_2 = -Cxe^{-x^2/2} + x + Cxe^{-x^2/2} = x;$ y_2 is a solution.

3. $y_1'(x) = \dfrac{-e^x}{(e^x + 1)^2};$ $y_1' + y_1 = \dfrac{-e^x}{(e^x + 1)^2} + \dfrac{1}{e^x + 1} = \dfrac{1}{(e^x + 1)^2} = y_1^2;$ y_1 is a solution.

$y_2'(x) = \dfrac{-Ce^x}{(Ce^x + 1)^2};$ $y_2' + y_2 = \dfrac{-Ce^x}{(Ce^x + 1)^2} + \dfrac{1}{Ce^x + 1} = \dfrac{1}{(Ce^x + 1)^2} = y_2^2;$

y_2 is a solution.

4. $y_1'' + 4y_1 = -8\sin 2x + 8\sin 2x = 0;$ y_1 is a solution.

$y_2'' + 4y_2 = -2\cos x + 8\cos x = 6\cos x;$ not a solution.

5. $y_1'(x) = 2e^{2x},$ $y_1'' = 4e^{2x};$ $y_1'' - 4y_1 = 4e^{2x} - 4e^{2x} = 0;$ y_1 is a solution.

$y_2'(x) = 2C\cosh 2x,$ $y_2'' = 4C\sinh 2x;$ $y_2'' - 4y_2 = 4C\sinh 2x - 4C\sinh 2x = 0;$

y_2 is a solution.

6. $y_1'' - 2y_1' - 3y_1 = e^{-x} + 18e^{3x} - 2(-e^{-x} + 6e^{3x}) - 3(e^{-x} + 2e^{3x}) = 0;$ not a solution

$y_2'' - 2y_2' - 3y_2 = \dfrac{7}{4}\left[(6 + 9x)e^{3x} - 2(1 + 3x)e^{3x} - 3xe^{3x}\right] = 7e^{3x};$ y_2 is a solution.

7. $y' - 2y = 1;$ $H(x) = \displaystyle\int (-2)\,dx = -2x,$ integrating factor: e^{-2x}

$$e^{-2x}y' - 2e^{-2x}y = e^{-2x}$$

$$\frac{d}{dx}\left[e^{-2x}y\right] = e^{-2x}$$

$$e^{-2x}y = -\frac{1}{2}e^{-2x} + C$$

$$y = -\frac{1}{2} + Ce^{2x}$$

8. $y' - \dfrac{2}{x}y = -1;$ $H(x) = \displaystyle\int -\frac{2}{x}\,dx,$ integrating factor: x^{-2}

$$x^{-2}y' - \frac{2}{x^3}y = -x^{-2}$$

$$\frac{d}{dx}(x^{-2}y) = -x^{-2}$$

$$x^{-2}y = \frac{1}{x} + C$$

$$y = x + Cx^2$$

9. $y' + \dfrac{5}{2}y = 1;$ $H(x) = \displaystyle\int \left(\dfrac{5}{2}\right) dx = \dfrac{5}{2}x,$ integrating factor: $e^{5x/2}$

$$e^{5x/2}y' + \dfrac{5}{2}e^{5x/2}y = e^{5x/2}$$
$$\dfrac{d}{dx}\left[e^{5x/2}y\right] = e^{5x/2}$$
$$e^{5x/2}y = \dfrac{2}{5}e^{5x/2} + C$$
$$y = \dfrac{2}{5} + Ce^{-5x/2}$$

10. $y' - y = -2e^{-x};$ $H(x) = \displaystyle\int -\,dx,$ integrating factor: e^{-x}

$$e^{-x}y' - e^{-x}y = -2e^{-2x}$$
$$\dfrac{d}{dx}\left(e^{-x}y\right) = -2e^{-2x}$$
$$e^{-x}y = e^{-2x} + C$$
$$y = e^{-x} + Ce^{x}$$

11. $y' - 2y = 1 - 2x;$ $H(x) = \displaystyle\int(-2)\,dx = -2x,$ integrating factor: e^{-2x}

$$e^{-2x}y' - 2e^{-2x}y = e^{-2x} - 2xe^{-2x}$$
$$\dfrac{d}{dx}\left[e^{-2x}y\right] = e^{-2x} - 2xe^{-2x}$$
$$e^{-2x}y = -\dfrac{1}{2}e^{-2x} + x\,e^{-2x} + \dfrac{1}{2}e^{-2x} + C = x\,e^{-2x} + C$$
$$y = x + Ce^{2x}$$

12. $y' + \dfrac{2}{x}y = \dfrac{\cos x}{x^2};$ $H(x) = \displaystyle\int \dfrac{2}{x}\,dx = 2\ln|x|,$ integrating factor: x^2

$$x^2y' + 2xy = \cos x$$
$$\dfrac{d}{dx}[x^2y] = \cos x$$
$$x^2y = \sin x + C$$
$$y = \dfrac{\sin x}{x^2} + \dfrac{C}{x^2}$$

13. $y' - \dfrac{4}{x}y = -2n;$ $H(x) = \displaystyle\int \left(-\dfrac{4}{x}\right) dx = -4\ln x = \ln x^{-4},$ integrating factor: $e^{\ln x^{-4}} = x^{-4}$

$$x^{-4}y' - \dfrac{4}{x}x^{-4}y = -2nx^{-4}$$
$$\dfrac{d}{dx}\left[x^{-4}y\right] = -2nx^{-4}$$
$$x^{-4}y = \dfrac{2}{3}nx^{-3} + C$$
$$y = \dfrac{2}{3}nx + Cx^4$$

14. $y' + y = 2 + 2x$; $H(x) = \int dx$,   integrating factor: e^x

$$e^x y' + e^x y = (2 + 2x)e^x$$

$$\frac{d}{dx}(e^x y) = 2(1 + x)e^x$$

$$e^x y = 2xe^x + C$$

$$y = 2x + Ce^{-x}$$

15. $y' - e^x\, y = 0$; $H(x) = \int -e^x\, dx = -e^x$,   integrating factor: e^{-e^x}

$$e^{-e^x} y' - e^x e^{-e^x} y = 0$$

$$\frac{d}{dx}\left[e^{-e^x} y\right] = 0$$

$$e^{-e^x} y = C$$

$$y = Ce^{e^x}$$

16. $y' - y = e^x$; $H(x) = \int - dx$,   integrating factor: e^{-x}

$$e^{-x} y' - e^{-x} y = 1$$

$$\frac{d}{dx}(e^{-x} y) = 1$$

$$e^{-x} y = x + C$$

$$y = xe^x + Ce^x$$

17. $y' + \dfrac{1}{1 + e^x}\, y = \dfrac{1}{1 + e^x}$; $H(x) = \displaystyle\int \frac{1}{1 + e^x}\, dx = \ln\frac{e^x}{1 + e^x}$,

integrating factor: $e^{H(x)} = \dfrac{e^x}{1 + e^x}$

$$\frac{e^x}{1 + e^x}\, y' + \frac{1}{1 + e^x} \cdot \frac{e^x}{1 + e^x}\, y = \frac{1}{1 + e^x} \cdot \frac{e^x}{1 + e^x}$$

$$\frac{d}{dx}\left[\frac{e^x}{1 + e^x}\, y\right] = \frac{e^x}{(1 + e^x)^2}$$

$$\frac{e^x}{1 + e^x}\, y = -\frac{1}{1 + e^x} + C$$

$$y = -e^{-x} + C\left(1 + e^{-x}\right)$$

This solution can also be written: $y = 1 + K\left(e^{-x} + 1\right)$,   where K is an arbitrary constant.

18. $y' + \dfrac{1}{x}y = \dfrac{1 + x}{x}e^x$; $H(x) = \displaystyle\int \frac{1}{x}\, dx$,   integrating factor: x

$$xy' + y = (1 + x)e^x$$

$$\frac{d}{dx}(xy) = (1 + x)e^x$$

$$xy = xe^x + C$$

$$y = e^x + \frac{C}{x}$$

19. $y' + 2xy = xe^{-x^2}$; $H(x) = \displaystyle\int 2x\,dx = x^2$,   integrating factor: e^{x^2}

$$e^{x^2}\,y' + 2xe^{x^2}\,y = x$$
$$\frac{d}{dx}\left[e^{x^2}\,y\right] = x$$
$$e^{x^2}\,y = \frac{1}{2}\,x^2 + C$$
$$y = e^{-x^2}\left(\tfrac{1}{2}\,x^2 + C\right)$$

20. $y' - \dfrac{1}{x}y = 2\ln x$; $H(x) = \displaystyle\int -\frac{1}{x}\,dx$,   integrating factor: $\dfrac{1}{x}$

$$\frac{1}{x}y' - \frac{1}{x^2}y = \frac{2}{x}\ln x$$
$$\frac{d}{dx}\left(\frac{1}{x}y\right) = \frac{2}{x}\ln x$$
$$\frac{1}{x}y = (\ln x)^2 + C$$
$$y = x(\ln x)^2 + Cx$$

21. $y' + \dfrac{2}{x+1}y = 0$; $H(x) = \displaystyle\int \frac{2}{x+1}\,dx = 2\ln(x+1) = \ln(x+1)^2$,

integrating factor: $e^{\ln(x+1)^2} = (x+1)^2$

$$(x+1)^2\,y' + 2(x+1)\,y = 0$$
$$\frac{d}{dx}\left[(x+1)^2\,y\right] = 0$$
$$(x+1)^2\,y = C$$
$$y = \frac{C}{(x+1)^2}$$

22. $y' + \dfrac{2}{x+1}y = (x+1)^{5/2}$; $H(x) = \displaystyle\int \frac{2}{x+1}\,dx$,   integrating factor: $(x+1)^2$

$$(x+1)^2y' + 2(x+1)y = (x+1)^{9/2}$$
$$\frac{d}{dx}\left[(x+1)^2y\right] = (x+1)^{9/2}$$
$$(x+1)^2y = \frac{2}{11}(x+1)^{11/2} + C$$
$$y = \frac{2}{11}(x+1)^{7/2} + C(x+1)^{-2}$$

23. $y' + y = x$; $H(x) = \displaystyle\int 1\,dx = x$,   integrating factor : e^x

$$e^x\,y' + e^x\,y = xe^x$$
$$\frac{d}{dx}\left[e^x\,y\right] = xe^x$$
$$e^x\,y = xe^x - e^x + C$$
$$y = (x-1) + Ce^{-x}$$

$y(0) = -1 + C = 1 \implies C = 2$. Therefore, $y = 2e^{-x} + x - 1$ is the solution which satisfies the initial condition.

24. $y' - y = e^{2x}$; $\qquad H(x) = \int -\,dx$, integrating factor: e^{-x}

$$\frac{d}{dx}(e^{-x}y) = e^x$$
$$e^{-x}y = e^x + C$$
$$y = e^{2x} + Ce^x$$

$1 = y(1) = e^2 + Ce \implies C = \dfrac{1 - e^2}{e}$ and $y = e^{2x} + \dfrac{1 - e^2}{e}e^x$ is the solution which satisfies the initial condition.

25. $y' + y = \dfrac{1}{1 + e^x}$; $\qquad H(x) = \int 1\,dx = x$, integrating factor : e^x

$$e^x\,y' + e^x\,y = \frac{e^x}{1 + e^x}$$
$$\frac{d}{dx}\left[e^x\,y\right] = \frac{e^x}{1 + e^x}$$
$$e^x\,y = \ln\left(1 + e^x\right) + C$$
$$y = e^{-x}\left[\ln\left(1 + e^x\right) + C\right]$$

$y(0) = \ln 2 + C = e \implies C = e - \ln 2$. Therefore, $y = e^{-x}\left[\ln\left(1 + e^x\right) + e - \ln 2\right]$ is the solution which satisfies the initial condition.

26. $y' + y = \dfrac{1}{1 + 2e^x}$; $\qquad H(x) = \int dx$, integrating factor: e^x

$$\frac{d}{dx}(e^x y) = \frac{e^x}{1 + 2e^x}$$
$$e^x y = \frac{1}{2}\ln(1 + 2e^x) + C$$
$$y = e^{-x}\left[\frac{1}{2}\ln(1 + 2e^x) + C\right]$$

$e = y(0) = \frac{1}{2}\ln 3 + C \implies C = e - \dfrac{1}{2}\ln 3$ and $y = e^{-x}\left[\frac{1}{2}\ln(1 + 2e^x) + e - \frac{1}{2}\ln 3\right]$ is the solution which satisfies the initial condition.

27. $y' - \dfrac{2}{x}y = x^2 e^x$; $\qquad H(x) = \int\left(-\dfrac{2}{x}\right)dx = -2\ln x = \ln x^{-2}$,

integrating factor: $e^{\ln x^{-2}} = x^{-2}$

$$x^{-2}\,y' - 2x^{-3}\,y = e^x$$
$$\frac{d}{dx}\left[x^{-2}\,y\right] = e^x$$
$$x^{-2}\,y = e^x + C$$
$$y = x^2\left(e^x + C\right)$$

$y(1) = e + C = 0 \implies C = -e.$ Therefore, $y = x^2(e^x - e)$ is the solution which satisfies the initial condition.

28. $y' + \dfrac{2}{x}y = e^{-x};$ $H(x) = \displaystyle\int \dfrac{2}{x}\,dx,$ integrating factor: x^2

$$\frac{d}{dx}(x^2 y) = x^2 e^{-x}$$
$$x^2 y = -e^{-x}(x^2 + 2x + 2) + C$$
$$y = -\frac{e^{-x}}{x^2}(x^2 + 2x + 2) + \frac{C}{x^2}$$

$-1 = y(1) = -5e^{-1} + C \implies C = 5e^{-1} - 1$ and $y = -\dfrac{e^{-x}}{x^2}(x^2 + 2x + 2) + \dfrac{5e^{-1} - 1}{x^2}$ is the solution which satisfies the initial condition.

29. Set $z = y' - y.$ Then $z' = y'' - y'.$

$$y' - y = y'' - y' \implies z = z' \implies z = C_1 e^x$$

Now,

$$z = y' - y = C_1 e^x \implies e^{-x}y' - e^{-x}y = C_1 \implies (e^{-x}y)' = C_1$$
$$\implies e^{-x}y = C_1 x + C_2 \implies y = C_1 x e^x + C_2 e^x$$

30. General solution: $y = Ce^{-rx}.$ (a) $y(a) = 0 = Ce^{-ra} \implies C = 0 \implies y(x) = 0$ for all $x.$

(b) $r < 0$ and $y \neq 0 \implies y \to \infty$ as $x \to \infty$

(c) $r > 0$ and $y \neq 0 \implies y \to 0$ as $x \to \infty$

(d) If $r = 0,$ then $y(x) = C,$ constant.

31. (a) Let y_1 and y_2 be solutions of $y' + p(x)y = 0,$ and let $u = y_1 + y_2.$ Then

$$u' + pu = (y_1 + y_2)' + p(y_1 + y_2)$$
$$= y_1' + y_2' + py_1 + py_2$$
$$= y_1' + py_1 + y_2' + py_2 = 0 + 0 = 0$$

Therefore u is a solution.

(b) Let $u = Cy$ where y is a solution of $y' + p(x)y = 0.$ Then

$$u' + pu = (Cy)' + p(Cy) = Cy' + Cpy = C(y' + py) = C \times 0 = 0$$

Therefore u is a solution.

32. (a)
$$y' + p(x)y = 0$$
$$e^{\int_a^x p(t)\,dt}y' + p(x)e^{\int_a^x p(t)\,dt}y = 0$$
$$\left[e^{\int_a^x p(t)\,dt}y\right]' = 0$$
$$e^{\int_a^x p(t)\,dt}y = C$$
$$y(x) = Ce^{-\int_a^x p(t)\,dt}$$

(b) $y(b) = 0 \implies Ce^{-\int_a^b p(t)\,dt} = 0 \implies C = 0 \implies y(x) = 0$ for all x.

(c) Let $z = y_1 - y_2$. Then z is a solution of $y' + p(x)y = 0$. If $y_1(b) = y_2(b)$, then $z(b) = 0 \implies$ $z(x) = 0$ for all x.

33. Let $y(x) = e^{-H(x)}\int_a^x q(t)\,e^{H(t)}\,dt$.

Note first that $y(a) = e^{-H(a)}\int_a^a q(t)\,e^{H(t)}\,dt = 0$ so y satisfies the initial condition.

Now,

$$y' + p(x)y = \left[e^{-H(x)}\int_a^x q(t)\,e^{H(t)}\,dt\right]' + p(x)\,e^{-H(x)}\int_a^x q(t)\,e^{H(t)}\,dt$$

$$= e^{-H(x)}q(x)\,e^{H(x)} + e^{-H(x)}[-p(x)]\int_a^x q(t)\,e^{H(t)}\,dt + p(x)\,e^{-H(x)}\int_a^x q(t)\,e^{H(t)}\,dt$$

$$= q(x)$$

Thus, $y(x) = e^{-H(x)}\int_a^x q(t)\,e^{H(t)}\,dt$ is the solution of the initial value problem.

34. Let $z = y_1 - y_2$. Then

$$z' = y_1' - y_2' = q - py_1 - (q - py_2) = -p\,(y_1 - y_2) = -pz \implies z' + pz = 0$$

35. According to Newton's Law of Cooling, the temperature T at any time t is given by

$$T(t) = 32 + [72 - 32]e^{-kt}$$

We can determine k by applying the condition $T(1/2) = 50°$:

$$50 = 32 + 40\,e^{-k/2}$$
$$e^{-k/2} = \frac{18}{40} = \frac{9}{20}$$
$$-\tfrac{1}{2}k = \ln(9/20)$$
$$k = -2\,\ln(9/20) \cong 1.5970$$

Therefore, $T(t) \cong 32 + 40\,e^{-1.5970t}$.

Now, $T(1) \cong 32 + 40\,e^{-1.5970} \cong 40.100;$ the temperature after 1 minute is (approx.) $40.10°$.

To find how long it will take for the temperature to reach $35°$, we solve

$$32 + 40\,e^{-1.5970t} = 35$$

for t:

$$32 + 40\,e^{-1.5970t} = 35$$

$$40\,e^{-1.5970t} = 3$$

$$-1.5970t = \ln(3/40)$$

$$t = \frac{\ln(3/40)}{-1.5970} \cong 1.62$$

It will take approximately 1.62 minutes for the thermometer to read $35°$.

36. By (9.1.4) $T(t) = 100 - 80e^{-kt}$

$$T(2) = 22 \implies 100 - 80e^{-2k} = 22 \implies k = \frac{\ln(39/40)}{-2} \cong 0.01266$$

$T(6) \cong 100 - 80e^{-0.01266(6)} \cong 25.85°\,C;$ $T(t) = 90 \implies -80e^{-0.01266t} = -10 \implies t \cong 164.25\,\text{secs}.$

37. (a) The solution of the initial value problem $v' = 32 - kv,\ (k > 0)\ v(0) = 0$ is:

$$v(t) = \frac{32}{k}\left(1 - e^{-kt}\right).$$

(b) At each time t, $1 - e^{-kt} < 1$. Therefore

$$v(t) = \frac{32}{k}\left(1 - e^{-kt}\right) < \frac{32}{k} \quad \text{and} \quad \lim_{t \to \infty} v(t) = \frac{32}{k}$$

(c)

38. (a) $\dfrac{dP}{dt} + (b - a)P = 0;$ $H(t) = \int (b - a)\,dt = (b - a)t,$ integrating factor : $e^{(b-a)t}$

$$e^{(b-a)t}\frac{dP}{dt} + (b - a)e^{(b-a)t}\,P = 0$$

$$\frac{d}{dt}\left[e^{(b-a)t}\,P\right] = 0$$

$$e^{(b-a)t}P = C$$

$$P = Ce^{(a-b)t}$$

$P(0) = P_0 \implies P(t) = P_0 e^{(a-b)t}.$

(b) (i) $a > b \Longrightarrow P_0 e^{(a-b)t}$ is increasing.

$\quad\quad P(t) \to \infty$ as $t \to \infty$.

$\quad$ (ii) $a = b \Longrightarrow P(t) = P_0$ is a constant.

$\quad$ (iii) $a < b \Longrightarrow P_0 e^{(a-b)t}$ is decreasing.

$\quad\quad P(t) \to 0$ as $t \to \infty$.

39. (a) $\dfrac{di}{dt} + \dfrac{R}{L} i = \dfrac{E}{L};$ $H(t) = \displaystyle\int \dfrac{R}{L} \, dt = \dfrac{R}{L} t,$ integrating factor : $e^{\frac{R}{L} t}$

$$e^{\frac{R}{L} t} \dfrac{di}{dt} + \dfrac{R}{L} e^{\frac{R}{L} t} i = \dfrac{E}{L} e^{\frac{R}{L} t}$$

$$\dfrac{d}{dt} \left[e^{\frac{R}{L} t} i \right] = \dfrac{E}{L} e^{\frac{R}{L} t}$$

$$e^{\frac{R}{L} t} i = \dfrac{E}{R} e^{\frac{R}{L} t} + C$$

$$i(t) = \dfrac{E}{R} + C e^{-\frac{R}{L} t}$$

$i(0) = 0 \implies C = -\frac{E}{R}, \quad$ so $\quad i(t) = \dfrac{E}{R} \left[1 - e^{-(R/L)\,t} \right].$

(b) $\displaystyle\lim_{t\to\infty} i(t) = \lim_{t\to\infty} \dfrac{E}{R} \left(1 - e^{-(R/L)\,t} \right) = \dfrac{E}{R}$ amps

(c) $i(t) = 0.9 \dfrac{E}{R} \implies e^{-(R/L)\,t} = \dfrac{1}{10} \implies -\dfrac{R}{L} t = -\ln 10 \implies t = \dfrac{L}{R} \ln 10$ seconds.

40. (a) $\dfrac{di}{dt} + \dfrac{R}{L} i = \dfrac{E}{L} \sin \omega t;$ $H(t) = \displaystyle\int \dfrac{R}{L} \, dt = \dfrac{R}{L} t,$ integrating factor : $e^{\frac{R}{L} t}$

$$e^{\frac{R}{L} t} \dfrac{di}{dt} + \dfrac{R}{L} e^{\frac{R}{L} t} i = \dfrac{E}{L} e^{\frac{R}{L} t} \sin \omega t$$

$$\dfrac{d}{dt} \left[e^{\frac{R}{L} t} i \right] = \dfrac{E}{L} e^{\frac{R}{L} t} \sin \omega t$$

$$e^{\frac{R}{L} t} i = \dfrac{E}{L} e^{\frac{R}{L} t} \dfrac{L^2}{R^2 + \omega^2 L^2} \left[\dfrac{R}{L} \sin \omega t - \omega \cos \omega t \right] + C$$

$$i(t) = \dfrac{EL}{R^2 + \omega^2 L^2} \left[\dfrac{R}{L} \sin \omega t - \omega \cos \omega t \right] + C e^{-\frac{R}{L} t}$$

$i(0) = i_0 \implies$ i(t)$= \dfrac{EL}{R^2 + \omega^2 L^2} \left[\dfrac{R}{L} \sin \omega t - \omega \cos \omega t \right] + \left[i_0 + \omega \dfrac{EL}{R^2 + \omega^2 L^2} \right] e^{-\frac{R}{L} t}.$

(b) $\displaystyle\lim_{t\to\infty}$ does not exist because the trigonometric functions continue to oscillate.

(c) A sample graph is:

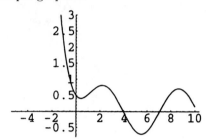

41. (a) $V'(t) = kV(t) \implies V(t) = V_0 e^{kt}$

Loses 20% in 5 minutes, so $V(5) = V_0 e^{5k} = 0.8V_0 \implies k = \frac{1}{5}\ln 0.8$

$\implies V(t) = V_0 e^{\frac{1}{5}(\ln 0.8)t} = V_0 \left(e^{\ln 0.8}\right)^{t/5} = V_0 (0.8)^{t/5} = V_0 \left(\frac{4}{5}\right)^{t/5}$.

Since $V_0 = 200$ liters, we get $V(t) = 200 \left(\frac{4}{5}\right)^{t/5}$

(b)

$$V'(t) = ktV(t)$$

$$V'(t) - ktV(t) = 0$$

$$e^{-kt^2/2}V'(t) - kte^{-kt^2/2} V(t) = 0$$

$$\frac{d}{dt}\left[e^{-kt^2/2} V(t)\right] = 0$$

$$e^{-kt^2/2} V(t) = C$$

$$V(t) = Ce^{kt^2/2}.$$

$V(0) = C = 200 \implies V(t) = 200e^{kt^2/2}$.

$V(5) = 160 \implies 200e^{k(25/2)} = 160, \quad e^{k(25/2)} = \frac{4}{5}, \quad e^k = \left(\frac{4}{5}\right)^{2/25}$.

Therefore $V(t) = 200 \left(\frac{4}{5}\right)^{t^2/25}$ liters.

42. Let $s(t)$ be the number of pounds of salt present after t minutes. Since

$$s'(t) = \text{ rate in } - \text{ rate out } = 3\,(0.2) - 3\left(\frac{s(t)}{100}\right),$$

we have

$$s'(t) + 0.03s(t) = 0.6.$$

Multiply by $e^{\int 0.03dt} = e^{0.03t}$:

$$e^{0.03t}s'(t) + 0.03e^{0.03t}s(t) = 0.6e^{0.03t}$$

$$\frac{d}{dt}\left[e^{0.03t}s(t)\right] = 0.6e^{0.03t}$$

$$e^{0.03t}s(t) = 20e^{0.03t} + C$$

$$s(t) = 20 + Ce^{-0.03t}.$$

Use the initial condition $s(0) = 100(0.25) = 25$ to determine C: $\quad 25 = 20 + Ce^0 \quad$ so $\quad C = 5$.

Thus, $s(t) = 20 + 5e^{-0.03t}$ lb.

43. (a) $\dfrac{dP}{dt} = k(M - P)$

(b) $\dfrac{dP}{dt} + kP = kM;$ $H(t) = \displaystyle\int k\,dt = kt,$ integrating factor : e^{kt}

$$e^{kt}\dfrac{dP}{dt} + ke^{kt}\,P = kM\,e^{kt}$$
$$\dfrac{d}{dt}\left[e^{kt}\,P\right] = kM\,e^{kt}$$
$$e^{kt}P = M\,e^{kt} + C$$
$$P = M + Ce^{-kt}$$

$P(0) = M + C = 0 \implies C = -M$ and $P(t) = M\left(1 - e^{-kt}\right)$

$P(10) = M\left(1 - e^{-10k}\right) = 0.3M \implies k \cong 0.0357$ and $P(t) = M\left(1 - e^{-0.0357t}\right)$

(c) $P(t) = M\left(1 - e^{-0.0357t}\right) = 0.9M \implies e^{-0.0357t} = 0.1 \implies t \cong 65$

Therefore, it will take approximately 65 days for 90 % of the population to be aware of the product.

44. (a) $\dfrac{dQ}{dt} = $ rate in $-$ rate out $= r - kQ,$ $k > 0$

(b) $\dfrac{dQ}{dt} + kQ = r,$ $Q(0) = 0 \implies Q(t) = \dfrac{r}{k}\left(1 - e^{-kt}\right)$

(c) $\displaystyle\lim_{t\to\infty} Q(t) = \dfrac{r}{k}.$

45. (a)

$\dfrac{dP}{dt} - 2\cos(2\pi t)P = 0 \implies P = Ce^{\frac{1}{\pi}\sin 2\pi t}.$

$P(0) = C = 1000 \implies P = 1000e^{\frac{1}{\pi}\sin 2\pi t}.$

(b)

$\dfrac{dP}{dt} - 2\cos(2\pi t)P = 2000\cos 2\pi t \implies P = Ce^{\frac{1}{\pi}\sin 2\pi t} - 1000.$

$P(0) = 1000 \implies C = 2000 \implies P = 2000e^{\frac{1}{\pi}\sin 2\pi t} - 1000.$

46. (a) Let $Q = \ln P.$ Then $\dfrac{dQ}{dt} = \dfrac{1}{P}\dfrac{dP}{dt} = a - bQ.$

Solving the differential equation $\dfrac{dQ}{dt} + bQ = a \implies Q = \dfrac{a}{b} + Ce^{-bt},$ so $P = e^{\frac{a}{b} + Ce^{-bt}}.$

$P(0) = P_0 \implies e^c = P_0 e^{-\frac{a}{b}}.$ Thus $P = e^{\frac{a}{b}}\left[P_0 e^{-\frac{a}{b}}\right]^{e^{-bt}}.$

(b) $e^{-bt} \to 0$ as $t \to \infty$, so $P \to e^{\frac{a}{b}}$.

(c) $P' = P(a - b\ln P) \implies P'' = P\left(\dfrac{-b}{P}\right) P' + P'(a - b\ln P) = P(a - b\ln P)(a - b - b\ln P)$.

If $0 < P < e^{a/b-1}$, then P is increasing and the graph is concave up; if $e^{a/b-1} < P < e^{a/b}$, then P is increasing and the graph is concave down; if $e^{a/b} < P$, then P is decreasing and the graph is concave down.

(d)

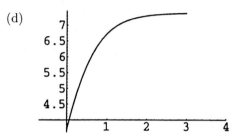

SECTION 9.2

1.
$$y' = y\sin(2x + 3)$$
$$\frac{1}{y}\,dy = \sin(2x + 3)\,dx$$
$$\int \frac{1}{y}\,dy = \int \sin(2x + 3)\,dx$$
$$\ln|y| = -\frac{1}{2}\cos(2x + 3) + C$$

This solution can also be written: $y = Ce^{-(1/2)\cos(2x+3)}$.

2.
$$y' = (x^2 + 1)(y^2 + y)$$
$$\int \frac{dy}{y^2 + y} = \int (x^2 + 1)\,dx$$
$$\ln\left|\frac{y}{y + 1}\right| = \frac{x^3}{3} + x + C$$

This solution can also be written: $y = \dfrac{1}{Ke^{-x-x^3/3} - 1}\,(K = e^C)$.

3.
$$y' = (xy)^3$$
$$\frac{1}{y^3}\,dy = x^3\,dx, \qquad y \neq 0$$
$$\int \frac{1}{y^3}\,dy = \int x^3\,dx$$
$$-\frac{1}{2}y^{-2} = \frac{1}{4}x^4 + C$$

This solution can also be written: $x^4 + \dfrac{2}{y^2} = C, \quad$ or $\quad y^2 = \dfrac{2}{C - x^4}$;

4.

$$y' = 3x^2(1 + y^2)$$

$$\int \frac{dy}{1 + y^2} = \int 3x^2 \, dx$$

$$\tan^{-1} y = x^3 + C$$

$$y = \tan(x^3 + C)$$

5.

$$y' = \frac{\sin(1/x)}{x^2 y \cos y}$$

$$y \cos y \, dy = \frac{1}{x^2} \sin(1/x) \, dx$$

$$\int y \cos y \, dy = \int \frac{1}{x^2} \sin(1/x) \, dx$$

$$y \sin y + \cos y = \cos(1/x) + C$$

6.

$$y' = \frac{y^2 + 1}{y + yx}$$

$$\int \frac{y}{1 + y^2} \, dy = \int \frac{1}{1 + x} \, dx$$

$$\ln \sqrt{1 + y^2} = \ln |1 + x| + C$$

$$1 + y^2 = K(1 + x)^2 \quad (K = \ln C)$$

7.

$$y' = x \, e^{y+x}$$

$$e^{-y} \, dy = x \, e^x \, dx$$

$$\int e^{-y} \, dy = \int x \, e^x \, dx$$

$$-e^{-y} = x \, e^x - e^x + C$$

$$e^{-y} = e^x - x \, e^x + C$$

This solution can also be written: $y = -\ln(e^x - xe^x + C)$.

8.

$$y' = xy^2 - x - y^2 + 1 = (x - 1)(y^2 - 1)$$

$$\int \frac{dy}{y^2 - 1} = \int \frac{dx}{x - 1}$$

$$\frac{1}{2} \ln \left| \frac{y - 1}{y + 1} \right| = \ln |x - 1| + C$$

This solution can also be written: $y = \dfrac{1 + Ke^{x^2 - 2x}}{1 - Ke^{x^2 - 2x}} \quad (K = e^C)$.

9.

$$(y \ln x)y' = \frac{(y + 1)^2}{x}$$

$$\frac{y}{(y + 1)^2} \, dy = \frac{1}{x \ln x} \, dx$$

$$\int \frac{y}{(y + 1)^2} \, dy = \int \frac{1}{x \ln x} \, dx$$

$$\ln |y + 1| + \frac{1}{y + 1} = \ln |\ln x| + C$$

10.
$$e^y \sin 2x \, dx + \cos x (e^{2y} - y) \, dy = 0$$
$$\int \frac{\sin 2x}{\cos x} \, dx + \int (e^y - ye^{-y}) \, dy = C$$
$$-2 \cos x + e^y + e^{-y}(1+y) = C$$

11.
$$(y \ln x) y' = \frac{y^2 + 1}{x}$$
$$\frac{y}{y^2 + 1} \, dy = \frac{1}{x \ln x} \, dx$$
$$\int \frac{y}{y^2 + 1} \, dy = \int \frac{1}{x \ln x} \, dx$$
$$\tfrac{1}{2} \ln (y^2 + 1) = \ln |\ln x| + K = \ln |C \ln x| \quad (K = \ln |C|)$$
$$\ln (y^2 + 1) = 2 \ln |C \ln x| = \ln (C \ln x)^2$$
$$y^2 = C(\ln x)^2 - 1$$

12.
$$y' = \frac{1 + 2y^2}{y \sin x}$$
$$\frac{y}{1 + 2y^2} \, dy = \csc x \, dx$$
$$\tfrac{1}{4} \ln(1 + 2y^2) = \ln |\csc x - \cot x| + C$$

the integral curves can be written as: $\ln(1 + 2y^2) = \ln \left[C(\csc x - \cot x)^4 \right]$, or as
$y^2 = K(\csc x - \cot x)^4 - \tfrac{1}{2}$.

13.
$$y' = x \sqrt{\frac{1 - y^2}{1 - x^2}}, \qquad y(0) = 0$$
$$\frac{1}{\sqrt{1 - y^2}} \, dy = \frac{x}{\sqrt{1 - x^2}} \, dx$$
$$\int \frac{1}{\sqrt{1 - y^2}} \, dy = \int \frac{x}{\sqrt{1 - x^2}} \, dx$$
$$\sin^{-1} y = -\sqrt{1 - x^2} + C$$
$$y(0) = 0 \quad \Longrightarrow \quad \arcsin 0 = -1 + C \quad \Longrightarrow \quad C = 1$$

Thus, $\arcsin y = 1 - \sqrt{1 - x^2}$.

14.
$$y' = \frac{e^{x-y}}{1 + e^x}$$
$$\int e^y \, dy = \int \frac{e^x}{1 + e^x} \, dx$$
$$e^y = \ln(1 + e^x) + C$$
$$y(1) = 0 \implies 1 = \ln(1 + e) + C \implies C = 1 - \ln(1 + e) \quad \text{and} \quad e^y = \ln(1 + e^x) + 1 - \ln(1 + e)$$

15.

$$y' = \frac{x^2 y - y}{y + 1}, \qquad y(3) = 1$$

$$\frac{y+1}{y}\, dy = (x^2 - 1)\, dx, \qquad y \neq 0$$

$$\int \frac{y+1}{y}\, dy = \int (x^2 - 1)\, dx$$

$$y + \ln|y| = \frac{1}{3} x^3 - x + C$$

$$y(3) = 1 \quad \Longrightarrow \quad 1 + \ln 1 = \frac{1}{3}(3)^3 - 3 + C \quad \Longrightarrow \quad C = -5.$$

Thus, $\quad y + \ln|y| = \frac{1}{3}x^3 - x - 5.$

16.

$$x^2 y' = y - xy$$

$$\int \frac{1}{y}\, dy = \int (1-x)x^{-2}\, dx$$

$$\ln|y| = -\frac{1}{x} - \ln|x| + C$$

$$-1 = y(-1) \implies C = -1 \quad \text{and} \quad \ln|xy| + \frac{1}{x} = -1$$

17. $\qquad (xy^2 + y^2 + x + 1)\, dx + (y-1)\, dy = 0, \qquad y(2) = 0$

$$(x+1)(y^2+1)\, dx + (y-1)\, dy = 0$$

$$(x+1)\, dx + \frac{y-1}{y^2+1}\, dy = 0$$

$$\int (x+1)\, dx + \int \frac{y-1}{y^2+1}\, dy = C$$

$$\frac{x^2}{2} + x + \frac{1}{2}\ln(y^2+1) - \tan^{-1} y = C$$

$y(2) = 0 \quad \Longrightarrow \quad C = 4.$ Thus, $\frac{1}{2}x^2 + x + \frac{1}{2}\ln(y^2+1) - \tan^{-1} y = 4$

18.

$$\cos y\, dx + (1 + e^{-x})\sin y\, dy = 0$$

$$\int \frac{dx}{1+e^{-x}} + \int \frac{\sin y}{\cos y}\, dy = C$$

$$\ln(e^x + 1) + \ln|\sec y| = C;$$

$$\frac{\pi}{4} = y(0) \implies \ln 2 + \ln\sqrt{2} = C$$

$$\ln(e^x + 1) + \ln|\sec y| = \tfrac{3}{2}\ln 2$$

19. $\qquad y' = 6\, e^{2x - y}, \qquad y(0) = 0$

$$y' = 6\, e^{2x - y}$$

$$e^y\, dy = 6\, e^{2x}\, dx$$

$$e^y = 3\, e^{2x} + C$$

$$y(0) = 0 \quad \Longrightarrow \quad 1 = 3 + C \quad \Longrightarrow \quad C = -2$$

Thus, $\quad e^y = 3\, e^{2x} - 2 \quad \Longrightarrow \quad y = \ln\left[3\, e^{2x} - 2\right]$

20. $xy' - y = 2x^2 y,$ $y(1) = 1$

$$xy' = y\left(1 + 2x^2\right)$$

$$\frac{1}{y}\, dy = \frac{1 + 2x^2}{x}\, dx = \left(\frac{1}{x} + 2x\right)\, dx$$

$$\ln|y| = \ln|x| + x^2 + C$$

$y(1) = 1$ $\implies$ $0 = 1 + C$ $\implies$ $C = -1$

Thus, $\ln|y| = \ln|x| + x^2 - 1$ or $y = xe^{x^2 - 1}$

21.   We assume that $C = 0$ at time $t = 0$.   (a) Let $A_0 = B_0$. Then

$$\frac{dC}{dt} = k(A_0 - C)^2 \quad \text{and} \quad \frac{dC}{(A_0 - C)^2} = k\, dt.$$

Integrating, we get

$$\int \frac{1}{(A_0 - C)^2}\, dC = \int k\, dt$$

$$\frac{1}{A_0 - C} = kt + M \qquad M \text{ a constant.}$$

Since $C(0) = 0$, $M = \dfrac{1}{A_0}$ and

$$\frac{1}{A_0 - C} = kt + \frac{1}{A_0}.$$

Solving this equation for C gives

$$C(t) = \frac{kA_0^2 t}{1 + kA_0 t}.$$

(b) Suppose that $A_0 \neq B_0$. Then

$$\frac{dC}{dt} = k(A_0 - C)(B_0 - C) \quad \text{and} \quad \frac{dC}{(A_0 - C)(B_0 - C)} = k\, dt.$$

Integrating, we get

$$\int \frac{1}{(A_0 - C)(B_0 - C)}\, dC = \int k\, dt$$

$$\frac{1}{B_0 - A_0} \int \left(\frac{1}{A_0 - C} - \frac{1}{B_0 - C}\right) dC = \int k\, dt$$

$$\frac{1}{B_0 - A_0} \left[- \ln(A_0 - C) + \ln(B_0 - C)\right] = kt + M$$

$$\frac{1}{B_0 - A_0} \ln\left(\frac{B_0 - C}{A_0 - C}\right) = kt + M \qquad M \text{ an arbitrary constant}$$

Since $C(0) = 0$, $M = \frac{1}{B_0 - A_0} \ln\left(\frac{B_0}{A_0}\right)$ and

$$\frac{1}{B_0 - A_0} \ln\left(\frac{B_0 - C}{A_0 - C}\right) = kt + \frac{\ln(B_0/A_0)}{B_0 - A_0}.$$

Solving this equation for C, gives

$$C(t) = \frac{A_0 B_0 \left(e^{kA_0 t} - e^{kB_0 t}\right)}{A_0 e^{kA_0 t} - B_0 e^{kB_0 t}}.$$

22. (a) From (9.2.4) with $\quad K = 0.0020, \quad M = 800, \quad R = P(0) = 100, \quad$ we have

$$P(t) = \frac{80,000}{100 + 700e^{-1.6t}}$$

(b)

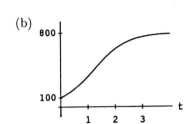

(c) $\dfrac{dP}{dt}$ is maximal at $t \cong 1.2162$

Maximum value $= 320$

23. (a)

$$m\frac{dv}{dt} = -\alpha v - \beta v^2$$

$$\frac{dv}{v(\alpha + \beta v)} = -\frac{1}{m}\,dt$$

$$\int \frac{1}{v(\alpha + \beta v)}\,dv = -\int \frac{1}{m}\,dt$$

$$\frac{1}{\alpha}\int \frac{1}{v}\,dv - \frac{\beta}{\alpha}\int \frac{1}{\alpha + \beta v}\,dv = -\int \frac{1}{m}\,dt$$

$$\frac{1}{\alpha}\ln v - \frac{1}{\alpha}\ln(\alpha + \beta v) = -\frac{1}{m}t + M, \quad M \text{ a constant}$$

$$\ln\left(\frac{v}{\alpha + \beta v}\right) = -\frac{\alpha}{m}t + \alpha M$$

$$\frac{v}{\alpha + \beta v} = Ke^{-\alpha t/m} \quad \left[K = e^{\alpha M}\right]$$

Solving this equation for v we get $\quad v(t) = \dfrac{\alpha K}{e^{\alpha t/m} - \beta K} = \dfrac{\alpha}{Ce^{\alpha t/m} - \beta} \quad [C = 1/K]$.

(b) Setting $v(0) = v_0$, we get

$$C = \frac{\alpha + \beta v_0}{v_0} \quad \text{and}$$

$$v(t) = \frac{\alpha v_0}{(\alpha + \beta v_0)e^{\alpha t/m} - \beta v_0}$$

(c) $\displaystyle\lim_{t\to\infty} v(t) = 0$

24. $\quad F = ma = m\dfrac{dv}{dt}$

(a)

$$m\frac{dv}{dt} = mg - \beta v^2$$

$$dt = \frac{m\,dv}{mg - \beta v^2} = \frac{m}{\beta}\left(\frac{dv}{v_c{}^2 - v^2}\right)$$

$$t = \frac{m}{\beta}\int \frac{1}{v_c{}^2 - v^2}\,dv = \frac{m}{2v_c\beta}\int \left(\frac{1}{v_c + v} + \frac{1}{v_c - v}\right)dv$$

$$= \frac{m}{2v_c\beta}\left[\ln\left(\frac{v_c + v}{v_c - v}\right)\right] + C$$

At $t = 0$, $v(0) = v_0$. Therefore

$$C = -\frac{m}{2v_c\beta}\left[\ln\left(\frac{v_c + v_0}{v_c - v_0}\right)\right] = \frac{m}{2v_c\beta}\left[\ln\left(\frac{v_c - v_0}{v_c + v_0}\right)\right].$$

Thus

$$t = \frac{m}{2v_c\beta}\left[\ln\left(\frac{v_c + v}{v_c + v_0}\cdot\frac{v_c - v_0}{v_c - v}\right)\right] = \frac{v_c}{2g}\left[\ln\left(\frac{v_c + v}{v_c + v_0}\cdot\frac{v_c - v_0}{v_c - v}\right)\right].$$

$$v_c = \sqrt{mg/\beta}$$

(b) $\dfrac{v_c + v}{v_c + v_0}\cdot\dfrac{v_c - v_0}{v_c - v} = e^{2tg/v_c}$

$$v_c + v = \frac{v_c + v_0}{v_c - v_0}e^{2tg/v_c}(v_c - v)$$

$$v\left[1 + \left(\frac{v_c + v_0}{v_c - v_0}\right)e^{2tg/v_c}\right] = v_c\left[\left(\frac{v_c + v_0}{v_c - v_0}\right)e^{2tg/v_c} - 1\right]$$

$$v = v_c\left[\frac{(v_c + v_0)e^{2tg/v_c} - (v_c - v_0)}{(v_c + v_0)e^{2tg/v_c} + (v_c - v_0)}\right].$$

We can bring the hyperbolic functions into play by writing

$$v = v_c\left[\frac{(v_c + v_0)e^{gt/v_c} - (v_c - v_0)e^{-gt/v_c}}{(v_c + v_0)e^{gt/v_c} - (v_c - v_0)e^{-gt/v_c}}\right]$$

$$= v_c\left[\frac{v_0\cosh(gt/v_c) + v_c\sinh(gt/v_c)}{v_0\sinh(gt/v_c) + v_c\cosh(gt/v_c)}\right]$$

(c) $a = g\left\{\dfrac{v_c{}^2 - v_0{}^2}{[v_0\sinh(gt/v_c) + v_c\cosh(gt/v_c)]^2}\right\}$

The acceleration can not change sign since the denominator is always positive and the numerator is constant. As $t \to \infty$, the denominator $\to \infty$, and the fraction $\to 0$.

(d) We can write

$$v = v_c\left[\frac{(v_c + v_0) - (v_c - v_0)e^{-2tg/v_c}}{(v_c + v_0) + (v_c - v_0)e^{-2tg/v_c}}\right].$$

As $t \to \infty$, $-2gt/v_c \to -\infty$ and $e^{-2tg/v_c} \to 0$. Thus $v \to v_c$

25. (a) Let $P = P(t)$ denote the number of people who have the disease at time t. Then, substituting into (9.2.4) with $M = 25,000$ and $R = 100$, we get

$$P(t) = \frac{25,000(100)}{100 + (249,00)e^{-25,000kt}} = \frac{25,000}{1 + 249e^{-25,000kt}}.$$

$$P(10)\frac{25,000}{1 + 249e^{-25,000(10k)}} = 400 \implies -25,000k \cong -0.1398.$$

Therefore, $P(t) = \dfrac{25,000}{1 + 249e^{-0.1398t}}.$

(b) $\dfrac{25,000}{1 + 249e^{-0.1398\,t}} = 12,500 \implies t \cong 40;$ (c)

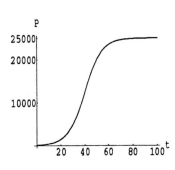

It will take 40 days for half the
population to have the disease.

26. $\dfrac{dy}{dt} = ky(M - y) = kMy - ky^2 \implies \dfrac{d^2y}{dt^2} = (kM - 2ky)\dfrac{dy}{dt} = k^2(M - 2y)(M - y)y$

$\dfrac{d^2y}{dt^2} > 0$ for $0 < y < \dfrac{M}{2},$ so $\dfrac{dy}{dt}$ is increasing

$\dfrac{d^2y}{dt^2} < 0$ for $\dfrac{M}{2} < y < M,$ so $\dfrac{dy}{dt}$ is decreasing

Therefore $\dfrac{dy}{dt}$ is maximal at $y = \dfrac{M}{2}.$ The disease is spreading fastest when half the population is
infected.

27.   Assume that the package is dropped from rest.

(a) Let $v = v(t)$ be the velocity at time $t,\ 0 \le t \le 10.$ Then

$$100\dfrac{dv}{dt} = 100g - 2v \quad \text{or} \quad \dfrac{dv}{dt} + \dfrac{1}{50}v = g \quad (g = 9.8 \text{ m/sec}^2)$$

This is a linear differential equation; $e^{t/50}$ is an integrating factor.

$$e^{t/50}\dfrac{dv}{dt} + \dfrac{1}{50}e^{t/50}v = g\,e^{t/50}$$

$$\dfrac{d}{dt}\left[e^{t/50}v\right] = g\,e^{t/50}$$

$$e^{t/50}v = 50g\,e^{t/50} + C$$

$$v = 50g + Ce^{-t/50}$$

Now, $v(0) = 0 \implies C = -50g$ and $v(t) = 50g\left(1 - e^{-t/50}\right).$

At the instant the parachute opens, $v(10) = 50g\left(1 - e^{-1/5}\right) \cong 50g(0.1813) \cong 88.82 \text{ m/sec}.$

(b) Now let $v = v(t)$ denote the velocity of the package t seconds after the parachute opens. Then

$$100\dfrac{dv}{dt} = 100g - 4v^2 \quad \text{or} \quad \dfrac{dv}{dt} = g - \dfrac{1}{25}v^2$$

This is a separable differential equation:

$$\frac{dv}{dt} = g - \frac{1}{25}v^2 \quad \text{set } u = v/5, \ du = (1/5)dv$$

$$\frac{du}{g - u^2} = \frac{1}{5}dt$$

$$\frac{1}{2\sqrt{g}}\ln\left|\frac{u + \sqrt{g}}{u - \sqrt{g}}\right| = \frac{t}{5} + K$$

$$\ln\left|\frac{u + \sqrt{g}}{u - \sqrt{g}}\right| = \frac{2\sqrt{g}}{5}t + M$$

$$\frac{u + \sqrt{g}}{u - \sqrt{g}} = Ce^{2\sqrt{g}t/5} \cong Ce^{1.25\,t}$$

$$u = \sqrt{g}\,\frac{Ce^{1.25\,t} + 1}{Ce^{1.25\,t} - 1}$$

$$v = 5\sqrt{g}\,\frac{Ce^{1.25\,t} + 1}{Ce^{1.25\,t} - 1}$$

Now, $v(0) = 88.82 \implies 5\sqrt{g}\,\dfrac{C + 1}{C - 1} = 88.82 \implies C \cong 1.43.$

Therefore, $\quad v(t) = 5\sqrt{g}\,\dfrac{1.43e^{1.25\,t} + 1}{1.43e^{1.25\,t} - 1} = \dfrac{15.65\left(1 + 0.70e^{-1.25\,t}\right)}{1 - 0.70e^{-1.25\,t}}$

(c) From part (b), $\quad \lim\limits_{t\to\infty} v(t) = 15.65$ m/sec.

28. (a) By the hint

$$\int \frac{dC}{\left(A_0 - \frac{1}{2}C\right)^2} = \int k\,dt$$

$$\frac{2}{A_0 - \frac{1}{2}C} = kt + K.$$

First, $\quad C(0) = 0 \implies K = 2/A_0.$ Then, $\quad C(1) = A_0 \implies k = 2/A_0.$ Thus,

$$\frac{2}{A_0 - \frac{1}{2}C} = \frac{2}{A_0}(t + 1), \quad \text{which gives} \quad C(t) = 2A_0\left(\frac{t}{t + 1}\right).$$

(b) By the hint

$$\int \frac{dC}{\left(A_0 - \frac{1}{2}C\right)\left(2A_0 - \frac{1}{2}C\right)} = \int k\,dt$$

$$\frac{1}{A_0}\int \left[\frac{1}{A_0 - \frac{1}{2}C} - \frac{1}{2A_0 - \frac{1}{2}C}\right]dC = \int k\,dt$$

$$\frac{1}{A_0}\left[-2\ln\left|A_0-\frac{1}{2}C\right|+2\ln\left|2A_0-\frac{1}{2}C\right|\right]=kt+K$$

$$\frac{2}{A_0}\ln\left|\frac{2A_0-\frac{1}{2}C}{A_0-\frac{1}{2}C}\right|=kt+K.$$

First, $C(0)=0\implies K=\frac{2}{A_0}\ln 2.$ Then,

$$C(1)=A_0\implies\frac{2}{A_0}\ln 3=k+\frac{2}{A_0}\ln 2\implies k=\frac{2}{A_0}\ln\frac{3}{2}.$$

Thus,

$$\frac{2}{A_0}\ln\left|\frac{2A_0-\frac{1}{2}C}{A_0-\frac{1}{2}C}\right|=\frac{2}{A_0}t\ln\frac{3}{2}+\frac{2}{A_0}\ln 2=\frac{2}{A_0}\ln\left[2\left(\frac{3}{2}\right)^t\right]$$

so that

$$\frac{2A_0-\frac{1}{2}C}{A_0-\frac{1}{2}C}=2\left(\frac{3}{2}\right)^t\quad\text{and therefore}\quad C(t)=4A_0\frac{3^t-2^t}{2(3^t)-2^t}.$$

(c) By the hint

$$\int\frac{dC}{\left(A_0-\frac{m}{m+n}C\right)\left(A_0-\frac{n}{m+n}C\right)}=\int k\,dt$$

$$\int\frac{1}{A_0(m-n)}\left[\frac{m}{A_0-\frac{m}{m+n}C}-\frac{n}{A_0-\frac{n}{m+n}C}\right]dC=\int k\,dt$$

$$\frac{1}{A_0(m-n)}\left[-(m+n)\ln\left|A_0-\frac{m}{m+n}C\right|+(m+n)\ln\left|A_0-\frac{n}{m+n}C\right|\right]=kt+K$$

$$\frac{m+n}{A_0(m-n)}\ln\left|\frac{A_0-\frac{n}{m+n}C}{A_0-\frac{m}{m+n}C}\right|=kt+K.$$

First, $C(0)=0\implies K=\frac{m+n}{A_0(m-n)}\ln\left|\frac{A_0}{A_0}\right|=0.$ Then,

$$C(1)=A_0\implies k=\frac{m+n}{A_0(m-n)}\ln\left|\frac{A_0-\frac{n}{m+n}A_0}{A_0-\frac{m}{m+n}A_0}\right|=\frac{m+n}{A_0(m-n)}\ln\frac{m}{n}.$$

Thus,

$$\frac{m+n}{A_0(m-n)} \ln \left| \frac{A_0 - \dfrac{n}{m+n}C}{A_0 - \dfrac{m}{m+n}C} \right| = \frac{m+n}{A_0(m-n)} \ln \left(\frac{m}{n}\right)(t) + 0$$

so that

$$\frac{A_0 - \dfrac{n}{m+n}C}{A_0 - \dfrac{m}{m+n}C} = \left(\frac{m}{n}\right)^t \quad \text{and therefore} \quad C(t) = A_0(m+n)\left[\frac{m^t - n^t}{m^{t+1} - n^{t+1}}\right].$$

PROJECT 9.2

1. (a) $2x + 3y = C \implies 2 + 3y' = 0 \implies y' = -\dfrac{2}{3}$

    The orthogonal trajectories are the solutions of:

    $$y' = \frac{3}{2}.$$

    $y' = \frac{3}{2} \implies y = \frac{3}{2}x + C$

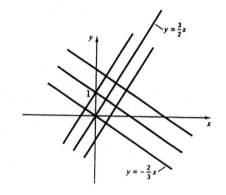

    (b) Curves: $y = Cx, \quad y' = C = \dfrac{y}{x}$

    orthogonal trajectories: $y' = -\dfrac{x}{y}$

    $$\int y\,dy + \int x\,dx = K_1; \quad x^2 + y^2 = K \ (= 2K_1)$$

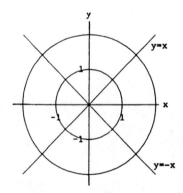

(c) $xy = C \implies y + xy' = 0 \implies y' = -\dfrac{y}{x}$

The orthogonal trajectories are the solutions of:

$$y' = \frac{x}{y}.$$

$$y' = \frac{x}{y}$$

$$\int x\,dx = \int y\,dy$$

$$\tfrac{1}{2}x^2 = \tfrac{1}{2}y^2 + C$$

$$\text{or} \quad x^2 - y^2 = C$$

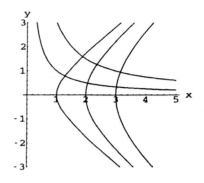

(d) $y = Cx^3, \quad y' = 3Cx^2 = \dfrac{3y}{x}$

orthogonal trajectories: $\quad y' = -\dfrac{x}{3y}$

$$\int 3y\,dy + \int x\,dx = K_1; \quad 3y^2 + x^2 = K \ \ (= 2K_1)$$

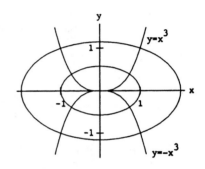

(e) $y = C\,e^x \implies y' = C\,e^x = y$

The orthogonal trajectories are the solutions of:

$$y' = -\frac{1}{y}.$$

$$y' = -\frac{1}{y}$$

$$\int y\,dy = -\int dx$$

$$\tfrac{1}{2}y^2 = -x + K$$

$$\text{or} \quad y^2 = -2x + C$$

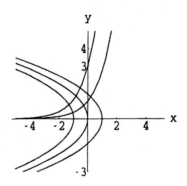

(f) $x = Cy^4, \quad 1 = 4Cy^3 y'; \quad y' = \dfrac{y}{4x}$

orthogonal trajectories: $\quad \dfrac{dy}{dx} = -\dfrac{4x}{y}$

$$\int y\,dy + \int 4x\,dx = K_1; \quad y^2 + 4x^2 = K \ \ (= 2K_1)$$

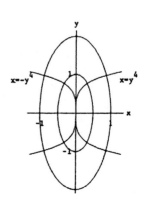

2. (a) Curves: $y^2 - x^2 = C$; $2yy' - 2x = 0 \implies y' = \dfrac{x}{y}$

orthogonal trajectories: $y' = -\dfrac{y}{x}$; $\displaystyle\int \frac{1}{y}\,dy = -\int \frac{1}{x}\,dx$; $y = \dfrac{C}{x}$

(b) Curves: $y^2 = Cx^3$, $2yy' = 3Cx^2$; $y^2 = \dfrac{2xyy'}{3} \implies y' = \dfrac{3y}{2x}$

orthogonal trajectories: $y' = -\dfrac{2x}{3y}$; $\displaystyle\int 3y\,dy + \int 2x\,dx = C$; $x^2 + \dfrac{3}{2}y^2 = C$ or $2x^2 + 3y^2 = C$

(c) Curves: $y = \dfrac{Ce^x}{x}$, $xy = Ce^x$; $xy' + y = Ce^x \implies y' = \dfrac{y(x-1)}{x}$

orthogonal trajectories: $y' = \dfrac{x}{y(1-x)}$; $\displaystyle\int y\,dy = \int \frac{x}{1-x}\,dx = \int \left(\frac{1}{1-x} - 1\right) dx$;

$\tfrac{1}{2}y^2 = -\ln|1-x| - x + C$ or $y^2 + 2x + \ln(1-x)^2 = C$

(d) Curves: $e^x \sin y = C$; $e^x \cos y\, y' + e^x \sin y = 0 \implies y' = -\dfrac{\sin y}{\cos y}$

orthogonal trajectories: $y' = \dfrac{\cos y}{\sin y}$; $\displaystyle\int \frac{\sin y}{\cos y}\,dy = \int dx$; $\ln|\sec y| = x + C$ or $\sec y = Ce^x$

3. (a) A differential equation for the given family is:

$$y^2 = 2xyy' + y^2(y')^2$$

A differential equation for the family of orthogonal trajectories is found by replacing y' by $-1/y'$. The result is:

$$y^2 = -\frac{2xy}{y'} + \frac{y^2}{(y')^2} \qquad \text{which simplifies to} \qquad y^2 = 2xyy' + y^2(y')^2$$

Thus, the given family is self-orthogonal.

(b) $\dfrac{x^2}{C^2} + \dfrac{y^2}{C^2 - 4} = 1 \implies \dfrac{2x}{C^2} + \dfrac{2yy'}{C^2 - 4} = 0 \implies C^2 = \dfrac{4x}{x + yy'}$

A differential equation for the given family is:

$$x^2 + xyy' - \frac{xy}{y'} - y^2 = 4$$

A differential equation for the family of orthogonal trajectories is found by replacing y' by $-1/y'$. The result is:

$$x^2 - xy\frac{1}{y'} + xyy' - y^2 = 4$$

Thus, the given family is self-orthogonal.

SECTION 9.3

1. The characteristic equation is:

$$r^2 + 2r - 8 = 0 \quad \text{or} \quad (r+4)(r-2) = 0.$$

The roots are: $r = -4, \; 2.$ The general solution is:

$$y = C_1 e^{-4x} + C_2 e^{2x}.$$

2. $r^2 - 13r + 42 = 0 \implies r = 6, \; 7; \quad y = C_1 e^{6x} + C_2 e^{7x}$

3. The characteristic equation is:

$$r^2 + 8r + 16 = 0 \quad \text{or} \quad (r+4)^2 = 0.$$

There is only one root: $r = -4.$ By Theorem 9.3.6 II, the general solution is:

$$y = C_1 e^{-4x} + C_2 x e^{-4x}.$$

4. $r^2 + 7r + 3 = 0 \implies r = -\dfrac{7}{2} \pm \dfrac{\sqrt{37}}{2}; \quad y = C_1 e^{\frac{-7+\sqrt{37}}{2}x} + C_2 e^{\frac{-7-\sqrt{37}}{2}x}.$

5. The characteristic equation is: $r^2 + 2r + 5 = 0.$

The roots are complex: $r = -1 \pm 2i.$ By Theorem 9.3.6 III, the general solution is:

$$y = e^{-x}\left(C_1 \cos 2x + C_2 \sin 2x\right).$$

6. $r^2 - 3r + 8 = 0 \implies r = \dfrac{3}{2} \pm \dfrac{\sqrt{23}}{2}i; \quad y = e^{3x/2}\left(C_1 \cos \dfrac{\sqrt{23}}{2}x + C_2 \sin \dfrac{\sqrt{23}}{2}x\right)$

7. The characteristic equation is:

$$2r^2 + 5r - 3 = 0 \quad \text{or} \quad (2r-1)(r+3) = 0.$$

The roots are: $r = \frac{1}{2}, \; -3.$ The general solution is:

$$y = C_1 e^{x/2} + C_2 e^{-3x}.$$

8. $r^2 - 12 = 0 \implies r = \pm 2\sqrt{3}; \quad y = C_1 e^{2\sqrt{3}x} + C_2 e^{-2\sqrt{3}x}.$

9. The characteristic equation is:

$$r^2 + 12 = 0.$$

The roots are complex: $r = \pm 2\sqrt{3}\,i.$ The general solution is:

$$y = C_1 \cos 2\sqrt{3}\,x + C_2 \sin 2\sqrt{3}\,x.$$

10. $r^2 - 3r + \dfrac{9}{4} = 0 \implies r = \dfrac{3}{2}; \quad y = C_1 e^{\frac{3}{2}x} + C_2 x e^{\frac{3}{2}x}$.

11. The characteristic equation is:

$$5r^2 + \tfrac{11}{4}r - \tfrac{3}{4} = 0 \quad \text{or} \quad 20r^2 + 11r - 3 = (5r - 1)(4r + 3) = 0.$$

The roots are: $r = \tfrac{1}{5}, \; -\tfrac{3}{4}$.   The general solution is:

$$y = C_1 e^{x/5} + C_2 e^{-3x/4}.$$

12. $2r^2 + 3r = 0 \implies r = 0, -\dfrac{3}{2}; \quad y = C_1 + C_2 e^{-\frac{3}{2}x}.$

13. The characteristic equation is:

$$r^2 + 9 = 0.$$

The roots are complex: $r = \pm 3i$.   The general solution is:

$$y = C_1 \cos 3x + C_2 \sin 3x.$$

14. $r^2 - r - 30 = 0 \implies r = 6, -5; \quad y = C_1 e^{6x} + C_2 e^{-5x}.$

15. The characteristic equation is:

$$2r^2 + 2r + 1 = 0.$$

The roots are complex: $r = -\tfrac{1}{2} \pm \tfrac{1}{2} i$.   The general solution is:

$$y = e^{-x/2} \left[C_1 \cos(x/2) + C_2 \sin(x/2) \right].$$

16. $r^2 - 4r + 4 = 0 \implies r = 2; \quad y = C_1 e^{2x} + C_2 x e^{2x}.$

17. The characteristic equation is:

$$8r^2 + 2r - 1 = 0 \quad \text{or} \quad (4r - 1)(2r + 1) = 0.$$

The roots are: $r = \tfrac{1}{4}, \; -\tfrac{1}{2}$.   The general solution is:

$$y = C_1 e^{x/4} + C_2 e^{-x/2}.$$

18. $5r^2 - 2r + 1 = 0 \implies r = \dfrac{1}{5} \pm \dfrac{2}{5} i; \quad y = e^{x/5} \left(C_1 \cos \dfrac{2x}{5} + C_2 \sin \dfrac{2x}{5} \right).$

19. The characteristic equation is:

$$r^2 - 5r + 6 = 0 \quad \text{or} \quad (r - 3)(r - 2) = 0.$$

The roots are: $r = 3, \, 2$.   The general solution and its derivative are:

$$y = C_1 e^{3x} + C_2 e^{2x}, \qquad y' = 3C_1 e^{3x} + 2C_2 e^{2x}.$$

The conditions: $y(0) = 1, \quad y'(0) = 1$ require that

$$C_1 + C_2 = 1 \quad \text{and} \quad 3C_1 + 2C_2 = 1.$$

Solving these equations simultaneously gives $C_1 = -1, \; C_2 = 2$.

The solution of the initial value problem is: $y = 2e^{2x} - e^{3x}.$

20. $r^2 + 2r + 1 = 0 \implies r = -1; \quad y = C_1 e^{-x} + C_2 x e^{-x}$

$1 = y(2) = C_1 e^{-2} + 2C_2 e^{-2}, \quad 2 = y'(2) = -C_1 e^{-2} - C_2 e^{-2}$

$\implies C_1 = -5e^2, \ C_2 = 3e^2 \implies y = -5e^{2-x} + 3xe^{2-x}.$

21. The characteristic equation is:

$$r^2 + \tfrac{1}{4} = 0.$$

The roots are: $r = \pm \tfrac{1}{2} i.$ The general solution and its derivative are:

$$y = C_1 \cos(x/2) + C_2 \sin(x/2) \qquad y' = -\tfrac{1}{2} C_1 \sin(x/2) + \tfrac{1}{2} C_2 \cos(x/2).$$

The conditions: $y(\pi) = 1, \quad y'(\pi) = -1$ require that

$$C_2 = 1 \qquad \text{and} \qquad C_1 = 2.$$

The solution of the initial value problem is: $y = 2\cos(x/2) + \sin(x/2).$

22. $r^2 - 2r + 2 = 0 \implies r = 1 \pm i; \quad y = e^x(C_1 \cos x + C_2 \sin x).$

$-1 = y(0) = C_1, \quad -1 = y'(0) = C_1 + C_2 \implies C_2 = 0, \quad y = -e^x \cos x$

23. The characteristic equation is:

$$r^2 + 4r + 4 = 0 \qquad \text{or} \qquad (r+2)^2 = 0.$$

There is only one root: $r = -2.$ The general solution and its derivative are:

$$y = C_1 e^{-2x} + C_2 x e^{-2x} \qquad y' = -2C_1 e^{-2x} + C_2 e^{-2x} - 2C_2 x e^{-2x}.$$

The conditions: $y(-1) = 2, \quad y'(-1) = 1$ require that

$$C_1 e^2 - C_2 e^2 = 2 \qquad \text{and} \qquad -2C_1 e^2 + 3C_2 e^2 = 1.$$

Solving these equations simultaneously gives $C_1 = 7e^{-2}, \ C_2 = 5e^{-2}.$

The solution of the initial value problem is: $y = 7e^{-2}e^{-2x} + 5e^{-2}xe^{-2x} = 7e^{-2(x+1)} + 5xe^{-2(x+1)}.$

24. $r^2 - 2r + 5 = 0 \implies r = 1 \pm 2i; \quad y = e^x(C_1 \cos 2x + C_2 \sin 2x).$

$0 = y(\pi/2) = e^{\pi/2}(-C_1) \implies C_1 = 0; \quad 2 = y'(\pi/2) = e^{\pi/2}(-2C_2) \implies C_2 = -e^{-\pi/2}$

$\implies y = -e^{x-\pi/2} \sin 2x.$

25. The characteristic equation is:

$$r^2 - r - 2 = 0 \qquad \text{or} \qquad (r-2)(r+1) = 0.$$

The roots are: $r = 2, \ -1.$ The general solution and its derivative are:

$$y = C_1 e^{2x} + C_2 e^{-x} \qquad y' = 2C_1 e^{2x} - C_2 e^{-x}.$$

(a) $y(0) = 1 \implies C_1 + C_2 = 1 \implies C_2 = 1 - C_1.$

Thus, the solutions that satisfy $y(0) = 1$ are: $y = Ce^{2x} + (1-C)e^{-x}.$

(b) $y'(0) = 1 \implies 2C_1 - C_2 = 1 \implies C_2 = 2C_1 - 1.$

Thus, the solutions that satisfy $y'(0) = 1$ are: $y = Ce^{2x} + (2C - 1)e^{-x}.$

(c)   To satisfy both conditions, we must have $2C - 1 = 1 - C \implies C = \frac{2}{3}.$

The solution that satisfies $y(0) = 1$, $y'(0) = 1$ is:

$$y = \tfrac{2}{3}e^{2x} + \tfrac{1}{3}e^{-x}.$$

26. $r^2 - \omega^2 = 0 \implies r = \pm\omega; \quad y = A_1 e^{\omega x} + A_2 e^{-\omega x}$

Since $e^{\omega x} = \cosh \omega x + \sinh \omega x$ and $e^{-\omega x} = \cosh \omega x - \sinh \omega x$,   we can write

$y = C_1 \cosh \omega x + C_2 \sinh \omega x$ (with $C_1 = A_1 + A_2$, $C_2 = A_1 - A_2$).

27. $\alpha = \dfrac{r_1 + r_2}{2}, \qquad \beta = \dfrac{r_1 - r_2}{2};$

$y = k_1 e^{r_1 x} + k_2 e^{r_2 x} = e^{\alpha x}\left(C_1 \cosh \beta x + C_2 \sinh \beta x\right), \quad \text{where} \quad k_1 = \dfrac{C_1 + C_2}{2}, \quad k_2 = \dfrac{C_1 - C_2}{2}.$

28. $r^2 + \omega^2 = 0 \implies r = \pm\omega i; \quad y = C_1 \cos \omega x + C_2 \sin \omega x.$

Assuming that $C_1^2 + C_2^2 > 0$, we have

$$C_1 \cos \omega x + C_2 \sin \omega x = \sqrt{C_1{}^2 + C_2{}^2}\left(\frac{C_1}{\sqrt{C_1{}^2 + C_2{}^2}}\cos \omega x + \frac{C_2}{\sqrt{C_1{}^2 + C_2{}^2}}\sin \omega x\right)$$
$$= A\left(\sin \phi_0 \cos \omega x + \cos \phi_0 \sin \omega x\right) = A \sin(\omega x + \phi_0),$$

where $A = \sqrt{C_1{}^2 + C_2{}^2}$ and ϕ_0, $\phi_0 \in [0, 2\pi)$,   is the angle such that

$$\sin \phi_0 = \frac{C_1}{\sqrt{C_1{}^2 + C_2{}^2}} \quad \text{and} \quad \cos \phi_0 = \frac{C_2}{\sqrt{C_1{}^2 + C_2{}^2}}$$

29. (a) Let $y_1 = e^{\alpha x}$, $y_2 = x e^{\alpha x}$.   Then

$$W(x) = y_1 y_2' - y_2 y_1' = e^{\alpha x}\left[e^{\alpha x} + \alpha x e^{\alpha x}\right] - x e^{\alpha x}\left[\alpha e^{\alpha x}\right] = e^{2\alpha x} \neq 0$$

(b) Let $y_1 = e^{\alpha x} \cos \beta x$, $y_2 = e^{\alpha x} \sin \beta x$, $\beta \neq 0$.   Then

$W(x) = y_1 y_2' - y_2 y_1'$

$= e^{\alpha x} \cos \beta x \left[\alpha e^{\alpha x} \sin \beta x + \beta e^{\alpha x} \cos \beta x\right] - e^{\alpha x} \sin \beta x \left[\alpha e^{\alpha x} \cos \beta x - \beta e^{\alpha x} \sin \beta x\right]$

$= \beta e^{2\alpha x} \neq 0$

30. Characteristic equation: $r^2 + 10^3 r + \dfrac{1}{C} = 0; \quad \text{roots:} \quad r = \dfrac{-10^3 \pm \sqrt{10^6 - 4/C}}{2}.$

(a) $r = 100(-5 \pm \sqrt{5}); \quad y = C_1 e^{100(-5+\sqrt{5})t} + C_2 e^{100(-5-\sqrt{5})t}$

(b) $r = -500; \quad y = C_1 e^{-500t} + C_2 t e^{-500t}$

(c) $r = 500(-1 \pm i); \quad y = e^{-500t}\left(C_1 \cos 500t + C_2 \sin 500t\right)$

31. (a) The solutions $y_1 = e^{2x}$, $y_2 = e^{-4x}$ imply that the roots of the characteristic equation
are $r_1 = 2$, $r_2 = -4$. Therefore, the characteristic equation is:
$$(r-2)(r+4) = r^2 + 2r - 8 = 0$$
and the differential equation is: $y'' + 2y' - 8y = 0$.

(b) The solutions $y_1 = 3e^{-x}$, $y_2 = 4e^{5x}$ imply that the roots of the characteristic equation
are $r_1 = -1$, $r_2 = 5$. Therefore, the characteristic equation is
$$(r+1)(r-5) = r^2 - 4r - 5 = 0$$
and the differential equation is: $y'' - 4y' - 5y = 0$.

(c) The solutions $y_1 = 2e^{3x}$, $y_2 = xe^{3x}$ imply that 3 is the only root of the characteristic
equation. Therefore, the characteristic equation is
$$(r-3)^2 = r^2 - 6r + 9 = 0$$
and the differential equation is: $y'' - 6y' + 9y = 0$.

32. (a) We want $r = \pm 2i$, so $r^2 = -4$. Differential equation: $y'' + 4y = 0$

(b) We want $r = -2 \pm 3i$, so $(r+2)^2 = -9$. Differential equation: $y'' + 4y' + 13y = 0$

33. (a) Let $y = e^{\alpha x} u$. Then
$$y' = \alpha e^{\alpha x} u + e^{\alpha x} u' \quad \text{and} \quad y'' = \alpha^2 e^{\alpha x} u + 2\alpha e^{\alpha x} u' + e^{\alpha x} u''$$
Now,
$$y'' - 2\alpha y + \alpha^2 y = \left(\alpha^2 e^{\alpha x} u + 2\alpha e^{\alpha x} u' + e^{\alpha x} u''\right) - 2\alpha \left(\alpha e^{\alpha x} u + e^{\alpha x} u'\right) + \alpha^2 e^{\alpha x} u$$
$$= e^{\alpha x} u''$$
Therefore, $y'' - 2\alpha y + \alpha^2 y = 0 \implies e^{\alpha x} u'' \implies u'' = 0$.

(b) $y'' - 2\alpha y' + \left(\alpha^2 + \beta^2\right) y = y'' - 2\alpha y' + \alpha^2 y + \beta^2 y$.

From part (a) $y = e^{\alpha x} u \implies y'' - 2\alpha y + \alpha^2 y = e^{\alpha x} u''$. Therefore,
$$y'' - 2\alpha y' + \left(\alpha^2 + \beta^2\right) y = 0 \implies e^{\alpha x} u'' + \beta^2 e^{\alpha x} u = 0 \implies u'' + \beta^2 u = 0.$$

34. $r^2 + ar + b = 0 \implies r_1, r_2 = \dfrac{-a \pm \sqrt{a^2 - 4b}}{2}$.

If $a^2 - 4b > 0$, then $\sqrt{a^2 - 4b} < a$, so $\dfrac{-a \pm \sqrt{a^2 - 4b}}{2}$ is negative, and the solutions:

$y = C_1 e^{r_1 x} + C_2 e^{r_2 x} \to 0$ as $x \to \infty$.

If $a^2 - 4b = 0$, then $r = r_1 = r_2 = -a/2 < 0$, and the solutions:

$y = C_1 e^{rx} + C_2 x e^{rx} \to 0$ as $x \to \infty$.

If $a^2 - 4b < 0$, then $y = e^{-ax/2}\left(C_1 \cos \frac{1}{2}\sqrt{b^2 - 4a}\, x + C_2 \sin \frac{1}{2}\sqrt{b^2 - 4a}\, x\right)$ satisfies

$|y| < e^{-ax/2} \implies y \to 0$ as $x \to \infty$.

35. (a) If $a = 0$, $b > 0$, then the general solution of the differential equation is:

$$y = C_1 \cos \sqrt{b}\, x + C_2 \sin \sqrt{b}\, x = A \cos\left(\sqrt{b}\, x + \phi\right)$$

where A and ϕ are constants. Clearly $|y(x)| \le |A|$ for all x.

(b) If $a > 0$, $b = 0$, then the general solution of the differential equation is:

$$y = C_1 + C_2 e^{-ax} \qquad \text{and} \qquad \lim_{x \to \infty} y(x) = C_1.$$

The solution which satisfies the conditions: $y(0) = y_0$, $y'(0) = y_1$ is:

$$y = y_0 + \frac{y_1}{a} - \frac{y_1}{a} e^{-ax} \quad \text{and} \quad \lim_{x \to \infty} y(x) = y_0 + \frac{y_1}{a}; \qquad k = y_0 + \frac{y_1}{a}.$$

36. Let y_1 and y_2 be solutions of the homogeneous equation.

Suppose that $y_2 = k\, y_1$ for some scalar k. Then

$$W(y_1, y_2) = \begin{vmatrix} y_1 & k\, y_1 \\ y_1' & k\, y_1' \end{vmatrix} = 0$$

Now suppose that $W(y_1, y_2) = 0$, and suppose that y_1 is not identically 0. Let I be an interval on which $y_1(x) \ne 0$. Then,

$$\left(\frac{y_2}{y_1}\right)' = \frac{y_1 y_2' - y_2 y_1'}{(y_1)^2} = \frac{W(y_1, y_2)}{(y_1)^2} = 0$$

Therefore, $\dfrac{y_2}{y_1} = k$ constant on I. Finally, $y_2 = ky_1$ on I implies $y_2 = ky_1$ for all x by the uniqueness theorem.

37. Let W be the Wronskian of y_1 and y_2. Then

$$W(a) = \begin{vmatrix} 0 & 0 \\ y_1'(a) & y_2'(a) \end{vmatrix} = 0$$

Therefore one of the solutions is a multiple of the other (see the Supplement to this Section).

38. From the hint, $\dfrac{dy}{dx} = \dfrac{dy}{dz}\dfrac{1}{x}$. Differentiating with respect to x again, we have

$$\frac{d^2y}{dx^2} = \frac{d^2y}{dz^2}\frac{dz}{dx}\frac{1}{x} + \frac{dy}{dz}\left(-\frac{1}{x^2}\right) = \frac{1}{x^2}\left(\frac{d^2y}{dz^2} - \frac{dy}{dz}\right).$$

Substituting into the differential equation $x^2 y'' + \alpha x y' + \beta y = 0$, we get

$$\left(\frac{d^2 y}{dz^2} - \frac{dy}{dz} \right) + \alpha \frac{dy}{dz} + \beta y = 0, \quad \text{or} \quad \frac{d^2 y}{dz^2} + a \frac{dy}{dz} + by = 0,$$

where $\quad a = \alpha - 1, \quad b = \beta.$

39. From Exercise 38, the change of variable $z = \ln x$ transforms the equation

$$x^2 y'' - x y' - 8y = 0$$

into the differential equation with constant coefficients

$$\frac{d^2 y}{dz^2} - 2\frac{dy}{dz} - 8y = 0.$$

The characteristic equation is:

$$r^2 - 2r - 8 = 0 \quad \text{or} \quad (r - 4)(r + 2) = 0$$

The roots are: $\quad r = 4, \ r = -2, \quad$ and the general solution (in terms of z) is:

$$y = C_1 e^{4z} + C_2 e^{-2z}.$$

Replacing z by $\ln x$ we get

$$y = C_1 e^{4 \ln x} + C_2 e^{-2 \ln x} = C_1 x^4 + C_2 x^{-2}.$$

40. Using the result of Exercise 38, we get

$$\frac{d^2 y}{dz^2} - 3\frac{dy}{dz} + 2y = 0, \quad \text{so} \quad r^2 - 3r + 2 = 0 \implies r = 1, 2.$$

$\implies y = C_1 e^z + C_2 e^{2z}.$ Substituting $z = \ln x,$ we get $y = C_1 x + C_2 x^2.$

41. From Exercise 38, the change of variable $z = \ln x$ transforms the equation

$$x^2 y'' - 3x y' + 4y = 0$$

into the differential equation with constant coefficients

$$\frac{d^2 y}{dz^2} - 4\frac{dy}{dz} + 4y = 0.$$

The characteristic equation is:

$$r^2 - 4r + 4 = 0 \quad \text{or} \quad (r - 2)^2 = 0.$$

The only root is: $\quad r = 2, \quad$ and the general solution (in terms of z) is:

$$y = C_1 e^{2z} + C_2 z e^{2z}.$$

Replacing z by $\ln x$ we get

$$y = C_1 e^{2 \ln x} + C_2 \ln x \, e^{2 \ln x} = C_1 x^2 + C_2 x^2 \ln x.$$

42. From Exercise 38, we get $\quad \dfrac{d^2 y}{dz^2} - 2\dfrac{dy}{dz} + 5y = 0$

$r^2 - 2r + 5 = 0 \implies r = 1 \pm 2i; \quad$ and $\quad y = e^z \left(C_1 \cos 2z + C_2 \sin 2z \right).$

Substituting $z = \ln x$ we get: $\quad y = x \left[C_1 \cos(2 \ln x) + C_2 \sin(2 \ln x) \right].$

REVIEW EXERCISES

1. First calculate the integrating factor $e^{H(x)}$:

$$H(x) = \int 1 dx = x \quad \text{and} \quad e^{H(x)} = e^x$$

Multiplication by e^x gives

$$e^x y' + e^x y = 2e^{-x} \quad \text{which is} \quad \frac{d}{dx}(e^x y) = 2e^{-x}$$

Integrating this equation, we get

$$e^x y = -2e^{-x} + C$$

and

$$y = -2e^{-2x} + Ce^{-x}$$

2. The equation can be written

$$-2\cos(2x) + \frac{2y^2 + 1}{y} y' = 0.$$

The equation is separable:

$$\int -2\cos(2x) dx + \int \frac{2y^2 + 1}{y} dy = C \quad \text{and} \quad -\sin(2x) + y^2 + \ln|y| = C$$

3. The equation can be written

$$\cos^2 x\, dx - \frac{y}{y^2 + 1} dy = 0.$$

The equation is separable:

$$\int \cos^2 x\, dx - \int \frac{y}{y^2 + 1} dy = C \quad \text{and} \quad \frac{1}{4}\sin(2x) + \frac{1}{2}x - \frac{1}{2}\ln(y^2 + 1) = C$$

or $\sin(2x) + 2x - 2\ln(y^2 + 1) = C.$

4. The equation can be written

$$xe^x - (y \ln y)y' = 0.$$

The equation is separable:

$$\int xe^x\, dx - \int y \ln y\, dy = C \quad \text{and} \quad e^x(x - 1) - \frac{1}{2}y^2 \ln y + \frac{1}{4}y^2 + C$$

5. The equation can be written

$$y' + \frac{3}{x}y = \frac{\sin 2x}{x^2}.$$

Calculate the integrating factor $e^{H(x)}$:

$$H(x) = \int \frac{3}{x}dx = \ln x^3 \quad \text{and} \quad e^{H(x)} = x^3.$$

Multiplying by x^3 gives

$$x^3 y' + 3x^2 y = x \sin 2x \quad \text{which is} \quad \frac{d}{dx}(x^3 y) = x \sin 2x.$$

Integrating this equation, we get

$$x^3 y = \int x \sin 2x \, dx + C = -\frac{1}{2}x \cos 2x + \frac{1}{4}\sin 2x + C.$$

and

$$y = -\frac{1}{2x^2}\cos 2x + \frac{1}{4x^3}\sin 2x + \frac{C}{x^3}.$$

6. The equation can be written

$$y' + \frac{2}{x}y = \frac{2}{x}e^{x^2}.$$

Calculate the integrating factor $e^{H(x)}$:

$$H(x) = \int \frac{2}{x}dx = \ln x^2 \quad \text{and} \quad e^{H(x)} = x^2.$$

Multiplication by x^2 gives

$$x^2 y' + 2xy = 2xe^{x^2} \quad \text{which is} \quad \frac{d}{dx}(x^2 y) = 2xe^{x^2}.$$

Integrating the equation, we get

$$x^2 y = e^{x^2} + C.$$

and

$$y = \frac{1}{x^2}(e^{x^2} + C).$$

7. The equation can be written

$$1 + x^2 - \frac{1}{1+y^2}y' = 0.$$

The equation is separable:

$$\int (1+x^2)dx - \int \frac{1}{1+y^2}dy = C \quad \text{and} \quad x + \frac{x^3}{3} - \arctan y = C$$

or $\arctan y = x + \frac{x^3}{3} + C.$

8. The equation can be written

$$x^2 - 1 - \frac{y+1}{y}y' = 0$$

The equation is separable:

$$\int (x^2 - 1)dx - \int \frac{y+1}{y} dy = C \quad \text{and} \quad \frac{1}{3}x^3 - x - y - \ln|y| = C.$$

9. The equation can be written

$$y' + \frac{2}{x}y = x^2.$$

Calculate the integrating factor $e^{H(x)}$:

$$H(x) = \int \frac{2}{x} dx = \ln x^2 \quad \text{and} \quad e^{H(x)} = x^2.$$

Multiplication by x^2 gives

$$x^2 y' + 2xy = x^4 \quad \text{which is} \quad \frac{d}{dx}(x^2 y) = x^4.$$

Integrating this equation, we get

$$x^2 y = \frac{1}{5}x^5 + C \quad \text{and} \quad y = \frac{1}{5}x^3 + \frac{C}{x^2}.$$

10. The equation can be written

$$x\sqrt{1 + x^2} - \frac{1}{y^2}y' = 0$$

The equation is separable:

$$\int x\sqrt{1 + x^2}dx - \int \frac{1}{y^2}dy = C \quad \text{and} \quad \frac{1}{3}(1 + x^2)^{3/2} + \frac{1}{y} = C.$$

Solving for y, we have

$$y = \frac{-3}{(1 + x^2)^{3/2} + C}.$$

11. The equation can be written

$$y' + \frac{1}{x}y = \frac{2}{x^2} + 1$$

The integrating factor is

$$e^{H(x)} = e^{\ln x} = x.$$

Multiplication by x gives

$$xy' + y = \frac{2}{x} + x \quad \text{which is} \quad \frac{d}{dx}(xy) = \frac{2}{x} + x.$$

Integrating this equation, we get

$$xy = \ln x^2 + \frac{1}{2}x^2 + C \quad \text{and} \quad y = \frac{1}{x}(\ln x^2 + \frac{1}{2}x^2 + C).$$

Applying the initial condition $y(1) = 2$, we have

$$\ln 1 + \frac{1}{2} + C = 2 \quad \text{and} \quad C = \frac{3}{2}.$$

Therefore

$$y = \frac{1}{x}\left(\ln x^2 + \frac{1}{2}x^2 + \frac{3}{2}\right).$$

12. The equation can be written

$$4x - \frac{y}{\sqrt{y^2+1}}y' = 0.$$

The equation is separable:

$$\int 4x\,dx - \int \frac{y}{\sqrt{y^2+1}}\,dy = C \quad \text{and} \quad 2x^2 - (y^2+1)^{1/2} = C.$$

To find the solution that satisfies $y(0) = 1$, we set $x = 0$, $y = 1$ and solve for C:

$$C = -(1+1)^{\frac{1}{2}} = -\sqrt{2}.$$

Therefore $2x^2 - (y^2+1)^{1/2} + \sqrt{2} = 0$ is the solution.

13. The equation can be written

$$e^{2x} + \frac{1}{2y-1}y' = 0.$$

The equation is separable:

$$\int e^{2x}\,dx + \int \frac{1}{2y-1}\,dy = C \quad \text{and} \quad \frac{1}{2}e^{2x} + \frac{1}{2}\ln|2y-1| = C.$$

Solving for y, we get

$$y = \frac{1}{2} + Ce^{-e^{2x}}.$$

To find the solution that satisfies $y(0) = \frac{1}{2} + \frac{1}{e}$, we set $x = 0$, $y = \frac{1}{2} + \frac{1}{e}$ and solve for C.

We have $C = 1$. Therefore $y = \frac{1}{2} + e^{-e^{2x}}$.

14. The equation can be written

$$\tan x - \cos y\, y' = 0.$$

The equation is separable:

$$\int \tan x\,dx - \int \cos y\,dy = C \quad \text{and} \quad \ln|\sec x| - \sin y = C.$$

Applying the initial condition: $y(0) = \frac{\pi}{2}$, we have $C = -1$. Therefore $\sin y = \ln|\sec x| + 1$.

15. The characteristic equation is

$$r^2 - 2r + 2 = 0.$$

The roots are: $r = 1 \pm i$.

The general solution is

$$y = C_1 e^x \cos x + C_2 e^x \sin x$$

16.   The characteristic equation is

$$r^2 + r + \frac{1}{4} = 0.$$

The roots are: $r = -\frac{1}{2}$ with multiplicity 2.

The general solution is

$$y = C_1 e^{-x/2} + C_2 x e^{-x/2}$$

17.   The characteristic equation is

$$r^2 - r - 2 = 0.$$

The roots are: $r = 2, -1$.

The general solution is

$$y = C_1 e^{2x} + C_2 e^{-x}$$

18.   The characteristic equation is

$$r^2 - 4r = 0.$$

The roots are: $r = 0, 4$.   The general solution is

$$y = C_1 + C_2 e^{4x}$$

19.   The characteristic equation is

$$r^2 - 6r + 9 = 0.$$

The roots are: $r = 3$ with multiplicity 2.

 The general solution is

$$y = C_1 e^{3x} + C_2 x e^{3x}$$

20.   The characteristic equation is

$$r^2 + 4 = 0$$

The roots are: $r = \pm 2i$.

The general solution is

$$y = C_1 \cos 2x + C_2 \sin 2x$$

21. The characteristic equation is

$$r^2 + 4r + 13 = 0$$

The roots are: $r = -2 \pm 3i$.

The general solution is

$$y = e^{-2x}(C_1 \cos 3x + C_2 \sin 3x)$$

22. The characteristic equation is

$$3r^2 - 5r - 2 = 0.$$

The roots are: $r = 2, -\frac{1}{3}$.

The general solution is

$$y = C_1 e^{2x} + C_2 e^{-x/3}.$$

23. The characteristic equation is

$$r^2 - r = 0.$$

The roots are: $r = 0, 1$.

The general solution is

$$y = C_1 + C_2 e^x.$$

Applying the initial conditions $y(0) = 1$ and $y'(0) = 0$, we have

$$C_1 + C_2 = 1, \quad C_2 = 0 \quad \Longrightarrow \quad C_1 = 1.$$

The solution of the initial-value problem is: $y = 1$.

24. The characteristic equation is

$$r^2 + 7r + 12 = 0.$$

The roots are: $r = -3, -4$.

The general solution is

$$y = C_1 e^{-3x} + C_2 e^{-4x}.$$

Applying the initial conditions $y(0) = 2, \quad y'(0) = 8$, we have

$$C_1 + C_2 = 2, \quad -3C_1 - 4C_2 = 8 \quad \Longrightarrow \quad C_1 = 16, \; C_2 = -14.$$

The solution of the initial-value problem is: $y = 16e^{-3x} - 14e^{-4x}$.

25. The characteristic equation is

$$r^2 - 6r + 13 = 0.$$

The roots are: $r = 3 \pm 2i$.

The general solution is

$$y = e^{3x}(C_1 \cos 2x + C_2 \sin 2x).$$

Applying the initial conditions $y(0) = 2, \quad y'(0) = 2$, we have

$$C_1 = 2, \quad 3C_1 + 2C_2 = 2 \quad \Longrightarrow \quad C_1 = 2, \; C_2 = -2.$$

The solution of the initial-value problem is: $y = e^{3x}(2 \cos 2x - 2 \sin 2x)$.

26. The characteristic equation is

$$r^2 + 4r + 4 = 0.$$

The roots are: $r = -2$ with multiplicity 2.

The general solution is

$$y = C_1 e^{-2x} + C_2 x e^{-2x}.$$

Applying the initial conditions $y(-1) = 2, \; y'(-1) = 1$, we have

$$C_1 e^2 - C_2 e^2 = 2, \quad -2C_2 e^2 + 3C_2 e^2 = 1 \quad \Longrightarrow \quad C_1 = 7e^{-2}, C_2 = 5e^{-2}.$$

The solution of the initial-value problem is: $y = 7e^{-2x-2} + 5xe^{-2x-2}$.

27. Curves: $y = Ce^{2x}; \quad y' = 2Ce^{2x} \quad \Longrightarrow \quad y' = 2y$

Orthogonal trajectories: $y' = -\dfrac{1}{2y}; \quad \displaystyle\int 2y \, dy = -\int dx; \quad y^2 = C - x.$

28. Curves: $y = \dfrac{C}{1+x^2}; \quad y' = -\dfrac{2xC}{(1+x^2)^2} \quad \Longrightarrow \quad y' = -\dfrac{2xy}{1+x^2}$

Orthogonal trajectories: $y' = \dfrac{1+x^2}{2xy}; \quad \displaystyle\int 2y \, dy = \int \dfrac{1+x^2}{x} \, dx = \int \left(\dfrac{1}{x} + x\right) dx;$

and

$$y^2 = \ln|x| + \frac{1}{2}x^2 + C$$

29. Substituting $y = x^r$ into the equation, we get

$$r(r-1)x^r + 4rx^r + 2x^r = 0 \quad \text{or} \quad x^r(r^2 + 3r + 2) = 0 \quad \Longrightarrow \quad r^2 + 3r + 2 = (r+2)(r+1) = 0.$$

The solutions are: $r = -1, -2$.

30. Substituting $y = x^r$ into the equation, we get

$$r(r-1)x^r - rx^r - 8x^r = 0 \quad \text{or} \quad x^r(r^2 - 2r - 8) = 0 \quad \Longrightarrow \quad r^2 - 2r - 8 = (r+2)(r-4) = 0.$$

The solutions are: $r = 4, -2$.

31. Let $y(t)$ be the value of the business (measured in millions) at time t. Then $y(t)$ satisfies

$$\frac{dy}{dt} = ky^2$$

The general solution of this equation is

$$y(t) = \frac{1}{-kt + C}$$

Applying the given conditions, $y(0) = 1$ and $y(1) = 1.5$, to find C and k, we get $C = 1$ and $k = 1/3$.

1 year from now the business will be worth:

$$y(2) = \frac{1}{-\frac{1}{3} + 1} = 3 \, \text{million.}$$

1.5 years from now the business will be worth:

$$y(2.5) = \frac{1}{-\frac{1}{3}\left(\frac{5}{2}\right) + 1} = 6 \, \text{million.}$$

2 years from now the business will be worth:

$$y(3) = \frac{1}{-\frac{1}{3}(3) + 1} = \infty.$$

Obviously the business cannot continue to grow at a rate proportional to its value squared.

32. Let $y(t)$ be the value of the business (measured in millions) at the time t. Then $y(t)$ satisfies

$$\frac{dy}{dt} = k\sqrt{y}$$

The general solution of this equation is

$$y(t) = \left(\frac{k}{2}t + C\right)^2$$

Applying the given conditions, $y(0) = 1$ and $y(2) = 1.44$, to find C and k, we get $C = 1$ and $k = 0.2$ so

$$y(t) = \left(\tfrac{1}{10}t + 1\right)^2.$$

5 years from now the business will be worth:

$$y(7) = (0.7 + 1)^2 = 2.89 \text{ million.}$$

Solve

$$y(t) = \left(\frac{t}{10} + 1\right)^2 = 4$$

$t = 10$. The business will be worth 4 million 8 years from now.

33. (a) The general solution of the differential equation is

$$y = \frac{a}{b} + Ce^{-bt}.$$

Applying the initial condition $y(0) = 0$, we get $C = -\frac{a}{b}$, and

$$y = \frac{a}{b}(1 - e^{-bt})$$

(b) $\displaystyle\lim_{t \to \infty} y(t) = \frac{a}{b}$

(c) Setting $y = 0.9\dfrac{a}{b}$, we have

$$0.9\frac{a}{b} = \frac{a}{b}(1 - e^{-bt}).$$

The solution to this equation is $t = \dfrac{\ln 10}{b}$ hours.

34. Let $T(t)$ be the temperature of the bar at time t. It follows from Newton's law of cooling that

$$T(t) = \tau + Ce^{-kt}.$$

By the conditions given in the problem, we have

$$\tau = 0, \quad T(0) = 100, \quad T(20) = 50.$$

Applying these conditions, we get $\quad C = 100$ and $k = \dfrac{\ln 2}{20}$. Therefore

$$T(t) = 100e^{-(t/20)\ln 2} = 100(2)^{-t/20}.$$

(a) $T(30) = 100(2)^{-3/2} \cong 35.4°$.

(b) Solve $25 = 100(2)^{-t/20}$ for t:

$$25 = 100(2)^{-t/20}, \quad -\frac{t}{20}\ln 2 = \ln(1/4), \quad t = \frac{20\ln 4}{\ln 2} = 40.$$

It will take 40 minutes for the bar to reach $25°$.

35. Let $T(t)$ be the temperature of the object at time t. It follows from Newton's law of cooling that

$$T(t) = \tau + Ce^{-kt}.$$

By the conditions given in the problem, we have

$$\tau = 70, \quad T(10) = 20, \quad T(20) = 35.$$

Applying these conditions, we get

$$C = -\frac{500}{7} \quad \text{and} \quad k = \frac{1}{10}\ln(10/7).$$

(a) The temperature of the object at time t is

$$T = 70 - \frac{500}{7}e^{-(t/10)\ln(10/7)} = 70 - \frac{500}{7}\left(\frac{7}{10}\right)^{t/10}$$

(b) $T(0) = 70 - \dfrac{500}{7} = -\dfrac{10}{7}$

36. (a) Let T be the length of time needed to fill the tank. Then

$$600 + (6-4)T = 1200$$

and $T = 300$ minutes.

(b) Let $S(t)$ be the amount of salt dissolved in the tank at time t. Then

$$\frac{dS}{dt} = \text{rate in} - \text{rate out} = \frac{1}{2} \times 6 - \frac{S}{600+2t} \times 4 = 3 - \frac{2S}{300+t}$$

The equation can be rewritten

$$S' + \frac{2S}{300+t} = 3.$$

The general solution to this equation is

$$S = 300 + t + \frac{C}{(300+t)^2}.$$

Since $S(0) = 40$, we get $C = -300^2 \cdot 260$. Therefore, the amount of salt in the tank at time t is:

$$S(t) = 300 + t - \frac{300^2 \cdot 260}{(300+t)^2}$$

(c) $S(T) = 300 + 300 - \dfrac{300^2 \cdot 260}{(300+t)^2} = 535$ pounds.

37. (a) Let T be the length of time needed to empty the tank. Then

$$80 - (8-4)T = 0 \quad \text{and} \quad T = 20 \text{ minutes.}$$

(b) Let $S(t)$ be the amount of salt in the tank at the time t. Then

$$\frac{dS}{dt} = 1 \times 4 - \frac{S}{80-4t} \times 8 = 4 - \frac{2S}{20-t}, \quad S(0) = \frac{1}{8} \times 80 = 10$$

The solution to this initial-value problem is

$$S(t) = 4(20-t) - \frac{7}{40}(20-t)^2 \tag{1}$$

(c) Let t_0 be the time that the tank contains exactly 40 gallons. Then

$$80 - 4t_0 = 40 \quad \text{and} \quad t_0 = 10.$$

Substituting $t = 10$ into (1), we get $S(10) = 22.5$ pounds.

38. (a) The differential equation is separable and can be written as

$$\frac{dP}{P(10^{-1} - 10^{-5}P)} = dt.$$

With the initial condition $P(0) = 2000$, the solution is

$$P(t) = \frac{2500e^{0.1t}}{1 + 0.25e^{0.1t}}.$$

Then

$$\lim_{t \to \infty} P(t) = 10^4.$$

(b) Setting $P = 0.9 \times 10^4$, we get

$$0.9 \times 10^4 = \frac{2500e^{0.1t}}{1 + 0.25e^{0.1t}}.$$

The solution to this equation is $t = 35.835 \cong 36$ months.

39. Let $P(t)$ be the number of people that have heard the rumor at time t. Then $P(t)$ satisfies:

$$\frac{dP}{dt} = kP(20,000 - P)$$

The general solution of this equation is

$$P(t) = \frac{20,000}{1 + Ce^{-20,000kt}}.$$

Now, $P(0) = \dfrac{20,000}{1 + C} = 500 \implies C = 39;$

$P(10) = \dfrac{20,000}{1 + 39e^{-20,000(10k)}} = 1200 \implies 20,000k \cong -.0912.$

Therefore, $P(t) = \dfrac{20,000}{1 + 39e^{-0.09120t}}.$

(a) $P(20) = \dfrac{20,000}{1 + 39e^{-0.0912(20)}} \cong 2742$

(b) The rumor will be spreading fastest when the number of people who have heard it is equal to the number of people who have not heard it:

$$\frac{20,000}{1 + 39e^{-0.0912t}} = 10,000.$$

The solution of this equation is: $t \cong 40$ days.

40. (a) From (2),

$$\frac{du}{\sqrt{1+u^2}} = \frac{1}{a}dx$$

Integrating, we get

$$\ln\left|u + \sqrt{1+u^2}\right| = \frac{x}{a} + C.$$

Applying the initial condition $y'(0) = u(0) = 0 \implies C = 0$. Thus, $\ln\left|u + \sqrt{1+u^2}\right| = \frac{x}{a}$.

(b) Set $u = y'$:

$$\ln\left|y' + \sqrt{1+(y')^2}\right| = \frac{x}{a}, \quad y' + \sqrt{1+(y')^2} = e^{x/a} \quad \text{and} \quad \sqrt{1+(y')^2} = e^{x/a} - y'.$$

Squaring both sides and simplyfying, we get

$$y' = \frac{e^{x/a} - e^{-x/a}}{2} = \sinh(x/a) \implies y = a\cosh(x/a) + C.$$

Applying the initial condition $y(0) = a$, we get $C = 0$.

Therefore $y(x) = a\cosh(x/a)$, a catenary.

CHAPTER 10

SECTION 10.1

1. $y = \frac{1}{2}x^2$

vertex $(0,0)$

focus $(0, \frac{1}{2})$

axis $x = 0$

directrix $y = -\frac{1}{2}$

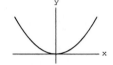

2. $y = -\frac{1}{2}x^2$

vertex $(0,0)$

focus $(0, -\frac{1}{2})$

axis $x = 0$

directrix $y = \frac{1}{2}$

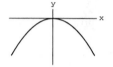

3. $y = \frac{1}{2}(x-1)^2$

vertex $(1,0)$

focus $(1, \frac{1}{2})$

axis $x = 1$

directrix $y = -\frac{1}{2}$

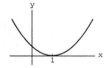

4. $y = -\frac{1}{2}(x-1)^2$

vertex $(1,0)$

focus $(1, -\frac{1}{2})$

axis $x = 1$

directrix $y = \frac{1}{2}$

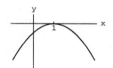

5. $y + 2 = \frac{1}{4}(x-2)^2$

vertex $(2,-2)$

focus $(2,-1)$

axis $x = 2$

directrix $y = -3$

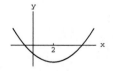

6. $y - 2 = \frac{1}{4}(x+2)^2$

vertex $(-2,2)$

focus $(-2,3)$

axis $x = -2$

directrix $y = 1$

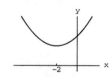

7. $y = x^2 - 4x$

vertex $(2, -4)$

focus $(2, -\frac{15}{4})$

axis $x = 2$

directrix $y = -\frac{17}{4}$

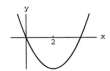

8. $y = x^2 + x + 1$

vertex $(-\frac{1}{2}, \frac{3}{4})$

focus $(-\frac{1}{2}, 1)$

axis $x = -\frac{1}{2}$

directrix $y = \frac{1}{2}$

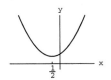

9. $\dfrac{x^2}{9} + \dfrac{y^2}{4} = 1$

center $(0, 0)$

foci $(\pm\sqrt{5}, 0)$

length of major axis 6

length of minor axis 4

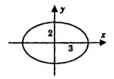

10. $\dfrac{x^2}{4} + \dfrac{y^2}{9} = 1$

center $(0, 0)$

foci $(0, \pm\sqrt{5})$

length of major axis 6

length of minor axis 4

11. $\dfrac{x^2}{4} + \dfrac{y^2}{6} = 1$

center $(0, 0)$

foci $(0, \pm\sqrt{2})$

length of major axis $2\sqrt{6}$

length of minor axis 4

12. $\dfrac{x^2}{4} + \dfrac{y^2}{3} = 1$

center $(0, 0)$

foci $(\pm 1, 0)$

length of major axis 4

length of minor axis $2\sqrt{3}$

13. $\dfrac{x^2}{9} + \dfrac{(y-1)^2}{4} = 1$

center $(0, 1)$

foci $(\pm\sqrt{5}, 1)$

length of major axis 6

length of minor axis 4

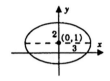

14. $x^2 + \dfrac{(y-3)^2}{4} = 1$

center $(0, 3)$

foci $(0, 3 \pm \sqrt{3})$

length of major axis 4

length of minor axis 2

15. $\dfrac{(x-1)^2}{16} + \dfrac{y^2}{64} = 1$

center $(1, 0)$

foci $(1, \pm 4\sqrt{3})$

length of major axis 16

length of minor axis 8

16. $\dfrac{(x-2)^2}{25} + \dfrac{(y-3)^2}{16} = 1$

center $(2, 3)$

foci $(5, 3)\ (-1, 3)$

length of major axis 10

length of minor axis 8

17. $x^2 - y^2 = 1$

center $(0, 0)$

transverse axis 2

vertices $(\pm 1, 0)$

foci $(\pm\sqrt{2}, 0)$

asymptotes $y = \pm x$

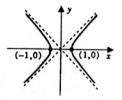

18. $y^2 - x^2 = 1$

center $(0, 0)$

transverse axis 2

vertices $(0, \pm 1)$

foci $(0, \pm\sqrt{2})$

asymptotes $y = \pm x$

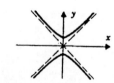

19. $\dfrac{x^2}{9} - \dfrac{y^2}{16} = 1$

center $(0,0)$

transverse axis 6

vertices $(\pm 3,\, 0)$

foci $(\pm 5,\, 0)$

asymptotes $y = \pm\frac{4}{3}x$

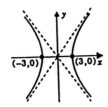

20. $\dfrac{x^2}{16} - \dfrac{y^2}{9} = 1$

center $(0,0)$

transverse axis 8

vertices $(\pm 4,\, 0)$

foci $(\pm 5,\, 0)$

asymptotes $y = \pm\frac{3}{4}x$

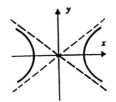

21. $\dfrac{y^2}{16} - \dfrac{x^2}{9} = 1$

center $(0,0)$

transverse axis 8

vertices $(0,\, \pm 4)$

foci $(0,\, \pm 5)$

asymptotes $y = \pm\frac{4}{3}x$

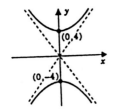

22. $\dfrac{y^2}{9} - \dfrac{x^2}{16} = 1$

center $(0,0)$

transverse axis 6

vertices $(0,\, \pm 3)$

foci $(0,\, \pm 5)$

asymptotes $y = \pm\frac{3}{4}x$

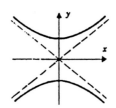

23. $\dfrac{(x-1)^2}{9} - \dfrac{(y-3)^2}{16} = 1$

center $(1,3)$

transverse axis 6

vertices $(4, 3)$ and $(-2, 3)$

foci $(6, 3)$ and $(-4, 3)$

asymptotes $y - 3 = \pm\frac{4}{3}(x - 1)$

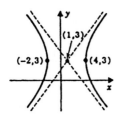

24. $\dfrac{(x-1)^2}{16} - \dfrac{(y-3)^2}{9} = 1$

center $(1,3)$

transverse axis 8

vertices $(5, 3)$ and $(-3, 3)$

foci $(6, 3)$ and $(-4, 3)$

asymptotes $y = \pm\frac{3}{4}(x - 1) + 3$

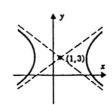

25. $\dfrac{(y-3)^2}{4} - \dfrac{(x-1)^2}{1} = 1$

center $(1, 3)$

transverse axis 4

vertices $(1, 5)$ and $(1, 1)$

foci $(1, 3 \pm \sqrt{5})$

asymptotes $y - 3 = \pm 2(x - 1)$

26. $(x+1)^2 - \dfrac{y^2}{3} = 1$

center $(-1, 0)$

transverse axis 2

vertices $(0, 0)$ and $(-2, 0)$

foci $(1, 0)$ and $(-3, 0)$

asymptotes $y = \pm \sqrt{3}(x + 1)$

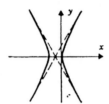

27. We can choose the coordinate system so that the parabola has an equation of the form $y = \alpha x^2, \alpha > 0$. One of the points of intersection is then the origin and the other is of the form $(c, \alpha c^2)$. We will assume that $c > 0$.

$$\text{area of } R_1 = \int_0^c \alpha x^2 dx = \frac{1}{3}\alpha c^3 = \frac{1}{3}A,$$

$$\text{area of } R_2 = A - \frac{1}{3}A = \frac{2}{3}A.$$

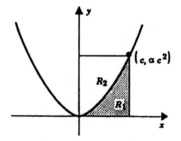

28. $A = A_1, \quad B = B_1$ by the reflection property of parabola;

$A = A_2, \quad B = B_2$ by simple geometry. Therefore

$A_1 = A_2, \quad B_1 = B_2$ and $C = D = \frac{1}{2}\pi$

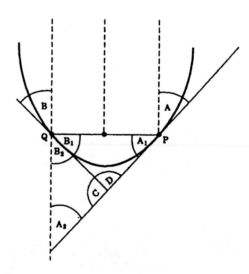

29. The equation of every such parabola takes the form

$$(x - x_0)^2 = 4c(y - y_0)$$

This equation can be written

$$y = \left(\frac{1}{4c}\right) x^2 - \left(\frac{x_0}{2c}\right) x + \left(y_0 + \frac{x_0{}^2}{4c}\right)$$

vertex $\left(-\dfrac{B}{2A}, \dfrac{4AC - B^2}{4A}\right)$, focus $\left(-\dfrac{B}{2A}, \dfrac{4AC - B^2 + 1}{4A}\right)$, directrix $y = \dfrac{4AC - B^2 - 1}{4A}$

30. Use $y = \dfrac{b}{a}\sqrt{a^2 - x^2}$.

$$A = \int_0^a \frac{b}{a}\sqrt{a^2 - x^2}\, dx = \frac{b}{a}\left[\frac{x}{2}\sqrt{a^2 - x^2} + \frac{a^2}{2}\sin^{-1}\frac{x}{a}\right]_0^a = \frac{ba\pi}{4}.$$

Also,

$$\overline{x}A = \int_0^a \frac{b}{a}x\sqrt{a^2 - x^2}\, dx = -\frac{b}{3a}\left[(a^2 - x^2)^{\frac{3}{2}}\right]_0^a = \frac{ba^2}{3},$$

and

$$\overline{y}A = \int_0^a \frac{b^2}{2a^2}(a^2 - x^2)\, dx = \frac{b^2}{2a^2}\left[a^2 x - \frac{x^3}{3}\right]_0^a = \frac{b^2 a}{3}.$$

Thus, $\overline{x} = \dfrac{4a}{3\pi}$, and $\overline{y} = \dfrac{4b}{3\pi}$.

31. By the hint, $xy = X^2 - Y^2 = 1$. In the XY-system $a = 1$, $b = 1$, $c = \sqrt{2}$. We have center $(0, 0)$, vertices $(\pm 1, 0)$, foci $(\pm\sqrt{2}, 0)$ and asymptotes $Y = \pm X$. Using

$$x = X + Y \quad \text{and} \quad y = X - Y$$

to convert to the xy-system, we find center $(0, 0)$, vertices $(1, 1)$ and $(-1, -1)$, foci $(\sqrt{2}, \sqrt{2})$ and $(-\sqrt{2}, -\sqrt{2})$, asymptotes $y = 0$ and $x = 0$, transverse axis $2\sqrt{2}$.

32. The points lie on the ellipse: $b^2 x^2 + a^2 y^2 = a^2 b^2$

$$b^2(a^2 \cos^2 t) + a^2(b^2 \sin^2 t) = a^2 b^2 (\cos^2 t + \sin^2 t) = a^2 b^2$$

33. $2\sqrt{\pi^2 a^4 - A^2}\,/\pi a$

34. Placing the parabola with its vertex at the origin, the equation becomes $x^2 = 4cy$.
When $y = 2$, $x = 2.5$, so $c = \dfrac{25}{32}$.

Thus the distance from the focus to the center of the mirror is $\dfrac{25}{32}$ ft.

35. In this case the length of the latus rectum is the width of the parabola at height $y = c$. With $y = c$, $4c^2 = x^2$, and $x = \pm 2c$. The length of the latus rectum is thus $4c$.

36. $y = \dfrac{1}{4c}x^2 \implies \dfrac{dy}{dx} = \dfrac{2x}{4c} = \dfrac{x}{2c}$

At $x = \pm 2c,$ we thus get $\dfrac{dy}{dx} = \pm 1$

37. $A = \displaystyle\int_{-2c}^{2c} \left(c - \dfrac{x^2}{4c} \right) dx = 2\int_0^{2c} \left(c - \dfrac{x^2}{4c} \right) dx = 2\left[cx - \dfrac{x^3}{12c} \right]_0^{2c} = \dfrac{8}{3}c^2$

$\overline{x} = 0$ by symmetry

$\overline{y}A = \displaystyle\int_{-2c}^{2c} \dfrac{1}{2}\left(c^2 - \dfrac{x^4}{16c^2} \right) dx = \int_0^{2c} \left(c^2 - \dfrac{x^4}{16c^2} \right) dx = \left[c^2 x - \dfrac{x^5}{80c^2} \right]_0^{2c} = \dfrac{8}{5}c^3$

$\overline{y} = \left(\dfrac{8}{5}c^3 \right)/\left(\dfrac{8}{3}c^2 \right) = \dfrac{3}{5}c$

38. $V = \displaystyle\int_0^{2c} 2\pi x \left(c - \dfrac{x^2}{4c} \right) dx = 2\pi \left[\dfrac{cx^2}{2} - \dfrac{x^4}{16c} \right]_0^{2c} = 2\pi c^3$

$\overline{y}V = \displaystyle\int_0^{2c} \pi x \left(c^2 - \dfrac{x^4}{16c^2} \right) dx = \pi \left[\dfrac{c^2 x^2}{2} - \dfrac{x^6}{6 \cdot 16c^2} \right]_0^{2c} = \dfrac{4}{3}\pi c^4$

$\implies \overline{y} = \dfrac{2}{3}c, \quad \overline{x} = 0$ by symmetry

39. $\dfrac{kx}{p(0)} = \tan\theta = \dfrac{dy}{dx}, \quad y = \dfrac{k}{2\,p(0)}x^2 + C$

In our figure $C = y(0) = 0$. Thus the equation of the cable is $y = kx^2/2p(0)$, the equation of a parabola.

40. We'll work in two dimensions. The directrix of the parabola is perpendicular to the undisturbed light rays. If the parabola were not there to intercept the rays, the rays would reach the directrix in paths of the same length (this is an assumption we are making). Also true upon reflection by the parabola since length of $\overline{PF}$=length of $\overline{PQ}$.

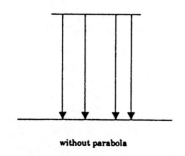

directrix

without parabola

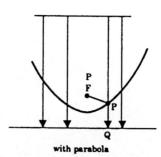

with parabola

41. Start with any two parabolas γ_1, γ_2. By moving them we can see to it that they have equations of the following form:

$$\gamma_1: x^2 = 4c_1 y, \quad c_1 > 0; \qquad \gamma_2: x^2 = 4c_2 y, \quad c_2 > 0.$$

Now we change the scale for γ_2 so that the equation for γ_2 will look exactly like the equation for γ_1. Set $X = (c_1/c_2)\, x, \quad Y = (c_1/c_2)\, y.$ Then

$$x^2 = 4c_2 y \quad \Longrightarrow \quad (c_2/c_1)^2\, X^2 = 4c_2\, (c_2/c_1)\, Y \quad \Longrightarrow \quad X^2 = 4c_1 Y.$$

Now γ_2 has exactly the same equation as γ_1; only the scale, the units by which we measure distance, has changed.

42. (a) Take $x(t) = a\cosh t, \quad y(t) = b\sinh t$ Then $\dfrac{x^2}{a^2} - \dfrac{y^2}{b^2} = \cosh^2 t - \sinh^2 t = 1$
   range$(x) = [a, \infty), \quad$ range$(y) = (-\infty, \infty)$

(b) Take $x(t) = -a\cosh t, \quad y(t) = b\sinh t$ Then $\dfrac{x^2}{a^2} - \dfrac{y^2}{b^2} = 1,$
   range$(x) = (-\infty, -a], \quad$ range $(y) = (-\infty, \infty)$

43. $A = \dfrac{2b}{a} \displaystyle\int_a^{2a} \sqrt{x^2 - a^2}\, dx = \dfrac{2b}{a} \left[\dfrac{x}{2}\sqrt{x^2 - a^2} - \dfrac{a^2}{2}\ln\left(x + \sqrt{x^2 - a^2}\right) \right]_a^{2a}$

$$= [2\sqrt{3} - \ln(2 + \sqrt{3})]ab$$

44. Let $P(x_0, y_0)$ be the point of tangency and let l be the tangent line.

$$\text{slope of } l = \frac{b^2 x_0}{a^2 y_0}, \quad \text{slope of } \overline{F_1 P} = \frac{y_0}{x_0 + c}, \quad \text{slope of } \overline{F_2 P} = \frac{y_0}{x_0 - c}$$

With θ_1 as the angle between l and $\overline{F_1 P}$ and with θ_2 as the angle between $\overline{F_2 P}$ and l, we have

$$\tan\theta_1 = \frac{\dfrac{b^2 x_0}{a^2 y_0} - \dfrac{y_0}{x_0 + c}}{1 + \dfrac{b^2 x_0}{a^2 y_0}\left(\dfrac{y_0}{x_0 + c}\right)} \quad \text{and} \quad \tan\theta_2 = \frac{\dfrac{y_0}{x_0 - c} - \dfrac{b^2 x_0}{a^2 y_0}}{1 + \left(\dfrac{y_0}{x_0 - c}\right)\left(\dfrac{b^2 x_0}{a^2 y_0}\right)}$$

From the fact that

$$b^2 x_0^2 - a^2 y_0^2 = a^2 b^2, \quad (P \text{ is on the hyperbola})$$

it follows readily that $\tan\theta_1 = \tan\theta_2$.

45. $e = \dfrac{\sqrt{25 - 16}}{\sqrt{25}} = \dfrac{3}{5}$

46. $e = \dfrac{\sqrt{25 - 16}}{\sqrt{25}} = \dfrac{3}{5}$

47. $e = \dfrac{\sqrt{25 - 9}}{\sqrt{25}} = \dfrac{4}{5}$

48. $e = \dfrac{\sqrt{169 - 144}}{\sqrt{169}} = \dfrac{5}{13}$

49. E_1 is fatter than E_2,   more like a circle.

50. The ellipse tends to a circle of radius a (Since b approaches a).

51. The ellipse tends to a line segment of length $2a$.

52. $a = 3, \quad c = ea = \dfrac{1}{3} \cdot 3 = 1, \quad b = \sqrt{a^2 - c^2} = \sqrt{8}$

   Center $(0,0)$,   so $\dfrac{x^2}{9} + \dfrac{y^2}{8} = 1$

53. $a = 3, \quad c = ea = \frac{2}{3}\sqrt{2} \cdot 3 = 2\sqrt{2}, \quad b = \sqrt{9 - 8} = 1; \quad x^2/9 + y^2 = 1$

54. The basic equation reads $\sqrt{x^2 + y^2} = e|x - k|$.   Square and arrange to

$$\left(x + \frac{e^2 k}{1 - e^2} \right)^2 + \frac{y^2}{1 - e^2} = \frac{e^2 k^2}{(1 - e^2)^2}$$

   Now set $a = ek/(1 - e^2)$ and $c = ea$.   This reduces the equation to

$$\frac{(x + c)^2}{a^2} + \frac{y^2}{a^2 - c^2} = 1$$

   This equation represents an ellipse of eccentricity $e = c/a$.

55. $e = \frac{5}{3}$ **56.** $e = \frac{5}{4}$

57. $e = \sqrt{2}$ **58.** $e = \frac{13}{5}$

59. The branches of H_1 open up less quickly than the branches of H_2.

60. The hyperbola tends to the union of two oppositely-directed half lines that begin at the ends of the transverse axis.

61. The hyperbola tends to a pair of parallel lines separated by the transverse axis.

62. The basic equation reads $\sqrt{x^2 + y^2} = e|x - k|$.   Square and rearrange to

$$\left(x - \frac{e^2 k}{e^2 - a} \right)^2 - \frac{y^2}{e^2 - 1} = \frac{e^2 k^2}{(e^2 - 1)^2}$$

   Now set $a = ek/(e^2 - 1)$ and $c = ea$.   This reduces the equation to

$$\frac{(x - c)^2}{a^2} - \frac{y^2}{c^2 - a^2} = 1$$

   This equation represents a hyperbola of eccentricity $e = c/a$

63. $P(x, y)$ is on the parabola with directrix $l: Ax + By + C = 0$ and focus $F(a, b)$ iff $d(P, l) = d(P, F)$

   which happens iff $\dfrac{|Ax + By + C|}{\sqrt{A^2 + B^2}} = \sqrt{(x - a)^2 + (y - b)^2}.$

   Squaring both sides of this equation and simplifying, we obtain

$$(Ay - Bx)^2 = (2aS + 2AC)x + (2bS + 2BC)y + c^2 - (a^2 + b^2)S$$

   with $S = A^2 + B^2 \neq 0$.

SECTION 10.2

1–8.

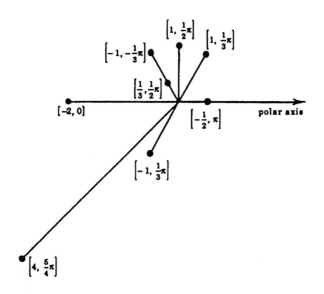

$\left[1, \frac{1}{2}\pi\right]$ $\left[1, \frac{1}{3}\pi\right]$ $\left[-1, -\frac{1}{3}\pi\right]$ $\left[\frac{1}{3}, \frac{1}{2}\pi\right]$ $[-2, 0]$ $\left[-\frac{1}{2}, \pi\right]$ polar axis $\left[-1, \frac{1}{3}\pi\right]$ $\left[4, \frac{5}{4}\pi\right]$

9. $x = 3\cos\frac{1}{2}\pi = 0$

$y = 3\sin\frac{1}{2}\pi = 3$

$(0, 3)$

10. $x = 4\cos\dfrac{\pi}{6} = 2\sqrt{3}$

$y = 4\sin\dfrac{\pi}{6} = 2$

$(2\sqrt{3}, 2)$

11. $x = -\cos(-\pi) = 1$

$y = -\sin(-\pi) = 0$

$(1, 0)$

12. $x = -1\cos\dfrac{\pi}{4} = -\sqrt{2}/2$

$y = -1\sin\dfrac{\pi}{4} = -\sqrt{2}/2$

$(-2\sqrt{2}, -2\sqrt{2})$

13. $x = -3\cos\left(-\frac{1}{3}\pi\right) = -\frac{3}{2}$

$y = -3\sin\left(-\frac{1}{3}\pi\right) = \frac{3}{2}\sqrt{3}$

$\left(-\frac{3}{2}, \frac{3}{2}\sqrt{3}\right)$

14. $x = 2\cos 0 = 2$

$y = 2\sin 0 = 0$

$(2, 0)$

15. $x = 3\cos\left(-\frac{1}{2}\pi\right) = 0$

$y = 3\sin\left(-\frac{1}{2}\pi\right) = -3$

$(0, -3)$

16. $x = 2\cos 3\pi = -2$

$y = 2\sin 3\pi = 0$

$(-2, 0)$

17. $r^2 = 0^2 + 1^2, \quad r = \pm 1$

$r = 1: \quad \cos\theta = 0$ and $\sin\theta = 1$

$\theta = \frac{1}{2}\pi$

$\left[1, \frac{1}{2}\pi + 2n\pi\right], \quad \left[-1, \frac{3}{2}\pi + 2n\pi\right]$

18. $r^2 = 1^2 + 0^2, \quad r = \pm 1$

$r = 1: \quad \cos\theta = 1$ and $\sin\theta = 0$

$\theta = 2\pi$

$\left[1, 2n\pi\right], \quad \left[-1, \pi + 2n\pi\right]$

19. $r^2 = (-3)^2 + 0^2 = 9, \quad r = \pm 3$

$r = 3: \quad \cos\theta = -1$ and $\sin\theta = 0$

$\theta = \pi$

$\left[3, \pi + 2n\pi\right], \quad \left[-3, 2n\pi\right]$

20. $r^2 = 4^2 + 4^2, \quad r = \pm 4\sqrt{2}$

$r = 4\sqrt{2}: \quad \cos\theta = \frac{1}{\sqrt{2}}$ and $\sin\theta = \frac{1}{\sqrt{2}}$

$\theta = \frac{1}{4}\pi$

$\left[4\sqrt{2}, \frac{1}{4}\pi + 2n\pi\right], \quad \left[-4\sqrt{2}, \frac{5}{4}\pi + 2n\pi\right]$

21. $r^2 = 2^2 + (-2)^2 = 8, \quad r = \pm 2\sqrt{2}$
$r = 2\sqrt{2}: \quad \cos\theta = \frac{1}{2}\sqrt{2}, \quad \sin\theta = -\frac{1}{2}\sqrt{2}$ $\Bigg\}$ $\left[2\sqrt{2}, \frac{7}{4}\pi + 2n\pi\right], \quad \left[-2\sqrt{2}, \frac{3}{4}\pi + 2n\pi\right]$
$\theta = \frac{7}{4}\pi$

22. $r^2 = 3^2 + (3\sqrt{3})^2, \quad r = \pm 6$
$r = 6: \quad \cos\theta = \frac{1}{2} \text{ and } \sin\theta = -\frac{\sqrt{3}}{2}$ $\Bigg\}$ $\left[6, \frac{5}{3}\pi + 2n\pi\right], \quad \left[-6, \frac{2}{3}\pi + 2n\pi\right]$
$\theta = \frac{5}{3}\pi$

23. $r^2 = \left(4\sqrt{3}\right)^2 + 4^2 = 64, \quad r \pm 8$
$r = 8: \quad \cos\theta = \frac{1}{2}\sqrt{3}, \quad \sin\theta = \frac{1}{2}$ $\Bigg\}$ $\left[8, \frac{1}{6}\pi + 2n\pi\right], \quad \left[-8, \frac{7}{6}\pi + 2n\pi\right]$
$r = \frac{1}{6}\pi$

24. $r^2 = (\sqrt{3})^2 + 1^2, \quad r = \pm 2$
$r = 2: \quad \cos\theta = \frac{\sqrt{3}}{2} \text{ and } \sin\theta = -\frac{1}{2}$ $\Bigg\}$ $\left[2, \frac{11}{6}\pi + 2n\pi\right], \quad \left[-2, \frac{5}{6}\pi + 2n\pi\right]$
$\theta = \frac{11}{6}\pi$

25. $d^2 = (x_1 - x_2)^2 + (y_1 - y_2)^2 = (r_1\cos\theta_1 - r_2\cos\theta_2)^2 + (r_1\sin\theta_1 - r_2\sin\theta_2)^2$
$$= r_1{}^2\cos^2\theta_1 - 2r_1 r_2\cos\theta_1\cos\theta_2 + r_2{}^2\cos^2\theta_2$$
$$+ r_1{}^2\sin^2\theta_1 - 2r_1 r_2\sin\theta_1\sin\theta_2 + r_2{}^2\sin^2\theta_2$$
$$= r_1{}^2 + r_2{}^2 - 2r_1 r_2\left(\cos\theta_1\cos\theta_2 + \sin\theta_1\sin\theta_2\right)$$
$$= r_1{}^2 + r_2{}^2 - 2r_1 r_2\cos\left(\theta_1 - \theta_2\right)$$
$$d = \sqrt{r_1{}^2 + r_2{}^2 - 2r_1 r_2\cos\left(\theta_1 - \theta_2\right)}$$

26. Draw a figure; the result is clear.

27. (a) $\left[\frac{1}{2}, \frac{11}{6}\pi\right]$ (b) $\left[\frac{1}{2}, \frac{5}{6}\pi\right]$ (c) $\left[\frac{1}{2}, \frac{7}{6}\pi\right]$

28. (a) $\left[3, \frac{5}{4}\pi\right]$ (b) $\left[3, \frac{1}{4}\pi\right]$ (c) $\left[3, \frac{7}{4}\pi\right]$

29. (a) $\left[2, \frac{2}{3}\pi\right]$ (b) $\left[2, \frac{5}{3}\pi\right]$ (c) $\left[2, \frac{1}{3}\pi\right]$

30. (a) $\left[3, \frac{3}{4}\pi\right]$ (b) $\left[3, \frac{7}{4}\pi\right]$ (c) $\left[3, \frac{1}{4}\pi\right]$

31. about the x-axis?: $\quad r = 2 + \cos(-\theta) \quad \Longrightarrow \quad r = 2 + \cos\theta, \quad$ yes.
about the y-axis?: $\quad r = 2 + \cos(\pi - \theta) \quad \Longrightarrow \quad r = 2 - \cos\theta, \quad$ no.
about the origin?: $\quad r = 2 + \cos(\pi + \theta) \quad \Longrightarrow \quad r = 2 - \cos\theta, \quad$ no.

32. about the x-axis?: $\quad r = \cos[2(-\theta)] \quad \Longrightarrow \quad r = \cos 2\theta, \quad$ yes.
about the y-axis?: $\quad r = \cos[2(\pi - \theta)] \quad \Longrightarrow \quad r = \cos 2\theta, \quad$ yes.
about the origin?: $\quad r = \cos[2(\pi + \theta)] \quad \Longrightarrow \quad r = \cos 2\theta, \quad$ yes.

33. about the x-axis?: $\quad r(\sin(-\theta) + \cos(-\theta)) = 1 \quad \Longrightarrow \quad r(-\sin\theta + \cos\theta) = 1, \quad$ no.
about the y-axis?: $\quad r(\sin(\pi - \theta) + \cos(\pi - \theta)) = 1 \quad \Longrightarrow \quad r(\sin\theta - \cos\theta) = 1, \quad$ no.
about the origin?: $\quad r(\sin(\pi + \theta) + \cos(\pi + \theta)) = 1 \quad \Longrightarrow \quad r(-\sin\theta - \cos\theta) = 1, \quad$ no.

34. about the x-axis?: $r\sin(-\theta) = 1 \implies -r\sin\theta = 1,$ no.

about the y-axis?: $r\sin(\pi - \theta) = 1 \implies r\sin\theta = 1$ yes.

about the origin?: $r\sin(\pi + \theta) = 1 \implies -r\sin\theta = 1,$ no.

35. about the x-axis?: $r^2\sin(-2\theta) = 1 \implies -r^2\sin 2\theta = 1,$ no.

about the y-axis?: $r^2\sin(2(\pi - \theta)) = 1 \implies -r^2\sin 2\theta = 1,$ no.

about the origin?: $r^2\sin(2(\pi + \theta)) = 1 \implies r^2\sin 2\theta = 1,$ yes.

36. about the x-axis?: $r^2\cos[2(-\theta)] = 1 \implies r^2\cos 2\theta = 1,$ yes.

about the y-axis?: $r^2\cos[2(\pi - \theta)] = 1 \implies r^2\cos 2\theta = 1,$ yes.

about the origin?: $r^2\cos[2(\pi + \theta)] = 1 \implies r^2\cos 2\theta = 1,$ yes.

37. $x = 2$

$r\cos\theta = 2$

38. $r\sin\theta = 3$

39. $2xy = 1$

$2(r\cos\theta)(r\sin\theta) = 1$

$r^2\sin 2\theta = 1$

40. $r^2 = 9$

$r = 3$

41. $x^2 + (y-2)^2 = 4$

$x^2 + y^2 - 4y = 0$

$r^2 - 4r\sin\theta = 0$

$r = 4\sin\theta$

42. $(x - a)^2 + y^2 = a^2$

$x^2 - 2ax + y^2 = 0$

$r^2 = 2ar\cos\theta$

$r = 2a\cos\theta$

43. $y = x$

$r\sin\theta = r\cos\theta$

$\tan\theta = 1$

$\theta = \pi/4$

[note: division by r okay
since $[0,0]$ is on the curve]

44. $x^2 - y^2 = 4$

$r^2(\cos^2\theta - \sin^2\theta) = 4$

$r^2\cos 2\theta = 4$

45. $x^2 + y^2 + x = \sqrt{x^2 + y^2}$

$r^2 + r\cos\theta = r$

$r = 1 - \cos\theta$

46. $y = mx$

$r\sin\theta = mr\cos\theta$

$\theta = a$ constant

47. $(x^2 + y^2)^2 = 2xy$

$r^4 = 2(r\cos\theta)(r\sin\theta)$

$r^2 = \sin 2\theta$

48. $(x^2 + y^2)^2 = x^2 - y^2$

$r^4 = r^2(\cos^2\theta - \sin^2\theta)$

$r^2 = \cos 2\theta$

49. The horizontal line $y = 4$

50. The vertical line $x = 4$

51. The line $y = \sqrt{3}x$

52. $\theta = \pm\frac{\pi}{3}:$ the lines $y = \pm\sqrt{3}x$

53.
$$r = 2(1 - \cos\theta)^{-1}$$
$$r - r\cos\theta = 2$$
$$\sqrt{x^2 + y^2} - x = 2$$
$$x^2 + y^2 = (x + 2)^2$$
$$y^2 = 4(x + 1)$$
a parabola

54. Circle $\quad x^2 + y^2 = 2y$

55.
$$r = 6\cos\theta$$
$$r^2 = 6r\cos\theta$$
$$x^2 + y^2 = 6x$$

56. The y-axis; $\quad x = 0$

57. The line $y = 2x$

58.
$$r = 4\sin(\theta + \pi) = -4\sin\theta$$
$$r^2 = -4r\sin\theta$$
$$x^2 + y^2 = -4y$$
$$x^2 + (y + 2)^2 = 4$$
a circle

59.
$$r = \frac{4}{2 - \cos\theta}$$
$$2r - r\cos\theta = 4$$
$$2\sqrt{x^2 + y^2} - x = 4$$
$$4(x^2 + y^2) = (x + 4)^2$$
$$3x^2 + 4y^2 - 8x = 16$$
an ellipse

60.
$$\frac{6}{1 + 2\sin\theta} = r$$
$$6 = r + 2r\sin\theta$$
$$\sqrt{x^2 + y^2} = 6 - 2y$$
$$x^2 + y^2 = 36 - 24y + 4y^2$$
$$x^2 - 3y^2 + 24y = 36$$
a hyperbola

61.
$$r = \frac{4}{1 - \cos\theta}$$
$$r - r\cos\theta = 4$$
$$\sqrt{x^2 + y^2} - x = 4$$
$$x^2 + y^2 = (x + 4)^2$$
$$y^2 = 8x + 16$$
a parabola

62.
$$r = \frac{2}{3 + 2\sin\theta}$$
$$3r + 2r\sin\theta = 2$$
$$3\sqrt{x^2 + y^2} = 2 - 2y$$
$$9x^2 + 5y^2 = 4 - 8y$$
an ellipse

63.
$$r = 2a\sin\theta + 2b\cos\theta$$
$$r^2 = 2ar\sin\theta + 2br\cos\theta$$
$$x^2 + y^2 = 2ay + 2bx$$
$$(x - b)^2 + (y - a)^2 = a^2 + b^2$$
center: (b, a); radius: $\sqrt{a^2 + b^2}$

64. $r = d + r \cos \theta$

65. $\frac{1}{2} (r \cos \theta + d) = r$

$r = \dfrac{d}{2 - \cos \theta}$

66. $r = 2(d + r \cos \theta)$

$= 2d + 2r \cos \theta$

SECTION 10.3

1.
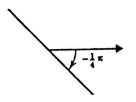

2.
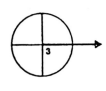

3.
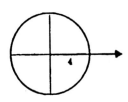

4.
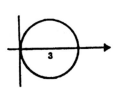

5.
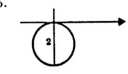

6.
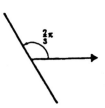

7.
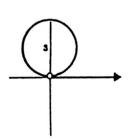

8.
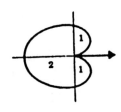

9.
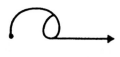

10.
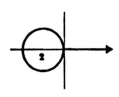

11.
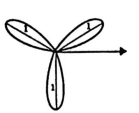

12.
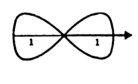

13.
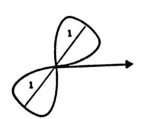

14.
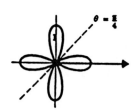

15.
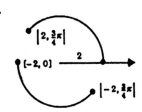

16.

17.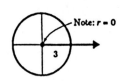

18.

19.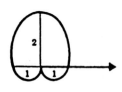

20.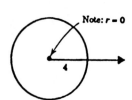

21.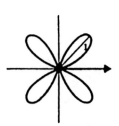

22.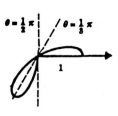

23.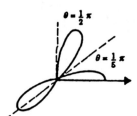

24.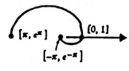

25.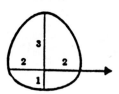

26.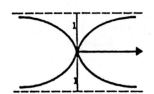

27.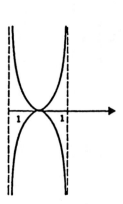

28.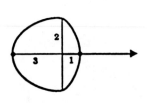

29.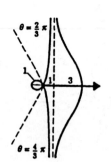

30.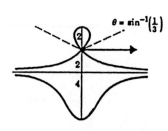

31.

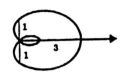

32.

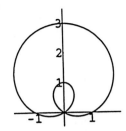

33. yes; $[1, \pi] = [-1, 0]$ and the pair $r = -1, \theta = 0$ satisfies the equation

34. No.   **35.** yes;   the pair $r = \frac{1}{2}, \theta = \frac{1}{2}\pi$ satisfies the equation

36. yes; $[1, -\frac{5}{6}\pi] = [-1, \frac{1}{6}\pi]$,   which lies on the curve.

37. $[2, \pi] = [-2, 0]$. The coordinates $[-2, 0]$ satisfy the equation $r^2 = 4\cos\theta$, and the coordinates $[2, \pi]$ satisfy the equation $r = 3 + \cos\theta$.

38. $[2, \pi/2] = [-2, -\pi/2]$ The coordinates $[2, \pi/2]$ satisfy $r^2 \sin\theta = 4$, and the coordinates $[-2, -\pi/2]$ satisfy $r = 2\cos 2\theta$.

39. $(0, 0), \left(-\frac{1}{2}, \frac{1}{2}\right)$

40. $(0, 1)$

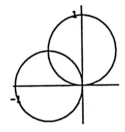

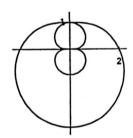

41. $(1, 0), (-1, 0)$

42. $(0, 0), (1, 1)$

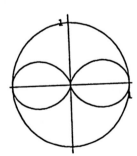

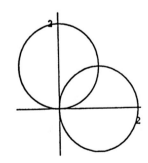

43. $(0,0)$, $\left(\frac{1}{4}, \frac{1}{4}\sqrt{3}\right)$, $\left(\frac{1}{4}, -\frac{1}{4}\sqrt{3}\right)$

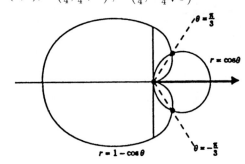

44. $(0,0)$, $(0,1)$

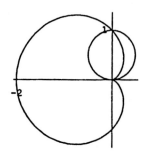

45. $(0,0)$, $\left(\pm\frac{\sqrt{3}}{4}, \frac{3}{4}\right)$

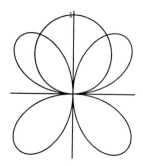

46. $(0,0)$, $\left(\frac{2+\sqrt{2}}{2}, \frac{2+\sqrt{2}}{2}\right)$, $\left(\frac{2-\sqrt{2}}{2}, \frac{2-\sqrt{2}}{2}\right)$

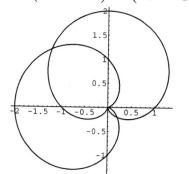

47. (a) The graph of $r = 1 + \cos\left(\theta - \frac{\pi}{3}\right)$ is the graph of $r = 1 + \cos\theta$ rotated counterclockwise $\pi/3$ radians; The graph of $r = 1 + \cos\left(\theta + \frac{\pi}{6}\right)$ is the graph of $r = 1 + \cos\theta$ rotated clockwise $\pi/6$ radians.

(b) The graph of $r = f(\theta - \alpha)$ is the graph of $r = f(\theta)$ rotated counterclockwise α radians.

48. (a)

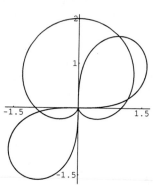

(b) The curves intersect at the pole and at:
$[1.172, 0.173]$, $[1.86, 1.036]$, $[0.90, 3.245]$

49. (a)

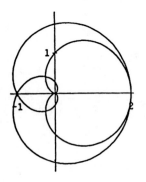

(b) The curves intersect at:
the pole and $(2,0)$, $(-1,0)$, $(-0.25, \pm 0.4330)$

50. (a)

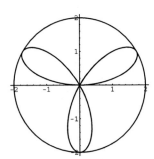

(b) The curves intersect at:

$(2, \pi/6), \ (2, 5\pi/6), \ (2, 3\pi/2)$

51. (a)

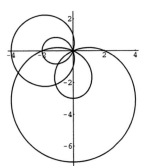

(b) The curves intersect at the pole and at:

$r = 1 - 3\cos\theta$	$r = 2 - 5\sin\theta$
$[-2, 0]$	$[2, \pi]$
$[3.800, 3.510]$	$[3.800, 3.510]$
$[2.412, 4.223]$	$[-2.412, 1.081]$
$[-1.267, 0.713]$	$[-1.267, 0.713]$

52. (a)

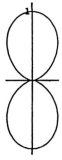

(b)

(c) $(\pm 0.7484, \pm 0.8651)$

53. "Butterfly" curves. The graph for the case $k = 2$ is:

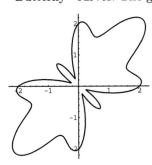

54. An insect? Fly? Mosquito?

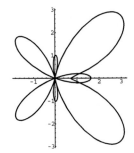

55. (a) $k = \frac{3}{2}$ (b) $k = \frac{5}{2}$

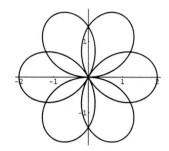

A petal curve with $2m$ petals.

56. $r = 2\cos\left(\frac{4}{3}\theta\right)$ $r = 2\sin\left(\frac{5}{3}\theta\right)$

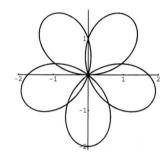

PROJECT 10.3

1. $e = \dfrac{r}{d - r\cos\theta} \implies r = ed - er\cos\theta \implies r(1 + e\cos\theta) = ed \implies r = \dfrac{ed}{1 + e\cos\theta}$

2. (a) Suppose $r = \dfrac{ed}{1 + e\cos\theta}$. Then

$$r + er\cos\theta = ed$$

$$r = ed - er\cos\theta$$

$$r^2 = e^2 d^2 - 2e^2 dr\cos\theta + e^2 r^2 \cos^2\theta$$

$$x^2 + y^2 = e^2 d^2 - 2e^2 dx + e^2 x^2$$

$$(1 - e^2)x^2 + 2e^2 dx + y^2 = e^2 d^2$$

$$\left(x + \frac{e^2 d}{1 - e^2}\right)^2 + \frac{y^2}{1 - e^2} = \frac{e^2 d^2}{(1 - e^2)^2}$$

$$(x + c)^2 + \frac{y^2}{1 - e^2} = a^2 \quad \left(a = \frac{ed}{1 - e^2}, \quad c = ea\right)$$

$$\frac{(x + c)^2}{a^2} + \frac{y^2}{(1 - e^2)a^2} = 1$$

$$\frac{(x + c)^2}{a^2} + \frac{y^2}{a^2 - c^2} = 1$$

(b) From part (a),

$$x^2 + y^2 = e^2 d^2 - 2e^2 dx + e^2 x^2$$

$$\text{becomes} \quad y^2 = d^2 - 2dx$$

$$\text{so} \quad y^2 = -4\frac{d}{2}\left(x - \frac{d}{2}\right)$$

(c) From part (a),

$$\left(x + \frac{e^2 d}{1 - e^2}\right)^2 + \frac{y^2}{1 - e^2} = \frac{e^2 d^2}{(1 - e^2)^2}$$

$$(x - c)^2 - \frac{y^2}{e^2 - 1} = a^2 \quad \left(a = \frac{ed}{e^2 - 1}, \quad c = ea\right)$$

$$\frac{(x - c)^2}{a^2} - \frac{y^2}{(e^2 - 1)a^2} = 1$$

$$\frac{(x - c)^2}{a^2} - \frac{y^2}{c^2 - a^2} = 1$$

3. (a) ellipse: $r = \dfrac{8}{4 + 3\cos\theta} = \dfrac{2}{1 + \frac{3}{4}\cos\theta}$.

Thus $e = \dfrac{3}{4}$ and $\dfrac{3}{4}d = 2 \implies d = \dfrac{8}{3}$.

Rectangular equation:

$$a = \frac{32}{7}, \quad c = \frac{24}{7}, \quad \text{so} \quad \frac{\left(x + \frac{24}{7}\right)^2}{\left(\frac{32}{7}\right)^2} + \frac{y^2}{\left(\frac{24}{7}\right)^2} = 1$$

(b) hyperbola: $r = \dfrac{6}{1 + 2\cos\theta}$

Thus $e = 2$ and $2d = 6 \implies d = 3$.

Rectangular equation:

$$a = 2, \quad c = 4, \quad \text{so} \quad \frac{(x - 4)^2}{4} - \frac{y^2}{12} = 1$$

(c) parabola: $r = \dfrac{6}{2 + 2\cos\theta} = \dfrac{3}{1 + \cos\theta}$

Thus $e = 1$ and $d = 3$.

Rectangular equation:

$$y^2 = -4\left(\frac{3}{2}\right)\left(x - \frac{3}{2}\right) = -6\left(x - \frac{3}{2}\right)$$

4. $r = \dfrac{\alpha}{1 - \beta\cos\theta}$ is the conic section $r = \dfrac{\alpha}{1 + \beta\cos\theta}$ rotated π radians in the counter clockwise direction:

$$\frac{\alpha}{1 - \beta\cos\theta} = \frac{\alpha}{1 + \beta\cos(\theta - \pi)}$$

$$r = \frac{\alpha}{1 - \beta \sin \theta} \text{ is the conic section } \quad r = \frac{\alpha}{1 + \beta \cos \theta} \text{ rotated } \frac{\pi}{2} \text{ radians in the clockwise direction:}$$

$$\frac{\alpha}{1 - \beta \sin \theta} = \frac{\alpha}{1 + \beta \cos \left(\theta + \frac{\pi}{2}\right)}$$

SECTION 10.4

1.

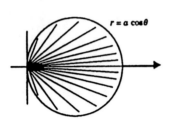

$$A = \int_{-\pi/2}^{\pi/2} \frac{1}{2} [a \cos \theta]^2 \, d\theta$$

$$= a^2 \int_0^{\pi/2} \frac{1 + \cos 2\theta}{2} \, d\theta$$

$$= a^2 \left[\frac{\theta}{2} + \frac{\sin 2\theta}{4} \right]_0^{\pi/2} = \frac{1}{4} \pi a^2$$

2.

$$A = \int_{-\pi/6}^{\pi/6} \frac{1}{2} a^2 \cos^2 3\theta \, d\theta$$

$$= \frac{a^2}{2} \left[\frac{\theta}{2} + \frac{\sin 6\theta}{12} \right]_{-\pi/6}^{\pi/6} = \frac{1}{12} \pi a^2$$

3.

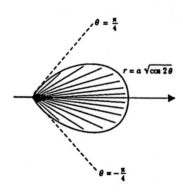

$$A = \int_{-\pi/4}^{\pi/4} \frac{1}{2} \left[a\sqrt{\cos 2\theta} \right]^2 \, d\theta$$

$$= a^2 \int_0^{\pi/4} \cos 2\theta \, d\theta$$

$$= a^2 \left[\frac{\sin 2\theta}{2} \right]_0^{\pi/4} = \frac{1}{2} a^2$$

4.

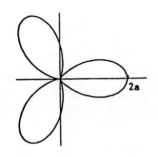

$$A = \int_{-\pi/3}^{\pi/3} \frac{1}{2} a^2 (1 + 2\cos 3\theta + \cos^2 3\theta) \, d\theta$$

$$= \frac{a^2}{2} \left[\theta + \frac{2}{3} \sin 3\theta + \frac{\theta}{2} + \frac{\sin 6\theta}{12} \right]_{-\pi/3}^{\pi/3} = \frac{1}{2} \pi a^2$$

5.

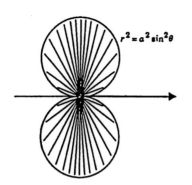

$$A = 2 \int_0^\pi \frac{1}{2} \left(a^2 \sin^2 \theta \right) d\theta$$

$$= a^2 \int_0^\pi \frac{1 - \cos 2\theta}{2} \, d\theta$$

$$= a^2 \left[\frac{\theta}{2} - \frac{\sin 2\theta}{4} \right]_0^\pi = \frac{1}{2} \pi a^2$$

6.

$$A = 4 \int_0^{\pi/2} \frac{1}{2} a^2 \sin^2 2\theta \, d\theta$$

$$= 2a^2 \left[\frac{\theta}{2} - \frac{\sin 4\theta}{8} \right]_0^{\pi/2} = \frac{1}{2} \pi a^2$$

7.

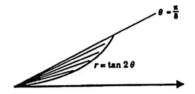

$$A = \int_0^{\pi/8} \frac{1}{2} \left[\tan 2\theta \right]^2 d\theta$$

$$= \frac{1}{2} \int_0^{\pi/8} \left(\sec^2 2\theta - 1 \right) d\theta$$

$$= \frac{1}{2} \left[\frac{1}{2} \tan 2\theta - \theta \right]_0^{\pi/8} = \frac{1}{4} - \frac{\pi}{16}$$

8.

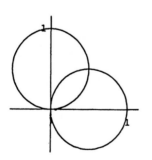

$$A = \int_0^{\pi/4} \frac{1}{2} (\cos^2 \theta - \sin^2 \theta) \, d\theta$$

$$= \frac{1}{2} \int_0^{\pi/4} \cos 2\theta \, d\theta$$

$$= \left[\frac{\sin 2\theta}{4} \right]_0^{\pi/4} = \frac{1}{4}$$

9.

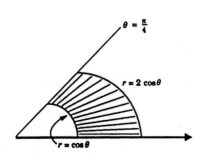

$$A = \int_0^{\pi/4} \frac{1}{2} \left([2 \cos \theta]^2 - [\cos \theta]^2 \right) d\theta$$

$$= \frac{3}{2} \int_0^{\pi/4} \frac{1 + \cos 2\theta}{2} \, d\theta$$

$$= \frac{3}{2} \left[\frac{\theta}{2} + \frac{\sin 2\theta}{4} \right]_0^{\pi/4} = \frac{3}{16} \pi + \frac{3}{8}$$

10.

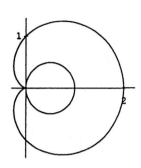

$$A = \int_0^{\pi/2} \frac{1}{2} (1 + 2\cos\theta + \cos^2\theta - \cos^2\theta) \, d\theta$$

$$= \frac{1}{2} \int_0^{\pi/2} (1 + 2\cos\theta) \, d\theta$$

$$= \frac{1}{2} [\theta + 2\sin\theta]_0^{\pi/2} = 1 + \frac{\pi}{4}$$

11.

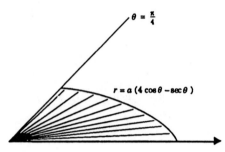

$$A = \int_0^{\pi/4} \frac{1}{2} [a(4\cos\theta - \sec\theta)]^2 \, d\theta$$

$$= \frac{a^2}{2} \int_0^{\pi/4} [16\cos^2\theta - 8 + \sec^2\theta] \, d\theta$$

$$= \frac{a^2}{2} \int_0^{\pi/4} [8(1 + \cos 2\theta) - 8 + \sec^2\theta] \, d\theta$$

$$= \frac{a^2}{2} [4\sin 2\theta + \tan\theta]_0^{\pi/4} = \frac{5}{2} a^2$$

12.

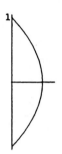

$$A = \int_{-\pi/2}^{\pi/2} \frac{1}{2} \cdot \frac{1}{4} \cdot \sec^4\frac{\theta}{2} \, d\theta$$

$$= \frac{1}{8} \int_{-\pi/2}^{\pi/2} (1 + \tan^2\frac{\theta}{2}) \sec^2\frac{\theta}{2} \, d\theta$$

$$= \frac{1}{8} \left[2\tan\frac{\theta}{2} + \frac{2}{3}\tan^3\frac{\theta}{2} \right]_{-\pi/2}^{\pi/2} = \frac{2}{3}$$

13.

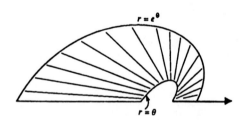

$$A = \int_0^{\pi} \frac{1}{2} \left([e^\theta]^2 - [\theta]^2 \right) d\theta$$

$$= \frac{1}{2} \int_0^{\pi} (e^{2\theta} - \theta^2) \, d\theta$$

$$= \frac{1}{2} [\frac{1}{2}e^{2\theta} - \frac{1}{3}\theta^3]_0^{\pi} = \frac{1}{12} (3e^{2\pi} - 3 - 2\pi^3)$$

14.

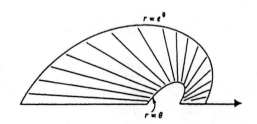

$$A = \int_0^{\pi} \frac{1}{2} \left[\left(e^{2\pi+\theta}\right)^2 - \theta^2 \right] d\theta$$

$$= \frac{1}{2} \left[\frac{e^{4\pi+2\theta}}{2} - \frac{\theta^3}{3} \right]_0^{\pi}$$

$$= \frac{1}{4}e^{6\pi} - \frac{1}{4}e^{4\pi} - \frac{1}{6}\pi^3$$

15.

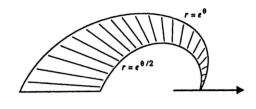

$$A = \int_0^\pi \frac{1}{2}\left(\left[e^\theta\right]^2 - \left[e^{\theta/2}\right]^2\right)d\theta$$

$$= \frac{1}{2}\int_0^\pi \left(e^{2\theta} - e^\theta\right)d\theta$$

$$= \frac{1}{2}\left[\frac{1}{2}e^{2\theta} - e^\theta\right]_0^\pi = \frac{1}{4}\left(e^{2\pi} + 1 - 2e^\pi\right)$$

16.

$$A = \int_0^\pi \frac{1}{2}\left[\left(e^{2\pi+\theta}\right)^2 - \left(e^\theta\right)^2\right]d\theta$$

$$= \frac{1}{2}\left[\frac{e^{4\pi+2\theta}}{2} - \frac{e^{2\theta}}{2}\right]_0^\pi$$

$$= \frac{1}{4}(e^{6\pi} - e^{4\pi} - e^{2\pi} + 1)$$

17.

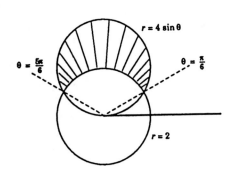

$$A = \int_{\pi/6}^{5\pi/6} \frac{1}{2}\left(\left[4\sin\theta\right]^2 - [2]^2\right)d\theta$$

18.

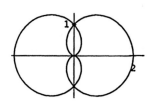

$$A = \int_{-\pi/2}^{\pi/2} \frac{1}{2}\left[(1 + \cos\theta)^2 - (1 - \cos\theta)^2\right]d\theta$$

19.

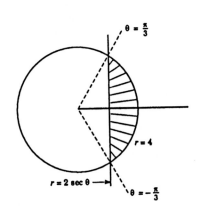

$$A = \int_{-\pi/3}^{\pi/3} \frac{1}{2}\left([4]^2 - [2\sec\theta]^2\right)d\theta$$

20.

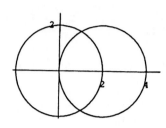

$$A = 2\int_{\pi/3}^{\pi/2} \frac{1}{2}\left[2^2 - (4\cos\theta)^2\right]d\theta + 2\int_{\pi/2}^{\pi} \frac{1}{2}\cdot 2^2\, d\theta$$

21.

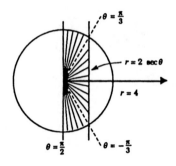

$$A = 2\left\{\int_0^{\pi/3} \frac{1}{2}(2\sec\theta)^2\, d\theta + \int_{\pi/3}^{\pi/2} \frac{1}{2}(4)^2\, d\theta\right\}$$

22.

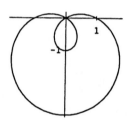

$$A = \int_{\pi/6}^{5\pi/6} \frac{1}{2}(1 - 2\sin\theta)^2\, d\theta$$

23.

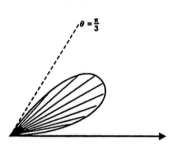

$$A = \int_0^{\pi/3} \frac{1}{2}(2\sin 3\theta)^2\, d\theta$$

24.

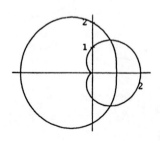

$$A = \int_{\pi/3}^{5\pi/3} \frac{1}{2}\left[(2 - \cos\theta)^2 - (1 + \cos\theta)^2\right]d\theta$$

25.

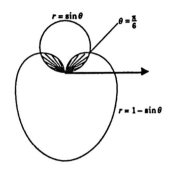

$$A = 2\left\{\int_0^{\pi/6} \frac{1}{2}\left(\sin\theta\right)^2 d\theta + \int_{\pi/6}^{\pi/2} \frac{1}{2}\left(1 - \sin\theta\right)^2 d\theta\right\}$$

26.

$$A = 2\int_0^{\pi/12} \frac{1}{2}(5\cos 6\theta)^2 d\theta$$

27.

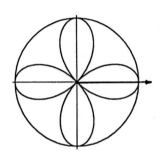

$$A = \pi - 8\int_0^{\pi/4} \frac{1}{2}\left(\cos 2\theta\right)^2 d\theta$$

28.

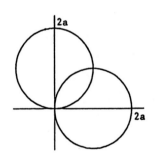

$$A = \int_0^{\pi/4} \frac{1}{2}(2a\sin\theta)^2 d\theta + \int_{\pi/4}^{\pi/2} \frac{1}{2}(2a\cos\theta)^2 d\theta$$

29.

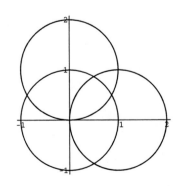

$$A = 2\int_0^{\pi/6} \frac{1}{2}\left[2\sin\theta\right]^2 d\theta + \frac{1}{12}\pi$$

$$= 4\int_0^{\pi/6} \sin^2\theta\, d\theta + \frac{1}{12}\pi$$

$$= 2\left[\theta - \frac{1}{2}\sin 2\theta\right]_0^{\pi/6} + \frac{1}{12}\pi = \frac{5}{12}\pi - \frac{1}{2}\sqrt{3}$$

30. The curves intersect at the point $[a, \pi/6]$ in the first quadrant. By symmetry:

$$A = 4 \int_0^{\pi/6} \tfrac{1}{2} \left[2a^2 \cos 2\theta - a^2 \right] \, d\theta = 2 \left[a^2 \sin 2\theta - a^2\theta \right]_0^{\pi/6} = a^2\sqrt{3} - \tfrac{1}{3}\pi a^2$$

31. The area of one petal of the curve $r = a\cos 2n\theta$ is given by:

$$2 \int_0^{\pi/4n} \tfrac{1}{2}(a\cos 2n\theta)^2 \, d\theta = a^2 \int_0^{\pi/4n} \cos^2 2n\theta \, d\theta$$

$$= a^2 \int_0^{\pi/4n} \left(\frac{1}{2} + \frac{\cos 4n\theta}{2} \right) \, d\theta$$

$$= a^2 \left[\frac{1}{2}\theta + \frac{\sin 4n\theta}{8n} \right]_0^{\pi/4n} = \frac{\pi a^2}{8n}$$

The total area enclosed by $r = a\cos 2n\theta$ is $\dfrac{\pi a^2}{2}$.

The area of one petal of the curve $r = a\sin 2n\theta$ is given by:

$$A = 2 \int_0^{\pi/4n} \tfrac{1}{2}(a\sin 2n\theta)^2 \, d\theta = a^2 \int_0^{\pi/4n} \left(\frac{1}{2} + \frac{\cos 4n\theta}{2} \right) \, d\theta = \frac{\pi a^2}{8n}$$

and the total area enclosed by the curve is $\dfrac{\pi a^2}{2}$.

32. Since there are $2n + 1$ petals, total area $= (2n + 1) \cdot$ (area of one petal). So

$$A = (2n + 1) \cdot 2 \int_0^{\pi/[2(2n+1)]} \frac{a^2}{2} \cos^2[(2n + 1)\theta] \, d\theta = (2n + 1)a^2 \left[\frac{\theta}{2} + \frac{\sin[2(2n + 1)\theta]}{4(2n + 1)} \right]_0^{\pi/[2(2n+1)]}$$

$$= (2n + 1) \cdot \frac{\pi a^2}{4(2n + 1)} = \frac{\pi}{4} a^2, \quad \text{independent of } n.$$

33. Let $P = \{\alpha = \theta_0, \theta_1, \theta_2, \ldots, \theta_n = \beta\}$ be a partition of the interval $[\alpha, \beta]$. Let θ_i^* be the midpoint of $[\theta_{i-1}, \theta_i]$ and let $r_i^* = f(\theta_i^*)$. The area of the ith "triangular" region is $\tfrac{1}{2}(r_i^*)\Delta\theta_i$, where $\Delta\theta_i = \theta_i - \theta_{i-1}$, and the rectangular coordinates of its centroid are(approximately) $\left(\tfrac{2}{3}r_i^* \cos\theta_i^*, \tfrac{2}{3}r_i^* \sin\theta_i^* \right)$.

The centroid $(\overline{x}_p, \overline{y}_p)$ of the union of the triangular regions satisfies the following equations

$$\overline{x}_p A_p = \frac{1}{3}(r_1^*)^3 \cos\theta_1 \Delta\theta_1 + \frac{1}{3}(r_2^*)^3 \cos\theta_2 \Delta\theta_2 + \cdots + \frac{1}{3}(r_n^*)^3 \cos\theta_n \Delta\theta_n$$

$$\overline{y}_p A_p = \frac{1}{3}(r_1^*)^3 \sin\theta_1 \Delta\theta_1 + \frac{1}{3}(r_2^*)^3 \sin\theta_2 \Delta\theta_2 + \cdots + \frac{1}{3}(r_n^*)^3 \sin\theta_n \Delta\theta_n$$

As $\|P\| \to 0$, the union of the triangular regions tends to the region Ω and the equations above tend to

$$\overline{x} A = \int_\alpha^\beta \frac{1}{3}r^3 \cos\theta \, d\theta$$

$$\overline{y} A = \int_\alpha^\beta \frac{1}{3}r^3 \sin\theta \, d\theta$$

The result follows from the fact that $A = \displaystyle\int_\alpha^\beta \frac{1}{2}r^2 \cos\theta \, d\theta$.

34. $A = \frac{1}{4}\pi r^2;$ $\rho(\theta) = r;$ by symmetry, $\bar{x} = \bar{y}.$

$$\bar{x}\,A = \frac{1}{3}\int_0^{\pi/2} r^3 \cos\theta\,d\theta = \frac{r^3}{3}\Big[\sin\theta\Big]_0^{\pi/2} = \frac{r^3}{3}; \quad \bar{x} = \bar{y} = \frac{4r}{3\pi}$$

35. Since the region enclosed by the cardioid $r = 1 + \cos\theta$ is symmetric with respect to the x-axis, $\bar{y} = 0.$
To find $\bar{x}$:

$$A = \int_0^{2\pi} r^2\,d\theta = \int_0^{2\pi} (1 + \cos\theta)^2\,d\theta$$

$$= \int_0^{2\pi} (1 + 2\cos\theta + \cos^2\theta)\,d\theta$$

$$= \int_0^{2\pi} \left(\frac{3}{2} + 2\cos\theta\frac{1}{2}\cos 2\theta\right)\,d\theta$$

$$= \left[\frac{3}{2} + 2\sin\theta + \frac{1}{4}\sin 2\theta\right]_0^{2\pi} = 3\pi$$

and

$$\frac{2}{3}\int_0^{2\pi} r^3 \cos\theta\,d\theta = \frac{2}{3}\int_0^{2\pi} (1 + \cos\theta)^3 \cos\theta\,d\theta$$

$$= \frac{2}{3}\int_0^{2\pi} \left(\cos\theta + 3\cos^2\theta + 3\cos^3\theta + \cos^4\theta\right)\,d\theta$$

$$= \frac{2}{3}\int_0^{2\pi} \left(\frac{15}{8} + 4\cos\theta + 2\cos 2\theta + \frac{1}{8}\cos 4\theta - 3\sin^2\theta\,\cos\theta\right)\,d\theta$$

$$= \frac{2}{3}\left[\frac{15}{8}\theta + 4\sin\theta + \sin 2\theta + \frac{1}{32}\sin 4\theta - \sin^3\theta\right]_0^{2\pi} = \frac{5}{2}\pi$$

Thus $\bar{x} = \dfrac{5\pi/2}{3\pi} = \dfrac{5}{6}.$

36. $\displaystyle\int_0^{2\pi} r^2\,d\theta = \int_0^{2\pi} (4 + 4\sin\theta + \sin^2\theta)\,d\theta = \left[4\theta - 4\cos\theta + \frac{\theta}{2} - \frac{\sin 2\theta}{4}\right]_0^{2\pi} = 9\pi$

$$\bar{x} = \frac{1}{9\pi}\cdot\frac{2}{3}\int_0^{2\pi} r^3 \cos\theta\,d\theta = \frac{2}{27\pi}\int_0^{2\pi} (2 + \sin\theta)^3 \cos\theta\,d\theta = 0$$

$$\bar{y} = \frac{2}{27\pi}\int_0^{2\pi} (2 + \sin\theta)^3 \sin\theta\,d\theta = \frac{2}{27\pi}\cdot\frac{51}{8}\cdot 2\pi = \frac{17}{18}$$

37. $A = \displaystyle\int_0^{2\pi} \frac{1}{2}[2 + \cos\theta]^2\,d\theta = \frac{1}{2}\int_0^{2\pi} (4 + 4\cos\theta + \cos^2\theta)\,d\theta = \frac{1}{2}\left[4\theta + 4\sin\theta + \frac{1}{2}\theta + \frac{1}{4}\sin 2\theta\right]_0^{2\pi} = \frac{9}{2}\pi$

38. $r = 2\cos 3\theta$ is a petal curve with 3 petals. The area of one petal is given by:

$$A = 2\int_0^{\pi/6} \frac{1}{2}(2\cos 3\theta)^2\,d\theta = 4\int_0^{\pi/6} \cos^2 3\theta\,d\theta = 2\int_0^{\pi/6}(1 + \cos 6\theta)\,d\theta = 2\left[\theta + \frac{1}{6}\sin 6\theta\right]_0^{\pi/6} = \frac{1}{3}\pi$$

The total area enclosed by the is $\pi.$

39.

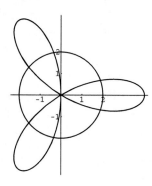

$$A = 6 \int_0^{\pi/9} \tfrac{1}{2} \left[(4 \cos 3\theta)^2 - 4 \right] \, d\theta$$

$$= 3 \int_0^{\pi/9} \left[16 \cos^2 3\theta - 4 \right] \, d\theta$$

$$= 3 \int_0^{\pi/9} (4 + 8 \cos 6\theta) \, d\theta$$

$$= 3 \left[4\theta + \tfrac{4}{3} \sin 6\theta \right]_0^{\pi/9} = \tfrac{4}{3} \pi + 2\sqrt{3}$$

40. The curves intersect at $[0.6667, 1.2310]$ in the first quadrant
By symmetry, the area is given by:

$$A = 2 \int_0^{1.2310} \tfrac{1}{2} \left[4 \cos^2 \theta - (1 - \cos \theta)^2 \right] \, d\theta$$

$$= \int_0^{1.2310} \left[3 \cos^2 \theta + 2 \cos \theta - 1 \right] \, d\theta$$

$$\cong 2.9725$$

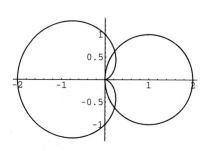

41. (a) $y^2 = x^2 \left(\dfrac{a - x}{a + x} \right)$

$r^2 \sin^2 \theta = r^2 \cos^2 \theta \left(\dfrac{a - r \cos \theta}{a + r \cos \theta} \right)$

$\sin^2 \theta (a + r \cos \theta) = \cos^2 \theta (a - r \cos \theta)$

$r \cos\theta = a \cos 2\theta$

$r = a \cos 2\theta \sec \theta$

(b) Let $a = 2$

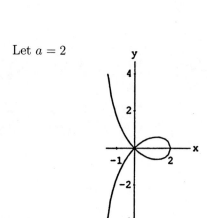

(c) $A = \displaystyle\int_{3\pi/4}^{5\pi/4} \dfrac{1}{2} a^2 \cos^2 2\theta \sec^2 \theta \, d\theta$

$$= 2 \int_{3\pi/4}^{5\pi/4} \cos^2 2\theta \sec^2 \theta \, d\theta \qquad (a = 2)$$

$$= 2 \int_{3\pi/4}^{5\pi/4} \dfrac{\left(2 \cos^2 \theta - 1 \right)^2}{\cos^2 \theta} \, d\theta$$

$$= 2 \int_{3\pi/4}^{5\pi/4} \left(4 \cos^2 \theta - 4 + \sec^2 \theta \right) \, d\theta$$

$$= 2 \int_{3\pi/4}^{5\pi/4} \left(-2 + 2 \cos 2\theta + \sec^2 \theta \right) \, d\theta$$

$$= 2 \left[-2\theta + \sin 2\theta + \tan \theta \right]_{3\pi/4}^{5\pi/4} = 8 - 2\pi$$

42. (a) $(x^2 + y^2)^2 = ax^2 y \implies r^4 = ar^2 \cos^2 \theta r \sin \theta \implies r = a \sin \theta \cos^2 \theta$

(b) same for all values of a,
    with different scale

(c) $A = \displaystyle\int_0^{\pi/2} \frac{1}{2} (2 \sin \theta)^2 \cos^4 \theta \, d\theta$

$= 2 \displaystyle\int_0^{\pi/2} (\cos^4 \theta - \cos^6 \theta) \, d\theta$

$= 2 \cdot \dfrac{\pi}{32} = \dfrac{\pi}{16}$

SECTION 10.5

1. $4x = (y - 1)^2$

2. $2x + 3y = 13$

3. $y = 4x^2 + 1, \quad x \geq 0$

4. $y = (x + 1)^3 - 5$

5. $9x^2 + 4y^2 = 36$

6. $x = (y - 2)^2 + 1$

7. $1 + x^2 = y^2$

8. $(x - 2)^2 + y^2 = 1$

9. $y = 2 - x^2, \quad -1 \leq x \leq 1$

10. $y = 4 - x^2, \quad x > 0$

11. $2y - 6 = x, \quad -4 \leq x \leq 4$

12. $1 + y^2 = x^2$

13. $y = x - 1$

14. $x = 6 - 3y$

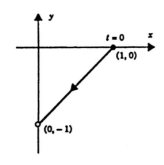

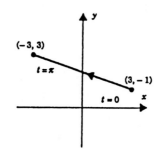

15. $xy = 1$

16. $y = x^2$

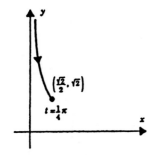

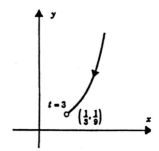

17. $2x + y = 11$

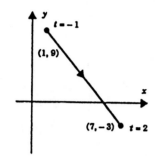

18. $1 + y^2 = x^2$

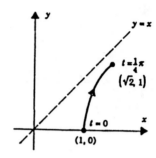

19. $x = \sin \frac{1}{2}\pi y$

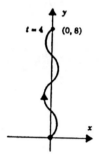

20. $x^2 + 4y^2 = 4$

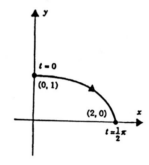

21. $1 + x^2 = y^2$

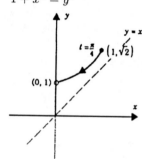

22. $x(t) = f(t)\cos t, \quad y(t) = f(t)\sin t, \quad t \in [\alpha, \beta]$

23.   (a) $x(t) = -\sin 2\pi t, \quad y(t) = \cos 2\pi t$ (b) $x(t) = \sin 4\pi t, \quad y(t) = \cos 4\pi t$

     (c) $x(t) = \cos \frac{1}{2}\pi t, \quad y(t) = \sin \frac{1}{2}\pi t$ (d) $x(t) = \cos \frac{3}{2}\pi t, \quad y(t) = -\sin \frac{3}{2}\pi t$

24.   (a) $x(t) = 3\cos 2\pi t, \quad y(t) = -4\sin 2\pi t$ (b) $x(t) = 3\sin 2\pi t, \quad y(t) = 4\cos 2\pi t$

     (c) $x(t) = -3\cos 4\pi t, \quad y(t) = -4\sin \pi t$ (d) $x(t) = 3\cos \frac{1}{2}\pi t, \quad y(t) = 4\sin \frac{1}{2}\pi t$

25. $x(t) = \tan \frac{1}{2}\pi t, \quad y(t) = 2$

26.   Any continuous function unbounded on $(0, 1)$ will do.   For example, $f(t) = \dfrac{1}{t}$

27. $x(t) = 3 + 5t, \quad y(t) = 7 - 2t$ **28.** $x(t) = 2 + 4t, \quad y(t) = 6 - 3t$

29. $x(t) = \sin^2 \pi t, \quad y(t) = -\cos \pi t$ **30.** $x(t) = 4(1-t)^2, \quad y(t) = 2(1-t)$

31. $x(t) = (2-t)^2, \quad y(t) = (2-t)^3$ **32.** $x(t) = (t+1)^3, \quad y(t) = (t+1)^2$

33. $\displaystyle \int_c^d y(t)x'(t)\,dt = \int_c^d f(x(t))x'(t)\,dt = \int_a^b f(x)\,dx = \text{area below } C$

34. $\displaystyle \int_c^d x(t)y(t)x'(t)\,dt = \int_c^d x(t)f(x(t))x'(t)\,dt = \int_a^b xf(x)\,dx = \overline{x}A$

     $\displaystyle \int_c^d \frac{1}{2}[y(t)]^2\, x'(t)\,dt = \int_c^d \frac{1}{2}[f(x(t))]^2\, x'(t)\,dt = \int_c^d \frac{1}{2}[f(x)]^2\,dx = \overline{y}A$

35. $\displaystyle \int_c^d \pi[y(t)]^2\, x'(t)\,dt = \int_c^d \pi[f(x(t))]^2\, x'(t)\,dt = \int_a^b \pi[f(x)]^2\,dx = V_x$

     $\displaystyle \int_c^d 2\pi x(t)y(t)x'(t)\,dt = \int_c^d 2\pi x(t)f(x(t))x'(t)\,dt = \int_a^b 2\pi x f(x)\,dx = V_y$

36. $\displaystyle \int_c^d \pi x(t)\,[y(t)]^2\, x'(t)\,dt = \int_c^d \pi x(t)\,[f(x(t))]^2\, x'(t)\,dt = \int_a^b \pi x\,[f(x)]^2\,dx = \overline{x}\,V_x = \overline{y}\,V_y$

37.

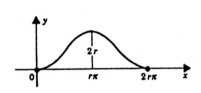

$$A = \int_0^{2\pi} x(t)\, y'(t)\, dt$$
$$= r^2 \int_0^{2\pi} (1 - \cos t)\, dt$$
$$= r^2\,[t - \sin t]_0^{2\pi} = 2\pi r^2$$

38. $\displaystyle \overline{x}A = \int_0^{2\pi} r\,t\,r(1-\cos t)\,r\,dt = r^3 \int_0^{2\pi} (t - t\cos t)\,dt = r^3\left[\frac{t^2}{2} - t\sin t - \cos t\right]_0^{2\pi} = 2r^3\pi^2$

    $\displaystyle \implies \overline{x} = \frac{2r^3\pi^2}{2\pi r^2} = \pi r$

    $\displaystyle \overline{y}A = \int_0^{2\pi} \frac{1}{2}\left[r^2(1-\cos t)^2\right]r\,dt = \frac{r^3}{2}\int_0^{2\pi}\left(1 - 2\cos t + \cos^2 t\right)\,dt = \frac{r^3}{2}\left[\frac{3t}{2} - 2\sin t + \frac{\sin 2t}{2}\right]_0^{2\pi}$

    $\displaystyle = \frac{3\pi r^3}{2} \implies \overline{y} = \frac{3\pi r^3/2}{2\pi r^2} = \frac{3}{4}r$

39.   (a) $V_x = 2\pi\overline{y}A = 2\pi\left(\frac{3}{4}r\right)\left(2\pi r^2\right) = 3\pi^2 r^3$

     (b) $V_y = 2\pi\overline{x}A = 2\pi\left(\pi r\right)2\pi r^2 = 4\pi^3 r^3$

40.   (a) $\displaystyle \overline{x}V_x = \int_0^{2\pi} \pi r\,t\,r^2(1-\cos t)^2 r\,dt = \pi r^4 \int_0^{2\pi}\left(t - 2t\cos t + t\cos^2 t\right)\,dt = 3\pi^3 r^4$

       $\implies \overline{x} = \pi r, \quad \overline{y} = 0$

     (b) $\displaystyle \overline{y}V_y = \int_0^{2\pi} \pi r\,t\,r^2(1-\cos t)^2 r\,dt = 3\pi^3 r^4 \implies \overline{y} = \frac{3}{4}r, \quad \overline{x} = 0$

41. $x(t) = -a \cos t, \quad y(t) = b \sin t \qquad t \in [0, \pi]$

42. (a) $A = 2 \int_0^\pi y(t) x'(t) \, dt = 2 \int_0^\pi (b \sin t)(a \sin t) \, dt = 2ab \left[\dfrac{t}{2} - \dfrac{\sin 2t}{4} \right]_0^\pi = \pi ab$

(b) $\bar{x} A = \int_0^\pi (-\cos t)(b \sin t)(a \sin t) \, dt = -a^2 b \int_0^\pi \sin^2 t \cos t \, dt = 0 \implies \bar{x} = 0$

$\bar{y} A = \int_0^\pi \dfrac{1}{2} b^2 \sin^2 t (a \sin t) \, dt = \dfrac{ab^2}{2} \int_0^\pi (1 - \cos^2 t) \sin t \, dt = \dfrac{2}{3} ab^2$

$\implies \bar{y} = \dfrac{2}{3} \dfrac{ab^2}{\pi ab/2} = \dfrac{4b}{3\pi}$

43. (a) Equation for the ray: $\quad y + 2x = 17, \quad x \geq 6$.

Equation for the circle: $\quad (x - 3)^2 + (y - 1)^2 = 25$.

Simultaneous solution of these equations gives the points of intersection: $(6, 5)$ and $(8, 1)$.

(b) The particle on the ray is at $(6, 5)$ when $t = 0$. However, when $t = 0$ the particle on the circle is at the point $(-2, 1)$. Thus, the intersection point $(6, 5)$ is not a collision point.

The particle on the ray is at $(8, 1)$ when $t = 1$. Since the particle on the circle is also at $(8, 1)$ when $t = 1$, the intersection point $(8, 1)$ is a collision point.

44. (a) Equation of ellipse: $\quad \dfrac{(x - 2)^2}{9} + \dfrac{(y - 3)^2}{49} = 1$

Equation of parabola: $\quad y = -\dfrac{7}{15}(x - 1)^2 + \dfrac{157}{15}$

Solving simultaneously gives the points of intersection $(2, 10) \quad$ and $\quad (5, 3)$

(b) Particle 2 is at $(2, 10)$ at $t = 0$, but particle 1 is at $(-1, 3)$ at $t = 0$, so no collision at $(2, 10)$.

Both particles are at $(5, 3)$ when $t = 1$, so $(5, 3)$ is a collision point.

45. If $x(r) = x(s)$ and $r \neq s$, then

$$r^2 - 2r = s^2 - 2s$$
$$r^2 - s^2 = 2r - 2s$$

(1) $\qquad\qquad\qquad\qquad\qquad r + s = 2.$

If $y(r) = y(s)$ and $r \neq s$, then

$$r^3 - 3r^2 + 2r = s^3 - 3s^2 + 2s$$
$$\left(r^3 - s^3 \right) - 3 \left(r^2 - s^2 \right) + 2 \left(r - s \right) = 0$$

(2) $\qquad\qquad \left(r^2 + rs + s^2 \right) - 3 \left(r + s \right) + 2 = 0.$

Simultaneous solution of (1) and (2) gives $r = 0$ and $r = 2$. Since $(x(0), y(0)) = (0, 0) = (x(2), y(2))$, the curve intersects itself at the origin.

46. $x(r) = x(s)$ and $r \neq s \implies \cos r(1 - 2\sin r) = \cos s(1 - 2\sin s)$

$y(r) = y(s)$ and $r \neq s \implies \sin r(1 - 2\sin r) = \sin s(1 - 2\sin s)$

Both equations can be satisfied simultaneously with $r \neq s$ in $[0, \pi]$ only for $r = \dfrac{\pi}{6}$, $s = \dfrac{5\pi}{6}$, or $r = \dfrac{5\pi}{6}$, $r = \dfrac{\pi}{6}$

So the curve intersects itself at $(0, 0)$.

47. Suppose that $r, s \in [0, 4]$ and $r \neq s$.

$$x(r) = x(s) \implies \sin 2\pi r = \sin 2\pi s.$$

$$y(r) = y(s) \implies 2r - r^2 = 2s - s^2 \implies 2(r - s) = r^2 - s^2 \implies 2 = r + s.$$

Now we solve the equations simultaneously:

$$\sin 2\pi r = \sin[2\pi(2 - r)] = -\sin 2\pi r$$

$$2\sin 2\pi r = 0$$

$$\sin 2\pi r = 0.$$

Since $r \in [0, 4]$, $r = 0, \frac{1}{2}, 1, \frac{3}{2}, 2, \frac{5}{2}, 3, \frac{7}{2}, 4$.

Since $s \in [0, 4]$ and $r \neq s$ and $r + s = 2$, we are left with $r = 0, \frac{1}{2}, \frac{3}{2}, 2$. Note that
$(x(0), y(0)) = (0, 0) = (x(2), y(2))$ and $\left(x\left(\frac{1}{2}\right), y\left(\frac{1}{2}\right)\right) = \left(0, \frac{3}{4}\right) = \left(x\left(\frac{3}{2}\right), y\left(\frac{3}{2}\right)\right)$.

The curve intersects itself at $(0, 0)$ and $\left(0, \frac{3}{4}\right)$.

48.

$$x(r) = x(s) \text{ and } r \neq s \implies \begin{cases} r^3 - 4r = s^3 - 4s \\ r^3 - s^3 = 4(r - s) \\ r^2 + rs + s^2 = 4 \end{cases}$$

$$y(r) = y(s) \text{ and } r \neq s \implies \begin{cases} r^3 - 3r^2 + 2r = s^3 - 3s^2 + 2s \\ r^3 - s^3 - 3(r^2 - s^2) + 2(r - s) = 0 \\ (r^2 + rs + s^2) - 3(r + s) + 2 = 0 \end{cases}$$

Solving simultaneously gives $r = 0$, $s = 2$, so curve intersect itself at $(0, 0)$

49. $x = 2t$, $y = 4t - t^2$; $0 \leq t \leq 6$: From the first equation, $t = \frac{1}{2}x$. Substituting this into the second equation, we get: $y = 2x - \frac{1}{4}x^2$, a parabola. The limits on t imply that $0 \leq x \leq 12$. The particle moves along the parabola $y = 2x - \frac{1}{4}x^2$ from the point $(0, 0)$ to the point $(12, -12)$.

50. The particle starts at the upper right and ends at the lower right.

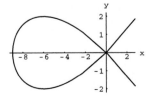

51. $x = \cos(t^2 + t)$, $y = \sin(t^2 + t)$ $\implies$ $x^2 = \cos^2(t^2 + t)$, $y^2 = \sin^2(t^2 + t)$.

Since $\cos^2(t^2 + t) + \sin^2(t^2 + t) = 1$, we have $x^2 + y^2 = 1$ the unit circle.

The particle starts at the point $(1, 0)$ and moves around the unit circle in the counterclockwise direction.

52. $x = \cos(\ln t)$, $y = \sin(\ln t)$ $\implies$ $x^2 + y^2 = 1$ the unit circle.

The particle starts at the point $(1, 0)$ and moves around the unit circle in the counterclockwise direction.

53. $x(\theta) = \cos\theta(a - b\sin\theta)$, $y(\theta) = \sin\theta(a - b\sin\theta)$

    (a) $a = 1$, $b = 2$ (b) $a = 2$, $b = 2$ (c) $a = 2$, $b = 1$

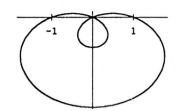 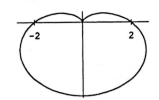 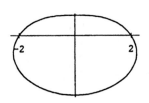

    (d) The curves are limaçons; the curve has an inner loop if $a < b$ and no loop if $a > b$.

PROJECT 10.5

1. $x''(t) = 0$ $\implies$ $x'(t) = C$; $x'(0) = v_0 \cos\theta$ $\implies$ $x'(t) = v_0 \cos\theta$.

Integrating again, $x(t) = (v_0 \cos\theta)t + x_0$ (since $x(0) = x_0$)

Similarly, since $y''(t) = -g$ $\implies$ $y'(t) = -gt + C$; $y'(0) = v_0 \sin\theta$ $\implies$ $y'(t) = -gt + v_0 \sin\theta$.

Integrating again, $y(t) = -\frac{1}{2}gt^2 + (v_0 \sin\theta)t + y_0$ (since $y(0) = y_0$)

2. From the first equation, we get $t = \dfrac{1}{v_0 \cos\theta}[x(t) - x_0]$.

Substituting this into the second equation gives the desired result.

3. (a) Set $x_0 = 0$, $y_0 = 0$, $g = 32$.

    parametric equations: $x = (v_0 \cos\theta)t$, $y = -16t^2 + (v_0 \sin\theta)t$

    rectangular equation: $y = -\dfrac{16}{v_0{}^2}(\sec^2\theta)x^2 + (\tan\theta)x$.

    (b) To find the range, set $y = 0$ and solve for x, $x \neq 0$:

$$-\frac{16}{v_0^2}\sec^2\theta\, x^2 + \tan\theta\, x = 0 \implies x = \frac{v_0^2}{16}\sin\theta\cos\theta$$

    (c) $y(t) = 0$ (and $t \neq 0$) when $t = \dfrac{v_0}{16}\sin\theta$

    (d) The range $\frac{1}{16}v_0{}^2 \sin\theta\cos\theta = \frac{1}{32}v_0{}^2 \sin 2\theta$ is maximal when $\theta = \frac{1}{4}\pi$ for then $\sin 2\theta = 1$.

    (e) Set $x = \dfrac{1}{32}v_0{}^2 \sin 2\theta = b$, $\theta = \dfrac{1}{2}\arcsin\left(\dfrac{32b}{v_0{}^2}\right)$.

4. (a) maximum height — approximately 9,000 ft.   range — approximately 61,000 ft.

SECTION 10.6

1. $x'(1) = 1$, $y'(1) = 3$,   slope 3,   point $(1,0)$;   tangent: $y = 3(x-1)$

2. $x'(2) = 4$, $y'(2) = 1$,   slope $= \frac{1}{4}$,   point $(4,7)$;   tangent: $y - 7 = \frac{1}{4}(x-4)$

3. $x'(0) = 2$, $y'(0) = 0$,   slope 0,   point $(0,1)$;   tangent: $y = 1$

4. $x'(1) = 2$, $y'(1) = 4$,   slope $= 2$,   point $(1,1)$;   tangent: $y - 1 = 2(x-1)$

5. $x'(1/2) = 1$, $y'(1/2) = -3$,   slope -3,   point $\left(\frac{1}{4}, \frac{9}{4}\right)$;   tangent: $y - \frac{9}{4} = -3\left(x - \frac{1}{4}\right)$

6. $x'(1) = -1$, $y'(1) = 2$,   slope $= -2$,   point $(1,2)$;   tangent: $y - 2 = -2(x-1)$

7. $x'\left(\frac{\pi}{4}\right) = -\frac{3}{4}\sqrt{2}$, $y'\left(\frac{\pi}{4}\right) = \frac{3}{4}\sqrt{2}$,   slope -1,   point $\left(\frac{1}{4}\sqrt{2}, \frac{1}{4}\sqrt{2}\right)$;
    tangent: $y - \frac{1}{4}\sqrt{2} = -\left(x - \frac{1}{4}\sqrt{2}\right)$

8. $x'(0) = 1$, $y'(0) = -3$,   slope $= -3$,   point $(1,3)$;   tangent: $y - 3 = -3(x-1)$

9. $x(\theta) = \cos\theta\,(4 - 2\sin\theta)$, $y(\theta) = \sin\theta\,(4 - 2\sin\theta)$,   point $(4,0)$
    $x'(\theta) = -4\sin\theta - 2\left(\cos^2\theta - \sin^2\theta\right)$, $y'(\theta) = 4\cos\theta - 4\sin\theta\cos\theta$
    $x'(0) = -2$, $y'(0) = 4$,   slope -2,   tangent: $y = -2\,(x - 4)$

10. $x(\theta) = 4\cos 2\theta\cos\theta$, $y(\theta) = 4\cos 2\theta\sin\theta$,   point $(0, -4)$
    $x'(\theta) = -8\sin 2\theta\cos\theta - 4\cos 2\theta\sin\theta$, $y'(\theta) = -8\sin 2\theta\sin\theta + 4\cos 2\theta\cos\theta$
    $x'(\frac{\pi}{2}) = 4$, $y'(\frac{\pi}{2}) = 0$, $m = 0$;   tangent: $y = -4$.

11. $x(\theta) = \dfrac{4\cos\theta}{5 - \cos\theta}$, $y(\theta) = \dfrac{4\sin\theta}{5 - \cos\theta}$,   point $\left(0, \dfrac{4}{5}\right)$
    $x'(\theta) = \dfrac{-20\sin\theta}{(5 - \cos\theta)^2}$, $y'(\theta) = \dfrac{4\,(5\cos\theta - 1)}{(5 - \cos\theta)^2}$
    $x'\left(\dfrac{\pi}{2}\right) = -\dfrac{4}{5}$, $y'\left(\dfrac{\pi}{2}\right) = -\dfrac{4}{25}$,   slope $\dfrac{1}{5}$,   tangent: $y - \dfrac{4}{5} = \dfrac{1}{5}x$

12. $x(\theta) = \dfrac{5\cos\theta}{4 - \cos\theta}$, $y(\theta) = \dfrac{5\sin\theta}{4 - \cos\theta}$,   point $\left(\dfrac{5\sqrt{3}}{8 - \sqrt{3}}, \dfrac{5}{8 - \sqrt{3}}\right)$
    $x'(\theta) = \dfrac{-20\sin\theta}{(4 - \cos\theta)^2}$, $y'(\theta) = \dfrac{5(4\cos\theta - 1)}{(4 - \cos\theta)^2}$,   slope $\dfrac{1 - 2\sqrt{3}}{2}$
    tangent: $y - \dfrac{5}{8 - \sqrt{3}} = \left(\dfrac{1 - 2\sqrt{3}}{2}\right)\left(x - \dfrac{5\sqrt{3}}{8 - \sqrt{3}}\right)$

13. $x(\theta) = \dfrac{\cos\theta\,(\sin\theta - \cos\theta)}{\sin\theta + \cos\theta}$, $y(\theta) = \dfrac{\sin\theta\,(\sin\theta - \cos\theta)}{\sin\theta + \cos\theta}$,   point $(-1, 0)$

$x'(\theta) = \dfrac{\sin\theta\,\cos 2\theta + 2\cos\theta}{(\sin\theta + \cos\theta)^2}$, $y'(\theta) = \dfrac{2\sin\theta - \cos\theta\,\cos 2\theta}{(\sin\theta + \cos\theta)^2}$

$x'(0) = 2$, $y'(0) = -1$,   slope $-\frac{1}{2}$,   tangent $y = -\frac{1}{2}(x+1)$

14. $x(\theta) = \dfrac{\sin\theta + \cos\theta}{\sin\theta - \cos\theta}\cos\theta$, $y(\theta) = \dfrac{\sin\theta + \cos\theta}{\sin\theta - \cos\theta}\sin\theta$,   point $(0, 1)$

$x'(\theta) = \dfrac{-2\cos\theta}{(\sin\theta - \cos\theta)^2} - \dfrac{\sin\theta + \cos\theta}{\sin\theta - \cos\theta}\sin\theta$, $y'(\theta) = \dfrac{-2\sin\theta}{(\sin\theta - \cos\theta)^2} + \dfrac{\sin\theta + \cos\theta}{\sin\theta - \cos\theta}\cos\theta$

$x'(\pi/2) = -1$, $y'(\pi/2) = -2$,   slope 2;   tangent: $y - 1 = 2x$

15. $x(t) = t$, $y(t) = t^3$

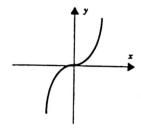

$x'(0) = 1$, $y'(0) = 0$,   slope 0
tangent $y = 0$

16. $x(t) = t^3$, $y(t) = t$

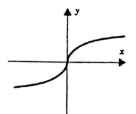

$x'(0) = 0$, $y'(0) = 1$
tangent $x = 0$

17. $x(t) = t^{5/3}$, $y(t) = t$

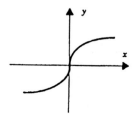

$x'(0) = 0$, $y'(0) = 1$,   slope undefined
tangent $x = 0$

18. $x(t) = t$, $y(t) = t^{5/3}$

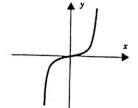

$x'(0) = 1$, $y'(0) = 0$
tangent $y = 0$

19. $x'(t) = 3 - 3t^2$, $y'(t) = 1$
$x'(t) = 0 \implies t = \pm 1$; $y'(t) \neq 0$
(a)   none
(b)   at $(2, 2)$ and $(-2, 0)$

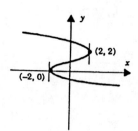

20. $x'(t) = 2t - 2, \quad y'(t) = 3t^2 + 12$

$x'(t) = 0 \quad \Longrightarrow \quad t = 1$

$y'(t) \neq 0 \quad$ for all t

(a)   no horizontal tangents

(b)   at $(-1, 13)$

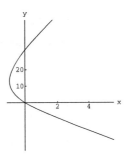

21.   curve traced once completely with $t \in [0, 2\pi)$

$x'(t) = -4\cos t, \quad y'(t) = -3\sin t$

$x'(t) = 0 \quad \Longrightarrow \quad t = \dfrac{\pi}{2}, \ \dfrac{3\pi}{2};$

$y'(t) = 0 \quad \Longrightarrow \quad t = 0, \pi$

(a)   at $(3, 7)$ and $(3, 1)$

(b)   at $(-1, 4)$ and $(7, 4)$

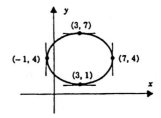

22. $x'(t) = 2\cos 2t, \quad y'(t) = \cos t$

$x'(t) = 0 \quad \Longrightarrow \quad t = \dfrac{\pi}{4}, \ \dfrac{3\pi}{4} \ \dfrac{5\pi}{4} \ \dfrac{7\pi}{4}$

$y'(t) = 0 \quad \Longrightarrow \quad t = \dfrac{\pi}{2}, \ \dfrac{3\pi}{2}$

(a)   at $(0, \pm 1)$

(b)   at $(\pm 1, \pm \frac{\sqrt{2}}{2})$

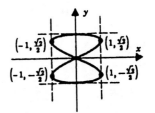

23. $x'(t) = 2t - 2, \quad y'(t) = 3t^2 - 6t + 2$

$x'(t) = 0 \quad \Longrightarrow \quad t = 1$

$y'(t) = 0 \quad \Longrightarrow \quad t = 1 \pm \frac{1}{3}\sqrt{3}$

(a)   at $\left(-\frac{2}{3}, \pm\frac{2}{9}\sqrt{3}\right)$

(b)   at $(-1, 0)$

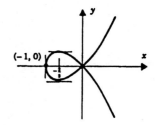

24. $x'(t) = 5\sin t, \quad y'(t) = \cos t$

$x'(t) = 0 \quad \Longrightarrow \quad t = 0, \ \pi$

$y'(t) = 0 \quad \Longrightarrow \quad t = \dfrac{\pi}{2}, \ \dfrac{3\pi}{2}$

(a)   at $(2, 2)$ and $(2, 4)$

(b)   at $(-3, 3)$ and $(7, 3)$

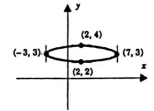

25.   curve traced completely with $t \in [0, 2\pi)$

$x'(t) = -\sin t, \quad y'(t) = 2\cos 2t$

$x'(t) = 0 \quad \Longrightarrow \quad t = 0, \pi$

$y'(t) = 0 \quad \Longrightarrow \quad t = \dfrac{\pi}{4}, \ \dfrac{3\pi}{4}, \ \dfrac{5\pi}{4}, \ \dfrac{7\pi}{4}$

(a)   at $\left(\pm\frac{1}{2}\sqrt{2}, \pm 1\right)$

(b)   at $(\pm 1, 0)$

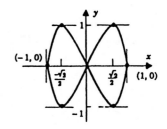

26. $x'(t) = 2\cos t, \quad y'(t) = 5\cos t$

$x'(t)$ and $y'(t)$ are near zero separately.

(a) none, (b) none.

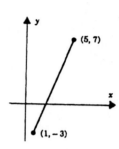

27. First, we find the values of t when the curve passes through $(2, 0)$.

$$y(t) = 0 \implies t^4 - 4t^2 = 0 \implies t = 0, \pm 2.$$
$$x(-2) = 2, \quad x(0) = 2, \quad x(2) = -2.$$

The curve passes through $(2, 0)$ at $t = -2$ and $t = 0$.

$$x'(t) = -1 - \frac{\pi}{2}\sin\frac{\pi t}{4}, \quad y'(t) = 4t^3 - 8t.$$

At $t = -2$, $x'(-2) = \frac{\pi}{2} - 1$, $y'(t) = -16$, tangent: $y = \dfrac{32}{2 - \pi}(x - 2)$.

At $t = 0$, $x'(0) = -1$, $y'(0) = 0$, tangent: $y = 0$.

28. Passes through $(0, 1)$ at $t = 1$ and $t = -1$

$$x'(t) = 3t^2 - 1, \quad y'(t) = \sin\frac{\pi}{2}t + \frac{\pi}{2}t\cos\frac{\pi}{2}t;$$

at $t = 1$: $x'(1) = 2$, $y'(1) = 1$, tangent: $y - 1 = \dfrac{x}{2}$

at $t = -1$: $x'(-1) = 2$, $y'(-1) = -1$, tangent: $y - 1 = -\dfrac{x}{2}$

29. The slope of $\overline{OP}$ is $\tan\theta_1$. The curve $r = f(\theta)$ can be parameterized by setting

$$x(\theta) = f(\theta)\cos\theta, \qquad y(\theta) = f(\theta)\sin\theta.$$

Differentiation gives

$$x'(\theta) = -f(\theta)\sin\theta + f'(\theta)\cos\theta, \qquad y'(\theta) = f(\theta)\cos\theta + f'(\theta)\sin\theta.$$

If $f'(\theta_1) = 0$, then

$$x'(\theta_1) = -f(\theta_1)\sin\theta_1, \qquad y'(\theta_1) = f(\theta_1)\cos\theta_1.$$

Since $f(\theta_1) \neq 0$, we have

$$m = \frac{y'(\theta_1)}{x'(\theta_1)} = -\cot\theta_1 = -\frac{1}{\text{slope of } \overline{OP}}.$$

30. $x(\theta) = (a - \cos\theta)\cos\theta, \quad y(\theta) = (a - \cos\theta)\sin\theta$ goes through $(0, 0)$ when $\theta = \pm\cos^{-1} a$

$x'(\theta) = -a\sin\theta + 2\cos\theta\sin\theta, \quad y'(\theta) = a\cos\theta + \sin^2\theta - \cos^2\theta$

At $\theta = \cos^{-1} a$, $\sin\theta = \sqrt{1 - a^2} \implies x'(\cos^{-1} a) = a\sqrt{1 - a^2}$, $y'(\cos^{-1} a) = 1 - a^2$

$$\implies m_1 = \frac{\sqrt{1 - a^2}}{a}$$

At $\theta = -\cos^{-1} a$, $\sin\theta = -\sqrt{1 - a^2} \implies x'(-\cos^{-1} a) = -a\sqrt{1 - a^2}$, $y'(-\cos^{-1} a) = 1 - a^2$

$$\implies m_2 = -\frac{\sqrt{1 - a^2}}{a}$$

We want $m_1 = -\dfrac{1}{m_2}$: $\dfrac{\sqrt{1 - a^2}}{a} = \dfrac{a}{\sqrt{1 - a^2}} \implies a = \dfrac{\sqrt{2}}{2}$

31. $x'(t) = 3t^2, \quad y'(t) = 2t$

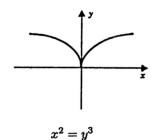

$$x^2 = y^3$$

32. $x'(t) = 3t^2, \quad y'(t) = 5t^4$

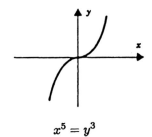

$$x^5 = y^3$$

33. $x'(t) = 5t^4, \quad y'(t) = 3t^2$

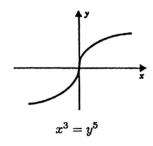

$$x^3 = y^5$$

34. $x'(t) = 3t^2, \quad y'(t) = 6t^2$

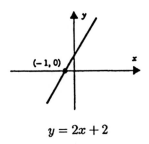

$$y = 2x + 2$$

35. $x'(t) = 2t, \quad y'(t) = 2t$

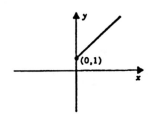

ray: $y = x + 1, \quad x \geq 0$

36. $\dfrac{d^2y}{dx^2} = \dfrac{d}{dt}\left(\dfrac{dy}{dx}\right) \cdot \dfrac{dt}{dx} = \dfrac{d}{dt}\left[\dfrac{y'(t)}{x'(t)}\right] \cdot \dfrac{1}{x'(t)} = \dfrac{x'(t)y''(t) - y'(t)x''(t)}{[x'(t)]^3}$

37. By (9.7.5), $\dfrac{d^2y}{dx^2} = \dfrac{(-\sin t)(-\sin t) - (\cos t)(-\cos t)}{(-\sin t)^3} = \dfrac{-1}{\sin^3 t}.$ At $t = \dfrac{\pi}{6}$, $\dfrac{d^2y}{dx^2} = -8.$

38. $\dfrac{d^2y}{dx^2} = \dfrac{3t^2 \cdot 0 - 1 \cdot 6t}{(3t^2)^3} = -\dfrac{2}{9t^5}.$ At $t = 1$, $\dfrac{d^2y}{dx^2} = -\dfrac{2}{9}$

39. By (9.7.5), $\dfrac{d^2y}{dx^2} = \dfrac{(e^t)(e^{-t}) - (-e^{-t})(e^t)}{(e^t)^3} = 2e^{-3t}.$ At $t = 0$, $\dfrac{d^2y}{dx^2} = 2.$

40. $\dfrac{d^2y}{dx^2} = \dfrac{-\sin 2t \, \cos t + 2\sin t \, \cos 2t}{\sin^3 2t}$ At $t = \dfrac{\pi}{4}$, $\dfrac{d^2y}{dx^2} = -\dfrac{\sqrt{2}}{2}$

41. $\dfrac{d^2y}{dx^2} = \cot^3 t$

42. the "curve" is the straight line $x + y = 1$

43.   tangent line: $y - 2 = -\dfrac{16}{3}\left(x - \dfrac{1}{8}\right)$

44. tangent line: $y - \dfrac{4\sqrt{3}}{4+\sqrt{3}} = \dfrac{-1}{1+\sqrt{3}}\left(x - \dfrac{4}{4+\sqrt{3}}\right)$

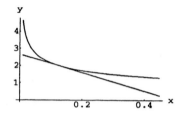

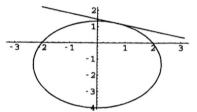

SECTION 10.7

1. $L = \displaystyle\int_0^1 \sqrt{1+2^2}\,dx = \sqrt{5}$

2. $L = \displaystyle\int_0^1 \sqrt{1+3^2}\,dx = \sqrt{10}$

3. $L = \displaystyle\int_1^4 \sqrt{1+\left[\dfrac{3}{2}\left(x-\dfrac{4}{9}\right)^{1/2}\right]^2}\,dx = \int_1^4 \dfrac{3}{2}\sqrt{x}\,dx = \left[x^{3/2}\right]_1^4 = 7$

4. $L = \displaystyle\int_0^{44} \sqrt{1+\left(\dfrac{3}{2}x^{1/2}\right)^2}\,dx = \int_0^{44}\sqrt{1+\dfrac{9}{4}x}\,dx = \dfrac{8}{27}\left[(1+\dfrac{9}{4}x)^{3/2}\right]_0^{44} = 296$

5. $L = \displaystyle\int_0^3 \sqrt{1+\left(\dfrac{1}{2}\sqrt{x}-\dfrac{1}{2\sqrt{x}}\right)^2}\,dx = \int_0^3\left(\dfrac{1}{2}\sqrt{x}+\dfrac{1}{2\sqrt{x}}\right)dx = \left[\dfrac{1}{3}x^{3/2}+x^{1/2}\right]_0^3 = 2\sqrt{3}$

6. $L = \displaystyle\int_1^2 \sqrt{1+(x-1)}\,dx = \dfrac{2}{3}\left[x^{3/2}\right]_1^2 = \dfrac{2}{3}(2\sqrt{2}-1)\cong 1.22$

7. $L = \displaystyle\int_0^1 \sqrt{1+\left[x\,(x^2+2)^{1/2}\right]^2}\,dx = \int_0^1 (x^2+1)\,dx = \left[\dfrac{1}{3}x^3+x\right]_0^1 = \dfrac{4}{3}$

8. $L = \displaystyle\int_2^4 \sqrt{1+x^2(x^2-2)}\,dx = \int_2^4 (x^2-1)\,dx = \left[\dfrac{x^3}{3}-x\right]_2^4 = \dfrac{50}{3}$

9. $L = \displaystyle\int_1^5 \sqrt{1+\left[\dfrac{1}{2}\left(x-\dfrac{1}{x}\right)\right]^2}\,dx = \int_1^5 \dfrac{1}{2}\left(x+\dfrac{1}{x}\right)dx = \left[\dfrac{1}{2}\left(\dfrac{1}{2}x^2+\ln x\right)\right]_1^5 = 6+\dfrac{1}{2}\ln 5 \cong 6.80$

10. $L = \displaystyle\int_1^4 \sqrt{1+\left(\dfrac{x}{4}-\dfrac{1}{x}\right)^2}\,dx = \int_1^4\left(\dfrac{x}{4}+\dfrac{1}{x}\right)dx = \left[\dfrac{x^2}{8}+\ln x\right]_1^4 = \dfrac{15}{8}+\ln 4 \cong 3.26$

11. $L = \int_1^8 \sqrt{1 + \left[\frac{1}{2}\left(x^{1/3} - x^{-1/3}\right)\right]^2}\, dx = \int_1^8 \frac{1}{2}\left(x^{1/3} + x^{-1/3}\right) dx = \frac{1}{2}\left[\frac{3}{4}x^{4/3} + \frac{3}{2}x^{2/3}\right]_1^8 = \frac{63}{8}$

12. $L = \int_1^2 \sqrt{1 + \left(\frac{x^4}{2} - \frac{1}{2}x^{-4}\right)^2}\, dx = \int_1^2 \left(\frac{x^4}{2} + \frac{x^{-4}}{2}\right) dx = \frac{1}{2}\left[\frac{x^5}{5} - \frac{1}{3x^3}\right]_1^2 = \frac{779}{240}$

13. $L = \int_0^{\pi/4} \sqrt{1 + \tan^2 x}\, dx = \int_0^{\pi/4} \sec x\, dx = [\ln|\sec x + \tan x|]_0^{\pi/4} = \ln\left(1 + \sqrt{2}\right) \cong 0.88$

14. $L = \int_0^1 \sqrt{1 + x^2}\, dx = \int_0^{\pi/4} \sec^3 u\, du = \frac{1}{2}\left[\sec u \tan u + \ln|\sec u + \tan u|\right]_0^{\pi/4}$

$= \frac{1}{2}\left[\sqrt{2} + \ln(1 + \sqrt{2})\right] \cong 1.15$

15. $L = \int_1^2 \sqrt{1 + \left(\sqrt{x^2 - 1}\right)^2}\, dx = \int_1^2 x\, dx = \left[\frac{1}{2}x^2\right]_1^2 = \frac{3}{2}$

16. $L = \int_0^{\ln 2} \sqrt{1 + \sinh^2 x}\, dx = \int_0^{\ln 2} \cosh x\, dx = [\sinh x]_0^{\ln 2} = \frac{3}{4}$

17. $L = \int_0^1 \sqrt{1 + \left[\sqrt{3 - x^2}\right]^2}\, dx = \int_0^1 \sqrt{4 - x^2}\, dx = \int_0^{\pi/6} 4\cos^2 u\, du$

$(x = 2\sin u)$

$= 2\int_0^{\pi/6}(1 + \cos 2u)\, du = 2\left[u + \frac{1}{2}\sin 2u\right]_0^{\pi/6} = \frac{1}{3}\pi + \frac{1}{2}\sqrt{3}$

18. $L = \int_{\pi/6}^{\pi/2} \sqrt{1 + \cot^2 x}\, dx = \int_{\pi/6}^{\pi/2} \csc x\, dx = [\ln|\csc x - \cot x|]_{\pi/6}^{\pi/2} = -\ln(2 - \sqrt{3}) \cong 1.32$

19. $v(t) = \sqrt{(2t)^2 + 2^2} = 2\sqrt{t^2 + 1}$

initial speed $= v(0) = 2$, terminal speed $= v\left(\sqrt{3}\right) = 4$

$s = \int_0^{\sqrt{3}} 2\sqrt{t^2 + 1}\, dt = 2\int_0^{\pi/3} \sec^3 u\, du = 2\left[\frac{1}{2}\sec u \tan u + \frac{1}{2}\ln|\sec u + \tan u|\right]_0^{\pi/3}$

$(t = \tan u)$ (by parts)

$= 2\sqrt{3} + \ln\left(2 + \sqrt{3}\right) \cong 4.78$

20. $v(t) = \sqrt{1 + t^2}$ initial speed $= v(0) = 1$,   terminal speed $= v(1) = \sqrt{2}$

$s = \int_0^1 \sqrt{1 + t^2}\, dt = \frac{1}{2}\left[\sqrt{2} + \ln(1 + \sqrt{2})\right] \cong 1.15$

21. $v(t) = \sqrt{(2t)^2 + (3t^2)^2}\, dt = t\left(4 + 9t^2\right)^{1/2}$

initial speed $= v(0) = 0$, terminal speed $= v(1) = \sqrt{13}$

$$s = \int_0^1 t\left(4 + 9t^2\right)^{1/2} dt = \left[\frac{1}{27}\left(4 + 9t^2\right)^{3/2}\right]_0^1 = \frac{1}{27}\left(13\sqrt{13} - 8\right)$$

22. $v(t) = \sqrt{9a^2\cos^4 t\sin^2 t + 9a^2\sin^4 t\cos^2 t} = 3a\cos t\sin t$

initial speed $= v(0) = 0$, terminal speed $= v(\pi/2) = 0$

$$s = \int_0^{\pi/2} 3a\cos t\sin t\, dt = \left[\frac{3a\sin^2 t}{2}\right]_0^{\pi/2} = \frac{3}{2}a$$

23. $v(t) = \sqrt{[e^t\cos t + e^t\sin t]^2 + [e^t\cos t - e^t\sin t]^2} = \sqrt{2}\,e^t$

initial speed $= v(0) = \sqrt{2}$, terminal speed $= \sqrt{2}\,e^\pi$

$$s = \int_0^\pi \sqrt{2}\,e^t\, dt = \left[\sqrt{2}\,e^t\right]_0^\pi = \sqrt{2}\left(e^\pi - 1\right)$$

24. $v(t) = \sqrt{(t\cos t)^2 + (t\sin t)^2} = t$ initial speed $= v(0) = 0$, terminal speed $= v(\pi) = \pi$

$$s = \int_0^\pi t\, dt = \left[\frac{t^2}{2}\right]_0^\pi = \frac{\pi^2}{2}$$

25. $L = \int_0^{2\pi} \sqrt{[x'(\theta)]^2 + [y'(\theta)]^2}\, d\theta = \int_0^{2\pi} \sqrt{a^2(1 - \cos\theta)^2 + a^2\sin^2\theta}\, d\theta$

$$= a\int_0^{2\pi} \sqrt{2(1 - \cos\theta)}\, d\theta = 2a\int_0^{2\pi}\sin\frac{\theta}{2}\, d\theta = -4a\left[\cos\frac{\theta}{2}\right]_0^{2\pi} = 8a$$

26. $\sqrt{[x'(\theta)]^2 + [y'(\theta)]^2} = \sqrt{(-2a\sin\theta + 2a\sin 2\theta)^2 + (2a\cos\theta - 2a\cos 2\theta)^2} = 2a\sqrt{2 - 2\cos\theta}$

$$L = \int_0^{2\pi} 2a\sqrt{2 - 2\cos\theta}\, d\theta = 2a\int_0^{2\pi} 2\sin\frac{\theta}{2}\, d\theta = 4a\left[-2\cos\frac{\theta}{2}\right]_0^{2\pi} = 16a$$

27.(a) $L = \int_0^{2\pi} \sqrt{(-3a\sin\theta - 3a\sin 3\theta)^2 + (3a\cos\theta - 3a\cos 3\theta)^2}\, d\theta$

$$= 3a\int_0^{2\pi} \sqrt{\sin^2\theta + 2\sin\theta\sin 3\theta + \sin^2 3\theta + \cos^2\theta - 2\cos\theta\cos 3\theta + \cos^2 3\theta}\, d\theta$$

$$= 3a\int_0^{2\pi} \sqrt{2(1 - \cos 4\theta)}\, d\theta = 6a\int_0^{2\pi} |\sin 2\theta|\, d\theta$$

$$= 24a\int_0^{\pi/2} \sin 2\theta\, d\theta = -12a\left[\cos 2\theta\right]_0^{\pi/2} = 24a$$

(b) The result follows from the identities: $\cos 3\theta = 4\cos^3\theta - 3\cos\theta$; $\sin 3\theta = 3\sin\theta - 4\sin^3\theta$

28. $\sqrt{[x'(\theta)]^2 + [y'(\theta)]^2} = \sqrt{(\cos\theta - \theta\sin\theta)^2 + (\sin\theta + \theta\cos\theta)^2} = \sqrt{1 + \theta^2}$

$L = \int_0^{2\pi} \sqrt{1 + \theta^2} \, d\theta = \left[\frac{1}{2}\theta\sqrt{1 + \theta^2} + \frac{1}{2}\ln|\theta + \sqrt{1 + \theta^2}|\right]_0^{2\pi} = \pi\sqrt{1 + 4\pi^2} + \frac{1}{2}\ln(2\pi + \sqrt{1 + 4\pi^2})$

29. $L = $ circumference of circle of radius $1 = 2\pi$

30. $L = $ half circumference of circle of radius $3 = 3\pi$

31. $L = \int_0^{4\pi} \sqrt{[e^\theta]^2 + [e^\theta]^2} \, d\theta = \int_0^{4\pi} \sqrt{2}\, e^\theta \, d\theta = \left[\sqrt{2}\, e^\theta\right]_0^{4\pi} = \sqrt{2}\left(e^{4\pi} - 1\right)$

32. $L = \int_{-2\pi}^{2\pi} \sqrt{(ae^\theta)^2 + (ae^\theta)^2} \, d\theta = \int_{-2\pi}^{2\pi} \sqrt{2}\, ae^\theta \, d\theta = a\sqrt{2}(e^{2\pi} - e^{-2\pi}) \quad (a > 0)$

33. $L = \int_0^{2\pi} \sqrt{[e^{2\theta}]^2 + [2e^{2\theta}]^2} \, d\theta = \int_0^{2\pi} \sqrt{5}\, e^{2\theta} \, d\theta = \left[\frac{1}{2}\sqrt{5}\, e^{2\theta}\right]_0^{2\pi} = \frac{1}{2}\sqrt{5}\left(e^{4\pi} - 1\right)$

34. $L = 2\int_0^{\pi} \sqrt{(1 + \cos\theta)^2 + (-\sin\theta)^2} \, d\theta = 2\int_0^{\pi} \sqrt{2 + 2\cos\theta} \, d\theta = 4\int_0^{\pi} \cos\frac{\theta}{2} \, d\theta = 8$

35. $L = \int_0^{\pi/2} \sqrt{(1 - \cos\theta)^2 + \sin^2\theta} \, d\theta = \int_0^{\pi/2} \sqrt{2 - 2\cos\theta} \, d\theta$

$= \int_0^{\pi/2} \left(2\sin\frac{1}{2}\theta\right) d\theta = \left[-4\cos\frac{1}{2}\theta\right]_0^{\pi/2} = 4 - 2\sqrt{2}$

36. $L = \int_0^{\pi/4} \sqrt{(2a\sec\theta)^2 + (2a\sec\theta\tan\theta)^2} \, d\theta = \int_0^{\pi/4} 2a\sec^2\theta \, d\theta = \left[2a\tan\theta\right]_0^{\pi/4} = 2a$

37. $s = \int_0^1 \sqrt{\left[\frac{1}{1 + t^2}\right]^2 + \left[\frac{-t}{1 + t^2}\right]^2} \, dt = \int_0^1 \frac{dt}{\sqrt{1 + t^2}}$

$= \int_0^{\pi/4} \sec u \, du = [\ln|\sec u + \tan u|]_0^{\pi/4} = \ln\left(1 + \sqrt{2}\right)$

$(t = \tan u)$

initial speed $= v(0) = 1$,   terminal speed $= v(1) = \frac{1}{2}\sqrt{2}$

38. $s = \int_0^{2\pi} \sqrt{(-\sin t)^2 + (1 - \cos t)^2} \, dt = \int_0^{2\pi} \sqrt{2 - 2\cos t} \, d\theta = \left[-4\cos\frac{t}{2}\right]_0^{2\pi} = 8$

initial speed $= v(0) = 0$,   terminal speed $= v(2\pi) = 0$

39. $c = 1$;   the curve $y = e^x$ is the curve $y = \ln x$ reflected in the line $y = x$

40. By the hint:
$$L = \int_1^4 \sqrt{1 + \left(\frac{3}{2}x^{1/2}\right)^2}\, dx = \int_1^4 \sqrt{1 + \frac{9}{4}x}\, dx = \frac{8}{27}\left[\left(1 + \frac{9}{4}x\right)^{3/2}\right]_1^4$$

$$= \frac{1}{27}\left[80\sqrt{10} - 13\sqrt{13}\right] \cong 7.63$$

41. coordinates of the midpoint: $\left(\frac{1}{2}, -\frac{7}{2}\right)$ **42.** coordinates of the midpoint (approx):
$$(3.2911, 5.9705)$$

43. $L = \int_a^b \sqrt{1 + \sinh^2 x}\, dx = \int_a^b \sqrt{\cosh^2 x}\, dx = \int_a^b \cosh x\, dx = A$

44. $L = \int_1^e \sqrt{1 + \frac{4}{x^2}}\, dx \cong 2.6625$

45.

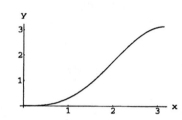

$$f(x) = \sin x - x \cos x$$
$$f'(x) = x \sin x$$
$$L = \int_0^\pi \sqrt{1 + x^2 \sin^2 x}\, dx \cong 4.6984$$

46. $L = \int_0^{\pi/3} \sqrt{\left(2e^{2t}\cos 2t - 2e^{2t}\sin 2t\right)^2 + \left(2e^{2t}\sin 2t + 2e^{2t}\cos 2t\right)^2}\, dt = \sqrt{8}\int_0^{\pi/3} e^{2t}\, dt \cong 10.0699$

47. $x = t^2, \quad y = t^3 - t$

$x' = 2t \quad y' = 3t^2 - 1$

$L = \int_{-1}^1 \sqrt{(2t)^2 + (3t^2 - 1)^2}\, dt \cong 2.7156$

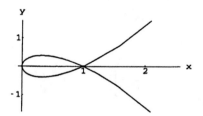

48. (a)

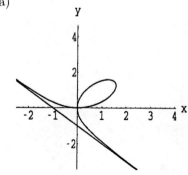

(b) $L = 2\int_0^1 \sqrt{(x'(t))^2 + (y'(t))^2}\, dt$

$$= 2\int_0^1 \sqrt{\left(\frac{3 - 6t^3}{(t^3 + 1)^2}\right)^2 + \left(\frac{6t - 3t^4}{(t^3 + 1)^2}\right)^2}\, dt$$

$$= 3.5485$$

49.

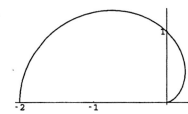

$r = 1 - \cos\theta, \quad r' = \sin\theta$

$$L = \int_0^\pi \sqrt{(1 - \cos\theta)^2 + \sin^2\theta} \; d\theta = 4$$

50. $r = \sin 5\theta$ is a petal curve with 5 petals. The length of one petal is given by:

$$L = \int_0^{\pi/5} \sqrt{\sin^2 5\theta + 25\cos^2 5\theta} \; d\theta \cong 2.101$$

The length of the entire curve is: 10.505 (approx.). The curve is traversed twice for $0 \le \theta \le 2\pi$. Thus the length of that curve is $2(10.505) \cong 21.01$.

51. (a) $L = \int_0^{2\pi} \sqrt{a^2\sin^2 t + b^2\cos^2 t} \; dt = 4\int_0^{\pi/2} \sqrt{a^2(1 - \cos^2 t) + b^2\cos^2 t} \; dt$

$\qquad = 4a\int_0^{\pi/2} \sqrt{1 - e^2\cos^2 t} \; dt, \quad$ where $\quad e = \dfrac{\sqrt{a^2 - b^2}}{a}$

$\quad$ (b) $L = 4\int_0^{\pi/2} \sqrt{25 - 9\cos^2 t} \; dt \cong 28.3617$

52. The line segment that joins P_{i-1} to P_i has length

$$\sqrt{(x_i - x_{i-1})^2 + [f(x_i) - f(x_{i-1})]^2}$$

This can be written

$$\sqrt{1 + \left[\frac{f(x_i) - f(x_{i-1})}{x_i - x_{i-1}}\right]^2} \; (x_i - x_{i-1}) = \sqrt{1 + \left[\frac{f(x_i) - f(x_{i-1})}{x_i - x_{i-1}}\right]^2} \; \Delta x_i$$

The mean value theorem gives the existence of x_i^* in $[x_{i-1}, x_i]$ such that

$$\frac{f(x_i) - f(x_{i-1})}{x_i - x_{i-1}} = f'(x_i^*).$$

The length of $\overline{P_{i-1}P_i}$ can then be written

$$\sqrt{1 + [f'(x_i^*)]^2} \; \Delta x_i$$

The length of the polygonal approximation is the sum of such expressions from $i = 1$ to $i = n$.

53. $\sqrt{1 + [f(x)]^2} = \sqrt{1 + \tan^2[\alpha(x)]} = |\sec[\alpha(x)]|$

54. Balancing the vertical forces we have

$$k\int_0^x \sqrt{1 + [f'(t)]^2} \; dt = p(x)\sin\theta. \quad \text{(weight=vertical pull at } x)$$

$\quad$ Balancing the horizontal forces we have

$$p(0) = p(x)\cos\theta. \quad \text{(pull at 0 = horizontal pull at } x)$$

Together these equations give

$$k \int_0^x \sqrt{1 + [f'(x)]^2} \, dt = p(0) \tan \theta = p(0) f'(\theta).$$

Differentiating with respect to x, we have

$$k \sqrt{1 + [f'(x)]^2} = p(0) f''(x)$$

$$\frac{f''(x)}{\sqrt{1 + [f'(x)]^2}} = b. \quad (\text{set} \quad k/p(0) = b)$$

Integration gives

$$\ln \left(f'(x) + \sqrt{1 + [f'(x)]^2} \right) = bx + C.$$

In the coordinate system of the figure, we have $f'(0) = 0$. Therefore $C = 0$ and

$$\ln \left(f'(x) + \sqrt{1 + [f'(x)]^2} \right) = bx$$

Exponentiating both sides we have

$$f'(x) + \sqrt{1 + [f'(x)]^2} = e^{bx}$$

$$\sqrt{1 + [f'(x)]^2} = e^{bx} - f'(x)$$

$$1 + [f'(x)]^2 = e^{bx} - 2e^{bx} f'(x) + [f'(x)]^2$$

$$2e^{bx} f'(x) = e^{2bx} - 1$$

$$f'(x) = \frac{1}{2}(e^{bx} - e^{-bx}) = \sinh bx$$

A final integration gives

$$f(x) = \frac{1}{b} \cosh bx + K.$$

Adjusting the coordinate system vertically so that $f(0) = 1/b$ we have

$$f(x) = \frac{1}{b} \cosh bx$$

Now set $a = 1/b$

SECTION 10.8

1. $L =$ length of the line segment $= 1$
    $(\overline{x}, \overline{y}) = \left(\frac{1}{2}, 4\right)$ (the midpoint of the line segment)
    $A_x =$ lateral surface area of cylinder of radius 4 and side $1 = 8\pi$.

2. $L = \displaystyle\int_0^1 \sqrt{1 + 2^2} \, dx = \sqrt{5}$

    $\overline{x}L = \displaystyle\int_0^1 x\sqrt{1 + 2^2} \, dx = \frac{\sqrt{5}}{2}, \quad \overline{x} = \frac{1}{2}$

    $\overline{y}L = \displaystyle\int_0^1 2x\sqrt{1 + 2^2} \, dx = \sqrt{5}, \quad \overline{y} = 1$

    $A_x = 2\pi \overline{y} L = 2\pi\sqrt{5}$

3.
$$L = \int_0^3 \sqrt{1 + \left(\frac{4}{3}\right)^2}\, dx = \left(\frac{5}{3}\right)^3 = 5$$

$$\overline{x}L = \int_0^3 x\sqrt{1 + \left(\frac{4}{3}\right)^2}\, dx = \frac{5}{3}\left[\frac{1}{2}x^2\right]_0^3 = \frac{15}{2}, \quad \overline{x} = \frac{3}{2}$$

$$\overline{y}L = \int_0^3 \frac{4}{3}x\sqrt{1 + \left(\frac{4}{3}\right)^2}\, dx = \left(\frac{4}{3}\right)\left(\frac{15}{2}\right) = 10, \quad \overline{y} = 2$$

$$A_x = 2\pi\overline{y}L = 2\pi(2)(5) = 20\pi$$

4.
$$L = \int_0^5 \sqrt{1 + \left(-\frac{12}{5}\right)^2}\, dx = 5\sqrt{1 + \frac{144}{25}} = 13$$

$$\overline{x}L = \int_0^5 x\sqrt{1 + \frac{144}{25}}\, dx = \frac{25}{2}\sqrt{1 + \frac{144}{25}}, \quad \overline{x} = \frac{5}{2}$$

$$\overline{y}L = \int_0^5 \left(12 - \frac{12}{5}x\right)\sqrt{1 + \frac{144}{25}}\, dx = \frac{12}{5}\cdot\frac{25}{2}\sqrt{1 + \frac{144}{25}}, \quad \overline{y} = 6$$

$$A_x = 2\pi\overline{y}L = 156\pi$$

5.
$$L = \int_0^2 \sqrt{(3)^2 + (4)^2}\, dt = (2)(5) = 10$$

$$\overline{x}L = \int_0^2 3t\sqrt{(3)^2 + (4)^2}\, dt = 15\left[\frac{1}{2}t^2\right]_0^2 = 30, \quad \overline{x} = 3$$

$$\overline{y}L = \int_0^2 4t\sqrt{(3)^2 + (4)^2}\, dt = 20\left[\frac{1}{2}t^2\right]_0^2 = 40, \quad \overline{y} = 4$$

$$A_x = 2\pi\overline{y}L = 2\pi(4)(10) = 80\pi$$

6.
$$L = \frac{1}{8} \text{ circumference of circle of radius } 5 = \frac{5}{4}\pi$$

$$\overline{x}L = \int_0^{\pi/4} x(\theta)\sqrt{x'(\theta)^2 + y'(\theta)^2}\, d\theta = \int_0^{\pi/4} 5\cos\theta \cdot 5\, d\theta = \frac{25\sqrt{2}}{2}, \quad \overline{x} = \frac{10\sqrt{2}}{\pi}$$

$$\overline{y}L = \int_0^{\pi/4} y(\theta)\sqrt{x'(\theta)^2 + y'(\theta)^2}\, d\theta = \int_0^{\pi/4} 5\sin\theta \cdot 5\, d\theta = 25\left(1 - \frac{\sqrt{2}}{2}\right), \quad \overline{y} = \frac{10}{\pi}(2 - \sqrt{2})$$

$$A_x = 2\pi\overline{y}L = 25\pi(2 - \sqrt{2})$$

7.
$$L = \int_0^{\pi/6} \sqrt{4\sin^2 t + 4\cos^2 t}\, dt = 2\left(\frac{\pi}{6}\right) = \frac{1}{3}\pi$$

$$\overline{x}L = \int_0^{\pi/6} 2\cos t\sqrt{4\sin^2 t + 4\cos^2 t}\, dt = 4\left[\sin t\right]_0^{\pi/6} = 2, \quad \overline{x} = \frac{6}{\pi}$$

$$\overline{y}L = \int_0^{\pi/6} 2\sin t\sqrt{4\sin^2 t + 4\cos^2 t}\, dt = 4\left[-\cos t\right]_0^{\pi/6} = 4 - 2\sqrt{3}, \quad \overline{y} = 6\left(2 - \sqrt{3}\right)/\pi$$

$$A_x = 2\pi\overline{y}L = 2\pi(6(2 - \sqrt{3})/\pi)\tfrac{1}{3}\pi = 4\pi(2 - \sqrt{3})$$

8.
$$L = \int_0^{\pi/2} \sqrt{(-3\cos^2 t \sin t)^2 + (3\sin^2 t \cos t)^2}\, dt = \int_0^{\pi/2} 3\cos t \sin t\, dt = \left[\frac{3\sin^2 t}{2}\right]_0^{\pi/2} = \frac{3}{2}$$

$$\bar{x}L = \int_0^{\pi/2} \cos^3 t \cdot 3\cos t \sin t\, dt = \left[-\frac{3}{5}\cos^5 t\right]_0^{\pi/2} = \frac{3}{5}, \quad \bar{x} = \frac{2}{5}$$

$$\bar{y}L = \int_0^{\pi/2} \sin^3 t \cdot 3\cos t \sin t\, dt = \left[\frac{3}{5}\sin^5 t\right]_0^{\pi/2} = \frac{3}{5}, \quad \bar{y} = \frac{2}{5}$$

$$A_x = 2\pi\bar{y}L = \frac{6}{5}\pi$$

9. $x(t) = a\cos t, \quad y = a\sin t; \quad t \in [\frac{1}{3}\pi, \frac{2}{3}\pi]$

$$L = \int_{\pi/3}^{2\pi/3} \sqrt{a^2\sin^2 t + a^2\cos^2 t}\, dt = \frac{1}{3}\pi a$$

by symmetry $\quad \bar{x} = 0$

$$\bar{y}L = \int_{\pi/3}^{2\pi/3} a\sin t \sqrt{a^2\sin^2 t + a^2\cos^2 t}\, dt = a^2 \int_{\pi/3}^{2\pi/3} \sin t\, dt$$

$$= a^2\left[-\cos t\right]_{\pi/3}^{2\pi/3} = a^2, \quad \bar{y} = 3a/\pi$$

$$A_x = 2\pi\bar{y}L = 2\pi a^2$$

10.
$$L = \int_0^\pi \sqrt{(1+\cos\theta)^2 + \sin^2\theta}\, d\theta = \int_0^\pi \sqrt{2 + 2\cos\theta}\, d\theta$$

$$= \int_0^\pi 2\sqrt{\frac{1+\cos\theta}{2}}\, d\theta = \int_0^\pi 2\cos\frac{1}{2}\theta\, d\theta = 4$$

$$\bar{x}L = \int_0^\pi \cos\theta(1+\cos\theta)\left(2\cos\frac{1}{2}\theta\right) d\theta = \int_0^\pi \left(\cos^2\frac{1}{2}\theta - \sin^2\frac{1}{2}\theta\right)\left(2\cos^2\frac{1}{2}\theta\right)\left(2\cos\frac{1}{2}\theta\right) d\theta$$

$$= 4\int_0^\pi \left(1 - 2\sin^2\frac{1}{2}\theta\right)\left(1 - \sin^2\frac{1}{2}\theta\right)\left(\cos\frac{1}{2}\theta\right) d\theta$$

$$= 4\int_0^\pi \left(\cos\frac{1}{2}\theta - 3\sin^2\frac{1}{2}\theta\cos\frac{1}{2}\theta + 2\sin^4\frac{1}{2}\theta\cos\frac{1}{2}\theta\right) d\theta = \frac{16}{5}; \quad \bar{x} = \frac{4}{5}$$

$$\bar{y}L = \int_0^\pi \sin\theta(1+\cos\theta)\left(2\cos\frac{1}{2}\theta\right) d\theta$$

$$= \int_0^\pi \left(2\sin\frac{1}{2}\theta\cos\frac{1}{2}\theta\right)\left(2\cos^2\frac{1}{2}\theta\right)\left(2\cos\frac{1}{2}\theta\right) d\theta$$

$$= 8\int_0^\pi \left(\cos^4\frac{1}{2}\theta\sin\frac{1}{2}\theta\right) d\theta = \frac{16}{5}; \quad \bar{y} = \frac{4}{5}$$

$$A_x = 2\pi\bar{y}L = \frac{32}{5}\pi$$

11. $A_x = \int_0^2 \frac{2}{3}\pi x^3 \sqrt{1+x^4}\, dx = \frac{1}{9}\pi \left[(1+x^4)^{3/2}\right]_0^2 = \frac{1}{9}\pi(17\sqrt{17} - 1) \cong 24.1179$

12. $A_x = \int_1^2 2\pi\sqrt{x}\sqrt{1 + \frac{1}{4x}}\, dx = \int_1^2 \pi\sqrt{4x+1}\, dx = \pi\left[\frac{1}{6}(4x+1)^{3/2}\right]_1^2 = \frac{1}{6}\pi(27 - 5\sqrt{5})$

13. $A_x = \displaystyle\int_0^1 \frac{1}{2}\pi x^3 \sqrt{1 + \frac{9}{16}x^4}\,dx = \frac{4}{27}\pi\left[\left(1 + \frac{9}{16}x^4\right)^{3/2}\right]_0^1 = \frac{61}{432}\pi$

14. $A_x = \displaystyle\int_0^4 2\pi \cdot 3\sqrt{x}\sqrt{1 + \frac{9}{4x}}\,dx = \int_0^4 3\pi\sqrt{4x+9}\,dx = \frac{\pi}{2}\left[(4x+9)^{3/2}\right]_0^4 = 49\pi$

15. $A_x = \displaystyle\int_0^{\pi/2} 2\pi \cos x \sqrt{1 + \sin^2 x}\,dx = \int_0^1 2\pi\sqrt{1+u^2}\,du$

$\qquad\qquad\qquad\qquad\qquad\qquad\qquad\qquad \uparrow\!\!\!\!\sqsubset u = \sin x$

$\qquad = 2\pi\left[\frac{1}{2}u\sqrt{1+u^2} + \frac{1}{2}\ln\left(u + \sqrt{1+u^2}\right)\right]_0^1 = \pi\left[\sqrt{2} + \ln\left(1+\sqrt{2}\right)\right]$

$\quad \uparrow\!\!\!\!\sqsubset (8.5.1)$

16. $A_x = \displaystyle\int_{-1}^0 4\pi\sqrt{1-x}\sqrt{1 + \frac{1}{1-x}}\,dx = 4\pi\int_{-1}^0 \sqrt{2-x}\,dx = -\frac{8}{3}\left[(2-x)^{3/2}\right]_{-1}^0 = \frac{8}{3}\pi(3\sqrt{3} - 2\sqrt{2})$

17. $A_x = \displaystyle\int_0^{\pi/2} 2\pi(e^\theta \sin\theta)\sqrt{[e^\theta\cos\theta - e^\theta\sin\theta]^2 + [e^\theta\sin\theta + e^\theta\cos\theta]^2}\,d\theta$

$\qquad = 2\pi\sqrt{2}\displaystyle\int_0^{\pi/2} e^{2\theta}\sin\theta\,d\theta$

$\qquad = 2\pi\sqrt{2}\left[\frac{1}{5}\left(2e^{2\theta}\sin\theta - e^{2\theta}\cos\theta\right)\right]_0^{\pi/2} = \frac{2}{5}\sqrt{2}\,\pi\,(2e^\pi + 1)$

$\qquad$ (by parts twice)

18. $A_x = \displaystyle\int_0^{\ln 2} 2\pi\cosh x\sqrt{1 + \sinh^2 x}\,dx$

$\qquad = 2\pi\displaystyle\int_0^{\ln 2} \cosh^2 x\,dx$

$\qquad = \frac{1}{2}\pi\displaystyle\int_0^{\ln 2}(e^{2x} + 2 + e^{-2x})\,dx = \frac{1}{2}\pi\left[\frac{1}{2}e^{2x} + 2x - \frac{1}{2}e^{-2x}\right]_0^{\ln 2} = \frac{1}{16}\pi(15 + 16\ln 2)$

19. (a) $A = \displaystyle\int_0^{2\pi} y(\theta)x'(\theta)\,d\theta \qquad$ [see (9.6.4)]

$\qquad = \displaystyle\int_0^{2\pi} a^2(1 - \cos\theta)^2\,d\theta$

$\qquad = a^2\displaystyle\int_0^{2\pi}(1 - 2\cos\theta + \cos^2\theta)\,d\theta$

$\qquad = a^2\displaystyle\int_0^{2\pi}\left(\frac{3}{2} - 2\cos\theta + \frac{1}{2}\cos 2\theta\right)d\theta$

$\qquad = a^2\left[\frac{3}{2}\theta - 2\sin\theta + \frac{1}{4}\sin 2\theta\right]_0^{2\pi}$

$\qquad = 3\pi a^2$

(b) $A = \displaystyle\int_0^{2\pi} 2\pi\, y(\theta) \sqrt{[x'(\theta)]^2 + [y'(\theta)]^2}\, d\theta$ (9.9.2)

$= \displaystyle\int_0^{2\pi} 2\pi\, a(1 - \cos\theta) \sqrt{a^2(1 - \cos\theta)^2 + a^2 \sin^2\theta}\, d\theta$

$= 2\pi\, a^2 \displaystyle\int_0^{2\pi} (1 - \cos\theta) \sqrt{2 - 2\cos\theta}\, d\theta$

$= 4\pi\, a^2 \displaystyle\int_0^{2\pi} (1 - \cos\theta) \sin\frac{\theta}{2}\, d\theta$

$= 4\pi\, a^2 \displaystyle\int_0^{2\pi} \left(2 \sin\frac{\theta}{2} - 2 \cos^2\frac{\theta}{2} \sin\frac{\theta}{2}\right) d\theta$

$= 4\pi\, a^2 \left[-4 \cos\frac{\theta}{2}\right]_0^{2\pi} + \dfrac{16\pi\, a^2}{3} \left[\cos^3(\theta/2)\right]_0^{2\pi} = \dfrac{64\pi\, a^2}{3}$

20. (a) The graph with $a = 1$;

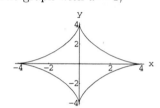

(b) $A = 2 \displaystyle\int_0^{\pi} y(\theta) x'(\theta)\, d\theta$

$= 2 \displaystyle\int_0^{\pi} (3\sin\theta - \sin 3\theta)(-3\sin\theta - 3\sin 3\theta)\, d\theta$

$= -6 \displaystyle\int_0^{\pi} (3\sin^2\theta + 2\sin\theta \sin 3\theta - \sin^2 3\theta)\, d\theta$

$= 6\pi$

(b) $A = \displaystyle\int_0^{\pi} 2\pi y(\theta) \sqrt{x'(\theta)^2 + y'(\theta)^2}\, d\theta = 2\pi \int_0^{\pi} (3a\sin\theta - a\sin 3\theta) 3a\sqrt{2 - 2\cos 4\theta}\, d\theta$

21. $A = \frac{1}{2}\theta s_2{}^2 - \frac{1}{2}\theta s_1{}^2$

$= \frac{1}{2}(\theta s_2 + \theta s_1)(s_2 - s_1)$

$= \frac{1}{2}(2\pi R + 2\pi r)s = \pi(R + r)s$

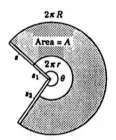

22. (a) $\overline{x} = \dfrac{1}{\frac{\pi}{4}r^2 - \frac{\pi}{4}a^2} \cdot \left(-\dfrac{4a}{3\pi} \cdot \dfrac{\pi}{4}a^2 + \dfrac{4r}{3\pi} \cdot \dfrac{\pi}{4}r^2\right) = \dfrac{4}{3\pi}\left(\dfrac{r^3 - a^3}{r^2 - a^2}\right) = \dfrac{4}{3\pi}\left(\dfrac{r^2 + ar + a^2}{r + a}\right)$

$\overline{y} = \overline{x}$

(b) $\displaystyle\lim_{a \to r} \dfrac{4}{3\pi}\left(\dfrac{r^2 + ar + a^2}{r + a}\right) = \dfrac{2r}{\pi};$ $= \overline{x} = \overline{y} = \dfrac{2r}{\pi}.$

23. (a) The centroids of the 3, 4, 5 sides are the midpoints $\left(\frac{3}{2}, 0\right)$, $(3, 2)$, $\left(\frac{3}{2}, 2\right)$.

(b) $\overline{x}(3 + 4 + 5) = \frac{3}{2}(3) + 3(4) + \frac{3}{2}(5)$, $12\overline{x} = 24$, $\overline{x} = 2$

$\overline{y}(3 + 4 + 5) = 0(3) + 2(4) + 2(5)$, $12\overline{y} = 18$, $\overline{y} = \dfrac{3}{2}$

(c) $A = \frac{1}{3}(3)(4) = 6$

$$\bar{x}A = \int_0^3 x\left(\frac{4}{3}x\right) dx = \int_0^3 \frac{4}{3}x^2 dx = \frac{4}{9}\left[x^3\right]_0^3 = 12, \quad \bar{x} = 2$$

$$\bar{y}A = \int_0^3 \frac{1}{2}\left(\frac{4}{3}x\right)^2 dx = \int_0^3 \frac{8}{9}x^2 dx = \frac{8}{27}\left[x^3\right]_0^3 = 8, \quad \bar{y} = \frac{4}{3}$$

(d) $\bar{x}(4+5) = 3(4) + \frac{3}{2}(5), \quad 9\bar{x} = \frac{39}{2}, \quad \bar{x} = \frac{13}{6}$

$\bar{y}(4+5) = 2(4) + 2(5), \quad 9\bar{y} = 18, \quad \bar{y} = 2$

(e) $A_x = 2\pi(2)(5) = 20\pi$

24. Set $y(t) = t, \quad x(t) = \frac{1}{2}\left[t\sqrt{t^2 - 1} + \ln|t - \sqrt{t^2 - 1}|\right], \quad t \in [2,5]$

$x'(t) = \sqrt{t^2 - 1} \quad y'(t) = 1$

$$A = \int_2^5 2\pi t\sqrt{(\sqrt{t^2 - 1})^2 + 1^2}\, dt = \int_2^5 2\pi t^2\, dt = \left[2\pi\frac{t^3}{3}\right]_2^5 = 78\pi$$

25. Set $y(t) = t, \quad x(t) = \frac{1}{6a^2}\left(t^3 + \frac{3a^4}{t}\right) \quad t \in [a, 3a]$

$x'(t) = \frac{1}{6a^2}\left(3t^2 - \frac{3a^4}{t^2}\right), \quad y'(t) = 1$

$$A = \int_a^{3a} 2\pi t\sqrt{\frac{1}{36a^4}\left(3t^2 - \frac{3a^4}{t^2}\right)^2 + 1}\, dt = 2\pi \int_a^{3a} \frac{t}{6a^2}\left(3t^2 + \frac{3a^4}{t^2}\right) dt$$

$$= \frac{\pi}{a^2}\int_a^{3a}\left(t^3 + \frac{a^4}{t}\right) dt = \frac{\pi}{a^2}\left[\frac{t^4}{4} + a^4 \ln 4\right]_a^{3a} = \pi a^2(20 + \ln 3)$$

26. $A_x = 2\pi\bar{y}L = 2\pi(b)(2\pi a) = 4\pi^2 ab$

27. (a) No: $f'(x) = -x/\sqrt{r^2 - x^2}$ is not defined at $x = \pm r$. We can however begin with a smaller

interval $[-r + \epsilon, r - \epsilon]$, integrate, and take the limit as $\epsilon \to 0$.

(b) $A = \int_0^{2\pi} 2\pi r\sqrt{(\sin t)^2}\, dt = 2\pi r\int_0^{2\pi} |\sin t|\, dt = 8\pi r$

C is not simple. The curve [in this case the line segment that joins $(1, r)$ to $(-1, r)$] is traced

out twice.

27. The band can be obtained by revolving about the x-axis the graph of the function

$$f(x) = \sqrt{r^2 - x^2}, \quad x \in [a, b].$$

A straightforward calculation shows that the surface area of the band is $2\pi r(b - a)$.

28. C revolved about the y-axis generates a surface of area

$$A_y = 2\pi\bar{x}L = 2\pi\bar{x}r(\theta_2 - \theta_1).$$

C revolved about the x-axis generates a surface of area

$$A_x = 2\pi x\bar{y}L = 2\pi\bar{y}r(\theta_2 - \theta_1).$$

By our solution to Exercise 27.

$$A_y = 2\pi r^2(\sin\theta_2 - \sin\theta_1)$$

and

$$A_x = 2\pi r^2(\cos\theta_1 - \cos\theta_2).$$

Therefore

$$\bar{x} = r\left(\frac{\sin\theta_2 - \sin\theta_1}{\theta_2 - \theta_1}\right), \quad \bar{y} = r\left(\frac{\cos\theta_1 - \cos\theta_2}{\theta_2 - \theta_1}\right).$$

29. (a) Parameterize the upper half of the ellipse by

$$x(t) = a\cos t, \quad y(t) = b\sin t; \qquad t \in [0, \pi].$$

Here

$$\sqrt{[x'(t)]^2 + [y'(t)]^2} = \sqrt{a^2\sin^2 t + b^2\cos^2 t} = \sqrt{a^2 - (a^2 - b^2)\cos^2 t},$$

which, with $c = \sqrt{a^2 - b^2}$, can be written $\sqrt{a^2 - c^2\cos^2 t}$.   Therefore,

$$A = \int_0^\pi 2\pi b\sin t\sqrt{a^2 - c^2\cos^2 t}\, dt = 4\pi b\int_0^{\pi/2}\sin t\sqrt{a^2 - c^2\cos^2 t}\, dt.$$

Setting $u = c\cos t$, we have $du = -c\sin t$ and

$$A = -\frac{4\pi b}{c}\int_c^0 \sqrt{a^2 - u^2}\, du = \frac{4\pi b}{c}\left[\frac{u}{2}\sqrt{a^2 - u^2} + \frac{a^2}{2}\sin^{-1}\left(\frac{u}{a}\right)\right]_0^c$$

$$= 2\pi b^2 + \frac{2\pi a^2 b}{c}\sin^{-1}\left(\frac{c}{a}\right) = 2\pi b^2 + \frac{2\pi ab}{e}\sin^{-1}e$$

where e is the eccentricity of ellipse: $e = c/a$.

(b) Parameterize the right half of the ellipse by

$$x(t) = a\cos t, \quad y(t) = b\sin t; \qquad t \in \left[-\frac{1}{2}\pi, \frac{1}{2}\pi\right].$$

Again $\sqrt{[x'(t)]^2 + [y'(t)]^2} = \sqrt{a^2 - c^2\cos^2 t}$ where $c = \sqrt{a^2 - b^2}$.

Therefore

$$A = \int_{-\pi/2}^{\pi/2} 2\pi a\cos t\sqrt{a^2 - c^2\cos^2 t}\, dt.$$

Set $u = c\sin t$.   Then $du = c\cos t\, dt$ and

$$A = \frac{2\pi a}{c}\int_{-c}^c \sqrt{b^2 + u^2}\, du = \frac{2\pi a}{c}\left[\frac{u}{2}\sqrt{b^2 + u^2} + \frac{b^2}{2}\ln\left|u + \sqrt{b^2 + u^2}\right|\right]_{-c}^c$$

Routine calculation gives

$$A = 2\pi a^2 + \frac{\pi b^2}{e} \ln \left| \frac{1+e}{1-e} \right|.$$

30. Set $s'(t) = \sqrt{[x'(t)]^2 + [y'(t)]^2}$.   Let $P = \{t_0, t_1, \cdots, t_n\}$ be a partition of $[c,d]$. The partition breaks up the surface into n surfaces of revolution with areas

$$A_i = \int_{t_{i-1}}^{t_i} 2\pi y(t) s'(t)\, dt$$

and centroids

$$\overline{x}_i = x(t_i^*) \quad \text{with} \quad t_i^* \in [t_{i-1}, t_i].$$

$$\overline{x}A = \overline{x}_1 A_1 + \cdots + \overline{x}_n A_n$$

$$= x(t_1^*) \int_{t_0}^{t_1} 2\pi y(t) s'(t)\, dt + \cdots + x(t_n^*) \int_{t_{n-1}}^{t_n} 2\pi y(t) s'(t)\, dt$$

$$\cong x(t_1^*)[2\pi y(t_1^*) s'(t_1^*) \Delta t_1] + \cdots + x(t_n^*)[2\pi y(t_n^*) s'(t_n^*) \Delta t_n]$$

$$= 2\pi x(t_1^*) y(t_1^*) s'(t_1^*) \Delta t_1 + \cdots + 2\pi x(t_n^*) y(t_n^*) s'(t_n^*) \Delta t_n$$

As $\|P\| \to 0$,   the expression on the right tends to

$$\int_c^d 2\pi x(t) y(t) s'(t)\, dt = \int_c^d 2\pi x(t) y(t) \sqrt{[x'(t)]^2 + [y'(t)]^2}\, dt.$$

31. Such a hemisphere can be obtained by revolving about the x-axis the curve

$$x(t) = r\cos t, \quad y(t) = r\sin t; \quad t \in [0, \tfrac{1}{2}\pi].$$

Therefore, $\displaystyle \overline{x}A = \int_0^{\pi/2} 2\pi(r\cos t)(r\sin t)\sqrt{r^2\sin^2 t + r^2\cos^2 t}\, dt$

$$= \int_0^{\pi/2} 2\pi r^3 \sin t \cos t\, dt = \pi r^3 \left[\sin^2 t\right]_0^{\pi/2} = \pi r^3.$$

$$A = 2\pi r^2; \quad \overline{x} = \overline{x}A/A = \tfrac{1}{2}r.$$

The centroid lies on the midpoint of the axis of the hemisphere.

32. The cone can be generated by revolving about the x-axis the graph of the function

$$f(x) = (r/h)x, \qquad x \in [0, h].$$

Formula 9.9.8 gives

$$\overline{x}A = \int_0^h 2\pi f(x)\sqrt{1 + [f'(x)]^2}\, dx = \frac{2\pi r}{h^2}\sqrt{h^2 + r^2} \int_0^h x^2\, dx = \frac{2}{3}\pi r h\sqrt{h^2 + r^2}.$$

$$A = \pi r\sqrt{h^2 + r^2}; \quad \overline{x} = \overline{x}A/A = \tfrac{2}{3}h$$

The centroid of the surface lies on the axis of the cone at a distance of $\tfrac{2}{3}h$ from the vertex of the cone.

33. Such a surface can be obtained by revolving about the x-axis the graph of the function

$$f(x) = \left(\frac{R-r}{h}\right)x + r, \quad x \in [0, h].$$

Formula (9.9.8) gives

$$\overline{x}A = \int_0^h 2\pi x f(x)\sqrt{1 + [f'(x)]^2}\, dx$$

$$= \frac{2\pi}{h}\sqrt{h^2 + (R-h)^2}\int_0^h \left[\left(\frac{R-r}{h}\right)x^2 + rx\right]dx$$

$$= \frac{\pi}{3}\sqrt{h^2 + (R-r)^2}\,(2R+r)h$$

$$A = \pi(R+r)s = \pi(R+r)\sqrt{h^2 + (R-r)^2} \quad\text{and}\quad \overline{x} = \frac{\overline{x}A}{A} = \left(\frac{2R+r}{R+r}\right)\frac{h}{3}.$$

The centroid of the surface lies on the axis of the cone $\left(\dfrac{2R+r}{R+r}\right)\dfrac{h}{3}$ units from the base of radius r.

PROJECT 10.8

1. Referring to the figure we have

$$x(\theta) = \overline{OB} - \overline{AB} = R\theta - R\sin\theta = R\,(\theta - \sin\theta)$$

$$y(\theta) = \overline{BQ} - \overline{QC} = R - R\cos\theta = R\,(1 - \cos\theta).$$

2. (a) $x'(\theta) = R(1 - \cos\theta), \quad y'(\theta) = R\sin\theta.$
  The arches end at $\theta = 2n\pi,$ and $x'(2n\pi) = y'(2n\pi) = 0$

(b) $A = \displaystyle\int_0^{2\pi} y(\theta)x'(\theta)\, d\theta = R^2\int_0^{2\pi}(1 - \cos\theta)^2\, d\theta = 3\pi R^2$

(c)

$$L = \int_0^{2\pi}\sqrt{[R\,(1-\cos\theta)]^2 + [R\sin\theta]^2}\, d\theta$$

$$= R\int_0^{2\pi}\sqrt{2 - 2\cos\theta}\, d\theta$$

$$= R\int_0^{2\pi}\sqrt{4\sin^2\left(\frac{\theta}{2}\right)}\, d\theta = 2R\int_0^{2\pi}\sin\frac{\theta}{2}\, d\theta = 4R\left[-\cos\frac{\theta}{2}\right]_0^{2\pi} = 8R$$

3. (a) $\bar{x} = \pi R$ by symmetry

$$\bar{y}A = \int_0^{2\pi} \frac{1}{2}[y(\theta)]^2 x'(\theta)\,d\theta$$

$$= \int_0^{2\pi} \frac{1}{2}R^2(1-\cos\theta)^2[R(1-\cos\theta)]\,d\theta$$

$$= \frac{1}{2}R^3 \int_0^{2\pi} \left(1 - 3\cos\theta + 3\cos^2\theta - \cos^3\theta\right)\,d\theta = \frac{5}{2}\pi R^3$$

$$A = 3\pi R^2 \quad \text{(by Exercise 2b)} \qquad \bar{y} = \left(\tfrac{5}{2}\pi R^3\right)/\left(3\pi R^2\right) = \tfrac{5}{6}R$$

(b) $V_x = 2\pi\bar{y}A = 2\pi\left(\dfrac{5}{6}R\right)\left(3\pi R^2\right) = 5\pi^2 R^3$

(c) $V_y = 2\pi\bar{x}A = 2\pi(\pi R)(3\pi R^2) = 6\pi^3 R^3$

4.

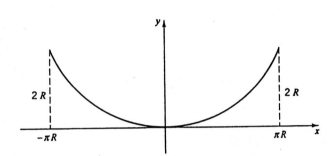

5. (a) $\dfrac{dy}{dx} = \dfrac{y'(\phi)}{x'(\phi)} = \dfrac{R\sin\phi}{R\,(1+\cos\phi)} = \dfrac{2\sin\frac{1}{2}\phi\cos\frac{1}{2}\phi}{2\cos^2\frac{1}{2}\phi} = \tan\dfrac{1}{2}\phi; \quad \alpha = \dfrac{1}{2}\phi$

(b)

$$s = \int_0^\phi \sqrt{[x'(t)]^2 + [y'(t)]^2}\,dt = \int_0^\phi \sqrt{[R(1+\cos t)]^2 + [R\sin t]^2}\,dt$$

$$= R \int_0^\phi \sqrt{2 + 2\cos t}\,dt = R \int_0^\phi 2\cos\frac{1}{2}t\,dt = 4R\left[\sin\frac{1}{2}t\right]_0^\phi$$

$$= 4R\sin\frac{1}{2}\phi = 4R\sin\alpha$$

Now note that the tangent at $(x(\phi), y(\phi))$ has slope

$$m = \frac{y'(\phi)}{x'(\phi)} = \frac{\sin\phi}{1+\cos\phi} = \frac{\sin\phi(1-\cos\phi)}{1-\cos^2\phi} = \frac{1-\cos\phi}{\sin\phi} = \frac{2\sin^2\frac{\phi}{2}}{2\sin\frac{\phi}{2}\cos\frac{\phi}{2}} = \tan\frac{\phi}{2}$$

So the inclination of the tangent is $\dfrac{\phi}{2} = \alpha$

6. (a) Already shown more generally in Example 6 of Section 9.8.

(b) Combining $d^2s/dt^2 = -g\sin\alpha$ with $s = 4R\sin\alpha$, we have

$$\frac{d^2s}{dt^2} = -\frac{g}{4R}s.$$

This is simple harmonic motion with angular frequency $\omega = \frac{1}{2}\sqrt{g/R}$ (see Section 18.6) and period $T = 2\pi/\omega = 4\pi\sqrt{R/g}$.

7. $\dfrac{d^2s}{dt^2} = -g\sin\alpha = -g \cdot \dfrac{2R}{\sqrt{\pi^2 R^2 + 4R^2}} = -\dfrac{2g}{\sqrt{\pi^2 + 4}}$

Since $\dfrac{ds}{dt} = 0$ and $s = R\sqrt{\pi^2 + 4}$ when $t = 0$, integrating twice gives

$$s = -\dfrac{g}{\sqrt{\pi^2 + 4}}\, t^2 + R\sqrt{\pi^2 + 4}$$

$$s = 0 \ \text{ at } \ t^2 = \dfrac{1}{g}R(\pi^2 + 4), \quad \text{so } \ t = \sqrt{R(\pi^2 + 4)/g}$$

REVIEW EXERCISES

1. The equation can be written $x^2 = 4(y+1)$, a parabola.

vertex $(0, -1)$

focus $(0, 0)$

axis y-axis

directrix $y = -2$

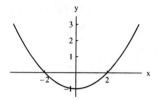

2. The equation can be written $\dfrac{y^2}{4} - \dfrac{(x-5)^2}{16} = 1$, a hyperbola.

center $(5, 0)$

foci $(5, \pm 2\sqrt{5})$

vertices $(5, \pm 2)$

asymptotes $x = 5 \pm 2y$

length of the transverse axis: 4.

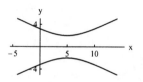

3. The equation can be written $\dfrac{(x+3)^2}{1} + \dfrac{y^2}{\frac{1}{4}} = 1$, an ellipse.

center $(-3, 0)$

foci $\left(-3 \pm \dfrac{3\sqrt{3}}{2}, 0\right)$

length of the major axis: 2

length of the minor axis: 1

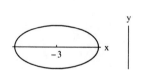

4. The equation can be written $\dfrac{(y-1)^2}{9} + \dfrac{(x-1)^2}{4} = 1$, an ellipse.

center $(1, 1)$

foci $(1, 1 \pm \sqrt{5})$

length of the major axis: 6

length of the minor axis: 4

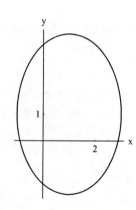

5. The equation can be written $\frac{(y+1)^2}{4} - \frac{(x-1)^2}{9} = 1$, a hyperbola.

center $(1, -1)$

foci $(1, -1 \pm \sqrt{13})$

vertices $(1, 1)$, $(1, -3)$

asymptotes: $y = -1 \pm \frac{2}{3}(x - 1)$

length of the transverse axis: 4

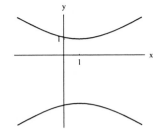

6. The equation can be written $(x - 5)^2 = 8(y - 2)$, a parabola.

vertex $(5, 2)$

focus $(5, 4)$

directrix: $y = 0$

axis: $x = 5$

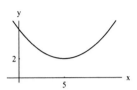

7. The equation can be written $\frac{(x-1)^2}{25} + \frac{(y+2)^2}{9} = 1$, an ellipse.

center $(1, -2)$

foci $(5, -2)$, $(-3, -2)$

length of the major axis: 10

length of the minor axis: 6

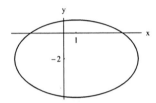

8. The equation can be written $\frac{(x+\sqrt{3})^2}{3} - \frac{(y+\sqrt{3})^2}{2} = 1$, a hyperbola.

center $(-\sqrt{3}, -\sqrt{3})$

foci $(\pm\sqrt{5} - \sqrt{3}, -\sqrt{3})$

vertices $(-2\sqrt{3}, -\sqrt{3})$ and $(0, -\sqrt{3})$

asymptotes: $y = -\sqrt{3} \pm \frac{\sqrt{2}}{\sqrt{3}}(x + \sqrt{3})$

length of the transverse axis: $2\sqrt{3}$

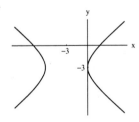

9. $x = -4\cos\frac{\pi}{3} = -2$, $y = -4\sin\frac{\pi}{3} = -2\sqrt{3}$; $(-2, -2\sqrt{3})$

10. $x = \sqrt{2}\cos\left(-\frac{5\pi}{4}\right) = -1$, $y = \sqrt{2}\sin\left(-\frac{5\pi}{4}\right) = 1$; $(-1, 1)$.

11. $r = 4$, $\theta = 3\pi/2$; $\left[4, \frac{3}{2}\pi + 2n\pi\right]$, $n = 0, \pm1, \pm2, \cdots$; $\left[-4, \frac{\pi}{2} + 2n\pi\right]$, $n = 0, \pm1, \pm2, \cdots$

12. $r = \sqrt{(1/4) + (1/4)} = \frac{\sqrt{2}}{2}$, $\theta = \frac{5\pi}{4}$; $\left[\frac{\sqrt{2}}{2}, \frac{5}{4}\pi + 2n\pi\right]$, $n = 0, \pm1, \pm2, \cdots$; $\left[-\frac{\sqrt{2}}{2}, \frac{1}{4}\pi + 2n\pi\right]$, $n = 0, \pm1, \pm2, \cdots$

13. $r = \sqrt{16} = 4$, $\theta = -\frac{\pi}{6}$; $\left[4, -\frac{\pi}{6} + 2n\pi\right]$, $n = 0, \pm1, \pm2, \cdots$; $\left[-4, \frac{5}{6}\pi + 2n\pi\right]$, $n = 0, \pm1, \pm2, \cdots$

14. $r = 2$, $\theta = \frac{2}{3}\pi$; $\left[2, \frac{2}{3}\pi + 2n\pi\right]$, $n = 0, \pm1, \pm2, \cdots$; $\left[-2, \frac{5}{3}\pi + 2n\pi\right]$, $n = 0, \pm1, \pm2, \cdots$

15. Set $y = r\sin\theta$ and $x = r\cos\theta$: $r\sin\theta = r^2\cos^2\theta \implies r = \sec\theta\tan\theta$.

16. Set $y = r \sin \theta$ and $x = r \cos \theta$: $\quad r^2 \cos^2 \theta + r^2 \sin^2 \theta - 2r \cos \theta = 0 \implies r = 2 \cos \theta$

17. Set $y = r \sin \theta$ and $x = r \cos \theta$:

$$r^2 \cos^2 \theta + r^2 \sin^2 \theta - 4r \cos \theta + 2r \sin \theta \implies r = 4 \cos \theta - 2 \sin \theta.$$

18. Set $y = r \sin \theta$ and $x = r \cos \theta$:

$$r^2 \sin^2 \theta (1 - r^2 \cos^2 \theta) = r^4 \cos^4 \theta$$

$$\sin^2 \theta - r^2 \cos^2 \theta \sin^2 \theta = r^2 \cos^4 \theta$$

$$\sin^2 \theta = r^2 \cos^2 \theta$$

$$r = \pm \tan \theta$$

19. $x = 5$

20. $r + 4 \cos \theta = 0$, $r^2 + 4r \cos \theta = 0 \implies x^2 + y^2 + 4x = 0$

21. Multiplication by r gives $r^2 = 3r \cos \theta + 4r \sin \theta \implies x^2 + y^2 - 3x - 4y = 0$.

22. The equation can be written $r^2 \cos 2\theta = 1$ Since $\cos 2\theta = \cos^2 \theta - \sin^2 \theta$, we have $x^2 - y^2 = 1$.

23.

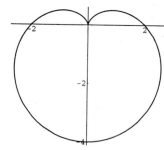

24.

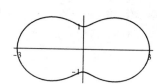

25.

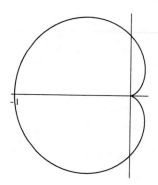

26.

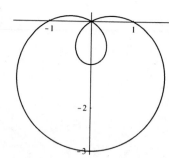

27.

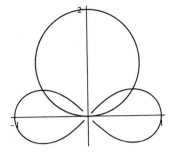

intersections: $[\frac{\sqrt{2}}{2}, \frac{1}{6}\pi], [\frac{\sqrt{2}}{2}, \frac{5}{6}\pi]$, the pole

28.

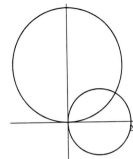

intersections: $[\sqrt{3}, \frac{\pi}{6}]$, the pole

29.

$$A = \int_0^{2\pi} \frac{1}{2}(2(1 - \cos\theta))^2 \, d\theta = 2\int_0^{2\pi} (1 - 2\cos\theta + \cos^2\theta) \, d\theta$$

$$= 2\int_0^{2\pi} \left(1 - 2\cos\theta + \frac{1 + \cos 2\theta}{2}\right) \theta$$

$$= 2\left[\frac{3}{2}\theta - 2\sin\theta + \frac{1}{4}\sin 2\theta\right]_0^{2\pi} = 6\pi$$

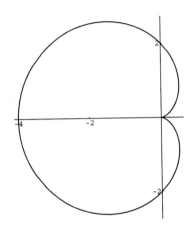

30. $A = 2\int_0^{\frac{\pi}{2}} \frac{1}{2}(4\sin 2\theta) \, d\theta$

$$= 4\left[-\frac{1}{2}\cos 2\theta\right]_0^{\pi/2} = 4$$

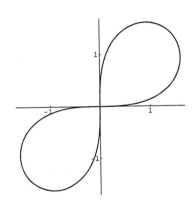

31.   One petal is formed when θ ranges from $-\pi/8$ to $\pi/8$. By symmetry, the area of one petal is

$$A = 2\int_0^{\pi/8} \frac{1}{2}(4\cos^2 4\theta) \, d\theta = 4\int_0^{\pi/8} \frac{1 + \cos 8\theta}{2} \, d\theta == 4\left[\frac{1}{2}\theta + \frac{1}{16}\sin 8\theta\right]_0^{\pi/8} = \frac{\pi}{4}$$

32. $A = \int_0^{\pi/2} \frac{1}{2}[\sin^2\theta - (1 - \cos\theta)^2] \, d\theta = \frac{1}{2}\int_0^{\pi/2} (2\cos\theta - 1 - \cos 2\theta) \, d\theta$

$$= \frac{1}{2}\left[2\sin\theta - \theta - \frac{1}{2}\sin 2\theta\right]_0^{\pi/2} = 1 - \frac{\pi}{4}$$

33. $\displaystyle\int_0^{\frac{\pi}{4}} \frac{1}{2}(2\sin\theta)^2 d\theta + \int_{\frac{\pi}{4}}^{\frac{3}{4}\pi} \frac{1}{2}(\sin\theta + \cos\theta)^2 d\theta = \int_0^{\frac{\pi}{4}}(1-\cos 2\theta)d\theta + \int_{\frac{\pi}{4}}^{\frac{3}{4}\pi} \frac{1}{2}(1+\sin 2\theta)d\theta$

$$= \frac{\pi}{4} - \frac{1}{2} + \frac{\pi}{4} = \frac{\pi}{2} - \frac{1}{2}$$

34. Since $x^2 = \dfrac{1}{t^2}$, we have

$$y = t^2 + 1 = \frac{1}{x^2} + 1 = \frac{x^2 + 1}{x^2}.$$

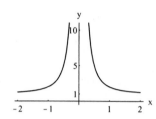

35. $y = \cos 2t = 1 - 2\sin^2\theta$

$y = 1 - 2x^2, \quad 0 \le x \le 1.$

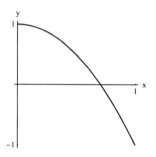

36. $\cosh^2 t - \sinh^2 = 1 \implies x^2 - y^2 = 1$, graph: the right branch of the hyperbola since $x > 0$.

37. $y = x^2 - 2x + 1 = (x-1)^2, \quad x \ge 1$, graph: the right half of the parabola.

38. Since $y = t^2 + 4t + 8 = (t+2)^2 + 4$,

$y = x^2 + 4, \quad x \le 2.$

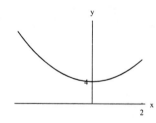

39. $x(t) = 1 + (5-1)t = 1 + 4t, \quad y(t) = 4 + (6-4)t = 4 + 2t, \quad t \in [0,1].$

40. $x(t) = 2 + (-2-2)t = 2 - 4t, \quad y(t) = -1 + (3 - [-1])t = -1 + 4t, \quad t \in [0,1].$

41. $x(t) = 3\cos(-t + \frac{\pi}{2}) = 3\sin t, \quad y(t) = 2\sin(-t + \frac{\pi}{2}) = 2\cos t.$

42. $x(t) = 3t - 1, \ x'(t) = 3; \quad y(t) = 9t^2 - 3t, \ y'(t) = 18t - 3.$

At $t = 1$: $x(1) = 2, \ y(1) = 6, \ x'(1) = 3, \ y'(1) = 15;$ tangent line: $y - 6 = 5(x - 2)$ or $y = 5x - 4.$

43. $x(t) = 3e^t, \ x'(t) = 3e^t; \quad y(t) = 5e^{-t}, \ y'(t) = -5e^{-t}.$

At $t = 0$: $x(0) = 3, \ y(0) = 5, \ x'(0) = 3, \ y'(0) = -5;$ tangent line: $y - 5 = -\frac{5}{3}(x - 3)$ or

$5x + 3y = 30.$

44. $x(\theta) = r\cos\theta = 2\sin 2\theta\,\cos\theta,\quad y(\theta) = r\sin\theta = 2\sin 2\theta\,\sin\theta;$

$x'(\theta) = 4\cos 2\theta\,\cos\theta - 2\sin 2\theta\,\sin\theta,\quad y'(\theta) = 4\cos 2\theta\,\sin\theta + 2\sin 2\theta\,\cos\theta;$

at $\theta = \pi/4:\quad x(\pi/4) = \sqrt{2},\ x'(\pi/4) = -\sqrt{2};\quad y(\pi/4) = \sqrt{2},\ y'(\pi/4) = \sqrt{2}$

tangent line: $y - \sqrt{2} = \dfrac{\sqrt{2}}{-\sqrt{2}}(x - \sqrt{2}) = -(x - \sqrt{2})\ $ or $\ x + y = 2\sqrt{2}.$

45. (a) For a horizontal tangent, set $y'(t) = 0:\ 3t^2 + 3t - 6 = 0 \implies t = -2,\ 1.$

The curve has a horizontal tangent at: $(x(-2), y(-2)) = (0, 10)$ and $(x(1), y(1)) = (3, -7/2).$

(b) For a vertical tangent, set $x'(t) = 0$ (provided $y'(t) \neq 0$ simultaneously):

$x'(t) = 2t + 2 \implies t = -1.$

Since $y'(-1) \neq 0$, the curve has a vertical tangent at $(x(-1), y(-1)) = (-1, 13/2).$

46. (a) For a horizontal tangent, set $y'(t) = 0:\ 2\cos t - 2\cos 2t = (2\cos t + 1)(\cos t - 1) = 0.$

The curve has a horizontal tangent at the points where $t = \frac{2\pi}{3} + 2n\pi,\ \frac{-2\pi}{3} + 2n\pi,\ 2n\pi,$

$n = 0, \pm 1, \pm 2, \cdots.$

(b) for a vertical tangent, set $x'(t) = 0:\ -2\sin t + 2\sin 2t = -2\sin t(1 - 2\cos t) = 0.$

the curve has a vertical tangent at the points where $t = \pm\frac{\pi}{3} + 2n\pi,\ n\pi,\ n = 0, \pm 1, \pm 2, \cdots.$

47. $x = 3t^2,\ x' = 6t,\ x'' = 6;\quad y = 4t^3,\ y' = 12t^2,\ y'' = 24t$

$$\frac{dy}{dx} = \frac{12t^2}{6t} = 2t;\quad \frac{d^2y}{dx^2} = \frac{6t(24t) - 12t^2(6)}{(6t)^3} = \frac{1}{3t}$$

48. Set $\dfrac{dx}{dt} = a,$ and $\dfrac{d^2y}{dt^2} = b.$ Then $x(t) = at + C_1$ and $y(t) = \frac{1}{2}bt^2 + C_2t + C_3$ where $C_1,\ C_2,\ C_3$ are arbitrary constants. In rectangular coordinates, the equation will have the form

$y = Ax^2 + Bx + C,$ a parabola

49. $L = \displaystyle\int_0^{5/9} \sqrt{1 + (y'(x))^2}\,dx = \int_0^{5/9} \sqrt{1 + \frac{9}{4}x}\,dx = \frac{19}{27}.$

50. $L = \displaystyle\int_0^{1/2} \sqrt{1 + [y'(x)]^2}\,dx = \int_0^{1/2} \sqrt{1 + \frac{4x^2}{(1 - x^2)^2}}\,dx = \int_0^{1/2} \frac{1 + x^2}{1 - x^2}\,dx$

$= \displaystyle\int_0^{1/2} \left[\frac{2}{1 - x^2} - 1\right] dx = \left[\ln\frac{(1 + x)}{(1 - x)} - x\right]_0^{1/2} = \ln 3 - \frac{1}{2}$

51. $L = \displaystyle\int_0^{\frac{\pi}{2}} \sqrt{\sin^2 t + 4\sin^2 t\,\cos^2 t}\,dt = \int_0^{\frac{\pi}{2}} 2\sin t\sqrt{\frac{1}{4} + \cos^2 t}\,dt$

$= 2\displaystyle\int_0^1 \sqrt{\frac{1}{4} + u^2}\,du$ (substitution $u = \cos t$)

$= 2\left[\dfrac{u}{2}\sqrt{\dfrac{1}{4} + u^2} + \dfrac{1}{8}\ln\left(u + \sqrt{\dfrac{1}{4} + u^2}\right)\right]_0^1 = \dfrac{\sqrt{5}}{2} + \dfrac{1}{4}\ln(\sqrt{5} + 2)$

52. $L = \int_0^4 \sqrt{t^2 + 6t + 9}\, dt = \int_0^4 (t + 3)\, dt = \left[\frac{1}{2}t^2 + 3t\right]_0^4 = 20.$

53. By symmetry $L = \int_0^{2\pi} \sqrt{(1 - \sin\theta)^2 + \cos^2\theta}\, d\theta = 2\int_0^\pi \sqrt{(1 - \sin\theta)^2 + \cos^2\theta}\, d\theta.$

$$L = \int_0^{2\pi} \sqrt{(1 - \sin\theta)^2 + \cos^2\theta}\, d\theta = 2\int_0^\pi \sqrt{2(1 - \sin\theta)}\, d\theta$$

$$= 2\int_0^\pi \sqrt{4\cos^2(\theta/2)}\, d\theta = 4\int_0^\pi \cos(\theta/2)\, d\theta = 8\Big[\sin(\theta/2)\Big]_0^\pi = 8$$

54. $L = \int_0^{\sqrt{5}} \sqrt{\theta^4 + 4\theta^2}\, d\theta = \int_0^{\sqrt{5}} \theta\sqrt{\theta^2 + 4}\, d\theta = \frac{1}{3}\Big[(\theta^2 + 4)^{3/2}\Big]_0^{\sqrt{5}} = \frac{19}{3}$

55. $\int_0^{24} 2\pi(2\sqrt{x})\sqrt{1 + \frac{1}{x}}\, dx = 4\pi \int_0^{24} \sqrt{1 + x}\, dx = \frac{8\pi}{3}\Big[(1 + x)^{3/2}\Big]_0^{24} = \frac{992\pi}{3}$

56. Let $u = \sqrt{t + 1}$. Then $\int_3^8 2\pi t\sqrt{t + 1}\, dt = \int_2^3 2\pi(u^2 - 1)2u^2\, du = 4\pi\int_2^3 (u^4 - u^2)\, du = \frac{2152}{15}\pi.$

57. $y = \dfrac{x^3}{6} + \dfrac{1}{2x}$

$$\int_1^3 2\pi\left(\frac{x^3}{6} + \frac{1}{2x}\right)\sqrt{1 + \left(\frac{x^2}{2} - \frac{1}{2x^2}\right)^2}\, dx = 2\pi\int_1^3 \left(\frac{x^3}{6} + \frac{1}{2x}\right)\left(\frac{x^2}{2} + \frac{1}{2x^2}\right)\, dx$$

$$= 2\pi\int_1^3 \left(\frac{x^5}{12} + \frac{x}{4} + \frac{x}{12} + \frac{1}{4x^3}\right)\, dx = \frac{2497\pi}{108}$$

58. $3x^2 + 4y^2 = 3a^2 \implies y = \dfrac{\sqrt{3}}{2}\sqrt{a^2 - x^2}$

$$A = 2\int_0^a 2\pi \cdot \frac{\sqrt{3}}{2}\sqrt{a^2 - x^2}\sqrt{1 + \frac{3}{4}\frac{x^2}{a^2 - x^2}}\, dx = \sqrt{3}\pi\int_0^a \sqrt{4a^2 - x^2}\, dx$$

$$= \sqrt{3}\pi\left[\frac{1}{2}x\sqrt{4a^2 - x^2} + 2a^2\arctan\frac{x}{\sqrt{4a^2 - x^2}}\right]_0^a = \frac{\sqrt{3}}{3}\pi^2 a^2 + \frac{3}{2}\pi a^2$$

59. Assume that $f'(x) > 0$. Then $(f^{-1})'(y) = \dfrac{1}{f'(x)}$, where $y = f(x)$, $dy = f'(x)\, dx$.
The length of the graph of f^{-1} is given by:

$$\int_{f(a)}^{f(b)} \sqrt{1 + [(f^{-1})'(y)]^2}\, dy = \int_{f(a)}^{f(b)} \sqrt{1 + 1/[f'(x)]^2}\, dy$$

$$= \int_a^b \frac{1}{f'(x)}\sqrt{(f'(x))^2 + 1}\, f'(x)\, dx$$

$$= \int_a^b \sqrt{1 + (f'(x))^2}\, dx$$

60. Solve the differential equations

$$x' = x \quad y' = 2y$$

with initial conditions $x(0) = 4$, $y(0) = 2$. We get $x(t) = 4e^t$, $y(t) = 2e^{2t}$.

Initial speed: $\nu_0 = 16\sqrt{2}$; terminal speed: $\nu_1 = \sqrt{16e^2 + 16e^4} = 4e\sqrt{1 + e^2}$.

$$s = \int_0^1 \sqrt{16e^{2t} + 16e^{4t}}\, dt = 4\int_0^1 e^t\sqrt{1 + e^{2t}}\, dt$$

$$= 2\left[e\sqrt{1 + e^2} + \ln\left(e + \sqrt{1 + e^2}\right) - 1 - \ln\left(1 + \sqrt{2}\right)\right]$$

61. If $\dfrac{x}{a} = \cos^3\theta$ and $\dfrac{y}{a} = \sin^3\theta$, then

$$\left(\frac{x}{a}\right)^{2/3} + \left(\frac{y}{a}\right)^{2/3} = \cos^2\theta + \sin^2\theta = 1 \quad \text{and} \quad x^{2/3} + y^{2/3} = a^2/3.$$

62. Assume $a \geq 0$. By symmetry,

$$L = 4\int_0^{\pi/2} \sqrt{(3a\cos^2\theta\sin\theta)^2 + (3a\sin^2\theta\cos\theta)^2}\, d\theta$$

$$= 4\int_0^{\pi/2} 3a\cos\theta\sin\theta\, d\theta = 12a\left[\frac{1}{2}\sin^2\theta\right]_0^{\pi/2} = 6a$$

63. By symmetry, $\bar{x} = \bar{y}$.

$$\bar{x}\frac{L}{4} = \int_0^{\pi/2} a\cos^3\theta(3a\cos\theta\sin\theta)\, d\theta = -\frac{3a^2}{5}\left[\cos^5\theta\right]_0^{\frac{\pi}{2}} = \frac{3a^2}{5}$$

Therefore, $(\bar{x}, \bar{y}) = \left(\dfrac{2a}{5}, \dfrac{2a}{5}\right)$

64. (a) The speed μ satisfies

$$|\mu|^2 = (x'(t))^2 + (y'(t))^2$$

$$= (4 - \pi\cos\pi t)^2 + (4 - \pi\sin\pi t)^2 = 32 + \pi - 8\pi(\cos\pi t + \sin\pi t)$$

$$= 32 + \pi - 8\pi\sqrt{2}\cos(\pi t - \frac{\pi}{4})$$

The particle has minimum speed $\sqrt{32 + \pi - 8\sqrt{2}\pi}$ at $t = 1/4$;

the particle has maximum speed $\sqrt{32 + 9\pi}$ at $t = 1$.

(b) $\dfrac{y'(1/4)}{x'(1/4)} = 1$.

CHAPTER 11

SECTION 11.1

1. lub $= 2$; glb $= 0$
2. lub $= 2$; glb $= 0$

3. no lub; glb $= 0$
4. lub $= 1$, no glb

5. lub $= 2$; glb $= -2$
6. lub $= 3$; glb $= -1$

7. no lub; glb $= 2$
8. lub $= 2$; glb $= -2$

9. lub $= 2\frac{1}{2}$; glb $= 2$
10. lub $= 0$; glb $= -1$

11. lub $= 1$; glb $= 0.9$
12. lub $= 2\frac{1}{9}$, glb $= -2\frac{1}{9}$

13. lub $= e$; glb $= 0$
14. no lub, glb $= 1$

15. lub $= \frac{1}{2}(-1 + \sqrt{5})$; glb $= \frac{1}{2}(-1 - \sqrt{5})$
16. no lub, no glb

17. no lub; no glb
18. no lub; no glb

19. no lub; no glb
20. lub $= 0$; no glb

21. glb $S = 0$, $0 \le \left(\frac{1}{11}\right)^3 < 0 + 0.001$
22. glb $S = 1$; $1 \le 1 < 1 + 0.0001$

23. glb $S = 0$, $0 \le \left(\frac{1}{10}\right)^{2n-1} < 0 + \left(\frac{1}{10}\right)^k$ $\left(n > \frac{k+1}{2}\right)$

24. glb $= 0$; $0 \le \left(\frac{1}{2}\right)^n < 0 + \left(\frac{1}{4}\right)^k$ for $n > 2k$

25. Let $\epsilon > 0$. The condition $m \le s$ is satisfied by all numbers s in S. All we have to show therefore is that there is some number s in S such that

$$s < m + \epsilon.$$

Suppose on the contrary that there is no such number in S. We then have

$$m + \epsilon \le x \quad \text{for all} \quad x \in S.$$

This makes $m + \epsilon$ a lower bound for S. But this cannot be, for then $m + \epsilon$ is a lower bound for S that is *greater* than m, and by assumption, m is the *greatest* lower bound.

26. (a) Let $M = |a_1| + \cdots + |a_n|$. Then for any i, $|a_i| < M$, so S is bounded

(b) lub $S = \max \{a_1, a_2, \ldots, a_n\} \in S$

   glb $S = \min \{a_1, a_2, \ldots, a_n\} \in S$

27. Let $c = \text{lub } S$. Since $b \in S$, $b \le c$. Since b is an upper bound for S, $c \le b$. Thus, $b = c$.

28. S consists of a single element, equal to lub S.

29. (a) Suppose that K is an upper bound for S and k is a lower bound. Let t be any element of T. Then $t \in S$ which implies that $k \le t \le K$. Thus K is an upper bound for T and k is a lower bound, and T is bounded.

    (b) Let $a = \text{glb } S$. Then $a \le t$ for all $t \in T$. Therefore, $a \le \text{glb } T$. Similarly, if $b = \text{lub } S$, then $t \le b$ for all $t \in T$, so lub $T \le b$. It now follows that glb $S \le \text{glb } T \le \text{lub } T \le \text{lub } S$.

30. (a) Let $S = \{r : r < \sqrt{2}, \ r \text{ rational}\}$.

    (b) Let $T = \{t : t < 2, \ t \text{ irrational}\}$.

31. Let c be a positive number and let $S = \{c, 2c, 3c, \ldots\}$. Choose any positive number M and consider the positive number M/c. Since the set of positive integers is not bounded above, there exists a positive integer k such that $k \ge M/c$. This implies that $kc \ge M$. Since $kc \in S$, it follows that S is not bounded above.

32. (a) If S is a set of negative numbers, then 0 is an upper bound for S. It follows that $\alpha = \text{lub } S \le 0$.

    (b) If T is a set of positive numbers, then 0 is a lower bound for T. It follows that $\beta = \text{glb } T \ge 0$.

33. (a) See Exercise 75 in Section 1.2.

    (b) Suppose $x_0^2 > 2$. Choose a positive integer n such that

$$\frac{2x_0}{n} - \frac{1}{n^2} < x_0^2 - 2.$$

Then,

$$\frac{2x_0}{n} - \frac{1}{n^2} < x_0^2 - 2 \implies 2 < x_0^2 - \frac{2x_0}{n} + \frac{1}{n^2} = \left(x_0 - \frac{1}{n}\right)^2$$

    (c) If $x_0^2 < 2$, then choose a positive integer n such that

$$\frac{2x_0}{n} + \frac{1}{n^2} < 2 - x_0^2.$$

Then

$$x_0^2 + \frac{2x_0}{n} + \frac{1}{n^2} < 2 \implies \left(x_0 + \frac{1}{n}\right)^2 < 2$$

34. Assume that there are only finitely many primes, $p_1, p_2, \cdots, p_n$ and let $Q = p_1 \cdot p_2 \cdots p_n + 1$. Q has a prime divisor p. But Q is not divisible by any of the p_i, so $p \ne p_1$ for all i a contradiction.

35. (a) $n = 5 : 2.48832$; $n = 10 : 2.59374$; $n = 100 : 2.70481$; $n = 1000 : 2.71692$; $n = 10,000 : 2.71815$

    (b) lub $= e$; glb $= 2$

36.

a_2	a_3	a_4	a_5	a_6	a_7	a_8	a_9	a_{10}
$\frac{9}{4}$	$\frac{5}{3}$	$\frac{6}{5}$	$\frac{1}{2}$	-3	4	$\frac{9}{4}$	$\frac{5}{3}$	$\frac{6}{5}$

(b) $a_{20} = \frac{9}{4}$, $a_{30} = -3$, $a_{40} = \frac{6}{5}$, $a_{50} = \frac{9}{4}$

(c) lub $S = 4$, glb $S = -3$

37. (a)

a_1	a_2	a_3	a_4	a_5
1.4142	1.6818	1.8340	1.9152	1.9571

a_6	a_7	a_8	a_9	a_{10}
1.9785	1.9892	1.9946	1.9973	1.9986

(b) Let S be the set of positive integers for which $a_n < 2$. Then $1 \in S$ since

$$a_1 = \sqrt{2} \cong 1.4142 < 2.$$

Assume that $k \in S$. Since $a_{k+1}^2 = 2a_k < 4$, it follows that $a_{k+1} < 2$. Thus $k+1 \in S$ and S is the set of positive integers.

(c) 2 is the least upper bound.

(d) Let c be a positive number. Then c is the least upper bound of the set

$$S = \left\{ \sqrt{c},\ \sqrt{c\sqrt{c}},\ \sqrt{c\sqrt{c\sqrt{c}}}, \ldots \right\}.$$

38. (a) $a_1 \cong 1.4142136$ $a_2 \cong 1.8477591$ $a_3 \cong 1.9615706$ $a_4 \cong 1.9903695$ $a_5 \cong 1.9975909$

$a_6 \cong 1.9993976$ $a_7 \cong 1.9998494$ $a_8 \cong 1.9999624$ $a_9 \cong 1.9999906$ $a_{10} \cong 1.9999976$

(b) $a_1 = \sqrt{2} < 2$.   Assume true for a_n. Then $a_{n+1} = \sqrt{2 + a_n} < \sqrt{2 + 2} = 2$.

(c) lub $S = 2$.

(d) For any positive number c, lub S is the positive number satisfying

$$x = \sqrt{c + x}, \quad \text{that is,} \quad x = (1 + \sqrt{1 + 4c})/2$$

SECTION 11.2

1. $a_n = 2 + 3(n-1) = 3n - 1$, $n = 1, 2, 3, \ldots$ **2.** $a_n = 1 - (-1)^n$, $n = 1, 2, 3, \ldots$

3. $a_n = \dfrac{(-1)^{n-1}}{2n - 1}$, $n = 1, 2, 3, \ldots$ **4.** $a_n = \dfrac{2^n - 1}{2^n}$, $n = 1, 2, 3, \ldots$

5. $a_n = \dfrac{n^2 + 1}{n}$, $n = 1, 2, 3, \ldots$ **6.** $a_n = (-1)^n \dfrac{n}{(n+1)^2}$, $n = 1, 2, 3, \ldots$

7. $a_n = \begin{cases} n & \text{if } n = 2k - 1 \\ 1/n, & \text{if } n = 2k, \end{cases}$ where $k = 1, 2, 3, \ldots$

8. $a_n = \begin{cases} n & \text{if } n = 2k \\ 1/n^2, & \text{if } n = 2k - 1, \end{cases}$ where $k = 1, 2, 3, \ldots$

9. decreasing;   bounded below by 0 and above by 2

10. not monotonic; bounded below by -1 and above by $\frac{1}{2}$.

11. $\dfrac{n + (-1)^n}{n} = 1 + (-1)^n \dfrac{1}{n}$: not monotonic;   bounded below by 0 and above by $\dfrac{3}{2}$

12. increasing; bounded below by 1.001 but not bounded above.

13. decreasing; bounded below by 0 and above by 0.9

14. increasing; bounded below by 0 and above by 1

15. $\dfrac{n^2}{n + 1} = n - 1 + \dfrac{1}{n + 1}$: increasing;   bounded below by $\dfrac{1}{2}$ but not bounded above

16. increasing; bounded below by $\sqrt{2}$, but not bounded above: $\sqrt{n^2 + 1} > n$

17. $\dfrac{4n}{\sqrt{4n^2 + 1}} = \dfrac{2}{\sqrt{1 + 1/4n^2}}$ and $\dfrac{1}{4n^2}$ decreases to 0: increasing;

    bounded below by $\frac{4}{5}\sqrt{5}$ and above by 2

18. $\dfrac{2^n}{4^n + 1} = \dfrac{2^n}{(2^n)^2 + 1} = \dfrac{1}{2^n + \frac{1}{2^n}}$.   decreasing; bounded below by 0 and above by $\frac{2}{5}$.

19. increasing;   bounded below by $\frac{2}{51}$ but not bounded above

20. $\dfrac{n^2}{\sqrt{n^3 + 1}} = \dfrac{1}{\sqrt{\frac{1}{n} + \frac{1}{n^4}}}$ and $\dfrac{1}{n} + \dfrac{1}{n^4}$ decreases to 0 $\implies$ increasing; bounded below by $\dfrac{\sqrt{2}}{2}$,

    but not bounded above.

21. $\dfrac{2n}{n + 1} = 2 - \dfrac{2}{n + 1}$ increases toward 2: increasing;   bounded below by 0 and above by $\ln 2$.

22. increasing $(n \geq 2)$;   bounded below by $\dfrac{2\sqrt{2}}{3^{10}}$, but not bounded above.

23. decreasing;   bounded below by 1 and above by 4

24. not monotonic; not bounded below and not bounded above

25. increasing;   bounded below by $\sqrt{3}$ and above by 2

26. decreasing (since $\dfrac{n + 1}{n}$ decreases to 1); bounded below by 0 and above by $\ln 2$.

27. $(-1)^{2n+1}\sqrt{n} = -\sqrt{n}$: decreasing; bounded above by -1 but not bounded below

28. $\dfrac{\sqrt{n+1}}{\sqrt{n}} = \sqrt{1 + \dfrac{1}{n}}$ decreasing; bounded below by 1 and above by $\sqrt{2}$

29. $\dfrac{2^n - 1}{2^n} = 1 - \dfrac{1}{2^n}$: increasing; bounded below by $\dfrac{1}{2}$ and above by 1

30. $\dfrac{1}{2n} - \dfrac{1}{2n + 3} = \dfrac{3}{2n(2n + 3)}$ decreasing; bounded below by 0 and above by $\frac{3}{10}$.

31. consider $\sin x$ as $x \to 0^+$: decreasing; bounded below by 0 and above by 1

32. not monotonic; bounded below by $-\frac{1}{2}$ and above by $\frac{1}{4}$

33. decreasing; bounded below by 0 and above by $\frac{5}{6}$

34. increasing (because $\dfrac{x}{\ln x}$ is an increasing function on $[4, \infty)$.); bounded below by $\dfrac{4}{\ln 4}$ but not bounded above.

35. $\dfrac{1}{n} - \dfrac{1}{n+1} = \dfrac{1}{n(n+1)}$: decreasing; bounded below by 0 and above by $\dfrac{1}{2}$

36. not monotonic; bounded below by -1 and above by 1

37. Set $f(x) = \dfrac{\ln x}{x}$. Then, $f'(x) = \dfrac{1 - \ln x}{x^2} < 0$ for $x > e$: decreasing;

bounded below by 0 and above by $\frac{1}{3}\ln 3$.

38. not monotonic; not bounded below nor above (because exponentials grow faster than polynomials).

39. Set $a_n = \dfrac{3^n}{(n+1)^2}$. Then, $\dfrac{a_{n+1}}{a_n} = 3\left(\dfrac{n+1}{n+2}\right)^2 > 1$: increasing;

bounded below by $\frac{3}{4}$ but not bounded above.

40. $\dfrac{1 - (\frac{1}{2})^n}{(\frac{1}{2})^n} = 2^n - 1$ increasing; bounded below by 1 but not bounded above.

41. For $n \geq 5$

$$\frac{a_{n+1}}{a_n} = \frac{5^{n+1}}{(n+1)!} \cdot \frac{n!}{5^n} = \frac{5}{n+1} < 1 \quad \text{and thus} \quad a_{n+1} < a_n.$$

Sequence is not nonincreasing: $a_1 = 5 < \frac{25}{2} = a_2$.

42. For $n \geq M$, $\dfrac{a_{n+1}}{a_n} = \dfrac{M^{n+1}}{(n+1)!} \cdot \dfrac{n!}{M^n} = \dfrac{M}{n+1} < 1$, so the sequence decreases for $n \geq M$.

43. boundedness: $0 < (c^n + d^n)^{1/n} < (2d^n)^{1/n} = 2^{1/n}d \le 2d$

monotonicity : $a_{n+1}^{n+1} = c^{n+1} + d^{n+1} = cc^n + dd^n$
$$< (c^n + d^n)^{1/n}c^n + (c^n + d^n)^{1/n}d^n$$
$$= (c^n + d^n)^{1+1/n}$$
$$= (c^n + d^n)^{(n+1)/n}$$
$$= a_n^{n+1}$$

Taking the $(n+1)$st root of each side we have $a_{n+1} < a_n$.   The sequence is decreasing.

44. If for all n

$$|a_n| \le M \quad \text{and} \quad |b_n| \le N,$$

then for all n

$$|\alpha a_n + \beta b_n| \le |\alpha||a_n| + |\beta||b_n| \le |\alpha|M + |\beta|N$$

and

$$|a_n b_n| = |a_n||b_n| \le MN$$

45. $a_1 = 1, \quad a_2 = \frac{1}{2}, \quad a_3 = \frac{1}{6}, \quad a_4 = \frac{1}{24}, \quad a_5 = \frac{1}{120}, \quad a_6 = \frac{1}{720}; \quad a_n = \frac{1}{n!}$

46. $a_1 = 1, \quad a_2 = 8, \quad a_3 = 27, \quad a_4 = 64, \quad a_5 = 125, \quad a_6 = 216; \quad a_n = n^3$

47. $a_1 = a_2 = a_3 = a_4 = a_5 = a_6 = 1; \quad a_n = 1$

48. $a_1 = 1, \quad a_2 = \frac{3}{2}, \quad a_3 = \frac{7}{4}, \quad a_4 = \frac{15}{8}, \quad a_5 = \frac{31}{16}, \quad a_6 = \frac{63}{32}; \quad a_n = (2^n - 1)/2^{n-1}$

49. $a_1 = 1, \quad a_2 = 3, \quad a_3 = 5, \quad a_4 = 7, \quad a_5 = 9, \quad a_6 = 11; \quad a_n = 2n - 1$

50. $a_1 = 1, \quad a_2 = \frac{1}{2}, \quad a_3 = \frac{1}{3}, \quad a_4 = \frac{1}{4}, \quad a_5 = \frac{1}{5}, \quad a_6 = \frac{1}{6}; \quad a_n = 1/n$

51. $a_1 = 1, \quad a_2 = 4, \quad a_3 = 9, \quad a_4 = 16, \quad a_5 = 25, \quad a_6 = 36; \quad a_n = n^2$

52. $a_1 = 1, \quad a_2 = 3, \quad a_3 = 7, \quad a_4 = 15, \quad a_5 = 31, \quad a_6 = 63; \quad a_n = 2^n - 1$

53. $a_1 = 1, \quad a_2 = 1, \quad a_3 = 2, \quad a_4 = 4, \quad a_5 = 8, \quad a_6 = 16; \quad a_n = 2^{n-2} \quad (n \ge 3)$

54. $a_1 = 3, \quad a_2 = 1, \quad a_3 = 3, \quad a_4 = 1, \quad a_5 = 3, \quad a_6 = 1; \quad a_n = 2 - (-1)^n$

55. $a_1 = 1, \quad a_2 = 3, \quad a_3 = 5, \quad a_4 = 7, \quad a_5 = 9, \quad a_6 = 11; \quad a_n = 2n - 1$

56. $a_1 = 1, \quad a_2 = 3, \quad a_3 = 4, \quad a_4 = 5, \quad a_5 = 6, \quad a_6 = 7; \quad a_n = n + 1 \quad (n \ge 2)$

57. First $a_1 = 2^1 - 1 = 1$.   Next suppose $a_k = 2^k - 1$ for some $k \ge 1$.   Then
$$a_{k+1} = 2a_k + 1 = 2(2^k - 1) + 1 = 2^{k+1} - 1.$$

58. True for $n = 1$. Assume true for n. Then $a_{n+1} = a_n + 5 = 5n - 2 + 5 = 5(n + 1) - 2$.

59. First $a_1 = \dfrac{1}{2^0} = 1$. Next suppose $a_k = \dfrac{k}{2^{k-1}}$ for some $k \geq 1$. Then

$$a_{k+1} = \frac{k+1}{2k} a_k = \frac{k+1}{2k} \frac{k}{2^{k-1}} = \frac{k+1}{2^k}.$$

60. True for $n = 1$. Assume true for n. Then $a_{n+1} = a_n - \dfrac{1}{n(n+1)} = \dfrac{1}{n} - \dfrac{1}{n(n+1)} = \dfrac{n+1-1}{n(n+1)} = \dfrac{1}{n+1}$

61. (a) If $r = 1$ then $S_n = n$ for $n = 1, 2, 3, \ldots$

(b)
$$\begin{aligned}
S_n &= 1 + r + r^2 + \cdots + r^{n-1} \\
rS_n &= r + r^2 + \cdots + r^n \\
S_n - rS_n &= 1 - r^n \\
S_n &= \frac{1 - r^n}{1 - r}, \qquad r \neq 1.
\end{aligned}$$

62. $\dfrac{1}{k(k+1)} = \dfrac{1}{k} - \dfrac{1}{k+1}$, so

$$\begin{aligned}
S_n &= a_1 + a_2 + \cdots + a_{n-1} + a_n \\
&= \frac{1}{1 \cdot 2} + \frac{1}{2 \cdot 3} + \cdots + \frac{1}{(n-1)n} + \frac{1}{n(n+1)} \\
&= \left(1 - \frac{1}{2}\right) + \left(\frac{1}{2} - \frac{1}{3}\right) + \cdots + \left(\frac{1}{n-1} - \frac{1}{n}\right) + \left(\frac{1}{n} - \frac{1}{n+1}\right) \\
&= 1 - \frac{1}{n+1} = \frac{n}{n+1} \quad \text{since all middle terms cancel out.}
\end{aligned}$$

63. (a) Let S_n denote the distance traveled between the nth and $(n+1)$st bounce. Then

$$S_1 = 75 + 75 = 150, \qquad S_2 = \tfrac{3}{4}(75) + \tfrac{3}{4}(75) = 150\left(\frac{3}{4}\right), \ldots, \qquad S_n = 150\left(\frac{3}{4}\right)^{n-1}.$$

(b) An object dropped from rest from a height h feet above the ground will hit the ground in $\frac{1}{4}\sqrt{h}$ seconds. Therefore it follows that the ball will be in the air

$$T_n = 2\left(\tfrac{1}{4}\right)\sqrt{\frac{S_n}{2}} = \frac{5\sqrt{3}}{2}\left(\frac{3}{4}\right)^{(n-1)/2} \qquad \text{seconds.}$$

64. $P_0 = 5$, $P_{12} = 5 \cdot 2$, $P_{24} = 5 \cdot 2^2$; so $P_n = 5 \, 2^{n/12}$

65. $\{a_n\}$ is an increasing sequence; $a_n \to \frac{1}{2}$ as $n \to \infty$.

66. $\{a_n\}$ is a decreasing sequence; $a_n \to 2$ as $n \to \infty$.

67. (a) Let S be the set of positive integers for which $a_{n+1} > a_n$. Since $a_2 = 1 + \sqrt{a_1} = 2 > 1$, $1 \in S$. Assume that $a_k = 1 + \sqrt{a_{k-1}} > a_{k-1}$. Then

$$a_{k+1} = 1 + \sqrt{a_k} > 1 + \sqrt{a_{k-1}} = a_k.$$

Thus, $k \in S$ implies $k + 1 \in S$. It now follows that $\{a_n\}$ is an increasing sequence.

(b) Since $\{a_n\}$ is an increasing sequence,

$$a_n = 1 + \sqrt{a_{n-1}} < 1 + \sqrt{a_n}, \quad \text{or} \quad a_n - \sqrt{a_n} - 1 < 0.$$

Rewriting the second inequality as

$$(\sqrt{a_n})^2 - \sqrt{a_n} - 1 < 0$$

and solving for $\sqrt{a_n}$ it follows that $\sqrt{a_n} < \frac{1}{2}(1 + \sqrt{5})$. Hence, $a_n < \frac{1}{2}(3 + \sqrt{5})$ for all n.

(c) $a_2 = 2, \quad a_3 \cong 2.4142, \quad a_4 \cong 2.5538, \quad a_5 \cong 2.5981, \ldots, \quad a_9 \cong 2.6179, \quad \ldots, \quad a_{15} \cong 2.6180$;

(e) $\text{lub}\,\{a_n\} = \frac{1}{2}(3 + \sqrt{5}) \cong 2.6180$

68. (a) We show that $a_n < a_{n+1}$ for all n.   True for $n = 1$ since $a_1 = 1 < \sqrt{3} = a_2$. Assume true for
  n, that is, $a_n < a_{n+1}$;   we need to show that $a_{n+1} < a_{n+2}$.
  But $a_{n+1} = \sqrt{3a_n} < \sqrt{3a_{n+1}} = a_{n+2}$,   as required.

(b) True since $a_1 < 3$, and $a_n < 3 \implies a_{n+1} < \sqrt{3 \cdot 3} = 3$

(c) $a_1 = 1, \quad a_2 = \sqrt{3}, \quad a_3 \cong 2.2795, \ldots, a_{14} \cong 2.9996, \quad a_{15} \cong 2.9998$

(d) $\text{lub} = 3$

SECTION 11.3

1. diverges

2. converges to 0

3. converges to 0

4. diverges

5. converges to 1: $\dfrac{n-1}{n} = 1 - \dfrac{1}{n} \to 1$

6. converges to 1: $\dfrac{n + (-1)^n}{n} = 1 + \dfrac{(-1)^n}{n} \to 1$

7. converges to 0: $\dfrac{n+1}{n^2} = \dfrac{1}{n} + \dfrac{1}{n^2} \to 0$

8. converges to 0: $\dfrac{\pi}{2n} \to 0, \quad \text{so} \quad \sin\left(\dfrac{\pi}{2n}\right) \to \sin 0 = 0$

9. converges to 0: $0 < \dfrac{2^n}{4^n + 1} < \dfrac{2^n}{4^n} = \dfrac{1}{2^n} \to 0$

10. diverges: $\dfrac{n^2}{n+1} \geq \dfrac{n^2}{2n} = \dfrac{n}{2}$

11. diverges

12. diverges: $\dfrac{4^n}{\sqrt{n^2 + 1}} \approx \dfrac{4^n}{n} \to \infty$

13. converges to 0

14. converges to 0: $\dfrac{4n}{2^n + 10^6} < \dfrac{4n}{2^n} = \dfrac{n}{2^{n-2}} \to 0$

15.   converges to 1: $\dfrac{n\pi}{4n+1} \to \dfrac{\pi}{4}$ so $\tan\dfrac{n\pi}{4n+1} \to \tan\dfrac{\pi}{4} = 1$

16.   converges to 0: $\dfrac{10^{10}\sqrt{n}}{n+1} = \dfrac{10^{10}}{\sqrt{n}+1/\sqrt{n}} \to 0$

17.   converges to $\dfrac{4}{9}$: $\dfrac{(2n+1)^2}{(3n-1)^2} = \dfrac{4+4/n+1/n^2}{9-6/n+1/n^2} \to \dfrac{4}{9}$

18.   converges to $\ln 2$: $\dfrac{2n}{n+1} \to 2,$ so $\ln\left(\dfrac{2n}{n+1}\right) \to \ln 2$

19.   converges to $\dfrac{1}{2}\sqrt{2}$: $\dfrac{n^2}{\sqrt{2n^4+1}} = \dfrac{1}{\sqrt{2+1/n^4}} \to \dfrac{1}{\sqrt{2}}$

20.   converges to 1: $\dfrac{n^4-1}{n^4+n-6} = \dfrac{1-\frac{1}{n^4}}{1+\frac{1}{n^3}-\frac{6}{n^4}} \to 1$

21.   diverges: $\cos n\pi = (-1)^n$ **22.**   diverges: $\dfrac{n^5}{17n^4+12} = n\left(\dfrac{1}{17+\frac{12}{n^4}}\right)$

23.   converges to 1: $\dfrac{1}{\sqrt{n}} \to 0$ so $e^{1/\sqrt{n}} \to e^0 = 1$

24.   converges to $\sqrt{4} = 2$

25.   converges to 0 : $\ln n - \ln(n+1) = \ln\left(\dfrac{n}{n+1}\right) \to \ln 1 = 0$

26.   converges to 1: $\dfrac{2^n-1}{2^n} = 1 - \dfrac{1}{2^n} \implies 1$

27.   converges to $\dfrac{1}{2}$: $\dfrac{\sqrt{n+1}}{2\sqrt{n}} = \dfrac{1}{2}\sqrt{1+\dfrac{1}{n}} \to \dfrac{1}{2}$ **28.**   converges to 0: $\dfrac{1}{n} - \dfrac{1}{n+1} = \dfrac{1}{n(n+1)} \to 0$

29.   converges to e^2: $\left(1+\dfrac{1}{n}\right)^{2n} = \left[\left(1+\dfrac{1}{n}\right)^n\right]^2 \to e^2$

30.   converges to $\sqrt{e}$: $\left(1+\dfrac{1}{n}\right)^{n/2} = \sqrt{\left(1+\dfrac{1}{n}\right)^n} \to \sqrt{e}$

31.   diverges;   since $2^n > n^3$ for $n \geq 10,$ $\dfrac{2^n}{n^2} > \dfrac{n^3}{n^2} = n$

32.   converges to $\ln 9$: $2\ln 3n - \ln(n^2+1) = \ln\left(\dfrac{9n^2}{n^2+1}\right) \to \ln 9$

33.   converges to 0: $\dfrac{|\sin n|}{\sqrt{n}} \leq \dfrac{1}{\sqrt{n}}$

34. converges to $\pi/4$: $\dfrac{n}{n+1} \to 1$, $\arctan 1 = \pi/4$

35. converges to $1/2$: $\sqrt{n^2+n} - n = \left(\sqrt{n^2+n} - n\right)\dfrac{\sqrt{n^2+n}+n}{\sqrt{n^2+n}+n} = \dfrac{n}{\sqrt{n^2+n}+n} \to \dfrac{1}{2}$

36. converges to 2: $\dfrac{\sqrt{4n^2+n}}{n} = \sqrt{4+(1/n)} \to \sqrt{4} = 2.$

37. converges to $\pi/2$

38. converges to $31/9$: let $s = 3.444\cdots$. Then $10s - s = 31$ and $s = 31/9$.

39. converges to $-\pi/2$: $\dfrac{1-n}{n} \to -1$; $\arcsin(-1) = \pi/2$.

40. converges to 1: $\dfrac{(n+1)(n+4)}{(n+2)(n+3)} = \dfrac{n^2+5n+5}{n^2+5n+6} \to 1.$

41. (a) $\sqrt[n]{n} \to 1$ \qquad\qquad (b) $\dfrac{3^n}{n!} \to 0$

42. (a) $\dfrac{n}{\sqrt[n]{n!}} \to e$ \qquad\qquad (b) does not converge

43. $b < \sqrt[n]{a^n + b^n} = b\sqrt[n]{(a/b)^n + 1} < b\sqrt[n]{2}$. Since $2^{1/n} \to 1$ as $n \to \infty$, it follows that $\sqrt[n]{a^n + b^n} \to b$ by the pinching theorem.

44. (a) $-1 < r \leq 1$ \qquad\qquad (b) $-1 < r < 1$

45. Set $\epsilon > 0$. Since $a_n \to L$, there exists N_1 such that

$$\text{if} \quad n \geq N_1, \quad \text{then} \quad |a_n - L| < \epsilon/2.$$

Since $b_n \to M$, there exists N_2 such that

$$\text{if} \quad n \geq N_2, \quad \text{then} \quad |b_n - M| < \epsilon/2.$$

Now set $N = \max\{N_1, N_2\}$. Then, for $n \geq N$,

$$|(a_n + b_n) - (L + M)| \leq |a_n - L| + |b_n - M| < \frac{\epsilon}{2} + \frac{\epsilon}{2} = \epsilon.$$

46. Let $\epsilon > 0$, choose k such that $n \geq k \implies |a_n - L| < \dfrac{\epsilon}{|\alpha| + 1}.$

Then for $n \geq k$, $|\alpha a_n - \alpha L| = |\alpha||a_n - L| \leq (|\alpha| + 1)|a_n - L| < \epsilon$

Therefore $\alpha a_n \to \alpha L$.

47. Since $\left(1 + \dfrac{1}{n}\right) \to 1$ and $\left(1 + \dfrac{1}{n}\right)^n \to e$,

$$\left(1 + \frac{1}{n}\right)^{n+1} = \left(1 + \frac{1}{n}\right)^n \left(1 + \frac{1}{n}\right) \to (e)(1) = e.$$

48. (a)   If $k = j$, $a_n = \dfrac{\alpha_k + \alpha_{k-1} \cdot \dfrac{1}{n} + \cdots + \alpha_0 \cdot \dfrac{1}{n^k}}{\beta_k + \beta_{k-1} \cdot \dfrac{1}{n} + \cdots + \beta_0 \cdot \dfrac{1}{n^k}} \to \dfrac{\alpha_k}{\beta_k}$

   (b)   If $k < j$, $a_n = \dfrac{\alpha_k + \alpha_{k-1} \cdot \dfrac{1}{n} + \cdots + \alpha_0 \cdot \dfrac{1}{n^k}}{\beta_j \cdot n^{j-k} + \beta_{j-1} \cdot n^{j-1-k} + \cdots + \beta_0 \cdot \dfrac{1}{n^k}} \to 0$

   (c)   If $k > j$, $a_n = \dfrac{\alpha_k \cdot n^{k-j} + \alpha_{k-1} \cdot n^{k-1-j} + \cdots + \alpha_0 \cdot \dfrac{1}{n^j}}{\beta_j + \beta_{j-1} \cdot \dfrac{1}{n} + \cdots + \beta_0 \cdot \dfrac{1}{n^j}}$ diverges

49. Suppose that $\{a_n\}$ is bounded and non-increasing. If L is the greatest lower bound of the range of this sequence, then $a_n \geq L$ for all n. Set $\epsilon > 0$. By Theorem 11.1.4 there exists a_k such that $a_k < L + \epsilon$. Since the sequence is non-increasing, $a_n \leq a_k$ for all $n \geq k$. Thus,

$$L \leq a_n < L + \epsilon \quad \text{or} \quad |a_n - L| < \epsilon \quad \text{for all} \quad n \geq k$$

and $a_n \to L$.

50. Let $\epsilon > 0$.   If $a_n \to L$,   then there exists a positive integer k such that

$$|a_n - L| < \epsilon \quad \text{for all} \quad n \geq k$$

If $n \geq k$,   then $2n \geq k$ and $2n - 1 \geq k$,   and thus

$$|e_n - L| = |a_{2n} - L| < \epsilon \quad \text{and} \quad |o_n - L| = |a_{2n-1} - L| < \epsilon$$

It follows that $e_n \to L$ and $o_n \to L$.   If $e_n \to L$ and $o_n \to L$, then there exist k_1 and k_2 such that

$$\text{if} \quad m \geq k_1, \quad \text{then} \quad |e_m - L| = |a_{2m} - L| < \epsilon$$

and

$$\text{if} \quad m \geq k_2, \quad \text{then} \quad |o_m - L| = |a_{2m-1} - L| < \epsilon$$

Let $k = max\{2k_1,\ 2k_2 - 1\}$.   If $n \geq k$ then

$$\text{either} \quad a_n = a_{2m} \ \text{with} \ m > k_1 \quad \text{or} \quad a_n = a_{2m-1} \ \text{with} \ m \geq k_2$$

In either case, $|a_n - L| < \epsilon$.   This shows that $a_n \to L$.

51. Let $\epsilon > 0$.   Choose k so that, for $n \geq k$,

$$L - \epsilon < a_n < L + \epsilon, \quad L - \epsilon < c_n < L + \epsilon \quad \text{and} \quad a_n \leq b_n \leq c_n.$$

   For such n,

$$L - \epsilon < b_n < L + \epsilon.$$

52. Let M be a bound for $\{b_n\}$.   Then $|a_n b_n| \leq |a_n| M$.

   Given $\epsilon > 0$,   choose k such that $|a_n| < \epsilon/M$ for $n \geq k$.   Then $|a_n b_n| < \epsilon$ for $n \geq k$.

53. Let $\epsilon > 0$. Since $a_n \to L$, there exists a positive integer N such that $L - \epsilon < a_n < L + \epsilon$ for all $n \geq N$. Now $a_n \leq M$ for all n, so $L - \epsilon < M$, or $L < M + \epsilon$. Since ϵ is arbitrary, $L \leq M$.

54. The converse is false. For example, let $a_n = (-1)^n$. Then $|a_n| \to 1$, but $\{a_n\}$ diverges.

55. Assume $a_n \to 0$ as $n \to \infty$. Let $\epsilon > 0$. There exists a positive integer N such that $|a_n - 0| < \epsilon$ for all $n \geq N$. Since $\big| |a_n| - 0 \big| \leq |a_n - 0|$, it follows that $|a_n| \to 0$. Now assume that $|a_n| \to 0$. Since $-|a_n| \leq a_n \leq |a_n|$, $a_n \to 0$ by the pinching theorem.

56. Let $\epsilon > 0$. There exists a positive integer N_1 such that $|a_n - L| < \epsilon$ for all $n > 2N_1 - 1$, and there exists a positive integer N_2 such that $|b_n - L| < \epsilon$ for all $n > 2N_2$. The sequence $a_1,\ b_1,\ a_2,\ b_2,\ \ldots$ can be represented by the sequence $c_1,\ c_2,\ c_3,\ \ldots,$ where

$$c_n = \begin{cases} a_{(n+1)/2}, & \text{if } n \text{ is odd} \\ b_{n/2}, & \text{if } n \text{ is even.} \end{cases}$$

Let $N = max\{2N_1 - 1,\ 2N_2\}$. Then $n > N \implies |c_n - L| < \epsilon \implies c_n \to L.$

57. By the continuity of f, $f(L) = f\left(\lim\limits_{n\to\infty} a_n\right) = \lim\limits_{n\to\infty} f(a_n) = \lim\limits_{n\to\infty} a_{n+1} = L.$

58. $\dfrac{2^n}{n!} = \dfrac{2}{1} \cdot \dfrac{2}{2} \cdot \dfrac{2}{3} \cdots \dfrac{2}{n} = 2 \cdot \dfrac{2}{n} \cdot (\text{terms that are } \leq 1) \leq \dfrac{4}{n}.$

Since $\dfrac{4}{n} \to 0$ and $0 < \dfrac{2^n}{n!} \leq \dfrac{4}{n}$, $\dfrac{2^n}{n!} \to 0$ as well.

59. Set $f(x) = x^{1/p}$. Since $\dfrac{1}{n} \to 0$ and f is continuous at 0, it follows by Theorem 11.3.12 that

$$\left(\dfrac{1}{n}\right)^{1/p} \to 0.$$

60. Since $|a_n - L| = |(a_n - L) - 0| = \big||a_n - L| - 0\big|,$

$|a_n - L| < \epsilon$ iff $|(a_n - L) - 0| < \epsilon$ iff $\big||a_n - L| - 0\big| < \epsilon,$

So $a_n \to L$ iff $a_n - L \to 0$ iff $|a_n - L| \to 0.$

61. $a_n = e^{1-n} \to 0$

62. diverges

63. $a_n = \dfrac{1}{n!} \to 0$

64. $a_n = 1 \cdot \dfrac{1}{2} \cdot \dfrac{2}{3} \cdots \dfrac{n-1}{n} = \dfrac{1}{n}$ converges to 0

65. $a_n = \dfrac{1}{2}[1 - (-1)^n]$ diverges

66. $a_n = \dfrac{2^n - 1}{2^{n-1}} \to 2$

67. $L = 0,\quad n = 32$

68. $\dfrac{1}{\sqrt{n}} \to 0.$ $\dfrac{1}{\sqrt{n}} < 0.001$ for $n \geq 1000^2 + 1$

69. $L = 0,\quad n = 4$

70. $\dfrac{n^{10}}{10^n} \to 0.$ $\dfrac{n^{10}}{10^n} < 0.001$ for $n \geq 15$

71. $L = 0,\quad n = 7$

72. $\dfrac{2^n}{n!} \to 0.\quad \dfrac{2^n}{n!} < 0.001$ for $n \geq 10$

73. $L = 0,\quad n = 65$

74. $\dfrac{\ln n}{n} \to 0.\quad \dfrac{\ln n}{n} < 0.001$ for $n \geq 9119$

75. (a) $a_{n+1} = 1 + \sqrt{a_n}$ Suppose that $a_n \to L$ as $n \to \infty$. Then $a_{n+1} \to L$ as $n \to \infty$. Therefore $L = 1 + \sqrt{L}$ which, since $L > 1$, implies $L = \frac{1}{2}(3 + \sqrt{5})$.

(b) $a_{n+1} = \sqrt{3a_n}$ Suppose that $a_n \to L$ as $n \to \infty$. Then $a_{n+1} \to L$ as $n \to \infty$. Therefore $L = \sqrt{3L}$ which, since $L > 1$, implies $L = 3$.

76. (a) $a_2 \cong 2.6458,\quad a_3 \cong 2.9404,\quad a_4 \cong 2.9900,\quad a_5 \cong 2.9983,\quad a_6 \cong 2.9997$

(b) True for $n = 1$.   Assume true for n.   Then $a_{n+1} = \sqrt{6 + a_{n-1}} \leq \sqrt{6 + 3} = 3$

(c) $a_{n+1}{}^2 - a_n{}^2 = 6 + a_n - a_n{}^2 = (3 - a_n)(2 + a_n) \geq 0$ since $0 \leq a_n \leq 3$.

   Since $a_k \geq 0$ for all k,   this implies $a_{n+1} \geq a_n$

(d) $a_n \to 3$

77. (a)

a_2	a_3	a_4	a_5	a_6	a_7	a_8	a_9	a_{10}
0.5403	0.8576	0.6543	0.7935	0.7014	0.7640	0.7221	0.7504	0.7314

(b) L is a fixed point of $f(x) = \cos x$, that is, $\cos L = L$; $L \cong 0.739085$.

78. (a) $a_2 \cong 1.5403,\quad a_3 \cong 1.5708,\quad a_4 \cong a_5 \cong \cdots \cong a_{10} \cong 1.5708$.

(b) $L \cong 1.570796$ Let $f(x) = x + \cos x$. L must satisfy $L = f(L)$,   so $L = L + \cos L$, and $\cos L = 0$.   Indeed, the L we found is just $\dfrac{\pi}{2} \cong 1.570796327$

PROJECT 11.3

1. (a)

a_2	a_3	a_4	a_5	a_6	a_7	a_8
2.000000	1.750000	1.732143	1.732051	1.732051	1.732051	1.732051

(b) $L = \dfrac{1}{2}\left(L + \dfrac{3}{L}\right)$ which implies $L^2 = 3$ or $L = \sqrt{3}$.

2. The Newton-Raphson method applied to the function $f(x) = x^2 - R$ gives

$$a_n = a_{n-1} - \frac{f(a_{n-1})}{f'(a_{n-1})} = a_{n-1} - \frac{a_{n-1}^2 - R}{2a_{n-1}}$$

$$= \frac{1}{2}\,a_{n-1} + \frac{1}{2}\frac{R}{a_{n-1}} = \frac{1}{2}\left(a_{n-1} + \frac{R}{a_{n-1}}\right), \quad n = 2, 3, \ldots .$$

3 & 4. (a) $f(x) = x^3 - 8$,   so $x_n \to 2$ (b) $f(x) = \sin x - \frac{1}{2}$,   so $x_n \to \frac{\pi}{6}$

(c) $f(x) = \ln x - 1$,   so $x_n \to e$

SECTION 11.4

1. converges to 1: $2^{2/n} = (2^{1/n})^2 \to 1^2 = 1$ **2.** converges to 1: $e^{-\alpha/n} \to e^0 = 1$

3. converges to 0: for $n > 3$, $0 < \left(\dfrac{2}{n}\right)^n < \left(\dfrac{2}{3}\right)^n \to 0$

4. converges to 0: $\dfrac{\log_{10} n}{n} = \dfrac{1}{\ln 10} \cdot \dfrac{\ln n}{n} \to 0$

5. converges to 0: $\dfrac{\ln(n+1)}{n} = \left[\dfrac{\ln(n+1)}{n+1}\right]\left(\dfrac{n+1}{n}\right) \to (0)(1) = 0$

6. converges to 0: $\dfrac{3^n}{4^n} = \left(\dfrac{3}{4}\right)^n \to 0$ **7.** converges to 0: $\dfrac{x^{100n}}{n!} = \dfrac{(x^{100})^n}{n!} \to 0$

8. converges to 1: $n^{1/(n+2)} = \left(n^{1/n}\right)^{n/(n+2)} \to 1$ **9.** converges to 1: $n^{\alpha/n} = (n^{1/n})^\alpha \to 1^\alpha = 1$

10. converges to 0: $\ln\left(\dfrac{n+1}{n}\right) \to \ln(1) = 0$

11. converges to 0: $\dfrac{3^{n+1}}{4^{n-1}} = 12\left(\dfrac{3^n}{4^n}\right) = 12\left(\dfrac{3}{4}\right)^n \to 12(0) = 0$

12. converges to $\dfrac{1}{2}$: $\displaystyle\int_{-n}^{0} e^{2x}\, dx = \dfrac{1}{2} - \dfrac{e^{-2n}}{2} \to \dfrac{1}{2}$

13. converges to 1: $(n+2)^{1/n} = e^{\frac{1}{n}\ln(n+2)}$ and, since

$$\frac{1}{n}\ln(n+2) = \left[\frac{\ln(n+2)}{n+2}\right]\left(\frac{n+2}{n}\right) \to (0)(1) = 0,$$

   it follows that $(n+2)^{1/n} \to e^0 = 1.$

14. converges to e^{-1} : $\left(1 - \dfrac{1}{n}\right)^n = \left(1 + \dfrac{(-1)}{n}\right)^n \to e^{-1}$ (by (11.4.7))

15.   converges to 1: $\displaystyle\int_0^n e^{-x}\, dx = 1 - \dfrac{1}{e^n} \to 1$ **16.**   diverges.

17. converges to π: integral $= 2\displaystyle\int_0^n \dfrac{dx}{1+x^2} = 2\tan^{-1} n \to 2\left(\dfrac{\pi}{2}\right) = \pi$

18. converges to 0: $\displaystyle\int_0^n e^{-nx}\, dx = -\dfrac{e^{-n^2}}{n} + \dfrac{1}{n} \to 0$

19.   converges to 1:   recall (11.4.6)   **20.**   converges to 0: $n^2 \sin n\pi = 0$ for all n

21. converges to 0: $\dfrac{\ln(n^2)}{n} = 2\dfrac{\ln n}{n} \to 2(0) = 0$

22. converges to π : $\displaystyle\int_{-1+1/n}^{1-1/n} \frac{dx}{\sqrt{1-x^2}} = \sin^{-1}\left(1 - \frac{1}{n}\right) - \sin^{-1}\left(-1 + \frac{1}{n}\right) \to \sin^{-1}(1) - \sin^{-1}(-1) = \pi$

23. diverges: Since $\displaystyle\lim_{x\to 0} \frac{\sin x}{x} = 1$, $\displaystyle\frac{n}{\pi}\sin\frac{\pi}{n} = \frac{\sin(\pi/n)}{\pi/n} \to 1$. Therefore,

$$n^2 \sin\frac{\pi}{n} = n\pi\left(\frac{n}{\pi}\sin\frac{\pi}{n}\right) \to n\pi.$$

24. diverges

25. converges to 0: $\displaystyle\frac{5^{n+1}}{4^{2n-1}} = 20\left(\frac{5}{16}\right)^n \to 0$

26. converges to e^{3x} : $\displaystyle\left(1 + \frac{x}{n}\right)^{3n} = \left[\left(1 + \frac{x}{n}\right)^n\right]^3 \to (e^x)^3 = e^{3x}$

27. converges to e^{-1}: $\displaystyle\left(\frac{n+1}{n+2}\right)^n = \left(1 - \frac{1}{n+2}\right)^n = \frac{\left(1 + \frac{(-1)}{n+2}\right)^{n+2}}{\left(1 + \frac{(-1)}{n+2}\right)^2} \to \frac{e^{-1}}{1} = e^{-1}$

28. converges to 2: $\displaystyle\int_{1/n}^1 \frac{dx}{\sqrt{x}} = 2 - \frac{2}{\sqrt{n}} \to 2.$

29. converges to 0: $\displaystyle 0 < \int_n^{n+1} e^{-x^2}\, dx \le e^{-n^2}[(n+1) - n] = e^{-n^2} \to 0$

30. converges to 1: $\displaystyle\left(1 + \frac{1}{n^2}\right)^n = \left[\left(1 + \frac{1}{n^2}\right)^{n^2}\right]^{1/n} \to (e^1)^0 = 1$

31. converges to 0: $\displaystyle\frac{n^n}{2^{n^2}} = \left(\frac{n}{2^n}\right)^n \to 0$ since $\displaystyle\frac{n}{2^n} \to 0$

32. converges to 0: $\displaystyle\int_0^{1/n} \cos e^x\, dx \quad\to\quad \int_0^0 \cos e^x\, dx = 0$

33. converges to e^x: use (11.4.7)

34. diverges: $\displaystyle\left(1 + \frac{1}{n}\right)^{n^2} = \left[\left(1 + \frac{1}{n}\right)^n\right]^n > 2^n, \quad \left[\left(1 + \frac{1}{n}\right)^n \approx e > 2\right]$

35. converges to 0: $\displaystyle\left|\int_{-1/n}^{1/n} \sin x^2\, dx\right| \le \int_{-1/n}^{1/n} |\sin x^2|\, dx \le \int_{-1/n}^{1/n} 1\, dx = \frac{2}{n} \to 0$

36. $\displaystyle\left(t + \frac{x}{n}\right)^n = t^n\left(1 + \frac{x/t}{n}\right)^n$; converges to 0 if $t < 1$, converges to e^x if $t = 1$, diverges if $t > 1$.

37. converges: $\displaystyle\frac{\sin(6/n)}{\sin(3/n)} \to 2$

38. converges: $\displaystyle\frac{\arctan n}{n} \to 0$

39. $\sqrt{n+1} - \sqrt{n} = \dfrac{\sqrt{n+1} - \sqrt{n}}{\sqrt{n+1} + \sqrt{n}}\left(\sqrt{n+1} + \sqrt{n}\right) = \dfrac{1}{\sqrt{n+1} + \sqrt{n}} \to 0$

40. $\sqrt{n^2+n} - n = \dfrac{\sqrt{n^2+n} - n}{\sqrt{n^2+n} + n}\left(\sqrt{n^2+n} + n\right) = \dfrac{n}{\sqrt{n^2+n} + n} = \dfrac{1}{1 + \sqrt{1 + 1/n}} \to \dfrac{1}{2}$

41. (a) The length of each side of the polygon is $2r\sin(\pi/n)$. Therefore the perimeter, p_n, of the polygon is given by: $p_n = 2rn\sin(\pi/n)$.

   (b) $2rn\sin(\pi/n) \to 2\pi r$ as $n \to \infty$: The number $2rn\sin(\pi/n)$ is the perimeter of a regular polygon of n sides inscribed in a circle of radius r. As n tends to ∞, the perimeter of the polygon tends to the circumference of the circle.

42. Since $0 < c < d,\quad d < (c^n + d^n)^{1/n} < (2d^n)^{1/n} = 2^{1/n}d \to d,\quad$ so by the pinching theorem

$$(c^n + d^n)^{1/n} \to d.$$

43. By the hint, $\quad \displaystyle\lim_{n\to\infty} \frac{1 + 2 + \ldots + n}{n^2} = \lim_{n\to\infty} \frac{n(n+1)}{2n^2} = \lim_{n\to\infty} \frac{1 + 1/n}{2} = \frac{1}{2}.$

44. diverges: $\dfrac{1^2 + 2^2 + \cdots + n^2}{(1+n)(2+n)} = \dfrac{n(n+1)(2n+1)}{6(1+n)(2+n)} = \dfrac{2n^3 + 3n^2 + n}{6n^2 + 18n + 12} \to \infty$

45. By the hint, $\quad \displaystyle\lim_{n\to\infty} \frac{1^3 + 2^3 + \ldots + n^3}{2n^4 + n - 1} = \lim_{n\to\infty} \frac{n^2(n+1)^2}{4(2n^4 + n - 1)} = \lim_{n\to\infty} \frac{1 + 2/n + 1/n^2}{8 + 4/n^3 - 4/n^4} = \frac{1}{8}.$

46. Here we show that every convergent sequence is a Cauchy sequence. Let $\epsilon > 0$. If $a_n \to L$, then there exists a positive integer k such that

$$|a_p - L| < \frac{\epsilon}{2} \quad \text{for all} \ \ p \geq k$$

With $\ m, n \geq k\ $ we have

$$|a_m - a_n| \leq |a_m - L| + |L - a_n| = |a_m - L| + |a_n - L| < \frac{\epsilon}{2} + \frac{\epsilon}{2} = \epsilon.$$

47. (a) $m_{n+1} - m_n = \dfrac{1}{n+1}(a_1 + \cdots + a_n + a_{n+1}) - \dfrac{1}{n}(a_1 + \cdots + a_n)$

$$= \frac{1}{n(n+1)}\left[\, na_{n+1} - \overbrace{(a_1 + \cdots + a_n)}^{n}\,\right]$$

$$> 0 \quad \text{since} \ \ \{a_n\} \text{ is increasing.}$$

   (b) We begin with the hint

$$m_n < \frac{|a_1 + \cdots + a_j|}{n} + \frac{\epsilon}{2}\left(\frac{n-j}{n}\right).$$

   Since j is fixed,

$$\frac{|a_1 + \cdots + a_j|}{n} \to 0$$

and therefore for n sufficiently large

$$\frac{|a_1 + \cdots + a_j|}{n} < \frac{\epsilon}{2}.$$

Since

$$\frac{\epsilon}{2}\left(\frac{n-j}{n}\right) < \frac{\epsilon}{2},$$

we see that, for n sufficiently large, $|m_n| < \epsilon$. This shows that $m_n \to 0$.

48. (a) Since $\{a_n\}$ converges, it is a Cauchy sequence (see Exercise 46), so given $\epsilon > 0$ we can find k
such that $|a_n - a_m| < \epsilon$ for $n, m \geq k$.

In particular, $|a_{n+1} - a_n| < \epsilon$, so $\lim_{n\to\infty}(a_n - a_{n-1}) = 0$.

(b) $\{a_n\}$ does not necessarily converge. For example let $a_n = \ln n$. This diverges, but

$$a_n - a_{n-1} = \ln n - \ln(n-1) = \ln\left(\frac{n}{n-1}\right) \to \ln 1 = 0$$

49. (a) Let S be the set of positive integers n $(n \geq 2)$ for which the inequalities hold. Since

$$\left(\sqrt{b}\right)^2 - 2\sqrt{ab} + \left(\sqrt{a}\right)^2 = \left(\sqrt{b} - \sqrt{a}\right)^2 > 0,$$

it follows that $\dfrac{a+b}{2} > \sqrt{ab}$ and so $a_1 > b_1$. Now,

$$a_2 = \frac{a_1 + b_1}{2} < a_1 \quad \text{and} \quad b_2 = \sqrt{a_1 b_1} > b_1.$$

Also, by the argument above,

$$a_2 = \frac{a_1 + b_1}{2} > \sqrt{a_1 b_1} = b_2,$$

and so $a_1 > a_2 > b_2 > b_1$. Thus $2 \in S$. Assume that $k \in S$. Then

$$a_{k+1} = \frac{a_k + b_k}{2} < \frac{a_k + a_k}{2} = a_k, \quad b_{k+1} = \sqrt{a_k b_k} > \sqrt{b_k^2} = b_k,$$

and

$$a_{k+1} = \frac{a_k + b_k}{2} > \sqrt{a_k b_k} = b_{k+1}.$$

Thus $k + 1 \in S$. Therefore, the inequalities hold for all $n \geq 2$.

(b) $\{a_n\}$ is a decreasing sequence which is bounded below.

$\{b_n\}$ is an increasing sequence which is bounded above.

Let $L_a = \lim_{n\to\infty} a_n$, $L_b = \lim_{n\to\infty} b_n$. Then

$$a_n = \frac{a_{n-1} + b_{n-1}}{2} \quad \text{implies} \quad L_a = \frac{L_a + L_b}{2} \quad \text{and} \quad L_a = L_b.$$

50. $\dfrac{e - \left(1 + \frac{1}{100}\right)^{100}}{e} \cong 0.004995:$ within 0.01%; $\quad \dfrac{e^5 - \left(1 + \frac{5}{100}\right)^{100}}{e^5} \cong 0.11395:$ within 12%

$\dfrac{e - \left(1 + \frac{1}{1000}\right)^{1000}}{e} \cong 0.0004995:$ within 0.05%; $\quad \dfrac{e^5 - \left(1 + \frac{5}{1000}\right)^{1000}}{e^5} \cong 0.01238:$ within 1.3%

51. The numerical work suggests $L \cong 1$. Justification: Set $f(x) = \sin x - x^2$. Note that $f(0) = 0$ and $f'(x) = \cos x - 2x > 0$ for x close to 0. Therefore $\sin x - x^2 > 0$ for x close to 0 and $\sin 1/n - 1/n^2 > 0$ for n large. Thus, for n large,

$$\frac{1}{n^2} < \sin \frac{1}{n} < \frac{1}{n} \qquad (|\sin x| \le |x| \quad \text{for all } x)$$

$$\left(\frac{1}{n^2}\right)^{1/n} < \left(\sin \frac{1}{n}\right)^{1/n} < \left(\frac{1}{n}\right)^{1/n}$$

$$\left(\frac{1}{n^{1/n}}\right)^2 < \left(\sin \frac{1}{n}\right)^{1/n} < \frac{1}{n^{1/n}}.$$

As $n \to \infty$ both bounds tend to 1 and therefore the middle term also tends to 1.

52. Numerical work suggests $L \cong 1/3$.   Conjecture: $L = 1/k$.

Proof: $(n^k + n^{k-1})^{1/k} - n = n(1 + 1/n)^{1/k} - n = \dfrac{(1 + 1/n)^{1/k} - 1}{1/n}$;

$$\lim_{n \to \infty} \frac{(1 + 1/n)^{1/k} - 1}{1/n} = \lim_{h \to 0} \frac{(1 + h)^{1/k} - 1}{h} = f'(1),$$

where $f(x) = x^{1/k}$.   Since $f'(x) = (1/k)x^{(1/k)-1}$, $f'(1) = 1/k$.

53. (a)

a_3	a_4	a_5	a_6	a_7	a_8	a_9	a_{10}
2	3	5	8	13	21	34	55

(b)

r_1	r_2	r_3	r_4	r_5	r_6
1	2	1.5	1.667	1.600	1.625

(c) Following the hint,

$$1 + \frac{1}{r_{n-1}} = 1 + \frac{1}{\dfrac{a_n}{a_{n-1}}} = 1 + \frac{a_{n-1}}{a_n} = \frac{a_n + a_{n-1}}{a_n} = \frac{a_{n+1}}{a_n} = r_n.$$

Now, if $r_n \to L$, then $r_{n-1} \to L$ and

$$1 + \frac{1}{L} = L \quad \text{which, since } L > 1, \text{ implies} \quad L = \frac{1 + \sqrt{5}}{2} \cong 1.618034.$$

54. With the partition $\{0, \frac{1}{n}, \frac{2}{n}, \dots, \frac{n}{n}\}$ and $f(x) = x$, we have

$$a_n = \frac{1}{n}\left(\frac{1}{n} + \frac{2}{n} + \cdots + \frac{n}{n}\right) = \frac{1}{n}\left[f\left(\frac{1}{n}\right) + f\left(\frac{2}{n}\right) + \cdots + f\left(\frac{n}{n}\right)\right] = \sum_{i=1}^{n} f(x_i)\Delta x_i,$$

so it is a Riemann sum for $\displaystyle\int_0^1 x \, dx$,   and therefore $\displaystyle\lim_{n \to \infty} a_n = \int_0^1 x \, dx = \frac{1}{2}$

SECTION 11.5

(We'll use $\star$ to indicate differentiation of numerator and denominator.)

1. $\displaystyle\lim_{x \to 0^+} \frac{\sin x}{\sqrt{x}} \overset{\star}{=} \lim_{x \to 0^+} 2\sqrt{x} \cos x = 0$

2. $\displaystyle\lim_{x \to 1} \frac{\ln x}{1 - x} \overset{\star}{=} \lim_{x \to 1} \frac{1/x}{-1} = -1$

3. $\lim\limits_{x\to 0} \dfrac{e^x - 1}{\ln(1+x)} \overset{\star}{=} \lim\limits_{x\to 0}(1+x)e^x = 1$

4. $\lim\limits_{x\to 4} \dfrac{\sqrt{x}-2}{x-4} \overset{\star}{=} \lim\limits_{x\to 4} \dfrac{\frac{1}{2\sqrt{x}}}{1} = \dfrac{1}{4}$

5. $\lim\limits_{x\to\pi/2} \dfrac{\cos x}{\sin 2x} \overset{\star}{=} \lim\limits_{x\to\pi/2} \dfrac{-\sin x}{2\cos 2x} = \dfrac{1}{2}$

6. $\lim\limits_{x\to a} \dfrac{x-a}{x^n - a^n} \overset{\star}{=} \lim\limits_{x\to a} \dfrac{1}{nx^{n-1}} = \dfrac{1}{na^{n-1}}$

7. $\lim\limits_{x\to 0} \dfrac{2^x - 1}{x} \overset{\star}{=} \lim\limits_{x\to 0} 2^x \ln 2 = \ln 2$

8. $\lim\limits_{x\to 0} \dfrac{\tan^{-1} x}{x} \overset{\star}{=} \lim\limits_{x\to 0} \dfrac{\frac{1}{1+x^2}}{1} = 1$

9. $\lim\limits_{x\to 1} \dfrac{x^{1/2} - x^{1/4}}{x-1} \overset{\star}{=} \lim\limits_{x\to 1}\left(\dfrac{1}{2}x^{-1/2} - \dfrac{1}{4}x^{-3/4}\right) = \dfrac{1}{4}$

10. $\lim\limits_{x\to 0} \dfrac{e^x - 1}{x(1+x)} \overset{\star}{=} \lim\limits_{x\to 0} \dfrac{e^x}{1+2x} = 1$

11. $\lim\limits_{x\to 0} \dfrac{e^x - e^{-x}}{\sin x} \overset{\star}{=} \lim\limits_{x\to 0} \dfrac{e^x + e^{-x}}{\cos x} = 2$

12. $\lim\limits_{x\to 0} \dfrac{1-\cos x}{3x} \overset{\star}{=} \lim\limits_{x\to 0} \dfrac{\sin x}{3} = 0$

13. $\lim\limits_{x\to 0} \dfrac{x + \sin \pi x}{x - \sin \pi x} \overset{\star}{=} \lim\limits_{x\to 0} \dfrac{1 + \pi\cos\pi x}{1 - \pi\cos\pi x} = \dfrac{1+\pi}{1-\pi}$

14. $\lim\limits_{x\to 0} \dfrac{a^x - (a+1)^x}{x} \overset{\star}{=} \lim\limits_{x\to 0} \dfrac{a^x \ln a - (a+1)^x \ln(a+1)}{1} = \ln\left(\dfrac{a}{a+1}\right)$

15. $\lim\limits_{x\to 0} \dfrac{e^x + e^{-x} - 2}{1 - \cos 2x} \overset{\star}{=} \lim\limits_{x\to 0} \dfrac{e^x - e^{-x}}{2\sin 2x} \overset{\star}{=} \lim\limits_{x\to 0} \dfrac{e^x + e^{-x}}{4\cos 2x} = \dfrac{1}{2}$

16. $\lim\limits_{x\to 0} \dfrac{x - \ln(x+1)}{1 - \cos 2x} \overset{\star}{=} \lim\limits_{x\to 0} \dfrac{1 - \frac{1}{x+1}}{2\sin 2x} \overset{\star}{=} \lim\limits_{x\to 0} \dfrac{\frac{1}{(x+1)^2}}{4\cos 2x} = \dfrac{1}{4}$

17. $\lim\limits_{x\to 0} \dfrac{\tan \pi x}{e^x - 1} \overset{\star}{=} \lim\limits_{x\to 0} \dfrac{\pi\sec^2 \pi x}{e^x} = \pi$

18. $\lim\limits_{x\to 0} \dfrac{\cos x - 1 + x^2/2}{x^4} \overset{\star}{=} \lim\limits_{x\to 0} \dfrac{-\sin x + x}{4x^3} \overset{\star}{=} \lim\limits_{x\to 0} \dfrac{-\cos x + 1}{12x^2} \overset{\star}{=} \lim\limits_{x\to 0} \dfrac{\sin x}{24x} = \dfrac{1}{24}$

19. $\lim\limits_{x\to 0} \dfrac{1 + x - e^x}{x(e^x - 1)} \overset{\star}{=} \lim\limits_{x\to 0} \dfrac{1 - e^x}{xe^x + e^x - 1} \overset{\star}{=} \lim\limits_{x\to 0} \dfrac{-e^x}{xe^x + 2e^x} = -\dfrac{1}{2}$

20. $\lim\limits_{x\to 0} \dfrac{\ln(\sec x)}{x^2} \overset{\star}{=} \lim\limits_{x\to 0} \dfrac{\tan x}{2x} \overset{\star}{=} \lim\limits_{x\to 0} \dfrac{\sec^2 x}{2} = \dfrac{1}{2}$

21. $\lim\limits_{x\to 0} \dfrac{x - \tan x}{x - \sin x} \overset{\star}{=} \lim\limits_{x\to 0} \dfrac{1 - \sec^2 x}{1 - \cos x} \overset{\star}{=} \lim\limits_{x\to 0} \dfrac{-2\sec^2 x \tan x}{\sin x} = \lim\limits_{x\to 0} \dfrac{-2\sec^2 x}{\cos x} = -2$

22. $\lim\limits_{x\to 0} \dfrac{xe^{nx} - x}{1 - \cos nx} \overset{\star}{=} \lim\limits_{x\to 0} \dfrac{e^{nx} + nxe^{nx} - 1}{n\sin nx} \overset{\star}{=} \lim\limits_{x\to 0} \dfrac{e^{nx}(2n + n^2 x)}{n^2 \cos nx} = \dfrac{2}{n}$

23. $\lim\limits_{x\to 1^-} \dfrac{\sqrt{1-x^2}}{\sqrt{1-x^3}} = \lim\limits_{x\to 1^-} \sqrt{\dfrac{1-x^2}{1-x^3}} = \sqrt{\dfrac{2}{3}} = \dfrac{1}{3}\sqrt{6}$ since $\lim\limits_{x\to 1^-} \dfrac{1-x^2}{1-x^3} \overset{\star}{=} \lim\limits_{x\to 1^-} \dfrac{2x}{3x^2} = \dfrac{2}{3}$

24. $\lim\limits_{x\to 0} \dfrac{2x - \sin \pi x}{4x^2 - 1} = 0$

25. $\lim\limits_{x\to \pi/2} \dfrac{\ln(\sin x)}{(\pi - 2x)^2} \overset{\star}{=} \lim\limits_{x\to \pi/2} \dfrac{-\cot x}{4(\pi - 2x)} \overset{\star}{=} \lim\limits_{x\to \pi/2} \dfrac{\csc^2 x}{-8} = -\dfrac{1}{8}$

26. $\lim\limits_{x\to 0^+} \dfrac{\sqrt{x}}{\sqrt{x} + \sin\sqrt{x}} \overset{\star}{=} \lim\limits_{x\to 0^+} \dfrac{\frac{1}{2\sqrt{x}}}{\frac{1}{2\sqrt{x}} + \frac{\cos\sqrt{x}}{2\sqrt{x}}} = \lim\limits_{x\to 0^+} \dfrac{1}{1 + \cos\sqrt{x}} = \dfrac{1}{2}$

27. $\lim\limits_{x\to 0} \dfrac{\cos x - \cos 3x}{\sin(x^2)} \overset{\star}{=} \lim\limits_{x\to 0} \dfrac{-\sin x + 3\sin 3x}{2x\cos(x^2)} \overset{\star}{=} \lim\limits_{x\to 0} \dfrac{-\cos x + 9\cos 3x}{2\cos(x^2) - 4x^2\sin(x^2)} = 4$

28. $\lim\limits_{x\to 0} \dfrac{\sqrt{a+x} - \sqrt{a-x}}{x} \overset{\star}{=} \lim\limits_{x\to 0} \dfrac{\frac{1}{2\sqrt{a+x}} + \frac{1}{2\sqrt{a-x}}}{1} = \dfrac{1}{\sqrt{a}} = \dfrac{\sqrt{a}}{a}$

29. $\lim\limits_{x\to \pi/4} \dfrac{\sec^2 x - 2\tan x}{1 + \cos 4x} \overset{\star}{=} \lim\limits_{x\to \pi/4} \dfrac{2\sec^2 x\,\tan x - 2\sec^2 x}{-4\sin 4x}$

$\overset{\star}{=} \lim\limits_{x\to \pi/4} \dfrac{2\sec^4 x + 4\sec^2 x\tan^2 x - 4\sec^2 x\tan x}{-16\cos\,4x} = \dfrac{1}{2}$

30. $\lim\limits_{x\to 0} \dfrac{x - \arcsin x}{\sin^3 x} \overset{\star}{=} \lim\limits_{x\to 0} \dfrac{1 - \frac{1}{\sqrt{1-x^2}}}{3\sin^2 x\cos x} \overset{\star}{=} \lim\limits_{x\to 0} \dfrac{\frac{-x}{(1-x^2)^{3/2}}}{6\sin x\cos^2 x - 3\sin^3 x}$

$\overset{\star}{=} \lim\limits_{x\to 0} \dfrac{\frac{-1-2x^2}{(1-x^2)^{5/2}}}{6\cos^3 x - 21\sin^2 x\cos x} = -\dfrac{1}{6}$

31. $\lim\limits_{x\to 0} \dfrac{\tan^{-1} x}{\tan^{-1} 2x} \overset{\star}{=} \lim\limits_{x\to 0} \dfrac{1}{1+x^2}\dfrac{1 + 4x^2}{2} = \dfrac{1}{2}$

32. $\lim\limits_{x\to 0} \dfrac{\sin^{-1} x}{x} \overset{\star}{=} \lim\limits_{x\to 0} \dfrac{\frac{1}{\sqrt{1-x^2}}}{1} = 1$

33. $1:\quad \lim\limits_{x\to \infty} \dfrac{\pi/2 - \tan^{-1} x}{1/x} \overset{\star}{=} \lim\limits_{x\to \infty} \dfrac{x^2}{1+x^2} = 1$

34. $-1:\quad \lim\limits_{n\to \infty} \dfrac{\ln(1 - \frac{1}{n})}{\sin(\frac{1}{n})} = \lim\limits_{x\to 0^+} \dfrac{\ln(1-x)}{\sin x} \overset{\star}{=} \lim\limits_{x\to 0^+} \dfrac{1}{(1-x)\cos x} = -1$

35. $1:\quad \lim\limits_{x\to \infty} \dfrac{1}{x[\ln(x+1) - \ln x]} = \lim\limits_{x\to \infty} \dfrac{1/x}{\ln(1 + 1/x)} = \lim\limits_{t\to 0^+} \dfrac{t}{\ln(1+t)} \overset{\star}{=} \lim\limits_{t\to 0^+}(1 + t) = 1$

36. $\dfrac{1}{3}:\quad \lim\limits_{n\to \infty} \dfrac{\sinh(\pi/n) - \sin(\pi/n)}{\sin^3(\pi/n)} = \lim\limits_{x\to \infty} \dfrac{\sinh(\pi/x) - \sin(\pi/x)}{\sin^3(\pi/x)} \overset{\star}{=} \lim\limits_{u\to 0^+} \dfrac{\sinh u - \sin u}{\sin^3 u}$

$\overset{\star}{=} \lim\limits_{u\to 0^+} \dfrac{\cosh u - \cos u}{3\sin^2 u\cos u} \overset{\star}{=} \lim\limits_{u\to 0^+} \dfrac{\sinh u + \sin u}{6\sin u\cos^2 u - 3\sin^3 u} \overset{\star}{=} \lim\limits_{u\to 0^+} \dfrac{\cosh u + \cos u}{6\cos^3 u - 21\sin^2 u\cos u} = \dfrac{2}{6} = \dfrac{1}{3}$

37. $\lim\limits_{x\to 0}\dfrac{x^3}{4^x-1}=0$

38. $\lim\limits_{x\to 0}\dfrac{4x}{\sin^2 x}$ does not exist

39. $\lim\limits_{x\to 0}\dfrac{x}{\frac{\pi}{2}-\arccos x}=1$

40. $\lim\limits_{x\to 2^+}\dfrac{\sqrt{2x}-2}{\sqrt{x-2}}=0$

41. $\lim\limits_{x\to 0}\dfrac{\tanh x}{x}=1$

42. $\lim\limits_{x\to \pi/2}\dfrac{1+\cos 2x}{1-\sin x}=4$

43. $\lim\limits_{x\to 0}(2+x+\sin x)\neq 0,\quad \lim\limits_{x\to 0}(x^3+x-\cos x)\neq 0$

44. $\lim\limits_{n\to\infty}n\left(a^{1/n}-1\right)=\lim\limits_{x\to\infty}\dfrac{a^{1/x}-1}{1/x}\overset{\star}{=}\lim\limits_{x\to\infty}\dfrac{a^{1/x}\ln a\left(-\frac{1}{x^2}\right)}{\left(-\frac{1}{x^2}\right)}=\ln a$

45. The limit does not exist if $b\neq 1$. Therefore, $b=1$.

$\lim\limits_{x\to 0}\dfrac{\cos ax-1}{2x^2}\overset{\star}{=}\lim\limits_{x\to 0}\dfrac{-a\sin ax}{4x}\overset{\star}{=}\lim\limits_{x\to 0}\dfrac{-a^2\cos ax}{4}=-\dfrac{a^2}{4}$

Now, $-\dfrac{a^2}{4}=-4$ implies $a=\pm 4$.

46. $\lim\limits_{x\to 0}\dfrac{\sin 2x+ax+bx^3}{x^3}\overset{\star}{=}\lim\limits_{x\to 0}\dfrac{2\cos 2x+a+3bx^2}{3x^2}$ need $a=-2$ to keep numerator 0

$\overset{\star}{=}\lim\limits_{x\to 0}\dfrac{-4\sin 2x+6bx}{6x}\overset{\star}{=}\lim\limits_{x\to 0}\dfrac{-8\cos 2x+6b}{6}=0$ if $6b=8$

$\Longrightarrow a=-2,\quad b=\dfrac{4}{3}$

47. Recall $\lim\limits_{x\to 0^+}(1+x)^{1/x}=e$:

$$\lim\limits_{x\to 0^+}\dfrac{(1+x)^{1/x}-e}{x}\overset{\star}{=}\lim\limits_{x\to 0^+}\left[(1+x)^{1/x}\right]\left[\dfrac{x-(1+x)\ln(1+x)}{x^2+x^3}\right]$$

$$=e\lim\limits_{x\to 0^+}\dfrac{x-(1+x)\ln(1+x)}{x^2+x^3}$$

$$\overset{\star}{=}e\lim\limits_{x\to 0^+}\dfrac{-\ln(1+x)}{2x+3x^2}$$

$$\overset{\star}{=}e\lim\limits_{x\to 0^+}\dfrac{-1/(1+x)}{2+6x}=-\dfrac{e}{2}$$

48. (a) $\lim\limits_{h\to 0}\dfrac{f(x+h)-f(x-h)}{2h}\overset{\star}{=}\lim\limits_{h\to 0}\dfrac{f'(x+h)-f'(x-h)(-1)}{2}=f'(x)$

(note that here we differentiated with respect to h, not x.)

(b) $\lim\limits_{h\to 0}\dfrac{f(x+h)-2f(x)+f(x-h)}{h^2}\overset{\star}{=}\lim\limits_{h\to 0}\dfrac{f'(x+h)-f'(x-h)}{2h}$

$\overset{\star}{=}\lim\limits_{h\to 0}\dfrac{f''(x+h)+f''(x-h)}{2}=f''(x)$

49. $\displaystyle\lim_{x\to 0}\frac{1}{x}\int_0^x f(t)\,dt \overset{\star}{=} \lim_{x\to 0}\frac{f(x)}{1} = f(0)$

50. (a) $\displaystyle\lim_{x\to 0}\frac{Si(x)}{x} \overset{\star}{=} \lim_{x\to 0}\frac{\sin x}{x} = 1$

(b)
$$\lim_{x\to 0}\frac{Si(x)-x}{x^3} \overset{\star}{=} \lim_{x\to 0}\frac{\sin x/x - 1}{3x^2} = \lim_{x\to 0}\frac{\sin x - x}{3x^3}$$
$$\overset{\star}{=} \lim_{x\to 0}\frac{\cos x - 1}{9x^2}$$
$$\overset{\star}{=} \lim_{x\to 0}\frac{-\sin x}{18x} = -\frac{1}{18}$$

51. (a) $\displaystyle\lim_{x\to 0}\frac{C(x)}{x} \overset{\star}{=} \lim_{x\to 0}\frac{\cos^2 x}{1} = 1$

(b) $\displaystyle\lim_{x\to 0}\frac{C(x)-x}{x^3} \overset{\star}{=} \lim_{x\to 0}\frac{\cos^2 x - 1}{3x^2} \overset{\star}{=} \lim_{x\to 0}\frac{-2\cos x \sin x}{6x} \overset{\star}{=} \lim_{x\to 0}\frac{-2\cos^2 x + 2\sin^2 x}{6} = -\frac{1}{3}$

52. (a) $\displaystyle\lim_{x\to a}\frac{\displaystyle\int_a^x f(t)\,dt}{f(x)} \overset{\star}{=} \lim_{x\to a}\frac{f(x)}{f'(x)} = 0$

(b) Similarly, $\displaystyle\lim_{x\to a}\frac{\displaystyle\int_a^x f(t)\,dt}{f(x)} \overset{\star}{=} \lim_{x\to a}\frac{f(x)}{f'(x)} \overset{\star}{=} \cdots \overset{\star}{=} \lim_{x\to a}\frac{f^{(k-1)}(x)}{f^{(k)}(x)} = 0$

53. $\displaystyle A(b) = 2\int_0^{\sqrt{b}}(b-x^2)\,dx = 2\left[bx - \frac{x^3}{3}\right]_0^{\sqrt{b}} = \frac{4}{3}\,b\sqrt{b}$ and $\displaystyle T(b) = \frac{1}{2}\left(2\sqrt{b}\right)b = b\sqrt{b}.$

Thus, $\displaystyle\lim_{b\to 0}\frac{T(b)}{A(b)} = \frac{b\sqrt{b}}{\frac{4}{3}\,b\sqrt{b}} = \frac{3}{4}.$

54. $\displaystyle T(\theta) = \frac{1}{2}(1-\cos\theta)\sin\theta; \quad S(\theta) = \frac{\theta}{2} - \frac{1}{2}\sin\theta:$
$$\lim_{\theta\to 0^+}\frac{T(\theta)}{S(\theta)} = \lim_{\theta\to 0^+}\frac{(1-\cos\theta)\sin\theta}{\theta - \sin\theta} \overset{\star}{=} \lim_{\theta\to 0^+}\frac{(1-\cos\theta)\cos\theta + \sin^2\theta}{1-\cos\theta} = \lim_{\theta\to 0^+}\frac{\cos\theta - \cos 2\theta}{1-\cos\theta}$$
$$\overset{\star}{=} \frac{-\sin\theta + 2\sin 2\theta}{\sin\theta} = \lim_{\theta\to 0^+}\frac{-\sin\theta + 4\sin\theta\cos\theta}{\sin\theta} = 3$$

55. (a) $f(x) \to \infty$ as $x \to \pm\infty$

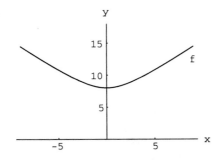

(b) $f(x) \to 10$ as $x \to 4$

Confirmation: $\displaystyle\lim_{x\to 4}\frac{x^2 - 16}{\sqrt{x^2+9}-5} \overset{\star}{=} \lim_{x\to 4}\frac{2x}{x\left(x^2+9\right)^{-1/2}} = \lim_{x\to 4}2\sqrt{x^2+9} = 10$

56. (a)

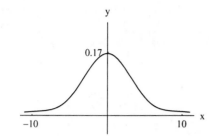

$$\lim_{x\to\pm\infty} f(x) = 0$$

(b) $f(x) \to \dfrac{1}{6}$ as $x \to 0$; $\displaystyle\lim_{x\to 0} \frac{x - \sin x}{x^3} \overset{\star}{=} \lim_{x\to 0} \frac{1 - \cos x}{3x^2} \overset{\star}{=} \lim_{x\to 0} \frac{\sin x}{6x} = \frac{1}{6}$

57. (a)

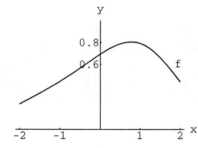

$f(x) \to 0.7$ as $x \to 0$

(b) Confirmation: $\displaystyle\lim_{x\to 0} \frac{2^{\sin x} - 1}{x} \overset{\star}{=} \lim_{x\to 0} \frac{\ln(2)\, 2^{\sin x} \cos x}{1} = \ln 2 \cong 0.6931$

58. (a)

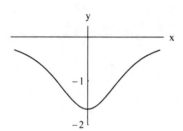

$g(x) \to -1.6$ as $x \to 0$

(b) Confirmation: $\displaystyle\lim_{x\to 0} \frac{3^{\cos x} - 3}{x^2} \overset{\star}{=} \lim_{x\to 0} \frac{3^{\cos x}(-\sin x)\ln 3}{2x} = -\lim_{x\to 0} 3^{\cos x}\, \frac{\sin x}{2x} \ln 3$

$$= -\frac{3\ln 3}{2} \cong -1.6479$$

SECTION 11.6

(We'll use $\star$ to indicate differentiation of numerator and denominator.)

1. $\displaystyle\lim_{x\to-\infty} \frac{x^2 + 1}{1 - x} \overset{\star}{=} \lim_{x\to-\infty} \frac{2x}{-1} = \infty$

2. $\displaystyle\lim_{x\to\infty} \frac{20x}{x^2 + 1} = 0$

3. $\displaystyle\lim_{x\to\infty} \frac{x^3}{1 - x^3} = \lim_{x\to\infty} \frac{1}{1/x^3 - 1} = -1$

4. $\displaystyle\lim_{x\to\infty} \frac{x^3 - 1}{2 - x} = -\infty$

5. $\displaystyle\lim_{x\to\infty} x^2 \sin \frac{1}{x} = \lim_{h\to 0^+} \left[\left(\frac{1}{h} \right) \left(\frac{\sin h}{h} \right) \right] = \infty$

6. $\lim\limits_{x\to\infty} \dfrac{\ln(x^k)}{x} \overset{\star}{=} \lim\limits_{x\to\infty} \dfrac{k/x}{1} = 0$

7. $\lim\limits_{x\to\frac{\pi}{2}^-} \dfrac{\tan 5x}{\tan x} = \lim\limits_{x\to\frac{\pi}{2}^-} \left[\left(\dfrac{\sin 5x}{\sin x}\right)\left(\dfrac{\cos x}{\cos 5x}\right)\right] = \dfrac{1}{5}$ since

$$\lim\limits_{x\to\frac{\pi}{2}^-} \dfrac{\sin 5x}{\sin x} = 1 \quad\text{and}\quad \lim\limits_{x\to\frac{\pi}{2}^-} \dfrac{\cos x}{\cos 5x} \overset{\star}{=} \lim\limits_{x\to\frac{\pi}{2}^-} \dfrac{\sin x}{5\sin 5x} = \dfrac{1}{5}$$

8. $\lim\limits_{x\to 0}(x\ln|\sin x|) = \lim\limits_{x\to 0} \dfrac{\ln|\sin x|}{1/x} \overset{\star}{=} \lim\limits_{x\to 0} \dfrac{\frac{\cos x}{\sin x}}{-\frac{1}{x^2}} = \lim\limits_{x\to 0}\left(-\dfrac{x}{\sin x}\right)(x\cos x) = 0$

9. $\lim\limits_{x\to 0^+} x^{2x} = \lim\limits_{x\to 0^+}(x^x)^2 = 1^2 = 1$ [see (11.6.4)]

10. $\lim\limits_{x\to\infty} x\sin\dfrac{\pi}{x} = \lim\limits_{t\to 0^+} \dfrac{\sin\pi t}{t} \overset{\star}{=} \lim\limits_{t\to 0^+} \dfrac{\pi\cos\pi t}{1} = \pi$

11. $\lim\limits_{x\to 0} x(\ln|x|)^2 = \lim\limits_{x\to 0} \dfrac{(\ln|x|)^2}{1/x} \overset{\star}{=} \lim\limits_{x\to 0} \dfrac{2\ln|x|}{-1/x} \overset{\star}{=} \lim\limits_{x\to 0} \dfrac{2}{1/x} = \lim\limits_{x\to 0} 2x = 0$

12. $\lim\limits_{x\to 0^+} \dfrac{\ln x}{\cot x} \overset{\star}{=} \lim\limits_{x\to 0^+} \dfrac{1/x}{-\csc^2 x} = \lim\limits_{x\to 0^+} -\dfrac{\sin^2 x}{x} = \lim\limits_{x\to 0^+} -\dfrac{\sin x}{x}\cdot\sin x = 0$

13. $\lim\limits_{x\to\infty} \dfrac{1}{x}\displaystyle\int_0^x e^{t^2}\,dt \overset{\star}{=} \lim\limits_{x\to\infty} \dfrac{e^{x^2}}{1} = \infty$

14. $\lim\limits_{x\to\infty} \dfrac{\sqrt{1+x^2}}{x} = \lim\limits_{x\to\infty} \sqrt{\dfrac{1}{x^2}+1} = 1$

15. $\lim\limits_{x\to 0}\left[\dfrac{1}{\sin^2 x} - \dfrac{1}{x^2}\right] = \lim\limits_{x\to 0} \dfrac{x^2 - \sin^2 x}{x^2\sin^2 x} \overset{\star}{=} \lim\limits_{x\to 0} \dfrac{2x - 2\sin x\cos x}{2x^2\sin x\cos x + 2x\sin^2 x}$

$$= \lim\limits_{x\to 0} \dfrac{2x - \sin 2x}{x^2\sin 2x + 2x\sin^2 x} \overset{\star}{=} \lim\limits_{x\to 0} \dfrac{2 - 2\cos 2x}{2x^2\cos 2x + 4x\sin 2x + 2\sin^2 x}$$

$$\overset{\star}{=} \lim\limits_{x\to 0} \dfrac{4\sin 2x}{-4x^2\sin 2x + 12x\cos 2x + 6\sin 2x}$$

$$\overset{\star}{=} \lim\limits_{x\to 0} \dfrac{8\cos 2x}{-8x^2\cos 2x - 32x\sin 2x + 24\cos 2x} = \dfrac{1}{3}$$

16. Since $\lim\limits_{x\to 0} \ln(|\sin x|^x) = \lim\limits_{x\to 0}(x\ln|\sin x|) = 0$ by Exercise 8, $\lim\limits_{x\to 0}|\sin x|^x = e^0 = 1$

17. $\lim\limits_{x\to 1} x^{1/(x-1)} = e$ since $\lim\limits_{x\to 1} \ln\left[x^{1/(x-1)}\right] = \lim\limits_{x\to 1} \dfrac{\ln x}{x-1} \overset{\star}{=} \lim\limits_{x\to 1} \dfrac{1}{x} = 1$

18. Take log:

$$\lim\limits_{x\to 0^+} \ln\left(x^{\sin x}\right) = \lim\limits_{x\to 0^+}(\sin x\ln x) = \lim\limits_{x\to 0^+}\left(\dfrac{\ln x}{\csc x}\right)$$

$$\overset{\star}{=} \lim\limits_{x\to 0^+} \dfrac{1/x}{-\csc x\cot x} = \lim\limits_{x\to 0^+} \dfrac{-\sin^2 x}{x\cos x} = 0, \quad\text{so}\quad \lim\limits_{x\to 0^+} x^{\sin x} = e^0 = 1$$

19.
$$\lim_{x \to \infty} \left(\cos \frac{1}{x} \right)^x = 1 \quad \text{since} \quad \lim_{x \to \infty} \ln \left[\left(\cos \frac{1}{x} \right)^x \right] = \lim_{x \to \infty} \frac{\ln \left(\cos \frac{1}{x} \right)}{(1/x)}$$

$$\overset{\star}{=} \lim_{x \to \infty} \left(-\frac{\sin (1/x)}{\cos (1/x)} \right) = 0$$

20. Take log:

$$\lim_{x \to \pi/2} \ln \left(| \sec x |^{\cos x} \right) = \lim_{x \to \pi/2} \cos x \ln | \sec x | = \lim_{x \to \pi/2} \frac{\ln | \sec x |}{\sec x}$$

$$\overset{\star}{=} \lim_{x \to \pi/2} \frac{\tan x}{\sec x \tan x} = \lim_{x \to \pi/2} \cos x = 0, \quad \text{so} \quad \lim_{x \to \pi/2} | \sec x |^{\cos x} = e^0 = 1$$

21.
$$\lim_{x \to 0} \left[\frac{1}{\ln (1 + x)} - \frac{1}{x} \right] = \lim_{x \to 0} \frac{x - \ln (1 + x)}{x \ln (1 + x)} \overset{\star}{=} \lim_{x \to 0} \frac{x}{x + (1 + x) \ln (1 + x)}$$

$$\overset{\star}{=} \lim_{x \to 0} \frac{1}{1 + 1 + \ln (1 + x)} = \frac{1}{2}$$

22. Take log: $\quad \lim_{x \to \infty} \ln(x^2 + a^2)^{(1/x)^2} = \lim_{x \to \infty} \frac{\ln(x^2 + a^2)}{x^2} \overset{\star}{=} \lim_{x \to \infty} \frac{\frac{2x}{x^2 + a^2}}{2x} = 0,$

so $\quad \lim_{x \to \infty} (x^2 + a^2)^{(1/x)^2} = e^0 = 1$

23.
$$\lim_{x \to 0} \left[\frac{1}{x} - \cot x \right] = \lim_{x \to 0} \frac{\sin x - x \cos x}{x \sin x} \overset{\star}{=} \lim_{x \to 0} \frac{x \sin x}{\sin x + x \cos x}$$

$$\overset{\star}{=} \lim_{x \to 0} \frac{\sin x + x \cos x}{2 \cos x - x \sin x} = 0$$

24. $\quad \lim_{x \to \infty} \ln \left(\frac{x^2 - 1}{x^2 + 1} \right)^3 = 3 \lim_{x \to \infty} \ln \left(\frac{x^2 - 1}{x^2 + 1} \right) = 0$

25. $\quad \lim_{x \to \infty} \left(\sqrt{x^2 + 2x} - x \right) = \lim_{x \to \infty} \left[\left(\sqrt{x^2 + 2x} - x \right) \left(\frac{\sqrt{x^2 + 2x} + x}{\sqrt{x^2 + 2x} + x} \right) \right]$

$$= \lim_{x \to \infty} \frac{2x}{\sqrt{x^2 + 2x} + x} = \lim_{x \to \infty} \frac{2}{\sqrt{1 + 2/x} + 1} = 1$$

26. $\quad \lim_{x \to \infty} \left(1 + \frac{a}{x} \right)^{bx} = \lim_{x \to \infty} \left[\left(1 + \frac{a}{x} \right)^x \right]^b = (e^a)^b = e^{ab}.$

27. $\quad \lim_{x \to \infty} \left(x^3 + 1 \right)^{1/\ln x} = e^3 \quad \text{since}$

$$\lim_{x \to \infty} \ln \left[\left(x^3 + 1 \right)^{1/\ln x} \right] = \lim_{x \to \infty} \frac{\ln \left(x^3 + 1 \right)}{\ln x} \overset{\star}{=} \lim_{x \to \infty} \frac{\left(\frac{3x^2}{x^3 + 1} \right)}{1/x} = \lim_{x \to \infty} \frac{3}{1 + 1/x^3} = 3.$$

28. Take log: $\quad \lim_{x \to \infty} \frac{\ln(e^x + 1)}{x} \overset{\star}{=} \lim_{x \to \infty} \frac{\frac{e^x}{e^x + 1}}{1} \overset{\star}{=} \lim_{x \to \infty} \frac{e^x}{e^x} = 1,$

so $\quad \lim_{x \to \infty} (e^x + 1)^{1/x} = e$

29. $\lim\limits_{x\to\infty}(\cosh x)^{1/x}=e$ since

$$\lim_{x\to\infty}\ln\left[(\cosh x)^{1/x}\right]=\lim_{x\to\infty}\frac{\ln(\cosh x)}{x}\overset{\star}{=}\lim_{x\to\infty}\frac{\sinh x}{\cosh x}=1.$$

30. Take log: $\lim\limits_{x\to\infty}3x\ln\left(1+\dfrac{1}{x}\right)=3\lim\limits_{x\to\infty}\dfrac{\ln\left(1+\frac{1}{x}\right)}{\frac{1}{x}}\overset{\star}{=}3\lim\limits_{x\to\infty}\dfrac{\frac{-1/x^2}{1+1/x}}{-\frac{1}{x^2}}=3,$

so $\lim\limits_{x\to\infty}\left(1+\dfrac{1}{x}\right)^{3x}=e^3$

31. $\lim\limits_{x\to0}\left(\dfrac{1}{\sin x}-\dfrac{1}{x}\right)=\lim\limits_{x\to0}\dfrac{x-\sin x}{x\sin x}\overset{\star}{=}\lim\limits_{x\to0}\dfrac{1-\cos x}{\sin x+x\cos x}$

$$\overset{\star}{=}\lim_{x\to0}\frac{\sin x}{2\cos x-x\sin x}=0$$

32. Take log: $\lim\limits_{x\to0}\dfrac{\ln(e^x+3x)}{x}\overset{\star}{=}\lim\limits_{x\to0}\dfrac{\frac{e^x+3}{e^x+3x}}{1}=4,$ so $\lim\limits_{x\to0}(e^x+3x)^{1/x}=e^4$

33. $$\lim_{x\to1}\left(\frac{1}{\ln x}-\frac{x}{x-1}\right)=\lim_{x\to1}\frac{x-1-x\ln x}{(x-1)\ln x}\overset{\star}{=}\lim_{x\to1}\frac{-\ln x}{(x-1)(1/x)+\ln x}$$

$$=\lim_{x\to1}\frac{-x\ln x}{x-1+x\ln x}\overset{\star}{=}\lim_{x\to1}\frac{-\ln x-1}{2+\ln x}=-\frac{1}{2}$$

34. $\sqrt{2}$:   take log: $\lim\limits_{x\to0}\dfrac{\ln(1+2^x)-\ln2}{x}\overset{\star}{=}\lim\limits_{x\to0}\dfrac{\frac{2^x\ln2}{1+2^x}}{1}=\dfrac{\ln2}{2};$ $\lim\limits_{x\to0}\left(\dfrac{1+2^x}{2}\right)^{1/x}=e^{\frac{1}{2}\ln2}=\sqrt{2}$

35. 0: $\dfrac{1}{n}\ln\dfrac{1}{n}=-\dfrac{\ln n}{n}\to0$

36. 0: $\lim\limits_{n\to\infty}\dfrac{n^k}{2^n}\to0$

37. 1: $\ln\left[(\ln n)^{1/n}\right]=\dfrac{1}{n}\ln(\ln n)\to0$

38. 0: $\lim\limits_{n\to\infty}\dfrac{\ln n}{n^p}\overset{\star}{=}\lim\limits_{n\to\infty}\dfrac{1/n}{p\,n^{p-1}}=\lim\limits_{n\to\infty}\dfrac{1}{p\,n^p}=0$

39. 1: $\ln\left[(n^2+n)^{1/n}\right]=\dfrac{1}{n}\ln[n(n+1)]=\dfrac{\ln n}{n}+\dfrac{\ln(n+1)}{n}\to0$

40. 1: $\lim\limits_{n\to\infty}\ln\left(n^{\sin(\pi/n)}\right)=\lim\limits_{n\to\infty}[\sin(\pi/n)\ln n]=\lim\limits_{n\to\infty}\left(\dfrac{\sin(\pi/n)}{1/n}\right)\left(\dfrac{\ln n}{n}\right)=0,$ so $n^{\sin(\pi/n)}\to1$

41. 0: $0\le\dfrac{n^2\ln n}{e^n}<\dfrac{n^3}{e^n},\quad\lim\limits_{x\to\infty}\dfrac{x^3}{e^x}=0$

42. 1:   take log: $\ln(\sqrt{n}-1)^{1/\sqrt{n}}=\dfrac{\ln(\sqrt{n}-1)}{\sqrt{n}}\to0,$ so $(\sqrt{n}-1)^{1/\sqrt{n}}\to1$

43. $\lim\limits_{x\to0}(\sin x)^x=1$

44. $\lim\limits_{x\to\pi/4}(\tan x)^{\tan2x}=\dfrac{1}{e}$

45. $\displaystyle\lim_{x\to 0}\left(\frac{1}{\sin x}-\frac{1}{\tan x}\right)=0$

46. $\displaystyle\lim_{x\to 0^+}(\sinh x)^{-x}=1$

47.

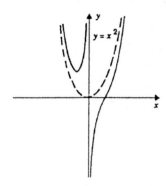

vertical asymptote y-axis

48.

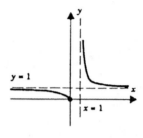

vertical asymptote $x=1$
horizontal asymptote $y=1$

49.

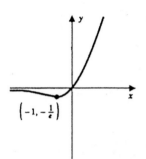

horizontal asymptote x-axis

50.

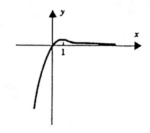

horizontal asymptote x-axis

51.

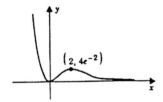

horizontal asymptote x-axis

52.

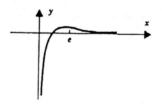

vertical asymptote y-axis
horizontal asymptote x-axis

53. $\displaystyle\frac{b}{a}\sqrt{x^2-a^2}-\frac{b}{a}x=\frac{\sqrt{x^2-a^2}+x}{\sqrt{x^2-a^2}+x}\left(\frac{b}{a}\right)\left(\sqrt{x^2-a^2}-x\right)=\frac{-ab}{\sqrt{x^2-a^2}+x}\to 0$ as $x\to\infty$

54. $\displaystyle\cosh x-\sinh x=\frac{1}{2}(e^x+e^{-x})-\frac{1}{2}(e^x-e^{-x})=e^{-x}\to 0,$ as $x\to\infty$

55. for instance, $f(x)=x^2+\dfrac{(x-1)(x-2)}{x^3}$

56. for instance, $F(x)=x+\dfrac{\sin x}{1+x^2}$

57. $\displaystyle\lim_{x\to 0^+}-\frac{2x}{\cos x}\neq\lim_{x\to 0^+}\frac{2}{-\sin x}.$ L'Hospital's rule does not apply here since $\displaystyle\lim_{x\to 0^+}\cos x=1.$

58. (a) Let S be the set of positive integers for which the statement is true. Since $\lim\limits_{x\to\infty} \dfrac{\ln x}{x} = 0$, $1 \in S$.

Assume that $k \in S$. By L'Hospital's rule,

$$\lim_{x\to\infty} \frac{(\ln x)^{k+1}}{x} \overset{\star}{=} \lim_{x\to\infty} \frac{(k+1)(\ln x)^k}{x} = 0 \quad (\text{since}\quad k \in S).$$

Thus $k + 1 \in S$, and S is the set of positive integers.

(b) Choose any positive number α. Choose a positive integer $k > \alpha$.

Then, for $x > e$,

$$0 < \frac{(\ln x)^\alpha}{x} < \frac{(\ln x)^k}{x}$$

and the result follows by the pinching theorem and part (a).

59. Let $y = \left(\dfrac{a^{1/x} + b^{1/x}}{2}\right)^x$. Then $\ln y = x \ln\left[\dfrac{a^{1/x} + b^{1/x}}{2}\right] = \dfrac{\ln\left(\dfrac{a^{1/x} + b^{1/x}}{2}\right)}{1/x}$.

Now,

$$\lim_{x\to\infty} \frac{\ln\left(\dfrac{a^{1/x} + b^{1/x}}{2}\right)}{1/x} \overset{\star}{=} \lim_{x\to\infty} \frac{\dfrac{2}{a^{1/x} + b^{1/x}} \cdot \dfrac{-a^{1/x}\ln a - b^{1/x}\ln b}{2x^2}}{-1/x^2}$$

$$= \lim_{x\to\infty} \frac{a^{1/x}\ln a + b^{1/x}\ln b}{a^{1/x} + b^{1/x}} = \tfrac{1}{2}\ln ab = \ln\sqrt{ab}$$

Thus, $\lim\limits_{x\to\infty} \ln y = \ln\sqrt{ab} \implies \lim\limits_{x\to\infty} y = \sqrt{ab}$.

60. (a) $\lim\limits_{k\to 0^+} v(t) = \lim\limits_{k\to 0^+} \dfrac{mg\left(1 - e^{-(k/m)t}\right)}{k} \overset{\star}{=} \lim\limits_{k\to 0^+} \dfrac{gte^{-(k/m)t}}{1} = gt$

(b) $\dfrac{dv}{dt} = g \implies v(t) = gt + C; \quad v(0) = 0 \implies C = 0 \quad\text{and}\quad v(t) = gt.$

61. (a) $A_b = 1 - (1 + b)e^{-b}$

(b) $\overline{x}_b = \dfrac{2 - \left(2 + 2b + b^2\right)e^{-b}}{1 - (1 + b)e^{-b}}; \qquad \overline{y}_b = \dfrac{\tfrac{1}{4} - \tfrac{1}{4}\left(1 + 2b + 2b^2\right)e^{-2b}}{2\left[1 - (1 + b)e^{-b}\right]}$

(c) $\lim\limits_{b\to\infty} A_b = 1; \quad \lim\limits_{b\to\infty} \overline{x}_b = 2; \quad \lim\limits_{b\to\infty} \overline{y}_b = \dfrac{1}{8}$

62. (a) $V_x = \pi\left[\dfrac{1}{4} - \dfrac{1}{4}\left(1 + 2b + 2b^2\right)e^{-2b}\right]$

(b) $V_y = 2\pi\left[2 - \left(2 + 2b + b^2\right)e^{-b}\right]$

(c) $\lim\limits_{b\to\infty} V_x = \dfrac{\pi}{4}, \qquad \lim\limits_{b\to\infty} V_y = 4\pi$

63. (a)
$$\lim_{x\to 0^+}\left(1+x^2\right)^{1/x}=1$$

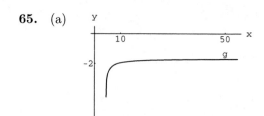

(b) $\displaystyle\lim_{x\to 0^+}\left(1+x^2\right)^{1/x}=1$ since

$$\lim_{x\to 0^+}\ln\left[\left(1+x^2\right)^{1/x}\right]=\lim_{x\to 0^+}\frac{\ln\left(1+x^2\right)}{x}\overset{\star}{=}\lim_{x\to 0^+}\frac{2x}{\left(1+x^2\right)}=0$$

64. (a)
$$\lim_{x\to\infty}f(x)\cong 1.5$$

(b) $\displaystyle\lim_{x\to\infty}f(x)=\lim_{x\to\infty}\left[\sqrt{x^2+3x+1}-x\right]$

$$=\lim_{x\to\infty}\frac{3x+1}{\sqrt{x^2+3x+1}+x}=\lim_{x\to\infty}\frac{3x+1}{x\sqrt{1+(3/x)+1/x^2}+x}=\frac{3}{2}$$

65. (a)
$$\lim_{x\to\infty}g(x)\cong -1.7.$$

(b) $\displaystyle\lim_{x\to\infty}g(x)=\lim_{x\to\infty}\left[\sqrt[3]{x^3-5x^2+2x+1}-x\right]$

$$=\lim_{x\to\infty}\frac{-5x^2+2x+1}{\left(\sqrt[3]{x^3-5x^2+2x+1}\right)^2+x\sqrt[3]{x^3-5x^2+2x+1}+x^2}$$

$$=-\frac{5}{3}\cong -1.667$$

66. $[P(x)]^{1/n}-x=\left([P(x)]^{1/n}-x\right)\cdot\dfrac{[P(x)]^{(n-1)/n}+x[P(x)]^{(n-2)/n}+\cdots+x^{n-2}[P(x)]^{1/n}+x^{n-1}}{[P(x)]^{(n-1)/n}+x[P(x)]^{(n-2)/n}+\cdots+x^{n-2}[P(x)]^{1/n}+x^{n-1}}$

$$=\frac{P(x)-x^n}{[P(x)]^{(n-1)/n}+x[P(x)]^{(n-2)/n}+\cdots+x^{n-2}[P(x)]^{1/n}+x^{n-1}}$$

$$=\frac{b_1x^{n-1}+b_2x^{n-2}+\cdots+b_n}{[P(x)]^{(n-1)/n}+x[P(x)]^{(n-2)/n}+\cdots+x^{n-2}[P(x)]^{1/n}+x^{n-1}}\to\frac{b_1}{n}\quad\text{as}\quad x\to\infty$$

SECTION 11.7

1. 1 : $\displaystyle\int_1^\infty \frac{dx}{x^2} = \lim_{b\to\infty} \int_1^b \frac{dx}{x^2} = \lim_{b\to\infty}\left[-\frac{1}{x}\right]_1^b = \lim_{b\to\infty}\left[1-\frac{1}{b}\right] = 1$

2. $\displaystyle\frac{\pi}{2}$: $\displaystyle\int_0^\infty \frac{dx}{1+x^2} = \lim_{b\to\infty} \tan^{-1} b = \frac{\pi}{2}$

3. $\displaystyle\frac{\pi}{4}$: $\displaystyle\int_0^\infty \frac{dx}{4+x^2} = \lim_{b\to\infty}\int_0^b \frac{dx}{4+x^2} = \lim_{b\to\infty}\left[\frac{1}{2}\tan^{-1}\frac{x}{2}\right]_0^b = \lim_{b\to\infty}\frac{1}{2}\tan^{-1}\left(\frac{b}{2}\right) = \frac{\pi}{4}$

4. $\displaystyle\frac{1}{p}$: $\displaystyle\int_0^\infty e^{-px}\,dx = \lim_{b\to 0}\left(-\frac{e^{-pb}}{p}+\frac{1}{p}\right) = \frac{1}{p}$

5. diverges: $\displaystyle\int_0^\infty e^{px}\,dx = \lim_{b\to\infty}\int_0^b e^{px}\,dx = \lim_{b\to\infty}\left[\frac{1}{p}e^{px}\right]_0^b = \lim_{b\to\infty}\frac{1}{p}\left(e^{pb}-1\right) = \infty$

6. 2 : $\displaystyle\int_0^1 \frac{dx}{\sqrt{x}} = \lim_{a\to 0^+}\int_a^1 \frac{dx}{\sqrt{x}} = \lim_{a\to 0^+}\left(2-2\sqrt{a}\right) = 2$

7. 6: $\displaystyle\int_0^8 \frac{dx}{x^{2/3}} = \lim_{a\to 0^+}\int_a^8 x^{-2/3}\,dx = \lim_{a\to 0^+}\left[3x^{1/3}\right]_a^8 = \lim_{a\to 0^+}\left[6-3a^{1/3}\right] = 6$

8. diverges: $\displaystyle\int_0^1 \frac{dx}{x^2} = \lim_{a\to 0^+}\int_a^1 \frac{dx}{x^2} = \lim_{a\to 0^+}\left(-1+\frac{1}{a}\right) = \infty$

9. $\displaystyle\frac{\pi}{2}$: $\displaystyle\int_0^1 \frac{dx}{\sqrt{1-x^2}} = \lim_{b\to 1^-}\int_0^b \frac{dx}{\sqrt{1-x^2}} = \lim_{b\to 1^-}\sin^{-1} b = \frac{\pi}{2}$

10. 2: $\displaystyle\int_0^1 \frac{dx}{\sqrt{1-x}} = \lim_{a\to 0^+}\int_a^1 \frac{dx}{\sqrt{1-x}} = \lim_{a\to 0^+}\left[-2\sqrt{1-x}\right]_a^1 = \lim_{a\to 0^+}\left(0+2\sqrt{1-a}\right) = 2$

11. 2: $\displaystyle\int_0^2 \frac{x}{\sqrt{4-x^2}}\,dx = \lim_{b\to 2^-}\int_0^b x\left(4-x^2\right)^{-1/2}\,dx = \lim_{b\to 2^-}\left[-\left(4-x^2\right)^{1/2}\right]_0^b$

$\displaystyle\qquad\qquad\qquad\qquad\qquad = \lim_{b\to 2^-}\left(2-\sqrt{4-b^2}\right) = 2$

12. $\displaystyle\frac{\pi}{2}$: $\displaystyle\int_0^a \frac{dx}{\sqrt{a^2-x^2}} = \lim_{b\to a^-}\int_0^b \frac{dx}{\sqrt{a^2-x^2}} = \lim_{b\to a^-}\left[\sin^{-1}\left(\frac{b}{a}\right)-\sin^{-1}(0)\right] = \frac{\pi}{2}$

13. diverges: $\displaystyle\int_e^\infty \frac{\ln x}{x}\,dx = \lim_{b\to\infty}\int_e^b \frac{\ln x}{x}\,dx = \lim_{b\to\infty}\left[\frac{1}{2}\left(\ln x\right)^2\right]_e^b = \lim_{b\to\infty}\left[\frac{1}{2}\left(\ln b\right)^2 - \frac{1}{2}\right] = \infty$

14. diverges: $\displaystyle\int_e^\infty \frac{dx}{x\ln x} = \lim_{b\to\infty}\int_e^b \frac{dx}{x\ln x} = \lim_{b\to\infty}\left[\ln(\ln x)\right]_e^b = \infty$

15. $-\dfrac{1}{4}$:

$$\int_0^1 x \ln x \, dx = \lim_{a \to 0^+} \int_a^1 x \ln x \, dx = \lim_{a \to 0^+} \left[\frac{1}{2} x^2 \ln x - \frac{1}{4} x^2 \right]_a^1$$

(by parts) ⤴

$$= \lim_{a \to 0^+} \left[\frac{1}{4} a^2 - \frac{1}{2} a^2 \ln a - \frac{1}{4} \right] = -\frac{1}{4}$$

Note: $\lim\limits_{t \to 0^+} t^2 \ln t = \lim\limits_{t \to 0^+} \dfrac{\ln t}{1/t^2} \overset{\star}{=} \lim\limits_{t \to 0^+} \dfrac{1/t}{-2/t^3} = -\dfrac{1}{2} \lim\limits_{t \to 0^+} t^2 = 0.$

16. 1: $\displaystyle\int_e^\infty \frac{dx}{x(\ln x)^2} = \lim_{b \to \infty} \int_e^b \frac{dx}{x(\ln x)^2} = \lim_{b \to \infty} \left[-\frac{1}{\ln x} \right]_e^b = \lim_{b \to \infty} \left[-\frac{1}{\ln b} + 1 \right] = 1$

17. π:

$$\int_{-\infty}^\infty \frac{dx}{1 + x^2} = \lim_{a \to -\infty} \int_a^0 \frac{dx}{1 + x^2} + \lim_{b \to \infty} \int_0^b \frac{dx}{1 + x^2}$$

$$= \lim_{a \to -\infty} \left[\tan^{-1} x \right]_a^0 + \lim_{b \to \infty} \left[\tan^{-1} x \right]_0^b = -\left(-\frac{\pi}{2} \right) + \frac{\pi}{2} = \pi$$

18. $\dfrac{\ln 3}{2}$:

$$\int_2^\infty \frac{dx}{x^2 - 1} = \lim_{b \to \infty} \int_2^b \frac{dx}{x^2 - 1} = \lim_{b \to \infty} \left[\frac{1}{2} \ln \left| \frac{x - 1}{x + 1} \right| \right]_2^b$$

$$= \lim_{b \to \infty} \left[\frac{1}{2} \ln \left(\frac{b - 1}{b + 1} \right) + \frac{1}{2} \ln 3 \right] = \frac{1}{2} \ln 3$$

19. diverges: $\displaystyle\int_{-\infty}^\infty \frac{dx}{x^2} = \lim_{a \to -\infty} \int_a^{-1} \frac{dx}{x^2} + \lim_{b \to 0^-} \int_{-1}^b \frac{dx}{x^2} + \lim_{c \to 0^+} \int_c^1 \frac{dx}{x^2} + \lim_{d \to \infty} \int_1^d \frac{dx}{x^2} ;$

and, $\displaystyle\lim_{c \to 0^+} \int_c^1 \frac{dx}{x^2} = \lim_{c \to 0^+} \left[-\frac{1}{x} \right]_c^1 = \lim_{c \to 0^+} \left[\frac{1}{c} - 1 \right] = \infty$

20. 2: $\displaystyle\int_{1/3}^3 \frac{dx}{\sqrt[3]{3x - 1}} = \lim_{a \to \frac{1}{3}^+} \int_a^3 \frac{dx}{(3x - 1)^{1/3}} = \lim_{a \to \frac{1}{3}^+} \left[\frac{3(3x - 1)^{2/3}}{2 \cdot 3} \right]_a^3$

$$= \lim_{a \to \frac{1}{3}^+} \left[\frac{8^{2/3}}{2} - \frac{(3a - 1)^{2/3}}{2} \right] = 2$$

21. $\ln 2$:

$$\int_1^\infty \frac{dx}{x(x + 1)} = \lim_{b \to \infty} \int_1^b \left[\frac{1}{x} - \frac{1}{x + 1} \right] dx$$

$$= \lim_{b \to \infty} \left[\ln \left(\frac{x}{x + 1} \right) \right]_1^b = \lim_{b \to \infty} \left[\ln \left(\frac{b}{b + 1} \right) - \ln \left(\frac{1}{2} \right) \right] = \ln 2$$

22. -1 : $\displaystyle\int_{-\infty}^0 x e^x \, dx = \lim_{a \to -\infty} \int_a^0 x e^x \, dx = \lim_{a \to -\infty} \left[x e^x - e^x \right]_a^0 = \lim_{a \to -\infty} \left[-1 - a e^a + e^a \right] = -1$

23. 4: $\displaystyle\int_3^5 \frac{x}{\sqrt{x^2 - 9}} \, dx = \lim_{a \to 3^-} \int_a^5 x \left(x^2 - 9 \right)^{-1/2} dx$

$$= \lim_{a \to 3^-} \left[\left(x^2 - 9 \right)^{1/2} \right]_a^5 = \lim_{a \to 3^-} \left[4 - \left(a^2 - 9 \right)^{1/2} \right] = 4$$

24.

$$\int_1^4 \frac{dx}{x^2-4} = \lim_{b\to 2^-}\int_1^b \frac{dx}{x^2-4} + \lim_{a\to 2^+}\int_a^4 \frac{dx}{x^2-4} = \lim_{b\to 2^-}\left[\frac14\ln\left|\frac{x-2}{x+2}\right|\right]_1^b + \lim_{a\to 2^+}\left[\frac14\ln\left|\frac{x-2}{x+2}\right|\right]_a^4$$

$$= \lim_{b\to 2^-}\left(\frac14\ln\left|\frac{b-2}{b+2}\right| - \frac14\ln\frac13\right) + \lim_{a\to 2^+}\left(\frac14\ln\frac26 - \frac14\ln\left|\frac{a-2}{a+2}\right|\right) = \infty \quad \text{diverges}$$

25. $\int_{-3}^3 \frac{dx}{x(x+1)}$ diverges since $\int_0^3 \frac{dx}{x(x+1)}$ diverges:

$$\int_0^3 \frac{dx}{x(x+1)} = \lim_{a\to 0^+}\int_a^3 \left(\frac1x - \frac1{x+1}\right)dx = \lim_{a\to 0^+}\left[\ln|x| - \ln|x+1|\right]_a^3$$

$$= \lim_{a\to 0^+}\left[\ln 3 - \ln 4 - \ln a + \ln(a+1)\right] = \infty.$$

26. $\frac14:$ $\int_1^\infty \frac{x}{(1+x^2)^2}\,dx = \lim_{b\to\infty}\int_1^b \frac{x}{(1+x^2)^2}\,dx = \lim_{b\to\infty}\left[\frac{-1}{2(1+x^2)}\right]_1^b = \lim_{b\to\infty}\left[\frac{-1}{2(1+b^2)}+\frac14\right] = \frac14$

27. $\int_{-3}^1 \frac{dx}{x^2-4}$ diverges since $\int_{-2}^1 \frac{dx}{x^2-4}$ diverges:

$$\int_{-2}^1 \frac{dx}{x^2-4} = \lim_{a\to -2^+}\int_a^1 \frac14\left[\frac1{x-2}-\frac1{x+2}\right]dx$$

$$= \lim_{a\to -2^+}\left[\frac14\left(\ln|x-2| - \ln|x+2|\right)\right]_a^1$$

$$= \lim_{a\to -2^+}\frac14\left[-\ln 3 - \ln|a-2| + \ln|a+2|\right] = -\infty.$$

28. $\frac{\pi}{2}:$

$$\int_0^\infty \frac1{e^x+e^{-x}}\,dx = \lim_{b\to\infty}\int_0^b \frac1{e^x+e^{-x}}\,dx = \lim_{b\to\infty}\left[\tan^{-1}e^x\right]_0^b = \frac{\pi}{2}-\frac{\pi}{4}$$

$$\int_{-\infty}^0 \frac1{e^x+e^{-x}}\,dx = \lim_{b\to-\infty}\int_b^0 \frac1{e^x+e^{-x}}\,dx = \lim_{b\to-\infty}\left[\tan^{-1}e^x\right]_b^0 = \frac{\pi}{4}$$

29. diverges: $\int_0^\infty \cosh x\,dx = \lim_{b\to\infty}\int_0^b \cosh x\,dx = \lim_{b\to\infty}\left[\sinh x\right]_0^b = \infty$

30. Since $\int_1^2 \frac{dx}{x^2-5x+6} = \lim_{b\to 2^-}\int_1^b \frac{dx}{(x-2)(x-3)} = \lim_{b\to 2^-}\left[\ln\left|\frac{x-3}{x-2}\right|\right]_1^b = \lim_{b\to 2^-}\left(\ln\left|\frac{b-3}{b-2}\right| - \ln 2\right)$

diverges, so does $\int_1^4 \frac{dx}{x^2-5x+6}$

31. $\frac12:$ $\int_0^\infty e^{-x}\sin x\,dx = \lim_{b\to\infty}\int_0^b e^{-x}\sin x\,dx = \lim_{b\to\infty}-\frac12\left[e^{-x}\cos x + e^{-x}\sin x\right]_0^b$

(by parts)⬆

$$= \lim_{b\to\infty}\frac12\left[1 - e^{-b}\cos b - e^{-b}\sin b\right] = \frac12$$

32. diverges: $\displaystyle\int_0^\infty \cos^2 x \, dx = \lim_{b\to\infty}\left[\frac{x}{2} + \frac{\sin 2x}{4}\right]_0^b = \lim_{b\to\infty}\left(\frac{b}{2} + \frac{\sin 2b}{4}\right) = \infty$

33. $2e - 2$: $\displaystyle\int_0^1 \frac{e^{\sqrt{x}}}{\sqrt{x}} \, dx = \lim_{a\to 0^+}\int_a^1 \frac{e^{\sqrt{x}}}{\sqrt{x}} \, dx = \lim_{a\to 0^+}\left[2\,e^{\sqrt{x}}\right]_a^1 = 2(e-1)$

34. 2 : $\displaystyle\int_0^{\pi/2} \frac{\cos x}{\sqrt{\sin x}} \, dx = \lim_{a\to 0^+}\int_a^{\pi/2} \frac{\cos x}{\sqrt{\sin x}} \, dx = \lim_{a\to 0^+}\left[2\sqrt{\sin x}\right]_a^{\pi/2} = 2$

35. (a) converges: $\displaystyle\int_0^\infty \frac{x}{(16+x^2)^2} \, dx = \frac{1}{32}$ (b) converges: $\displaystyle\int_0^\infty \frac{x^2}{(16+x^2)^2} \, dx = \frac{\pi}{16}$

    (c) converges: $\displaystyle\int_0^\infty \frac{x}{16+x^4} \, dx = \frac{\pi}{16}$ (d) diverges

36. (a) converges: $\displaystyle\int_0^2 \frac{x^3}{\sqrt[3]{2-x}} \, dx = \frac{243\sqrt[3]{4}}{55}$ (b) converges: $\displaystyle\int_0^2 \frac{1}{\sqrt{2-x}} \, dx = 2\sqrt{2}$

    (c) converges: $\displaystyle\int_0^2 \frac{x}{\sqrt{2-x}} \, dx = \frac{8\sqrt{2}}{3}$ (d) converges: $\displaystyle\int_0^2 \frac{1}{\sqrt{2x-x^2}} \, dx = \pi$

37. $\displaystyle\int_0^1 \sin^{-1} x \, dx = \left[x\sin^{-1}x\right]_0^1 - \int_0^1 \frac{x}{\sqrt{1-x^2}} \, dx = \frac{\pi}{2} - \lim_{a\to 1^-}\int_0^a \frac{x}{\sqrt{1-x^2}} \, dx$

    (by parts)

    Now, $\displaystyle\int_0^a \frac{x}{\sqrt{1-x^2}} \, dx = -\frac{1}{2}\int_1^{1-a^2} \frac{1}{\sqrt{u}} \, du = \left[-\sqrt{u}\right]_1^{1-a^2} = 1 - \sqrt{1-a^2}$

      $u = 1 - x^2$

    Thus, $\displaystyle\int_0^1 \sin^{-1} x \, dx = \frac{\pi}{2} - \lim_{a\to 1^-}\left(1 - \sqrt{1-a^2}\right) = \frac{\pi}{2} - 1.$

38. (a) $\displaystyle\int_0^\infty x^r e^{-x} \, dx$ diverges if $r \le -1$:

    $\displaystyle\int_0^\infty x^r e^{-x} \, dx = \int_0^1 x^r e^{-x} \, dx + \int_1^\infty x^r e^{-x} \, dx$ and $\displaystyle\int_0^1 x^r e^{-x} \, dx$ diverges.

    For any $r > -1$, we can find k such that $x^r < e^{x/2}$ for $x \ge k$ ($e^{x/2}$ grows faster than any power of x). Then $\displaystyle\int_0^\infty x^r e^{-x} \, dx < \int_0^k x^r e^{-x} \, dx + \int_k^\infty e^{-x/2} \, dx$, which converges. Thus $\displaystyle\int_0^\infty x^r e^{-x} \, dx$ converges for all $r > -1$.

    (b) For $n = 1$: $\displaystyle\int_0^\infty x e^{-x} \, dx = \lim_{b\to\infty}\left[-x e^{-x} - e^{-x}\right]_0^b = \lim_{b\to\infty}\left[-b e^{-b} - e^{-b} + 1\right] = 1$

      Assume true for n. $\displaystyle\int_0^\infty x^{n+1} e^{-x} \, dx = \lim_{b\to\infty}\left(\left[-x^{n+1} e^{-x}\right]_0^b + (n+1)\int_0^b x^n e^{-x} \, dx\right)$

      $= \lim_{b\to\infty}\left(-b^{n+1} e^{-b}\right) + (n+1)\int_0^\infty x^n e^{-x} \, dx = 0 + (n+1)n! = (n+1)!$

39.

$$\int_0^\infty \frac{1}{\sqrt{x}\,(1+x)}\,dx = \int_0^1 \frac{1}{\sqrt{x}\,(1+x)}\,dx + \int_1^\infty \frac{1}{\sqrt{x}\,(1+x)}\,dx$$

$$= \lim_{a\to 0^+} \int_a^1 \frac{1}{\sqrt{x}\,(1+x)}\,dx + \lim_{b\to\infty} \int_1^b \frac{1}{\sqrt{x}\,(1+x)}\,dx$$

Now, $\displaystyle\int \frac{1}{\sqrt{x}\,(1+x)}\,dx = \int \frac{2}{1+u^2}\,du = 2\arctan u + C = 2\arctan\sqrt{x} + C.$

$u = \sqrt{x}$

Therefore, $\displaystyle\lim_{a\to 0^+}\int_a^1 \frac{1}{\sqrt{x}\,(1+x)}\,dx = \lim_{a\to 0^+}\left[2\arctan\sqrt{x}\right]_a^1 = \lim_{a\to 0^+} 2\left[\pi/4 - \arctan\sqrt{a}\right] = \frac{\pi}{2}$

and $\displaystyle\lim_{b\to\infty}\int_1^b \frac{1}{\sqrt{x}\,(1+x)}\,dx = \lim_{b\to\infty}\left[2\arctan\sqrt{x}\right]_1^b = \lim_{b\to\infty} 2\left[\arctan\sqrt{b} - \pi/4\right] = \frac{\pi}{2}.$

Thus, $\displaystyle\int_0^\infty \frac{1}{\sqrt{x}\,(1+x)}\,dx = \pi.$

40.

$$\int_1^\infty \frac{1}{x\sqrt{x^2-1}}\,dx = \lim_{a\to 1^+}\int_a^2 \frac{1}{x\sqrt{x^2-1}}\,dx + \lim_{b\to\infty}\int_2^b \frac{1}{x\sqrt{x^2-1}}\,dx$$

$$= \lim_{a\to 1^+}\left[\sec^{-1}x\right]_a^2 + \lim_{b\to\infty}\left[\sec^{-1}x\right]_2^\infty$$

$$= \lim_{a\to 1^+}\left(\sec^{-1}2 - \sec^{-1}a\right) + \lim_{b\to\infty}\left(\sec^{-1}b - \sec^{-1}2\right)$$

$$= (\sec^{-1}2 - 0) + \left(\frac{\pi}{2} - \sec^{-1}2\right) = \frac{\pi}{2}$$

41. surface area $\displaystyle S = \int_1^\infty 2\pi\left(\frac{1}{x}\right)\sqrt{1+\frac{1}{x^4}}\,dx = 2\pi\int_1^\infty \frac{\sqrt{x^4+1}}{x^3}\,dx = \infty$ by comparison with $\displaystyle\int_1^\infty \frac{1}{x}\,dx$

42. $\displaystyle A = \int_0^{\pi/2}(\sec x - \tan x)\,dx = \lim_{b\to\pi/2^-}\int_0^b (\sec x - \tan x)\,dx = \lim_{b\to\pi/2^-}\left[\ln(\sec x - \tan x) - \ln\sec x\right]_0^b$

$$= \lim_{b\to\pi/2^-}\left[\ln(1+\sin x)\right]_0^b = \ln 2$$

43. (a)

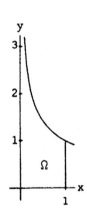

(b) $\displaystyle A = \int_0^1 \frac{1}{\sqrt{x}}\,dx = \lim_{a\to 0^+}\int_a^1 \frac{1}{\sqrt{x}}\,dx$

$$= \lim_{a\to 0^+}\left[2\sqrt{x}\right]_a^1 = 2$$

(c) $\displaystyle V = \int_0^1 \pi\left(\frac{1}{\sqrt{x}}\right)^2 dx = \pi\int_0^1 \frac{1}{x}\,dx = \pi\lim_{a\to 0^+}\int_a^1 \frac{1}{x}\,dx = \pi\lim_{a\to 0^+}\left[\ln x\right]_a^1$ diverges

44. **(a)**

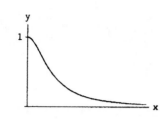

(b) $A = \int_0^\infty \frac{1}{1 + x^2}\, dx = \lim_{b\to\infty} \tan^{-1} b = \frac{\pi}{2}$

(c) $V_x = \int_0^\infty \pi \cdot \frac{1}{(1+x^2)^2}\, dx = \lim_{b\to\infty} \frac{\pi}{2}\left[\tan^{-1} x + \frac{x}{1+x^2}\right]_0^b$

$= \lim_{b\to\infty} \frac{\pi}{2}\left(\tan^{-1} b + \frac{b}{1+b^2} - 0\right) = \frac{\pi^2}{4}$

(d) $V_y = \int_0^\infty \frac{2\pi x}{1+x^2}\, dx = \lim_{b\to\infty} \pi \left[\ln(1+x^2)\right]_0^b = \infty$

45. **(a)**

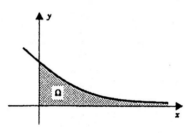

(b) $A = \int_0^\infty e^{-x}\, dx = 1$

(c) $V_x = \int_0^\infty \pi e^{-2x}\, dx = \pi/2$

(d)

$V_y = \int_0^\infty 2\pi x e^{-x}\, dx = \lim_{b\to\infty} \int_0^b 2\pi x e^{-x}\, dx = \lim_{b\to\infty}\left[2\pi(-x-1)e^{-x}\right]_0^b$

(by parts)

$= 2\pi\left(1 - \lim_{b\to\infty} \frac{b+1}{e^b}\right) = 2\pi(1-0) = 2\pi$

(e) $A = \int_0^\infty 2\pi e^{-x}\sqrt{1+e^{-2x}}\, dx = \lim_{b\to\infty}\int_0^b 2\pi e^{-x}\sqrt{1+e^{-2x}}\, dx$

$\int_0^b 2\pi e^{-x}\sqrt{1+e^{-2x}}\, dx = -2\pi\int_1^{e^{-b}} \sqrt{1+u^2}\, du$

$u = e^{-x}$

$= -\pi\left[u\sqrt{1+u^2} + \ln\left(u + \sqrt{1+u^2}\right)\right]_1^{e^{-b}}$

$= \pi\left[\sqrt{2} + \ln\left(1+\sqrt{2}\right) - e^{-b}\sqrt{1+e^{-2b}} - \ln\left(e^{-b} + \sqrt{1+e^{-2b}}\right)\right]$

Taking the limit of this last expression as $b \to \infty$, we have

$$A = \pi\left[\sqrt{2} + \ln\left(1+\sqrt{2}\right)\right].$$

46. $\overline{x}A = \int_0^\infty x e^{-x}\, dx = 1, \quad \overline{x} = \frac{\overline{x}A}{A} = 1$

$\overline{y}A = \int_0^\infty \frac{1}{2} e^{-2x}\, dx = \frac{1}{4}, \quad \overline{y} = \frac{\overline{y}A}{A} = \frac{1}{4};$ centroid: $(1, \frac{1}{4})$

Yes: $2\pi\overline{x}A = 2\pi = V_y, \quad 2\pi\overline{y}A = \frac{1}{2}\pi = V_x$

47. (a) The interval $[0, 1]$ causes no problem. For $x \geq 1$, $e^{-x^2} \leq e^{-x}$ and $\int_1^\infty e^{-x}\, dx$ is finite.

(b) $V_y = \int_0^\infty 2\pi x e^{-x^2}\, dx = \lim_{b\to\infty} \int_0^b 2\pi x e^{-x^2}\, dx = \lim_{b\to\infty} \pi \left[-e^{-x^2}\right]_0^b = \lim_{b\to\infty} \pi \left(1 - e^{-b^2}\right) = \pi$

48. (a) $A = \int_1^\infty \left(\frac{1}{x} - \frac{x}{x^2+1}\right) dx = \frac{1}{2}\ln 2$

(b) $V_x = \int_1^\infty \left[\left(\frac{1}{x}\right)^2 - \left(\frac{x}{x^2+1}\right)^2\right] dx < \int_1^\infty \frac{dx}{x^2}$ finite

(c) $V_y = \int_1^\infty 2\pi x \left(\frac{1}{x} - \frac{x}{x^2+1}\right) dx = 2\pi \int_1^\infty \frac{1}{x^2+1}\, dx = \frac{1}{2}\pi^2$

49. (a)

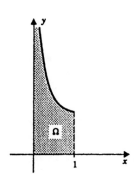

(b) $A = \lim_{a\to 0^+} \int_a^1 x^{-1/4}\, dx = \lim_{a\to 0^+} \left[\frac{4}{3}x^{3/4}\right]_a^1 = \frac{4}{3}$

(c) $V_x = \lim_{a\to 0^+} \int_a^1 \pi x^{-1/2}\, dx = \lim_{a\to 0^+} \left[2\pi x^{1/2}\right]_a^1 = 2\pi$

(d) $V_y = \lim_{a\to 0^+} \int_a^1 2\pi x^{3/4}\, dx = \lim_{a\to 0^+} \left[\frac{8\pi}{7}x^{7/4}\right]_a^1 = \frac{8}{7}\pi$

50. (i) Suppose that $\int_a^\infty g(x)\, dx = L$. Since $f(x) \geq 0$ for $x \in [a, \infty)$, $\int_a^x f(t)\, dt$ is increasing. Therefore it is sufficient to show that $\int_a^x f(t)\, dt$ is bounded above. For any number $M \geq a$, we have

$$\int_a^M f(x)\, dx \leq \int_a^M g(x)\, dx \leq \int_a^\infty g(x)\, dx = L$$

Therefore, $\int_a^x f(t)\, dt$ is bounded and $\int_a^\infty f(x)\, dx$ converges.

(ii) If $\int_0^\infty f(x)\, dx$ diverges, then $\int_0^\infty g(x)\, dx$ can not converge, by (i)

51. converges by comparison with $\int_1^\infty \frac{dx}{x^{3/2}}$

52. converges by comparison with $\int_2^\infty e^{-x}\, dx$ on $[2, \infty)$.

53. diverges since for x large the integrand is greater than $\frac{1}{x}$ and $\int_1^\infty \frac{dx}{x}$ diverges

54. Converges by comparison with $\int_\pi^\infty \frac{dx}{x^2}$

55. converges by comparison with $\int_1^\infty \frac{dx}{x^{3/2}}$

56. Diverges by comparison with $\int_e^\infty \frac{dx}{(x+1)\ln(x+1)}$

57. (a) $\displaystyle\lim_{b\to\infty}\int_0^b \frac{2x}{1+x^2}\,dx = \lim_{b\to\infty}\left[\ln\left(1+x^2\right)\right]_0^b = \infty$

Thus, the improper integral $\displaystyle\int_0^\infty \frac{2x}{1+x^2}\,dx$ diverges.

(b)
$$\lim_{b\to\infty}\int_{-b}^b \frac{2x}{1+x^2}\,dx = \lim_{b\to\infty}\left[\ln\left(1+x^2\right)\right]_{-b}^b$$
$$= \lim_{b\to\infty}\left(\ln\left[1+b^2\right] - \ln\left[1+(-b)^2\right]\right) = \lim_{b\to\infty}(0) = 0$$

58. $\displaystyle\int_0^\infty \sin x\,dx = \lim_{b\to\infty}\int_0^b \sin x\,dx = \lim_{b\to\infty}\left[-\cos x\right]_0^b = 1 - \lim_{b\to\infty}\cos b;$ the limit does not exist

$$\lim_{b\to\infty}\int_{-b}^b \sin x\,dx = \lim_{b\to\infty}\left[-\cos x\right]_{-b}^b = \lim_{b\to\infty}\left[-\cos b + \cos(-b)\right] = \lim_{b\to\infty}[0] = 0$$

59. $r(\theta) = ae^{c\theta}, \quad r'(\theta) = ace^{c\theta}$

$$L = \int_{-\infty}^{\theta_1} \sqrt{a^2 e^{2c\theta} + a^2 c^2 e^{2c\theta}}\,d\theta$$

$(10.7.3)\underset{\uparrow}{}$

$$= \left(a\sqrt{1+c^2}\right)\left(\lim_{b\to-\infty}\int_b^{\theta_1} e^{c\theta}\,d\theta\right)$$

$$= \left(a\sqrt{1+c^2}\right)\left(\lim_{b\to-\infty}\left[\frac{e^{c\theta}}{c}\right]_b^{\theta_1}\right)$$

$$= \left(\frac{a\sqrt{1+c^2}}{c}\right)\left(\lim_{b\to-\infty}\left[e^{c\theta_1} - e^{cb}\right]\right) = \left(\frac{a\sqrt{1+c^2}}{c}\right)e^{c\theta_1}$$

60. For all real t, $-\dfrac{t^2}{2} < t+1$. Therefore $\displaystyle\int_{-\infty}^x e^{-t^2/2}\,dt$ converges by comparison with $\displaystyle\int_{-\infty}^x e^{t+1}\,dt$.

61. $F(s) = \displaystyle\int_0^\infty e^{-sx}\cdot 1\,dx = \lim_{b\to\infty}\int_0^b e^{-sx}\,dx = \lim_{b\to\infty}\left[-\frac{1}{s}e^{-sx}\right]_0^b = \frac{1}{s}$ provided $s > 0$.

Thus, $F(s) = \dfrac{1}{s};$ $\mathrm{dom}(F) = (0,\infty)$.

62.
$$F(s) = \int_0^\infty xe^{-sx}\,dx = \lim_{b\to\infty}\left[\frac{-xe^{-sx}}{s} - \frac{e^{-sx}}{s^2}\right]_0^b = \lim_{b\to\infty}\left(\frac{-be^{-sb}}{s} - \frac{e^{-sb}}{s^2} + \frac{1}{s^2}\right)$$
$$= \frac{1}{s^2} \quad\text{if } s > 0, \quad\text{diverges for } s \le 0, \quad\text{so } \mathrm{dom}(F) = (0,\infty).$$

63. $F(s) = \displaystyle\int_0^\infty e^{-sx}\cos 2x\,dx = \lim_{b\to\infty}\int_0^b e^{-sx}\cos 2x\,dx$

Using integration by parts $\displaystyle\int e^{-sx}\cos 2x\,dx = \frac{4}{s^2+4}\left[\frac{1}{2}e^{-sx}\sin 2x - \frac{s}{4}e^{-sx}\cos 2x\right] + C.$

Therefore,

$$F(s) = \lim_{b \to \infty} \frac{4}{s^2 + 4} \left[\frac{1}{2} e^{-sx} \sin 2x - \frac{s}{4} e^{-sx} \cos 2x \right]_0^b$$

$$= \frac{4}{s^2 + 4} \lim_{b \to \infty} \left[\frac{1}{2} e^{-sb} \sin 2b - \frac{s}{4} e^{-sb} \cos 2b + \frac{s}{4} \right] = \frac{4}{s^2 + 4} \cdot \frac{s}{4} = \frac{s}{s^2 + 4} \qquad \text{provided} \quad s > 0.$$

Thus, $F(s) = \dfrac{s}{s^2 + 4}$; $\text{dom}(F) = (0, \infty)$.

64.
$$F(s) = \int_0^\infty e^{ax} e^{-sx} \, dx = \int_0^\infty e^{(a-s)x} \, dx = \lim_{b \to \infty} \left[\frac{e^{(a-s)b}}{a - s} - \frac{1}{a - s} \right]$$

$$= \frac{1}{s - a} \quad \text{if } s > a, \quad \text{diverges if } s \le a; \quad \text{so dom } F = (a, \infty)$$

65.   The function f is nonnegative on $(-\infty, \infty)$ and

$$\int_{-\infty}^\infty f(x) \, dx = \int_{-\infty}^0 0 \, dx + \int_0^\infty \frac{6x}{(1 + 3x^2)^2} \, dx = \int_0^\infty \frac{6x}{(1 + 3x^2)^2} \, dx$$

Now, $\displaystyle \int \frac{6x}{(1 + 3x^2)^2} \, dx = -\frac{1}{1 + 3x^2} + C.$

Therefore,

$$\int_{-\infty}^\infty f(x) \, dx = \lim_{b \to \infty} \left[-\frac{1}{1 + 3x^2} \right]_0^b = \lim_{b \to \infty} \left(1 - \frac{1}{1 + 3b^2} \right) = 1.$$

66. f is nonnegative, and

$$\int_{-\infty}^\infty f(x) \, dx = \int_0^\infty k e^{-kx} \, dx = \lim_{b \to \infty} \left[-e^{-kx} \right]_0^b = \lim_{b \to \infty} \left(-e^{-kb} + 1 \right) = 1$$

So, f is a probability density function.

67.
$$\mu = \int_{-\infty}^\infty x f(x) \, dx = \int_{-\infty}^0 0 \, dx + \int_0^\infty k x e^{-kx} \, dx = \lim_{b \to \infty} \int_0^b k x e^{-kx} \, dx$$

Using integration by parts, $\displaystyle \int k x e^{-kx} \, dx = -x e^{-kx} - \frac{1}{k} e^{-kx} + C.$

Therefore,

$$\mu = \int_{-\infty}^\infty x f(x) \, dx = \lim_{b \to \infty} \left[-x e^{-kx} - \frac{1}{k} e^{-kx} \right]_0^b = \lim_{b \to \infty} \left[-b e^{-kb} - \frac{1}{k} e^{-kb} + \frac{1}{k} \right] = \frac{1}{k}$$

68. $\displaystyle \sigma = \int_{-\infty}^\infty (x - \mu)^2 f(x) \, dx = \int_0^\infty \left(x - \frac{1}{k} \right)^2 k e^{-kx} \, dx$

$$= \int_0^\infty k x^2 e^{-kx} \, dx - 2 \int_0^\infty x e^{-kx} \, dx + \frac{1}{k} \int_0^\infty e^{-kx} \, dx = \frac{1}{k^2}$$

69.   Observe that $\displaystyle F(t) = \int_1^t f(x) \, dx$ is continuous and increasing, that $\displaystyle a_n = \int_1^n f(x) \, dx$
is increasing, and that $(*)$ $a_n \le \int_1^t f(x) \, dx \le a_{n+1}$ for $t \in [n, n+1]$.

If $\displaystyle\int_1^\infty f(x)\,dx$ converges, then F, being continuous, is bounded and, by $(*)$, $\{a_n\}$ is bounded and therefore convergent. If $\{a_n\}$ converges, then $\{a_n\}$ is bounded and, by $(*)$, F is bounded. Being increasing, F is also convergent; i.e., $\displaystyle\int_1^\infty f(x)\,dx$ converges.

REVIEW EXERCISES

1. $|x-2| \le 3 \implies -1 \le x \le 5:$ lub $= 5$, glb $= -1$.

2. $x^2 > 3 \implies x > \sqrt{3}$ or $x < -\sqrt{3};$ no lub, no glb.

3. $x^2 - x - 2 \le 0 \implies (x-2)(x+1) \le 0 \implies -1 \le x \le 2:$ lub $= 2$, glb $= -1$.

4. $\cos x \le 1$ for all $x;$ no lub, no glb.

5. Since $e^{-x^2} \le 1$ for all x, $e^{-x^2} \le 2$ for all $x;$ no lub, no glb.

6. $\ln x < e \implies 0 < x < e^e:$ lub $= e^e$, glb $= 0$.

7. increasing; bounded below by $\frac{1}{2}$ and above by $\frac{2}{3}$.

8. increasing; bounded below by 0 but not bounded above: $\dfrac{n^2-1}{n} = n - \dfrac{1}{n} \to \infty$ as $n \to \infty$.

9. bounded below by 0 and above by $\frac{3}{2}$;   not monotonic

10. increasing; bounded below by $\frac{4}{5}$ and above by 1.

11. $\left\{\dfrac{2^n}{n^2}\right\} = \{2, 1, \frac{8}{9}, 1, \frac{32}{25}, \dots\};$ the sequence is not monotonic.

    However, it is increasing from a_3 on. The sequence is bounded below by $\frac{8}{9}$; it is not bounded above.

12. $\left\{\dfrac{\sin(n\pi/2)}{n^2}\right\} = \{1, 0, -\frac{1}{9}, 0, \frac{1}{25}, \dots\};$ bounded below by $-\frac{1}{9}$ and above by 1;   not monotonic

13. the sequence does not converge; $n2^{1/n} \to \infty$ as $n \to \infty$

14. converges to 1: $\dfrac{n^2+3n+2}{n^2+7n+12} = \dfrac{1+\dfrac{3}{n}+\dfrac{2}{n^2}}{1+\dfrac{7}{n}+\dfrac{12}{n^2}} \to 1.$

15. converges to 1: $\displaystyle\lim_{n\to\infty}\frac{1}{n}\ln\left(\frac{n}{n+1}\right) = 0 \implies \lim_{n\to\infty}\left(\frac{n}{1+n}\right)^{1/n} = 1$

16. converges to 0: $\dfrac{4n^2 + 5n + 1}{n^3 + 1} = \dfrac{\frac{4}{n} + \frac{5}{n^2} + \frac{1}{n^3}}{1 + \frac{1}{n^3}} \to 0.$

17. converges to 0: $\cos\frac{\pi}{n}\sin\frac{\pi}{n} \to \cos 0 \sin 0 = 0$ as $n \to \infty$

18. diverges: $(2 + \frac{1}{n})^n > 2^n$ and 2^n diverges.

19. converges to 0: $0 = [\ln 1]^n \le [\ln(1 + \frac{1}{n})]^n \le [\ln 2]^n$; $[\ln 2]^n \to 0$ as $n \to \infty$.

20. converges to $\ln 8$: $3\ln 2n - \ln(n^3 + 1) = \ln\dfrac{8n^3}{n^3 + 1} \to \ln 8.$

21. converges to $\frac{3}{2}$: $\dfrac{3n^2 - 1}{\sqrt{4n^4 + 2n^2 + 3}} = \dfrac{3 - \frac{1}{n^2}}{\sqrt{4 + \frac{2}{n^2} + \frac{3}{n^4}}} \to \dfrac{3}{2}$ as $n \to \infty$.

22. converges to 0: $\dfrac{(n^2 + 4)^{\frac{1}{3}}}{2n + 1} = \dfrac{(1/n + 4/n^3)^{1/3}}{2 + 1/n} \to 0$ as $n \to \infty$.

23. converges to 0: $\dfrac{\pi}{n}\cos\dfrac{\pi}{n} \to 0\cos 0 = 0$ as $n \to \infty$.

24. converges 0: $(n/\pi)\sin(n\pi) = 0$ for all positive integers n.

25. converges to 0: $\displaystyle\int_n^{n+1} e^{-x}dx = \left[-e^{-x}\right]_n^{n+1} = e^{-n}(1 - \frac{1}{e}) \to 0$ as $n \to \infty$

26. diverges: $\displaystyle\int_1^n \dfrac{1}{\sqrt{x}}dx = \left[2\sqrt{x}\right]_1^n = 2\sqrt{n} - 2$ and $2\sqrt{n} - 2$ diverges.

27. Given $\epsilon > 0$. Since $a_n \to L$, there exists a positive integer K such that if $n \ge K$, then $|a_n - L| < \epsilon$. Now, if $n \ge K - 1$, then $n + 1 \ge K$ and $|a_{n+1} - L| < \epsilon$. Therefore, $a_{n+1} \to L$.

28. Let $\epsilon > 0$. Since $a_n \to L$, there is positive integer K such that if $n \ge K$,

$$|a_n - L| < \dfrac{\epsilon}{2}.$$

The set $\{|a_1 - L|, \cdots, |a_K - L|\}$ is a finite set so there is a positive integer N such that if $n > N$,

$$\dfrac{|a_i - L|}{n} < \dfrac{\epsilon}{2K}, i = 1, 2, \cdots, K.$$

Let $M = \max\{K, N\}$. Then, if $n \ge M$,

$$\left|\dfrac{a_1 + \cdots + a_n}{n} - L\right| \le \sum_{i=1}^K \dfrac{|a_i - L|}{n} + \sum_{j=K+1}^n \dfrac{|a_j - L|}{n} < K(\dfrac{\epsilon}{2K}) + n(\dfrac{\epsilon}{2n}) = \epsilon.$$

Therefore $m_n \to L$.

29. As an example, let $a = \frac{\pi}{3}$. Then

$$\cos \pi/3 = 0.5, \quad \cos\cos 0.5 \cong 0.87758, \cdots.$$

Using technology (graphing calculator, CAS), we get

$$\cos\cos\cdots\cos \pi/3 \to 0.73910.$$

and $\cos(0.73910) \cong 0.73910$.

Hence, numerically, this sequence converges to 0.73910.

30. Let $f(x) = \sin(\cos x)$ and let $a = \pi/3$. Then

$$f(\pi/3) \cong 0.4794, f(f(\pi/3)) \cong 0.7753, \cdots$$

After 14 steps, we get $f(f(\cdots f(\pi/3)) \cong 0.6948$ and $\sin(\cos 0.6948) \cong 0.6948$.

31. $\displaystyle\lim_{x\to\infty} \frac{5x + 2\ln x}{x + 3\ln x} = \lim_{x\to\infty} \frac{5 + 2\dfrac{\ln x}{x}}{1 + 3\dfrac{\ln x}{x}} = 5; \quad \left(\lim_{x\to\infty} \frac{\ln x}{x} = 0\right)$

32. $\displaystyle\lim_{x\to 0} \frac{e^x - 1}{\tan 2x} \overset{\star}{=} \lim_{x\to 0} \frac{e^x}{2\sec^2 2x} = \frac{1}{2}$

33. $\displaystyle\lim_{x\to 0} \frac{\ln(\cos x)}{x^2} \overset{\star}{=} \lim_{x\to 0} \frac{\dfrac{-\sin x}{\cos x}}{2x} = \lim_{x\to 0} \frac{-1}{2\cos x} \cdot \frac{\sin x}{x} = -\frac{1}{2}$

34. Set $y = x^{1/(x-1)}$. Then $\ln y = \dfrac{\ln x}{x - 1}$, and $\displaystyle\lim_{x\to 1} \frac{\ln x}{x - 1} \overset{\star}{=} \lim_{x\to 1} \frac{1}{x} = 1$

Therefore, $\displaystyle\lim_{x\to 1} x^{1/(x-1)} = e$.

35. $\displaystyle\lim_{x\to\infty} \left(1 + \frac{4}{x}\right)^{2x} = \lim_{x\to\infty} \left[\left(1 + \frac{4}{x}\right)^x\right]^2 = e^8$

36. $\displaystyle\lim_{x\to 0} \frac{e^{2x} - e^{-2x}}{\sin x} \overset{\star}{=} \lim_{x\to 0} \frac{2e^{2x} + 2e^{-2x}}{\cos x} = 4$

37. $\displaystyle\lim_{x\to 0^+} x^2 \ln x = \lim_{x\to 0^+} \frac{\ln x}{\dfrac{1}{x^2}} \overset{\star}{=} \lim_{x\to 0^+} \frac{\dfrac{1}{x}}{\dfrac{-2}{x^3}} = \lim_{x\to 0^+} \frac{-x^2}{2} = 0$

38. $\displaystyle\lim_{x\to\infty} \frac{10^x}{x^{10}} \overset{\star}{=} \lim_{x\to\infty} \frac{10^x \ln 10}{10x^9} \overset{\star}{=} \lim_{x\to\infty} \frac{10^x(\ln 10)^2}{(10)(9)x^8} \overset{\star}{=} \cdots = \lim_{x\to\infty} \frac{10^x(\ln 10)^{10}}{10!} \to \infty$

39. $\displaystyle\lim_{x\to 0} \frac{e^x + e^{-x} - x^2 - 2}{\sin^2 x - x^2} \overset{\star}{=} \lim_{x\to 0} \frac{e^x - e^{-x} - 2x}{2\sin x \cos x - 2x} \overset{\star}{=} \lim_{x\to 0} \frac{e^x - e^{-x} - 2}{\sin 2x - 2x} \overset{\star}{=} \lim_{x\to 0} \frac{e^x + e^{-x} - 2}{2\cos 2x - 2}$

$\displaystyle \overset{\star}{=} \lim_{x\to 0} \frac{e^x - e^{-x}}{-4\sin x 2x} \overset{\star}{=} \lim_{x\to 0} \frac{e^x + e^{-x}}{-8\cos 2x} = -\frac{1}{4}$

40. $\displaystyle \lim_{x\to 1} \csc(\pi x)\ln x = \lim_{x\to 1} \frac{\ln x}{\sin \pi x} \overset{\star}{=} \lim_{x\to 1} \frac{1/x}{\pi \cos \pi x} = -\frac{1}{\pi}$

41. $\displaystyle \lim_{x\to\infty} xe^{-x^2}\int_0^x e^{t^2}\,dt = \lim_{x\to\infty} \frac{\displaystyle\int_0^x e^{t^2}\,dt}{\dfrac{e^{x^2}}{x}} \overset{\star}{=} \lim_{x\to\infty} \frac{e^{x^2}}{\dfrac{2x^2 e^{x^2}-e^{x^2}}{x^2}} = \lim_{x\to\infty} \frac{x^2}{2x^2-1} = \frac{1}{2}$

42. Consider $\displaystyle \ln \frac{e^{-1/x^2}}{x^n} = \left(-\frac{1}{x^2}-n\ln|x|\right).$

$$\lim_{x\to 0}\left(-\frac{1}{x^2}-n\ln|x|\right) = \lim_{x\to 0}-\frac{1+nx^2\ln|x|}{x^2} = -\infty$$

Therefore $\displaystyle \lim_{x\to 0}\frac{e^{-1/x^2}}{x^n}\to 0.$

43. $\displaystyle \int_1^\infty \frac{e^{-\sqrt{x}}}{\sqrt{x}}\,dx = \lim_{b\to\infty}\int_1^b \frac{e^{-\sqrt{x}}}{\sqrt{x}}\,dx = \lim_{b\to\infty}\left[-2e^{-\sqrt{x}}\right]_1^b = \lim_{b\to\infty}(-2e^{-\sqrt{b}}+2e^{-1}) = 2e^{-1}$

44. $\displaystyle \int \frac{x}{\sqrt{1-x^2}}\,dx = -\sqrt{1-x^2}+C$

$$\int_0^1 \frac{x}{\sqrt{1-x^2}}\,dx = \lim_{b\to 1^-}\int_0^b \frac{x}{\sqrt{1-x^2}}\,dx = \lim_{b\to 1^-}\left[-\sqrt{1-x^2}\right]_0^b = \lim_{b\to 1^-}(-\sqrt{1-b^2}+1) = 1$$

45. $\displaystyle \int_0^1 \frac{1}{1-x^2}\,dx = \lim_{a\to 1^-}\int_0^a \frac{1}{1-x^2}\,dx = -\frac{1}{2}\lim_{a\to 1^-}\int_0^a\left(\frac{1}{x-1}-\frac{1}{x+1}\right)dx$ (partial fractions)

$$= \frac{1}{2}\lim_{a\to 1^-}\Big[\ln(x+1)-\ln(1-x)\Big]_0^a = \frac{1}{2}\lim_{a\to 1^-}\ln\left[\frac{a+1}{1-a}\right] = \infty;$$

the integral diverges.

46. $\displaystyle \int_0^{\pi/2}\sec x\,dx = \lim_{c\to\pi/2^-}\int_0^c \sec x\,dx = \lim_{c\to\pi/2^-}\Big[\ln(\sec x+\tan x)\Big]_0^c = \lim_{c\to\pi/2^-}\ln(\sec c+\tan c) = \infty;$

the integral diverges.

47. $\displaystyle \int_1^\infty \frac{\sin(\pi/x)}{x^2}\,dx = \lim_{b\to\infty}\int_1^b \frac{\sin(\pi/x)}{x^2}\,dx = \lim_{b\to\infty}\left[\frac{1}{\pi}\cos \pi/x\right]_1^b = \frac{2}{\pi}$

48. $\displaystyle \int_0^9 \frac{1}{(x-1)^{2/3}}\,dx = \lim_{c\to 1^-}\int_0^c \frac{1}{(x-1)^{2/3}}\,dx + \lim_{c\to 1^+}\int_c^9 \frac{1}{(x-1)^{2/3}}\,dx$

$$\lim_{c\to 1^-}\int_0^c \frac{1}{(x-1)^{2/3}}\,dx = \lim_{c\to 1^-}3\Big[(x-1)^{1/3}\Big]_0^c = 3$$

$$\lim_{c\to 1^+}\int_c^9 \frac{1}{(x-1)^{2/3}}\,dx = \lim_{c\to 1^+}3\Big[(x-1)^{1/3}\Big]_c^9 = 6$$

$$\int_0^9 \frac{1}{(x-1)^{2/3}} = 3+6 = 9.$$

49. $\displaystyle\int \frac{1}{e^x + e^{-x}}\,dx = \int \frac{e^x}{e^{2x}+1}\,dx = \arctan e^x + C$

$\displaystyle\int_0^\infty \frac{1}{e^x+e^{-x}}\,dx = \lim_{c\to\infty}\int_0^c \frac{1}{e^x+e^{-x}}\,dx = \lim_{c\to\infty}\arctan e^x\Big|_0^c = \frac{\pi}{2}$

50. Set $u = \ln x,\ du = \dfrac{1}{x}\,dx;\ u(2) = \ln 2$. Then

$$\int_2^\infty \frac{1}{x(\ln x)^k}\,dx = \int_{\ln 2}^\infty \frac{1}{u^k}\,du$$

The integral converges if $k > 1$ and diverges otherwise.

For $k > 1$,

$$\int_2^\infty \frac{1}{x(\ln x)^k}\,dx = \lim_{c\to\infty}\int_{\ln 2}^c \frac{1}{u^k}\,du = \lim_{c\to\infty}\frac{1}{1-k}u^{1-k}\Big|_{\ln 2}^c = \frac{1}{(k-1)(\ln 2)^{k-1}}$$

51.
$$\int_0^a \ln(1/x)\,dx = \int_0^a -\ln x\,dx = \lim_{c\to 0^+}\int_c^a -\ln x\,dx = \lim_{c\to 0^+}\Big[-x\ln x + x\Big]_c^a$$
$$= \lim_{c\to 0^+}\big[-a\ln a + a + c\ln c - c\big] = a\ln(1/a) + a$$

52. $y = (a^{2/3} - x^{2/3})^{3/2};\quad y' = -x^{-1/3}(a^{2/3} - x^{2/3})^{1/2}$

$$L = \int_0^a \sqrt{1+(y')^2}\,dx = \int_0^a a^{1/3}x^{-1/3}\,dx = \frac{3a}{2}$$

53. For any $a \in S + T,\ a = s + t$ for some $s \in S$ and $t \in T$. Hence $a \le \mathrm{lub}(S) + \mathrm{lub}(T)$.

Therefore, $S + T$ is bounded above and $\mathrm{lub}(S) + \mathrm{lub}(T)$ is an upper bound for $S + T$.

Let $M = \mathrm{lub}(S + T)$ and suppose $M < \mathrm{lub}(S) + \mathrm{lub}(T)$. Set $\epsilon = (\mathrm{lub}(S) + \mathrm{lub}(T)) - M$. There exist $s \in S$ and $t \in T$ such that

$$\mathrm{lub}(S) - s < \epsilon/2, \quad \text{and} \quad \mathrm{lub}(T) - t < \epsilon/2$$

Now,

$$\mathrm{lub}(S) + \mathrm{lub}(T) - (s+t) = \mathrm{lub}(S) - s + \mathrm{lub}(T) - t < \epsilon = \mathrm{lub}(S) + \mathrm{lub}(T) - M$$

which implies $s + t > M$, a contradiction. Therefore $\mathrm{lub}(S) + \mathrm{lub}(T) = \mathrm{lub}(S + T)$.

54. (a) Since S is bounded below, there is a number b such that $b \le s$ for every $s \in S$. Thus b is a lower bound for S and $B \ne \emptyset$.

(b) Choose any $s \in S$. Then, for any $b \in B,\ b \le s$. Therefore B is bounded above (each element $s \in S$ is an upper bound for B).

(c) We show first that $\mathrm{glb}(S)$ is an upper bound for B. For if not, there is $b \in B$ such that $b > \mathrm{glb}(S)$. Then there is an $s \in S$ such that $\mathrm{glb}(S) < s < b$, which contradicts the fact that b is a lower bound of S. It follows that $\mathrm{lub}(B) \le \mathrm{glb}(S)$. If $\mathrm{lub}(B) < \mathrm{glb}(S)$, then there exists a number a such that $\mathrm{lub}(B) < a < \mathrm{glb}(S)$ which implies that a is a lower bound for S and $a \in B$. Therefore $a \le \mathrm{lub}(B)$, a contradiction. Thus, $\mathrm{lub}(B) = \mathrm{glb}(S)$.

55. (a) If $\displaystyle\int_{-\infty}^{\infty} f(x)dx = L,$ then

$$\lim_{c\to\infty}\int_0^c f(x)\,dx, \quad \text{and} \quad \lim_{b\to-\infty}\int_b^0 f(x)\,dx$$

both exist and

$$\lim_{c\to\infty}\int_0^c f(x)\,dx + \lim_{b\to-\infty}\int_b^0 f(x)\,dx = L$$

Let $c = -b,$ then

$$\lim_{c\to\infty}\int_{-c}^c f(x)\,dx = L$$

(b) Set $f(x) = x.$ then $\displaystyle\lim_{c\to\infty}\int_{-c}^c x\,dx = 0,$ but $\displaystyle\int_{-\infty}^{\infty} x\,dx$ diverges.

56. (a) Assume that f is nonnegative on $(-\infty, \infty).$

By Exercise 55, $\displaystyle\int_{-\infty}^{\infty} f(x)\,dx = L \implies \lim_{c\to\infty}\int_{-c}^c f(x)\,dx = L$

Now assume that $\displaystyle\lim_{c\to\infty}\int_{-c}^c f(x)\,dx = L.$ Since f is nonnegative,

$$\int_0^x f(t)\,dt \le L \quad \text{on} \quad [0,\infty).$$

Therefore $\displaystyle\int_0^x f(t)\,dt$ is a bounded and nondecreasing function, which implies that

$\displaystyle\lim_{c\to\infty}\int_0^c f(x)\,dx$ exists. Similarly, $\displaystyle\lim_{c\to\infty}\int_{-c}^0 f(x)\,dx$ exists.

Therefore, $\displaystyle\int_{-\infty}^{\infty} f(x)\,dx$ exists, and, by the uniqueness of the limit, $\displaystyle\int_{-\infty}^{\infty} f(x)\,dx = L.$

57. Let S be a set of integers which is bounded above. Then there is an integer $k \in S$ such that $k \ge n$ for all $n \in S,$ for if not, S is not bounded above. Therefore, k is an upper bound for $S.$

Let $M = \mathrm{lub}(S).$ Then $M \ge k$ since $k \in S.$ Also $M \le k$ since k is an upper bound for $S.$ Therefore $M = k;$ the least upper bound of S is an element of $S.$

58. $\mathrm{lub}\,[L_f(P)] = \int_a^b f(x)\,dx; \quad \mathrm{glb}\,[U_f(P)] = \int_a^b f(x)\,dx.$

CHAPTER 12

SECTION 12.1

1. $1 + 4 + 7 = 12$ 2. $2 + 5 + 8 + 11 = 26$

3. $1 + 2 + 4 + 8 = 15$ 4. $\frac{1}{2} + \frac{1}{4} + \frac{1}{8} + \frac{1}{16} = \frac{15}{16}$

5. $1 - 2 + 4 - 8 = -5$ 6. $2 - 4 + 8 - 16 = -10$

7. $\frac{1}{3} + \frac{1}{9} + \frac{1}{27} = \frac{13}{27}$ 8. $-\frac{1}{6} + \frac{1}{24} - \frac{1}{120} = -\frac{2}{15}$

9. $1 + \frac{1}{4} + \frac{1}{16} + \frac{1}{64} = \frac{85}{64}$ 10. $1 - \frac{1}{4} + \frac{1}{16} - \frac{1}{64} = \frac{51}{64}$

11. $\displaystyle\sum_{n=1}^{11}(2n-1)$ 12. $\displaystyle\sum_{k=1}^{10}(-1)^{k+1}(2k-1)$ 13. $\displaystyle\sum_{k=1}^{35}k(k+1)$

14. $\displaystyle\sum_{k=1}^{n} m_k \, \Delta x_k$ 15. $\displaystyle\sum_{k=1}^{n} M_k \Delta x_k$ 16. $\displaystyle\sum_{k=1}^{n} f(x_k^*)\,\Delta x_k$

17. $\displaystyle\sum_{k=3}^{10}\frac{1}{2^k}, \quad \sum_{i=0}^{7}\frac{1}{2^{i+3}}$ 18. $\displaystyle\sum_{k=3}^{10}\frac{k^k}{k!}, \quad \sum_{i=0}^{7}\frac{(i+3)^{i+3}}{(i+3)!}$

19. $\displaystyle\sum_{k=3}^{10}(-1)^{k+1}\frac{k}{k+1}, \quad \sum_{i=0}^{7}(-1)^i\frac{i+3}{i+4}$ 20. $\displaystyle\sum_{k=3}^{10}\frac{1}{2k-3}, \quad \sum_{i=0}^{7}\frac{1}{2i+3}$

21. Set $k = n + 3$. Then $n = -1$ when $k = 2$ and $n = 7$ when $k = 10$.
$$\sum_{k=2}^{10}\frac{k}{k^2+1} = \sum_{n=-1}^{7}\frac{n+3}{(n+3)^2+1} = \sum_{n=-1}^{7}\frac{n+3}{n^2+6n+10}$$

22. $\displaystyle\sum_{n=2}^{12}\frac{(-1)^n}{n-1} = \sum_{k=1}^{11}\frac{(-1)^{k+1}}{k+1-1} = \sum_{k=1}^{11}\frac{(-1)^{k+1}}{k}$

23. Set $k = n - 3$. Then $n = 7$ when $k = 4$ and $n = 28$ when $k = 25$.
$$\sum_{k=4}^{25}\frac{1}{k^2-9} = \sum_{n=7}^{28}\frac{1}{(n-3)^2-9} = \sum_{n=7}^{28}\frac{1}{n^2-6n}$$

24. $\displaystyle\sum_{k=0}^{15}\frac{3^{2k}}{k!} = \sum_{n=-2}^{13}\frac{3^{2(n+2)}}{(n+2)!} = 81\sum_{n=-2}^{13}\frac{3^{2n}}{(n+2)!}$

25. $0.a_1 a_2 \cdots a_n = \dfrac{a_1}{10} + \dfrac{a_2}{10^2} + \cdots + \dfrac{a_n}{10^n} = \displaystyle\sum_{k=1}^{n}\frac{a_k}{10^k}$

26. $\displaystyle\sum_{k=1}^{n}\frac{1}{\sqrt{k}}=\frac{1}{\sqrt{1}}+\frac{1}{\sqrt{2}}+\frac{1}{\sqrt{3}}+\cdots+\frac{1}{\sqrt{n}}\geq\frac{1}{\sqrt{n}}+\frac{1}{\sqrt{n}}+\frac{1}{\sqrt{n}}+\cdots+\frac{1}{\sqrt{n}}=\frac{n}{\sqrt{n}}=\sqrt{n}.$

27. $\displaystyle\sum_{k=0}^{50}\frac{1}{4^k}=1.3333\cdots$

28. $\displaystyle\sum_{k=1}^{50}\frac{1}{k^2}\cong 1.62513$

29. $\displaystyle\sum_{k=0}^{50}\frac{1}{k!}=2.71828\cdots$

30. $\displaystyle\sum_{k=0}^{50}\left(\frac{2}{3}\right)^k\cong 3$

SECTION 12.2

1. $\dfrac{1}{2};$

$$s_n=\frac{1}{2}\left[\frac{1}{1\cdot 2}+\frac{1}{2\cdot 3}+\cdots+\frac{1}{(n)(n+1)}\right]$$

$$=\frac{1}{2}\left[\left(1-\frac{1}{2}\right)+\left(\frac{1}{2}-\frac{1}{3}\right)+\cdots+\left(\frac{1}{n}-\frac{1}{n+1}\right)\right]=\frac{1}{2}\left[1-\frac{1}{n+1}\right]\to\frac{1}{2}$$

2. $\dfrac{1}{2};$ $\displaystyle\sum_{k=3}^{\infty}\frac{1}{k^2-k}=\sum_{k=3}^{\infty}\left(\frac{1}{k-1}-\frac{1}{k}\right)=\lim_{n\to\infty}\left(\frac{1}{2}-\frac{1}{n}\right)=\frac{1}{2}$

3. $\dfrac{11}{18};$

$$s_n=\frac{1}{1\cdot 4}+\frac{1}{2\cdot 5}+\cdots+\frac{1}{n(n+3)}$$

$$=\frac{1}{3}\left[\left(1-\frac{1}{4}\right)+\left(\frac{1}{2}-\frac{1}{5}\right)+\cdots+\left(\frac{1}{n}-\frac{1}{n+3}\right)\right]$$

$$=\frac{1}{3}\left[1+\frac{1}{2}+\frac{1}{3}-\frac{1}{n+1}-\frac{1}{n+2}-\frac{1}{n+3}\right]\to\frac{1}{3}\left(1+\frac{1}{2}+\frac{1}{3}\right)=\frac{11}{18}$$

4. $\dfrac{3}{4};$ $\displaystyle\sum_{k=0}^{\infty}\frac{1}{(k+1)(k+3)}=\frac{1}{2}\sum_{k=0}^{\infty}\left(\frac{1}{k+1}-\frac{1}{k+3}\right)=\frac{1}{2}\lim_{n\to\infty}\left(1+\frac{1}{2}-\frac{1}{n+2}-\frac{1}{n+3}\right)=\frac{3}{4}$

5. $\dfrac{10}{3};$ $\displaystyle\sum_{k=0}^{\infty}\frac{3}{10^k}=3\sum_{k=0}^{\infty}\left(\frac{1}{10}\right)^k=3\left(\frac{1}{1-1/10}\right)=\frac{30}{9}=\frac{10}{3}$

6. $\dfrac{5}{6};$ $\displaystyle\sum_{k=0}^{\infty}\frac{(-1)^k}{5^k}=\sum_{k=0}^{\infty}\left(-\frac{1}{5}\right)^k=\frac{1}{1+\frac{1}{5}}=\frac{5}{6}$

7. $-\dfrac{3}{2};$ $\displaystyle\sum_{k=0}^{\infty}\frac{1-2^k}{3^k}=\sum_{k=0}^{\infty}\left(\frac{1}{3}\right)^k-\sum_{k=0}^{\infty}\left(\frac{2}{3}\right)^k=\frac{1}{1-1/3}-\frac{1}{1-2/3}=\frac{3}{2}-3=-\frac{3}{2}$

8. $\dfrac{1}{4};$ $\displaystyle\sum_{k=0}^{\infty}\frac{1}{2^{k+3}}=\frac{1}{8}\sum_{k=0}^{\infty}\frac{1}{2^k}=\frac{1}{8}\cdot\frac{1}{1-\frac{1}{2}}=\frac{1}{4}$

9. $24;$ geometric series with $a=8$ and $r=\dfrac{2}{3},$ sum $=\dfrac{a}{1-r}=24$

10. $\dfrac{3}{15,616}$; $\displaystyle\sum_{k=2}^{\infty}\dfrac{3^{k-1}}{4^{3k+1}}=\sum_{k=0}^{\infty}\dfrac{3^{k+1}}{4^{3k+7}}=\dfrac{3}{4^{7}}\sum_{k=0}^{\infty}\left(\dfrac{3}{4^{3}}\right)^{k}=\dfrac{3}{4^{7}}\cdot\dfrac{1}{1-\frac{3}{4^{3}}}=\dfrac{3}{15,616}$

11. Let $x=0.\overbrace{a_1a_2\cdots a_n}\overbrace{a_1a_2\cdots a_n}\cdots$. Then

$$x=\sum_{k=1}^{\infty}\dfrac{a_1a_2\cdots a_n}{(10^n)^k}=a_1a_2\cdots a_n\sum_{k=1}^{\infty}\left(\dfrac{1}{10^n}\right)^k$$

$$=a_1a_2\cdots a_n\left[\dfrac{1}{1-1/10^n}-1\right]=\dfrac{a_1a_2\cdots a_n}{10^n-1}.$$

12. (a) Denote the partial sums of the first series by s_n and those of the second series by t_n and observe that

$$s_n=(a_0+a_1+\cdots+a_j)+t_n.\quad\text{Obviously}\quad s_n\to L\quad\text{iff}$$

$$t_n=s_n-(a_0+a_1+\cdots+a_j)\to L-(a_0+a_1+\cdots+a_j).$$

Parts (b) and (c) follow from this equation.

13. $\dfrac{1}{1+x}=\dfrac{1}{1-(-x)}=\displaystyle\sum_{k=0}^{\infty}(-x)^k=\sum_{k=0}^{\infty}(-1)^k x^k$

14. $\dfrac{1}{1+x^2}=\dfrac{1}{1-(-x^2)}=\displaystyle\sum_{k=0}^{\infty}(-x^2)^k=\sum_{k=0}^{\infty}(-1)^k x^{2k}$

15. $\dfrac{x}{1-x}=x\left(\dfrac{1}{1-x}\right)=x\displaystyle\sum_{k=0}^{\infty}(x^k)=\sum_{k=0}^{\infty}x^{k+1}$

16. $\dfrac{x}{1+x}=x\cdot\dfrac{1}{1-(-x)}=x\displaystyle\sum_{k=0}^{\infty}(-x)^k=\sum_{k=0}^{\infty}(-1)^k x^{k+1}$

17. $\dfrac{x}{1+x^2}=x\left[\dfrac{1}{1-(-x^2)}\right]=x\displaystyle\sum_{k=0}^{\infty}(-x^2)^k=\sum_{k=0}^{\infty}(-1)^k x^{2k+1}$

18. $\dfrac{x}{1+4x^2}=\dfrac{x}{1-(-4x^2)}=x\displaystyle\sum_{k=0}^{\infty}(-4x^2)^k=\sum_{k=0}^{\infty}(-1)^k(2x)^{2k+1}$

19. $1+\dfrac{3}{2}+\dfrac{9}{4}+\dfrac{27}{8}+\dfrac{81}{16}+\cdots=\displaystyle\sum_{k=0}^{\infty}\left(\dfrac{3}{2}\right)^k$

This is a geometric series with $x=\frac{3}{2}>1$. Therefore the series diverges.

20. $a_k=\dfrac{1}{4}\left(\dfrac{-5}{4}\right)^k$ does not go to zero

21. $\displaystyle\lim_{k\to\infty}\left(\dfrac{k+1}{k}\right)^k=e\neq0$

22. $a_k = \dfrac{k^{k-2}}{3^k} = \left(\dfrac{k}{3}\right)^k \cdot \dfrac{1}{k^2} > \dfrac{2^k}{k^2}$ for $k > 6$, so $a_k \to \infty$

23. Rebounds to half its previous height:

$$s = 6 + 3 + 3 + \frac{3}{2} + \frac{3}{2} + \frac{3}{4} + \frac{3}{4} + \cdots = 6 + 6\sum_{k=0}^{\infty} \frac{1}{2^k} = 6 + \frac{6}{1 - \frac{1}{2}} = 18 \text{ ft.}$$

24. $s = 6 + 2h + 2h\left(\dfrac{h}{6}\right) + 2h\left(\dfrac{h}{6}\right)^2 + \cdots = 6 + 2h\sum_{k=0}^{\infty}\left(\dfrac{h}{6}\right)^k = 6 + \dfrac{11h}{6 - h} = 21$

$\implies 11h = 15(6 - h) \implies h = \frac{45}{13}$

25. A principal x deposited now at $r\%$ interest compounded annually will grow in k years to

$$x\left(1 + \frac{r}{100}\right)^k.$$

This means that in order to be able to withdraw n_k dollars after k years one must place

$$n_k\left(1 + \frac{r}{100}\right)^{-k}$$

dollars on deposit today. To extend this process in perpetuity as described in the text, the total deposit must be

$$\sum_{k=1}^{\infty} n_k\left(1 + \frac{r}{100}\right)^{-k}.$$

26. (a) $\displaystyle\sum_{k=1}^{\infty} 5000\left(\frac{1}{2}\right)^{k-1}(1.05)^{-k} = \frac{5000}{1.05}\sum_{k=1}^{\infty}\left[\frac{1}{2(1.05)}\right]^{k-1} = \frac{5000}{1.05}\cdot\frac{1}{1 - \frac{1}{2\cdot 1}} \cong \$9090.91$

(b) $800\,\dfrac{\frac{0.8}{1.06}}{1 - \frac{0.8}{1.06}} \cong \$2461.54$

(c) $\dfrac{N}{1.05}\cdot\dfrac{1}{1 - \frac{1}{1.05}} = 20N$

27. $\displaystyle\sum_{n=1}^{\infty}\left(\frac{9}{10}\right)^n = \frac{\frac{9}{10}}{1 - \frac{9}{10}} = 9$ or $\$9$

28. Total length removed $= \dfrac{1}{3} + \dfrac{2}{9} + \dfrac{4}{27} + \cdots = \dfrac{1}{3}\displaystyle\sum_{k=0}^{\infty}\left(\frac{2}{3}\right)^k = \dfrac{1}{3}\cdot\dfrac{1}{1 - \frac{2}{3}} = 1$

Some points: $0, 1, \frac{1}{3}, \frac{2}{3}, \frac{1}{9}, \frac{2}{9}, \frac{7}{9}, \frac{8}{9}.$

29.

$$A = 4^2 + (2\sqrt{2})^2 + 2^2 + (\sqrt{2})^2 + 1^2 + \cdots + \left[4\left(\frac{1}{\sqrt{2}}\right)^n\right]^2 + \cdots$$

$$= \sum_{n=0}^{\infty}\left[4\left(\frac{1}{\sqrt{2}}\right)^n\right]^2 = 16\sum_{n=0}^{\infty}\left(\frac{1}{2}\right)^n = 16\cdot\frac{1}{1 - \frac{1}{2}} = 32$$

30. (a) If $\sum(a_k + b_k)$ converges, then $\sum b_k = \sum(a_k + b_k) - \sum a_k$ would also converge.

(b) If $a_k = b_k = 2^k$, $\sum a_k$, $\sum b_k$ and $\sum(a_k + b_k)$ diverge, but $\sum(a_k - b_k) = \sum 0$ converges.

(c) If $a_k = 2^k, b_k = -2^k$, $\sum a_k, \sum b_k$ and $\sum(a_k - b_k)$ diverge, but $\sum(a_k + b_k) = \sum 0$ converges.

31. Let $L = \sum_{k=0}^{\infty} a_k$. Then

$$L = \sum_{k=0}^{\infty} a_k = \sum_{k=0}^{n} a_k + \sum_{k=n+1}^{\infty} a_k = s_n + R_n.$$

Therefore, $R_n = L - s_n$ and since $s_n \to L$ as $n \to \infty$, it follows that $R_n \to 0$ as $n \to \infty$.

32. (a) By convergence, $a_k \to 0$, so $\dfrac{1}{a_k}$ diverges, hence $\sum \dfrac{1}{a_k}$ diverges.

(b) If $a_k = \sqrt{k}$, then $\sum a_k$ diverges and $\sum \dfrac{1}{a_k}$ diverges (Example 5)

If $a_k = 2^k$, then $\sum a_k$ diverges and $\sum \dfrac{1}{a_k} = \sum \dfrac{1}{2^k}$ converges.

33. $s_n = \sum_{k=1}^{n} \ln\left(\dfrac{k+1}{k}\right) = [\ln(n+1) - \ln(n)] + [\ln n - \ln(n-1)] + \cdots + [\ln 2 - \ln 1] = \ln(n+1) \to \infty$

34. $a_k = \left(\dfrac{k}{k+1}\right)^k = \left(\dfrac{1}{\frac{k+1}{k}}\right)^k \to \dfrac{1}{e} \neq 0$

35. (a) $s_n = \sum_{k=1}^{n} (d_k - d_{k+1}) = d_1 - d_{n+1} \to d_1$

(b) We use part (a).

(i) $\sum_{k=1}^{\infty} \dfrac{\sqrt{k+1} - \sqrt{k}}{\sqrt{k(k+1)}} = \sum_{k=1}^{\infty} \left[\dfrac{1}{\sqrt{k}} - \dfrac{1}{\sqrt{k+1}}\right] = 1$

(ii) $\sum_{k=1}^{\infty} \dfrac{2k+1}{2k^2(k+1)^2} = \sum_{k=1}^{\infty} \dfrac{1}{2}\left[\dfrac{1}{k^2} - \dfrac{1}{(k+1)^2}\right] = \dfrac{1}{2}$

36. Use induction to verify the hint. Then

$$s_n = \dfrac{1 - (n+1)x^n + nx^{n+1}}{(1-x)^2} \to \dfrac{1}{(1-x)^2}$$

Since $-(n+1)x^n + nx^{n+1} \to 0$ for $|x| < 1$. This last statement follows from observing that $nx^n \to 0$.

To see this, choose $\epsilon > 0$ so that $(1 + \epsilon)|x| < 1$. Since $n^{1/n} \to 1$, there exists k so that

$$n^{1/n} < 1 + \epsilon \quad \text{for} \quad n \geq k.$$

Then for $n \geq k$

$$|nx^n| = |n^{1/n}x|^n \leq ((1 + \epsilon)|x|)^n \to 0$$

37. $R_n = \displaystyle\sum_{k=n+1}^{\infty} \frac{1}{4^k} = \dfrac{\left(\frac{1}{4}\right)^{n+1}}{1 - \frac{1}{4}} = \dfrac{1}{3 \cdot 4^n};$

$\dfrac{1}{3 \cdot 4^n} < 0.0001 \implies 4^n > 3333.33 \implies n > \dfrac{\ln 3333.33}{\ln 4} \cong 5.85$

Take $N = 6$.

38. $R_n = \displaystyle\sum_{k=n+1}^{\infty} (0.9)^k = (0.9)^{n+1}\dfrac{1}{1 - 0.9} = 9(0.9)^n < 0.0001 \implies n \geq \dfrac{\ln\left(\frac{0.0001}{9}\right)}{\ln 0.9} \cong 109$

39. $R_n = \displaystyle\sum_{k=n+1}^{\infty} \frac{1}{k(k+2)} = \frac{1}{2}\sum_{k=n+1}^{\infty}\left(\frac{1}{k} - \frac{1}{k+2}\right) = \frac{1}{2}\left(\frac{1}{n+1} + \frac{1}{n+2}\right);$

$\dfrac{1}{2}\left(\dfrac{1}{n+1} + \dfrac{1}{n+2}\right) < 0.0001 \implies n \geq 9999$. Take $N = 9999$.

40. $R_n = \displaystyle\sum_{k=n+1}^{\infty} \left(\frac{2}{3}\right)^k = \left(\frac{2}{3}\right)^{n+1}\dfrac{1}{1 - \frac{2}{3}} = 2\left(\frac{2}{3}\right)^n < 0.0001 \implies n \geq \dfrac{\ln\left(\frac{0.0001}{2}\right)}{\ln\left(\frac{2}{3}\right)} \cong 25$

41. $|R_n| = \left|\displaystyle\sum_{k=n+1}^{\infty} x^k\right| = \left|\dfrac{x^{n+1}}{1-x}\right| = \dfrac{|x|^{n+1}}{1-x};$

$$\frac{|x|^{n+1}}{1-x} < \epsilon$$

$$|x|^{n+1} < \epsilon(1-x)$$

$$(n+1)\ln|x| < \ln\epsilon(1-x)$$

$$n + 1 > \frac{\ln\epsilon(1-x)}{\ln|x|} \qquad [\text{recall} \quad \ln|x| < 0]$$

$$n > \frac{\ln\epsilon(1-x)}{\ln|x|} - 1$$

Take N to be smallest integer which is greater than $\dfrac{\ln\epsilon(1-x)}{\ln|x|}$.

42. $s_n = a_n - a_1$. Thus $\{s_n\}$ converges iff $\{a_n\}$ converges.

Hence $\displaystyle\sum_{k=1}^{\infty}(a_{k+1} - a_k)$ converges iff $\{a_n\}$ converges.

SECTION 12.3

1. converges; basic comparison with $\sum \frac{1}{k^2}$

2. diverges; limit comparison with $\sum \frac{1}{k}$

3. converges; basic comparison with $\sum \frac{1}{k^2}$

4. diverges; basic comparison with $\sum \frac{1}{k}$

5. diverges; basic comparison with $\sum \frac{1}{k+1}$

6. converges; basic comparison with $\sum \frac{1}{k^2}$

7. diverges; limit comparison with $\sum \frac{1}{k}$

8. converges; geometric with $x = \frac{2}{5}$

9. converges; integral test, $\int_1^\infty \frac{\tan^{-1} x}{1+x^2}\,dx = \lim_{b\to\infty}\left[\frac{1}{2}(\tan^{-1} x)^2\right]_1^b = \frac{3\pi^2}{32}$

10. converges; basic comparison with $\sum \frac{1}{k^2}$

11. diverges; p-series with $p = \frac{2}{3} \le 1$

12. converges; basic comparison with $\sum \frac{1}{k^3}$

13. diverges; divergence test, $\left(\frac{3}{4}\right)^{-k} \not\to 0$

14. diverges; basic comparison with $\sum \frac{1}{1+2k}$

15. diverges; basic comparison with $\sum \frac{1}{k}$

16. converges; integral test, $\int_2^\infty \frac{dx}{x(\ln x)^2} = \lim_{b\to\infty}\left[\frac{-1}{\ln x}\right]_2^\infty = \frac{1}{\ln 2}$

17. diverges; divergence test, $\frac{1}{2+3^{-k}} \to \frac{1}{2} \ne 0$

18. converges; limit comparison with $\sum \frac{1}{k^4}$

19. converges; limit comparison with $\sum \frac{1}{k^2}$

20. diverges; $a_k \not\to 0$.

21. diverges; integral test, $\int_2^\infty \frac{dx}{x\ln x} = \lim_{b\to\infty}\left[\ln(\ln x)\right]_2^b = \infty$

22. converges; limit comparison with $\sum \frac{1}{2^k}$

23. converges; limit comparison with $\sum \frac{2^k}{5^k}$

24. diverges; limit comparison with $\sum \frac{1}{k}$

25. diverges; limit comparison with $\sum \frac{1}{k}$

26. diverges; limit comparison with $\sum \frac{1}{\sqrt{k}}$

27. converges; limit comparison with $\sum \frac{1}{k^{3/2}}$

28. diverges; limit comparison with $\sum \frac{1}{k}$

29. converges; integral test, $\displaystyle\int_1^\infty xe^{-x^2}\,dx = \lim_{b\to\infty}\left[-\frac{1}{2}e^{-x^2}\right]_1^b = \frac{1}{2e}$

30. converges; integral test: $\displaystyle\int_1^\infty x^2 2^{-x^3}\,dx = \lim_{b\to\infty}\left[\frac{2^{-x^3}}{-3\ln 2}\right]_1^\infty = \frac{1}{6\ln 2}$

31. converges; basic comparison with $\displaystyle\sum \frac{3}{k^2}$, $2 + \sin k \le 3$ for all k.

32. diverges; basic comparison with $\displaystyle\sum \frac{1}{\sqrt{k}}$, $\dfrac{2 + \cos k}{\sqrt{k+1}} > \dfrac{1}{\sqrt{k}}$

33. Recall that $1 + 2 + 3 + \cdots + k = \dfrac{k(k+1)}{2}$. Therefore

$$\sum \frac{1}{1+2+3+\cdots+k} = \sum \frac{2}{k(k+1)}.$$ This series converges; direct comparison with $\displaystyle\sum \frac{2}{k^2}$

34. $\displaystyle\sum \frac{n}{1 + 2^2 + 3^2 + \cdots + n^2} = \sum \frac{n}{\frac{1}{6}n(n+1)(n+2)}$ converges: limit comparison with $\displaystyle\sum \frac{1}{n^2}$

35. converges; basic comparison with $\displaystyle\sum \frac{1}{k^2}$: $\displaystyle\sum \frac{2k}{(2k)!} = \sum \frac{1}{(2k-1)(2k-2)\cdots 3 \cdot 2 \cdot 1} < \sum \frac{1}{k^2}$

36. converges; basic comparison with $\displaystyle\sum \frac{1}{k^2}$: $\displaystyle\sum \frac{2k!}{(2k)!} = \sum \frac{1}{k(2k-1)(2k-2)\cdots(k+1)} < \sum \frac{1}{k^2}$

37. Use the integral test:

Let $u = \ln x$, $du = \dfrac{1}{x}\,dx$: $\displaystyle\int \frac{1}{x(\ln x)^p}\,dx = \int u^{-p}\,du = \frac{u^{1-p}}{1-p} + C.$

$$\int_1^\infty \frac{1}{x(\ln x)^p}\,dx = \lim_{b\to\infty}\int_1^b \frac{1}{x(\ln x)^p}\,dx = \lim_{b\to\infty}\frac{1}{1-p}(\ln a)^{1-p}$$

The series converges for $p > 1$.

38. If $p \le 1$, $\displaystyle\sum \frac{\ln k}{k^p} > \sum \frac{1}{k^p}$ diverges.

If $p > 1$,   then $\dfrac{p-1}{2} > 0$,   so for large k, $\ln k < k^{\frac{p-1}{2}}$

Then $\dfrac{\ln k}{k^p} < \dfrac{k^{\frac{p-1}{2}}}{k^p} = \dfrac{1}{k^{\frac{p+1}{2}}}$.   Since $\dfrac{p+1}{2} > 1, \displaystyle\sum \frac{1}{k^{\frac{p+1}{2}}}$ converges

Hence so does $\displaystyle\sum \frac{\ln k}{k^p}$.   so converges iff $p > 1$.

39. (a) Use the integral test: $\displaystyle\int_0^\infty e^{-\alpha x}\,dx = \lim_{b\to\infty}\left[-\frac{1}{\alpha}e^{-\alpha x}\right]_0^b = \frac{1}{\alpha}$ converges.

(b) Use the integral test: $\displaystyle\int_0^\infty xe^{-\alpha x}\,dx = \lim_{b\to\infty}\left[-\frac{x}{\alpha e^{-\alpha x}} - \frac{1}{\alpha^2}e^{-\alpha x}\right]_0^b = \frac{1}{\alpha^2}$ converges.

(c) The proof follows by induction using parts (a) and (b) and the reduction formula

$$\int x^n e^{ax}\,dx = \frac{x^n e^{ax}}{a} - \frac{n}{a}\int x^{n-1}e^{ax}\,dx \quad \text{[see Exercise 67, Section 8.2]}$$

40. $\displaystyle\int_{n+1}^{\infty}\frac{dx}{x^p} < \sum_{k=n+1}^{\infty}\frac{1}{k^p} < \int_{n}^{\infty}\frac{dx}{x^p}$

$$\Longrightarrow \quad \frac{1}{(p-1)(n+1)^{p-1}} < \sum_{k=1}^{\infty}\frac{1}{k^p} - \sum_{k=1}^{n}\frac{1}{k^p} < \frac{1}{(p-1)n^{p-1}}$$

41. (a) $\displaystyle\sum_{k=1}^{4}\frac{1}{k^3} \cong 1.1777$
         (b) $\displaystyle\frac{1}{2\cdot 5^2} < R_4 < \frac{1}{2\cdot 4^2}$

$$0.02 < R_4 < 0.0313$$

(c) $\displaystyle 1.1777 + 0.02 = 1.1977 < \sum_{k=1}^{\infty}\frac{1}{k^3} < 1.1777 + 0.0313 = 1.2090$

42. (a) $\displaystyle\sum_{k=1}^{4}\frac{1}{k^4} = 1 + \frac{1}{2^4} + \frac{1}{3^4} + \frac{1}{4^4} \cong 1.0788$

(b) $\displaystyle\frac{1}{3(5)^3} < R_4 < \frac{1}{3(4)^3} \quad \Longrightarrow \quad 0.0027 < R_4 < 0.0052$

(c) $\displaystyle 1.0815 < \sum_{k=1}^{\infty}\frac{1}{k^4} < 1.0840$

43. (a) Put $p = 2$ and $n = 100$ in the estimates in Exercise 38. The result is: $\displaystyle\frac{1}{101} < R_{100} < \frac{1}{100}$.

(b) $\displaystyle R_n < \frac{1}{(2-1)n^{2-1}} < 0.0001 \quad \Longrightarrow \quad n > 10,000 \quad \text{Take } n = 10,001.$

44. (a) $\displaystyle\frac{1}{2(101)^2} < R_{100} < \frac{1}{2(100)^2} \quad \Longrightarrow \quad 0.000049 < R_{100} < 0.00005$

(b) $\displaystyle R_n < \frac{1}{2n^2} < 0.0001 \quad \Longrightarrow \quad n \geq 71$

(c) $\displaystyle\sum_{k=1}^{\infty}\frac{1}{k^3} \cong \sum_{k=1}^{71}\frac{1}{k^3} \cong 1.20196$

45. (a) $\displaystyle R_n < \frac{1}{(4-1)n^{4-1}} < 0.0001 \quad \Longrightarrow \quad n^3 > 3333 \quad \Longrightarrow \quad n > 14.94 : \text{Take } n = 15.$

(b) $\displaystyle R_n < \frac{1}{(4-1)n^{4-1}} < 0.001 \quad \Longrightarrow \quad n^3 > 333.33 \quad \Longrightarrow \quad n > 6.93 : \text{Take } n = 7.$

(c) $\displaystyle\sum_{k=1}^{\infty}\frac{1}{k^4} \cong \sum_{k=1}^{7}\frac{1}{k^4} \cong 1.082$

46. (a) $R_n < \dfrac{1}{4n^4} < 0.0001 \implies n \geq 8$

   (b) $R_n < \dfrac{1}{(4)n^4} < 0.001 \implies n^3 > 333.33 \implies n > 3.97 :$ Take $n = 4$.

   (c) $\displaystyle\sum_{k=1}^{\infty} \frac{1}{k^5} \cong \sum_{k=1}^{4} \frac{1}{k^5} \cong 1.0363$

47. (a) If $a_k/b_k \to 0$, then $a_k/b_k < 1$ for all $k \geq K$ for some K. But then $a_k < b_k$ for all $k \geq K$

   and, since $\sum b_k$ converges, $\sum a_k$ converges. [The Basic Comparison Theorem 12.3.6.]

   (b) Similar to (a) except that this time we appeal to part (ii) of Theorem 12.3.6.

   (c) $\displaystyle\sum a_k = \sum \frac{1}{k^2}$ converges, $\displaystyle\sum b_k = \sum \frac{1}{k^{3/2}}$ converges, $\dfrac{1/k^2}{1/k^{3/2}} = \dfrac{1}{\sqrt{k}} \to 0$

   $\displaystyle\sum a_k = \sum \frac{1}{k^2}$ converges, $\displaystyle\sum b_k = \sum \frac{1}{\sqrt{k}}$ diverges, $\dfrac{1/k^2}{1/\sqrt{k}} = \dfrac{1}{k^{3/2}} \to 0$

   (d) $\displaystyle\sum b_k = \sum \frac{1}{\sqrt{k}}$ diverges, $\displaystyle\sum a_k = \sum \frac{1}{k^2}$ converges, $\dfrac{1/k^2}{1/\sqrt{k}} = \dfrac{1}{k^{3/2}} \to 0$

   $\displaystyle\sum b_k = \sum \frac{1}{\sqrt{k}}$ diverges, $\displaystyle\sum a_k = \sum \frac{1}{k}$ diverges, $\dfrac{1/k}{1/\sqrt{k}} = \dfrac{1}{\sqrt{k}} \to 0$

48. (a) Since $a_k/b_k \to \infty$, $b_k/a_k \to 0$,   so this follows from Exercise 45(b)

   (b) Follows from Exercise 45(a)

   (c) $\displaystyle\sum a_k = \sum \frac{1}{\sqrt{k}}$ diverges, $\displaystyle\sum b_k = \sum \frac{1}{k^2}$ converges, $\dfrac{1/\sqrt{k}}{1/(k^2)} = k^{3/2} \to \infty$

   $\displaystyle\sum a_k = \sum \frac{1}{\sqrt{k}}$ diverges, $\displaystyle\sum b_k = \sum \frac{1}{k}$ diverges, $\dfrac{1/\sqrt{k}}{1/k} = \sqrt{k} \to \infty$

   (d) $\displaystyle\sum b_k = \sum \frac{1}{k^2}$ converges, $\displaystyle\sum a_k = \sum \frac{1}{k^{3/2}}$ converges, $\dfrac{1/k^{3/2}}{1/(k^2)} = \sqrt{k} \to \infty$

   $\displaystyle\sum b_k = \sum \frac{1}{k^2}$ converges, $\displaystyle\sum a_k = \sum \frac{1}{\sqrt{k}}$ diverges, $\dfrac{1/\sqrt{k}}{1/(k^2)} = k^{3/2} \to \infty$

49. (a) Since $\sum a_k$ converges, $a_k \to 0$. Therefore there exists a positive integer N such that $0 < a_k < 1$

   for $k \geq N$. Thus, for $k \geq N$, $a_k^2 < a_k$ and so $\sum a_k^2$ converges by the comparison test.

   (b) $\sum a_k$ may either converge or diverge: $\sum 1/k^4$ and $\sum 1/k^2$ both converge; $\sum 1/k^2$ converges

   and $\sum 1/k$ diverges.

50. Since $0 < \left(a_k - \dfrac{1}{k}\right)^2 < a_k{}^2 + \dfrac{1}{k^2}$, $\displaystyle\sum \left(a_k - \frac{1}{k}\right)^2$ converges by comparison with

   $\displaystyle\sum a_k{}^2 + \sum \frac{1}{k^2}$ But $\displaystyle\sum \left(a_k - \frac{1}{k}\right)^2 = \sum a_k^2 - 2 \sum \frac{a_k}{k} + \sum \frac{1}{k^2}$,

   so $\displaystyle\sum \frac{a_k}{k}$ must converge.

51. $0 < L - \sum\limits_{k=1}^{n} f(k) = L - s_n = \sum\limits_{k=n+1}^{\infty} f(k) < \int_{n}^{\infty} f(x)\, dx$ [see the proof of the integral test]

52. $0 < L - S_n < \int_{n}^{\infty} \dfrac{1}{x^2+1}\, dx = \dfrac{\pi}{2} - \arctan n < 0.001 \implies n > \tan\left(\dfrac{\pi}{2} - 0.001\right) \cong 1000$

53. $L - s_n < \int_{n}^{\infty} xe^{-x^2}\, dx = \lim\limits_{b\to\infty} \int_{n}^{b} xe^{-x^2}\, dx$

$$= \lim\limits_{b\to\infty}\left[-\dfrac{1}{2}e^{-x^2}\right]_{n}^{b} = \dfrac{1}{2}e^{-n^2}$$

$\dfrac{1}{2}e^{-n^2} < 0.001 \implies e^{n^2} > 500 \implies n > 2.49;$ take N = 3.

54. (a) Set $f(x) = 1/x$ in the proof of the integral test.

(b) $\sum\limits_{1}^{n} \dfrac{1}{k} > 100$ if $\ln(n+1) > 100 \implies n+1 > e^{100} \cong 2.7 \times 10^{43}$

55. Set $f(x) = x^{1/4} - \ln x$. Then

$$f'(x) = \dfrac{1}{4}x^{-3/4} - \dfrac{1}{x} = \dfrac{1}{4x}(x^{1/4} - 4).$$

Since $f(e^{12}) = e^3 - 12 > 0$ and $f'(x) > 0$ for $x > e^{12}$, we have that

$$n^{1/4} > \ln n \quad \text{and therefore} \quad \dfrac{1}{n^{5/4}} > \dfrac{\ln n}{n^{3/2}}$$

for sufficiently large n. Since $\sum \dfrac{1}{n^{5/4}}$ is a convergent p-series, $\sum \dfrac{\ln n}{n^{3/2}}$ converges by the basic comparison test.

56. The series converges if $\deg q \geq \deg p + 2$ and diverges otherwise.

SECTION 12.4

1. converges; ratio test: $\dfrac{a_{k+1}}{a_k} = \dfrac{10}{k+1} \to 0$ **2.** converges; root test: $\left(\dfrac{1}{k2^k}\right)^{1/k} = \dfrac{1}{2k^{1/k}} \to \dfrac{1}{2}$

3. converges; root test: $(a_k)^{1/k} = \dfrac{1}{k} \to 0$ **4.** converges; root test: $a_k^{1/k} = \dfrac{k}{2k+1} \to \dfrac{1}{2}$

5. diverges; ratio test: $\dfrac{a_{k+1}}{a_k} = \dfrac{k+1}{100} \to \infty$ **6.** diverges; comparison with $\sum \dfrac{1}{k}$

7. diverges; limit comparison with $\sum \dfrac{1}{k}$ **8.** converges; root test $(a_k)^{1/k} = \dfrac{1}{\ln k} \to 0$

9. converges; root test: $(a_k)^{1/k} = \dfrac{2}{3}k^{1/k} \to \dfrac{2}{3}$ **10.** diverges; comparison with $\sum \dfrac{1}{k}$

11. diverges; limit comparison with $\sum \dfrac{1}{\sqrt{k}}$ **12.** converges; limit comparison with $\sum \dfrac{1}{k^2}$

13. diverges; ratio test: $\dfrac{a_{k+1}}{a_k} = \dfrac{k+1}{10^4} \to \infty$ **14.** converges; root test: $(a_k)^{1/k} = \dfrac{k^{2/k}}{e} \to \dfrac{1}{e}$

15. converges; basic comparison with $\sum \dfrac{1}{k^{3/2}}$

16. converges; ratio test, $\dfrac{a_{k+1}}{a_k} = \dfrac{2^{k+1}(k+1)!}{(k+1)^{k+1}} \cdot \dfrac{k^k}{2^k k!} = 2\left(\dfrac{k}{k+1}\right)^k = \dfrac{2}{\left(\frac{k+1}{k}\right)^k} \to \dfrac{2}{e}$

17. converges; basic comparison with $\sum \dfrac{1}{k^2}$

18. converges; integral test $\displaystyle\int_2^\infty \dfrac{dx}{x(\ln x)^{3/2}} = \dfrac{2}{\sqrt{\ln 2}}$

19. diverges; integral test: $\displaystyle\int_2^\infty \dfrac{1}{x}(\ln x)^{-1/2} dx = \lim_{b \to \infty} \left[2(\ln x)^{1/2}\right]_2^b = \infty$

20. converges; limit comparison with $\sum \dfrac{1}{k^{3/2}}$

21. diverges; divergence test: $\left(\dfrac{k}{k+100}\right)^k = \left(1 + \dfrac{100}{k}\right)^{-k} \to e^{-100} \neq 0$

22. converges; ratio test: $\dfrac{[(k+1)!]^2}{(2k+2)!} \cdot \dfrac{(2k)!}{(k!)^2} = \dfrac{(k+1)^2}{(2k+1)(2k+2)} \to \dfrac{1}{4}$

23. diverges; limit comparison with $\sum \dfrac{1}{k}$ **24.** diverges; $a_k \not\to 0$

25. converges; ratio test: $\dfrac{a_{k+1}}{a_k} = \dfrac{\ln(k+1)}{e \ln k} \to \dfrac{1}{e}$

26. converges; ratio test: $\dfrac{(k+1)!}{(k+1)^{k+1}} \cdot \dfrac{k^k}{k!} = \dfrac{1}{\left(\frac{k+1}{k}\right)^k} \to \dfrac{1}{e}$

27. converges; basic comparison with $\sum \dfrac{1}{k^{3/2}}$

28. converges: ratio test $\dfrac{(k+1)!}{1 \cdot 3 \cdot \ldots \cdot (2k+1)} \cdot \dfrac{1 \cdot 3 \ldots \cdot (2k-1)}{k!} = \dfrac{k+1}{2k+1} \to \dfrac{1}{2}$

29. converges; ratio test: $\dfrac{a_{k+1}}{a_k} = \dfrac{2(k+1)}{(2k+1)(2k+2)} \to 0$

30. converges; root test: $(a_k)^{1/k} = \dfrac{(2k+1)^2}{5k^2+1} \to \dfrac{4}{5}$

31. converges; ratio test: $\dfrac{a_{k+1}}{a_k} = \dfrac{(k+1)(2k+1)(2k+2)}{(3k+1)(3k+2)(3k+3)} \to \dfrac{4}{27}$

32. converges by Exercise 38, section 11.2

33. converges;   ratio test: $\dfrac{a_{k+1}}{a_k} = \dfrac{1}{(k+1)^{1/2}}\left(\dfrac{k+1}{k}\right)^{k/2} \to 0 \cdot \sqrt{e} = 0$

34. diverges: $a_k = \left(\dfrac{k}{9}\right)^k \not\to 0$

35. converges;   root test: $(a_k)^{1/k} = \sqrt{k} - \sqrt{k-1} = \dfrac{1}{\sqrt{k}+\sqrt{k+1}} \to 0$

36. converges;   root test: $(a_k)^{1/k} = \dfrac{k}{3^k} \to 0$

37. $\dfrac{1}{2} + \dfrac{2}{3^2} + \dfrac{4}{4^3} + \dfrac{8}{5^4} + \cdots = \displaystyle\sum_{k=0}^{\infty} \dfrac{2^k}{(k+2)^{k+1}}$

   converges;   root test: $(a_k)^{1/k} = \dfrac{2}{(k+2)^{1+1/k}} \to 0$

38. converges: ratio test (see Exercise 28)

39. $\dfrac{1}{4} + \dfrac{1 \cdot 3}{4 \cdot 7} + \dfrac{1 \cdot 3 \cdot 5}{4 \cdot 7 \cdot 10} + \cdots = \displaystyle\sum_{k=0}^{\infty} \dfrac{1 \cdot 3 \cdots (1+2k)}{4 \cdot 7 \cdots (4+3k)}$

   converges;   ratio test: $\dfrac{a_{k+1}}{a_k} = \dfrac{3+2k}{7+3k} \to \dfrac{2}{3}$

40. converges; ratio test : $\dfrac{a_{k+1}}{a_k} = \dfrac{2 \cdot 4 \cdot 6 \cdot \ldots \cdot 2(k+1)}{3 \cdot 7 \cdot 11 \cdot \ldots \cdot [4(k+1)-1]} \cdot \dfrac{3 \cdot 7 \cdot 11 \cdot \ldots \cdot (4k-1)}{2 \cdot 4 \cdot 6 \cdot \ldots \cdot 2k} = \dfrac{2k+2}{4k+3} \to \dfrac{1}{2}$

41. By the hint

$$\sum_{k=1}^{\infty} k \left(\dfrac{1}{10}\right)^k = \dfrac{1}{10} \sum_{k=1}^{\infty} k \left(\dfrac{1}{10}\right)^{k-1} = \dfrac{1}{10}\left[\dfrac{1}{1-1/10}\right]^2 = \dfrac{10}{81}.$$

42. (a) If $\lambda > 1$,   then for k sufficiently large

$$\dfrac{a_{k+1}}{a_k} > 1 \quad \text{and thus} \quad a_{k+1} > a_k$$

   This shows that the kth term cannot tend to 0 and thus the series cannot converge.

   (b) $\displaystyle\sum \dfrac{1}{k}$ diverges, $\dfrac{a_{k+1}}{a_k} = \dfrac{k}{k+1} \to 1$

   $\displaystyle\sum \dfrac{1}{k^2}$ converges, $\dfrac{a_{k+1}}{a_k} = \dfrac{k^2}{(k+1)^2} \to 1$

43. The series $\displaystyle\sum_{k=0}^{\infty} \dfrac{k!}{k^k}$ converges (see Exercise 26). Therefore, $\displaystyle\lim_{k\to\infty} \dfrac{k!}{k^k} = 0$ by Theorem 12.2.5.

44. $\dfrac{r^{n+1}}{(n+1)!} \cdot \dfrac{n!}{r^n} = \dfrac{r}{n+1} \to 0,$ so by ratio test $\sum \dfrac{r^n}{n!}$ converges, and therefore $\dfrac{r^n}{n!} \to 0$

45. Use the ratio test:

$$\frac{a_{k+1}}{a_k} = \frac{\dfrac{[(k+1)!]^2}{[p(k+1)]!}}{\dfrac{(k!)^2}{(pk)!}} = (k+1)^2 \, \frac{(pk)!}{(pk)!(pk+1)\cdots(pk+p)} = \frac{(k+1)^2}{(pk+1)\cdots(pk+p)}$$

Thus

$$\frac{a_{k+1}}{a_k} \to \begin{cases} \dfrac{1}{4}, & \text{if } p = 2 \\[2mm] 0, & \text{if } p > 2. \end{cases}$$

The series converges for all $p \geq 2$.

46. By root test: $(a_k)^{1/k} = \dfrac{r}{k^{r/k}} = \dfrac{r}{\left(k^{1/k}\right)^r} \to r$ converges if $r < 1$,   diverges if $r > 1$.

If $r = 1$, we get $\sum \dfrac{1}{k}$, which diverges.

47. Set $b_k = a_k r^k$.   If $(a_k)^{1/k} \to \rho$ and $\rho < \dfrac{1}{r}$,   then

$$(b_k)^{1/k} = (a_k r^k)^{1/k} = (a_k)^{1/k} r \to \rho r < 1$$

and thus, by the root test, $\Sigma b_k = \Sigma a_k r^k$ converges.

48. (a)
$$a_k = \begin{cases} (\tfrac{1}{2})^k, & k \text{ is odd} \\[2mm] (\tfrac{1}{2})^{k-2}, & k \text{ is even} \end{cases}$$
Clearly, $(a_k)^{1/k} \to \tfrac{1}{2} < 1$.

(b) $\displaystyle\lim_{k\to\infty} \frac{a_{k+1}}{a_k}$ does not exist since

$$\frac{a_{k+1}}{a_k} = \begin{cases} \dfrac{1}{8}, & k \text{ is even} \\[2mm] 2, & k \text{ is odd} \end{cases}$$

SECTION 12.5

1. diverges; $a_k \not\to 0$

2. (a) $\sum |a_k| = \sum \dfrac{1}{2k}$ diverges, so not absolutely convergent.

    (b) $\dfrac{1}{2(k+1)} < \dfrac{1}{2k},$ $a_k \to 0:$ converges conditionally;   Theorem 12.5.3.

3. diverges; $\dfrac{k}{k+1} \to 1 \neq 0$

4. (a) $\sum |a_k| = \sum \dfrac{1}{k \ln k},$ does not converge absolutely.

    (b) converges conditionally;   Theorem 12.5.3.

5. (a) does not converge absolutely;   integral test,

$$\int_1^\infty \frac{\ln x}{x}\,dx = \lim_{b\to\infty}\left[\frac{1}{2}(\ln x)^2\right]_1^b = \infty$$

(b) converges conditionally;   Theorem 12.5.3

6. diverges; $a_k \not\to 0$

7. diverges;   limit comparison with $\displaystyle\sum \frac{1}{k}$

another approach: $\displaystyle\sum\left(\frac{1}{k} - \frac{1}{k!}\right) = \sum\frac{1}{k} - \sum\frac{1}{k!}$ diverges since $\displaystyle\sum\frac{1}{k}$ diverges and

$\displaystyle\sum\frac{1}{k!}$ converges

8. converges absolutely (terms already positive): ratio test, $\dfrac{a_{k+1}}{a_k} = \dfrac{(k+1)^3}{2^{k+1}}\cdot\dfrac{2^k}{k^3} = \left(\dfrac{k+1}{k}\right)^3\cdot\dfrac{1}{2} \to \dfrac{1}{2}$

9. (a) does not converge absolutely;   limit comparison with $\displaystyle\sum\frac{1}{k}$

(b) converges conditionally;   Theorem 12.5.3

10. converges absolutely by ratio test.

11. diverges; $a_k \not\to 0$ 　　　　　　　　　　　　**12.**   diverges: $a_k \not\to 0$

13. (a) does not converge absolutely;

$$(\sqrt{k+1} - \sqrt{k})\cdot\frac{(\sqrt{k+1}+\sqrt{k})}{(\sqrt{k+1}+\sqrt{k})} = \frac{1}{\sqrt{k+1}+\sqrt{k}}$$

and

$$\sum\frac{1}{\sqrt{k}+\sqrt{k+1}} > \sum\frac{1}{2\sqrt{k+1}} = \frac{1}{2}\sum\frac{1}{\sqrt{k+1}} \qquad \text{(a p-series with $p<1$)}$$

(b) converges conditionally;   Theorem 12.5.3

14. (a) does not converge absolutely: $\dfrac{k}{k^2+1} > \dfrac{k}{2k^2} = \dfrac{1}{2k}$,   comparison with $\displaystyle\sum\frac{1}{2k}$

(b) $\dfrac{k+1}{(k+1)^2+1} < \dfrac{k}{k^2+1}$;   converges conditionally;   Theorem 12.5.3.

15. converges absolutely (terms already positive); basic comparison,

$$\sum \sin\left(\frac{\pi}{4k^2}\right) \le \sum\frac{\pi}{4k^2} = \frac{\pi}{4}\sum\frac{1}{k^2} \qquad (|\sin x| \le |x|)$$

16. (a) does not converges absolutely: $\displaystyle\sum\frac{1}{\sqrt{k(k+1)}} > \sum\frac{1}{k+1}$

(b) converges conditionally by Theorem 12.5.3

17. converges absolutely;   ratio test, $\dfrac{a_{k+1}}{a_k} = \dfrac{k+1}{2k} \to \dfrac{1}{2}$

18. terms all positive, converges absolutely: $a_k = \dfrac{1}{\sqrt{k}\sqrt{k+1}(\sqrt{k}+\sqrt{k+1})}$,   comparison with $\sum \dfrac{1}{k^{3/2}}$

19. (a) does not converge absolutely;   limit comparison with $\sum \dfrac{1}{k}$

(b) converges conditionally;   Theorem 12.5.3

20. (a) does not converge absolutely: $\dfrac{k+2}{k^2+k} > \dfrac{k}{2k^2} = \dfrac{1}{2k}$,   comparison with $\sum \dfrac{1}{2k}$

(b) converges conditionally;   Theorem 12.5.3.

21. diverges; $a_k = \dfrac{4^{k-2}}{e^k} = \dfrac{1}{16}\left(\dfrac{4}{e}\right)^k \not\to 0$

22. converges absolutely by integral test: $\displaystyle\int_1^\infty x^2 2^{-x}\,dx$ converges

23. diverges; $a_k = k\sin(1/k) = \dfrac{\sin(1/k)}{1/k} \to 1 \neq 0$

24. diverges: $\left|\dfrac{a_{k+1}}{a_k}\right| = \dfrac{(k+1)^{k+1}}{(k+1)!}\cdot\dfrac{k!}{k^k} = \left(\dfrac{k+1}{k}\right)^k > 1$,   so $a_k \not\to 0$

25. converges absolutely;   ratio test, $\dfrac{a_{k+1}}{a_k} = \dfrac{(k+1)e^{-(k+1)}}{k\,e^{-k}} = \dfrac{k+1}{k}\dfrac{1}{e} \to \dfrac{1}{e}$

26. (a) $\sum \dfrac{\cos\pi k}{k} = \sum \dfrac{(-1)^k}{k}$ does not converge absolutely.

(b) converges conditionally;   Theorem 12.5.3.

27. diverges; $\sum(-1)^k \dfrac{\cos\pi k}{k} = \sum(-1)^k\dfrac{(-1)^k}{k} = \sum\dfrac{1}{k}$

28. Converges absolutely; $|a_k| = \left|\dfrac{\sin(\pi k/2)}{k\sqrt{k}}\right| < \dfrac{1}{k^{3/2}}$

29. converges absolutely;   basic comparison

$$\sum \left|\dfrac{\sin(\pi k/4)}{k^2}\right| \leq \sum \dfrac{1}{k^2}$$

30. The series $\sum\left(\dfrac{1}{3k+2} - \dfrac{1}{3k+3}\right) = \sum \dfrac{1}{(3k+2)(3k+3)}$ converges by comparison with $\sum \dfrac{1}{k^2}$.

If $\sum\left(\dfrac{1}{3k+2} - \dfrac{1}{3k+3} - \dfrac{1}{3k+4}\right)$ converged, then

$$\sum \dfrac{1}{3k+4} = \sum\left(\dfrac{1}{3k+2} - \dfrac{1}{3k+3}\right) - \sum\left(\dfrac{1}{3k+2} - \dfrac{1}{3k+3} - \dfrac{1}{3k+4}\right)$$ would converge, which is

not the case.

31.   diverges; $a_k \not\to 0$ **32.**   error $< a_{21} = \dfrac{1}{21} \cong 0.04762$

33. Use (12.5.4); $|s - s_{80}| < a_{81} = \dfrac{1}{\sqrt{82}} \cong 0.1104$ **34.** error $< a_5 = \dfrac{1}{10^5} = 0.00001$

35. Use (12.5.4); $|s - s_9| < a_{10} = \dfrac{1}{10^3} = 0.001$

36. error $< a_{n+1} = \dfrac{1}{10^{n+1}}$ (a) $\dfrac{1}{10^{n+1}} < 10^{-3} \implies n \geq 3$ (b) $\dfrac{1}{10^{n+1}} < 10^{-4} \implies n \geq 4$

37. $\dfrac{10}{11}$;   geometric series with $a = 1$ and $r = -\dfrac{1}{10}$,   sum $= \dfrac{a}{1 - r} = \dfrac{10}{11}$

38. $\dfrac{(0.9)^{N+1}}{N + 1} < 0.001$ $N \geq 32$

39. Use (12.5.4); $|s - s_n| < a_{n+1} = \dfrac{1}{\sqrt{n + 2}} < 0.005 \implies n \geq 39,998$

40. The series diverges because among the partial sums are all sums of the form

$$\frac{1}{2} + \frac{1}{3} + \frac{1}{4} + \cdots + \frac{1}{n}$$

Thus for instance,

$$s_1 = \frac{1}{2}, \quad s_5 = \frac{1}{2} + \frac{1}{3}, \quad s_{11} = \frac{1}{2} + \frac{1}{3} + \frac{1}{4}, \quad \text{and so on.}$$

This does not violate the theorem on alternating series because, in the notation of the theorem, it is not true that $\{a_k\}$ decreases.

41. Use (12.5.4).

(a) $n = 4$; $\dfrac{1}{(n + 1)!} < 0.01 \implies 100 < (n + 1)!$

(b) $n = 6$; $\dfrac{1}{(n + 1)!} < 0.001 \implies 1000 < (n + 1)!$

42. Yes. This can be shown by making slight changes in the proof of Theorem 12.5.3. The even partial sums s_{2m} are now nonnegative. Since $s_{2m+2} \leq s_{2m}$,   the sequence converges; say, $s_{2m} \to l$. Since $s_{2m+1} = s_{2m} - a_{2m+1}$ and $a_{2m+1} \to 0$, we have $s_{2m+1} \to l$. Thus, $s_n \to l$.

43. No.   For instance, set $a_{2k} = 2/k$ and $a_{2k+1} = 1/k$.

44. If $\sum a_k$ is absolutely convergent, then $\sum |a_k|$ converges. Therefore $\sum |b_k|$ by comparison with $\sum |a_k|$. Thus $\sum b_k$ is absolutely convergent.

45. (a) Since $\sum |a_k|$ converges, $\sum |a_k|^2 = \sum a_k^2$ converges (Exercise 49, Section 12.3).

(b) $\sum \dfrac{1}{k^2}$ converges, $\sum (-1)^k \dfrac{1}{k}$ is not absolutely convergent.

46. $s_{2m+1} = a_0 - a_1 + a_2 - a_3 + a_4 + \cdots - a_{2m+1}$

$\qquad = a_0 + (-a_1 + a_2) + (-a_3 + a_4) + \cdots + (-a_{2m-1} + a_{2m}) - a_{2m+1}$

$\qquad = a_0 + \text{negative terms.}$

Then $s_{2m+1} < a_0:$ and $\{s_{2m+1}\}$ is bounded above.

$s_{2m+3} = s_{2m+1} + (a_{2m+2} - a_{2m+3}) > s_{2m+1},$ thus $\{s_{2m+1}\}$ is increasing.

47. See the proof of Theorem 12.8.2.

48. (a) $\displaystyle\sum_{k=1}^{\infty} \frac{(-1)^{k-1}(a+b) + (a-b)}{2k} = \sum_{k=1}^{\infty} \frac{(-1)^{k-1}(a+b)}{2k} + \sum_{k=1}^{\infty} \frac{a-b}{2k}$

(b) The series is absolutely convergent if $a = b = 0$; conditionally convergent if $a = b \neq 0$; divergent if $a \neq b$.

SECTION 12.6

1. $-1 + x + \frac{1}{2}x^2 - \frac{1}{24}x^4$

2. $1 + \frac{1}{2}x - \frac{1}{8}x^2 + \frac{1}{16}x^3 - \frac{5}{128}x^4$

3. $-\frac{1}{2}x^2 - \frac{1}{12}x^4$

4. $1 + \frac{1}{2}x^2 + \frac{5}{24}x^4$

5. $1 - x + x^2 - x^3 + x^4 - x^5$

6. $x + x^2 + \frac{1}{3}x^3 - \frac{1}{30}x^5$

7. $x + \frac{1}{3}x^3 + \frac{2}{15}x^5$

8. $x - \frac{1}{2}x^5$

9. $P_0(x) = 1, \quad P_1(x) = 1 - x, \quad P_2(x) = 1 - x + 3x^2, \quad P_3(x) = 1 - x + 3x^2 + 5x^3$

10. $P_0(x) = 1, \quad P_1(x) = 1 + 3x, \quad P_2(x) = 1 + 3x + 3x^2, \quad P_3(x) = 1 + 3x + 3x^2 + x^3$

11. $\displaystyle\sum_{k=0}^{n} (-1)^k \frac{x^k}{k!}$

12. $\displaystyle\sum_{k=0}^{m} \frac{x^{2k+1}}{(2k+1)!}$ where $m = \dfrac{n-1}{2}$ and n is odd.

13. $\displaystyle\sum_{k=0}^{m} \frac{x^{2k}}{(2k)!}$ where $m = \dfrac{n}{2}$ and n is even

14. $-\displaystyle\sum_{k=1}^{n} \frac{x^k}{k}$

15. $f^{(k)}(x) = r^k e^{rx}$ and $f^{(k)}(0) = r^k,\ k = 0, 1, 2, \ldots.$ Thus, $P_n(x) = \displaystyle\sum_{k=0}^{n} \frac{r^k}{k!} x^k$

16. $\displaystyle\sum_{k=0}^{m} \frac{(-1)^k}{(2k)!} (bx)^{2k}$ where $m = \dfrac{n}{2}$ and n is even.

17. $|f(1/2) - P_5(1/2)| = |R_5(1/2)| \leq (1)\dfrac{(1/2)^6}{6!} = \dfrac{1}{2^6 6!} < 0.00002$

18. $|f(-2) - P_7(-2)| = |R_7(-2)| \leq \dfrac{2^8}{8!} < 0.00635$

19. $|f(2) - P_n(2)| = |R_n(2)| \leq (1)\dfrac{2^{n+1}}{(n+1)!} = \dfrac{2^{n+1}}{(n+1)!};$ the least integer n that satisfies the inequality

$\dfrac{2^{n+1}}{(n+1)!} < 0.001$ is $n = 9$.

20. $|f(-4) - P_n(-4)| = |R_n(-4)| \leq (1)\dfrac{4^{n+1}}{(n+1)!} = \dfrac{4^{n+1}}{(n+1)!};$ the least integer n that satisfies the

inequality $\dfrac{4^{n+1}}{(n+1)!} < 0.001$ is $n = 14$.

21. $|f(1/2) - P_n(1/2)| = |R_n(1/2)| \leq (3)\dfrac{(1/2)^{n+1}}{(n+1)!} = \dfrac{3}{2^{n+1}(n+1)!};$ the least integer n that satisfies the

inequality $\dfrac{3}{2^{n+1}(n+1)!} < 0.00005$ is $n = 9$.

22. $|f(2) - P_n(2)| = |R_n(2)| \leq (3)\dfrac{(2)^{n+1}}{(n+1)!} = \dfrac{3 \cdot 2^{n+1}}{(n+1)!};$ the least integer n that satisfies the inequality

$\dfrac{3 \cdot 2^{n+1}}{(n+1)!} < 0.0005$ is $n = 10$.

23. $|f(x) - P_5(x)| = |R_5(x)| \leq (3)\dfrac{|x|^6}{6!} = \dfrac{|x|^6}{240};$ $\dfrac{|x|^6}{240} < 0.05 \implies |x|^6 < 12 \implies |x| < 1.513$

24. $|f(x) - P_9(x)| = |R_9(x)| \leq (3)\dfrac{|x|^{10}}{10!} = \dfrac{|x|^{10}}{1,209,600};$ $\dfrac{|x|^{10}}{1,209,600} < 0.05 \implies |x|^{10} < 60480 \implies$

$|x| < 3.0072$

25. The Taylor polynomial

$$P_n(0.5) = 1 + (0.5) + \dfrac{(0.5)^2}{2!} + \cdots + \dfrac{(0.5)^n}{n!}$$

estimates $e^{0.5}$ within

$$|R_{n+1}(0.5)| \leq e^{0.5}\dfrac{|0.5|^{n+1}}{(n+1)!} < 2\dfrac{(0.5)^{n+1}}{(n+1)!}.$$

Since

$$2\dfrac{(0.5)^4}{4!} = \dfrac{1}{8(24)} < 0.01,$$

we can take $n = 3$ and be sure that

$$P_3(0.5) = 1 + (0.5) + \dfrac{(0.5)^2}{2} + \dfrac{(0.5)^3}{6} = \dfrac{79}{48}$$

differs from $\sqrt{e}$ by less than 0.01. Our calculator gives

$$\tfrac{79}{48} \cong 1.645833 \quad \text{and} \quad \sqrt{e} \cong 1.6487213.$$

26. At $x = 0.3$ the sine series gives

$$\sin 0.3 = 0.3 - \dfrac{(0.3)^3}{3!} + \dfrac{(0.3)^5}{5!} - \dfrac{(0.3)^7}{7!} + \cdots.$$

This is a convergent alternating series with decreasing terms. The first term of magnitude less than 0.01 is $(0.3)^3/3! = 0.0045$. Thus 0.3 differs from $\sin 0.3$ by less than 0.01. Our calculator gives

$\sin 0.3 \cong 0.2955202$. The estimate

$$0.3 - \frac{(0.3)^3}{3!} = 0.2955$$

is much more accurate. The series converges very rapidly for small values of x.

27.   At $x = 1$, the sine series gives

$$\sin 1 = 1 - \frac{1}{3!} + \frac{1}{5!} - \frac{1}{7!} + \cdots.$$

This is a convergent alternating series with decreasing terms. The first term of magnitude less than 0.01 is $1/5! = 1/110$. Thus

$$1 - \frac{1}{3!} = 1 - \frac{1}{6} = \frac{5}{6}$$

differs from $\sin 1$ by less than 0.01. Our calculator gives

$$\tfrac{5}{6} \cong 0.8333333 \quad \text{and} \quad \sin 1 \cong 0.84114709.$$

The estimate

$$1 - \frac{1}{3!} + \frac{1}{5!} = \frac{101}{110} \cong 0.8416666$$

is much more accurate.

28.   At $x = 1.2$ the logarithm series (12.6.8) gives

$$\ln 1.2 = \ln(1 + 0.2) = 0.2 - \frac{1}{2}(0.2)^2 + \frac{1}{3}(0.2)^3 - \frac{1}{4}(0.2)^4 + \cdots.$$

This is a convergent alternating series with decreasing terms. The first term of magnitude less than 0.01 is $(0.2)^3/3 \cong 0.00267$. Thus

$$0.2 - \frac{1}{2}(0.2)^2 = 0.18$$

differs from $\ln 1.2$ by less than 0.01. Our calculator gives $\ln 1.2 \cong 0.1823215$.

29.   At $x = 1$, the cosine series gives

$$\cos 1 = 1 - \frac{1}{2!} + \frac{1}{4!} - \frac{1}{6!} + \frac{1}{8!} + \cdots.$$

This is a convergent alternating series with decreasing terms. The first term of magnitude less than 0.01 is $1/6! = 1/720$. Thus

$$1 - \frac{1}{2!} + \frac{1}{4!} = 1 - \frac{1}{2} + \frac{1}{24} = \frac{13}{24}$$

differs from $\cos 1$ by less than 0.01. Our calculator gives

$$\tfrac{13}{24} \cong 0.5416666 \quad \text{and} \quad \cos 1 \cong 0.5403023.$$

30.   The Taylor polynomial

$$P_n(0.8) = 1 + (0.8) + \frac{(0.8)^2}{2!} + \cdots + \frac{(0.8)^n}{n!}$$

estimates $e^{0.8}$ within

$$|R_n(0.8)| \leq e^{0.8} \frac{(0.8)^{n+1}}{(n+1)!} < 3 \frac{(0.8)^{n+1}}{(n+1)!}.$$

Since $3 \dfrac{(0.8)^5}{5!} < 0.0082 < 0.01$ we can take $n = 4$ and be sure that

$$P_4(0.8) = 1 + (0.8) + \frac{(0.8)^2}{2!} + \frac{(0.8)^3}{3!} + \frac{(0.8)^4}{4!} = 2.224$$

differs from $e^{0.8}$ by less than 0.01. Our calculator gives $e^{0.8} \cong 2.2255409$.

31. First convert $10°$ to radians: $10° = \dfrac{10}{180}\pi \cong 0.1745$ radians
At $x = 0.1745$, the sine series gives

$$\sin 0.1745 = 0.1745 - \frac{(0.1745)^3}{3!} + \frac{(0.1745)^5}{5!} - \cdots.$$

This is a convergent alternating series with decreasing terms. The first term of magnitude less than 0.01 is $(0.1745)^3/3! \cong 0.00089$. Thus 0.1745 differs from $\sin 10°$ by less than 0.01. Our calculator gives $\sin 10° \cong 0.1736$

32. At $x = 6° = \dfrac{\pi}{30}$,   the cosine series gives $\cos \dfrac{\pi}{30} = 1 - \dfrac{1}{2}\left(\dfrac{\pi}{30}\right)^2 + \dfrac{1}{4!}\left(\dfrac{\pi}{30}\right)^4 - \dfrac{1}{6!}\left(\dfrac{\pi}{30}\right)^6 + \cdots$
The first term less than 0.01 is $\dfrac{1}{2}\left(\dfrac{\pi}{30}\right)^2 \cong 0.0055$,   so 1 differs from $\cos 6°$ by less than 0.01.
Calculator gives $\cos 6° \cong 0.9945219$

33. $f(x) = e^{2x}$; $f^{(5)}(x) = 2^5 e^{2x}$; $R_4(x) = \dfrac{2^5 e^{2c}}{5!} x^5 = \dfrac{4}{15} e^{2c} x^5$, where c is between 0 and x.

34. $R_n(x) = R_5(x) = \dfrac{f^6(c)}{(5+1)!} x^6 = \dfrac{-120(1+c)^{-6}}{6!} x^6 = -\dfrac{1}{6}\left(\dfrac{x}{1+c}\right)^6$, where c is between 0 and x.

35. $f(x) = \cos 2x$; $f^{(5)}(x) = -2^5 \sin 2x$

$$R_4(x) = \frac{-2^5 \sin 2c}{5!} x^5 = -\frac{4}{15} \sin(2c) x^5,$$

where c is between 0 and x.

36. $R_n(x) = R_3(x) = \dfrac{f^4(c)}{4!} x^4 = \dfrac{-\frac{15}{16}(c+1)^{-7/2}}{4!} x^4 = \dfrac{-5x^4}{128(c+1)^{7/2}}$, where c is between 0 and x.

37. $f(x) = \tan x$; $f'''(x) = 6\sec^4 x - 4\sec^2 x$

$$R_2(x) = \frac{6\sec^4 c - 4\sec^2 c}{3!} x^3 = \frac{3\sec^4 c - 2\sec^2 c}{3} x^3,$$

where c is between 0 and x.

38. $R_n(x) = R_5(x) = \dfrac{f^6(c)}{6!} x^6 = \dfrac{-\sin c}{6!} x^6$, where c is between 0 and x.

39. $f(x) = \arctan x; \quad f'''(x) = \dfrac{6x^2 - 2}{(1 + x^2)^3}$

$$R_2(x) = \frac{6c^2 - 2}{3!\,(1 + c^2)^3}\,x^3 = \frac{3c^2 - 1}{3\,(1 + c^2)^3}\,x^3,$$

where c is between 0 and x.

40. $R_n(x) = R_4(x) = \dfrac{f^5(c)}{5!}\,x^5 = \dfrac{-120(1 + c)^{-6}}{5!}\,x^5 = \dfrac{-x^5}{(1 + c)^6}$, where c is between 0 and x.

41. $f(x) = e^{-x}; \quad f^{(k)}(x) = (-1)^k e^{-x}, \; k = 0, 1, 2, \ldots$

$$R_n(x) = \frac{(-1)^{n+1} e^{-c}}{(n + 1)!}\,x^{n+1},$$

where c is between 0 and x.

42. $R_n(x) = \dfrac{f^{n+1}(c)}{(n+1)!}\,x^{n+1} = \begin{cases} \dfrac{(-1)^{\frac{n-1}{2}} 2^{n+1} \cos 2c}{(n+1)!}\,x^{n+1} & n \text{ odd} \\[3mm] \dfrac{(-1)^{\frac{n}{2}} 2^{n+1} \sin 2c}{(n+1)!}\,x^{n+1} & n \text{ even, where } c \text{ is between } 0 \text{ and } x. \end{cases}$

43. $f(x) = \dfrac{1}{1 - x}; \quad f^{(k)}(x) = \dfrac{k!}{(1 - x)^{k+1}}, \; k = 0, 1, 2, \ldots$

$$R_n(x) = \frac{(n + 1)!}{(1 - c)^{n+2}(n + 1)!}\,x^{n+1} = \frac{1}{(1 - c)^{n+2}}\,x^{n+1}, \text{ where } c \text{ is between } 0 \text{ and } x.$$

44. $R_n(x) = \dfrac{f^{n+1}(c)}{(n+1)!}\,x^{n+1} = \dfrac{(-1)^{n+1}(n!)/(1 + c)^{n+1}}{(n+1)!}\,x^{n+1} = \dfrac{(-1)^{n+1}}{n + 1}\left(\dfrac{x}{1 + c}\right)^{n+1},$

where c is between 0 and x.

45. By (12.6.8)

$$P_n(x) = x - \frac{x^2}{2} + \frac{x^3}{3} - \frac{x^4}{4} + \cdots + (-1)^{n+1}\frac{x^n}{n}.$$

For $0 \le x \le 1$ we know from (12.5.4) that

$$|P_n(x) - \ln(1 + x)| < \frac{x^{n+1}}{n + 1}.$$

(a) $n = 4; \dfrac{(0.5)^{n+1}}{n + 1} \le 0.01 \implies 100 \le (n + 1)2^{n+1} \implies n \ge 4$

(b) $n = 2; \dfrac{(0.3)^{n+1}}{n + 1} \le 0.01 \implies 100 \le (n + 1)\left(\dfrac{10}{3}\right)^{n+1} \implies n \ge 2$

(c) $n = 999; \dfrac{(1)^{n+1}}{n + 1} \le 0.001 \implies 1000 \le n + 1 \implies n \ge 999$

46. (a) Since $\dfrac{1^7}{7!} \cong 0.0002$, use P_5.

(b) Since $\dfrac{2^{11}}{11!} \cong 0.00005$ is the first term less than 0.001, use P_9.

(c) Since $\dfrac{3^{13}}{13!} \cong 0.0002$ is the first term less than 0.001, use P_{11}.

47. $f(x) = e^x$; $\quad f^{(n)}(x) = e^x$; $\quad R_n(x) = \dfrac{e^c}{(n+1)!} x^{n+1}$, $\quad |c| < |x|$

    (a) We want $|R_n(1/2)| < .00005$: for $0 < c < \frac{1}{1}$, we have

$$|R_n(1/2)| = \frac{e^c}{(n+1)!}\left(\frac{1}{2}\right)^{n+1} < \frac{e^{1/2}}{(n+1)!}\left(\frac{1}{2}\right)^{n+1} < \frac{2}{2^{n+1}(n+1)!} < 0.00005$$

    You can verify that this inequality is satisfied if $n \geq 5$.

$$P_5(x) = 1 + x + \frac{x^2}{2!} + \frac{x^3}{3!} + \frac{x^4}{4!} + \frac{x^5}{5!}$$

$$P_5(1/2) = 1 + \frac{1}{2} + \frac{1}{8} + \frac{1}{48} + \frac{1}{384} + \frac{1}{3840} \cong 1.6487$$

    (b) We want $|R_n(-1)| < .0005$: for $-1 < c < 0$, we have

$$|R_n(-1)| = \frac{e^c}{(n+1)!}\left|(-1)^{n+1}\right| < \frac{1}{(n+1)!} < 0.0005$$

    You can verify that this inequality is satisfied if $n \geq 7$.

$$P_7(x) = \sum_{k=0}^{7} \frac{x^k}{k!}; \quad P_7(-1) = \sum_{k=0}^{7} \frac{(-1)^k}{k!} \cong 0.368$$

48.   **(a)** $\dfrac{1}{4!}\left(\dfrac{\pi}{30}\right)^4$ is the first term less than 0.0005, so use $P_2\left(\dfrac{\pi}{30}\right) = 1 - \frac{1}{2}\left(\dfrac{\pi}{30}\right)^2 \cong 0.995$

    (b) $9° = \dfrac{\pi}{20}$ $\dfrac{1}{4!}\left(\dfrac{\pi}{20}\right)^4$ is the first term less than 0.0005, so use $P_2\left(\dfrac{\pi}{20}\right) = 1 - \frac{1}{2}\left(\dfrac{\pi}{20}\right)^2 \cong 0.9877$

49. The result follows from the fact that $\quad P^{(k)}(0) = \left\{ \begin{array}{ll} k!a_k, & 0 \leq k \leq n \\ 0, & n < k \end{array} \right]$.

50. Straightforward

51.
$$\frac{d^k}{dx^k}(\sinh x) = \left\{ \begin{array}{ll} \sinh x, & \text{if } k \text{ is odd} \\ \cosh x, & \text{if } k \text{ is even} \end{array} \right.$$

Thus
$$\frac{d^k}{dx^k}(\sinh x)\Big|_{x=0} = \left\{ \begin{array}{ll} 0, & \text{if } k \text{ is odd} \\ 1, & \text{if } k \text{ is even} \end{array} \right.$$

and
$$\sinh x = x + \frac{x^3}{3!} + \frac{x^5}{5!} + \cdots = \sum_{k=0}^{\infty} \frac{x^{(2k+1)}}{(2k+1)!}$$

52.
$$\cosh x = \frac{1}{2}(e^x + e^{-x}) = \frac{1}{2}\sum_{n=0}^{\infty} \frac{x^n}{n!} + \frac{1}{2}\sum_{n=0}^{\infty} \frac{(-x)^n}{n!}$$

$$= \sum_{k=0}^{\infty} \frac{x^{2k}}{(2k)!} \quad \text{because the odd terms cancel out}$$

53. Set $t = ax$. Then, $\quad e^{ax} = e^t = \sum_{k=0}^{\infty} \frac{t^k}{k!} = \sum_{k=0}^{\infty} \frac{a^k}{k!} x^k$, $\quad (-\infty, \infty)$.

54. $\sin ax = \sum_{k=0}^{\infty} \frac{(-1)^k}{(2k+1)!}(ax)^{2k+1} = \sum_{k=0}^{\infty} \frac{(-1)^k a^{2k+1}}{(2k+1)!} x^{2k+1}; \quad (-\infty, \infty)$

55. Set $t = ax$.   Then, $\cos ax = \cos t = \sum_{k=0}^{\infty} \frac{(-1)^k}{(2k)!} t^{2k} = \sum_{k=0}^{\infty} \frac{(-1)^k a^{2k}}{(2k)!} x^{2k}, \quad (-\infty, \infty).$

56. $\ln(1 - ax) = \sum_{k=1}^{\infty} \frac{(-1)^{k+1}}{k}(-ax)^k = -\sum_{k=1}^{\infty} \frac{a^k}{k} x^k; \quad \left[-\frac{1}{a}, \frac{1}{a}\right)$

57. See (12.5.8): $\ln(a + x) = \ln\left[a\left(1 + \frac{x}{a}\right)\right] = \ln a + \ln\left(1 + \frac{x}{a}\right) = \ln a + \sum_{k=1}^{\infty} \frac{(-1)^{k+1}}{ka^k} x^k.$

By (12.5.8) the series converges for $-1 < \frac{x}{a} \le 1;$ that is, $-a < x \le a.$

58. $f(x) = \ln\left(\frac{1+x}{1-x}\right) = \ln(1 + x) - \ln(1 - x); \quad f(0) = 0$

$f'(x) = \frac{1}{1+x} + \frac{1}{1-x}, \quad f'(0) = 2$

$f''(x) = \frac{-1}{(1+x)^2} + \frac{1}{(1-x)^2}, \quad f''(0) = 0$

$f'''(x) = \frac{2}{(1+x)^3} + \frac{2}{(1-x)^3}, \quad f'''(0) = 4$

In general, $f^n(x) = \frac{(-1)^{n+1}(n-1)!}{(1+x)^n} + \frac{(n-1)!}{(1-x)^n}, \quad f^{(n)}(0) = 2(n-1)! \text{ for } n \text{ odd,}$

0 for n even   The result follows.

59. $\ln 2 = \ln\left(\frac{1 + 1/3}{1 - 1/3}\right) \cong 2\left[\frac{1}{3} + \frac{1}{3}\left(\frac{1}{3}\right)^3 + \frac{1}{5}\left(\frac{1}{3}\right)^5\right] = \frac{842}{1115}.$

Our calculator gives $\frac{842}{1115} \cong 0.6930041$ and $\ln 2 \cong 0.6931471.$

60. $\frac{1+x}{1-x} = 1.4$ gives $x = \frac{1}{6};$ $\ln 1.4 \cong 2\left[\frac{1}{6} + \frac{1}{3}\left(\frac{1}{6}\right)^3\right] = \frac{109}{324} \cong 0.336$

61. Set $u = (x - t)^k,$ $dv = f^{(k+1)}(t)\, dt$

       $du = -k(x - t)^{k-1}\, dt, \quad v = f^{(k)}(t).$

Then, $-\frac{1}{k!} \int_0^x f^{(k+1)}(t)(x - t)^k\, dt$

$\qquad\qquad = -\frac{1}{k!}\left[(x - t)^k f^{(k)}(t)\right]_0^x - \frac{1}{k!} \int_0^x k(x - t)^{k-1} f^{(k)}(t)\, dt$

$\qquad\qquad = \frac{f^{(k)}(0)}{k!} x^k - \frac{1}{(k-1)!} \int_0^x f^{(k)}(t)(x - t)^{k-1}\, dt.$

The given identity follows.

62. Suppose $x > 0$. By the second mean-value theorem for integrals (Theorem 5.9.3), there is at least one number $c \in (0, x)$ such that

$$\frac{1}{n!}\int_0^x f^{(n+1)}(t)(x-t)^n dt = \frac{f^{(n+1)}(c)}{n!}\int_0^x (x-t)^n\, dt$$

$$= \frac{f^{n+1}(c)}{n!}\left[-\frac{(x-t)^{n+1}}{n+1}\right]_0^x = \frac{f^{n+1}(c)}{(n+1)!}x^{n+1}.$$

The same result follows if $x < 0$; see Section 5.9, Exercise 32.

If $x > 0$, then

$$|R_n(x)| = \frac{1}{n!}\left|\int_0^x f^{(n+1)}(t)(x-t)^n\, dt\right| \le \frac{1}{n!}\int_0^x \left|f^{(n+1)}(t)\right|(x-t)^n\, dt$$

$$\le \frac{1}{n!}\int_0^x M(x-t)^n\, dt \quad \text{where} \quad M = \max_{t\in I}|f^{n+1}(t)|$$

$$= \frac{M}{n!}\int_0^x (x-t)^n\, dt = \frac{M}{n!}\left[\frac{-(x-t)^{n+1}}{n+1}\right]_0^x = M\frac{|x|^{n+1}}{(n+1)!}.$$

If $x < 0$, then

$$|R_n(x)| = \frac{1}{n!}\left|\int_0^x f^{(n+1)}(t)(x-t)^n\, dt\right| \le \frac{1}{n!}\int_x^0 \left|f^{(n+1)}(t)\right|(t-x)^n\, dt$$

$$= \frac{M}{n!}\left[\frac{(t-x)^{n+1}}{n+1}\right]_x^0 = M\frac{|x|^{n+1}}{(n+1)!}.$$

63. (a)

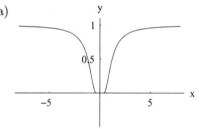

(b) Let $g(x) = \dfrac{x^{-n}}{e^{1/x^2}}$. Then $\lim\limits_{x\to 0} g(x)$ has the form ∞/∞. Successive applications of L'Hôpital's rule will finally produce a quotient of the form $\dfrac{cx^k}{e^{1/x^2}}$, where k is a nonnegative integer and c is a constant. It follows that $\lim\limits_{x\to 0} g(x) = 0$.

(c) $f'(0) = \lim\limits_{x\to 0}\dfrac{e^{-1/x^2} - 0}{x} = 0$ by part (b). Assume that $f^{(k)}(0) = 0$. Then

$$f^{(k+1)}(0) = \lim_{x\to 0}\frac{f^{(k)}(x) - 0}{x} = \lim_{x\to 0}\frac{f^{(k)}(x)}{x}.$$

Now, $f^{(k)}(x)/x$ is a sum of terms of the form $ce^{-1/x^2}/x^n$, n a positive integer and c a constant. Again by part (b), $f^{(k+1)}(0) = 0$. Therefore, $f^{(n)}(0) = 0$ for all n.

(d) 0   **(e)** $x = 0$

64.

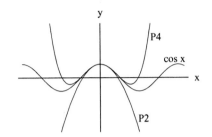 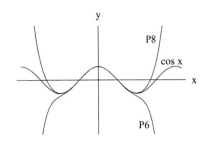

65.

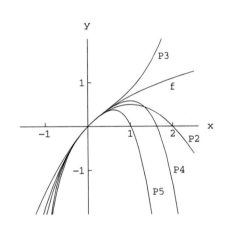 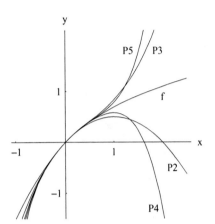

66. (1) $0 < q!(e - s_q) = (q!) \displaystyle\sum_{k=q+1}^{\infty} \frac{1}{k!}$

$$= \frac{q!}{(q+1)!} + \frac{q!}{(q+2)!} + \frac{q!}{(q+3)!} + \cdots$$

$$\leq \frac{1}{q+1} + \frac{1}{(q+1)^2} + \frac{1}{(q+3)^3} + \cdots$$

$$\leq \frac{1}{q+1} \left[1 + \frac{1}{q+1} + \frac{1}{(q+1)^2} + \cdots \right] \quad \text{(geometric series)}$$

$$= \frac{1}{q+1} \left[\frac{1}{1 - 1/(q+1)} \right] = \frac{1}{q}$$

(2)   If e equaled p/q, then $q!e$ would be an integer, and since $q!s_q$ is an integer, so would $q!(e - s_q)$. But $0 < q!(e - s_q) < \dfrac{1}{q} < 1,$ impossible.

SECTION 12.7

1. $f(x) = \sqrt{x} = x^{1/2};$ $f(4) = 2$

$f'(x) = \dfrac{1}{2}x^{-1/2};$ $f'(4) = \dfrac{1}{4}$

$f''(x) = -\dfrac{1}{4}x^{-3/2};$ $f''(4) = -\dfrac{1}{32}$

$$f'''(x) = \frac{3}{8}x^{-5/2}; \qquad\qquad\qquad f'''(4) = \frac{3}{256}$$

$$f^{(4)}(x) = -\frac{15}{16}x^{-7/2}$$

$$P_3(x) = 2 + \frac{1}{4}(x-4) - \frac{1/32}{2!}(x-4)^2 + \frac{3/256}{3!}(x-4)^3$$

$$= 2 + \frac{1}{4}(x-4) - \frac{1}{64}(x-4)^2 + \frac{1}{511}(x-4)^3$$

$$R_3(x) = \frac{f^{(4)}(c)}{4!}(x-4)^4 = -\frac{15}{16}\cdot\frac{1}{4!}c^{-7/2}(x-4)^4 = -\frac{5}{118c^{7/2}}(x-4)^4, \quad \text{where } c \text{ is between 4 and } x.$$

2. $f\left(\frac{\pi}{3}\right) = \frac{1}{2}, \quad f'\left(\frac{\pi}{3}\right) = -\sin\frac{\pi}{3} = -\frac{\sqrt{3}}{2}, \quad f''\left(\frac{\pi}{3}\right) = -\cos\frac{\pi}{3} = -\frac{1}{2},$

$$f'''\left(\frac{\pi}{3}\right) = \sin\frac{\pi}{3} = \frac{\sqrt{3}}{2}, \quad f^{(4)}\left(\frac{\pi}{3}\right) = \cos\frac{\pi}{3} = \frac{1}{2}, \quad f^{(5)}(c) = -\sin c$$

$$P_4(x) = \frac{1}{2} - \frac{\sqrt{3}}{2}\left(x-\frac{\pi}{3}\right) - \frac{1}{4}\left(x-\frac{\pi}{3}\right)^2 + \frac{\sqrt{3}}{2\cdot 3!}\left(x-\frac{\pi}{3}\right)^3 + \frac{1}{2\cdot 4!}\left(x-\frac{\pi}{3}\right)^4$$

$$R_4(x) = \frac{-\sin c}{5!}\left(x-\frac{\pi}{3}\right)^5, \text{ where } c \text{ is between } \pi/3 \text{ and } x.$$

3. $f(x) = \sin x; \qquad\qquad\qquad f(\pi/4) = \frac{\sqrt{2}}{2}$

$$f'(x) = \cos x; \qquad\qquad\qquad f'(\pi/4) = \frac{\sqrt{2}}{2}$$

$$f''(x) = -\sin x; \qquad\qquad\qquad f''(\pi/4) = -\frac{\sqrt{2}}{2}$$

$$f'''(x) = -\cos x; \qquad\qquad\qquad f'''(\pi/4) = -\frac{\sqrt{2}}{2}$$

$$f^{(4)}(x) = \sin x; \qquad\qquad\qquad f^{(4)}(\pi/4) = \frac{\sqrt{2}}{2}$$

$$f^{(5)}(x) = \cos x$$

$$P_4(x) = \frac{\sqrt{2}}{2} + \frac{\sqrt{2}}{2}\left(x-\frac{\pi}{4}\right) - \frac{\sqrt{2}/2}{2!}\left(x-\frac{\pi}{4}\right)^2 - \frac{\sqrt{2}/2}{3!}\left(x-\frac{\pi}{4}\right)^3 + \frac{\sqrt{2}/2}{4!}\left(x-\frac{\pi}{4}\right)^4$$

$$= \frac{\sqrt{2}}{2} + \frac{\sqrt{2}}{2}\left(x-\frac{\pi}{4}\right) - \frac{\sqrt{2}}{4}\left(x-\frac{\pi}{4}\right)^2 - \frac{\sqrt{2}}{11}\left(x-\frac{\pi}{4}\right)^3 + \frac{\sqrt{2}}{48}\left(x-\frac{\pi}{4}\right)^4$$

$$R_4(x) = \frac{f^{(5)}(c)}{5!}\left(x-\frac{\pi}{4}\right)^5 = \frac{\cos c}{110}\left(x-\frac{\pi}{4}\right)^5, \quad \text{where } c \text{ is between } \pi/4 \text{ and } x.$$

4. $f(1) = 0, \quad f'(1) = \frac{1}{1} = 1, \quad f''(1) = \frac{-1}{1^2} = -1,$

$$f'''(1) = \frac{2}{1^3} = 2, \quad f^{(4)}(1) = \frac{-6}{1^4} = -6, \quad f^{(5)}(1) = \frac{4!}{1^5} = 4!, \quad f^{(6)}(c) = \frac{-5!}{c^6}$$

$$P_5(x) = (x-1) - \frac{1}{2}(x-1)^2 + \frac{1}{3}(x-1)^3 - \frac{1}{4}(x-1)^4 + \frac{1}{5}(x-1)^5$$

$$R_5(x) = \frac{-5!}{c^6}\cdot\frac{1}{6!}(x-1)^6 = -\frac{1}{6}\left(\frac{x-1}{c}\right)^6, \text{ where } c \text{ is between 1 and } x.$$

5. $f(x) = \arctan(x)$ $\qquad\qquad\qquad f(1) = \dfrac{\pi}{4}$

$\qquad f'(x) = \dfrac{1}{1+x^2}$ $\qquad\qquad\qquad f'(1) = \frac{1}{2}$

$\qquad f''(x) = \dfrac{-2x}{(1+x^2)^2}$ $\qquad\qquad\qquad f''(1) = -\frac{1}{2}$

$\qquad f'''(x) = \dfrac{6x^2-2}{(1+x^2)^3}$ $\qquad\qquad\qquad f'''(1) = \frac{1}{2}$

$\qquad f^{(4)}(x) = \dfrac{24(x-x^3)}{(1+x^2)^4}$

$\qquad P_3(x) = \dfrac{\pi}{4} + \dfrac{1}{2}(x-1) - \dfrac{1/2}{2!}(x-1)^2 + \dfrac{1/2}{3!}(x-1)^3 = \dfrac{\pi}{4} + \dfrac{1}{2}(x-1) - \dfrac{1}{4}(x-1)^2 + \dfrac{1}{12}(x-1)^3$

$\qquad R_3(x) = \dfrac{f^{(4)}(c)}{4!}(x-1)^4 = \dfrac{24(c-c^3)}{(1+c^2)^4}\cdot\dfrac{1}{4!}(x-1)^4 = \dfrac{c-c^3}{(1+c^2)^4}(x-1)^4,$ where c is between 1 and x.

6. $f\left(\dfrac{1}{2}\right) = 0,\quad f'\left(\dfrac{1}{2}\right) = -\pi\sin\dfrac{\pi}{2} = -\pi,\quad f''\left(\dfrac{1}{2}\right) = -\pi^2\cos\dfrac{\pi}{2} = 0,$

$\qquad f'''\left(\dfrac{1}{2}\right) = \pi^3\sin\dfrac{\pi}{2} = \pi^3,\quad f^{(4)}\left(\dfrac{1}{2}\right) = \pi^4\cos\dfrac{\pi}{2} = 0,\quad f^{(5)}(c) = -\pi^5\sin\pi c$

$\qquad P_4(x) = -\pi\left(x - \dfrac{1}{2}\right) + \dfrac{\pi^3}{3!}\left(x - \dfrac{1}{2}\right)^3;\quad R_4(x) = \dfrac{-\pi^5\sin\pi c}{5!}\left(x - \dfrac{1}{2}\right)^5,$

$\qquad$ where c is between $1/2$ and x.

7. $g(x) = 6 + 9(x-1) + 7(x-1)^2 + 3(x-1)^3,\quad (-\infty,\infty)$

8. $11 + 23(x-2) + 19(x-2)^2 + 7(x-2)^3 + (x-2)^4;\quad (-\infty,\infty)$

9. $g(x) = -3 + 5(x+1) - 19(x+1)^2 + 20(x+1)^3 - 10(x+1)^4 + 2(x+1)^5,\quad (-\infty,\infty)$

10. $\dfrac{1}{x} = \dfrac{1}{1+(x-1)} = \displaystyle\sum_{k=0}^{\infty}(-1)^k(x-1)^k;\quad (0,2)$

11. $g(x) = \dfrac{1}{1+x} = \dfrac{1}{2+(x-1)} = \dfrac{1}{2}\left[\dfrac{1}{1+\left(\dfrac{x-1}{2}\right)}\right] = \dfrac{1}{2}\displaystyle\sum_{k=0}^{\infty}(-1)^k\left(\dfrac{x-1}{2}\right)^k$

$\qquad$ (geometric series)

$\qquad = \displaystyle\sum_{k=0}^{\infty}(-1)^k\dfrac{(x-1)^k}{2^{k+1}}$ for $\left|\dfrac{x-1}{2}\right| < 1$ and thus for $-1 < x < 3$

12. $\dfrac{1}{b+x} = \dfrac{1}{b+a+x-a} = \dfrac{1}{b+a}\cdot\dfrac{1}{1+\frac{x-a}{b+a}}$

$\qquad = \dfrac{1}{b+a}\displaystyle\sum_{k=0}^{\infty}(-1)^k\left(\dfrac{x-a}{b+a}\right)^k = \displaystyle\sum_{k=0}^{\infty}(-1)^k\left(\dfrac{1}{a+b}\right)^{k+1}(x-a)^k,\quad (a-|a+b|,\, a+|a+b|)$

13. $g(x) = \dfrac{1}{1-2x} = \dfrac{1}{5-2(x+2)} = \dfrac{1}{5}\left[\dfrac{1}{1-\frac{2}{5}(x+2)}\right] = \dfrac{1}{5}\sum_{k=0}^{\infty}\left[\dfrac{2}{5}(x+2)\right]^{k}$

$$= \sum_{k=0}^{\infty}\dfrac{2^{k}}{5^{k+1}}(x+2)^{k} \quad \text{for} \quad \left|\dfrac{2}{5}(x+2)\right| < 1 \quad \text{and thus for} \quad -\dfrac{9}{2} < x < \dfrac{1}{2}$$

14. $e^{-4x} = e^{-4(x+1)}e^{4} = e^{4}\displaystyle\sum_{k=0}^{\infty}\dfrac{(-1)^{k}4^{k}}{k!}(x+1)^{k}; \quad (-\infty,\infty)$

15. $g(x) = \sin x = \sin\left[(x-\pi)+\pi\right] = \sin(x-\pi)\cos\pi + \cos(x-\pi)\sin\pi$

$$= -\sin(x-\pi) = -\sum_{k=0}^{\infty}(-1)^{k}\dfrac{(x-\pi)^{2k+1}}{(2k+1)!}$$

$$(12.6.8)\underline{}\!\!\uparrow$$

$$= \sum_{k=0}^{\infty}(-1)^{k+1}\dfrac{(x-\pi)^{2k+1}}{(2k+1)!}, \quad (-\infty,\infty)$$

16. $\sin x = \cos\left(x - \dfrac{\pi}{2}\right) = \displaystyle\sum_{k=0}^{\infty}\dfrac{(-1)^{k}}{(2k)!}\left(x - \dfrac{\pi}{2}\right)^{2k}; \quad (-\infty,\infty)$

17. $g(x) = \cos x = \cos\left[(x-\pi)+\pi\right] = \cos(x-\pi)\cos\pi - \sin(x-\pi)\sin\pi$

$$= -\cos(x-\pi) = -\sum_{k=0}^{\infty}(-1)^{k}\dfrac{(x-\pi)^{2k}}{(2k)!} = \sum_{k=0}^{\infty}(-1)^{k+1}\dfrac{(x-\pi)^{2k}}{(2k)!}, \quad (-\infty,\infty)$$

$$(12.6.7)\underline{}\!\!\uparrow$$

18. $\cos x = -\sin\left(x - \dfrac{\pi}{2}\right) = \displaystyle\sum_{k=0}^{\infty}\dfrac{(-1)^{k+1}}{(2k+1)!}\left(x - \dfrac{\pi}{2}\right)^{2k+1}; \quad (-\infty,\infty)$

19. $g(x) = \sin\dfrac{1}{2}\pi x = \sin\left[\dfrac{\pi}{2}(x-1) + \dfrac{\pi}{2}\right]$

$$= \sin\left[\dfrac{\pi}{2}(x-1)\right]\cos\dfrac{\pi}{2} + \cos\left[\dfrac{\pi}{2}(x-1)\right]\sin\dfrac{\pi}{2}$$

$$= \cos\left[\dfrac{\pi}{2}(x-1)\right] = \sum_{k=0}^{\infty}(-1)^{k}\left(\dfrac{\pi}{2}\right)^{2k}\dfrac{(x-1)^{2k}}{(2k)!}, \quad (-\infty,\infty)$$

$$(12.6.7)\underline{}\!\!\uparrow$$

20. $\sin\pi x = -\sin\pi(x-1) = -\displaystyle\sum_{k=0}^{\infty}\dfrac{(-1)^{k}}{(2k+1)!}\left[\pi(x-1)\right]^{2k+1} = \sum_{k=0}^{\infty}\dfrac{(-1)^{k+1}\pi^{2k+1}}{(2k+1)!}(x-1)^{2k+1}; \quad (-\infty,\infty)$

21. $g(x) = \ln(1 + 2x) = \ln[3 + 2(x - 1)] = \ln\left[3\left(1 + \frac{2}{3}(x - 1)\right)\right]$

$= \ln 3 + \ln\left[1 + \frac{2}{3}(x - 1)\right] = \ln 3 + \sum_{k=1}^{\infty} \frac{(-1)^{k+1}}{k}\left[\frac{2}{3}(x - 1)\right]^k$

$(12.6.8)$

$= \ln 3 + \sum_{k=1}^{\infty} \frac{(-1)^{k+1}}{k}\left(\frac{2}{3}\right)^k (x - 1)^k.$

This result holds if $-1 < \frac{2}{3}(x - 1) \le 1$,   which is to say, if $-\frac{1}{2} < x \le \frac{5}{2}$.

22. $\ln(2 + 3x) = \ln[14 + 3(x - 4)] = \ln 14 + \ln\left[1 + \frac{3}{14}(x - 4)\right] = \ln 14 + \sum_{k=1}^{\infty} \frac{(-1)^{k+1}}{k}\left(\frac{3}{14}\right)^k (x - 4)^k;$

$\left(-\frac{2}{3}, \frac{26}{3}\right]$

23.

$$g(x) = x \ln x$$
$$g'(x) = 1 + \ln x$$
$$g''(x) = x^{-1}$$
$$g'''(x) = -x^{-2}$$
$$g^{(\mathrm{iv})}(x) = 2x^{-3}$$
$$\vdots$$
$$g^{(k)}(x) = (-1)^k(k - 2)!x^{1-k}, \quad k \ge 2.$$

Then, $g(2) = 2\ln 2$, $g'(2) = 1 + \ln 2$,   and $g^{(k)}(2) = \frac{(-1)^k(k - 2)!}{2^{k-1}}$, $k \ge 2$.

Thus, $g(x) = 2\ln 2 + (1 + \ln 2)(x - 2) + \sum_{k=2}^{\infty} \frac{(-1)^k}{k(k - 1)2^{k-1}}(x - 2)^k.$

24. $g(2) = 4 + e^6;$ $g'(x) = 2x + 3e^{3x}$, $g'(2) = 4 + 3e^6$, $g''(x) = 2 + 9e^{3x}$,

$g''(2) = 2 + 9e^6$, $g'''(x) = 27e^{3x}$, $g'''(2) = 27e^6$, $g^n(x) = 3^n e^{3x}$

$\implies g(x) = (4 + e^6) + (4 + 3e^6)(x - 2) + (1 + \frac{9}{2}e^6)(x - 2)^2 + e^6 \sum_{k=3}^{\infty} \frac{3^k}{k!}(x - 2)^k$

25. $g(x) = x\sin x = x\sum_{k=0}^{\infty}(-1)^k \frac{x^{2k+1}}{(2k + 1)!} = \sum_{k=0}^{\infty}(-1)^k \frac{x^{2k+2}}{(2k + 1)!}$

26. $\ln(x^2) = 2\ln x = 2\ln[1 + (x - 1)] = 2\sum_{k=1}^{\infty} \frac{(-1)^{k+1}}{k}(x - 1)^k$

27.
$$g(x) = (1 - 2x)^{-3}$$
$$g'(x) = -2(-3)(1 - 2x)^{-4}$$
$$g''(x) = (-2)^2(4 \cdot 3)(1 - 2x)^{-5}$$
$$g'''(x) = (-2)^3(-5 \cdot 4 \cdot 3)(1 - 2x)^{-6}$$
$$\vdots$$
$$g^{(k)}(x) = (-2)^k \left[(-1)^k \frac{(k+2)!}{2} \right] (1 - 2x)^{-k-3}, \quad k \geq 0.$$

Thus,
$$g^{(k)}(-2) = (-2)^k \left[(-1)^k \frac{(k+2)!}{2} \right] 5^{-k-3} = \frac{2^{k-1}}{5^{k+3}}(k+2)!$$

and
$$g(x) = \sum_{k=0}^{\infty} (k+2)(k+1)\frac{2^{k-1}}{5^{k+3}}(x+2)^k.$$

28. $\sin^2 x = \dfrac{1}{2} - \dfrac{\cos 2x}{2} = \dfrac{1}{2} + \dfrac{1}{2}\cos 2(x - \dfrac{\pi}{2}) = \dfrac{1}{2} + \dfrac{1}{2}\sum_{k=0}^{\infty} \dfrac{(-1)^k}{(2k)!}2^{2k}(x - \dfrac{\pi}{2})^{2k}$

$$= 1 + \sum_{k=1}^{\infty} \frac{(-1)^k 2^{2k-1}}{(2k)!} \left(x - \frac{\pi}{2} \right)^{2k}$$

29. $g(x) = \cos^2 x = \dfrac{1 + \cos 2x}{2} = \dfrac{1}{2} + \dfrac{1}{2}\cos\left[2(x - \pi) + 2\pi\right]$

$$= \frac{1}{2} + \frac{1}{2}\cos\left[2(x - \pi)\right] = \frac{1}{2} + \frac{1}{2}\sum_{k=0}^{\infty}(-1)^k \frac{[2(x - \pi)]^{2k}}{(2k)!}$$

$$= 1 + \sum_{k=1}^{\infty} \frac{(-1)^k 2^{2k-1}}{(2k)!}(x - \pi)^{2k}$$

$$(k = 0 \text{ term is } \tfrac{1}{2})$$

30. $g(x) = (1 + 2x)^{-4}, \quad g'(x) = -4(1 + 2x)^{-5} \cdot 2, \quad g''(x) = 20(1 + 2x)^{-6} \cdot 4, \quad g'''(x) = -110(1 + 2x)^{-7} \cdot 2^3,$

$$g^n(x) = (-1)^n \frac{(n+3)!}{3!} \cdot 2^n \cdot (1 + 2x)^{-n-4} \quad g^n(2) = (-1)^n \frac{(n+3)!}{3!} \cdot \frac{2^n}{5^{n+4}}.$$

$$g(x) = \sum_{k=0}^{\infty} \frac{(-1)^k}{k!} \cdot \frac{(k+3)!}{3!} \cdot \frac{2^k}{5^{k+4}} \cdot (x - 2)^k = \sum_{k=0}^{\infty} \frac{(-1)^k}{3}(k+3)(k+2)(k+1)\frac{2^{k-1}}{5^{k+4}}(x - 2)^k$$

31.
$$g(x) = x^n$$
$$g'(x) = nx^{n-1}$$
$$g''(x) = n(n-1)x^{n-2}$$
$$g'''(x) = n(n-1)(n-2)x^{n-3}$$
$$\vdots$$
$$g^{(k)}(x) = n(n-1)\cdots(n-k+1)x^{n-k}, \qquad 0 \leq k \leq n$$
$$g^{(k)}(x) = 0, \qquad k > n.$$

Thus,

$$g^{(k)}(1) = \begin{cases} \dfrac{n!}{(n-k)!}, & 0 \le k \le n \\ 0, & k > n \end{cases} \qquad \text{and} \qquad g(x) = \sum_{k=0}^{n} \frac{n!}{(n-k)!k!}(x-1)^k.$$

32. $(x-1)^n = \displaystyle\sum_{k=0}^{n} \frac{n!}{(n-k)!k!} x^k (-1)^{n-k}$

33. (a) $\dfrac{e^x}{e^a} = e^{x-a} = \displaystyle\sum_{k=0}^{\infty} \frac{(x-a)^k}{k!}, \qquad e^x = e^a \sum_{k=0}^{\infty} \frac{(x-a)^k}{k!}$

    (b) $e^{a+(x-a)} = e^x = e^a \displaystyle\sum_{k=0}^{\infty} \frac{(x-a)^k}{k!}, \qquad e^{x_1+x_2} = e^{x_1} \sum_{k=0}^{\infty} \frac{x_2^k}{k!} = e^{x_1} e^{x_2}$

    (c) $e^{-a} \displaystyle\sum_{k=0}^{\infty} (-1)^k \frac{(x-a)^k}{k!}$

34. (a) $\sin x = \sin a + (x-a)\cos a - \dfrac{(x-a)^2}{2!}\sin a + \dfrac{(x-a)^3}{3!}\cos a + \cdots$

       $\cos x = \cos a - (x-a)\sin a - \dfrac{(x-a)^2}{2!}\cos a + \dfrac{(x-a)^3}{3!}\sin a + \cdots$

    (b) in both instances $\displaystyle\sum_{k=0}^{\infty} |a_k| \le \sum_{k=0}^{\infty} \frac{|x-a|^k}{k!}$

    (c)

$$\sin(x_1 + x_2)$$

$$= \sin x_1 + x_2 \cos x_1 - \frac{(x_2)^2}{2!}\sin x_1 - \frac{(x_2)^3}{3!}\cos x_1 + \cdots$$

$$= \left(\sin x_1 - \frac{x_2^2}{2!}\sin x_1 + \frac{x_2^4}{4!}\sin x_1 - \cdots \right) + \left(x_2 \cos x_1 - \frac{x_2^2}{3!}\cos x_1 + \frac{x_2^5}{5!}\cos x_1 - \cdots \right)$$

$$= \sin x_1 \left(1 - \frac{x_2^2}{2!} + \frac{x_2^4}{4!} - \cdots \right) + \cos x_1 \left(x_2 - \frac{x_2}{3!} + \frac{x_2^5}{5!} - \cdots \right)$$

$$= \sin x_1 \cos x_2 + \cos x_1 \sin x_2$$

The other formula can be derived in a similar manner.

35. $P_6(x) = \dfrac{\pi}{4} + \dfrac{1}{2}(x-1) - \dfrac{1}{4}(x-1)^2 + \dfrac{1}{12}(x-1)^3 - \dfrac{1}{40}(x-1)^5 + \dfrac{1}{48}(x-1)^6$

36. $P_8(x) = \cosh 4 + 2\sinh 4(x-2) + 2\cosh 4(x-2)^2 + \dfrac{4}{3}\sinh 4(x-2)^3 + \cdots + \dfrac{2}{315}\cosh 4(x-2)^8$

SECTION 12.8

1. (a) converges   (b) absolutely converges   (c) ?   (d) diverges

2. (a) absolutely converges   (b) diverges   (c) ?   (d) convergent

3. $(-1, 1)$;   ratio test: $\dfrac{b_{k+1}}{b_k} = \dfrac{k+1}{k}|x| \to |x|$,   series converges for $|x| < 1$.

   At the endpoints $x = 1$ and $x = -1$ the series diverges since at those points $b_k \not\to 0$.

4. $[-1, 1)$;   ratio test: $\left|\dfrac{x^{k+1}}{k+1} \cdot \dfrac{k}{x^k}\right| = \dfrac{k}{k+1}|x| \to |x| \implies r = 1$

   At $x = 1 : \displaystyle\sum \frac{1}{k}$, diverges;   at $x = -1, \displaystyle\sum \frac{(-1)^k}{k}$ converges.

5. $(-\infty, \infty)$;   ratio test: $\dfrac{b_{k+1}}{b_k} = \dfrac{|x|}{(2k+1)(2k+2)} \to 0$, series converges all x.

6. $\left[-\dfrac{1}{2}, \dfrac{1}{2}\right]$;   root test: $\left|\dfrac{2^k}{k^2}x^k\right|^{1/k} = \dfrac{2 \cdot |x|}{k^{2/k}} \to 2|x| \implies r = \dfrac{1}{2}$

   At $x = \dfrac{1}{2} : \displaystyle\sum \frac{1}{k^2}$, converges;   at $x = -\dfrac{1}{2} : \displaystyle\sum \frac{(-1)^k}{k^2}$ converges.

7. Converges only at 0;   divergence test: $(-k)^{2k}x^{2k} \to 0$ only if $x = 0$, and series clearly converges at $x = 0$.

8. $(-1, 1]$;   root test: $\left|\dfrac{x^k}{\sqrt{k}}\right|^{1/k} = \dfrac{|x|}{k^{k/2}} \to |x| \implies r = 1$

   At $x = 1 : \displaystyle\sum \frac{(-1)^k}{\sqrt{k}}$, converges;   at $x = -1 : \displaystyle\sum \frac{1}{\sqrt{k}}$ diverges.

9. $[-2, 2)$;   root test: $(b_k)^{1/k} = \dfrac{|x|}{2k^{1/k}} \to \dfrac{|x|}{2}$,   series converges for $|x| < 2$.

   At $x = 2$ series becomes $\displaystyle\sum \frac{1}{k}$, the divergent harmonic series.

   At $x = -2$ series becomes $\displaystyle\sum (-1)^k \frac{1}{k}$, a convergent alternating series.

10. $[-2, 2]$;   root test: $\left|\dfrac{x^k}{k^2 2^k}\right|^{1/k} = \dfrac{|x|}{k^{2/k}2} \to \dfrac{|x|}{2} \implies r = 2$

    At $x = 2 : \displaystyle\sum \frac{1}{k^2}$, converges;   at $x = -2 : \displaystyle\sum \frac{(-1)^k}{k^2}$ converges.

11. Converges only at 0; divergence test: $\left(\dfrac{k}{100}\right)^k x^k \to 0$ only if $x = 0$, and series clearly converges at $x = 0$.

12. $(-1, 1)$;   ratio test: $\dfrac{(k+1)^2|x|^{k+1}}{1+(k+1)^2} \cdot \dfrac{1+k^2}{k^2|x|^k} \to |x| \implies r = 1$

At $x = 1$: $\sum \dfrac{k^2}{1+k^2}$, diverges;   at $x = -1$: $\sum \dfrac{(-1)^k k^2}{1+k^2}$ diverges.

13. $\left[-\dfrac{1}{2}, \dfrac{1}{2}\right)$;   root test: $(b_k)^{1/k} = \dfrac{2|x|}{\sqrt{k^{1/k}}} \to 2|x|$, series converges for $|x| < \dfrac{1}{2}$.

At $x = \dfrac{1}{2}$ series becomes $\sum \dfrac{1}{\sqrt{k}}$, a divergent p-series.

At $x = -\dfrac{1}{2}$ series becomes $\sum (-1)^k \dfrac{1}{\sqrt{k}}$, a convergent alternating series.

14. $[-1, 1)$;   ratio test: $\dfrac{|x|^{k+1}}{\ln(k+1)} \cdot \dfrac{\ln k}{|x|^k} = |x| \dfrac{\ln k}{\ln(k+1)} \to |x| \implies r = 1$

At $x = 1$: $\sum \dfrac{1}{\ln k}$, diverges;   at $x = -1$: $\sum \dfrac{(-1)^k}{\ln k}$ converges.

15. $(-1, 1)$;   ratio test: $\dfrac{b_{k+1}}{b_k} = \dfrac{k^2}{(k+1)(k-1)}|x| \to |x|$, series converges for $|x| < 1$.

At the endpoints $x = 1$ and $x = -1$ the series diverges since there $b_k \nrightarrow 0$.

16. $\left(-\dfrac{1}{|a|}, \dfrac{1}{|a|}\right)$;   root test: $\left|ka^k x^k\right|^{1/k} = k^{1/k}|a| \cdot |x| \to |a| \cdot |x| \implies r = \dfrac{1}{|a|}$

At $x = \dfrac{1}{|a|}$: $\sum ka^k \dfrac{1}{|a|^k}$, diverges, similarly at $x = -\dfrac{1}{|a|}$.

17. $(-10, 10)$;   root test: $(b_k)^{1/k} = \dfrac{k^{1/k}}{10}|x| \to \dfrac{|x|}{10}$, series converges for $|x| < 10$.

At the endpoints $x = 10$ and $x = -10$ the series diverges since there $b_k \nrightarrow 0$.

18. $(-e, e)$;   root test: $\left|\dfrac{3k^2 x^k}{e^k}\right|^{1/k} = \dfrac{3^{1/k} k^{2/k}}{e}|x| \to \dfrac{|x|}{e} \implies r = e$

Diverges at $x = \pm e$.

19. $(-\infty, \infty)$;   root test: $(b_k)^{1/k} = \dfrac{|x|}{k} \to 0$, series converges for all x.

20. $(-\infty, \infty)$;   ratio test: $\dfrac{7^{k+1}|x|^{k+1}}{(k+1)!} \cdot \dfrac{k!}{7^k|x|^k} = \dfrac{7|x|}{k+1} \to 0, \implies r = \infty$

21. $(-\infty, \infty)$;   root test: $(b_k)^{1/k} = \dfrac{|x-2|}{k} \to 0$, series converges all x.

22. converges only at 0, ratio test: $\dfrac{(k+1)!|x|^{k+1}}{k!|x|^k} = (k+1)|x| \to \infty \implies r = 0$

23. $\left(-\dfrac{3}{2}, \dfrac{3}{2}\right)$;   ratio test: $\dfrac{b_{k+1}}{b_k} = \dfrac{\dfrac{2^{k+1}}{3^{k+2}}|x|}{\dfrac{2^k}{3^{k+1}}} = \dfrac{2}{3}|x|$, series converges for $|x| < \dfrac{3}{2}$.

At the endpoints $x = 3/2$ and $x = -3/2$, the series diverges since there $b_k \nrightarrow 0$.

24. $(-\infty, \infty)$;   ratio test: $\dfrac{2^{k+1}|x|^{k+1}}{(2k+2)!} \cdot \dfrac{(2k)!}{2^k|x|^k} = \dfrac{2|x|}{(2k+1)(2k+2)} \to 0 \implies r = \infty$

25. Converges only at $x = 1$;   ratio test: $\dfrac{b_{k+1}}{b_k} = \dfrac{k^3}{(k+1)^2}|x - 1| \to \infty$ if $x \ne 1$

The series clearly converges at $x = 1$; otherwise it diverges.

26. $\left[-\dfrac{1}{e}, \dfrac{1}{e}\right]$ root test: $\left|\dfrac{(-e)^k}{k^2} x^k\right|^{1/k} = \dfrac{e|x|}{k^{2/k}} \to e|x| \implies r = \dfrac{1}{e}$.

Converges at $\ \ x = \pm\dfrac{1}{e}$;

27. $(-4, 0)$;   ratio test: $\dfrac{b_{k+1}}{b_k} = \dfrac{k^2 - 1}{2k^2}|x + 2| \to \dfrac{|x + 2|}{2}$, series converges for $|x + 2| < 2$.

At the endpoints $x = 0$ and $x = -4$, the series diverges since there $b_k \nrightarrow 0$.

28. $[-2, 0)$;   ratio test: $\dfrac{\ln(k+1)}{k+1}|x+1|^{k+1} \cdot \dfrac{k}{\ln k |x+1|^k} = \dfrac{\ln(k+1)}{\ln k} \cdot \dfrac{k}{k+1}|x+1| \to |x+1| \implies r = 1$

At $x = 0$, $\sum \dfrac{\ln k}{k}$ diverges,   at $x = -2$, $\sum \dfrac{\ln k}{k}(-1)^k$ converges.

29. $(-\infty, \infty)$;   ratio test: $\dfrac{b_{k+1}}{b_k} = \dfrac{(k+1)^2}{k^2(k+2)}|x+3| \to 0$, series converges for all x.

30. $(4 - e, 4 + e)$;   root test: $\left|\dfrac{k^3}{e^k}(x-4)^k\right|^{1/k} = \dfrac{k^{3/k}}{e}|x-4| \to \dfrac{|x-4|}{e} \implies r = e$

At $x = 4 + e$, $\sum k^3$ diverges,   at $x = 4 - e$, $\sum(-1)^k k^3$ diverges.

31. $(-1, 1)$;   root test: $(b_k)^{1/k} = \left(1 + \dfrac{1}{k}\right)|x| \to |x|$, series converges for $|x| < 1$.

At the endpoints $x = 1$ and $x = -1$, the series diverges since there $b_k \nrightarrow 0$

[recall $\left(1 + \dfrac{1}{k}\right)^k \to e$]

32. $\left[a - \dfrac{1}{|a|}, a + \dfrac{1}{|a|}\right]$;   root test: $\left|\dfrac{(-1)^k a^k}{k^2}(x-a)^k\right|^{1/k} = \dfrac{|a|}{k^{2/k}}|x-a| \to |a| \cdot |x-a| \implies r = \dfrac{1}{|a|}$

Converges at both $\ a - \dfrac{1}{|a|}, a + \dfrac{1}{|a|}$ $\left(\text{compare with } \sum \dfrac{1}{k^2}\right)$.

33. $(0, 4)$;   ratio test: $\dfrac{b_{k+1}}{b_k} = \dfrac{\ln(k+1)}{\ln k}\dfrac{|x-2|}{2} \to \dfrac{|x-2|}{2}$,   series converges for $|x - 2| < 2$.

At the endpoints $x = 0$ and $x = 4$ the series diverges since there $b_k \not\to 0$.

34. $(-\infty, \infty)$;   root test: $\dfrac{|x-1|}{\ln k} \to 0 \implies r = \infty$

35. $\left(-\dfrac{5}{2}, \dfrac{1}{2}\right)$;   root test: $(b_k)^{1/k} = \dfrac{2}{3}|x+1| \to \dfrac{2}{3}|x+1|$,   series converges for $|x + 1| < \dfrac{3}{2}$.

At the endpoints $x = -\dfrac{5}{2}$ and $x = \dfrac{1}{2}$ the series diverges since there $b_k \not\to 0$.

36. $\left[2 - \dfrac{1}{\pi}, 2 + \dfrac{1}{\pi}\right]$;   ratio test: $\dfrac{2^{\frac{1}{k+1}}\pi^{k+1}|x-2|^{k+1}}{(k+1)(k+2)(k+3)} \cdot \dfrac{k(k+1)(k+2)}{2^{\frac{1}{k}}\pi^k|x-2|^k} \to \pi|x-2| \implies r = \dfrac{1}{\pi}$

Converges at $x = 2 \pm \dfrac{1}{\pi}$ $\left(\text{compare with } \sum \dfrac{1}{k^3}\right)$.

37. $1 - \dfrac{x}{2} + \dfrac{2x^2}{4} - \dfrac{3x^3}{8} + \dfrac{4x^4}{16} - \cdots = 1 + \displaystyle\sum_{k=1}^{\infty}(-1)^k\dfrac{kx^k}{2^k}$

$(-2, 2)$;   ratio test: $\dfrac{b_{k+1}}{b_k} = \dfrac{k+1}{2k}|x| \to \dfrac{|x|}{2}$,   series converges for $|x| < 2$.

At the endpoints $x = 2$ and $x = -2$ the series diverges since there $b_k \not\to 0$.

38. $\dfrac{1}{5^2}(x-1) + \dfrac{4}{5^4}(x-1)^2 + \dfrac{9}{5^6}(x-1)^3 + \cdots = \displaystyle\sum_{k=1}^{\infty}\dfrac{k^2}{5^{2k}}(x-1)^k$.

$(-24, 26)$;   root test: $\left|\dfrac{k^2}{5^{2k}}(x-1)^k\right|^{1/k} = \dfrac{k^{2/k}}{5^2}|x-1| \to \dfrac{|x-1|}{25} \implies r = 25$

At $x = -24$, $\sum(-1)^k k^2$ diverges,   at $x = 26$, $\sum k^2$ diverges.

39. $\dfrac{3x^2}{4} + \dfrac{9x^4}{9} + \dfrac{27x^6}{16} + \dfrac{81x^8}{25} + \cdots = \displaystyle\sum_{k=1}^{\infty}\dfrac{3^k}{(k+1)^2}x^{2k}$

$\left[-\dfrac{1}{\sqrt{3}}, \dfrac{1}{\sqrt{3}}\right]$;   ratio test: $\dfrac{b_{k+1}}{b_k} = \dfrac{3(k+1)^2}{(k+2)^2}x^2 \to 3x^2$, series converges for $x^2 < \dfrac{1}{3}$

or $|x| < \dfrac{1}{\sqrt{3}}$.

At $x = \pm\dfrac{1}{\sqrt{3}}$, the series becomes $\displaystyle\sum\dfrac{1}{(k+1)^2} \cong \sum\dfrac{1}{n^2}$, a convergent series p-series.

40. $\dfrac{1}{16}(x+1) - \dfrac{2}{25}(x+1)^2 + \dfrac{3}{36}(x+1)^3 + \cdots = \displaystyle\sum_{k=1}^{\infty}\dfrac{(-1)^{k+1}k}{(k+3)^2}(x+1)^k$.

$(-2, 0]$;   ratio test: $\dfrac{(k+1)|x+1|^{k+1}}{(k+4)^2} \cdot \dfrac{(k+3)^2}{k|x+1|^k} \to |x+1| \implies r = 1$

At $x = 0$, $\displaystyle\sum\dfrac{(-1)^{k+1}k}{(k+3)^2}$ converges;   at $x = -2$, $-\displaystyle\sum\dfrac{k}{(k+3)^2}$ diverges.

41. $\sum a_k(x-1)^k$ convergent at $x = 3$ $\implies$ $\sum a_k(x-1)^k$ is absolutely convergent on $(-1, 3)$.

(a) $\sum a_k = \sum a_k(2-1)^k$; absolutely convergent

(b) $\sum(-1)^k a_k = \sum a_k(0-1)^k$; absolutely convergent

(c) $\sum(-1)^k a_k 2^k = \sum a_k(-1-1)^k$; ??

42. It must converge absolutely for $-8 < x < 4$.

43. (a) Suppose that $\sum a_k r^k$ is absolutely convergent. Then,

$$\sum \left|a_k(-r)^k\right| = \sum |a_k| \left|(-r)^k\right| = \sum |a_k| \left|r^k\right| = \sum \left|a_k(-r)^k\right|.$$

Therefore, $\sum \left|a_k(-r)^k\right|$ is absolutely convergent.

(b) If $\sum \left|a_k r^k\right|$ converged, then, from part (a), $\sum \left|a_k(-r)^k\right|$ would also converge.

44. $\displaystyle\sum_{k=0}^{\infty} \frac{x^k}{r^k}$

45. $$\sum_{k=0}^{\infty} = a_0 + a_1 x + a_2 x^2 + a_0 x^3 + a_1 x^4 + a_2 x^5 + a_0 x^6 + \cdots$$

$$= \sum_{k=0}^{\infty} \left(a_0 + a_1 x + a_2 x^2\right) x^{3k}$$

$$= \sum_{k=0}^{\infty} \left(a_0 + a_1 x + a_2 x^2\right) \left(x^3\right)^k$$

(a) $\left|\dfrac{\left(a_0 + a_1 x + a_2 x^2\right)\left(x^3\right)^{k+1}}{\left(a_0 + a_1 x + a_2 x^2\right)\left(x^3\right)^k}\right| = \left|x^3\right| \implies r = 1.$

(b) $\displaystyle\sum_{k=0}^{\infty} \left(a_0 + a_1 x + a_2 x^2\right)\left(x^3\right)^k = \left(a_0 + a_1 x + a_2 x^2\right)\sum_{k=0}^{\infty}\left(x^3\right)^k = \left(a_0 + a_1 x + a_2 x^2\right)\dfrac{1}{1-x^3}.$

46. $(-1, 1)$; $s_k \le k$ and $\sum k x^k$ converges for $|x| < 1$;   for $|x| \ge 1$, $s_k x^k \not\to 0$.

47. Examine the convergence of $\sum |a_k x^k|$; for (a) use the root test and for (b) use the ratio rest.

48. By ratio test: $\left|\dfrac{a_{k+1}}{a_k}\right| \cdot |x| \to \dfrac{|x|}{r}$,   so $\left|\dfrac{a_{k+1} x^{2(k+1)}}{a_k x^{2k}}\right| = \left|\dfrac{a_{k+1}}{a_k}\right| |x|^2 \to \dfrac{|x|^2}{r}$

$\implies$ radius of convergence is $\sqrt{r}$.

SECTION 12.9

1. Use the fact that $\dfrac{d}{dx}\left(\dfrac{1}{1-x}\right) = \dfrac{1}{(1-x)^2}$:

$$\frac{1}{(1-x)^2} = \frac{d}{dx}(1 + x + x^2 + x^3 + \cdots + x^n + \cdots) = 1 + 2x + 3x^2 + \cdots + nx^{n-1} + \cdots.$$

2.
$$\frac{1}{(1-x)^3} = \frac{1}{2}\frac{d^2}{dx^2}\left[\frac{1}{1-x}\right] = \frac{1}{2}\frac{d^2}{dx^2}\left[1 + x + x^2 + \cdots + x^n + \cdots\right]$$

$$= \frac{1}{2}\left[2 + 6x + 12x^2 + \cdots + n(n-1)x^{n-2} + \cdots\right]$$

$$= 1 + 3x + 6x^2 + \cdots + \frac{n(n-1)}{2}x^{n-2} + \cdots$$

3. Use the fact that $\dfrac{d^{(k-1)}}{dx^{(k-1)}}\left[\dfrac{1}{1-x}\right] = \dfrac{(k-1)!}{(1-x)^k}$:

$$\frac{1}{(1-x)^k} = \frac{1}{(k-1)!}\frac{d^{(k-1)}}{dx^{(k-1)}}\left[1 + x + \cdots + x^{k-1} + x^k + x^{k+1} + \cdots + x^{n+k-1} + \cdots\right]$$

$$= \frac{1}{(k-1)!}\frac{d^{(k-1)}}{dx^{(k-1)}}\left[x^{k-1} + x^k + x^{k+1} + \cdots + x^{n+k-1} + \cdots\right]$$

$$= 1 + kx + \frac{(k+1)k}{2}x^2 + \cdots + \frac{(n+k-1)(n+k-2)\cdots(n+1)}{(k-1)!}x^n + \cdots$$

$$= 1 + kx + \frac{(k+1)k}{2!}x^2 + \cdots + \frac{(n+k-1)!}{n!(k-1)!}x^n + \cdots.$$

4. $\ln(1-x) = -\displaystyle\int \frac{dx}{1-x} = -\left[x + \frac{x^2}{2} + \frac{x^3}{3} + \cdots + \frac{x^{n+1}}{n+1} + \cdots\right] + C; \quad \ln 1 = 0 \implies C = 0$

$$\implies \ln(1-x) = -x - \frac{x^2}{2} - \frac{x^3}{3} - \cdots - \frac{x^{n+1}}{n+1} - \cdots$$

5. Use the fact that $\dfrac{d}{dx}[\ln(1-x^2)] = \dfrac{-2x}{1-x^2}$:

$$\frac{1}{1-x^2} = 1 + x^2 + x^4 + \cdots + x^{2n} + \cdots$$

$$\frac{-2x}{1-x^2} = -2x - 2x^3 - 2x^5 - \cdots - 2x^{2n+1} - \cdots.$$

By integration

$$\ln(1-x^2) = \left(-x^2 - \frac{1}{2}x^4 - \frac{1}{3}x^6 - \cdots - \frac{x^{2n+2}}{n+1} - \cdots\right) + C.$$

At $x = 0$, both $\ln(1-x^2)$ and the series are 0. Thus, $C = 0$ and

$$\ln(1-x^2) = -x^2 - \frac{1}{2}x^4 - \frac{1}{3}x^6 - \cdots - \frac{1}{n+1}x^{2n+2} - \cdots.$$

6. $\ln(2-3x) = \ln 2 + \ln\left(1 - \dfrac{3}{2}x\right) = \ln 2 - \dfrac{3}{2}x - \dfrac{1}{2}\left(\dfrac{3}{2}\right)^2 x^2 - \dfrac{1}{3}\left(\dfrac{3}{2}\right)^3 x^3 - \cdots - \dfrac{1}{n+1}\left(\dfrac{3}{2}\right)^{n+1} x^{n+1} - \cdots$

7. $\sec^2 x = \dfrac{d}{dx}(\tan x) = \dfrac{d}{dx}\left(x + \dfrac{1}{3}x^3 + \dfrac{2}{15}x^5 + \dfrac{17}{315}x^7 + \cdots\right) = 1 + x^2 + \dfrac{2}{3}x^4 + \dfrac{17}{45}x^6 + \cdots$

8. $\ln \cos x = -\displaystyle\int \dfrac{\sin x}{\cos x}\,dx = -\int \tan x\,dx = -\dfrac{x^2}{2} - \dfrac{x^4}{12} - \dfrac{x^6}{45} - \dfrac{17}{2520}x^8 - \cdots + C$

$\ln \cos 0 = 0 \implies C = 0$

9. On its interval of convergence a power series is the Taylor series of its sum. Thus,

$$f(x) = x^2 \sin^2 x = x^2\left(x - \dfrac{x^3}{3!} + \dfrac{x^5}{5!} - \dfrac{x^7}{7!} + \cdots\right)$$

$$= x^3 - \dfrac{x^5}{3!} + \dfrac{x^7}{5!} - \dfrac{x^9}{7!} + \cdots = \sum_{n=0}^{\infty} f^{(n)}(0)\dfrac{x^n}{n!}$$

implies $f^{(9)}(0) = -9!/7! = -72.$

10. $f(x) = x\cos x^2 = x\left(1 - \dfrac{(x^2)^2}{2!} + \dfrac{(x^2)^4}{4!} - \dfrac{(x^2)^6}{6!} + \cdots\right)$

$\implies \dfrac{f^{(9)}(0)}{9!}x^9 = \dfrac{x^9}{4!} \implies f^{(9)}(0) = \dfrac{9!}{4!} = 15120.$

11. $\sin x^2 = \displaystyle\sum_{k=0}^{\infty}(-1)^k \dfrac{(x^2)^{2k+1}}{(2k+1)!} = \sum_{k=0}^{\infty}(-1)^k \dfrac{x^{4k+2}}{(2k+1)!}$

12. $x^2 \arctan x = x^2 \displaystyle\int \dfrac{1}{1+x^2}\,dx = x^2\int\left(\sum_{k=0}^{\infty}(-1)^k x^{2k}\right)dx = x^2\left[\sum_{k=0}^{\infty}\dfrac{(-1)^k x^{2k+1}}{2k+1} + C\right]$

$= \displaystyle\sum_{k=0}^{\infty}\dfrac{(-1)^k}{2k+1}x^{2k+3} \quad (\arctan 0 = 0 \implies C = 0)$

13. $e^{3x^3} = \displaystyle\sum_{k=0}^{\infty}\dfrac{(3x^3)^k}{k!} = \sum_{k=0}^{\infty}\dfrac{3^k}{k!}x^{3k}$

14. $\dfrac{1-x}{1+x} = \dfrac{1}{1+x} - \dfrac{x}{1+x} = \displaystyle\sum_{k=0}^{\infty}(-1)^k x^k - \sum_{k=0}^{\infty}(-1)^k x^{k+1} = 1 + 2\sum_{k=0}^{\infty}(-1)^{k+1}x^{k+1}$

15. $\dfrac{2x}{1-x^2} = 2x\left(\dfrac{1}{1-x^2}\right) = 2x\displaystyle\sum_{k=0}^{\infty}(x^2)^k = \sum_{k=0}^{\infty}2x^{2k+1}$

16. $x\sinh x^2 = \dfrac{x}{2}\left(e^{x^2} - e^{-x^2}\right) = \dfrac{x}{2}\left[\displaystyle\sum_{k=0}^{\infty}\dfrac{x^{2k}}{k!} - \sum_{k=0}^{\infty}\dfrac{(-1)^k x^{2k}}{k!}\right] = \dfrac{x}{2}\left[2\sum_{k=0}^{\infty}\dfrac{x^{4k+2}}{(2k+1)!}\right]$

$= \displaystyle\sum_{k=0}^{\infty}\dfrac{x^{4k+3}}{(2k+1)!}$

17. $\dfrac{1}{1-x} + e^x = \displaystyle\sum_{k=0}^{\infty} x^k + \sum_{k=0}^{\infty} \dfrac{x^k}{k!} = \sum_{k=0}^{\infty} \dfrac{(k!+1)}{k!} x^k$

18. $\cosh x \sinh x = \dfrac{1}{2}\sinh 2x = \dfrac{1}{2}\displaystyle\sum_{k=0}^{\infty} \dfrac{(2x)^{2k+1}}{(2k+1)!} = \sum_{k=0}^{\infty} \dfrac{4^k}{(2k+1)!} x^{2k+1}$

19. $x\ln(1+x^3) = x\displaystyle\sum_{k=1}^{\infty} \dfrac{(-1)^{k+1}}{k}(x^3)^k = \sum_{k=1}^{\infty} \dfrac{(-1)^{k+1}}{k} x^{3k+1}$

$(12.6.8)\text{───}\uparrow$

20. $(x^2+x)\ln(1+x) = (x^2+x)\displaystyle\sum_{k=1}^{\infty} \dfrac{(-1)^{k+1}}{k} x^k = x^2 + \sum_{k=3}^{\infty} \dfrac{(-1)^{k+1}}{(k-1)(k-2)} x^k$

21. $x^3 e^{-x^3} = x^3\displaystyle\sum_{k=0}^{\infty} \dfrac{(-x^3)^k}{k!} = \sum_{k=0}^{\infty} \dfrac{(-1)^k}{k!} x^{3k+3}$

22.
$$x^5(\sin x + \cos 2x) = x^5\left[\sum_{k=0}^{\infty} \dfrac{(-1)^k}{(2k+1)!} x^{2k+1} + \sum_{k=0}^{\infty} \dfrac{(-1)^k}{(2k)!} 2^{2k} x^{2k}\right]$$
$$= \sum_{k=0}^{\infty} \dfrac{(-1)^k}{(2k+1)!}\left[(2k+1)4^k x^{2k+5} + x^{2k+6}\right]$$

23. (a) $\displaystyle\lim_{x\to 0} \dfrac{1-\cos x}{x^2} \overset{\star}{=} \lim_{x\to 0} \dfrac{\sin x}{2x} = \dfrac{1}{2}$ ($\star$ indicates differentiation of numerator and denominator).

   (b) $\displaystyle\lim_{x\to 0} \dfrac{1-\cos x}{x^2} = \lim_{x\to 0} \dfrac{\dfrac{x^2}{2!} - \dfrac{x^4}{4!} + \dfrac{x^6}{6!} - \cdots}{x^2} = \lim_{x\to 0}\left(\dfrac{1}{2} - \dfrac{x^2}{4!} + \dfrac{x^4}{6!} - \cdots\right) = \dfrac{1}{2}$

24. (a) $\displaystyle\lim_{x\to 0} \dfrac{\sin x - x}{x^2} \overset{\star}{=} \lim_{x\to 0} \dfrac{\cos x - 1}{2x} \overset{\star}{=} \lim_{x\to 0} -\dfrac{\sin x}{2} = 0$

   (b) $\dfrac{\sin x - x}{x^2} = \dfrac{-\dfrac{x^3}{3!} + \dfrac{x^5}{5!} - \cdots}{x^2} \to 0$ as $x \to 0$

25. (a) $\displaystyle\lim_{x\to 0} \dfrac{\cos x - 1}{x\sin x} \overset{\star}{=} \lim_{x\to 0} \dfrac{-\sin x}{\sin x + x\cos x} \overset{\star}{=} \lim_{x\to 0} \dfrac{-\cos x}{2\cos x - x\sin x} = -\dfrac{1}{2}$

   (b)
$$\lim_{x\to 0} \dfrac{\cos x - 1}{x\sin x} = \dfrac{-\dfrac{x^2}{2!} + \dfrac{x^4}{4!} - \dfrac{x^6}{6!} + \cdots}{x^2 - \dfrac{x^4}{3!} + \dfrac{x^6}{5!}\cdots}$$
$$= \dfrac{-\dfrac{1}{2!} + \dfrac{x^2}{4!} - \dfrac{x^4}{6!} + \cdots}{1 - \dfrac{x^2}{3!} + \dfrac{x^4}{5!}\cdots} = -\dfrac{1}{2}$$

26. (a) $\quad \lim_{x \to 0} \dfrac{e^x - 1 - x}{x \arctan x} \overset{\ast}{=} \lim_{x \to 0} \dfrac{e^x - 1}{\arctan x + x/(1 + x^2)} \overset{\ast}{=} \lim_{x \to 0} \dfrac{e^x}{\dfrac{1}{1 + x^2} + \dfrac{1 - x^2}{(1 + x^2)^2}} = \dfrac{1}{2}$

(b) $\quad \dfrac{e^x - 1 - x}{x \arctan x} = \dfrac{\dfrac{x^2}{2} + \dfrac{x^3}{3!} + \dfrac{x^4}{4!} + \cdots}{x^2 - \dfrac{x^4}{3} + \dfrac{x^6}{5} - \cdots} \to \dfrac{1}{2}.$

27. $\quad \displaystyle\int_0^x \dfrac{\ln(1 + t)}{t}\, dt = \int_0^x \dfrac{1}{t}\left(\sum_{k=1}^\infty \dfrac{(-1)^{k-1}}{k} t^k \right) dt = \int_0^x \left(\sum_{k=1}^\infty \dfrac{(-1)^{k-1}}{k} t^{k-1} \right) dt$

$$= \sum_{k=1}^\infty \dfrac{(-1)^{k-1}}{k} \int_0^x t^{k-1}\, dt = \sum_{k=1}^\infty \dfrac{(-1)^{k-1}}{k^2} x^k, \quad -1 \le x \le 1$$

28. $\quad \displaystyle\int_0^x \dfrac{1 - \cos t}{t^2}\, dt = \int_0^x \dfrac{1}{t^2}\left[\sum_{k=1}^\infty \dfrac{(-1)^{k+1}}{(2k)!} t^{2k} \right] dt$

$$= \int_0^x \left[\sum_{k=1}^\infty \dfrac{(-1)^{k+1}}{(2k)!} t^{2k-2} \right] dt = \sum_{k=1}^\infty \dfrac{(-1)^{k+1}}{(2k)!} \cdot \dfrac{x^{2k-1}}{2k - 1}$$

29. $\quad \displaystyle\int_0^x \dfrac{\arctan t}{t}\, dt = \int_0^x \dfrac{1}{t}\left(\sum_{k=0}^\infty \dfrac{(-1)^k}{2k + 1} t^{2k+1} \right) dt = \int_0^x \left(\sum_{k=0}^\infty \dfrac{(-1)^k}{2k + 1} t^{2k} \right) dt$

$$= \sum_{k=0}^\infty \dfrac{(-1)^k}{2k + 1} \int_0^x t^{2k}\, dt$$

$$= \sum_{k=0}^\infty \dfrac{(-1)^k}{(2k + 1)^2} x^{2k+1}, \quad -1 \le x \le 1$$

30. $\quad \displaystyle\int_0^x \dfrac{\sinh t}{t}\, dt = \int_0^x \dfrac{1}{t}\left[\sum_{k=0}^\infty \dfrac{t^{2k+1}}{(2k + 1)!} \right] dt = \int_0^x \sum_{k=0}^\infty \dfrac{t^{2k}}{(2k + 1)!}\, dt = \sum_{k=0}^\infty \dfrac{x^{2k+1}}{(2k + 1)^2 (2k)!}$

31. $\quad 0.804 \le I \le 0.808; \qquad I = \displaystyle\int_0^1 \left(1 - x^3 + \dfrac{x^6}{2!} - \dfrac{x^9}{3!} + \cdots \right) dx$

$$= \left[x - \dfrac{x^4}{4} + \dfrac{x^7}{14} - \dfrac{x^{10}}{60} + \dfrac{x^{13}}{(13)(24)} - \cdots \right]_0^1$$

$$= 1 - \dfrac{1}{4} + \dfrac{1}{14} - \dfrac{1}{60} + \dfrac{1}{311} - \cdots.$$

Since $\quad \dfrac{1}{311} < 0.01, \quad$ we can stop there:

$$1 - \dfrac{1}{4} + \dfrac{1}{14} - \dfrac{1}{60} \le I \le 1 - \dfrac{1}{4} + \dfrac{1}{14} - \dfrac{1}{60} + \dfrac{1}{311} \quad \text{gives} \quad 0.804 \le I \le 0.808.$$

32. $I = \int_0^1 \left(x^2 - \frac{x^6}{3!} + \frac{x^{10}}{5!} - \frac{x^{14}}{7!} + \cdots \right) dx = \frac{1}{3} - \frac{1}{7(3!)} + \frac{1}{11(5!)} - \frac{1}{15(7!)} + \cdots.$

Since $\frac{1}{11(5!)} = \frac{1}{840} < 0.01,$ we can stop there:

$$\frac{1}{3} - \frac{1}{7(3!)} \le I \le \frac{1}{3} - \frac{1}{7(3!)} + \frac{1}{11(5!)} \quad \text{gives} \quad 0.309 \le I \le 0.311.$$

33. $0.600 \le I \le 0.603;$ $\qquad I = \int_0^1 \left(x^{1/2} - \frac{x^{3/2}}{3!} + \frac{x^{5/2}}{5!} - \cdots \right) dx$

$$= \left[\frac{2}{3}x^{3/2} - \frac{1}{15}x^{5/2} + \frac{1}{420}x^{7/2} - \cdots \right]_0^1$$

$$= \frac{2}{3} - \frac{1}{15} + \frac{1}{420} - \cdots.$$

Since $\frac{1}{420} < 0.01,$ we can stop there:

$$\frac{2}{3} - \frac{1}{15} \le I \le \frac{2}{3} - \frac{1}{15} + \frac{1}{420} \quad \text{gives} \quad 0.600 \le I \le 0.603.$$

34. $I = \int_0^1 \left(x^4 - x^6 + \frac{x^8}{2!} - \frac{x^{10}}{3!} + \frac{x^{11}}{4!} - \cdots \right) dx = \frac{1}{5} - \frac{1}{7} + \frac{1}{18} - \frac{1}{66} + \frac{1}{288} - \cdots.$

Since $\frac{1}{288} < 0.01,$ we can stop there:

$$\frac{1}{5} - \frac{1}{7} + \frac{1}{18} - \frac{1}{66} \le I \le \frac{1}{5} - \frac{1}{7} + \frac{1}{18} - \frac{1}{66} + \frac{1}{288} \quad \text{gives} \quad 0.097 \le I \le 0.101.$$

35. $0.294 \le I \le 0.304;$ $\qquad I = \int_0^1 \left(x^2 - \frac{x^6}{3} + \frac{x^{10}}{5} - \frac{x^{14}}{7} + \cdots \right) dx$

$(12.9.6)\underline{\qquad}\Bigg\uparrow$

$$= \left[\tfrac{1}{3}x^3 - \tfrac{1}{21}x^7 + \tfrac{1}{55}x^{11} - \tfrac{1}{105}x^{15} + \cdots \right]_0^1$$

$$= \tfrac{1}{3} - \tfrac{1}{21} + \tfrac{1}{55} - \tfrac{1}{105} + \cdots.$$

Since $\tfrac{1}{105} < 0.01,$ we can stop there:

$$\tfrac{1}{3} - \tfrac{1}{21} + \tfrac{1}{55} - \tfrac{1}{105} \le I \le \tfrac{1}{3} - \tfrac{1}{21} + \tfrac{1}{55} \quad \text{gives} \quad 0.294 \le I \le 0.304.$$

36. $I = \int_1^2 \left(\frac{x}{2!} - \frac{x^3}{4!} + \frac{x^5}{6!} - \frac{x^7}{8!} + \cdots \right) dx$

$$= \frac{3}{2(2!)} - \frac{15}{4(4!)} + \frac{63}{6(6!)} - \frac{255}{8(8!)} + \cdots = \frac{3}{4} - \frac{15}{96} + \frac{63}{4320} - \frac{255}{322560} + \cdots.$$

Since $\frac{255}{322560} < 0.01,$ we can stop there:

$$\frac{3}{4} - \frac{15}{96} + \frac{63}{4320} - \frac{255}{322560} \le I \le \frac{3}{4} - \frac{15}{96} + \frac{63}{4320} \quad \text{gives} \quad 0.607 \le I \le 0.609.$$

37. $I \cong 0.9461;$ $I = \int_0^1 \left(1 - \frac{x^2}{3!} + \frac{x^4}{5!} - \cdots \right) dx$

$$= \left[x - \frac{x^3}{3 \cdot 3!} + \frac{x^5}{5 \cdot 5!} - \cdots \right]_0^1$$

$$= 1 - \frac{1}{3 \cdot 3!} + \frac{1}{5 \cdot 5!} - \frac{1}{7 \cdot 7!} \cdots.$$

Since $\frac{1}{7 \cdot 7!} = \frac{1}{35,280} \cong 0.000028 < 0.0001,$ we can stop there:

$$1 - \frac{1}{3 \cdot 3!} + \frac{1}{5 \cdot 5!} - \frac{1}{7 \cdot 7!} < I < 1 - \frac{1}{3 \cdot 3!} + \frac{1}{5 \cdot 5!}; \quad I \cong 0.9461$$

38. $I = \int_0^{0.5} \left(\frac{1}{2!} - \frac{x^2}{4!} + \frac{x^4}{6!} - \frac{x^7}{8!} + \cdots \right) dx = \frac{0.5}{2!} - \frac{(0.5)^3}{3 \cdot 4!} + \frac{(0.5)^5}{5 \cdot 6!} - \frac{(0.5)^7}{7 \cdot 8!} + \cdots$

$\frac{(0.5)^5}{5 \cdot 6!} < 0.0001;$ $I \cong 0.2483$

39. $I \cong 0.4485;$ $I = \int_0^{0.5} \left(1 - \frac{x}{2} + \frac{x^2}{3} - \frac{x^3}{4} + \cdots \right) dx$

$$= \left[x - \frac{x^2}{2^2} + \frac{x^3}{3^2} - \frac{x^4}{4^2} + \cdots \right]_0^{1/2}$$

$$= \frac{1}{2} - \frac{1}{2^2 \cdot 2^2} + \frac{1}{3^2 \cdot 2^3} - \frac{1}{4^2 \cdot 2^4} + \cdots = \sum_{k=1}^{\infty} \frac{(-1)^{k-1}}{k^2 \cdot 2^k}$$

Now, $\frac{1}{8^2 \cdot 2^8} = \frac{1}{16,384} \cong 0.000061$ is the first term which is less than 0.0001. Thus

$$\sum_{k=1}^{7} \frac{(-1)^{k-1}}{k^2 \cdot 2^k} < I < \sum_{k=1}^{8} \frac{(-1)^{k-1}}{k^2 \cdot 2^k}; \quad I \cong 0.4485$$

40. $I = \int_0^{0.2} \left(x^2 - \frac{x^4}{3!} + \frac{x^6}{5!} - \frac{x^8}{7!} + \cdots \right) dx = \frac{(0.2)^3}{3} - \frac{(0.2)^5}{5 \cdot 3!} + \frac{(0.2)^7}{7 \cdot 5!} - \cdots$

$\frac{(0.2)^5}{5 \cdot 3!} < 0.0001,$ so $I \cong 0.0027$

41. $e^{x^3};$ by (12.6.5)   **42** $\sum_{k=0}^{\infty} \frac{1}{k!} x^{3k+1} = x \sum_{k=0}^{\infty} \frac{1}{k!} (x^3)^k = x e^{x^3}$

43. $3x^2 e^{x^3} = \frac{d}{dx} (e^{x^3})$

44. (a) $f(x) = \frac{e^x - 1}{x} = \frac{1}{x} \left(x + \frac{x^2}{2!} + \frac{x^3}{3!} + \cdots \right) = 1 + \frac{x}{2!} + \frac{x^2}{3!} + \frac{x^3}{4!} + \cdots$

(b) $f'(x) = \frac{xe^x - e^x + 1}{x^2} = \frac{1}{2} + \frac{2x}{3!} + \frac{3x^2}{4!} + \cdots + \frac{nx^{n-1}}{(n+1)!} + \cdots$

$f'(1) = 1 = \sum_{k=1}^{\infty} \frac{k}{(k+1)!}$

45. (a) $f(x) = xe^x = x \sum_{k=0}^{\infty} \dfrac{x^k}{k!} = \sum_{k=0}^{\infty} \dfrac{x^{k+1}}{k!}$

(b) Using integration by parts: $\displaystyle\int_0^1 xe^x\,dx = [xe^x - e^x]_0^1 = e - e + 1 = 1.$

Using the power series representation:

$$\int_0^1 xe^x\,dx = \int_0^1 \left(\sum_{k=0}^{\infty} \frac{x^{k+1}}{k!}\right) dx = \sum_{k=0}^{\infty} \int_0^1 \left(\frac{x^{k+1}}{k!}\right) dx$$

$$= \sum_{k=0}^{\infty} \frac{1}{k!}\left[\frac{x^{k+2}}{k+2}\right]_0^1$$

$$= \sum_{k=0}^{\infty} \frac{1}{k!(k+2)}$$

$$= \frac{1}{2} + \sum_{k=1}^{\infty} \frac{1}{k!(k+2)}$$

Thus, $\quad 1 = \dfrac{1}{2} + \sum_{k=1}^{\infty} \dfrac{1}{k!(k+2)}$ and $\displaystyle\sum_{k=1}^{\infty} \frac{1}{k!(k+2)} = \frac{1}{2}.$

46. $\dfrac{d}{dx}(\sinh x) = \dfrac{d}{dx}\left[\sum_{k=0}^{\infty} \dfrac{x^{2k+1}}{(2k+1)!}\right] = \sum_{k=0}^{\infty} \dfrac{(2k+1)x^{2k}}{(2k+1)!} = \sum_{k=0}^{\infty} \dfrac{x^{2k}}{(2k)!} = \cosh x$

$\dfrac{d}{dx}(\cosh x) = \dfrac{d}{dx}\left[\sum_{k=0}^{\infty} \dfrac{x^{2k}}{(2k)!}\right] = \sum_{k=1}^{\infty} \dfrac{2kx^{2k-1}}{(2k)!} = \sum_{k=1}^{\infty} \dfrac{x^{2k-1}}{(2k-1)!} = \sinh x.$

47. Let $f(x)$ be the sum of these series; $\quad a_k$ and b_k are both $\dfrac{f^{(k)}(0)}{k!}.$

48. As $k \to \infty$,

$$k^{1/k}|x|^{(k-1)/k} \to |x|.$$

Thus, for k sufficiently large,

$$k^{1/k}|x|^{(k-1)/k} < |x| + \epsilon \quad \text{and} \quad |kx^{k-1}| = k|x|^{k-1} < (|x| + \epsilon)^k.$$

49. (a) If f is even, then the odd ordered derivatives $f^{(2k-1)}$, $k = 1, 2, \ldots$ are odd. This implies that $f^{(2k-1)}(0) = 0$ for all k and so $a_{2k-1} = f^{(2k-1)}(0)/(2k-1)! = 0$ for all k.

(b) If f is odd, then all the even ordered derivatives $f^{(2k)}$, $k = 1, 2, \ldots$ are odd. This implies that $f^{(2k)}(0) = 0$ for all k and so $a_{2k} = f^{(2k)}(0)/(2k)! = 0$ for all k.

50. $f(0) = 1, \quad f'(0) = -2f(0) = -2, \quad f''(x) = -2f'(x) = 4f(x), \quad f''(0) = 4$

$f^{(n)}(x) = (-2)^n f(x), \quad f^{(n)}(0) = (-2)^n, \quad f(x) = \displaystyle\sum_{k=0}^{\infty} \frac{(-2)^k}{k!}x^k = \sum_{k=0}^{\infty} \frac{(-2x)^k}{k!} = e^{-2x}$

51. $f''(x) = -2f(x);$ $f(0) = 0,$ $f'(0) = 1$

$f''(x) = -2f(x);$ $f''(0) = 0$

$f'''(x) = -2f'(x);$ $f'''(0) = -2$

$f^{(4)}(x) = -2f''(x);$ $f^{(4)}(0) = 0$

$f^{(5)}(x) = -2f'''(x);$ $f^{(5)}(0) = 4$

$f^{(6)}(x) = -2f^{(4)}(x);$ $f^{(6)}(0) = 0$

$f^{(7)}(x) = -2f^{(5)}(x);$ $f^{(7)}(0) = -8$

$\vdots$

$$f(x) = x - \frac{2}{3!}x^3 + \frac{4}{5!}x^5 - \frac{8}{7!}x^7 + \cdots = \sum_{k=0}^{\infty} \frac{(-1)^k 2^k}{(2k+1)!}x^{2k+1} = \frac{1}{\sqrt{2}}\sin\left(x\sqrt{2}\right)$$

52.   (a) $f(x) = xe^{-x^2 \ln 2} = x - \ln 2\, x^3 + \frac{(\ln 2)^2}{2}x^5 - \frac{(\ln 2)^3}{6}x^7 + \frac{(\ln 2)^4}{24}x^9 + \cdots .$

$$f'(x) = 1 - 3\ln 2\, x^2 + \frac{5(\ln 2)^2}{2}x^4 - \frac{7(\ln 2)^3}{6}x^6 + \frac{3(\ln 2)^4}{8}x^8 + \cdots .$$

$$\int f(x)\, dx = \frac{1}{2}x^2 - \frac{\ln 2}{4}x^4 + \frac{(\ln 2)^2}{12}x^6 - \frac{(\ln 2)^3}{48}x^8 + \frac{(\ln 2)^4}{240}x^{10} + \cdots .$$

(b) $f(x) = x \arctan x = x^2 - \frac{1}{3}x^4 + \frac{1}{5}x^6 - \frac{1}{7}x^8 + \frac{1}{9}x^{10} \cdots .$

$$f'(x) = 2x - \tfrac{4}{3}x^3 + \tfrac{6}{5}x^5 - \tfrac{8}{7}x^7 + \tfrac{10}{9}x^9 \cdots .$$

$$\int f(x)\, dx = \frac{1}{3}x^3 - \frac{1}{15}x^5 + \frac{1}{35}x^7 - \frac{1}{63}x^9 + \frac{1}{99}x^{11} \cdots .$$

53. $0.0352 \leq I \leq 0.0359;$ $I = \displaystyle\int_0^{1/2}\left(x^2 - \frac{x^3}{2} + \frac{x^4}{3} - \frac{x^5}{4} + \cdots\right) dx$

$$= \left[\frac{x^3}{3} - \frac{x^4}{8} + \frac{x^5}{15} - \frac{x^6}{24} + \cdots\right]_0^{1/2}$$

$$= \frac{1}{3(2^3)} - \frac{1}{8(2^4)} + \frac{1}{15(2^5)} - \frac{1}{24(2^6)} + \cdots .$$

Since $\dfrac{1}{24(2^6)} = \dfrac{1}{1536} < 0.001,$ we can stop there:

$$\frac{1}{3(2^3)} - \frac{1}{8(2^4)} + \frac{1}{15(2^5)} - \frac{1}{24(2^6)} \leq I \leq \frac{1}{3(2^3)} - \frac{1}{8(2^4)} + \frac{1}{15(2^5)}$$

gives $0.0352 \leq I \leq 0.0359.$ Direct integration gives

$$I = \int_0^{1/2} x \ln(1+x)\, dx = \left[\frac{1}{2}(x^2 - 1)\ln(1+x) - \frac{1}{4}x^2 + \frac{1}{2}x\right]_0^{1/2} = \frac{3}{16} - \frac{3}{8}\ln 1.5 \cong 0.0354505.$$

54. $I = \displaystyle\int_0^1 \left(x^2 - \frac{x^4}{3!} + \frac{x^6}{5!} - \frac{x^8}{7!} + \cdots \right) = \frac{1}{3} - \frac{1}{5(3!)} + \frac{1}{7(5!)} - \frac{1}{9(7!)} + \cdots.$

Since $\dfrac{1}{9(7!)} = \dfrac{1}{5040} < 0.001,$ we can stop there:

$$\frac{1}{3} - \frac{1}{5(3!)} + \frac{1}{7(5!)} - \frac{1}{9(7!)} \le I \le \frac{1}{3} - \frac{1}{5(3!)} + \frac{1}{7(5!)} \quad \text{gives} \quad 0.3009 \le I \le 0.3011.$$

Direct integration gives

$$I = \int_0^1 x \sin x \, dx = [-x\cos x + \sin x]_0^1 = \sin 1 - \cos 1 \cong 0.3011686.$$

55. $0.2640 \le I \le 0.2643;$

$$I = \int_0^1 \left(x - x^2 + \frac{x^3}{2!} - \frac{x^4}{3!} + \frac{x^5}{4!} - \frac{x^6}{5!} + \frac{x^7}{6!} - \cdots \right) dx$$

$$= \left[\frac{x^2}{2} - \frac{x^3}{3} + \frac{x^4}{4(2!)} - \frac{x^5}{5(3!)} + \frac{x^6}{6(4!)} - \frac{x^7}{7(5!)} + \frac{x^8}{8(6!)} - \cdots \right]_0^1$$

$$= \frac{1}{2} - \frac{1}{3} + \frac{1}{4(2!)} - \frac{1}{5(3!)} + \frac{1}{6(4!)} - \frac{1}{7(5!)} + \frac{1}{8(6!)} - \cdots.$$

Note that $\dfrac{1}{8(6!)} = \dfrac{1}{5760} < 0.001.$ The integral lies between

$$\frac{1}{2} - \frac{1}{3} + \frac{1}{4(2!)} - \frac{1}{5(3!)} + \frac{1}{6(4!)} - \frac{1}{7(5!)}$$

and

$$\frac{1}{2} - \frac{1}{3} + \frac{1}{4(2!)} - \frac{1}{5(3!)} + \frac{1}{6(4!)} - \frac{1}{7(5!)} + \frac{1}{8(6!)}.$$

The first sum is greater than 0.2640 and the second sum is less than 0.2643.

Direct integration gives

$$\int_0^1 xe^{-x}\, dx = \left[-xe^{-x} - e^{-x} \right]_0^1 = 1 - 2/e \cong 0.2642411.$$

56. For $x \in [0, 4]$

$$0 \le e^x - \left(1 + x + \frac{x^2}{2!} + \cdots + \frac{x^n}{n!} \right) \le \frac{e^4 4^{n+1}}{(n+1)!}$$

$$(12.6.3) \overset{\uparrow}{\underline{\quad}}$$

Thus for $x \in [0, 2]$

$$0 \le e^{x^2} - \left(1 + x^2 + \frac{x^4}{2!} + \cdots + \frac{x^{2n}}{n!} \right) \le \frac{e^4 4^{n+1}}{(n+1)!}$$

It follows that

$$0 \le \int_0^2 e^{x^2}\, dx - \int_0^2 \left(1 + x^2 + \frac{x^4}{2!} + \cdots + \frac{x^{2n}}{n!} \right) dx \le \int_0^2 \frac{e^4 4^{n+1}}{(n+1)!}\, dx$$

$$0 \le \int_0^2 e^{x^2}\, dx - \left[x + \frac{x^3}{3} + \frac{x^5}{5(2!)} + \cdots + \frac{x^{2n+1}}{(2n+1)n!} \right] \le \frac{2e^4 4^{n+1}}{(n+1)!}$$

$$0 \le \int_0^2 e^{x^2}\, dx - \left(2 + \frac{2^3}{3} + \frac{2^5}{5(2!)} + \cdots + \frac{2^{2n+1}}{(2n+1)n!} \right) \le \frac{e^4 2^{2n+3}}{(n+1)!}$$

PROJECT 12.9A

1. $f(x) = (1+x)^\alpha$ $f(0) = 1$

    $f'(x) = \alpha(1+x)^{\alpha-1}$ $f'(0) = \alpha$

    $f''(x) = \alpha(\alpha-1)(1+x)^{\alpha-2}$ $f''(0) = \alpha(\alpha-1)$

    and so on

    $$f(x) = 1 + \alpha x + \frac{\alpha(\alpha-1)}{2!}x^2 + \frac{\alpha(\alpha-1)(\alpha-2)}{3!}x^3 + \cdots.$$

2.

    $$\left| \frac{\dfrac{\alpha(\alpha-1)(\alpha-2)\cdots(\alpha-k)}{(k+1)!}}{\dfrac{\alpha(\alpha-1)(\alpha-2)\cdots(\alpha-[k+1])}{k!}} \right| = \left| \frac{\alpha-k}{k+1} \right| \to 1 \text{ as } k \to \infty.$$

3. $\phi(x) = 1 + \alpha x + \frac{\alpha(\alpha-1)}{2!}x^2 + \frac{\alpha(\alpha-1)(\alpha-2)}{3!}x^3 + \cdots$ $\phi'(x) = \alpha + \alpha(\alpha-1)x + \frac{1}{2}\alpha(\alpha-1)(\alpha-2)x^2 + \cdots$

    $(1+x)\phi'(x) = \phi'(x) + x\phi'(x)$

    $$= \alpha + \alpha(\alpha-1)x + \frac{1}{2}\alpha(\alpha-1)(\alpha-2)x^2 + \cdots + \alpha x + \alpha(\alpha-1)x^2 + \frac{1}{2}\alpha(\alpha-1)(\alpha-2)x^3 + \cdots$$

    $$= \alpha + \alpha^2 x + \frac{\alpha^2(\alpha-1)}{2!}x^2 + \frac{\alpha^3(\alpha-1)(\alpha-2)}{3!}x^3 + \cdots = \alpha\phi(x)$$

4. $g(x) = \dfrac{\phi(x)}{(1+x)^\alpha};$ $g'(x) = \dfrac{(1+x)^\alpha\phi'(x) - \alpha\phi(x)(1+x)^{\alpha-1}}{(1+x)^{2\alpha}} = \dfrac{(1+x)\phi'(x) - \alpha\phi(x)}{(1+x)^{\alpha+1}} = 0.$

    Therefore, $g(x) \equiv C$, constant. Since $g(0) = 1$, $C = 1$; $\phi(x) = (1+x)^\alpha$.

5. (a) Take $\alpha = 1/2$ in (12.9.7) to obtain $1 + \dfrac{1}{2}x - \dfrac{1}{8}x^2 + \dfrac{1}{16}x^3 - \dfrac{5}{128}x^4$.

    (b) $\sqrt{1-x} = [1 + (-x)]^{1/2} = 1 - \dfrac{x}{2} + \dfrac{\frac{1}{2}(-\frac{1}{2})}{2!}x^2 - \dfrac{\frac{1}{2}(-\frac{1}{2})(-\frac{3}{2})}{3!}x^3 + \dfrac{\frac{1}{2}(-\frac{1}{2})(-\frac{3}{2})(-\frac{5}{2})}{4!}x^4$

    $$= 1 - \frac{1}{2}x - \frac{1}{8}x^2 - \frac{1}{16}x^3 - \frac{5}{128}x^4 - \cdots$$

    (c) Replace x by x^2 and take $\alpha = 1/2$ to obtain $\sqrt{1-x^2} \cong 1 + \frac{1}{2}x^2 - \frac{1}{8}x^4$.

    (d) Replace x by $-x^2$ and take $\alpha = 1/2$: $\sqrt{1-x^2} \cong 1 - \dfrac{x^2}{2} - \dfrac{1}{8}x^4$

    (e) Take $\alpha = -1/2$: $1 - \dfrac{1}{2}x + \dfrac{3}{8}x^2 - \dfrac{5}{16}x^3 + \dfrac{35}{128}x^4$.

    (f) Take $\alpha = -1/4$: $\dfrac{1}{\sqrt[4]{1+x}} = 1 - \dfrac{1}{4}x + \dfrac{5}{32}x^2 - \dfrac{15}{128}x^3 + \dfrac{195}{2048}x^4 + \cdots$

6. (a) $f(x) = \dfrac{1}{\sqrt{1-x^2}} = \left(1-x^2\right)^{-1/2}$

In 12.9.7, replace x by x^2 and take $\alpha = -1/2$ to obtain

$$\frac{1}{\sqrt{1-x^2}} = 1 - \frac{1}{2}x^2 + \frac{(-1/2)(-3/2)}{2!}x^4 - \frac{(-1/2)(-3/2)(-5/2)}{3!}x^6 + \cdots$$

$$= 1 - \frac{1}{2}x^2 + \frac{3}{8}x^4 + \frac{5}{16}x^6 + \cdots$$

By Problem 2, this series has radius of convergence $r = 1$.

(b)

$$\arcsin x = \int_0^x \frac{1}{\sqrt{1-x^2}}\, dt = \int_0^x \left(1 - \frac{1}{2}t^2 + \frac{3}{8}t^4 + \frac{5}{16}t^6 + \cdots\right) dt$$

$$= x - \frac{1}{6}x^3 + \frac{3}{40}x^5 + \frac{5}{112}x^7 + \cdots$$

By Theorem 12.9.3, the radius of convergence of this series is $r = 1$.

7. (a) $\qquad \alpha = -\dfrac{1}{2}: \quad \dfrac{1}{\sqrt{1+x^2}} = 1 - \dfrac{x^2}{2} + \dfrac{\left(-\frac{1}{2}\right)\left(-\frac{3}{2}\right)}{2!}x^4 + \dfrac{\left(-\frac{1}{2}\right)\left(-\frac{3}{2}\right)\left(-\frac{5}{2}\right)}{3!}x^6 + \cdots$

$$= 1 - \frac{1}{2}x^2 + \frac{3}{8}x^4 - \frac{5}{16}x^6 + \cdots$$

(b)

$$\sinh^{-1} x = \int_0^x \frac{1}{\sqrt{1+t^2}}\, dt = \int_0^x \left(1 - \frac{1}{2}t^2 + \frac{3}{8}t^4 - \frac{5}{16}t^6 + \cdots\right) dt$$

$$= x - \frac{1}{6}x^3 + \frac{3}{40}x^4 - \frac{5}{112}x^7 + \cdots; \quad r = 1$$

PROJECT 12.9B

1. $\tan\left[2\arctan\left(\frac{1}{5}\right)\right] = \dfrac{2\tan\left[\arctan\left(\frac{1}{5}\right)\right]}{1-\tan^2\left[\arctan\left(\frac{1}{5}\right)\right]} = \dfrac{\frac{2}{5}}{1-\frac{1}{25}} = \dfrac{5}{12}$

$2\arctan\left(\frac{1}{5}\right) = \arctan\left(\frac{5}{12}\right)$

$4\arctan\left(\frac{1}{5}\right) = 2\arctan\left(\frac{5}{12}\right)$

$\tan\left(\left[\arctan\left(\frac{1}{5}\right)\right]\right) = \tan\left[2\arctan\left(\frac{1}{5}\right)\right] = \dfrac{\frac{10}{12}}{1-\frac{25}{144}} = \dfrac{120}{119}$

$\tan\left[4\arctan\left(\frac{1}{5}\right) - \arctan\left(\frac{1}{239}\right)\right] = \dfrac{\frac{120}{119}-\frac{1}{239}}{1+\frac{120}{119}\cdot\frac{1}{239}} = \dfrac{120(239)-119}{119(239)+120} = 1$

Thus $\quad 4\arctan\left(\frac{1}{5}\right) - \arctan\left(\frac{1}{239}\right) = \dfrac{\pi}{4}.$

2. $4\arctan\left(\frac{1}{5}\right) - \arctan\left(\frac{1}{239}\right) < 4\displaystyle\sum_{k=1}^{5}\frac{(-1)^{k-1}}{2k-1}\left(\frac{1}{5}\right)^{2k-1} - \left[\frac{1}{239} - \frac{1}{3}\left(\frac{1}{239}\right)^3\right]$

$\qquad = 0.789582246 - 0.004184076 = 0.78539817.$

$4\arctan\frac{1}{5} - \arctan\frac{1}{239} > 4\displaystyle\sum_{k=1}^{5}\frac{(-1)^{k-1}}{2k-1}\left(\frac{1}{5}\right)^{2k-1} - \frac{1}{239}$

$\qquad = 0.789582238 - 0.0041841 = 0.785398138.$

These inequalities imply $\quad 3.14159255 < \pi < 3.14159268.$

3. $4\arctan\frac{1}{5} - \arctan\frac{1}{239} < 4\sum_{k=1}^{15}\frac{(-1)^{k-1}}{2k-1}\left(\frac{1}{5}\right)^{2k-1} - \left[\sum_{k=1}^{4}\frac{(-1)^{k-1}}{2k-1}\left(\frac{1}{239}\right)^{2k-1}\right]$

$= 0.785398163397448309616$

$4\arctan\frac{1}{5} - \arctan\frac{1}{239} > 4\sum_{k=1}^{14}\frac{(-1)^{k-1}}{2k-1}\left(\frac{1}{5}\right)^{2k-1}\left[\sum_{k=1}^{3}\frac{(-1)^{k-1}}{2k-1}\left(\frac{1}{239}\right)^{2k-1}\right]$

$= 0.785398163397448306408$

These inequalities imply $3.14159265358979322563 < \pi < 3.14159265358979323846.$

REVIEW EXERCISES

1. $\sum_{k=0}^{\infty}\left(\frac{3}{4}\right)^k = \frac{1}{1-\frac{3}{4}} = 4,$ a geometric series with $r = \frac{3}{4}$.

2. $\sum_{k=0}^{\infty}(-1)^k\left(\frac{1}{2}\right)^k = \sum_{0}^{\infty}\left(-\frac{1}{2}\right)^k = \frac{1}{1-\left(-\frac{1}{2}\right)} = \frac{2}{3},$ a geometric series with $r = -\frac{1}{2}$

3. Since $e^x = \sum_{k=0}^{\infty}\frac{x^k}{k!},$ $\sum_{k=0}^{\infty}\frac{(\ln 2)^k}{k!} = e^{\ln 2} = 2$

4. $\sum_{k=1}^{\infty}\frac{1}{k(k+1)} = \sum_{k=1}^{\infty}\left(\frac{1}{k} - \frac{1}{k+1}\right) = \lim_{n\to\infty}\left(1 - \frac{1}{n+1}\right) = 1$

5. diverges; limit comparison with $\sum\frac{1}{k}$

6. converges; limit comparison with $\sum\frac{1}{k^2}$

7. absolutely convergent; basic comparison

$$\sum\left|\frac{(-1)^k}{(k+1)(k+2)}\right| \le \sum\frac{1}{k^2}$$

8. conditionally convergent; $\sum\frac{(-1)^k}{2k+1}$ converges by Theorem 11.5.3;

$\sum\left|\frac{(-1)^k}{2k+1}\right| = \sum\frac{1}{2k+1}$ diverges.

9. absolutely convergent; $\sum_{k=0}^{\infty}\left|\frac{(-1)^k(100)^k}{k!}\right| = \sum_{k=0}^{\infty}\frac{100^k}{k!}$ which converges by the ratio test.

10. converges; root test: $\left(\frac{k+1}{3^k}\right)^{1/k} \to \frac{1}{3}$ as $k \to \infty,$ or ratio test: $\frac{k+2}{3^{k+1}}\frac{3^k}{k+1} \to \frac{1}{3}$ as $k \to \infty$

11. diverges; ratio test:

$$\frac{(k+1)!}{(k+1)^{(k+1)/2}} \cdot \frac{k^{k/2}}{k!} = \frac{k^{k/2}}{(k+1)^{(k-1)/2}} = \left(\frac{k}{k+1}\right)^{k/2} \sqrt{k+1} \to \infty.$$

12. converges; limit comparison with: $\displaystyle\sum \frac{1}{k^2}$

13. conditionally convergent: $\displaystyle\sum \frac{(-1)^k}{\sqrt{(k+1)(k+2)}}$ converges by Theorem 11.5.3;

$$\sum \left|\frac{(-1)^k}{\sqrt{(k+1)(k+2)}}\right| = \sum \frac{1}{\sqrt{(k+1)(k+2)}} \quad \text{diverges -- limit comparison with } \sum \frac{1}{k}.$$

14. converges; root test: $\displaystyle \left[k\left(\frac{3}{4}\right)^k\right]^{1/k} \to \frac{3}{4}$ as $k \to \infty$

15. converges; ratio test: $\displaystyle \frac{a_{k+1}}{a_k} = \left(\frac{k+1}{k}\right)^e \cdot \frac{1}{e} \to \frac{1}{e} < 1$

16. conditionally convergent; Theorem 12.5.3: $\displaystyle\sum \frac{\ln k}{\sqrt{k}}$ diverges by the integral test.

17. diverges; ratio test: $\displaystyle \frac{a_{k+1}}{a_k} = \frac{[2(k+1)]!}{2^{k+1}(k+1)!} \cdot \frac{2^k k!}{(2k)!} = 2k+1 \to \infty$

18. absolutely convergent: limit comparison with $\displaystyle\sum \frac{1}{k^{3/2}}$

19. converges; basic comparison:

$$\sum \frac{(\arctan k)^2}{1+k^2} \le \frac{\pi^2}{4} \sum \frac{1}{1+k^2} \le \frac{\pi^2}{4} \sum \frac{1}{k^2};$$

or the integral test: $\displaystyle \int_0^\infty \frac{(\arctan x)^2}{1+x^2}\,dx$ converges.

20. converges; $\displaystyle \frac{2^k + k^4}{3^k} = \frac{2^k}{3^k} + \frac{k^4}{3^k}$, and each of the series $\displaystyle\sum \frac{2^k}{3^k}$ and $\displaystyle\sum \frac{k^4}{3^k}$ is convergent.

21. $\displaystyle e^x = \sum_{k=0}^\infty \frac{x^k}{k!}$. Therefore,

$$xe^{2x^2} = x\sum_{k=0}^\infty \frac{(2x^2)^k}{k!} = \sum_{k=0}^\infty \frac{2^k}{k!} x^{2k+1}$$

22. $\displaystyle \ln(1+x) = \sum_{k=1}^\infty \frac{(-1)^{k+1}}{k} x^k$. Therefore,

$$\ln(1+x^2) = \sum_{k=1}^\infty \frac{(-1)^{k+1}}{k} x^{2k}$$

23. $\displaystyle \arctan x = \sum_{k=0}^\infty \frac{(-1)^k x^{2k+1}}{2k+1}$. Therefore,

$$\sqrt{x}\arctan\left(\sqrt{x}\right) = x^{1/2}\sum_0^\infty \frac{(-1)^k x^{(2k+1)/2}}{2k+1} = x^{\frac{1}{2}}\sum_0^\infty \frac{(-1)^k x^{k+\frac{1}{2}}}{2k+1} = \sum_0^\infty (-1)^k \frac{x^{k+1}}{2k+1}$$

24. $a^x = e^{x \ln a}$. Therefore,

$$a^x = \sum_{k=0}^{\infty} \frac{(\ln a)^k}{k!} x^k$$

25. $\ln(1+x) = \sum_{k=1}^{\infty} \frac{(-1)^{k+1}}{k} x^k$. Therefore,

$$\ln(1+x^2) = \sum_{k=1}^{\infty} \frac{(-1)^{k+1}}{k} (x^2)^k \quad \text{and} \quad \ln(1-x^2) = \sum_{k=1}^{\infty} \frac{(-1)^{k+1}}{k} (-x^2)^k.$$

$$f(x) = x \ln \frac{1+x^2}{1-x^2} = x[\ln(1+x^2) - \ln(1-x^2)] = x \left(\sum_{k=1}^{\infty} \frac{(-1)^{k+1}}{k} (x^2)^k - \sum_{k=1}^{\infty} \frac{(-1)^{k+1}}{k} (-x^2)^k \right)$$

$$= 2 \sum_{k=0}^{\infty} \frac{x^{4k+3}}{2k+1}$$

26. $\sin x = \sum_{k=0}^{\infty} \frac{(-1)^k}{(2k+1)!} x^{2k+1}$. Therefore,

$$(x+x^2)\sin x^2 = (x+x^2) \sum_{k=0}^{\infty} \frac{(-1)^k}{(2k+1)!} (x^2)^{2k+1} = \sum_{k=0}^{\infty} \frac{(-1)^k}{(2k+1)!} \left(x^{4k+3} + x^{4k+4} \right).$$

27. $f(x) = (1-x)^{1/3}$ $f(0) = 1$

    $f'(x) = -\frac{1}{3}(1-x)^{-\frac{2}{3}}$ $f'(0) = -\frac{1}{3}$

    $f''(x) = -\frac{2}{9}(1-x)^{-\frac{5}{3}}$ $f''(0) = -\frac{2}{9}$

    $f'''(x) = -\frac{10}{27}(1-x)^{-\frac{7}{3}}$ $f'''(0) = -\frac{10}{27}$

    $P_3(x) = 1 - \frac{1}{3}x - \frac{1}{9}x^2 - \frac{5}{81}x^3$

28. $f(x) = \arcsin x$ $f(0) = 0$

    $f'(x) = \dfrac{1}{\sqrt{1-x^2}}$ $f'(0) = 1$

    $f''(x) = \dfrac{x}{(1-x^2)^{3/2}}$ $f''(0) = 0$

    $f'''(x) = \dfrac{3x^2}{(1-x^2)^{5/2}} + \dfrac{1}{(1-x^2)^{3/2}}$ $f'''(0) = 1$

    $f^{(4)}(x) = \dfrac{15x^3}{(1-x^2)^{7/2}} + \dfrac{9x}{(1-x^2)^{5/2}}$ $f^{(4)}(0) = 0$

    $P_4(x) = x + \frac{1}{6}x^3$

29. $\left[-\frac{1}{5}, \frac{1}{5}\right)$;   ratio test: $\dfrac{b_{k+1}}{b_k} = \dfrac{5k}{k+1}|x| \to 5|x| \implies r = \frac{1}{5}$

    At $x = -\frac{1}{5}$, $\sum \dfrac{(-1)^k}{k}$ converges;   at $x = \frac{1}{5}$, $\sum \dfrac{1}{k}$ diverges.

30. $(-3,3)$;   ratio test: $\dfrac{b_{k+1}}{b_k} = \dfrac{1}{3}|x| \implies r = 3$

at $x = -3$, $\displaystyle\sum \dfrac{(-1)^k}{3^k}(-3)^{k+1} = \sum 3(-1)^{2k+1}$ diverges;

at $x = 3$, $\displaystyle\sum \dfrac{(-1)^k}{3^k} 3^{k+1} = \sum 3(-1)^k$ diverges

31. $(-\infty, \infty)$;   ratio test: $\dfrac{b_{k+1}}{b_k} = \dfrac{2|x-1|^2}{(2k+2)(2k+1)} \to 0 \implies r = \infty$

32. $(0,4)$;   ratio test: $\dfrac{b_{k+1}}{b_k} = \dfrac{1}{2}|x-2| \implies r = 2$

at $x = 0$, $\displaystyle\sum \dfrac{1}{2^k}(-2)^k = \sum(-1)^k$ diverges;

at $x = 4$, $\displaystyle\sum \dfrac{1}{2^k}2^k = \sum 1$ diverges

33. $(-9,9)$;   ratio test: $\dfrac{b_{k+1}}{b_k} = \dfrac{k+1}{9k}|x| \to \dfrac{1}{9}|x| \implies r = 9$

At $x = -9$, $\sum k$ diverges;   at $x = 9$, $\sum(-1)^k k$ diverges

34. $(-1,1)$;   ratio test: $\dfrac{b_{k+1}}{b_k} = \dfrac{(k+1)(2k+1)}{k(2k+3)}|x|^2 \to |x|^2 \implies r = 1$

at $x = 1$, $\displaystyle\sum \dfrac{k}{2k+1}$ diverges $\left(\dfrac{k}{2k+1} \to \dfrac{1}{2} \text{ as } k \to \infty\right)$;

at $x = -1$, $\displaystyle\sum \dfrac{-k}{2k+1}$ diverges for the same reason.

35. $(-4,-2]$;   ratio test: $\dfrac{b_{k+1}}{b_k} = \dfrac{\sqrt{k}}{\sqrt{k+1}}|x+3| \to |x+3| \implies r = 1$

at $x = -2$, $\displaystyle\sum \dfrac{(-1)^k}{\sqrt{k}}$ converges;

at $x = -4$, $\displaystyle\sum \dfrac{1}{\sqrt{k}}$ diverges

36.   diverges except at $x = -1$: $\dfrac{b_{k+1}}{b_k} = (k+1)|x+1| \to \infty \implies r = 0$

37. $f(x) = e^{-2x} = e^{-2(x+1)+2} = e^2 \cdot e^{-2(x+1)} = e^2 \displaystyle\sum_0^\infty \dfrac{[-2(x+1)]^k}{k!} = e^2 \sum_0^\infty \dfrac{(-1)^k 2^k}{k!}(x+1)^k$; $r = \infty$.

38. $\sin 2x = \displaystyle\sum_0^\infty \dfrac{(-1)^k 2^{2k}}{(2k)!}\left(x - \dfrac{\pi}{4}\right)^{2k}$; $r = \infty$.

39. $f(x) = \ln x = \ln[1 + (x-1)] = \displaystyle\sum_1^\infty \dfrac{(-1)^{k+1}}{k}(x-1)^k$; $r = 1$

40. $\sqrt{x+1} = 1 + \dfrac{1}{2}x - \dfrac{1}{8}x^2 + \sum_{3}^{\infty}(-1)^{k+1}\dfrac{1\cdot3\cdot5\cdots(2k-3)}{2^k\,k!}x^k, \quad r = 1$

41. $\dfrac{1}{1+x^4} = \sum_{k=0}^{\infty}(-1)^k x^{4k}$

$$\int_0^{1/2}\frac{1}{1+x^4}\,dx = \sum_{k=0}^{\infty}(-1)^k\int_0^{1/2}x^{4k}\,dx = \sum_{k=0}^{\infty}(-1)^k\frac{1}{4k+1}\frac{1}{2^{4k+1}}$$

This is an alternating series with decreasing terms and the third term $\dfrac{1}{9(2^9)} \approx 0.0002$. Hence

$$\int_0^{1/2}\frac{1}{1+x^4}dx \approx \frac{1}{2} - \frac{1}{5(2^5)} \approx 0.4938$$

42. $e^x = \sum_{k=0}^{n}\dfrac{x^k}{k!}$ with remainder $|R_n(x)| \le \max|f^{(n+1)}(t)|\dfrac{|x|^{n+1}}{(n+1)!}$.

$$R_n(2/3) < 3\frac{(2/3)^{n+1}}{(n+1)!} < 0.01 \quad \text{iff} \quad \left(\frac{3}{2}\right)^{n+1}(n+1)! > 300 \implies n = 4.$$

Therefore $e^{2/3} \approx 1 + \frac{2}{3} + \frac{1}{2}(2/3)^2 + \frac{1}{6}(2/3)^3 + \frac{1}{24}(2/3)^4 \approx 1.9465$

43. Set $f(x) = x^{1/3}$ and take $a = 64$. Then

$$x^{1/3} = f(x) = f(64) + f'(64)(x-64) + \frac{f''(64)}{2!}(x-64)^2 + \cdots.$$

$f(x) = x^{1/3}$ $\qquad\qquad\qquad\qquad\qquad$ $f(64) = 4$

$f'(x) = \dfrac{1}{3}x^{-2/3}$ $\qquad\qquad\qquad\qquad$ $f'(64) = \dfrac{1}{48}$

$f''(x) = -\dfrac{2}{9}x^{-5/3}$ $\qquad\qquad\qquad\quad$ $f''(64) = -\dfrac{2}{9\cdot4^5}$

$\cdots$

$$|R_2(68)| \le \frac{\max|f'''(t)|}{3!}4^3 < 0.005. \quad \text{Therefore,}$$

$$\sqrt[3]{68} = f(68) \cong 4 + \frac{1}{12} - \frac{1}{9\cdot4^3} \cong 4.0816.$$

44. $\sin x = \sum_{k=0}^{\infty}\dfrac{(-1)^k}{(2k+1)!}x^{2k+1}; \quad x\sin x^4 = \sum_{k=0}^{\infty}\dfrac{(-1)^{k+1}}{(2k+1)!}x^{8k+5}.$

$$\int_0^1 x\sin x^4\,dx = \sum_{k=0}^{\infty}\frac{(-1)^k}{(2k+1)!}\frac{1}{8k+6}$$

This is an alternating series with decreasing terms and the 3th term $\dfrac{1}{5!(24)} < 0.01$. Therefore

$$\int_0^1 x\sin x^4\,dx \approx \frac{1}{6} - \frac{1}{3!(14)} \approx 0.155$$

45. Let $g(x) = \sin x$ and $a = \pi/4$. Then $\sin x = \dfrac{\sqrt{2}}{2} + \dfrac{\sqrt{2}}{2}(x - \pi/4) - \dfrac{\sqrt{2}}{2(2!)}(x - \pi/4)^2 - \dfrac{\sqrt{2}}{2(3!)}$
$(x - \pi/4)^3 + \cdots$

$$|R_n(x)| = \frac{|g^{(n+1)}(c)|}{(n+1)!}\left|\left(x - \frac{\pi}{4}\right)^{n+1}\right|$$

$$\leq \frac{|(x - \pi/4)|^{n+1}}{(n+1)!} \qquad (g^{(n+1)}(c) = \pm \sin c \text{ or } \pm \cos c)$$

Now, $48° = \dfrac{48\pi}{180}$ radians. We want to find the smallest positive integer n such that
$|R_n(48\pi/180 - \pi/4)| < 0.0001$.

$$|R_n(48\pi/180 - \pi/4)| \leq \left(\frac{\pi}{60}\right)^{n+1}\frac{1}{(n+1)!} \cong \frac{(0.05236)^{n+1}}{(n+1)!} < 0.0001 \implies n \geq 2.$$

$$\sin x \cong P_2(x) = \frac{\sqrt{2}}{2} + \frac{\sqrt{2}}{2}\left(x - \frac{\pi}{4}\right) - \frac{\sqrt{2}}{4}\left(x - \frac{\pi}{4}\right)^2; \quad \sin 48° \cong \frac{\sqrt{2}}{2}\left[1 + \frac{\pi}{60} - \frac{1}{2}\left(\frac{\pi}{60}\right)^2\right] \cong 0.7432$$

46. $\displaystyle\int_0^1 x^2 e^{-x^2}\, dx = \sum_{k=0}^{\infty}\int_0^1 \frac{(-1)^k}{k!}x^{2k+2}\, dx = \sum_{k=0}^{\infty}\frac{(-1)^k}{k!}\frac{1}{2k+3}$

This is an alternating series with decreasing terms and the 6th term $\dfrac{1}{5!(13)} < 0.001$. Therefore

$$\int_0^1 x^2 e^{-x^2}\, dx \approx \sum_{k=0}^{4}\frac{(-1)^k}{k!}\frac{1}{2k+3} \approx 0.1900$$

47. For the sine function, $x - \frac{1}{6}x^3 + \frac{1}{120}x^5 = P_5 = P_6$. Therefore, for $x \in [0, \pi/4]$ we have

$$|\sin x - P_5(x)| = \left|\frac{f^{(7)}(c)}{7!}x^7\right| \leq \frac{1}{7!}\left(\frac{\pi}{4}\right)^7 < 0.000037$$

$$(|f^{(7)}(c)| = \cos c \leq 1)$$

48. For the cosine function, $1 - \frac{1}{2}x^2 + \frac{1}{24}x^4 - \frac{1}{720}x^6 = P_6 = P_7$. Therefore, for $x \in [0, \pi/4]$ we have

$$|\cos x - P_6(x)| = \left|\frac{f^{(8)}(c)}{8!}x^8\right| \leq \frac{1}{8!}\left(\frac{\pi}{4}\right)^8 < 0.0000036$$

$$(|f^{(8)}(c)| = \cos c \leq 1)$$

49. $\displaystyle\sum_{k=1}^{\infty} a_k = \int_1^{\infty} xe^{-x}\, dx = \frac{2}{e}$

50. Let $\epsilon > 0$. For each positive integer n, set $a_n = x_n - \dfrac{\epsilon}{2^{n+1}}$ and $b_n = x_n + \dfrac{\epsilon}{2^{n+1}}$. Then

$$b_n - a_n = \frac{\epsilon}{2^{n+1}} \quad \text{and} \quad \sum_{n=1}^{\infty}(b_n - a_n) = \sum_{n=1}^{\infty}\frac{\epsilon}{2^{n+1}} = \frac{\epsilon}{2} < \epsilon$$

51. If $\displaystyle\sum_{k=1}^{\infty}(a_{k+1} - a_k)$ converges, then the sequence of partial sums $s_n = a_{n+1} - a_1$ converges.

Therefore, the sequence a_k converges.

If the sequence a_k converges, then the sequence $s_n = a_{n+1} - a_1$ converges which implies that the

series $\displaystyle\sum_{k=1}^{\infty}(a_{k+1} - a_k)$ converges.

52. (a) $a_k = \displaystyle\sum_{n=0}^{\infty}(\frac{1}{k})^n = \frac{1}{1 - 1/k} = \frac{k}{k-1}$ and $\displaystyle\sum_{k=2}^{\infty} a_k = \sum_{k=2}^{\infty}\frac{k}{k-1}$.

The series diverges because $a_k = \dfrac{k}{k-1} \not\to 0$.

(b) $a_k = \displaystyle\sum_{n=1}^{\infty}(\frac{1}{k})^n = \frac{1/k}{1 - 1/k} = \frac{1}{k-1}$ and $\displaystyle\sum_{k=2}^{\infty} a_k = \sum_{k=2}^{\infty}\frac{1}{k-1}$.

The series diverges; limit comparison with $\displaystyle\sum\frac{1}{k}$.

(c) $a_k = \displaystyle\sum_{n=2}^{\infty}(\frac{1}{k})^n = \frac{1/k^2}{1 - 1/k} = \frac{1}{k(k-1)}$ and $\displaystyle\sum_{k=2}^{\infty} a_k = \sum_{k=2}^{\infty}\frac{1}{k(k-1)}$.

The series converges; limit comparison with $\displaystyle\sum\frac{1}{k^2}$.

Printed in the United States
81347LV00003B/3-10